T0317664

Hybrid Electric Vehicles

Automotive Series

Series Editor: Thomas Kurfess

Hybrid Electric Vehicles: Principles and Applications with Practical Perspectives, 2nd Edition	Mi and Masrur	October 2017
Hybrid Electric Vehicle System Modeling and Control, 2nd Edition	Liu	April 2017
Thermal Management of Electric Vehicle Battery Systems	Dincer, Hamut and Javani	March 2017
Automotive Aerodynamics	Katz	April 2016
The Global Automotive Industry	Nieuwenhuis and Wells	September 2015
Vehicle Dynamics	Meywerk	May 2015
Vehicle Gearbox Noise and Vibration: Measurement, Signal Analysis, Signal Processing and Noise Reduction Measures	Tůma	April 2014
Modeling and Control of Engines and Drivelines	Eriksson and Nielsen	April 2014
Modelling, Simulation and Control of Two-Wheeled Vehicles	Tanelli, Corno and Savaresi	March 2014
Advanced Composite Materials for Automotive Applications: Structural Integrity and Crashworthiness	Elmarakbi	December 2013
Guide to Load Analysis for Durability in Vehicle Engineering	Johannesson and Speckert	November 2013

Hybrid Electric Vehicles

Principles and Applications with Practical Perspectives

Second Edition

Chris Mi

San Diego State University
USA

M. Abul Masrur

University of Detroit Mercy
USA

This edition first published 2018

Registered Office(s)
John Wiley & Sons, Inc., 111 River Street, Hoboken, NJ 07030, USA
John Wiley & Sons Ltd, The Atrium, Southern Gate, Chichester, West Sussex, PO19 8SQ, UK

Editorial Office
The Atrium, Southern Gate, Chichester, West Sussex, PO19 8SQ, UK

For details of our global editorial offices, customer services, and more information about Wiley products visit us at www.wiley.com.

Wiley also publishes its books in a variety of electronic formats and by print-on-demand. Some content that appears in standard print versions of this book may not be available in other formats.

Library of Congress Cataloging-in-Publication Data

Names: Mi, Chris, author. | Masrur, Abul, author.
Title: Hybrid electric vehicles : principles and applications with practical perspectives / Chris Mi, San Diego State University, US, M. Abul Masrur, University of Detroit-Mercy, US.
Description: Second edition. | Hoboken, NJ, USA : Wiley, 1918. | Series: Automotive series | Includes bibliographical references and index. | Identifiers: LCCN 2017019753 (print) | LCCN 2017022859 (ebook) | ISBN 9781118970539 (pdf) | ISBN 9781118970546 (epub) | ISBN 9781118970560 (cloth)
Subjects: LCSH: Hybrid electric vehicles.
Classification: LCC TL221.15 (ebook) | LCC TL221.15 .M545 2018 (print) | DDC 629.22/93–dc23
LC record available at https://lccn.loc.gov/2017019753

Cover Design: Wiley
Cover Images: © Taina Sohlman/Shutterstock; © J.D.S/Shutterstock; © Sjo/iStockphoto; © Monty Rakusen/Gettyimages

Set in 10/12pt Warnock by SPi Global, Pondicherry, India

10 9 8 7 6 5 4 3 2 1

Contents

About the Authors

Chris Mi is a Fellow of the IEEE, Professor and Chair of the Department of Electrical and Computer Engineering, and Director of the US DOE funded GATE Center for Electric Drive Transportation at San Diego State University, California, USA. He was previously a professor at the University of Michigan-Dearborn from 2001 to 2015. He received the BSc and MSc degrees from Northwestern Polytechnical University, Xi'an, China, and the PhD degree from the University of Toronto, Canada, all in electrical engineering. Previously he was an electrical engineer with General Electric, Canada Inc. He was the President and the Chief Technical Officer of 1Power Solutions, Inc. from 2008 to 2011.

His research interests are in electric and hybrid vehicles. He has taught tutorials and seminars on the subject of HEVs/PHEVs for the Society of Automotive Engineers (SAE), the IEEE, workshops sponsored by the National Science Foundation (NSF), and the National Society of Professional Engineers. He has delivered courses to major automotive OEMs and suppliers, including GM, Ford, Chrysler, Honda, Hyundai, Tyco Electronics, A&D Technology, Johnson Controls, Quantum Technology, Delphi, and the European PhD School. He has offered tutorials in many countries, including the USA, China, Korea, Singapore, Italy, France, and Mexico. He has published more than 200 articles and delivered 30 invited talks and keynote speeches. He has also served as a panelist in major IEEE and SAE conferences.

Dr Mi is a recipient of the "Distinguished Teaching Award" and the "Distinguished Research Award" of University of Michigan-Dearborn. He is a recipient of the 2007 IEEE Region 4 "Outstanding Engineer Award," "IEEE Southeastern Michigan Section Outstanding Professional Award," and the "SAE Environmental Excellence in Transportation (E2T) Award." He was also a recipient of the National Innovation Award and the Government Special Allowance Award from the China Central Government. In December 2007, he became a Member of Eta Kappa Nu, which is the Electrical and Computer Engineering Honor Society, for being "a leader in education and an example of good moral character."

Dr Mi was the chair (2008–2009) and vice-chair (2006–2007) of the IEEE Southeastern Michigan Section, and was the general chair of the 5th IEEE Vehicle Power and Propulsion Conference held in Dearborn, Michigan, USA in September 2009. Dr Mi is one of the three Area Editors of the IEEE Transactions on Vehicular Technology, associate editor of IEEE Transactions on Power Electronics, and Associate Editor of IEEE Transactions on Industry Applications. He served on the review panel for the NSF, the US Department of Energy, the Natural Sciences and Engineering Research Council of

Canada, Hong Kong Research Grants Council, French Centre National de la Recherche Scientifique, Agency for Innovation by Science and Technology in Flanders (Belgium), and the Danish Research Council. He is the topic chair of the 2011 IEEE International Future Energy Challenge, and the general chair of the 2013 IEEE International Future Energy Challenge. He is a Distinguished Lecturer (DL) of the IEEE Vehicular Technology Society.

Dr Mi is also the general co-chair of the IEEE Workshop on Wireless Power Transfer, guest editor-in-chief of the IEEE Journal of Emerging and Selected Topics in Power Electronics – Special Issue on WPT, guest co-editor-in-chief of IEEE Transactions on Power Electronics Special Issue on WPT, guest editor of IEEE Transactions on Industrial Electronics – Special Issue on dynamic wireless power transfer, and steering committee member of the IEEE Transportation Electrification Conference (ITEC, Asian). He is the program chair for the 2014 IEEE International Electric Vehicle Conference (IEVC) in Florence Italy December 2014 and is also the chair for the IEEE Future Direction's Transportation Electrification Initiative (TEI) e-Learning Committee and developed an e-learning module on wireless power transfer.

M. Abul Masrur received his PhD in Electrical Engineering from the Texas A & M University, College Station, TX, USA in 1984. Prior to that he received BSc and MSc degrees in Electrical Engineering. He also has a Master's degree in Computer Engineering. Dr Masrur is an Adjunct Professor at the University of Detroit Mercy, where he has been teaching various courses since 2003, which include Advanced Electric and Hybrid Vehicles, Vehicular Power Systems, Electric Drives and Power Electronics. He has also been instructing graduate level courses at the University of Michigan-Dearborn since 2014, related to vehicular electronics and electrical systems. He was with the Scientific Research Labs, Ford Motor Co., between 1984 and 2001 and was involved in research and development related to electric drives and power electronics, advanced automotive power system architectures, electric active suspension systems for automobiles, electric power assist steering, and standalone UPS protection design, among other things.

Since April 2001, Dr Masrur has been with the US Army RDECOM-TARDEC (R&D) where he has been involved in vehicular electric power system architecture concept design and development, advanced vehicular propulsion, microgrid and vehicle to vehicle (V2V) and vehicle to grid (V2G) systems, wireless power transfer, electric power management, and artificial intelligence-based fault diagnostics in electric drives. He has authored/co-authored over 90 publications, many of which are in public domain international journals and conference proceedings. He is also the co-inventor of eight US patents (and one additional US patent is pending), two of the patents are also patented in Europe and one in Japan. He received the Best Automotive Electronics Paper Award from the IEEE Vehicular Technology Society in 1998 for his papers proposing novel vehicular power system architectures published in the *IEEE Transactions on Vehicular Technology*, and in 2006 was a joint recipient of the SAE Environmental Excellence in Transportation Award – Education, Training, & Public Awareness (or E2T) for a tutorial course he had been jointly presenting on hybrid vehicles.

Dr Masrur is a Fellow of the IEEE, cited for "Contributions to fault diagnostics in electric motor drives and automotive electric power systems". From 1999–2007 he served as an associate editor (Vehicular Electronics Section) of the *IEEE Transactions on Vehicular Technology*. He also served as chair of the Motor-Subcommittee of the IEEE Power & Energy Society – Electric Machinery Committee for two years ending in December 2010. As a member of this motor subcommittee he also participated in the development of the IEEE Draft Trial-Use Guide for Testing Permanent Magnet Machines (P1812), which was recently released, and was cited as an outstanding contribution by the IEEE.

Preface To the First Edition

It is well recognized today that, technologies of hybrid electric vehicles (HEVs) and electric vehicles (EVs) are vital to the overall automotive industry and also to the user, in terms of both better fuel economy and a better effect on the environment. Over the past decade, these technologies have taken a significant leap forward. As they have developed, the literature in the public domain has also grown accordingly, in the form of publications in conference proceedings and journals, and also in the form of textbooks and reference books. Why then was the effort made to write this book? The question is legitimate. The authors observed that existing textbooks have topics like drive cycle, fuel economy, and drive technology as their main focus. In addition, the authors felt that the main focus of such textbooks was on regular passenger automobiles. It is against this backdrop that the authors felt a wider look at the technology was necessary. By this, it is meant that HEV technology is one which is applicable not just to regular automobiles, but also to other vehicles such as locomotives, off-road vehicles (construction and mining vehicles), ships, and even to some extent to aircraft. The authors believe that the information probably exists, but not specifically in textbook form where the overall viewpoint is included. In fact, HEV technology is not new – a slightly different variant of it was present many years ago in diesel–electric locomotives. However, the availability of high-power electronics and the development of better materials for motor technology have made it possible to give a real boost to HEV technology during the past decade or so, making it viable for wider applications.

A textbook, unlike a journal paper, has to be reasonably self-contained. Hence the authors decided to review the basics, including power electronics, electric motors, and storage elements like batteries, capacitors, flywheels, and so on. All these are the main constituent elements of HEV technology. Also included is a discussion on the system-level architecture of the vehicles, modeling and simulation methods, transmission and coupling. Drive cycles and their meaning, and optimization of the vehicular power usage strategy (and power management), have also been included. The issue of dividing power between multiple sources lies within the domain of power management, which is an extremely important matter in any power system where more than one source of power is used. These sources may be similar or diverse in nature – that is, they could be electrical, mechanical, chemical, and so on – and even if they could all be similar, they might potentially have different characteristics. Optimization involves a decision on resource allocation in such situations. Some of these optimization methods actually exist in and are used by the utility industry, but they have lately attracted significant interest in vehicular applications. To make the book relatively complete and more holistic in nature, the topics of applications to off-road vehicles, locomotives, ships, and aircraft have also been included. In the recent past, the interface between a vehicle and the utility grid for plug-in capabilities has become important, hence the inclusion of topics on plug-in hybrids and vehicle-to-grid or vehicle-to-vehicle power transfer. Also presented is a discussion on diagnostics and prognostics, the reliability of the HEV from a system-level perspective, electromechanical vibration and noise vibration harshness (NVH), electromagnetic compatibility and electromagnetic interference (EMC/EMI), and overall life cycle issues. These topics are almost non-existent in the textbooks on HEVs known to the authors. In fact, some of the topics have not been discussed much in the research literature either, but they are all very important issues. The success of a technology is ultimately manifested in the form of user acceptance and is intimately connected with the mass manufacture of the product. It is not sufficient for a technology to be good; unless a technology, particularly the ones meant for ordinary consumers, can be mass produced in a relatively inexpensive manner, it may not have much of an impact on society. This is very much valid for HEVs as well. The book therefore concludes with a chapter on commercialization issues in HEVs.

The authors have significant industrial experience in many of the technical areas covered in the book, as reflected in the material and presentation. They have also been involved in teaching both academic and industrial professional courses in the area of HEV and EV systems and components. The book evolved to some extent from the notes used in these courses. However, significant amounts of extra material have been added, which is not covered in those courses.

It is expected that the book will fill some of the gaps in the existing literature and in the areas of HEV and EV technologies for both regular and off-road vehicles. It will also help the reader to get a better system-level perspective of these.

There are 15 chapters, the writing of which was shared between the three authors. Chris Mi is the main author of Chapters 1, 4, 5, 9, and 10. M. Abul Masrur is the main author of Chapters 2, 6, 7, 8, 14, and 15. David Wenzhong Gao is the main author of Chapters 3, 11, 12, and 13.

Since this is the first edition of the book, the authors very much welcome any input and comments from readers, and will ensure that any corrections or amendments, as needed, are incorporated into future editions.

The authors are grateful to all those who helped to complete the book. In particular, a large portion of the material presented is the result of many years of work by the authors as well as other members of their research groups at the University of Michigan-Dearborn, Tennessee Technological University, and University of Denver. Thanks are due to the many dedicated staff and graduate students who made enormous contributions and provided supporting material for this book.

The authors also owe a debt of gratitude to their families, who gave tremendous support and made sacrifices during the process of writing this book.

Sincere acknowledgment is made to various sources that granted permission to use certain materials or pictures in this book. Acknowledgments are included where those materials appear. The authors used their best efforts to get approval to use those materials that are in the public domain and on open Internet web sites. Sometimes the original sources of the materials (in some web sites in particular) no longer exist or could not be traced. In these cases, the authors have noted where they found the materials and expressed their acknowledgment. If any of these sources were missed, the authors apologize for that oversight, and will rectify this in future editions of the book if brought to the attention of the publisher. The names of any product or supplier referred to in this book are provided for information only and are not in any way to be construed as an endorsement (or lack thereof) of such product or supplier by the publisher or the authors.

Finally, the authors are extremely grateful to John Wiley & Sons, Ltd and its editorial staff for giving them the opportunity to publish this book and helping in all possible ways. Finally, the authors acknowledge with great appreciation the efforts of the late Ms. Nicky Skinner of John Wiley & Sons, who initiated this book project on behalf of the publisher, but passed away in an untimely way very recently, and so did not see her efforts come to successful fruition.

Preface To the Second Edition

Although the first edition of this book was very well received by individuals, academic institutions, and others, the authors felt and the publisher also agreed that it would enrich the book and help the readership if we revised some of the materials in the first edition and also added some new items due to the introduction of new technologies in the vehicle electrification technology which has taken place over the past few years. With that in mind, the authors pursued the following activities.

In Chapters 1–11, we revised certain things overall, which included correcting a few relatively minor errors which we noticed. Chapter 6 has been significantly updated with important materials on off-road vehicles, with emphasis on excavators, which are relatively more complex in terms of architecture. Chapter 8 has also been updated to some extent. Chapter 11 on energy storage has been completely reorganized and rewritten to make it more application oriented. Chapter 12 in this edition is a new chapter, with focus on battery modeling. Chapter 13 is also a new chapter, related to battery charger design, which is an important issue in EV and PHEV. Chapters 12 and 13 from first edition have now become chapters 14 and 15, with minor changes incorporated. Chapter 16 is a completely new addition, related to wireless charging. Since wireless power transfer is a new technology and is under serious consideration in the automotive industry for charging of EV and PHEV, the authors felt that it is important to include it in this edition. Previous chapters 14 and 15 from the first edition have now become chapters 17 and 18 with some modifications. Finally, a new Chapter 19 has been added, which takes a holistic perspective on HEV and EV and discusses various viewpoints and pros and cons of introduction of HEV and EV. This chapter also discusses situations where EV and HEV may not necessarily be a good idea, as indicated by various researchers.

This second edition has been written by only the first two authors (Chris Mi and M. Abul Masrur) of the first edition, primarily due to various preoccupations of the third author (David Gao) since writing of the first edition of this book. The authors (Chris Mi and M. Abul Masrur) most sincerely appreciate the contribution of David Gao to the first edition which was very helpful in initiating the undertaking of this book writing project. The authors are also grateful to John Wiley Publishers (UK) who invited us to produce this second edition.

Finally, as is understandable, any text or reference book of this nature may have some inadvertent errors, which could be of typographic, grammatical, or of a technical nature. The authors would be most grateful if readers were to bring those to the notice of the publisher and/or the authors.

Chris Mi & M. Abul Masrur

Series Preface

Hybrid electric vehicles (HEVs) have been in existence for many years. One can see numerous HEVs on the road today, as they are quite commonplace. However, their presence extends well beyond the roads of the world. HEVs are seen on rails, on and beneath our seas, and in the air. The need for ever-increasing efficiency and reduced emissions continues to spur the growth of the HEV market sector as well as ever-improving and complex technologies in support of the expanding demands placed on HEV systems. Thus, the need to fully understand HEVs from an integrated systems perspective is critical for those who design next generation systems, not only in the automotive industry, but across all transportation sectors.

Modern Hybrid Electric Vehicles is a second-generation text that presents the hybrid electric vehicle from an integrated systems perspective. It is a well-balanced text that presents a system-level architecture of HEV, that includes design concepts, hardware, and critical aspects of HEV implementation including power usage and management strategies. The text is designed as part of an advanced engineering course in HEV systems and is part of the *Automotive Series* whose primary goal is to publish practical and topical books for researchers and practitioners in industry, and for postgraduates and advanced undergraduates in automotive engineering. The series addresses new and emerging technologies in automotive engineering, supporting the development of more fuel-efficient, safer and more environmentally friendly vehicles. It covers a wide range of topics, including design, manufacture, and operation, and the intention is to provide a source of relevant information that will be of use to leading professionals in the field.

Modern Hybrid Electric Vehicles provides a thorough technical foundation for HEV design, analysis, operation, and control. It also, incorporates a number of real-world concepts that are useful to the practicing engineer, resulting in a text that is an excellent blend of analytical concepts and pragmatic applications. The text goes beyond discussions of automobiles and extends the technical discussions to off-road vehicles, locomotives, ships and aircraft, making it an excellent reference for a wide spectrum of transportation systems designers. It also provides thorough insight into HEV system diagnostics, prognostics, and reliability from a traditional mechanical noise vibration harshness (NVH) viewpoint, and it also integrates issues related to electromechanical vibration and to electromagnetic compatibility and electromagnetic interference (EMC/EMI). Such topics are critical in HEV design, and are not typically covered in textbooks. Thus this text provides significantly new insights into HEVs. It is a well-written text, authored by recognized industrial and academic experts in a field that is critical to the transportation sector providing a thorough understanding of HEV systems from both design and implementation perspectives, and it is a welcome addition to the *Automotive Series.*

November 2016 *Thomas Kurfess*

1 Introduction

Modern society relies heavily on fossil fuel based transportation for economic and social development – freely moving goods and people. There are about 800 million cars in the world and about 260 million motor vehicles on the road in the United States in 2014 according to the US Department of Transportation's estimate [1]. In 2009, China overtook the United States to become the world's largest auto maker and auto market, with output and sales respectively hitting 13.79 and 13.64 million units in that year [2]. With further urbanization, industrialization, and globalization, the trend of rapid increase in the number of personal automobiles worldwide is inevitable. The issues related to this trend become evident because transportation relies heavily on oil. Not only are the oil resources on Earth limited, but also the emissions from burning oil products have led to climate change, poor urban air quality, and political conflict. Thus, global energy system and environmental problems have emerged, which can be attributed to a large extent to personal transportation.

Personal transportation offers people the freedom to go wherever and whenever they want. However, this freedom of choice creates a conflict, leading to growing concerns about the environment and concerns about the sustainability of human use of natural resources.

First, the world faces a serious challenge in energy demand and supply. The world consumes approximately 85 million barrels of oil every day but there are only 1300 billion barrels of proven reserves of oil. At the current rate of consumption, the world will run out of oil in 40 years [3]. New discoveries of oil reserves are at a slower pace than the increase in demand. Of the oil consumed, 60% is used for transportation [4]. The United States consumes approximately 25% of the world's total oil [5]. Reducing oil consumption in the personal transportation sector is essential for achieving energy and environmental sustainability.

Second, the world faces a great challenge from global climate change. The emissions from burning fossil fuels increase the carbon dioxide (CO_2) concentration (also referred to as greenhouse gas or GHG emissions) in the Earth's atmosphere. The increase in CO_2 concentration leads to excessive heat being captured on the Earth's surface, which leads to a global temperature increase and extreme weather conditions in many parts of the world. The long-term consequences of global warming can lead to rising sea levels and instability of ecosystems.

Hybrid Electric Vehicles: Principles and Applications with Practical Perspectives,
Second Edition. Chris Mi and M. Abul Masrur.

Gasoline and diesel powered vehicles are among the major contributors to CO_2 emissions. In addition, there are other emissions from conventional fossil fuel powered vehicles, including carbon monoxide (CO) and nitrogen oxides (NO and NO_2, or NO_X) from burning gasoline, hydrocarbons or volatile organic compounds (VOCs) from evaporated, unburned fuel, and sulfur oxide and particulate matter (soot) from burning diesel fuel. These emissions cause air pollution and ultimately affect human and animal health.

Third, society needs sustainability, but the current model is far from it. Cutting fossil fuel usage and reducing carbon emissions are part of the collective effort to retain human uses of natural resources within sustainable limits. Therefore, future personal transportation should provide enhanced freedom, sustainable mobility, and sustainable economic growth and prosperity for society. In order to achieve these, vehicles driven by electricity from clean, secure, and smart energy are essential.

Electrically driven vehicles have many advantages and challenges. Electricity is more efficient than the combustion process in a car. Well-to-wheel studies show that, even if the electricity is generated from petroleum, the equivalent miles that can be driven by 1 gallon (3.8 l) of gasoline is 108 miles (173 km) in an electric car, compared to 33 miles (53 km) in an internal combustion engine (ICE) car [6–8]. In a simpler comparison, it costs 2 cents per mile to use electricity (at US $0.12 per kWh) but 10 cents per mile to use gasoline (at $3.30 per gallon) for a compact car.

Electricity can be generated through renewable sources, such as hydroelectric, wind, solar, and biomass. On the other hand, the current electricity grid has extra capacity available at night when usage of electricity is off-peak. It is ideal to charge electric vehicles (EVs) at night when the grid has the extra energy capacity.

High cost, limited driving range, and long charging time are the main challenges for battery-powered EVs. Hybrid electric vehicles (HEVs), which use both an ICE and an electric motor to drive the vehicle, overcome the cost and range issues of a pure EV without the need to plug in to charge. The fuel consumption of HEVs can be significantly reduced compared to conventional gasoline engine-powered vehicles. However, the vehicle still operates on gasoline/diesel fuel.

Plug-in hybrid electric vehicles (PHEVs) are equipped with a larger battery pack and a larger-sized motor compared to HEVs. PHEVs can be charged from the grid and driven a limited distance (20–40 miles) using electricity, referred to as charge-depletion (CD) mode operation. Once the battery energy has been depleted, PHEVs operate similar to a regular HEV, referred to as charge-sustain (CS) mode operation, or extended range operation. Since most of the personal vehicles are for commuting and 75% of them are driven only 40 miles or less daily [9], a significant amount of fossil fuel can be displaced by deploying PHEVs capable of a range of 40 miles of purely electricity-based propulsion. In the extended range operation, a PHEV works similar to an HEV by using the onboard electric motor and battery to optimize the engine and vehicle system operation to achieve a higher fuel efficiency. Thanks to the larger battery power and energy capacity, the PHEV can recover more kinetic energy during braking, thereby further increasing fuel efficiency.

1.1 Sustainable Transportation

The current model of the personal transportation system is not sustainable in the long run because the Earth has limited reserves of fossil fuel, which provide 97% of all transportation energy needs at the present time [10]. To understand how sustainable

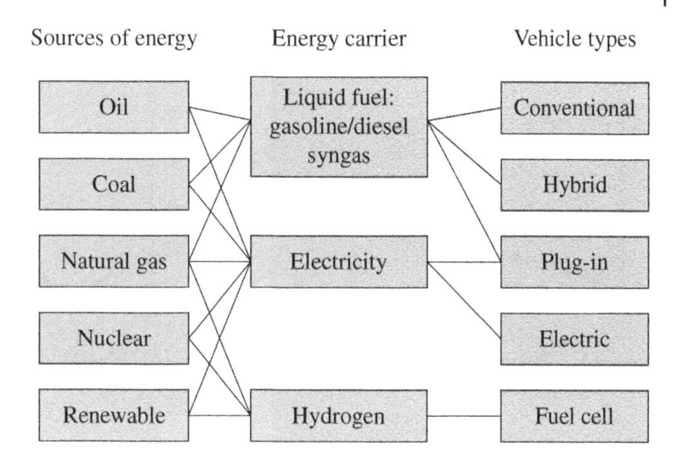

Figure 1.1 A sustainable.

transportation can be achieved, let us look at the ways energy can be derived and the ways vehicles are powered.

The energy available to us can be divided into three categories: renewable energy, fossil fuel-based non-renewable energy, and nuclear energy. Renewable energy includes hydropower, solar, wind, ocean, geothermal, biomass, and so on. Non-renewable energy includes coal, oil, and natural gas. Nuclear energy, though abundant, is not renewable since there are limited resources of uranium and other radioactive elements on Earth. In addition, there is concern on nuclear safety (such as the accident in Japan due to earthquake and tsunami) and nuclear waste processing in the long term. Biomass energy is renewable because it can be derived from wood, crops, cellulose, garbage, and landfill. Electricity and hydrogen are secondary forms of energy. They can be generated by using a variety of sources of original energy, including renewable and non-renewable energy. Gasoline, diesel, and syngas are energy carriers derived from fossil fuel.

Figure 1.1 shows the different types of sources of energy, energy carriers, and vehicles. Conventional gasoline/diesel-powered vehicles rely on liquid fuel which can only be derived from fossil fuel. HEVs, though more efficient and consuming less fuel than conventional vehicles, still rely on fossil fuel as the primary energy. Therefore, both conventional cars and HEVs are not sustainable. EVs and fuel cell vehicles rely on electricity and hydrogen, respectively. Both electricity and hydrogen can be generated from renewable energy sources, therefore they are sustainable as long as only renewable energy sources are used for the purpose. PHEVs, though not totally sustainable, offer the advantages of both conventional vehicles and EVs at the same time. PHEVs can displace fossil fuel usage by using grid electricity. They are not the ultimate solution for sustainability but they build a pathway to future sustainability.

1.1.1 Population, Energy, and Transportation

The world's population is growing at a rapid pace, as shown in Figure 1.2a [11]. At the same time, personal vehicle sales are also growing at a rapid pace, as shown in Figure 1.2b (www.dot.gov, also http://en.wikipedia.org/wiki/Passenger_vehicles_in_the_United_States). There is a clear correlation between population growth and the number of vehicles sold every year.

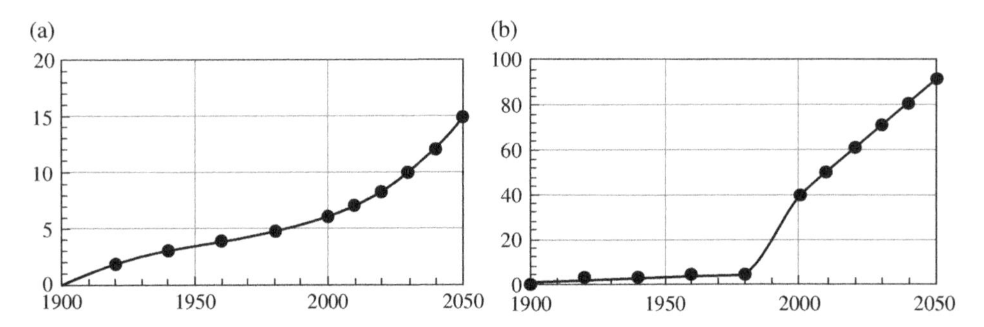

Figure 1.2 Trends of world population and vehicles sold per year. (a) World population, in billion. (b) Passenger cars sold per year, in millions.

Fuel economy, as used in the United States, evaluates how many miles can be driven with 1 gallon of gas, or miles per gallon (MPG). Fuel consumption, as used in most countries in the world, evaluates the gasoline (or diesel) consumption in liters for every 100 km the car is driven (l per 100 km). The US Corporate Average Fuel Economy Standard, known as the CAFÉ standard, sets the fuel economy for passenger cars at 27.5 MPG from 1989 to 2008 [12]. With an average 27.5 MPG fuel economy, an average 15,000 miles driven per year, and 250 million cars on the road, the United States would consume 136 billion gallons of gasoline per year. This is equivalent to 7 billion barrels of oil, or 0.5% of all the proven oil reserves on Earth.

China surpassed the United States in 2009 to become the largest vehicle market in the world, with more than 13 million motor vehicles sold in 2009. Growth in China has been in double digits for five consecutive years. In 2009, overall vehicle sales dropped 20% worldwide due to the global financial crisis, but China's car market still grew by more than 6%, along with its sustained economic growth of close to 10%. In 2016, China sold more than 27 million vehicles. China used to be self-sufficient in oil supplies, but is now estimated to import 50% of its oil consumption (http://data.chinaoilweb.com/crudeoil-import-data/index.html).

In addition to industrialized countries such as Japan and Germany which have high demand for oil imports, developing countries such as India and Brazil have also seen tremendous growth in car sales recently. These countries face the same challenges in oil demand and environmental aspects. Figure 1.3 shows liquid energy consumption and demand per day by country [13].

Figure 1.4 shows the history and projections of oil demand and production (http://www.eia.doe.gov/steo/contents.html). Many analysts believe in the theory of peak oil at the present time, which predicts that oil production is at its peak in history, and will soon be below oil demand. The gap generated by demand and production can most likely cause another energy crisis in the absence of careful planning.

1.1.2 Environment

Carbon emissions from burning fossil fuel are the primary source of GHG emissions that lead to global environment and climate change. Figure 1.5 shows the fossil carbon emissions from 1900 to the present time [14]. The most dramatic increase of GHG emissions has happened in the past 100 years. Associated with the increase of GHG emissions is the global temperature increase. Figure 1.6 shows the global mean

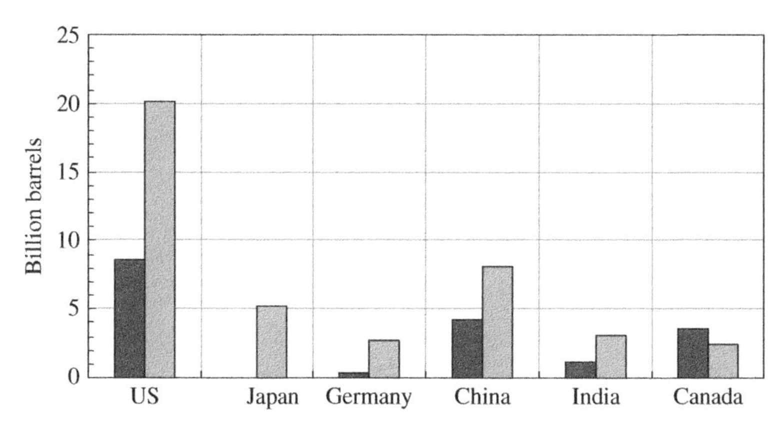

Figure 1.3 Average crude oil consumption per day by country in 2014, in million barrels. The left column for each country is the production and the right column is the consumption [13].

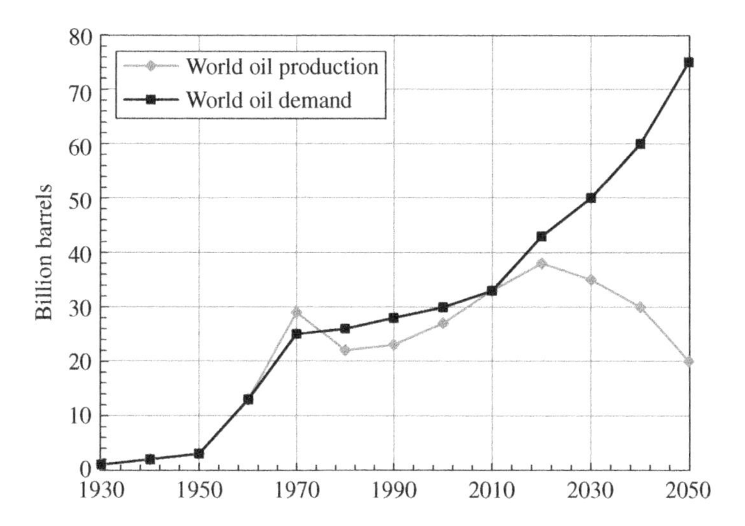

Figure 1.4 World oil demand and depletion history and projections.

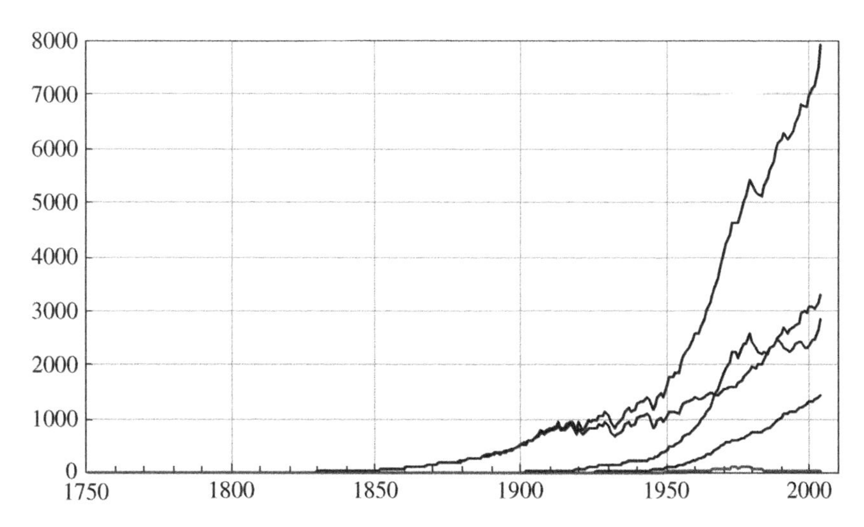

Figure 1.5 Global fossil carbon emissions from 1800 to 2004 [14]. On the right tip points, from top to bottom: total CO_2, oil, coal, cement production, and other. *Source:* ONRL.

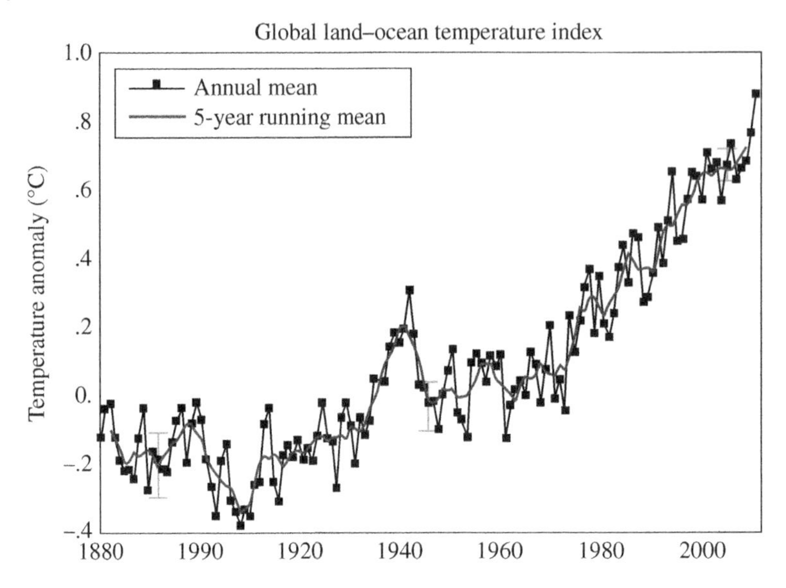

Figure 1.6 Global annual mean surface air temperature change. Data from http://data.giss.nasa.gov/gistemp/graphs/. Courtesy NASA.

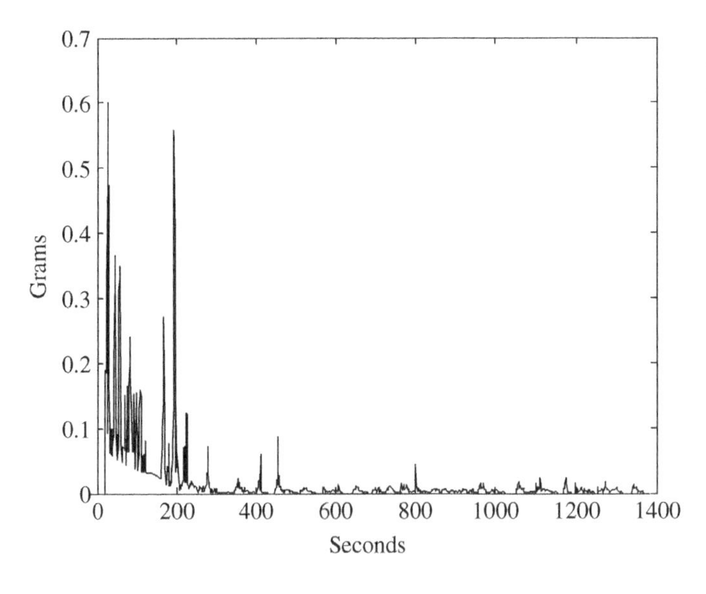

Figure 1.7 Typical emissions of a passenger car during cold starting (showing the total emissions in grams, made up of hydrocarbons, carbon monoxide, nitrogen oxide, and particulate matter).

land–ocean temperature change from 1880 to 2015, using the period of 1951–1980 temperature as the basis for comparison (http://data.giss.nasa.gov/gistemp/graphs/).

As an example of how car emissions contribute to GHG emissions, Figure 1.7 shows the emissions of a typical passenger car during a cold start. Modern cars are equipped with catalytic converters to reduce emissions from the car tailpipes/exhausts. But the catalytic converter needs to heat up to approximately 350°C in order to function efficiently. It has been estimated that 70–80% of the total emissions occur during the first two minutes after a cold start during a standard driving cycle.

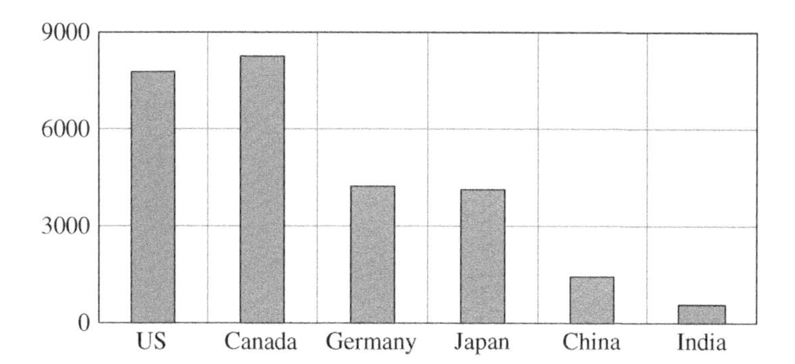

Figure 1.8 Energy consumption per capita in 2014 in kilograms of oil equivalent. (http://data.worldbank.org/indicator/EG.USE.PCAP.KG.OE?order=wbapi_data_value_2014+wbapi_data_value+wbapi_data_value-last&sort=desc)

1.1.3 Economic Growth

Economic growth relies heavily on energy supply. For example, from 1999 to 2015, China's economy attained an average growth rate of nearly 10%. In the same period, energy demand increased by more than 15% per year. In the early 1990s, China's oil production was sufficient to support its own economy, but by 2009, China imported a large portion of its oil consumption, estimated at 40% (http://data.chinaoilweb.com/crude-oil-import-data/index.html). China imports more than 50% of its liquid fuel consumption.

Figure 1.8 shows the energy consumption per capita, in kilograms of oil equivalent [13]. It is evident that developing countries are still well below the level of the developed countries. To reach sustainability, the global economy must embrace a new model.

1.1.4 New Fuel Economy Requirement

In 2009, the US government announced its new CAFÉ standard, requiring that all car manufacturers achieve an average fuel economy of 35 MPG by 2020 and 54.5 by 2030. This is equivalent to 6.7 l/100 km. The new requirement is a major increase in fuel economy in the United States in 20 years, and represents approximately a 40% increase from the current standard as shown in Figure 1.9. This new legislation is a major step forward to effectively reduce energy consumption and GHG emissions. To achieve this goal, a mixed portfolio is necessary for all car manufacturers.

First, auto makers must shift from large cars and pickup trucks to smaller vehicles to balance the portfolio. Second, they must continue to develop technologies that support fuel efficiency improvements in conventional gasoline engines. Lastly and most importantly, they have to increase HEV and PHEV production.

1.2 A Brief History of HEVs

EVs were invented in 1834, that is, about 60 years earlier than gasoline-powered cars, which were invented in 1895. By 1900, there were 4200 automobiles sold in the United States, of which 40% were electric cars (http://sites.google.com/site/petroleumhistoryresources/Home/cantankerous-combustion).

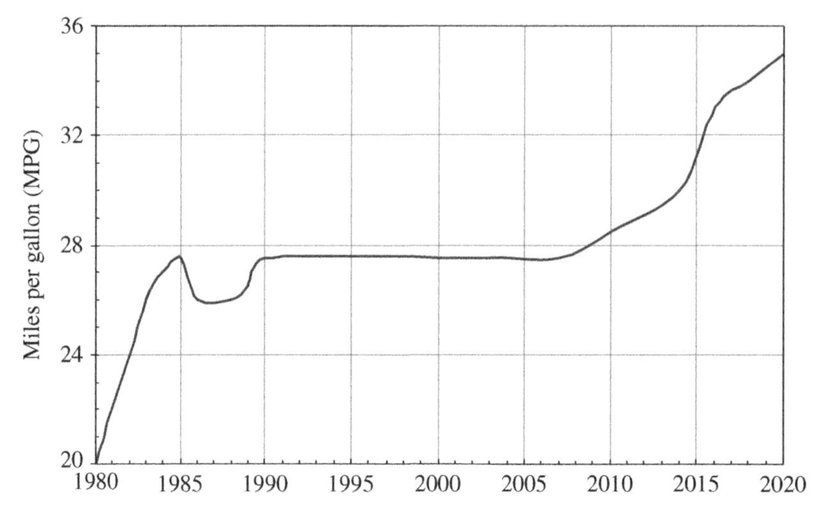

Figure 1.9 Fuel economy evolution in the United States (CAFÉ requirements).

Dr Ferdinand Porsche in Germany built probably the world's first HEV in 1898, using an ICE to spin a generator that provided power to electric motors located in the wheel hubs (http://aoghs.org/editors-picks/first-auto-show/). Another hybrid vehicle, made by the Krieger Company in 1903, used a gasoline engine to supplement the power of the electric motor which used electricity from a battery pack (http://www.hybridcars.com/history/history-of-hybrid-vehicles.html). Both hybrids are similar to the modern series HEV.

Also in the 1900s, a Belgian car maker, Pieper, introduced a 3.5 hp Voiturette in which the small gasoline engine was mated to an electric motor under the seat (http://en.wikipedia.org/wiki/Voiturette). When the car was cruising, its electric motor was used as a generator to charge the batteries. When the car was climbing a grade, the electric motor, mounted coaxially with the gas engine, helped the engine to drive the vehicle. In 1905, a US engineer, H. Piper, filed a patent for a petrol–electric hybrid vehicle. His idea was to use an electric motor to assist an ICE, enabling the vehicle to achieve 25 mph. Both hybrid designs are similar to the modern parallel HEV.

In the United States, there were a number of electric car companies in the 1920s, with two of them dominating the EV markets – Baker of Cleveland and Woods of Chicago. Both car companies offered hybrid electric cars. However, the hybrid cars were more expensive than gasoline cars, and sold poorly.

HEVs, together with EVs, faded away by 1930 and the electric car companies all failed. There were many reasons leading to the disappearance of the EV and HEV. When compared to gasoline-powered cars, EVs and HEVs:

- were more expensive than gasoline cars due to the large battery packs used
- were less powerful than gasoline cars due to the limited power from the onboard battery
- had limited range between each charge
- needed many hours to recharge the onboard battery.

Figure 1.10 Ford Electric Ranger.

In addition, urban and rural areas lacked accessibility to electricity for charging electric and hybrid cars.

The major progress in gasoline-powered cars also hastened the disappearance of the EV and HEV. The invention of starters made the starting of gasoline engines easier, and assembly line production of gasoline-powered vehicles, such as the Model-T by Henry Ford, made these vehicles a lot more affordable than electric and hybrid vehicles.

It was not until the Arab oil embargo in 1973 that the soaring price of gasoline sparked new interest in EVs. The US Congress introduced the Electric and Hybrid Vehicle Research, Development, and Demonstration Act in 1976 recommending the use of EVs as a means of reducing oil dependency and air pollution. In 1990, the California Air Resource Board (CARB), in consideration of the smog affecting Southern California, passed the zero emission vehicle (ZEV) mandate, which required 2% of vehicles sold in California to have no emissions by 1998 and 10% by 2003. California car sales have approximately a 10% share of the total car sales in the United States. Major car manufacturers were afraid that they might lose the California car market without a ZEV. Hence, every major auto maker developed EVs and HEVs. Fuel cell vehicles were also developed in this period. Many EVs were made, such as GM's EV1, Ford's Ranger pickup EV (Figure 1.10), Honda's EV Plus, Nissan's Altra EV, and Toyota's RAV4 EV.

In 1993, the US Department of Energy set up the Partnership for Next Generation Vehicle (PNGV) program to stimulate the development of EVs and HEVs. The partnership was a cooperative research program between the US government and major auto corporations, aimed at enhancing vehicle efficiency dramatically. Under this program, the three US car companies demonstrated the feasibility of a variety of new automotive technologies, including an HEV that can achieve 70 MPG. This program was cancelled in 2001 and was transitioned to the Freedom CAR (Cooperative Automotive Research), which is responsible for the HEV, PHEV, and battery research programs under the US Department of Energy.

Unfortunately, the EV program faded again away by 2000, with thousands of EV programs terminated by the auto companies. This is due partly to the fact that consumer acceptance was not overwhelming, and partly to the fact that the CARB relaxed its ZEV mandate.

The world's automotive history turned to a new page in 1997 when the first modern hybrid electric car, the Toyota Prius, was sold in Japan. This car, along with Honda's Insight and Civic HEVs, has been available in the United States since 2000. These early HEVs marked a radical change in the types of cars offered to the public: vehicles that take advantage of the benefits of both battery EVs and conventional gasoline-powered vehicles. At the time of writing, there are more than 40 models of HEVs available in the marketplace from more than 10 major car companies.

1.3 Why EVs Emerged and Failed in the 1990s, and What We Can Learn

During the 1990s, California had a tremendous smog and pollution problem that needed to be addressed. The CARB passed a ZEV mandate that required car manufacturers to sell ZEVs if they wanted to sell cars in California. This led to the development of electric cars by all major car manufacturers. Within a few years, there were more than 10 production EVs available to consumers, such as the GM EV1, the Toyota RAV4, and the Ford Ranger.

Unfortunately, the EV market collapsed in the late 1990s. What caused the EV industry to fail? The reasons were mixed, depending on how one looks at it, but the following were the main contributors to the collapse of EVs in the 1990s:

- **Limitations of EVs:** These concerned the limited range (most EVs provided 60–100 miles, compared to 300 or more miles from gasoline-powered vehicles); long charging time (eight or more hours); high cost (40% more expensive than gasoline cars); and limited cargo space in many of the EVs.
- **Cheap gasoline:** The operating cost (fuel cost) of cars is insignificant in comparison to the investment that an EV owner makes in buying an EV.
- **Consumers:** Consumers believed that large sports utility vehicles (SUVs) and pickup trucks were safer to drive and more convenient for many other functions, such as towing. Therefore, consumers preferred large SUVs to smaller efficient vehicles (partly due to the low gasoline prices).
- **Car companies:** Automobile manufacturers spent billions of dollars in research, development, and deployment of EVs, but the market did not respond very well. They were losing money in selling EVs at that time. Maintenance and servicing of EVs were additional burdens on the car dealerships. Liability was a major concern, though there was no evidence that EVs were less safe than gasoline vehicles.
- **Gas companies:** EVs were seen as a threat to gas companies and the oil industry. Lobbying by the car and gasoline companies of the federal government and the California government to drop the mandate was one of the key factors leading to the disappearance of EVs in the 1990s.
- **Government:** The CARB switched at the last minute from a mandate for EVs to hydrogen vehicles.
- **Battery technology:** Lead acid batteries were used in most EVs in the 1990s. The batteries were large and heavy, and needed a long time to charge.
- **Infrastructure:** There was limited infrastructure for recharging the EVs.

As we strive for a way toward sustainable transportation, lessons from history will help us avoid the same mistakes. In the current context of HEV and PHEV development, we must overcome many barriers in order to succeed:

- **Key technology:** That is, batteries, power electronics, and electric motors. In particular, without significant breakthroughs in batteries and with gasoline prices continuing at low levels, there will be significant obstacles to large-scale deployment of EVs and PHEVs.
- **Cost:** HEVs and PHEVs cost significantly more than their gasoline counterparts. Efforts need to be made to cut component and system cost. When savings in fuel can quickly recover the investment in the HEV, consumers will rapidly switch to HEVs and PHEVs.
- **Infrastructure:** This needs to be ready for the large deployment of PHEVs, including electricity generation for increased demand by PHEVs and increased renewable energy generation, and for rapid and convenient charging of grid PHEVs.
- **Policy:** Government policy has a significant impact on the deployment of many new technologies. Favorable policies including taxation, standards, consumer incentives, investment in research, development, and manufacturing of advanced technology products will all have a positive impact on the deployment of HEV and PHEV.
- **Approach:** An integrated approach that combines high-efficiency engines, vehicle safety, and smarter roadways will ultimately help form a sustainable future for personal transportation.

1.4 Architectures of HEVs

A HEV is a combination of a conventional ICE-powered vehicle and an EV. It uses both an ICE and an electric motor/generator for propulsion. The two power devices, the ICE and the electric motor, can be connected in series or in parallel from the power flow point of view. When the ICE and motor are connected in series, the HEV is a series hybrid in which only the electric motor is providing mechanical power to the wheels. When the ICE and the electric motor are connected in parallel, the HEV is a parallel hybrid in which both the electric motor and the ICE can deliver mechanical power to the wheels.

In an HEV, liquid fuel is still the source of energy. The ICE is the main power converter that provides all the energy for the vehicle. The electric motor increases system efficiency and reduces fuel consumption by recovering kinetic energy during regenerative braking, and optimizes the operation of the ICE during normal driving by adjusting the engine torque and speed. The ICE provides the vehicle with an extended driving range therefore overcoming the disadvantages of a pure EV.

In a PHEV, in addition to the liquid fuel available on the vehicle, there is also electricity stored in the battery, which can be recharged from the electric grid. Therefore, fuel usage can be further reduced.

In a series HEV or PHEV, the ICE drives a generator (referred to as the I/G set). The ICE converts energy in the liquid fuel to mechanical energy, and the generator converts the mechanical energy of the engine output to electricity. An electric motor will propel the vehicle using electricity generated by the I/G set. This electric motor is also used to capture the kinetic energy during braking. There will be a battery between the generator and the electric motor to buffer the electric energy between the I/G set and the motor.

In a parallel HEV or PHEV, both the ICE and the electric motor are coupled to the final drive shaft through a mechanical coupling mechanism, such as clutchs, gears, belts, or pulleys. This parallel configuration allows both the ICE and the electric motor to drive

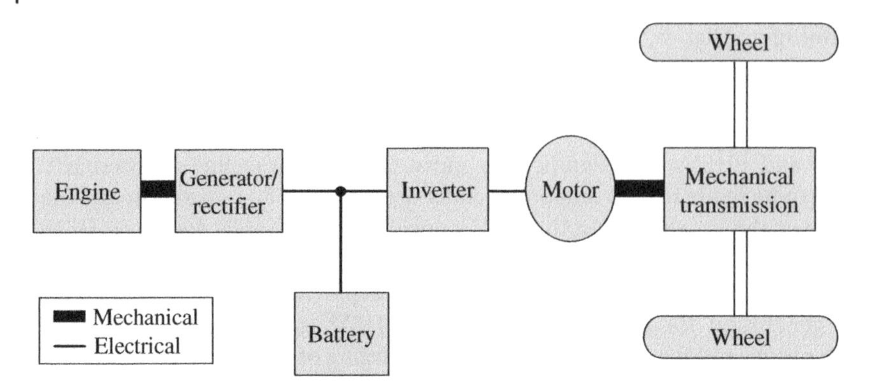

Figure 1.11 The architecture of a series HEV.

the vehicle, either in combined mode or separately. The electric motor is also used for regenerative braking and for capturing the excess energy from the ICE during coasting.

HEVs and PHEVs can also have either the series–parallel configuration or a more complex configuration which usually contains more than one electric machine. These configurations can generally further improve the performance and fuel economy of the vehicle with added component cost.

1.4.1 Series HEVs

Figure 1.11 shows the configuration of a series HEV. In this HEV, the ICE is the main energy converter that converts the original energy in gasoline to mechanical power. The mechanical output of the ICE is then converted to electricity using a generator. The electric motor moves the final drive using electricity generated by the generator or electricity stored in the battery. The electric motor can receive electricity directly from the engine, or from the battery, or both. Since the engine is decoupled from the wheels, the engine speed can be controlled independently of vehicle speed. This not only simplifies the control of the engine, but, more importantly, can allow the operation of the engine at its optimum speed to achieve the best fuel economy. It also provides flexibility in locating the engine on the vehicle. There is no need for the traditional mechanical transmission in a series HEV. Based on the vehicle operating conditions, the propulsion components on a series HEV can operate with different combinations:

- **Battery alone:** When the battery has sufficient energy, and the vehicle power demand is low, the I/G set is turned off, and the vehicle is powered by the battery only.
- **Combined power:** At high power demands, the I/G set is turned on and the battery also supplies power to the electric motor.
- **Engine alone:** During highway cruising and at moderately high power demands, the I/G set is turned on. The battery is neither charged nor discharged. This is mostly due to the fact that the battery's state of charge (SOC) is already at a high level but the power demand of the vehicle prevents the engine from off or it may not be efficient to turn the engine off.
- **Power split:** When the I/G is turned on, the vehicle power demand is below the I/G optimum power, and the battery SOC is low, then a portion of the I/G power is used to charge the battery.

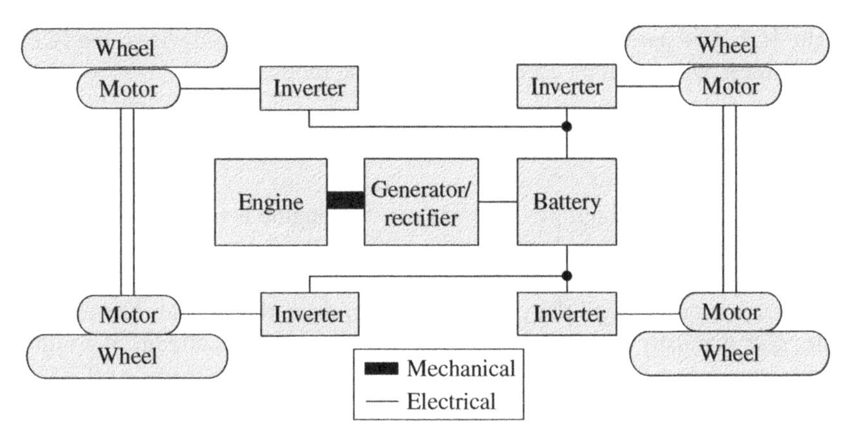

Figure 1.12 Hub motor configuration of a series HEV.

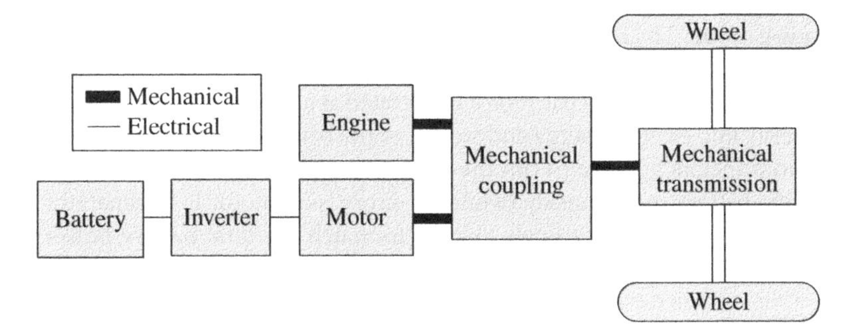

Figure 1.13 The architecture of a parallel HEV.

- **Stationary charging:** The battery is charged from the I/G power without the vehicle being driven.
- **Regenerative braking:** The electric motor is operated as a generator to convert the vehicle's kinetic energy into electric energy and charge the battery.

A series HEV can be configured in the same way that conventional vehicles are configured, that is, the electric motor in place of the engine as shown in Figure 1.11. Other choices are also available, such as in-wheel hub motors. In this case, as shown in Figure 1.12, there are four electric motors, one installed inside each wheel. Due to the elimination of transmission and final drive, the efficiency of the vehicle system can be significantly increased. The vehicle will also have all-wheel drive (AWD) capability. However, controlling the four electric motors independently can be a challenge.

1.4.2 Parallel HEVs

Figure 1.13 shows the configuration of a parallel hybrid. In this configuration, the ICE and the electric motor are coupled to the final drive through a mechanism such as clutchs, belts, pulleys, and gears. Both the ICE and the motor can deliver power to the final drive, either in combined mode, or each separately. The electric motor can be used as a generator to recover the kinetic energy during braking or by absorbing a portion of power from the ICE. The parallel hybrid needs only two propulsion devices, the ICE and the electric motor, which can be used in the following modes:

- **Motor-alone mode:** When the battery has sufficient energy, and the vehicle power demand is low, then the engine is turned off and the vehicle is powered by the motor and battery only.
- **Combined power mode:** At high power demands, the engine is turned on and the motor also supplies power to the wheels.
- **Engine-alone mode:** During highway cruising and at moderately high power demands, the engine provides all the power needed to drive the vehicle. The motor remains idle. This is mostly due to the fact that the battery SOC is already at a high level but the power demand of the vehicle prevents the engine from turning off, or it may not be efficient to turn the engine off.
- **Power split mode:** When the engine is on, but the vehicle power demand is low and the battery SOC is also low, then a portion of the engine power is converted to electricity by the motor to charge the battery.
- **Stationary charging mode:** The battery is charged by running the motor as a generator and driven by the engine, without the vehicle being driven.
- **Regenerative braking mode:** The electric motor is operated as a generator to convert the vehicle's kinetic energy into electric energy and store it in the battery. Note that in regenerative mode it is in principle possible to run the engine as well, and provide additional current to charge the battery more quickly (while the propulsion motor is in generator mode) and command its torque accordingly, that is, to match the total battery power input. In this case, the engine and motor controllers have to be properly coordinated.

1.4.3 Series–Parallel HEVs

The series–parallel HEV shown in Figure 1.14 incorporates the features of both a series and a parallel HEV. Therefore, it can be operated as a series or parallel HEV. In comparison to a series HEV, the series–parallel HEV adds a mechanical link between the engine and the final drive, so the engine can drive the wheels directly. When compared to a parallel HEV, the series–parallel HEV adds a second electric machine that serves primarily as a generator.

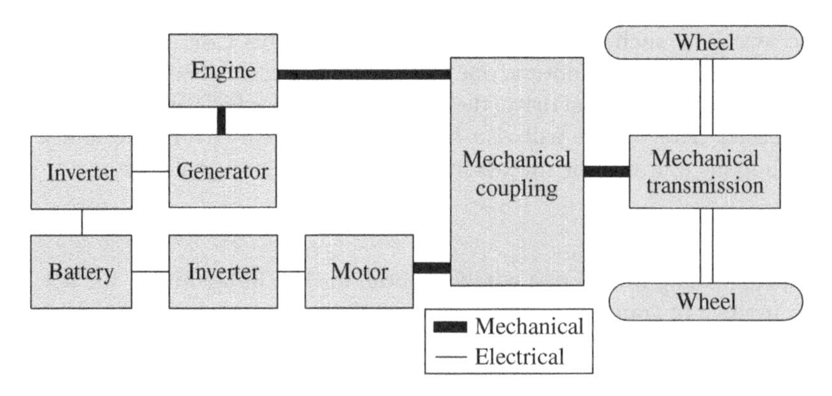

Figure 1.14 The architecture of a series–parallel HEV.

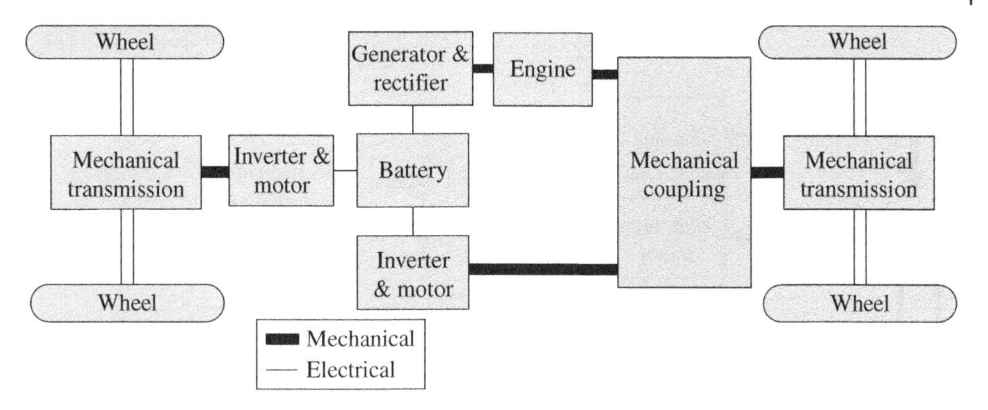

Figure 1.15 The electrical four-wheel drive system using a complex architecture.

Because a series–parallel HEV can operate in both parallel and series modes, the fuel efficiency and drivability can be optimized based on the vehicle's operating condition. The increased degree of freedom in control makes the series–parallel HEV a popular choice. However, due to increased components and complexity, a series–parallel HEV is generally more expensive than a series or a parallel HEV.

1.4.4 Complex HEVs

Complex HEVs usually involve the use of planetary gear systems and multiple electric motors (in the case of four/all-wheel drive). One typical example is a four-wheel drive (4WD) system that is realized through the use of separate drive axles, as shown in Figure 1.15. The generator in this system is used to realize the series operation as well as to control the engine operating condition for maximum efficiency. The two electric motors are used to realize all-wheel drive, and to provide better performance in regenerative braking. They may also enhance vehicle stability control and antilock braking control by their use.

1.4.5 Diesel and other Hybrids

HEVs can also be built around diesel vehicles. All topologies explained earlier, such as series, parallel, series–parallel, and complex HEVs, are applicable to diesel hybrids. Due to the fact that diesel vehicles can generally achieve a higher fuel economy, the fuel efficiency of hybridized diesel vehicles can be even better when compared to their gasoline counterparts.

Vehicles such as delivery trucks and buses have unique driving patterns and relatively low fuel economy. When hybridized, these vehicles can provide significant fuel savings. Hybrid trucks and buses can be series, parallel, series–parallel, or complex structured and may run on gasoline or diesel.

Diesel locomotives are a special type of hybrid. A diesel locomotive uses a diesel engine and generator set to generate electricity. It uses electric motors to drive the train. Even though a diesel locomotive can be referred to as a series hybrid, in some architectures there is no battery for the main drive system to buffer energy between the I/G set and the electric motor. This special configuration is sometimes referred to as

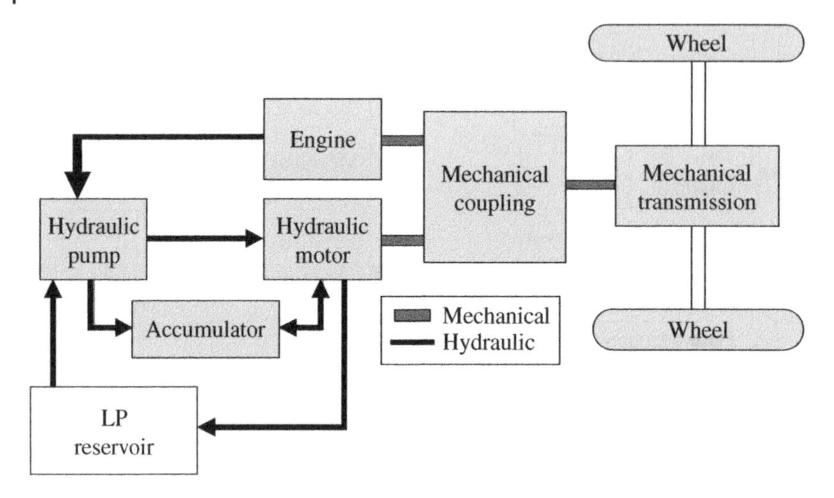

Figure 1.16 A parallel hydraulic hybrid vehicle (LP, Low Pressure).

simple hybrid. In other architectures, batteries are used and can help reduce the size of the generator, and can also be used for regenerative energy capture. The batteries, in this case, can also be utilized for short-term high current due to torque needs, without resorting to a larger generator.

1.4.6 Other Approaches to Vehicle Hybridization

The main focus of this book is on HEVs, that is, electric–gasoline or electric–diesel hybrids. However, there exist other types of hybridization methods that involve other types of energy storage and propulsion, such as compressed air, flywheels, and hydraulic systems. A typical hydraulic hybrid is shown in Figure 1.16. Hydraulic systems can provide a large amount of torque, but due to the complexity of the hydraulic system, a hydraulic hybrid is considered only for large trucks and utility vehicles where frequent and extended period of stops of the engine are necessary.

1.4.7 Hybridization Ratio

Some new concepts have also emerged in the past few years, including full hybrid, mild hybrid, and micro hybrid. These concepts are usually related to the power rating of the main electric motor in an HEV. For example, if the HEV contains a fairly large electric motor and associated batteries, it can be considered as a full hybrid. But if the size of the electric motor is relatively small, then it may be considered as a micro hybrid.

Typically, a full hybrid should be able to operate the vehicle using the electric motor and battery up to a certain speed limit and drive the vehicle for a certain amount of time. The speed threshold is typically the speed limit in a residential area. The typical power rating of an electric motor in a full hybrid passenger car is 50–75 kW.

The micro hybrid, on the other hand, does not offer the capability to drive the vehicle with the electric motor only. The electric motor is merely for starting and stopping the engine. The typical rating of electric motors used in micro hybrids is less than 10 kW. A mild hybrid is in between a full hybrid and a micro hybrid.

An effective approach for evaluating HEVs is to use a hybridization ratio to reflect the degree of hybridization of an HEV. In a parallel hybrid, the hybridization ratio is defined as

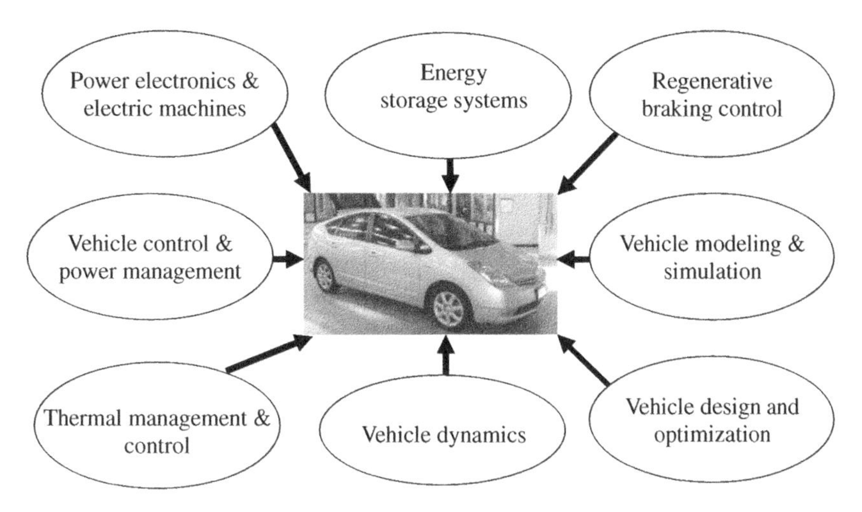

Figure 1.17 The general nature and required engineering field by HEVs.

the ratio of electric power to the total powertrain power. For example, an HEV with a motor rated at 50 kW and an engine rated at 75 kW will have a hybridization ratio of 50/(50 + 75) kW = 40%. A conventional gasoline-powered vehicle has a 0% hybridization ratio and a battery EV has a hybridization ratio of 100%. A series HEV will also have a hybridization ratio of 100% due to the fact that the vehicle is capable of being driven in EV mode.

1.5 Interdisciplinary Nature of HEVs

HEVs involve the use of electric machines, power electronics converters, and batteries, in addition to conventional ICEs and mechanical and hydraulic systems. The interdisciplinary nature of HEV systems can be summarized in Figure 1.17. The HEV field involves engineering subjects beyond traditional automotive engineering, which was mechanical engineering oriented. Power electronics, electric machines, energy storage systems, and control systems are now integral parts of the engineering of HEVs and PHEVs.

In addition, thermal management is also important in HEVs and PHEVs because the power electronics, electric machines, and batteries all require a much lower temperature to operate properly, compared to a non-hybrid vehicle's powertrain components. Modeling and simulation, vehicle dynamics, and vehicle design and optimization also pose challenges to the traditional automotive engineering field due to the increased difficulties in packaging the components and associated thermal management systems, as well as the changes in vehicle weight, shape, and weight distribution.

1.6 State of the Art of HEVs

In the past 20 years, many HEVs have been deployed by the major automotive manufacturers. Figure 1.18 shows HEV sales in the United States from 2000 to 2016, and predictions (http://electricdrive.org/ht/d/sp/i/20952/pid/20952). Figure 1.19

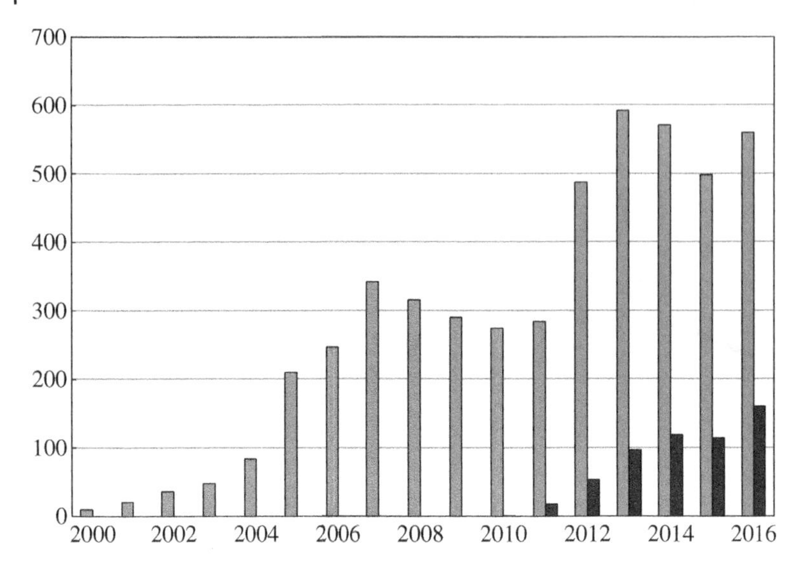

Figure 1.18 Total numbers of HEVs sold in the United States from 2000 to 2016 (in thousands): left bar, actual sales number; right bar, predicted.

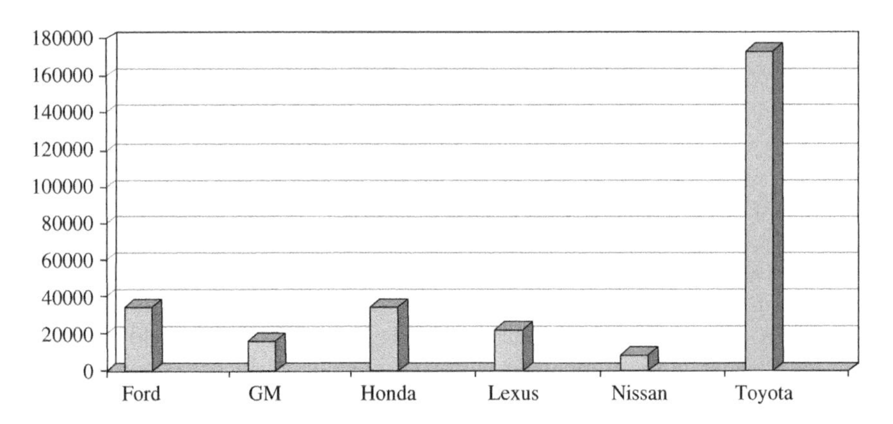

Figure 1.19 Breakdown of HEV sales by manufacturer in the United States in 2009 (in thousands).

shows the US HEV sales breakdown by manufacturer (https://www.google.com/url?sa=t&rct=j&q=&esrc=s&source=web&cd=2&ved=0ahUKEwjOkp-wtZvNAhVDF2MKHSWGAMQQFggmMAE&url=http%3A%2F%2Fwww.afdc.energy.gov%2Fuploads%2Fdata%2Fdata_source%2F10301%2F10301_hev_sales.xlsx&usg=AFQjCNEpgbwPbD7Y-swdSbwDq14QJHVCDg&sig2=ofMr8WxcjNxIC4d9mgc5Mw). It is clear that HEV sales have grown significantly over the past 20 years. In 2008, these sales had a downturn, which is consistent with conventional car sales which dropped more than 20% in 2008 from the previous year. Another observation is that most HEV sales belong to Toyota, which manufactured the earliest modern HEV, the Prius, and also makes most of the models available (including the Lexus).

Table 1.1 Hybrid Electric Vehicle (HEV) Sales by Model.

Vehicle	Pre 2010	2010	2011	2012	2013	2014	2015	Total
Toyota Prius	814,173	140,928	136,463	164,618	145,172	122,776	113,829	1,637,959
Toyota Camry	154,977	14,587	9,241	45,656	44,448	39,515	30,640	339,064
Honda Civic	197,177	7,336	4,703	7,156	7,719	5,070	4,887	234,048
Ford Fusion	15,554	20,816	11,286	14,100	37,270	35,405	24,681	159,112
Lexus RX400h	87,790	15,119	10,723	12,223	11,307	9,351	7,722	154,235
Toyota Prius C	–			30,838	41,979	40,570	38,484	151,871
Toyota Highlander	102,053	7,456	4,549	5,921	5,070	3,621	4,015	132,685
Toyota Prius V	–			28,450	34,989	30,762	28,290	122,491
Ford Escape	95,285	11,182	10,089	1,440				117,996
Hyundai Sonata	–		19,673	20,754	21,559	21,052	19,908	102,946
Honda Insight	34,490	20,962	15,549	5,846	4,802	3,562	1,458	86,669
Lexus CT 200 h	–		14,381	17,831	15,071	17,673	14,657	79,613
Ford C-Max Hybrid	–			10,935	28,056	19,162	14,177	72,330
Honda Accord	27,086				996	13,977	11,065	53,124
Lexus ES Hybrid	–			7,027	16,562	14,837	11,241	49,667
Kia Optima Hybrid	–			10,245	13,919	13,776	11,492	49,432
Toyota Avalon Hybrid	–			747	16,468	17,048	11,956	46,219
Ford Lincoln MKZ	–	1,192	5,739	6,067	7,469	10,033	8,403	38,903
Chevy Malibu	6,255	405	24	16,664	13,779	1,018	59	38,204
Nissan Altima	26,564	6,710	3,236	103				36,613
Buick Lacrosse	–		1,801	12,010	7,133	7,353	4,042	32,339
Honda CR-Z	–	5,249	11,330	4,192	4,550	3,562	3,073	31,956
Lexus HS 250 h	6,699	10,663	2,864	650	4			20,880
Mercury Mariner	11,916	890						12,806
Saturn Vue	9,979	50						10,029
Chevy Tahoe	7,045	1,426	519	533	376	65	8	9,972
Volkswagen Jetta Hybrid	–			162	5,655	1,939	740	8,496
Nissan Infiniti Q50	–				307	3,456	4,012	7,775
Lexus GS 450 h	4,576	305	282	615	522	183	91	6,574
Buick Regal	–		123	2,564	2,893	662	186	6,428
GMC Yukon	3,543	1,221	598	560	288	31	10	6,251
Cadillac Escalade	2,759	1,210	819	708	372	41	7	5,916
Chevrolet Sierra/ Silverado	1,598	2,393	1,165	471	104	24	2	5,757

(*Continued*)

Table 1.1 (Continued)

Vehicle	Pre 2010	2010	2011	2012	2013	2014	2015	Total
Subaru XV Crosstrek	–						5,589	5,589
Nissan Pathfinder Hybrid	–				334	2,480	2,245	5,059
Nissan Infiniti QX60	–				676	1,678	2,356	4,710
Porsche Cayenne	–	206	1,571	1,180	615	650		4,222
Lexus NX Hybrid	–					354	2,573	2,927
Mercury Milan	1,468	1,416						2,884
Acura ILX Hybrid	–			972	1,461	379	22	2,834
Lexus LS600hL	2,102	129	84	54	115	65	47	2,596
Nissan Infiniti M35h	–		378	691	475	180	176	1,900
Saturn Aura	1,584	54						1,638
Audi Q5 Hybrid	–			270	854	283	97	1,504
Toyota RAV4	–						1,494	1,494
BMW ActiveHybrid 3 (335ih)	–			402	905	151	30	1,488
Mercedes S400	–	801	309	121	64	10	1	1,306
Mazda Tribute		570	484	90				1,144
BMW ActiveHybrid 5 (535ih)	–			403	520	112	12	1,047
Chevy Impala Hybrid	–				56	565	272	893
VW Touareg Hybrid	–		390	250	118	30	16	804
BMW ActiveHybrid 7	–	102	338	230	31	45	25	771
Porsche Panamera S	–		52	570	113			735
Mercedes ML450h	–	627	1	22	11	20	10	691
Mercedes E400H	–				282	158	53	493
Acura RLX	–					133	250	383
BMW X6	–	205	43	3				251
Chrysler Aspen	79							79
GMC Sierra	–				65	6	1	72
Dodge Durango	9							9
TOTAL	1,614,761	274,210	268,807	434,344	495,534	443,823	384,404	3,915,883

(Data Sources: Worksheet and notes available at www.afdc.energy.gov/data/)

Table 1.2 shows the current HEVs available in the United States, along with a comparison to the base model of gasoline-powered cars (www.toyota.com, www.ford.com, www.gm.com, http://www.nissanusa.com/, www.honda.com, www.fca.com). In the case of the Toyota Prius, the comparison is made to the Toyota Corolla. It can be seen that the price of HEVs is generally 40% more than that of their base models. The increase in fuel economy in HEVs is also significant, in particular for city driving.

1.6.1 Toyota Prius

Toyota produced the world's first mass-marketed modern HEV in 1997, the Prius, as shown in Figure 1.20. The worldwide sales of the Prius exceeded 1 million units in 2009. The Toyota Prius uses a planetary gear set to realize continuous variable transmission (CVT). Therefore, the conventional transmission is not needed in this system. As shown in Figure 1.21, the engine is connected to the carrier of the planetary gear, while the generator is connected to the sun gear. The ring gear is coupled to the final drive, as is the electric motor. The planetary gear set also acts as a power/torque splitting device. During normal operations, the ring gear speed is determined by the vehicle speed, while the generator speed can be controlled such that the engine speed is in its optimum efficiency range.

The 6.5 Ah, 2.1 kW nickel metal hydride battery pack is charged by the generator during coasting and by the propulsion motor (in generation mode) during regenerative braking. The engine is shut off during low-speed driving.

The same technology is used in the Camry hybrid, the Highlander hybrid, and the Lexus brand hybrids. However, the Highlander and the Lexus hybrids add a third motor at the rear wheel. The drive performance, such as for acceleration and braking, can thus be further improved.

1.6.2 The Honda Civic

The Honda Civic hybrid has an electric motor mounted between the ICE and the CVT, as shown in Figure 1.22. The electric motor either provides assistance to the engine during high power demand, or splits the engine power during low power demand.

1.6.3 The Ford Escape

The Escape hybrid from the Ford Motor Company (Figure 1.23) is the first hybrid in the SUV category. The Escape hybrid uses the same planetary gear concept as the Toyota system.

1.6.4 The Two-Mode Hybrid

The GM two-mode hybrid transmission was initially developed by GM (Alison) in 1996, and later advanced by GM, Chrysler, BMW, and Mercedes-Benz through a joint venture named the Global Hybrid Cooperation in 2005. The GM two-mode hybrids (Figure 1.24) use two planetary gear sets and two electric machines to realize two

Table 1.2 Partial list of HEVs available in the United States (data from 2011).

Manufacturer	Model	HEV price (US $)	Base model price (US $)	Price increase (%)	HEV MPG		Base MPG		Increase in MPG (%)	
					City	Hwy	City	Hwy	City	Hwy
Toyota	Prius[a]	22 800	15 450	47.6	51	48	26	35	96	37
	Camry	26 400	19 595	34.7	33	34	22	33	50	3
	Highlander	34 900	25 855	35.0	27	25	20	27	35	12
Ford/Mercury	Fusion	27 950	19 695	41.9	41	36	22	34	86	6
	Escape	29 860	21 020	42.1	34	31	22	28	55	11
	Mariner	30 105	23 560	27.8	34	31	21	28	62	11
	Milan	31 915	21 860	46.0	41	26	23	34	78	−24
Honda	Insight[b]	19 800	15 655	26.5	40	43	26	34	54	26
	Civic	23 800	15 655	52.0	40	45	26	34	54	32
Nissan	Altima	26 780	19 900	34.6	35	33	23	32	52	3
Lexus	RX 450h	42 685	37 625	13.4	32	28	18	25	78	12
	GS 450h	57 450	54 070	6.3	22	25	17	24	29	4
	LS 600h	108 800	74 450	46.1	20	22	16	23	25	−4
GM GMC, Chevrolet, and Cadillac Saturn[c]	Tahoe	50 720	37 280	36.1	21	22	15	21	40	5
	Yukon	51 185	38 020	34.6	21	22	15	21	40	5
	Sierra	38 710	20 850	85.7	21	22	15	22	40	0
	Malibu	22 800	21 825	4.5	26	34	22	33	18	3
	Escalade	73 425	62 495	17.5	21	22	13	20	62	10
	Silverado	38 340	29 400	30.4	22	21	13	17	69	24
Chrysler Dodge	Aspen[d]	44 700	40 000	11.8	18	19	15	20	38	6
	Durango[d]	45 900	40 365	13.7	18	19	15	20	38	6

a) Comparison to Corolla
b) Comparison to Civic
c) Saturn Brand vehicle including Vue and Saturn Aura are not offered
d) Chrysler Aspen uses similar platform to Durango but is no longer offered

Figure 1.20 The Toyota Prius (2010 model).

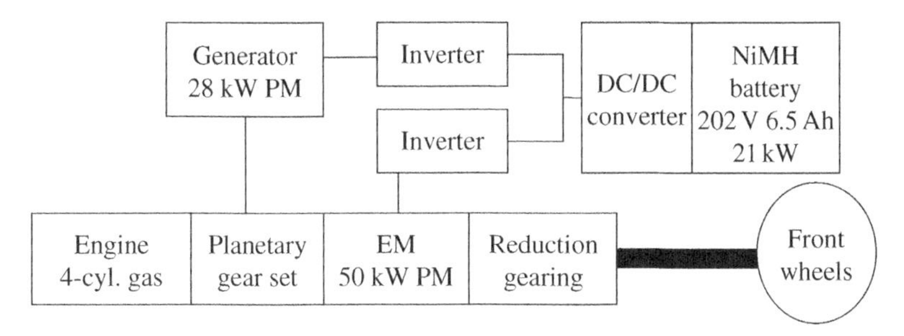

Figure 1.21 The powertrain layout of the Toyota Prius (EM, electric machine; PM, permanent magnet).

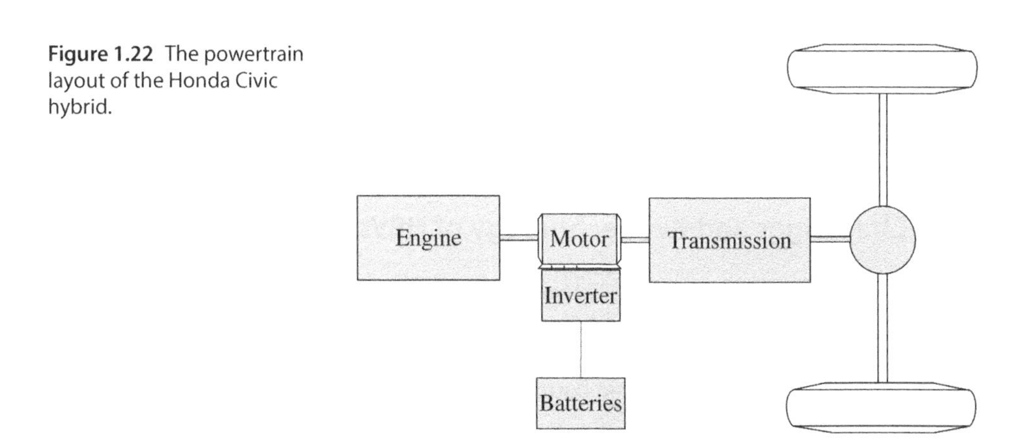

Figure 1.22 The powertrain layout of the Honda Civic hybrid.

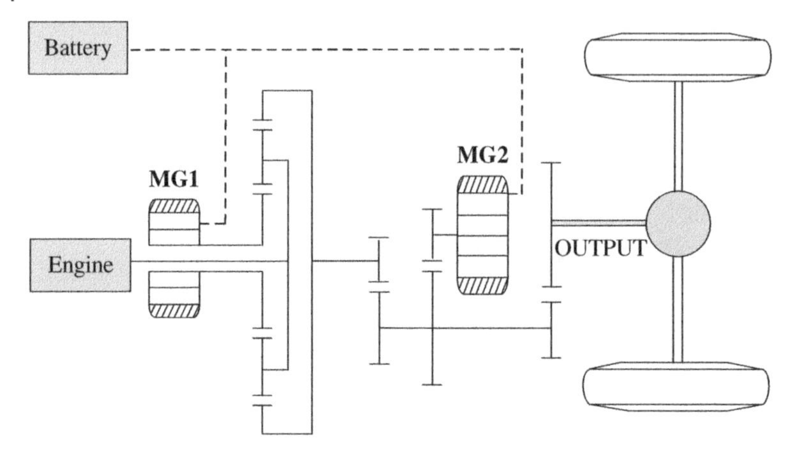

Figure 1.23 The Ford Escape hybrid SUV.

Figure 1.24 The Chrysler Aspen two-mode hybrid.

different operating modes, namely, high-speed mode and low-speed mode. Detailed operation of the two-mode hybrid will be discussed in Chapter 4.

1.7 Challenges and Key Technology of HEVs

HEVs can overcome some of the disadvantages of battery-powered pure EVs and gasoline-powered conventional vehicles. These advantages include optimized fuel economy and reduced emissions when compared to conventional vehicles, and increased range, reduced charging time, and reduced battery size (hence reduced cost) when compared to pure EVs.

However, HEVs and PHEVs still face many challenges, including higher cost when compared to conventional vehicles, electromagnetic interference caused by high-power components, and safety and reliability concerns due to increased components and complexity, packaging of the system, vehicle control, and power management.

- **Power electronics and electric machines:** The subject of power electronics and electric motors is not new. However, the use of power electronics in a vehicle environment poses significant challenges. Environmental conditions, such as extreme high and low temperatures, vibration, shock, and transient behavior are very different from what usually affects electric motors and power electronic converters. Challenges in power electronics in an HEV include packaging, size, cost, and thermal management.
- **Electromagnetic interference:** High-frequency switching and high-power operation of power electronics and electric motors will generate abundant electromagnetic noise that will interfere with the rest of the vehicle system if not dealt with properly.
- **Energy storage systems:** These systems are a major challenge for HEVs and PHEVs. The pulsed power behavior and energy content required for the best performance are typically difficult for conventional batteries to satisfy. Life cycle and abuse tolerance are also critical for vehicle applications. Currently, nickel metal hydride batteries are used by most HEVs and lithium-ion batteries are targeted by PHEVs. Ultracapacitors have also been considered in some special applications where power demand is a major concern. Flywheels have also been investigated. The limitations of the current energy storage systems are unsatisfactory power density and energy density, limited life cycle, high cost, and potential safety issues.
- **Regenerative braking control:** Recovering the kinetic energy during braking is a key feature of HEVs and PHEVs. However, coordinating regenerative braking with the hydraulic/frictional braking system presents a major challenge as far as safety and braking performance are concerned.
- **Power management and vehicle control:** HEVs involve the use of multiple propulsion components that require harmonious coordination. Hence, power management is a critical aspect of vehicle control functions in an HEV. An optimized vehicle controller can help achieve better fuel efficiency in an HEV.
- **Thermal management:** Power electronics, electric machines, and batteries all require a much lower operating temperature than a gasoline engine. A separate cooling loop is necessary in an HEV.
- **Modeling and simulation, vehicle dynamics, vehicle design, and optimization:** Due to the increased number of components in an HEV, packaging of the components in the same space is a challenge. Associated vehicle dynamics, vehicle design, and modeling and simulation all involve major challenges.

1.8 The Invisible Hand–Government Support

Without the government support, the HEV and PHEV may take a longer time to succeed in the marketplace due to their high cost and other limitations. As far as consumers are concerned, there are two kinds of buyers of HEVs. One kind expects to save money over time by saving fuel consumption; the other buys the hybrid because of environmental concerns. The payback period, or the time it takes for the owner to recover the investment in an HEV due to fuel savings, depends greatly on the price of gasoline.

Take the Toyota Prius as an example. This HEV is priced at $22,800 while the conventional similar model, the Corolla, is only $15,450. The MPG increases by 96% in city driving and 37% in highway driving. Assume that an owner annually drives an average of 10,000 miles in the city and 5000 miles on the highway. Then the annual fuel consumption for the base model and for the hybrid will be as follows:

Conventional:	10,000 miles/26 MPG + 5000/35 MPG = 527 gallons
HEV:	10,000 miles/51 MPG + 5000/48 MPG = 300 gallons

At $2.50 per gallon, the fuel savings are $568 per year. In other words, it would take 13 years to make up the cost difference between the two vehicles. However, if gasoline were to cost $6.5 per gallon, it would only take 5 years to recover the cost differences.

Consider the Ford Escape SUV as another example. This HEV is priced at $29,860 while the conventional model is priced at $21,020. Assume an owner drives an average of 10,000 miles in the city and 5000 miles on the highway annually. Then the annual fuel consumption for the base model and for the hybrid will be as follows:

Conventional:	10,000 miles/22 MPG + 5000/28 MPG = 633 gallons
HEV:	10,000 miles/34 MPG + 5000/31 MPG = 455 gallons

The total fuel saved is 178 gallons. At $2.50 per gal, the fuel savings are $455 per year, so it would take 19 years to make up the difference. However, if gasoline were to cost $6.5 per gallon, it would take 7.5 years to recover the cost difference.

Government incentives can make a difference in this scenario. Table 1.3 lists the tax credits available on certain PHEVs by the US federal government for the years 2011–2016. If we take the Ford Escape HEV as an example, when the tax credit is considered

Table 1.3 Available tax credits for selected PHEVs sold in the United States. (https://www.irs.gov/businesses/qualified-vehicles-acquired-after-12-31-2009).

Model year	Make	Model	Credit amount ($)
2011	BMW	I3	$7,500
2011–2017	GM	Chevrolet Volt	$7,500
		Cadillac ELR	$7,500
		Chevrolet Spark EV	$7,500
2013–2016	Mercedes-Benz	smart Coupe/Cabrio EV	$7,500
2011–2016	Nissan	Leaf	$7,500
2015	Porsche	Caynee S E-Hybrid	$5,336
2012–2015	Toyota	Prius PHEV	$2,500
		RAV4 EV	$7,500

the recovery time of investment would be 5 and 13 years, respectively, for gasoline prices of $6.5 and $2.5 per gallon.

A few states and metropolitan cities in the United States, such as California and Washington DC, allow access of HEVs to the high-occupancy-vehicle (HOV) lane. Due to the fact that a significant amount of time can be saved by driving in the HOV lane, some consumers are motivated to buy an HEV or PHEV for their daily commuting to save time.

Some other states, such as Colorado and California, have separate incentives in addition to those from the federal government. A number of private companies such as Google, Bank of America, STMicroelectronics, and Hyperion reward their employees for buying or leasing an HEV. It is interesting to note that the motivation of the companies toward such investment by their employees may be to allow them to spend time in the office rather than in traffic.

In 2009, the US president announced funding of $2.4 billion for PHEV development that supported 48 projects being undertaken by major car manufacturers and automotive suppliers. The projects support the development of batteries and electric drive components as well as the deployment of electric drive vehicles. The president of the United States set the goal of reaching 1 million plug-in hybrids by 2015. Additional incentives are in place for PHEV developers and consumers.

At the time of writing the first edition of this book, China, Japan, the European Union, and many other countries around the world have initiated support for the development, demonstration, and deployment of EVs, PHEVs, and associated battery and electric drive components.

However, hybridization of vehicles is not the ultimate solution for sustainability, though it builds the pathway to a sustainable future. The technology developed along this pathway will allow the necessary transition from fossil fuel based transportation to ultimate electrification of the transportation sector.

Electrification alone will not provide sustainability. There will be a need for a coordinated effort along this critical path, including smarter urban planning, public transient systems, high-speed rail networks, and smarter and safer vehicles and roads. There will also be a need for a collective effort by multiple stakeholders. The power industry must increase its renewable energy and cleaner energy generation. Governments will need to develop fast rail transportation systems in and between dense metropolitan areas.

1.9 Latest Development in EV and HEV, China's Surge in EV Sales

At the time of writing the second edition of this book, there have been significant advances in the electrification of the automobile:

- HEV, EV, and PHEV sales are steadily increasing, with PHEV and EV sales significantly increased over the past five years. The US saw 114,022 EV and PHEV sales and 384,404 HEV sales in 2015.
- China sold 21.15 million passenger cars in 2015, including sedans, sport-utility vehicles, and minivans, a 7.3% increase from 2014; and sold 23.9 million cars in 2016, a 15.9% gain over 2015. This is less than the 10% and 16% gain in 2013 and 2014, respectively. It is predicted that China will see a 7.8% increase in auto sales in 2016, increasing to 22.76 million vehicles for 2016.

- China had the largest surge of EV sales due to its massive government incentives, and sold more than 300,000 in 2015 (188,700 passenger cars and 124,000 electric buses) or 223% of the previous year's sales, not including more than 300,000 low-speed vehicle sales. (http://cleantechnica.com/2016/03/08/china-electric-car-sales-increased-223-in-2015/, and http://www.wsj.com/articles/china-car-sales-growth-slows-further-1452587244). It sold 350,000 electric vehicle and plug-in electric vehicles in 2017, almost half of all EV/PHEVs sold in the world.
- A number of companies have gone bankrupt in the past 6 years, while EV sales have steadily increased, including EV maker Fisker Karma, lithium-ion battery maker A123, and battery swapping company Better Place.
- The oil price is maintained low, reaching below $30 per barrel in 2016, 80% down from its peak in 2008.

Historically, the high oil prices have spiked the interest in electric vehicles. The current low price of oil will generate uncertainties for the future of electric and hybrid electric vehicles, partly due to the low gasoline prices. Take the example in the previous section where the gasoline price was set at $3.30 per gallon, the cost to drive with gasoline was almost five times the cost to drive using electricity. With the gasoline price at $1.65 per gallon, the cost to drive a car using gasoline is only 2.5 times that of electricity. With an average of 15,000 miles per year, it will save $375 per year for a typical compact sized car. Therefore, if the incremental cost of EVs (including battery cost) is at $5000, it will take nearly 15 years to recover the initial investment. Hence, economically, EV and HEV becomes less attractive.

Government incentives can completely change the scenario. For example, in 2015, China sold more than 300,000 EVs. However, in the last three months of that year they saw 75% sales of EV compared to the total of EVs sold in the first nine months. This surge is due to the fact that some government subsidies were due to expire at the end of the year. In some cases, the subsidy will cover more than 90% of the selling price of the car. However, China will gradually reduce the government subsidies, at the rate of 20% per year starting in 2016, resulting in no or little subsidies for EVs by 2020. We will have to see how the market and consumers react with the reduction in subsidies.

On the other hand, Tesla, for example, has gained a lot of momentum in its EV sales, especially the much anticipated Model 3, which starts at just $35,000, compared to $75,000 for a Model S. The Model 3 has already had more than 400,000 orders placed as of June 2016. But even with the sales volume realized, Tesla lost $889 million in 2015 and $282 million in the first quarter of 2016. (https://finance.yahoo.com/q/is?s=TSLA&annual, http://gizmodo.com/tesla-is-losing-money-but-making-more-cars-1758351727, http://seekingalpha.com/article/3968035-tesla-lose-money-every-car-sold).

The battery is still the bottle neck for EV penetration. Cost, energy density, durability, reliability, and safety are the major concerns for the currently available lithium-ion batteries. Other technologies, such as graphene material, which could potentially help to significantly increase battery energy density, are still to be further developed and validated. If indeed the energy density could be tripled or quadrupled in the next few years, and the cost could come down to 1/3 of the current cost, and EVs would be more competitive than gasoline-powered cars.

References

1 Table 1060. State Motor Vehicle Registrations: 1990 to 2007, http://www.statista.com/statistics/183505/number-of-vehicles-in-the-united-states-since-1990/(accessed June 11, 2016).
2 China's Auto Sales Rebound in August after July's Monthly Decline, http://news.xinhuanet.com/english2010/business/2010-09/09/c_13487102.htm (accessed June 11, 2016).
3 Owen, N.A., Inderwildi, O.R., and King, D.A. (2010) The status of conventional world oil reserves – hype or cause for concern? *Energy Policy*, 38, 4743 http://dx.doi.org/10.1016%2Fj.enpol.2010.02.026 (accessed June 11, 2011).
4 US Oil Demand by End-Use Sector (1950–2004), http://www.eia.doe.gov/pub/oil_gas/petroleum/analysis_publications/oil_market_basics/dem_image_us_cons_sector.htm (accessed June 11, 2016).
5 International Energy Outlook. United States Energy Information Administration (2007) Petroleum and Other Liquid Fuels, May, http://www.eia.doe.gov/oiaf/archive/ieo07/pdf/oil.pdf (accessed June 11, 2016).
6 Williamson, S.S. and Emadi, A. (2005) Comparative assessment of hybrid electric and fuel cell vehicles based on comprehensive well-to-wheels efficiency analysis. *IEEE Transactions on Vehicular Technology*, 54 (3), 856–862.
7 Imai, S., Takeda, N., and Horii, Y. (1997) Total efficiency of a hybrid electric vehicle. Proceedings of the Power Conversion Conference, Nagaoka.
8 Rousseau, A. and Sharer, P. (2004) Comparing Apples to Apples: Well-to-wheel Analysis of Current ICE and Fuel Cell Vehicle Technologies. Argonne National Laboratory, http://www.autonomie.net/docs/6%20-%20Papers/WTW/apples_to_apples.pdf (accessed June 11, 2016).
9 Sanna, L. (2005) Driving the Solution – the Plug-in hybrid Vehicle, http://mydocs.epri.com/docs/CorporateDocuments/EPRI_Journal/2005-Fall/1012885_PHEV.pdf (accessed June 11, 2016).
10 The Energy Report, http://comptroller.texas.gov/specialrpt/energy/(accessed June 11, 2016).
11 World Population, http://www.google.com/publicdata?ds=wb-wdi&met=sp_pop_totl&tdim=true&dl=en&hl=en&q=world+population (accessed June 11, 2016).
12 Annual Update on the Automotive Fuel Economy Program, http://www.nhtsa.gov/cars/rules/CAFE/updates.htm (accessed June 11, 2016).
13 The World Factbook, Oil Consumption by Country, http://data.worldbank.org/indicator/EG.USE.PCAP.KG.OE?order=wbapi_data_value_2014+wbapi_data_value+wbapi_data_value-last&sort=desc (accessed June 11, 2016).
14 Global, Regional, and National Fossil Fuel CO_2 Emissions, http://cdiac.esd.ornl.gov/trends/emis/tre_glob.html (accessed January 27, 2011).
http://cdiac.esd.ornl.gov/trends/emis/tre_glob.html (accessed January 27, 2011).
http://web.archive.org/web/20080508060713/

2

Concept of Hybridization of the Automobile

2.1 Vehicle Basics

2.1.1 Constituents of a Conventional Vehicle

Present-day engine-propelled automobiles have evolved over many years. Automobiles initially started with steam propulsion and later transitioned into ones based on the internal combustion engine (ICE). The focus of this chapter will be on ICE vehicles. So, the vehicles we use nowadays have diesel or gasoline (or petrol, as it is called in countries outside North America) engines. The engine provides the power to drive the vehicle. An illustration of an ICE is shown in Figure 2.1.

The engine has a chamber where gasoline or diesel is ignited, and as a consequence, when the gaseous air–fuel mixture explodes, it creates a very high pressure to drive the pistons. A piston is connected through a reciprocating arm to a crankshaft, as shown in Figure 2.1. The crankshaft is connected to a flywheel which is then connected to a transmission system. The purpose of the transmission system is to match the torque speed profile of the engine to the torque speed profile of the load. Figure 2.2 shows a simplified diagram of a transmission system connected to an engine and a few intermediate devices.

The shaft from the transmission system is ultimately connected to the wheels through some additional mechanical interfaces such as differential gears. The overall vehicular system is shown in Figure 2.3 with the basic system-level constituent elements in a present-day vehicle.

Figure 2.3 shows the cutaway views of a hybrid and a regular vehicle which indicate the complexity and tight packaging of numerous components within the confines of a small space in present-day automobiles.

2.1.2 Vehicle and Propulsion Load

The power generated by the engine is ultimately used to drive a load. In an automobile, this load includes the road resistance due to friction, uphill or downhill drive related to the road profile, and the environmental effect of, for example, the wind, rain, and snow. In addition, some of the energy developed in the vehicle is wasted in overcoming the internal resistance within the vehicle's various components and subsystems, none of which is 100% efficient. Examples of such subsystems or components include the

Hybrid Electric Vehicles: Principles and Applications with Practical Perspectives,
Second Edition. Chris Mi and M. Abul Masrur.

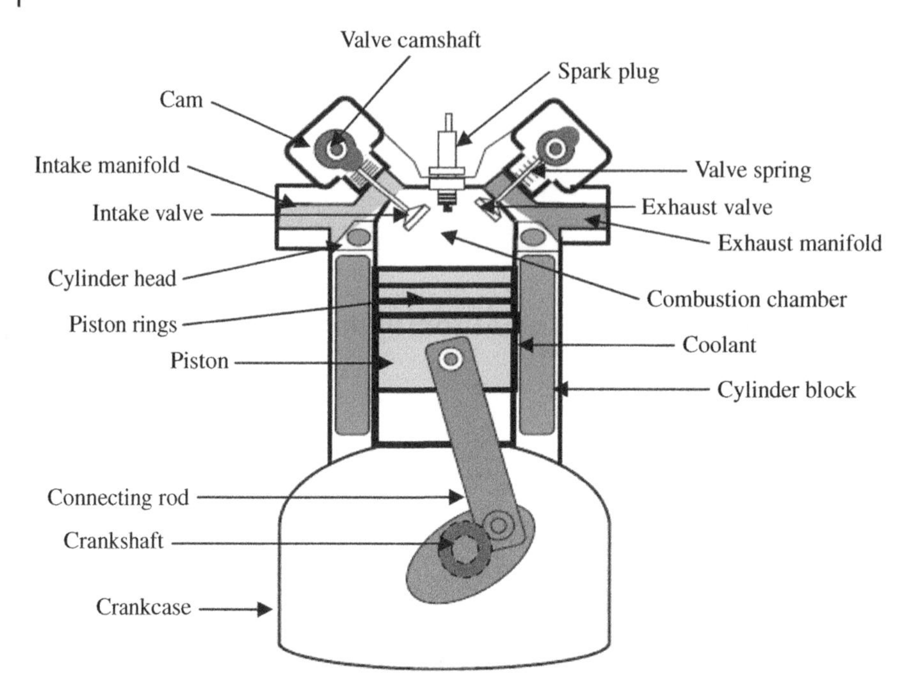

Figure 2.1 Cutaway view of an ICE.

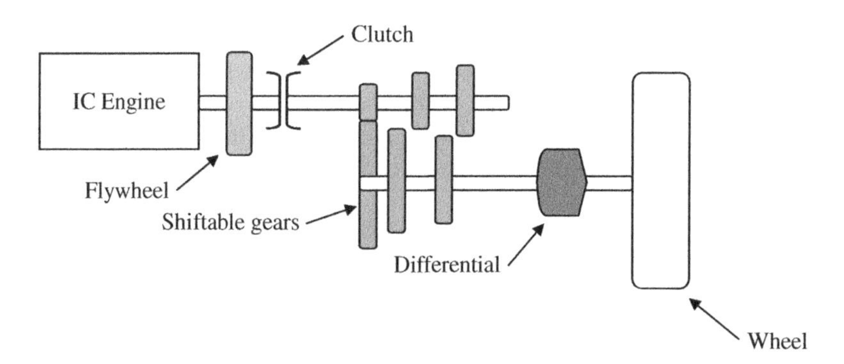

Figure 2.2 Transmission system and engine connected together.

radiator fan, various pumps, whether electrical or mechanical, motors for the wipers, and window lift. These items are just a few examples from of vehicular loads. The energy lost in these devices is released eventually as heat and expelled into the atmosphere.

Normally "load" can be related to the amount of opposing force or torque, but a more scientific definition of load comes from the fact that it is not defined by a single number or numerical value. Load is a collection of number defined by the speed–torque or speed–force characteristics in the form of a table or graph, that is, through a mathematical equation relating speed and torque. Similarly, the engine is also defined by speed–torque characteristics in the form of a table or graph, that is, through a mathematical

(a)

(b)

Figure 2.3 (a) Cutaway view of a Lexus RX 400h. (b) Cutaway view of a Lexus LS 400. Courtesy Wikimedia, http://en.wikipedia.org/wiki/File:Lexus_RX_400h_cutaway_model.jpg; http://commons.wikimedia.org/wiki/File:Lexus_Cutaway_LS_400.jpg

equation. The operating point of the combination of the engine and the load system together will then be at the intersection of these two characteristics. This situation is shown in Figure 2.4. It should also be noted that, for the engine, the torque speed curve is the relationship when a specific amount of fuel (corresponding to a specific accelerator pedal position) is going into the engine. Similarly, for the load, the torque speed curve is the relationship, for example under a specific brake pressure (i.e. brake pedal pressure in case it is a brake load), or a specific slope or gradient over which the vehicle is moving, and so on. In other words, the curve in Figure 2.4 is actually a single member out of a family of curves.

A complete vehicle or automotive system has various loads. Some of these are electrical devices, and others are mechanical devices. The electrical loads are normally run at a low voltage (nominally 12 V). The reason for running the non-propulsion loads at a low voltage is primarily related to safety issues and various standards that have to be followed. There is an existing manufacturing base for many of these non-propulsion loads (as indicated below), where it is easier to take advantage of the

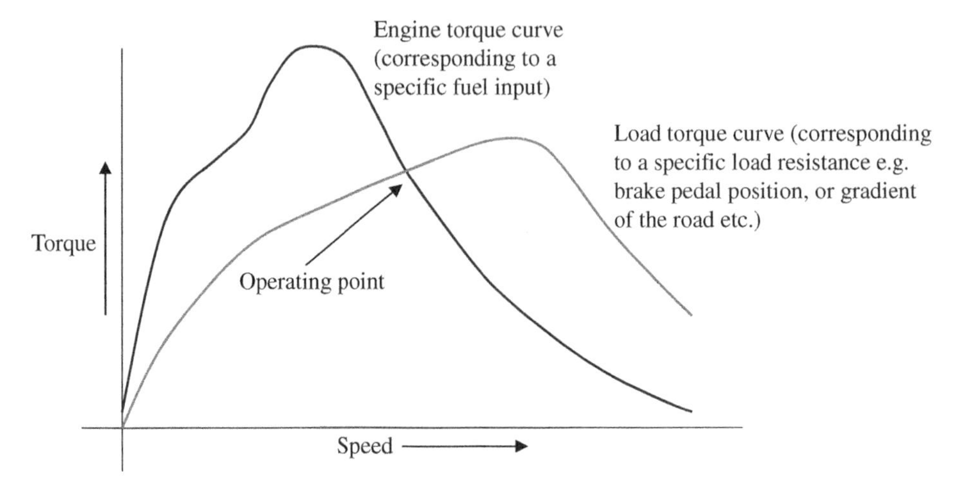

Figure 2.4 Load and engine characteristics of a vehicle.

situation, and use the existing low-voltage components, rather than transform the voltage system. Examples of these loads are:

- brakes – mechanical (hydraulic or low-voltage electrically assisted)
- air-conditioner – generally mechanical
- radiator fan – can be belt driven mechanically or driven by a low-voltage electric motor
- various pumps – can be mechanical or low-voltage electrical
- window lift – electrical
- door locks – electrical
- wipers – electrical
- various lights – non-motor load, low-voltage electrical
- radio, TV, GPS – non-motor load, low-voltage electrical
- various controllers – for example, engine controller, transmission controller, vehicle body controller
- various computational microprocessors – non-motor load, low-voltage electrical.

2.1.3 Drive Cycles and Drive Terrain

Since a vehicle will be driven through all kinds of road profiles and environmental conditions, to exactly know beforehand about which loads the vehicle will encounter under all circumstances is difficult. It is of course possible to perform experiments and place sensors to monitor the speed and torque of a vehicle, but to do so under all circumstances for all vehicle platforms is simply unrealistic, if not impossible. Hence, for the sake of engineering studies, a few limited situations have been developed, which more or less cover typical road profiles and the terrains normally encountered. Using a few of these profiles, we can create or synthesize various arbitrary road profiles. Such profiles can involve things like driving within a city, on a highway, across some special uphill or downhill terrain, to name a few. Drive cycles only provide time and corresponding speed fluctuations, with labels attached to these telling us what kind of drive cycle it is, for example, city or highway. Drive cycle data by itself does not reveal

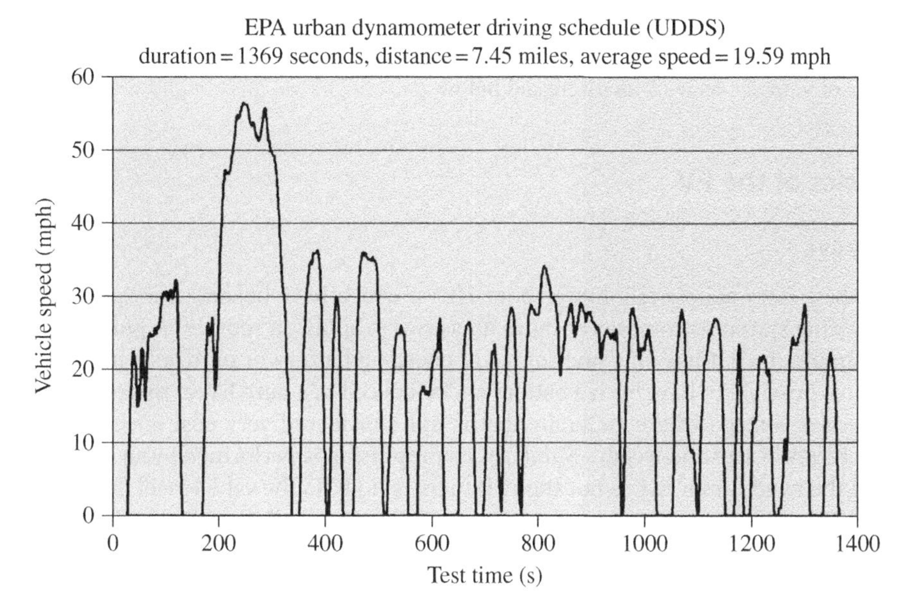

Figure 2.5 A typical automotive drive cycle. Courtesy US EPA, www.epa.gov/oms/standards/light-duty/udds.htm

the terrain; it is the label attached to the data that tells what the drive cycle is. One possible drive cycle is shown in Figure 2.5 (www.epa.gov/oms/standards/light-duty/udds.htm). It should be noted that the drive cycle data coupled with the nature of the terrain determine the torque (or the power) that the vehicle will need, to go through the particular terrain. For example, an identical torque speed curve on a flat road will need much lower power to drive through, compared to a road with a 30° slope with the same speed versus time profile.

Based on the above, it is easy to see that if a vehicle goes through different driving situations, partly city, partly highway, and so on, then one can obtain the cumulative speed vs. time data by synthesizing multiple typical drive cycles.

The question then arises about the ways to utilize the drive cycle information. Assume that we want to know about the fuel economy of a particular vehicle X. It is not sufficient to say that vehicle X achieves 25 MPG. We also need to say under what conditions this was obtained, that is, whether it was under a city drive cycle, or highway drive cycle, or something else. Only then can we make a fair comparison between vehicle X and another vehicle Y.

As there are different kinds of drive cycles, that of a passenger car cannot be compared with the drive cycle of a refuse truck or a postal mail vehicle, since they have very different kinds of stop and start driving. Similarly the drive cycle of a heavy mining vehicle is different again. We will say more about this in a later chapter on off-road vehicles. The bottom line is that drive cycles allow us to make a fair comparison between vehicles, in terms of fuel economy in particular, and also in terms of performance.

Finally, it should be noted that a drive cycle concerns the road profile through which a vehicle goes and hence is a situation external to the vehicle. However, the response of a vehicle to a given drive cycle, in terms of fuel economy, will be different depending on whether the vehicle is a regular ICE vehicle, fully electric vehicle (EV), hybrid electric vehicle (HEV), and so on. Hence the discussion on drive cycles above applies to the other types of vehicles as well, as discussed below.

2.2 Basics of the EV

2.2.1 Why EV?

Although these days people talk more about HEVs, which have become very popular, their underlying system is complex because it has two propulsion sources. A pure EV is relatively simple since it has only one source of energy, a battery or perhaps a fuel cell. Although it is possible to have both a battery and a fuel cell in a pure EV, at this time the technology is at a stage where generally that is not considered very cost effective, or efficient in terms of size and weight. Similarly, its propulsion is performed by an electric motor, and the need for an ICE is not there. If there is no ICE, the vehicle will not need any fuel injectors, various complicated engine controllers, and all the other peripherals associated with the engine and transmission. With a reduced parts count and a simpler system, it will also be more reliable [1].

In addition, an EV is virtually a zero-emission vehicle (only "virtually" since nothing has technically zero emissions in a true global sense). Of course, if we consider the ultimate source of energy, by tracing the path backward from the battery to the utility industry, it will be found that the location of pollution has been essentially shifted from the vehicle to elsewhere. Furthermore, an EV is very quiet. In fact, it can be so quiet that people have even talked about introducing artificial noise into the vehicle so that they can hear it, which is something important to know from a safety point of view. It should also be recognized that if the battery or the fuel cell technology – the single weak link in terms of technological maturity – were fully mature, that is, able to provide the necessary power and energy within a compact size, weight and cost, the HEV would probably not be necessary. Obviously, the EV is on everyone's wish list and can be considered to be the real culminating point of automotive technology.

From a technical viewpoint, the EV has another benefit. In the ICE, which is a reciprocating engine, the torque produced is pulsating in nature. The flywheel helps smooth the torque, which would otherwise cause vibration. In the EV, the motor can create a very smooth torque and, in fact, it is possible to do away with the flywheel, thus saving material and manufacturing cost, in addition to reducing weight. Finally, the efficiency of an ICE (gasoline to shaft torque) is very low. The engine itself has about 30–37% efficiency for gasoline and about 40% for diesel, but by the time the power arrives at the wheel, the efficiency is just 5–10%. But the efficiency of the electric motor is very high – around 90%. The battery and power electronics to drive the motor also have high efficiency. If each of these components has efficiencies on the order of 90%, by the time the battery energy leaves the motor shaft and reaches the wheel, the overall efficiency will be something like 70%, substantially higher than that of the ICE.

2.2.2 Constituents of an EV

As noted earlier, an EV is simpler than an HEV. Its basic system-level constituents are shown in Figure 2.6.

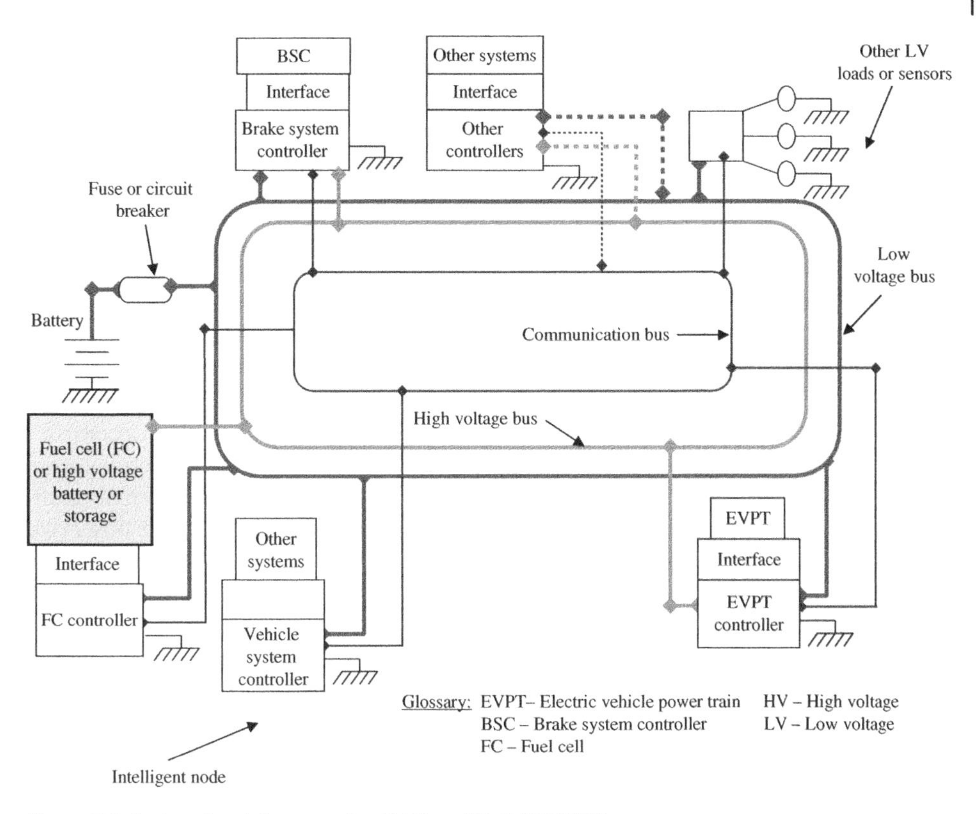

Figure 2.6 System-level diagram of an EV. From [1], © 2008 IEEE.

The complete EV consists of not only the electric drive and power electronics for propulsion, but also other subsystems to make the whole system work. In Figure 2.6, we need a battery (or a fuel cell) to provide the electrical energy. This is shown by the block on the left, which provides power to drive the electric motor. The motor is part of the EV powertrain, labeled as EVPT on the right. For each of these items, battery or EVPT, there is a controller. The battery controller controls the charging or discharging, and similarly the EVPT controller controls the speed or torque of the motor by controlling the power electronics. It should be realized that even though the blocks in the diagram are shown to be at quite different positions, in reality some of those could be physically in close proximity. This is due to packaging issues, and also, by positioning them nearby, it is possible to reduce the high-current and/or high-voltage cable lengths. Similarly, although the FC controller (for fuel cell controller, or it would be a battery controller if a battery is used) and the EVPT controller are shown separately to indicate separate functionality, in reality they could be part of the same physical box and could even use the same microprocessors to achieve their functions. These aspects are subtle design issues involving cost and packaging. In addition, there is a box shown as the interface. This is the controller box that receives signals and also power, both high voltage for propulsion and low voltage for certain specific devices which operate at low voltage, and then through the interface function channels them to the EVPT motor or the high-voltage battery. Again, these functionally separate blocks may be merged when physically integrating the system.

In addition to the above blocks, there are various other blocks, for example, the vehicle controller, which can receive signals such as the velocity of the vehicle, and driver pedal position, and make a decision whether or not additional torque is needed from the motor. Based on that information it can send a signal to the EVPT controller with the appropriate torque request. Similarly, the brake controller can receive signals corresponding to brake pedal position, vehicle velocity, and so on, and decide how much brake force is needed. It can also receive signals such as the battery's state of charge and figure out whether the opportunity of regenerative braking is present. If so, it can then send signals to the EVPT controller to carry this out. All the above illustrate the importance of continuous flow of information and signals between various control blocks and corresponding decision-making in each subsystem.

Information transmission between various blocks is normally done through a controller area network (CAN) bus. This is basically a type of computer network where a single wire (which could also be a twisted pair of wires) can contain a variety of information or communication signals multiplexed together. Some sort of protocol has to be observed when multiple signals are passed through and shared by a common medium. In other words, there is some sort of priority-based signal flow when trying to share the same physical medium. For relatively slow signals – for example, to turn the door lock switch – we can afford to wait, whereas for very important functions such as braking and steering, which are safety functions, the signals need to be transmitted immediately. There are some newer protocols that can allow such activities. In addition, for safety-critical functions, it may be necessary to have additional hardware-based backup communication mechanisms, so as to avoid failures.

2.2.3 Vehicle and Propulsion Loads

There is a significant amount of commonality between the loads in an EV and a regular automobile. Hence, just like a regular vehicle, some of these loads are electrical devices and others are mechanical devices. As noted earlier, those loads that are normally electrical run at a low voltage (nominally 12 V), with the exception of the propulsion load, that is, the propulsion motor, which runs at a high voltage (several hundred volts). These low voltage loads for non-propulsion purposes are also referred to as auxiliary loads. The reason for using low voltage for auxiliary loads is primarily safety related. And, of course, the existing manufacturing base for many of these non-propulsion loads can be an advantage, by using the existing low-voltage components, rather than transforming the voltage system. Examples of these auxiliary loads are the same as those noted in Section 2.1.2:

- brake motor (if a fully or partially electrical brake system is used) – low voltage
- air-conditioner motor (if electrical) – low voltage
- radiator fan (if electrical) – low voltage
- various pumps (if electrical) – low voltage
- window lift – low voltage
- door locks – electrical
- wipers – electrical
- various lights – non-motor load, low-voltage electrical
- radio, TV, GPS – non-motor, low-voltage electrical

- various controllers, such as engine controller, transmission controller, vehicle body controller
- various computational microprocessors, digital signal processors (DSPs) – non-motor, low-voltage electrical

The propulsion load can be several kilowatts for a mild hybrid vehicle regenerative braking system, up to say 50 kW, or a few hundred kilowatts for propulsion in a hybrid vehicle. The various pumps and fans can be only a few hundred watts or less, whereas some small motors such as door lock motors could be just a few tens of watts. Similarly the lights can range from a few tens to about a hundred watts.

The above loads are fed by the battery, the generator, or a combination of the two. We can also see from Figure 2.6 that the non-propulsion loads are fed by the low-voltage battery. This low voltage can be derived either by using part of the main high-voltage battery system, or through the transformation of the high-voltage system by a DC–DC down converter; it could even be a totally separate low-voltage battery with its own generator system to charge it. In other words, more than one architecture is possible for the low-voltage system.

2.3 Basics of the HEV

2.3.1 Why HEV?

In the previous section we discussed the architecture of a purely EV. As we saw, the EV propulsion uses an electric motor. The energy comes from the battery (or perhaps a fuel cell). The battery bank in a pure EV can be quite large if the vehicle has to go a few hundred miles on one charge. The reason for this is that battery technology, as it stands today, does not have a very high energy density in terms of weight and size, compared to a liquid fuel such as gasoline. Although new batteries technologies such as lithium-ion have a much higher energy density than lead acid or nickel metal hydride, it is still much lower than liquid fuel.

As noted earlier, the HEV is a complex system since it has at least two propulsion sources, whereas a pure EV has only one source of energy, a battery or fuel cell. In the EV, the propulsion is produced only by the electric motor and there is no ICE. This removes the need for fuel injectors, complicated engine controllers, and other peripherals. Hence, with a reduced parts count, the system is simpler and more reliable [1].

Of course, there is an efficiency improvement in the HEV compared to the ICE, but it will still be lower than in the EV. The overall efficiency will depend on the relative size of the ICE and the electric propulsion motor power.

A variant of the HEV is found in locomotives and in very high-powered off-road vehicles. In a number of variants of such systems there is no battery. The ICE is used to drive a generator which creates (uncontrolled) AC (alternating current) power. This power is translated to DC and then to another (controlled) AC power required to drive an electric motor. The problem with this system is that the engine has to be run continuously to produce the electricity. The advantage is that it does not need a battery. Furthermore, the ICE can be run at an optimal speed to achieve the best possible efficiency. One problem with this system is that it does not lend itself to regenerative energy recovery during braking. The battery helps regenerative energy recovery by allowing storage, and it can

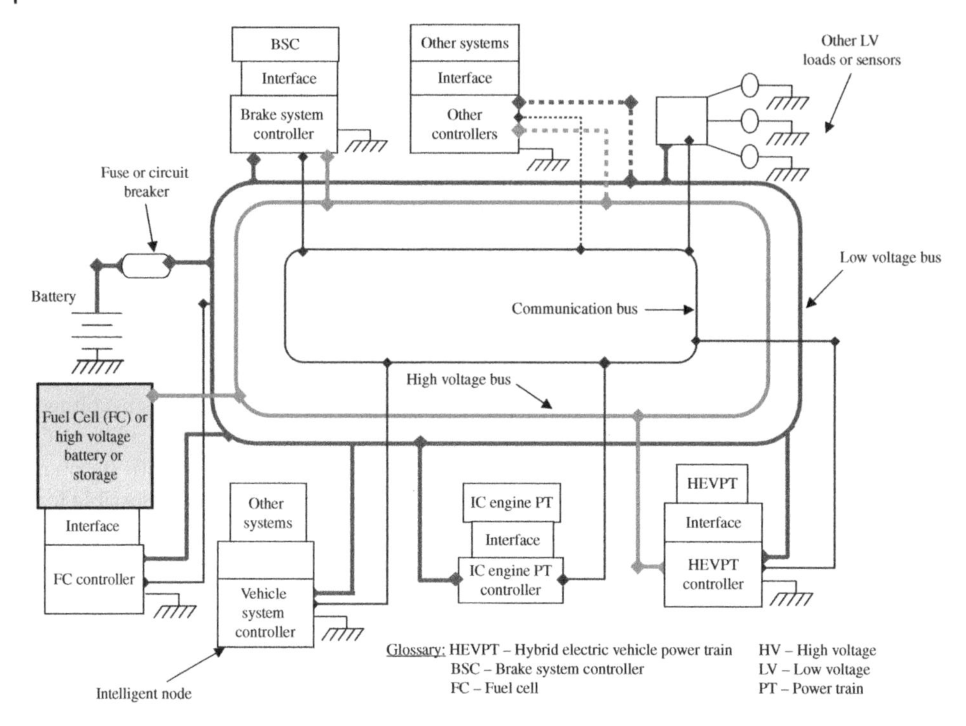

Figure 2.7 System-level diagram of an HEV. From [1], © 2008 IEEE.

also be coordinated more optimally in terms of when the ICE and/or the electric motor should be run. More on this will be presented in a later chapter.

2.3.2 Constituents of an HEV

As noted earlier, an EV is simpler than an HEV. The basic system-level constituents for an HEV are shown in Figure 2.7 [1].

As we can see, the only difference between this diagram (Figure 2.7) and the one for the EV (Figure 2.6) is that this one has an additional subsystem called IC engine, along with the necessary interface and the controller. Otherwise the two diagrams are identical.

All the other items and the discussion are the same as for the EV.

2.4 Basics of Plug-In Hybrid Electric Vehicle (PHEV)

2.4.1 Why PHEV?

The PHEV, as the name suggests, differs from an HEV only by the fact that it can be plugged in, via a cable, to a household utility wall socket or elsewhere, to charge the vehicle's battery. To extend the flexibility of the system, it is also possible in principle to use the engine and/or the battery system in the vehicle to generate AC power and feed it back to the utility grid. Since the plug-in allows a fair amount of external utility system energy to drive the vehicle, it is helpful to use a larger battery (in terms of energy content) than in a regular HEV. A larger battery is not a required part of the PHEV propulsion, but having one definitely benefits fuel economy and also increases the range of the vehicle. In an HEV, using a much larger battery may not necessarily be the optimal choice in terms of design, since the ICE is always capable of kicking in, when the battery needs to be charged. People sometimes think that a large battery is mandatory for a PHEV, which may not be the case. How large the battery can be, depends on the packaging space available in the vehicle. If the battery size is small, then the benefits of the PHEV will be merely incremental, whereas if it is too big, it can be very expensive and will take longer to recharge from the utility system. Note also that the household utility system may have some limitations on how much current it can sustain in charging a battery system, and hence some safeguards are necessary for the plug-in. Since the cost of utility energy at present is much lower than the price of gasoline, it makes sense to use the PHEV where possible.

2.4.2 Constituents of a PHEV

The same diagram as before, that is, Figure 2.6 (for the EV), applies to a PHEV. The only difference is that it now has an extra connecting socket in the vehicle, from where a lead can be pulled out and plugged into the wall utility outlet. Obviously, when the vehicle is connected to a utility outlet, its propulsion motor is not needed and neither is the ICE, as far as turning the wheels is concerned. However, the vehicle may still need to use auxiliary loads (normally low-voltage loads at 12 V), the air-conditioner (can be low voltage as well), or the heater and some lights. Hence it is appropriate to deliver those loads at low voltage. If fast charging of the battery is necessary, it will also be appropriate to run the ICE and use the propulsion motor as a generator, or have a separate generator for this purpose. Depending on the scheme used, changes in the gear train system are called for. Even though the whole process of interconnection between the utility and the PHEV system is simple in principle, there are quite a few considerations to be taken into account, as will be obvious from the possible architecture for such a vehicle shown in Figure 2.8.

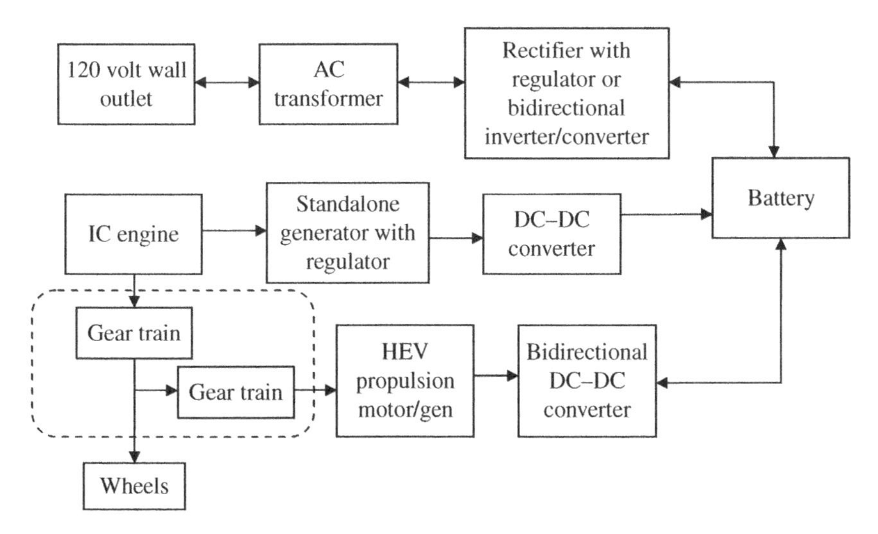

Figure 2.8 A possible architecture for the plug-in hybrid vehicle and home outlet interface.

From Figure 2.8 it is apparent that in order to charge the battery, one path goes directly from the wall outlet to the battery, through a transformer isolation and a rectifier or DC–DC converter combination. This situation is directly involved with the plug-in part of the system. The bottom part of the figure shows that the charging process is done by either driving a standalone alternator or using the propulsion motor itself run as a generator, ultimately charging the battery. Of course, it is understood that, when plugged in, the vehicle is stationary and the wheels are not moving. Even though the process indicated in the bottom part of Figure 2.8 is not involved directly with the plug-in, an overall power management process has to coordinate both the plug-in and the ICE, since there may be a situation when fast charging becomes necessary, and both the plug-in and the ICE (in generation mode) need to run concurrently. Finally, note the inclusion of a bidirectional converter in the plug-in part of the figure. This covers the possibility that in future the utility regulations may allow power to be fed back into the utility grid from the vehicle, assuming that it has enough power to do so. This issue is not an immediate consideration within the automotive industry at present. However, the possibility may in fact help the vehicle to be used as an emergency generator to light a home in case there is a utility power failure.

One final note here may be useful. Although in the above the charging of the battery by connecting directly to some outlet has been mentioned, nowadays the industry is also focusing on the use of wireless charging of batteries, where the charging power from the electrical source to the vehicle is passed wirelessly by an inductive process. This involves inductive coils located in the source and the vehicle, and when these coils are within close proximity – a few inches normally – a significant amount of power can flow wirelessly for charging the vehicle battery. The technology is equally applicable to HEV, EV, and PHEV, and more detailed discussion on this will be made in Chapter 16.

2.4.3 Comparison of HEV and PHEV

Fundamentally the only difference between the HEV and the PHEV is related to the upper part of Figure 2.8, where the wall outlet is used to charge the battery. The size of the battery may be substantially different. In addition, as indicated above, the plug-in system has to be properly coordinated with the rest of the charging process and the overall power management in the vehicular system.

2.5 Basics of Fuel Cell Vehicles (FCVs)

2.5.1 Why FCV?

The FCV architecture is most closely related to the pure EV. If the power source in the EV is replaced by a fuel cell system, then potentially it could replace the battery in the EV. Even in an HEV with an ICE, if the electrical energy source for propulsion were replaced by a fuel cell, it could still be classified as a FCV, though in reality it would be an HEV. Of course, as discussed in other parts of this book on the topic of the fuel cell, it needs chemical energy in the form of either hydrogen gas which can be directly used as fuel, or natural gas with appropriate reformers. It is also possible to have certain other kinds of liquid fuels along with reformers so that the fuel can be transformed into hydrogen. In such cases, it is possible to do the hydrogen reformation and creation, and the hydrogen consumption, by the fuel cell simultaneously, so that no additional storage of the hydrogen is necessary. Without reformers and conversion and use as it is produced, the hydrogen has to be stored in a tank. As long as sufficient fuel is present, the cell will run. Note that the fuel cell does not run on its own initially, and some warm-up is necessary. For that purpose a relatively small battery can be used, but as soon as the fuel cell is started, it will be self-sustaining and the initial energy drawn from the battery to warm up the fuel cell can soon be replenished. Advantages of the fuel cell are that it is virtually pollution free and quiet. But at present the technology has still to mature before it can replace existing ICE-based hybrid vehicles.

2.5.2 Constituents of a FCV

The same figure which was used for the EV applies here. It can be seen from Figure 2.6 that the high-voltage source is already labeled as either a battery or a fuel cell. The constituent elements are, therefore, identical to the EV.

2.5.3 Some Issues Related to Fuel Cells

One of the problems with the fuel cell is that it is a unidirectional device, that is, it can deliver power output but, unlike a battery or ultracapacitor, it cannot receive any power back. Obviously then, the fuel cell has to be ruled out for any regenerative efforts in a vehicle. This implies that a battery or an ultracapacitor has to be introduced in order to provide a regenerative capability. It is not just for regenerative braking that a storage battery or an ultracapacitor is necessary; such storage elements also serve as a mechanism by which the fuel cell can be started. This is important for both series and parallel hybrid vehicles. The battery or ultracapacitor has to be designed to meet such a starting current capability for at least half a minute, if not for longer. In addition, based on the typical drive cycle of the vehicle, an assessment has to be made about the regenerative needs of the vehicle during braking, and the size of the battery or ultracapacitor should be large enough for the worst-case scenario.

The second problem with the fuel cell is its sensitivity in terms of individual cell voltage. This gives an indication of the health of the fuel cell condition. If the cell voltages show a difference, that can indicate a problem. Fuel cells generally cannot handle large transients, hence a battery often helps reduce the size of fuel cell needed and protects it during large transients in the dynamic process.

The fuel cell – not the cell per se, rather the whole module – along with all the peripheral devices such as a compressor, water disposal mechanism, and warming system, together lead to a relatively low overall efficiency for the complete fuel cell system.

Reference

1 Masrur, M.A. (2008) Penalty for fuel economy – system level perspectives on the reliability of hybrid electric vehicles during normal and graceful degradation operation. *IEEE Systems Journal*, 2 (4), 476–483.

3

HEV Fundamentals

3.1 Introduction

Hybrid electric vehicles (HEVs) have been discussed along with their rationale in Chapter 2. As noted there, HEVs are vehicles that combine an internal combustion engine (ICE) with an electrical traction system. A hybrid vehicle should contain at least two sources of energy storage devices, such as battery, gasoline tank, or hydrogen tank. Based on the energy storages devices, there should also be corresponding power conversion sources, such as ICE or electric motor, within the vehicle to convert the energy into mechanical power. The end result is that the mechanical power to the wheels can potentially (but does not necessarily have to) come from multiple (at least two) sources. The main benefit of hybrid vehicle architecture is that it allows advantage to be taken of the best operating points of the sources (i.e. source characteristics) in terms of efficiency for delivering a certain amount of power, and matching the source characteristics to the load characteristics to achieve the best overall efficiency when delivering a particular load power. Advantages offered by HEVs are:

- operating the IC engine operation at its best operating point in the engine efficiency map
- efficiency improvement through regenerative braking, which allows braking energy to be returned to some energy storage element
- less engine idling
- improved fuel economy, automatically leading to reduced emission of greenhouse gases
- reduced fossil fuel consumption
- limited amount of gracefully degradable mode operation of the vehicle under partial failure conditions, depending on the HEV architecture.

Chapter 1 discussed various HEV architectures based on the configuration of the drivetrain: series hybrid, parallel hybrid, series–parallel hybrid, complex hybrid, and plug-in hybrid. In this chapter, some fundamentals will be discussed.

During the design phase of an HEV, it is necessary to select the ratings for the IC engine, traction electric motor/generator, generator (if there is a separate one), and energy storage elements based on the desired vehicle performance. It should be noted that in principle it is possible for the same electrical machine to function as either a motor in propulsion mode, or as a generator (which can be considered as the negative of propulsion mode) when its input comes from the IC engine, allowing acceptance

Hybrid Electric Vehicles: Principles and Applications with Practical Perspectives,
Second Edition. Chris Mi and M. Abul Masrur.

of mechanical power and converting the same to electrical energy. But sometimes, depending on the architecture, there can be a separate electrical machine which can predominantly work as a generator (but can also operate as a motor from time to time, depending on how the system architecture is designed). The design process of an HEV system nowadays involves a significant amount of modeling and simulation prior to actual development. To go through some of the above, this chapter will discuss vehicle road load modeling, specify requirements for vehicle performance and fuel economy, and include two design examples for series hybrid and parallel hybrid vehicles.

3.2 Vehicle Model

For the purpose of this section, the vehicle is modeled as a load, where the load resistance felt by the vehicle is due to the road profile. In other words, the engine, and the electric motor will basically see some speed and torque against which they have to work, in order to drive the vehicle under the particular road profile.

Consider the vehicle and the associated forces illustrated in Figure 3.1 [1]. Here, a vehicle of mass M_v is considered, moving at a velocity v, and moving up a slope of angle α (in degrees). The propulsion force or the tractive force (i.e. driving force) is F_{te}. This force has to overcome rolling resistance F_{rr}, aerodynamic drag F_{ad}, the climbing resistance force F_{rg} (the component of the vehicle's weight acting down the slope), and the force to accelerate the vehicle (the acceleration force), if the velocity is not constant. The total of the first three terms is the road load force F_{RL}. The road load, therefore, can be written as

$$F_{RL} = F_{rr} + F_{ad} + F_{rg} \tag{3.1}$$

The rolling resistance F_{rr} is really produced at the tire's internal material level due to the hysteresis of the tire, and ultimately arising from the interaction of the tire with the roadway. Rolling resistance depends on the coefficient of rolling friction between the tire and the road C_f, the normal force F_N due to the vehicle's weight $M_v g$, and the gravitational acceleration g. If the vehicle is at rest and the force applied to the road is not great enough to overcome the rolling resistance, then the rolling resistance must exactly

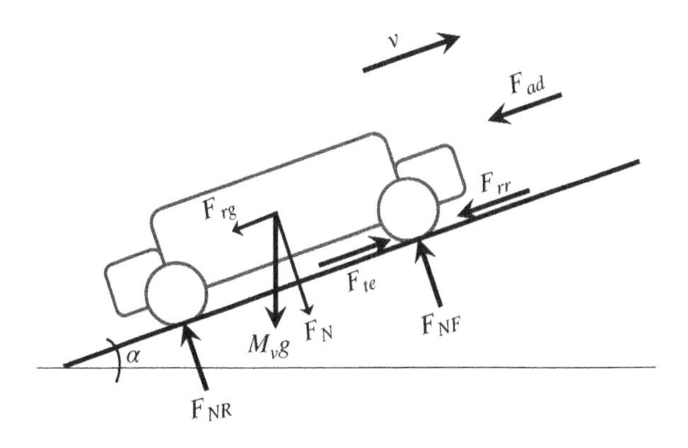

Figure 3.1 Vehicle model.

cancel out the applied tractive force to keep the vehicle from moving. Hence, the equation for rolling resistance can be written as

$$F_{rr} = -F_{te} \quad \text{if} \quad v = 0 \quad \text{and} \quad F_{te} < C_f \ M_v g \cos\left(\frac{\alpha\pi}{180°}\right)$$
$$F_{rr} = -C_f \ M_v g \cos\left(\frac{\alpha\pi}{180°}\right) \quad \text{otherwise} \tag{3.2}$$

The aerodynamic drag (i.e. the resistance encountered by the vehicle due to air resistance) depends on the air density ρ (kg/m^3), coefficient of drag C_d, frontal area of the vehicle A, and vehicle speed v. The equation for the aerodynamic drag is

$$F_{ad} = 0.5\rho C_d A v^2 \operatorname{sgn}(v) \tag{3.3}$$

where

$$\operatorname{sgn}(v) = +1 \quad \text{if} \quad v > 0$$
$$= -1 \quad \text{if} \quad v < 0$$

The force due to the road grade (slope of the road) depends on the mass of the vehicle M_v, slope angle in degrees α, and gravitational acceleration g. The equation for this force is

$$F_{rg} = -M_v g \sin\left(\frac{\alpha\pi}{180°}\right) \tag{3.4}$$

The road load curves of a vehicle corresponding to different road angles are shown in Figure 3.2. The vehicle parameters are given in Table 3.1. As can be anticipated, the road load (opposing force) increases with the velocity and with road angle. In this road load, the rolling resistance and air drag are related to velocity, whereas the force due to slope relates to the angle α.

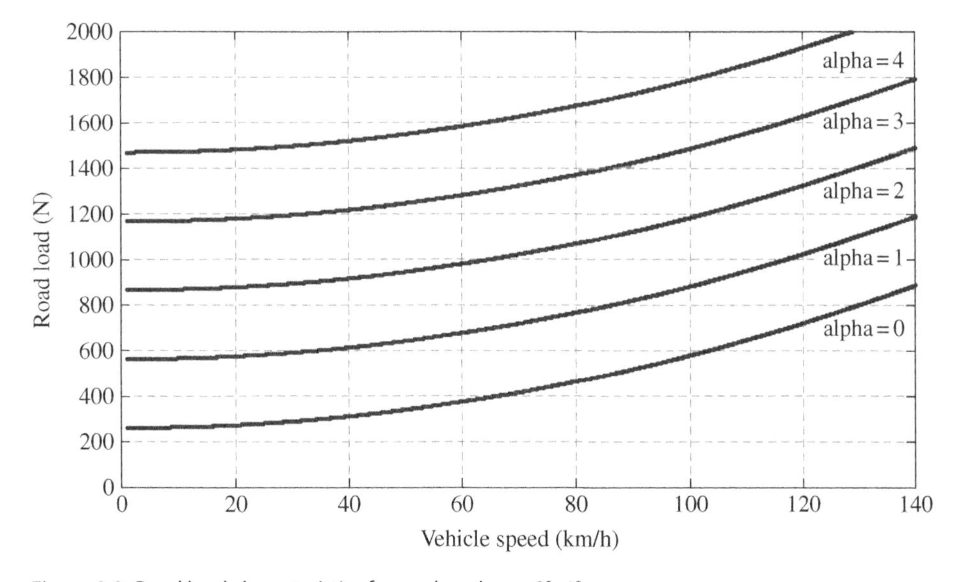

Figure 3.2 Road load characteristics for road angle $\alpha = 0°–4°$.

Table 3.1 Parameters for simulated vehicle.

Parameters	Value	Unit
Vehicle mass	1767	kg
Gravity	9.81	m/s^2
Rolling friction	0.015	—
Air density	1.225	kg/m^3
Aerodynamic drag coefficient	0.35	—
Frontal area	1.93	m^2
Wheel inertia	3.2639	kg/m^2
Wheel radius	0.2794	m
Headwind speed	0	m/s

The force to accelerate the vehicle is governed by Newton's second law, and leads to the linear acceleration of the vehicle. This is given by

$$F_{acc} = M_v a = M_v \frac{dv}{dt} \tag{3.5}$$

The total tractive effort is the sum of all the above forces:

$$F_{te} = F_{rr} + F_{ad} + F_{rg} + F_{acc} \tag{3.6}$$

The vehicle's velocity is calculated by integrating the vehicle's acceleration from time t = 0 seconds to a desired time t, with the initial velocity set to 0 km/h. This is given by

$$V = \frac{1}{M_v} \int_{t}^{t=0} \left(F_{te} - F_{rr} - F_{ad} - F_{rg}\right) dt \tag{3.7}$$

In the case of an ICE-driven vehicle, the vehicle tractive force comes from the engine shaft torque. The axle torque and engine torque are related as

$$T_{axle} = (T_{ICE})(GR_{trans})(GR_{diff})(\eta_{trans})(\eta_{diff}) \tag{3.8}$$

where T denotes torque, GR denotes gear ratio, η denotes efficiency, subscript "*trans*" denotes transmission, and subscript "*diff*" denotes differential. Thus, the tractive force is

$$F_{te} = \frac{T_{axle}}{\text{tire radius}} \tag{3.9}$$

In the case of series hybrid vehicles, the tractive (propulsion) force to the wheels comes from electric motor shaft torque; in the case of parallel hybrid vehicles, the propulsion torque can come either from the combination of the torques from the ICE and electric traction motor, or it can come from only one of these entities, depending on the algorithm used to make the decision.

3.3 Vehicle Performance

Prior to vehicle design, the performance constraints which are to be met must be defined. These constraints depend on the vehicle type and size. Typical performance specifications include initial acceleration, cruising speed, maximum speed, gradability, and drive range. Acceleration rate is the minimum time required to accelerate the vehicle from 0 to a specified speed such as 40, 60, or 80 mph. Sometimes acceleration rate from a non-zero lower speed to a higher speed is specified, such as from 40 to 60 mph. The maximum acceleration is limited by the maximum tractive power and the road condition. The gradability is the maximum grade that a vehicle can move along at a certain speed with the maximum tractive force available from the powertrain. Drive range refers to the distance (i.e. miles or kilometers) that a vehicle can travel with a full tank of fuel and/or fully charged batteries before refueling or recharging. The drive range of an EV or HEV is extremely important competing against their ICE counterparts.

The US PNGV's (Partnership for a New Generation of Vehicle) performance goal for midsize vehicles is:

- 0–60 mph: ≤12 seconds
- 40–60 mph: ≤5.3 seconds
- 0–85 mph: ≤23.4 seconds
- Maximum speed: 85 mph
- Maximum grade at 55 mph: 6.5%.
- Drive cycles are standard vehicle speed versus time profile for testing vehicle performance, fuel economy, and emissions. For example, the Federal Highway Driving Schedule (FHDS) and Federal Urban Driving Schedule (FUDS) are plotted in Figure 3.3.

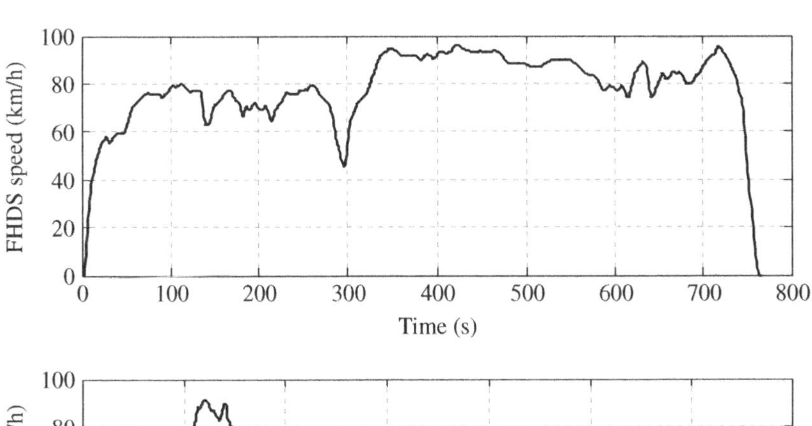

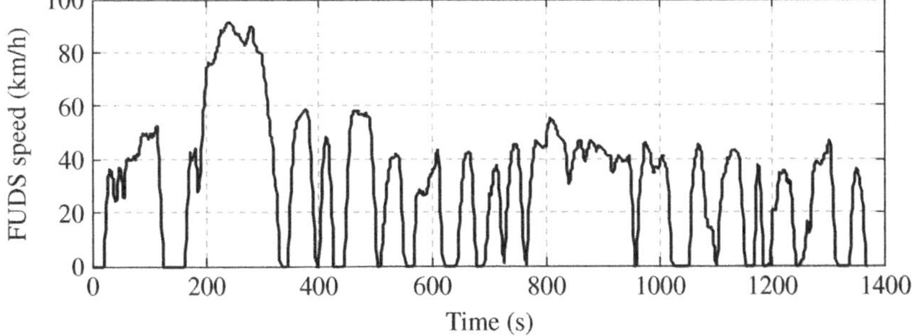

Figure 3.3 The FHDS and FUDS drive cycles.

(a)

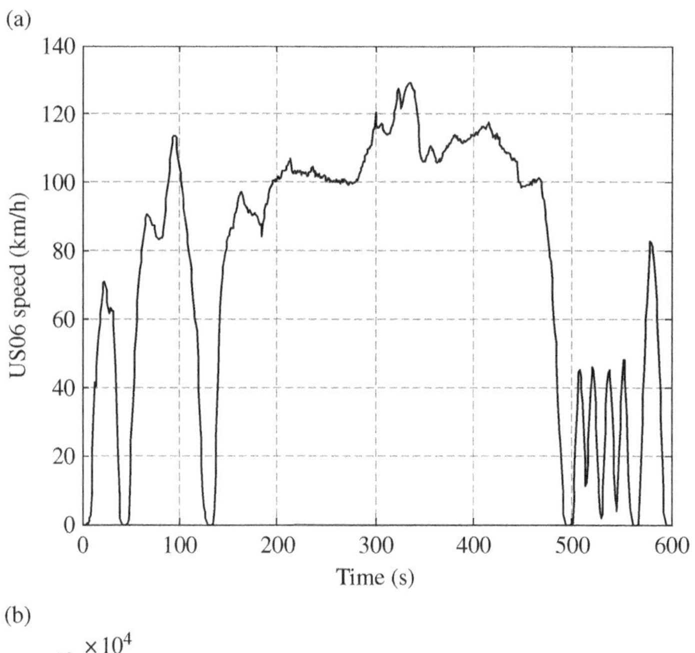

(b)

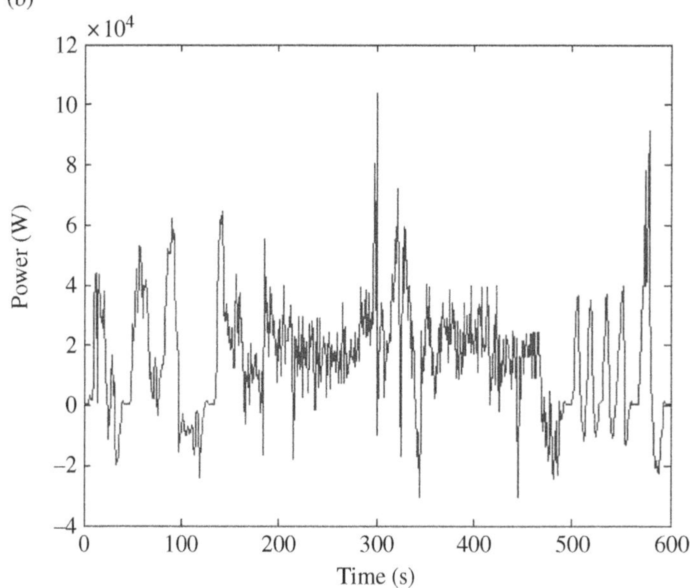

Figure 3.4 (a) The US06 drive cycle and (b) vehicle required electric power under US06 driving cycle.

The power necessary for operating a vehicle can be calculated from the drive cycles, and is dependent on the mass of the vehicle. For example, the power required for driving a vehicle with a total mass of 1380 kg under the US06 drive cycle is shown in Figure 3.4. US06, the aggressive drive cycle, came from the US Environmental Protection Agency (EPA) to measure fuel economy and emissions. The positive power is due to acceleration while the negative power is due to deceleration. Part of the negative power corresponding to braking can be recovered through regenerative braking of the HEV.

Table 3.2 The characteristics of example driving cycles.

Drive cycle	Idle time (percentage of total time)	Stops	Time percentage in specific speed zone
CONSTANT	4	0	≤16 km/h 100%
NYGCT	68.3	5	≤5 km/h 82%
NEWYORKBUS	67.3	11	≤6 km/h 75%
NYCTRUCK	52.1	20	≤12 km/h 70%
HWFET_HTN	0.7	1	≥70 km/h 85%
HWFET	0.7	1	≥70 km/h 85%
UDDS_HDV	33	14	≤20 km/h 80%
BEIJINGBUS	59.1	23	≤20 km/h 85%

Different drive cycles in different parts of the world are used to test the dynamic performance and fuel economy of different types of vehicles. At present, the drive cycles mostly come from Europe, North America, and Japan. China is also developing its own drive cycles based on Chinese road systems.

Table 3.2 shows eight example drive cycles [2]: CYC_HWFET, CYC_HWFET_MTN, CYC_NYCTRUCK, CYC_NYGCT, CYC_NEWYORKBUS, CYC_UDDS_HDV, CYC_BEIJINGBUS, and CYC_CONSTANT.

As can be seen in Table 3.2, each drive cycle has different characteristics. But in general the idle time percentage (spent), number of stops, and time percentage (spent) in a particular speed zone, are the predominant items affecting the fuel economy. For example, in the UDDS_HDV drive cycle, a vehicle has 14 stops and starts, 33% idle time, and runs at a lower speed of 20 km/h or less during 80% of the driving time period.

The fuel economy refers to the distance a vehicle can travel per unit of fuel (usually in miles or kilometers per gallon). For EVs or HEVs the definition can be somewhat complex. Here, miles per gallon gasoline equivalent (MPGGE) is used to measure how many miles a vehicle can travel with the consumption of energy equivalent to the amount of energy released from combustion of 1 gallon of gasoline. This is due to the fact that in an EV the source of energy is fully electrical and in an HEV the energy is due to the mix of two difference sources, i.e. gasoline and electricity from battery. Fuel economy of HEVs also depends on the drive cycles. Thus sometimes composite fuel economy or combined fuel economy is used. For example, composite fuel economy can be computed as the weighted average of the state of charge (SOC) balanced fuel economy values during the city drive cycle and highway drive cycle [3]:

$$\text{Composite fuel economy} = \frac{1}{\dfrac{0.55}{\text{City\_FE}} + \dfrac{0.45}{\text{Hwy\_FE}}} \tag{3.10}$$

where City_FE and Hwy_FE denote the city and highway fuel economy values, respectively.

It is possible to use the vehicle road load modeling in the previous section, in order to perform a parametric design study for vehicle design. This can help to show how vehicle performance is affected by changing the values for C_f, A, C_d, and M_v.

3.4 EV Powertrain Component Sizing

In this section, several examples are given to illustrate fundamental concepts and methods for EV component sizing.

As discussed earlier, when we size the powertrain of an EV, we must ensure sufficient tractive force for the vehicle to:

- accelerate from zero speed to a certain speed within a required time limit
- overcome wind resistance force if headwind speed is non-zero
- overcome aerodynamic force
- overcome rolling resistance
- climb a certain slope (grade).

Example 3.1 Let us find the approximate rating of an EV powertrain with a vehicle weight of 1364 kg (3000 lb). First, let us determine the forces needed to accelerate at 4.47 m/s^2, assuming that aerodynamic, rolling, and hill-climbing force counts for an extra 10% of the needed acceleration force.

$$\begin{aligned} F_{te} &= 1.1 \times \text{mass} \times \text{acceleration} \\ &= 1.1 \times 1364 \times 4.47 \\ &= 6704\,\text{N} \end{aligned}$$

Then, let us determine the average power needed to accelerate the vehicle from 0 to 96.5 km/h (60 mph):

$$\begin{aligned} \text{Energy required} &= \text{mass} \times V^2/2 \\ &= 1364 \times (26.8\,\text{m/s})^2/2 \\ &= 489839\,\text{J} \end{aligned}$$

Time required for the vehicle to accelerate from 0 to 96.5 km/h is

$$\begin{aligned} \text{Time} &= (26.8\,\text{m/s})/(4.47\,\text{m/s}^2) \\ &= 6s \end{aligned}$$

$$\begin{aligned} \text{Average power} &= \text{force} \times \text{distance/seconds} = \text{energy/time} \\ &\approx 81.7\,\text{kW} \left(\text{peak power } P_{\max} = F_{te} \times V \approx 180\,\text{kW}\right) \end{aligned}$$

Example 3.2 Determine the tractive force needed for a 1364 kg vehicle to accelerate to 96.5 km/h in 10 seconds, assuming a constant acceleration. The total power required for acceleration is

$$\text{Acceleration}\, a = V/t = 26.8/10 = 2.68\,\text{m/s}^2$$

$$\text{Force}\, F_{te} = \text{mass} \times \text{acceleration} = 1364 \times 2.68\,\text{m/s}^2 = 3657\,\text{N}$$

$$\text{Final power} = F_{te} \times V = 3657 \times 26.8 = 98\,\text{kW} \left(\text{at a speed of } 60\,\text{mph}\right)$$

In this example, a constant acceleration is assumed. In real life, the acceleration near 60 mph will be greatly reduced. Therefore, the actual power needed to accelerate the vehicle is much less than 98 kW:

$$\text{Average power} = \text{final power}/2 = 49\,\text{kW}$$

Example 3.3 Assume that the vehicle accelerates according to a sine wave (Figure 3.5, top plot) to a final speed of 60 mph and ignore all other resistances. Let us find the tractive force needed for the 3000 lb vehicle to accelerate.

$$V_{final} = 60\,\text{mph} = 60 \times 1608/3600 = 26.8\,\text{m/s}$$
$$a_m(1-\cos\pi)/\omega = 26.8 \qquad (\omega t = \pi \,\text{at}\, 10s, \omega = 0.314)$$
$$\therefore a_m = 0.314 \times 26.8/2 = 4.2\,\text{m/s}^2$$
$$a = 4.2 \sin 0.314t$$
$$F = \text{mass} \times a = 3000/2.2 \times 4.2 \sin 0.314t = 5727 \sin 0.314t$$
$$P = FV = 5727 \sin 0.314t \times 4.2(1-\cos 0.314t)/0.314$$
$$= 76 \sin 0.314t - 38 \sin 0.628t$$
$$P_{\max} \approx 90\,\text{kW}$$

P is also illustrated in Figure 3.5, bottom plot, where the two terms in P are plotted separately.

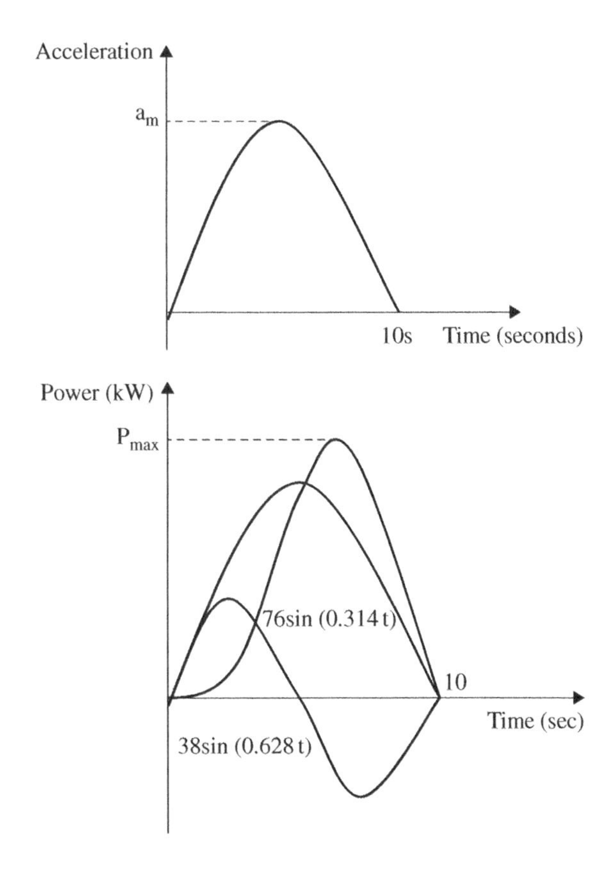

Figure 3.5 Vehicle acceleration and power.

Example 3.4 In this example, we will compare the resistive and the acceleration forces of a vehicle. The acceleration force was computed in Example 3.2 as 3657 N (accelerate to 60 mph in 10 seconds). The vehicle mass is 1364 kg.

1) Aerodynamic force: Consider the vehicle with the following:

   a) Aerodynamic drag coefficient 0.109
   b) Air density 1.16 kg/m^3
   c) Frontal area 1.543 m^2 (or 16.6 ft^2)

   Then the aerodynamic force at 60 mph (96 km/h) is

   $$F_{ad} = 0.5 \times 1.16 \times 0.109 \times 1.543 \times 26.82 = 70\,\text{N}$$

2) Rolling resistance: Consider the vehicle with a rolling coefficient of 0.02. Thus the rolling resistance force is

   $$F_{rr} = 0.02 \times 1364 \times 9.8 = 270\,\text{N}$$

3) Hill-climbing resistance: Consider the vehicle with a grade of 0.5°. Thus the hill-climbing force is

   $$F_{rg} = mg \times \sin\alpha = 1364 \times 9.8 \times \sin 0.5° = 117\,\text{N}$$

   The total resistance force on the highway is (except for acceleration)

   $$F_{RL} = F_{ad} + F_{rr} + F_{rg} = 70 + 270 + 117 = 457\,\text{N}$$
   $$P_r = F_{RL}V = 457 \times 26.8\,\text{m/s} = 12.3\,\text{kW}(\text{at top speed})$$

Note that the dragging force accounts only for P_r/P_{max} = 12.3 kW/98 kW = 12.5% of the total tractive force.

Further, note that the power required for a vehicle to cruise on a highway at 60 mph is only about 6% of the power needed to accelerate the vehicle from 0 to 60 mph in 10 seconds. Since most electric motors can be designed to overload for a short time, a motor can be designed at much lower ratings.

Note also that an electric motor can have an efficiency (including controller) of over 90%, while an ICE has an efficiency less than 30%. Furthermore, an ICE does not have the transient overload capability as does a motor. This is why the rated power of an ICE is usually much higher than required for highway cruising.

Now let us briefly look at the braking in an EV. Energy is wasted during braking in conventional vehicles, but the braking energy can be partially recovered in EVs and HEVs. Consequently, the performance of the antilock brake system (ABS) can be improved in HEVs/EVs, and traction control is easier to achieve in HEVs/EVs.

Let us compute the energy expected when bringing a 1364 kg vehicle to a halt from a speed of 60 mph in 10 seconds:

$$\text{Energy} = ½ \times \text{mass} \times V^2 = ½ \times 1364 \times \left(26.8\,\text{m/s}\right)^2$$
$$= 489{,}709\,\text{J} = 0.136\ \text{kWh}$$

Using an average speed of 30 mph, the vehicle will travel at 44 ft/s or 440 ft in 10 seconds. Assuming an average drag force of 100 lbf, the drag loss is

$$100 \times 4.455 \times 440/3.28 = 59{,}762\,\text{J} = 0.0166\,\text{kWh}$$

The energy that can potentially be recovered is 0.136 − 0.0166 = 0.1194 kWh.

The design of the complete propulsion system is a complex issue involving numerous variables, constraints, considerations, and judgment, which is beyond the scope of this book. The power and energy requirement from the powertrain is determined from a given set of vehicle cruising and acceleration specifications. EV/HEV design is an iterative process and requires many engineers from multiple disciplines to collaborate to meet design goals:

- Electrical and mechanical engineers design the electric motor for the EV or the combination of electric motor and ICE for HEVs.
- Power electronics engineers design the power conversion circuit which links the energy source with the electric motor.
- Control engineers, working in conjunction with the power electronics engineers, develop the propulsion control system.
- Electrochemists and chemical engineers design the energy source based on the energy requirement and guidelines of the vehicle manufacturer.

3.5 Series Hybrid Vehicle*

In the series hybrid powertrain, as illustrated in Figure 3.6, the mechanical output from the ICE is converted into electrical energy using a generator and the electrical energy is either used to charge the battery or is bypassed from the battery to the electric traction motor which propels the wheels. With respect to power electronics components, an AC–DC converter for charging the batteries and a DC–AC inverter for traction motor propulsion are required. In a series configuration, the engine is decoupled from the road load so the engine will not undergo abrupt changes in operating conditions and will have little idling time, thus emissions are reduced and this is better for the environment. Some other advantages of series hybrids are flexibility in the location of the engine–generator set and simplicity in design. However, three propulsion components are needed: the ICE, generator, and motor. This results in a longer chain of energy transmission and so the efficiency of series hybrids is generally lower than parallel hybrids. The motor must be designed for the maximum sustained power that the vehicle may require, such as when climbing a high grade. Nevertheless, the vehicle operates below the maximum power most of the time. All three drivetrain components need to be sized for maximum power for long-distance, sustained, and high-speed driving, otherwise the batteries will discharge fairly quickly, leaving the ICE to supply all the power through the generator. An example application for a series hybrid is in locomotive drives.

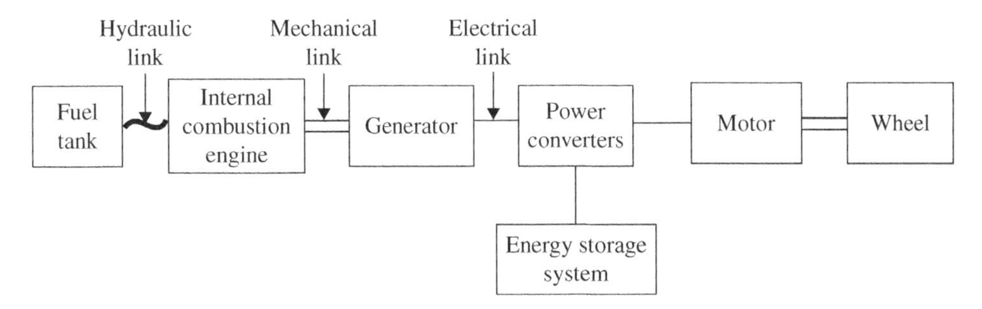

Figure 3.6 Series HEV powertrain.

* (This section © 2005 IEEE. Used, with permission.)

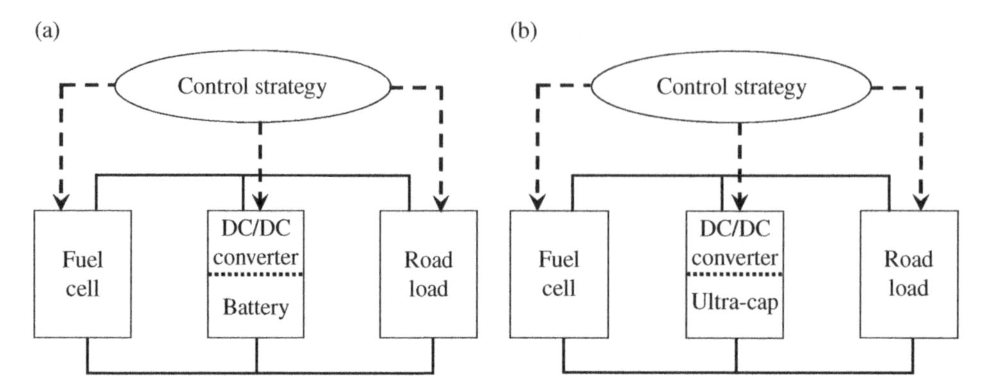

Figure 3.7 Configurations of fuel cell hybrid powertrains.

The fuel cell EV is one type of series hybrid and will be used as a design example for it [4]. In the fuel cell EV, the energy storage devices can be batteries or ultracapacitors or both. The fuel cell system supplies the base power while the battery/ultracapacitor supplies peak power for fast acceleration, and captures the braking energy for regeneration. The hybridization degree is defined as

$$HD = \frac{P_{ESS}}{P_{ESS} + P_{fc}} \tag{3.11}$$

where P_{ESS} is the power of the energy storage device and P_{fc} is the fuel cell power. An *HD* of 0 corresponds to a pure fuel cell vehicle, and an *HD* of 1 corresponds to a pure EV such as a battery electric vehicle (BEV). In a hybrid fuel cell powertrain, a good control strategy is necessary to manage the power flow considering the largely different characteristics of each component. The two types of hybrid fuel cell powertrains to be studied are illustrated in Figure 3.7. The operating modes of this series hybrid include: (1) battery- or ultracapacitor-alone mode in which the fuel cell is turned off; (2) fuel-cell-alone mode; (3) combined mode in which both the fuel cell and energy storage provide power to the road load; (4) fuel cell power split mode in which the fuel cell provides power for the vehicle road load and charging the battery or ultracapacitor; and (5) regenerative braking mode, in which energy storage recuperates part of the braking energy.

To compare the fuel economy and performance of the two types of fuel cell hybrid powertrains, four fuel cell hybrid vehicles are designed in ADVISOR (ADvanced VehIcle SimulatOR [3, 5]). The first design case in ADVISOR is the default fuel cell vehicle with a battery, and is used as a baseline vehicle (input file name: FUEL_CELL_defaults_in). The hypothetical small car is roughly based on a 1994 Saturn SL1 with the main data listed in Table 3.3. The fuel cell is the type of ambient pressure hydrogen fuel cell system. A one-speed gearbox is used for the transmission. The powertrain controller uses the hybrid with a thermostat control strategy as defined in ADVISOR [3, 5]. In brief, the fuel cell is controlled to be off when the energy storage SOC achieves a predetermined upper limit and is controlled to be on when the SOC drops below a predetermined lower limit. The designs of the fuel cell and energy storage are described in more detail below.

Table 3.3 Assumed parameters for the vehicle.

Parameters	Value	Unit
Vehicle glider mass, m	592	kg
Gravity, g	9.81	m/s^2
Rolling resistance coefficient, f_r	0.009	—
Transmission efficiency, η_t	0.92	—
Air density, ρ_a	1.2	kg/m^3
Aerodynamic drag coefficient, C_d	0.335	—
Frontal area, A_f	2.0	m^2
Wheel radius	0.282	m
Average electrical accessory load	700	W
Vehicle cargo mass	136	kg
Electric motor mass	91	kg
Fuel cell system mass	223	kg

In the hybrid powertrain, the fuel cell supplies the base power, P_e, to meet the vehicle power requirement for cruising and/or driving on a road with a grade [6, 7]:

$$P_e = \frac{(mgf_r + ½\rho_a\, C_d\, A_f\, V^2 + mgi)V}{1000\eta_t\, \eta_{em}} \tag{3.12}$$

where m is the vehicle total mass (assumed to be 1380 kg), i is the road grade, and η_{em} is the motor average efficiency (assumed to be 0.90). Then, by calculation, P_e is 13.1 kW at a cruising speed of 60 mph (96 km/h) on a flat road; P_e is 42.1 kW at a maximum speed of 95 mph (152 km/h) on a flat road; and P_e is 50.0 kW if the maximum grade at 55 mph is 10%. This determines the size of the fuel cell for steady driving.

The maximum total electric power required for vehicle acceleration from 0 to 60 mph in 11 seconds is [8]

$$P_{tot} = \frac{\left(mgf_r + \frac{1}{2}\rho_a\, C_d\, A_f\, V^2 + m\delta\frac{dv}{dt}\right)V}{1000\eta_t\, \eta_{em}} \tag{3.13}$$

where δ is the mass factor (assumed to be 1.035). The maximum P_{tot} is computed to be 124.8 kW.

Based on the above calculation, the net peak power of the fuel cell system is designed to be 50 kW and the battery size at least 75 kW (including electrical accessory load). A Westinghouse 75 kW (continuous) AC induction motor/inverter with a peak efficiency of 0.92 is used in the powertrain.

In summary, the four design cases are

- **Case 1 (baseline):** 25 modules of lead acid (Hawker Genesis 12 V, 26 Ah 10EP VRLA) batteries are used; battery peak power is 86.35 kW; total vehicle mass is 1380 kg.

- **Case 2:** 85 cells of Maxwell ultracapacitors are used; ultracapacitor peak power is 86.0 kW; total vehicle mass is 1140 kg (the same hybridization degree as Case 1).
- **Case 3:** 170 cells of ultracapacitors are used (twice those in Case 2); ultracapacitor peak power is 172.0 kW; total vehicle mass is 1174 kg.
- **Case 4:** 50 modules of lead acid battery are used (twice those in Case 1); battery peak power is 172.8 kW; total vehicle mass is 1655 kg.

As a double check, the total required electric power under the US06 driving cycle for Case 1 is given in Figure 3.4b. Note that the maximum total power required is about 105 kW at a time instant near 300 seconds. This is close to (but smaller than) the above maximum P_{tot} of 124.4 kW. So our design does satisfy the vehicle road load power requirement.

The assumptions for the battery and ultracapacitor are given in Table 3.4 [3,9]. The peak power, energy storage capacity, weight, and cost of the two types of fuel cell hybrid powertrains are compared in Table 3.5.

Note that the designed fuel cell vehicles have large power rating energy storage devices because the fuel cell vehicle is a series hybrid [10]. Thus, the fuel cell system can be considered as an auxiliary power unit, which is controlled to run at its optimum efficiency region.

In the simulation studies, energy management strategies (series thermostat control strategy [11]) remain the same in all cases. The highest desired SOC is 0.8; the lowest desired SOC is 0.4. The initial SOC for the battery and ultracapacitor is set to the same value (0.7) for all cases. The simulations are run under three different driving schedules: UDDS (urban dynamometer driving schedule), HWFET (highway fuel economy test), and US06 (aggressive driving cycle). The fuel economy in terms of MPGGE for the four cases is compared in Table 3.6. The MPGGE is calculated based on the lower heating value of gasoline (42.6 kJ/g), density of gasoline (749 g/l), and total energy consumption from the energy storage system. The detailed definition of MPGGE is given in [3].

Table 3.4 Energy storage system assumptions.

	Mass per cell (kg)	Specific power (W/kg)	Specific energy (Wh/kg)	Cost ($/kW)
Battery	11	314.6	28.4	10
Ultracapacitor	0.408	2500	6	15

Table 3.5 Energy storage system comparison.

	Peak power (kW)	HD	Energy storage capacity (kWh)	Mass (kg)	Cost ($)
Case 1 (battery)	86.4	0.63	7.81	275	864
Case 2 (ultracapacitor)	86.4	0.63	0.21	35	1296
Case 3 (ultracapacitor)	172	0.78	0.42	71	2580
Case 4	172.8	0.78	15.62	550	1728

Table 3.6 Fuel economy (MPGGE).

	Case 1	Case 2	Case 3	Case 4
UDDS (four cycles, 29.8 miles (47.7 km))	58.5	75.8	73.3	50.9
HWFET (four cycles, 41 miles (66 km))	78.7	84.4	83.2	74.7
US06 (four cycles, 32 miles (51 km))	58.7	64.1	63.1	56.8

Table 3.7 Simulated vehicle performance.

	0–60 mph (s)	40–60 mph (s)	0–85 mph (s)	Maximum speed (mph)	Maximum grade at 55 mph (%)
Case 1	11.1	5.8	22	97.6	9.8
Case 2	19.2	13.6	44.5	95.0	12.0
Case 3	6.9	3.2	36	95.1	11.7
Case 4	8.2	4.0	16	97.2	8.0
PNGV constraints	≤12	≤5.3	≤23.4	85	6.5

From the numbers in the table, Case 2 with an ultracapacitor has a higher fuel economy than Case 1. Increasing the modules of the ultracapacitor from Case 2 to Case 3 does not increase fuel economy (the difference is small). Increasing the modules of batteries from Case 1 to Case 4 decreases fuel economy.

The simulated vehicle performance is listed in Table 3.7. Specifically, the performance indices include the times for the vehicle to accelerate from 0 to 60 mph, from 40 to 60 mph, and from 0 to 85 mph, the maximum achievable speed, and the maximum sustainable grade at 55 mph. The last row also lists the PNGV performance constraints for a midsize car [9]. From the numbers given, Case 1 has a better performance than Case 2. Increasing the modules of the ultracapacitor from Case 2 to Case 3 greatly improves the vehicle performance. As a result, the fuel cell–ultracapacitor hybrid (Case 3) has a much better performance than the baseline fuel cell–battery hybrid (Case 1). Increasing the modules of batteries from Case 1 to Case 4 also improves the vehicle performance.

From the simulation results in Tables 3.6 and 3.7, with the same hybridization degree (0.63), the fuel cell–ultracapacitor hybrid vehicle's fuel economy is higher by about 30% than its battery counterpart, but the performance is worse. If more ultracapacitors are used with a hybridization degree of 0.78, the vehicle performance can be improved tremendously, while the fuel economy is maintained roughly at the same high level.

Overall, the fuel cell–ultracapacitor hybrid is better since ultracapacitors can more effectively assist the fuel cell to meet transient power demand. If more ultracapacitors are added, the performance is improved, while the fuel economy remains high. Although battery modules can also be increased for a fuel cell–battery hybrid, only the performance can be improved, while the fuel economy will be decreased as shown by Case 4. The battery has a higher specific energy than an ultracapacitor. But in a charge-sustaining hybrid powertrain, this is not necessarily an advantage, since the range is determined by the fuel tank volume. As demonstrated in this chapter, higher specific power characteristics of ultracapacitors are a big advantage for hybrid fuel cell powertrains, which can enhance both the fuel economy and the vehicle acceleration performance. In addition, the characteristics of allowing a wide variation of SOC also enable the ultracapacitor to more effectively assist the fuel cell to meet the vehicle power demand and to help achieve a better performance and higher fuel economy.

3.6 Parallel Hybrid Vehicle

In a parallel hybrid powertrain, more than one energy source can provide propulsion power. In this hybrid system, the ICE and the motor are coupled in many possible ways. The propulsion force may be supplied by the ICE alone, by the electric motor alone, or both. The electric motor can operate as an electric generator to charge the battery during regenerative braking or when the ICE output power is greater than the required power at the wheels. A smaller ICE and a smaller motor can be selected to obtain a similar performance in the non-hybrid counterpart. Parallel hybrids can offer lower cost as compared to series hybrids. However, one disadvantage is that the former need a complex control system. In a parallel hybrid vehicle, the ICE and the electric motor can separately provide the propulsion force. Figure 3.8 shows a block diagram for one possible parallel hybrid vehicle, which is a "through the road" four wheel drive hybrid.

Based on the location of the gearbox or transmission, the parallel hybrid can be further classified as:

1) **Pre-transmission parallel hybrid:** This is also known as the engine-transmission-motor system. In this system, the gearbox is located on the main drive shaft after the torque coupler. As a result, the gear speed ratios apply on both the engine and the electric motor. Power flow is summed at the gearbox. Also the torque from the motor is added to the torque of the engine at the gearbox input shaft.
2) **Post-transmission parallel hybrid:** This is also known as the engine-motor-transmission system. In this system, the gearbox is located before the torque coupler. As a result, gearbox speed ratios apply only on the engine. In this system, the torque from the motor is added to the torque of the engine at the gearbox output shaft.
3) **"Through the road" parallel hybrid:** This belongs to the class of parallel hybrids. The ICE-based powertrain propels one axle and the electric traction motor propels another axle. The design of the hybrid powertrain is simplified since the above two powertrains are decoupled.

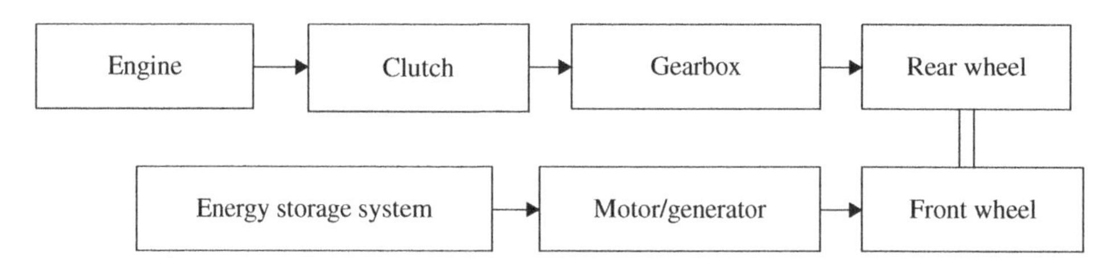

Figure 3.8 "Through the road" parallel hybrid powertrain.

3.6.1 Electrically Peaking Hybrid Concept

The electrically peaking hybrid (ELPH) concept was proposed by the HEV Research Group from Texas A&M University, and is a parallel control strategy. The electric traction motor is controlled to accommodate the acceleration and deceleration power while the ICE is controlled to provide the average power of the vehicle load. The main goal of this control strategy is to operate the ICE in the high-efficiency torque–speed region and thus increase ICE fuel economy while reducing harmful emissions. Also, the battery SOC is maintained within a predetermined range by this control strategy during the vehicle driving period [12].

In this section, the design of a parallel hybrid vehicle is given as an example. The design is done in a hierarchical manner starting at the system level and ending at the component level. The system design starts by defining the performance constraints to be met, such as the initial acceleration, cruise velocity, maximum velocity, and gradability. The system is designed so as to meet the above criteria with minimum power. The performance specification for the present vehicle is as follows:

- Acceleration: from 0 to 100 km/h in 16 seconds.
- Gradability: 5° at 100 km/h and maximum 25° at 60 km/h.
- Speed: 160 km/h (ICE only), 140 km/h (electric motor only).
- The parameters and constants used in the calculations are given in Table 3.1. In this design, a single gear ratio and ideal loss-free gears are chosen for simplicity. The primary energy source in the parallel hybrid is decided based on the energy management strategy used. In this design example, the ELPH control strategy is implemented. Component sizing is selected in such a way that the battery SOC can be controlled within a predetermined range.

In the first step of this design, the power ratings of the ICE and the electric motor are estimated based on vehicle performance requirements or specifications. First, an electric motor is designed based on the ELPH strategy, which meets both the acceleration and the road load requirements. The motor operates in three regions as shown in Figure 3.9, according to the different vehicle speed range. The motor is a variable frequency induction-type motor [13]. The first region, called the constant torque/force region, extends from 0 to the rated motor speed (v_{rm}), corresponding to the motor rated power. From the base speed up to the maximum speed (v_n), the motor runs in the constant power region. The motor runs in the natural mode if it is operated beyond v_n. The torque decreases roughly according to the inverse of speed squared in this high-speed region. Note that the natural mode is not shown in the figure and v_{rv} represents the vehicle's rated speed. Generally, the motor is operated in the constant torque region for rapid acceleration. The corresponding motor efficiency plot is shown in Figure 3.10. The following motor loss constants are assumed in plotting the motor efficiency: copper losses $k_c = 1.5$, iron losses $k_i = 0.1$, windage losses $k_w = 0.0001$, and motor constant losses = 20 [13].

The differential equation describing the performance of the hybrid vehicle is given by

$$a = \frac{dv}{dt} = \frac{F_{te} - F_{RL}}{K_m m} \tag{3.14}$$

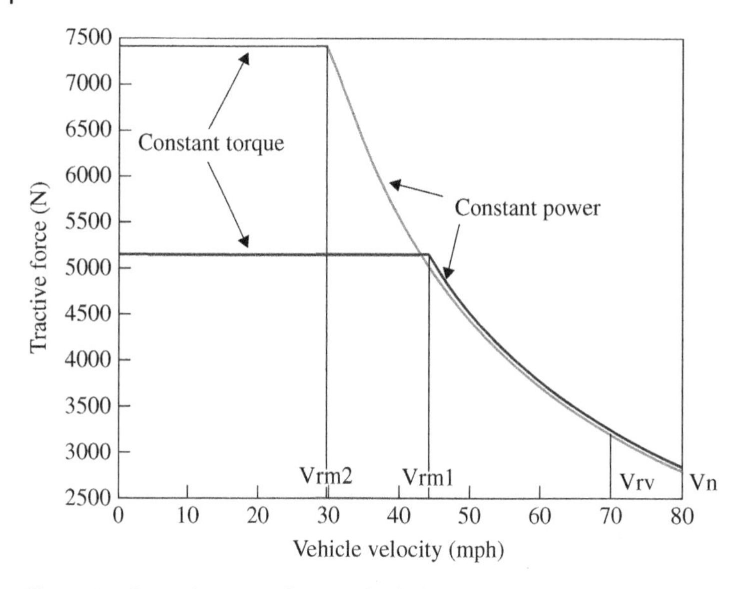

Figure 3.9 Example tractive force and vehicle velocity curve corresponding to electric traction motor.

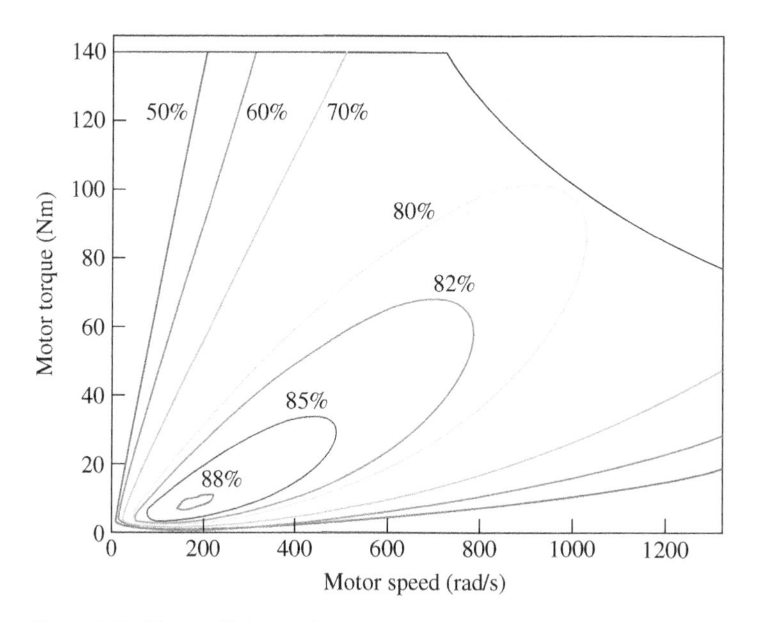

Figure 3.10 Motor efficiency plot.

where F and F_{RL} are the motive and road load forces, respectively. Assuming the road load to be 0 for simplicity and K_m = 1, then the above equation becomes

$$a = \frac{dv}{dt} = \frac{F}{m} \tag{3.15}$$

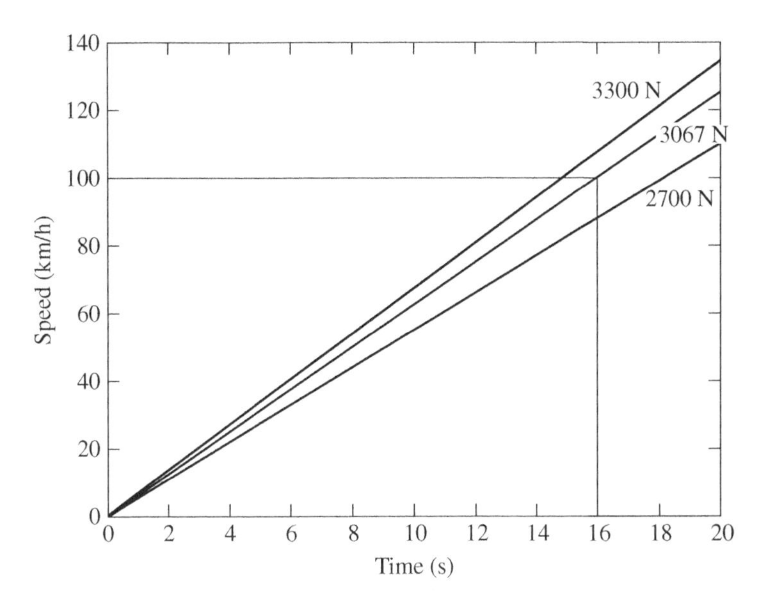

Figure 3.11 Acceleration performance with zero road load.

A vehicle with a mass of 1767 kg moving with an acceleration of 1.7361 m/s^2 requires an average force of 3067.7 N at the tire/road interface, approximately 857 Nm at the wheels. Note that this is the torque required to accelerate the vehicle, not to overcome the base load. It can be seen from Figure 3.11 that the vehicle meets the acceleration performance of 0–100 km/h in 16 seconds.

The power requirement to meet the acceleration performance is given in Figure 3.12, at approximately 95 kW.

Integrating the above acceleration equation for a time period of t_f and for a final velocity of v_{rv}, we get

$$m\int_{V_{rv}}^{0}\frac{dv}{F}=\int_{t_f}^{0}dt \tag{3.16}$$

The left integral can be separated into two integrals, one for a constant torque region and one for a constant power region:

$$m\int_{V_{rm}}^{0}\frac{dv}{Pm/v_{rm}}+m\int_{V_{rv}}^{V_{rm}}\frac{dv}{Pm/v}=t_f \tag{3.17}$$

Solving for Pm gives

$$Pm=\frac{m}{2t_f}\left(v_{rm}^2+v_{rv}^2\right) \tag{3.18}$$

The fact that the power requirement will be minimum if the motor is operated in the constant power region can be obtained from this equation. To find the minimum

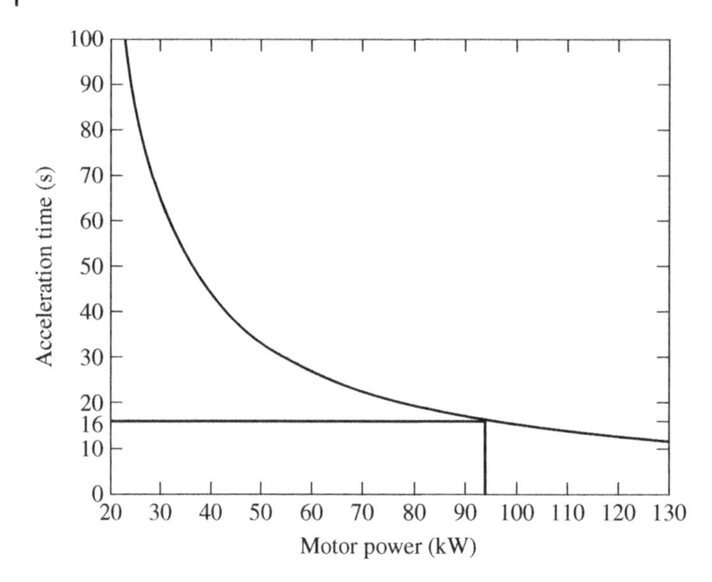

Figure 3.12 Motor power requirement to meet 0–100 km/h in 16 seconds.

power, we take the derivative of Pm with respect to v_{rm} and set the derivative to 0. This will yield the solution

$$v_{rm} = 0 \tag{3.19}$$

Therefore, if the motor is accelerating from 0 to v_{rv} in t_f seconds in the constant power region alone, the power requirement is minimum; in fact, this power is half of the required power when the motor is operated entirely in the constant torque region. The motor power dependence on the motor rated speed is shown in Figure 3.13. It can be seen that the power required decreases as the constant power region increases and it is minimum when the motor is completely operated in the constant power region. The effect of extending the constant power range is shown in Figure 3.14. Defining the constant power range ratio as v_{rm}/v_{rv}, it can be seen from Figure 3.14 that, with a ratio of about 1:4, the decrease in the required Pm for a certain t_f becomes less significant. Here, a ratio of 1:1 corresponds to the purely constant torque region. Note here that as the constant power range is increased, the torque required increases as shown in Figure 3.9. So, a tradeoff has to be made between motor power and motor torque, depending upon the requirements. Note also that the previous discussion for the required motor power is obtained based on neglecting the road load force. When this road load force is included, we have the following equation:

$$m\int_{V_{rm}}^{0}\frac{dv}{\left(Pm/v_{rm}\right)-F_{RL}}+m\int_{V_{rv}}^{V_{rm}}\frac{dv}{\left(Pm/v\right)-F_{RL}}=t_f \tag{3.20}$$

It can be seen from Figure 3.12 that a 95 kW motor is needed to meet the acceleration performance. But the power requirement should also meet the maximum vehicle velocity requirement. The motor power demand with vehicle speed for

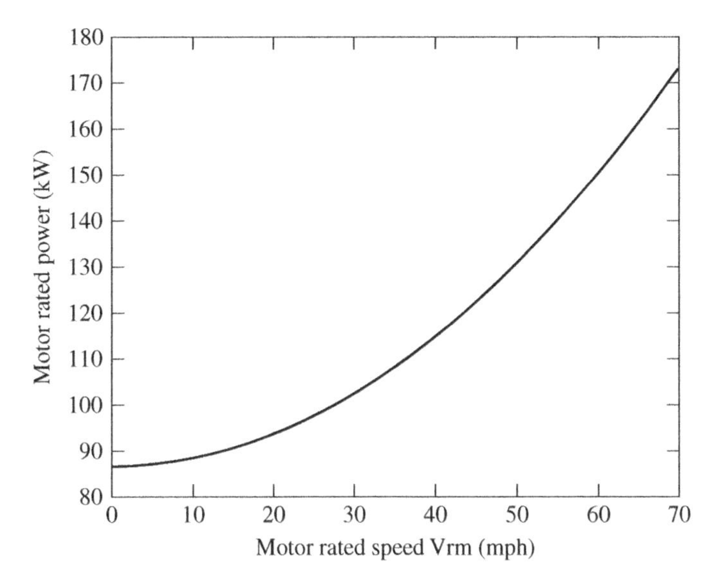

Figure 3.13 Motor rated power as a function of motor rated speed during acceleration.

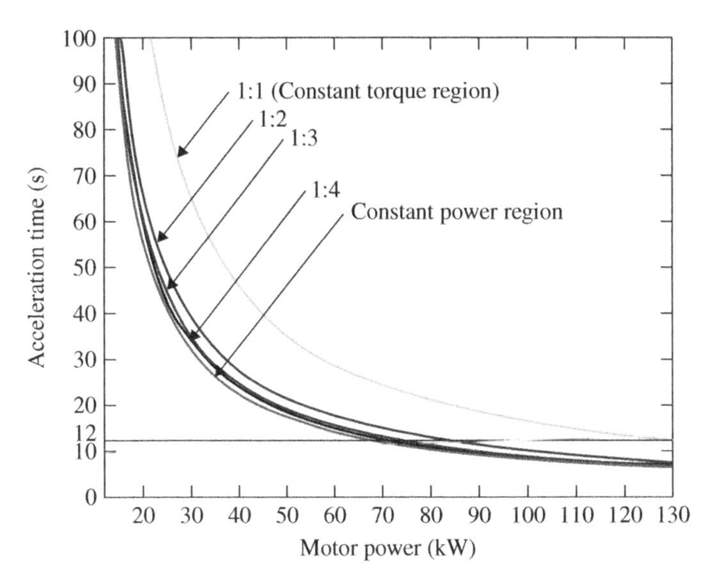

Figure 3.14 Motor power as a function of constant power range.

different grade angles is shown in Figure 3.15. The following equation is used in plotting Figure 3.15 with $\eta_t = 0.92$:

$$P_e = \frac{v}{1000\eta_t}\left(C_f mg + 0.5\rho C_d\ A_f\ v^2 + mg\sin\frac{\alpha\pi}{180°}\right) \tag{3.21}$$

where η_t is the transmission efficiency.

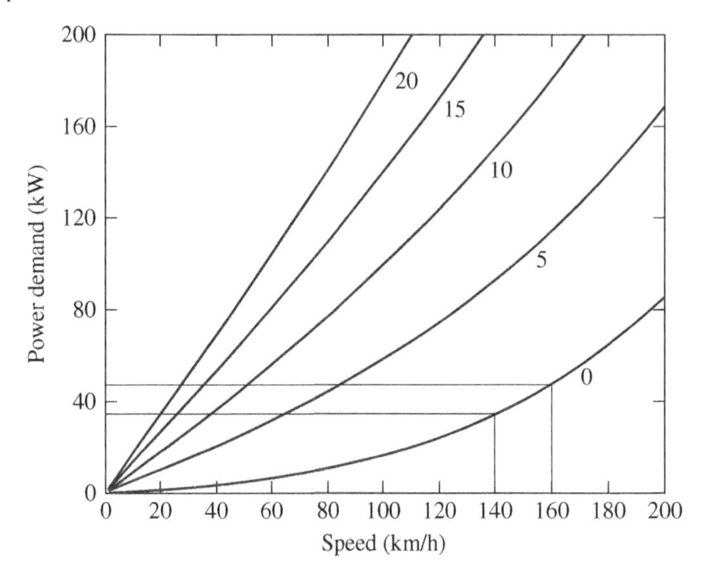

Figure 3.15 Vehicle power demand at zero acceleration with vehicle speed at different road grades.

The power demand required depends on the maximum vehicle speed and the maximum road angle to be climbed. In our case, the requirement is 140 km/h (ICE alone) and 160 km/h (electric motor alone); then from Figure 3.15 the power requirement of the motor is about 38 kW. Note here that the power required to meet the maximum speed of the vehicle is less than the power required to meet the acceleration performance. In order to meet both requirements, a motor power of 95 kW is chosen.

3.6.2 ICE Characteristics

A 55 kW ICE's torque–speed and power–speed characteristics with different throttle angles of 80° and 90° are shown in Figure 3.16. This plot is obtained by using a two-dimensional lookup table [14].

Figure 3.16 clearly indicates that, for high-torque outputs, the ICE must be operated at moderate speeds. Also, the efficiency of the ICE (not shown in the figure) is high at moderate speeds. In the ELPH control strategy, the ICE needs to be sized so as to provide the road load power and to recharge the batteries. From Figure 3.15, to achieve 160 km/h on ICE alone the power needed is 45 kW. If we assume an extra power demand of 10 kW for hotel loads such as air-conditioner, lighting, and other auxiliary loads, then the size of the ICE needs to be at least 55 kW.

3.6.3 Gradability Requirement

So far, two vehicle components, the motor and ICE, are designed to meet the acceleration and maximum speed requirements. Now, let us double check gradability requirements. Figure 3.17 shows the required vehicle power at grade angles of 0°, 5°, 10°, 15°, 20°, and 25°.

It can be seen from Figure 3.17 that approximately 60 kW and 140 kW are required to meet the gradability requirements of 100 km/h at 5° grade and 60 km/h at 25° grade, respectively. The available power from the vehicle is the sum of available powers from both the traction motor and the ICE. The available power of the vehicle is 150 kW (95 kW

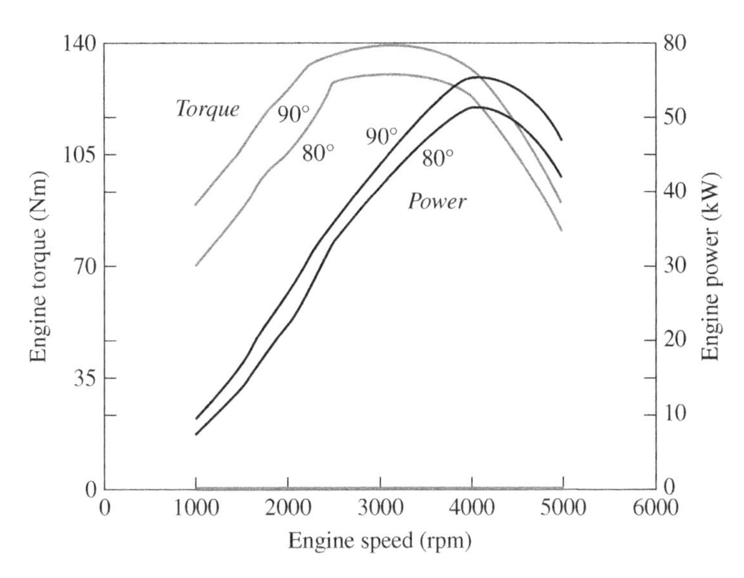

Figure 3.16 ICE torque–speed characteristics.

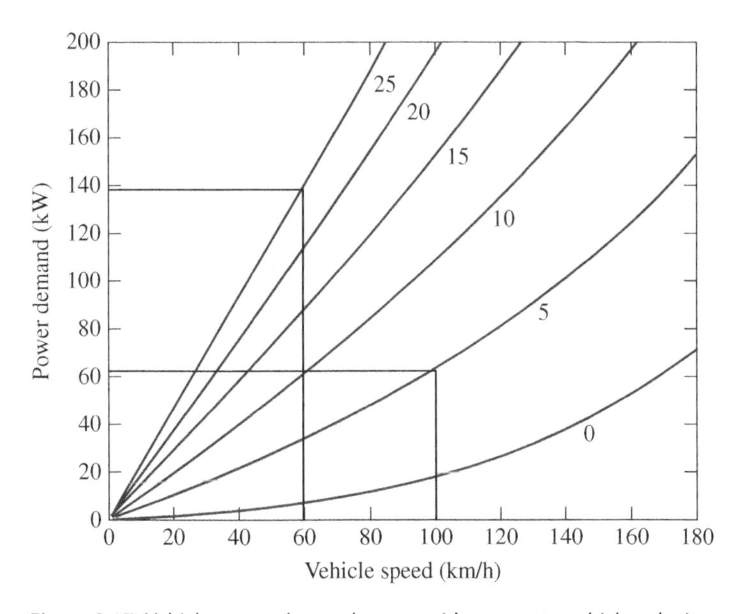

Figure 3.17 Vehicle power demand curves with respect to vehicle velocity and grade angle.

from the motor and 55 kW from the ICE). Thus, the available power is greater than the two gradability power requirements. Hence, these requirements are met by our design.

3.6.4 Selection of Gear Ratio from ICE to Wheel

In this design, a single gear ratio was assumed for simplicity. Here, we relax the single ratio specification and find out the required gear ratio from engine to drive wheel. Gears are mechanical devices used to gain a mechanical advantage through an increase in torque or reduction in speed. Gear ratio is the ratio of the ICE speed (rad/s)

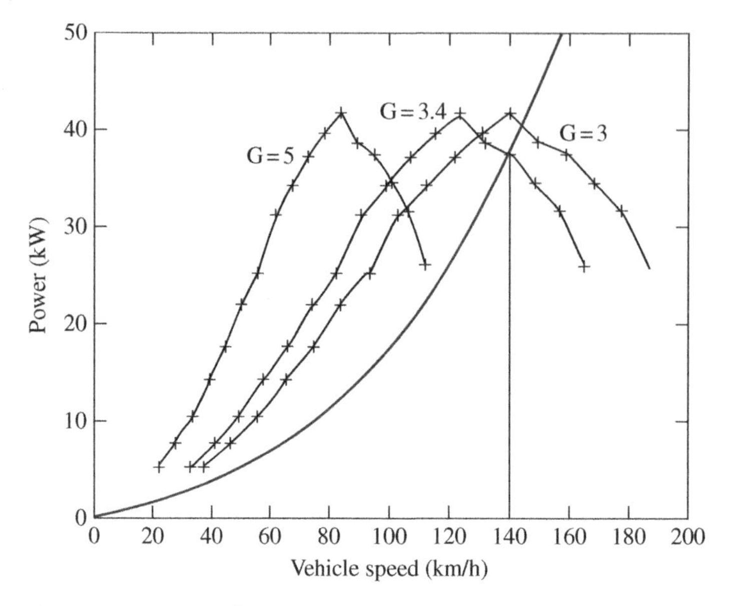

Figure 3.18 Gear ratio from ICE to drive wheel (the curve with '+' marks represents ICE power and the curve without '+' marks represents vehicle power).

multiplied by the tire radius (m) with respect to the vehicle speed (m/s). In Figure 3.18, the vehicle power demand curve (Figure 3.15) at zero road grade is plotted on the same graph as the engine power curves (Figure 3.16) at 90° throttle angle under different gear ratios [8]. The ratio curve on which these two curves touch each other, satisfying the maximum speed constraint and the available ICE power, corresponds to the minimum gear ratio required to meet the performance constraints. It can be seen from Figure 3.18 that a gear ratio of 3.4 is required to sustain a speed of 140 km/h in ICE-only operating mode.

In a similar fashion, the gear ratio of the electric motor to drive wheel can be determined.

3.7 Wheel Slip Dynamics

In this section, vehicle wheel slip dynamics is briefly discussed. As the power delivered to the wheels increases, vehicle acceleration eventually becomes limited by traction power. In power-limited acceleration, the vehicle reaches its maximum acceleration because the engine cannot deliver any more traction power. In acceleration with limited traction, the engine can and does deliver more power, but the traction force from the tires to the road surface cannot be further increased, resulting in limited acceleration. This is due to the nonlinear characteristics of the coefficient of friction between the tire and road surface. A schematic of a single-wheel braking model and the corresponding free body diagram is shown in Figure 3.19 [15].

In this figure, the normal force z balances the vehicle weight F_z. F_x is the tire reaction force, which is produced by the friction between the tire and road surface. This tire reaction force will cause a torque that drives the wheel forward with an angular

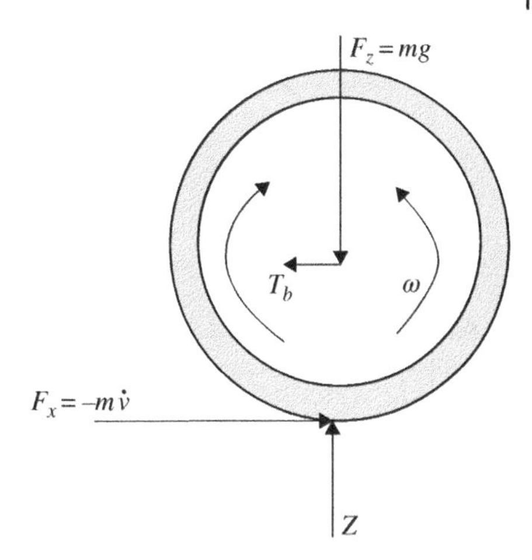

Figure 3.19 Model of quarter car forces and torques.

velocity ω. On the contrary, the braking torque T_b counteracts F_x and slows down the wheel's forward rotational motion. The dynamic motion equations can be written following Newton's laws:

$$m\frac{dv}{dt} = -F_x \tag{3.22}$$

$$J\frac{d\omega}{dt} = rF_x - T_b\,\mathrm{sgn}(\omega) \tag{3.23}$$

$$0 = Z - F_z \tag{3.24}$$

where

v = longitudinal speed at which the wheel moves along the vehicle
ω = angular velocity of the wheel
F_z = vehicle weight force
Z = normal force due to road reaction force
F_x = tire reaction force due to friction
r = wheel radius
J = wheel inertia
T_b = braking torque.

The tire friction force is given by

$$F_x = F_z\mu(\lambda, \mu_H, \alpha) \tag{3.25}$$

where the friction coefficient μ is a nonlinear function of λ, μ_H, and α; λ is longitudinal tire slip; μ_H is the friction coefficient between the tire and road; and α is the slip angle of the wheel.

If the tire friction force exceeds the force given by the above equation, the tire slips excessively and enters dynamic friction, where the coefficient of friction decreases dramatically, that is, it brakes traction.

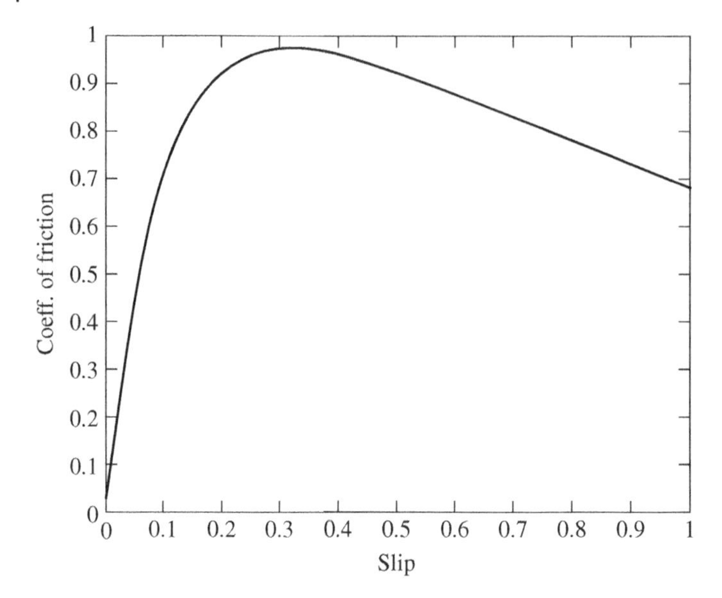

Figure 3.20 Friction coefficient (μ) and dependence on slip (λ).

The longitudinal slip is defined by

$$\lambda = \frac{v - \omega r}{v} \tag{3.26}$$

which quantifies the normalized difference between the vehicle speed v and the linear speed at the radius of the wheel corresponding to the wheel rotational speed ω. If $\lambda = 0$, the wheel is in free motion with no friction force between the tire and road surface. If $\lambda = 1$, the wheel is in a locked condition without any rotational motion ($\omega = 0$) and thus only sliding friction is present.

The longitudinal tire friction force is usually obtained by curve-fitting road test data. The so-called *magic formula* of Pacejka gives a good representation of the longitudinal friction characteristic [16, 17], as expressed by

$$y(x) = D\sin(C\arctan(Bx - E(Bx - \arctan(Bx)))) \tag{3.27}$$

The function parameters B, C, D, and E are determined based on curve-fitting. A more simple dependence of the friction coefficient as a function of slip can be given by unimodal friction characteristics [18]:

$$\mu(\lambda) = 1.18(1 - e^{-10\lambda}) - \lambda/2 \tag{3.28}$$

An example friction coefficient as a function of slip is shown in Figure 3.20. In the figure, the curve of the friction coefficient (μ) can be divided into two regions: to the left of 0.3, where the friction coefficient increases as slip value increases; and to the right of 0.3, where the friction coefficient decreases until the wheel locking condition ($\lambda = 1$) as slip value increases. Note that the curve in Figure 3.20 is different for different road conditions such as dry asphalt, wet asphalt, gravel, and packed snow [18]. Understandably, the friction coefficient is much higher for dry asphalt than for snow.

References

1 Gillespie, T.D. (1992) *Fundamentals of Vehicle Dynamics*, SAE, Warrendale, PA.
2 Fu, J. and Gao, W. (2009) Principal component analysis based on drive cycles for hybrid electric vehicle. 5th IEEE Vehicular Power and Propulsion Conference, September 7–11, Dearborn, MI.
3 Brooker, A., Haraldsson, K., Hendricks, T., et al. (2002) ADVISOR Documentation, Version 2002, National Renewable Energy Laboratory.
4 Gao, W. (2005) Performance comparison of a hybrid fuel cell – battery powertrain and a hybrid fuel cell – ultracapacitor powertrain. *IEEE Transactions on Vehicular Technology*, 54 (3), 846–855.
5 ADVISOR 2004 Documentation, http://www.avl.com.
6 Gao, Y. and Ehsani, M. (2001) Systematic design of fuel-cell powered hybrid vehicle drive train. Future Transportation Technology Conference, August 20–22, Costa Mesa, CA, paper 2001-01-2532.
7 Ehsani, M., Gao, Y., and Butler, K.L. (1999) Application of electrically peaking hybrid propulsion system to a full-size passenger car simulation design verification. *IEEE Transactions on Vehicular Technology*, 48 (6), 1779–1787.
8 Rahman, Z., Butler, K.L., and Ehsani, M. (1999) Designing parallel hybrid electric vehicles using V-ELPH 2.01. Proceedings of the American Control Conference, June 2–4, San Diego, CA.
9 Markel, T., Zolot, M., Wipke, K.B., and Pesaran, A.A. (2003) Energy storage requirements for hybrid fuel cell vehicles. Advanced Automotive Battery Conference, June 10–13, Nice, France.
10 Wipke, K.B., Markel, T., and Nelson, D. (2001) Optimizing energy management strategy and degree of hybridization for a hydrogen fuel cell SUV. 18th Electric Vehicle Symposium, EVS 18, Berlin, Germany.
11 Rahman, Z., Butler, K.L., and Ehsani, M. (1999) Design studies of a series hybrid heavy-duty transit bus using V-ELPH 2.01. Proceedings of the 49th IEEE Vehicular Technology Conference, May, Houston, TX, vol. 3, pp. 2268–2272.
12 Butler, K.L., Ehsani, M., and Kamath, P. (1999) A MATLAB-based modeling and simulation package for electric and hybrid electric vehicle design. *IEEE Transactions on Vehicular Technology*, 48 (6), 1770–1778.
13 Jefferson, C.M. and Barnard, R.H. (2002) *Hybrid Vehicle Propulsion*, WIT Press, Southampton.
14 Larminie, J. and Lowry, J. (2003) *Electric Vehicle Technology Explained*, John Wiley & Sons, Ltd, Chichester.
15 Johansen, T.A., Petersen, I., Kalkkuhl, J., and Ludemann, J. (2003) Gain-scheduled wheel slip control in automotive brake systems. *IEEE Transactions on Control Systems Technology*, 11 (6), 799–811.
16 Bakker, E., Pacejka, H., and Lidner, L. (1989) A New Tire Model with an Application in Vehicle Dynamics Studies. SAE Paper no. 890087, pp. 101–113.
17 Bakker, E., Nyborg, L., and Pacejka, H. (1987) Tyre Modeling for Use in Vehicle Dynamics Studies. SAE Paper no. 870421, pp. 190–204.
18 Olson, B.J., Shaw, S.W., and Stépán, G.S. (2003) Nonlinear dynamics of vehicle traction. *Vehicle System Dynamics*, 40 (6), 377–399.

4

Advanced HEV Architectures and Dynamics of HEV Powertrain

There are various hybrid powertrain architectures which are in use. The general goals of a hybrid transmission design are to realize the different operating modes of a hybrid vehicle system, such as the capability to run motor-alone mode, engine-alone mode, combined mode, power split mode, regenerative braking mode, and stationary charging operations. In addition, it is important to be able to control engine power and speed during hybrid operations so that better fuel economy and lower emissions can be achieved for all ranges of vehicle speeds and power demands. Lastly, the system design should be easy to implement and control, and bear low overall cost. This chapter discusses the principles of a few advanced hybrid electric powertrain architectures. These architectures include the popular GM two-mode hybrid and its variations, dual-clutch-based hybrid, Tsai's hybrid, Zhang's hybrid, Renault hybrid, and Timken hybrid.

The steady state operating modes and torque/speed relationships are presented first. Then, the HEV powertrain dynamics will be briefly discussed for the Toyota hybrid transmission.

4.1 Principle of Planetary Gears

Many hybrids in the marketplace today replace the traditional automatic transmission with a planetary gear train which can present functionalities of a continuous variable transmission (CVT). The engine, motor, and generator together form a type of transmission that can provide electric continuous variable transmission, known as e-CVT.

A planetary gear train has one or more planetary gears orbiting around a sun gear or central axis of the train. Therefore, there is a moving axis in planetary gear trains (carrier). A pictorial representation of a planetary train is shown in Figure 4.1. As can be seen, there are three axes in total: sun axis, planet carrier axis, and ring axis. The possible relative motions of these axes make planetary gear trains very interesting.

Of the three shafts of a planetary gear train, any shaft can be treated as the input shaft or output shaft. Hence, a combination will be either two input shafts and one output shaft, or one input shaft and two output shafts. In some applications, one of the gear sets is grounded so that only one input and one output is available.

Hybrid Electric Vehicles: Principles and Applications with Practical Perspectives,
Second Edition. Chris Mi and M. Abul Masrur.

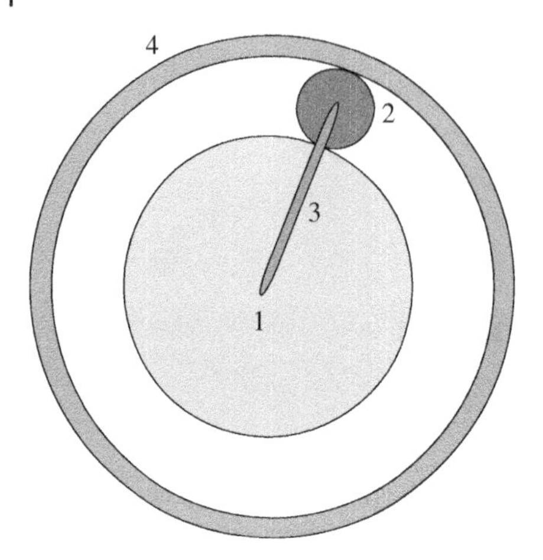

Figure 4.1 A planetary gear train: 1, the sun gear; 2, the planetary gear; 3, the arm or planet carrier; 4, the ring gear.

To understand the speed/torque relationships inside a planetary gear train, we can first look at the relationship between a simple set of gears. The linear velocity of the edge of a wheel relative to its center is defined as

$$V = \omega r \tag{4.1}$$

In simple gears, two wheels are put in contact with each other. The two wheels will travel at the same linear speed. Hence, the speed relationship of the two wheels is

$$\frac{\omega_1}{\omega_2} = \frac{r_2}{r_1} \tag{4.2}$$

The relationship between a gear's radius and a gear's tooth number can be written as $2\pi r_i = pN_i$, where N is the tooth number and p is the pitch (arc distance between two adjunct teeth). If gear i and gear j are in contact, then they must have the same circular pitch p in order to transfer the movement.

Therefore, we have

$$\frac{r_i}{r_j} = \frac{N_i}{N_j} \tag{4.3}$$

This expression is valid for any gear with an angular movement with respect to one point. For the planetary gear train shown in Figure 4.2, there are two contact points: point A between the sun gear and the planet gear, and point B between the planet gear and the ring gear. The linear velocities of the contact points can be obtained using two different paths for each point:

For point A:

$$v_a = \omega_s \; r_s \tag{4.4}$$

$$v_a = \omega_p \; r_p - \omega_c \; r_a \tag{4.5}$$

For point B:

$$v_b = \omega_r \; r_r \tag{4.6}$$

$$v_b = \omega_p \; r_p + \omega_c \; r_a \tag{4.7}$$

Figure 4.2 The operation of a planetary gear train.

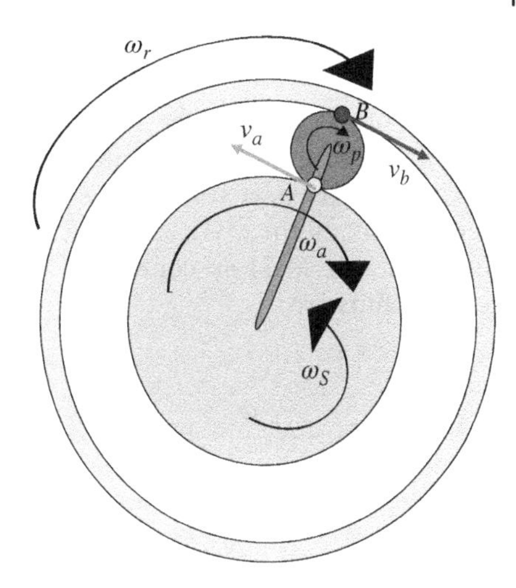

The directions of the various components' rotations have been chosen so that the movement of the planetary train is feasible. Thus, we have the following two equations:

$$\omega_s \ r_s = \omega_p \ r_p - \omega_c \ r_a \tag{4.8}$$

$$\omega_r \ r_r = \omega_p \ r_p + \omega_c \ r_a \tag{4.9}$$

After manipulation, we get

$$\omega_r \ r_r = \omega_s \ r_s + 2\omega_c \ r_a$$

But $r_a = r_r - r_p$ and $2r_p + r_s = r_r$, so $r_a = (r_r + r_s)/2$. Therefore, we get

$$\omega_r \ r_r = \omega_s \ r_s + \omega_c \left(r_s + r_r\right) \tag{4.10}$$

In order to make use of the expression easier, the clockwise direction will be considered positive and the anticlockwise direction negative. As can be seen in Figure 4.2, not all the angular velocities have the same direction. So, we can rewrite the above equation by including the correct reference direction:

$$\omega_r \ r_r = -\omega_s \ r_s + \omega_c \left(r_s + r_r\right) \tag{4.11}$$

Since

$$2\pi r_i = pN_i \tag{4.12}$$

we have

$$\omega_r \ N_r + \omega_s \ N_s = \omega_c \left(N_s + N_r\right) \tag{4.13}$$

or

$$\frac{N_r}{N_s + N_r}\omega_r + \frac{N_s}{N_s + N_r}\omega_s = \omega_c \tag{4.14}$$

As mentioned earlier, two inputs are required before a planetary gear set can be uniquely analyzed. If a gear is grounded, its velocity is zero; nevertheless this zero velocity constitutes one of the input values.

4.2 Toyota Prius and Ford Escape Hybrid Powertrain

The Toyota Prius and the Ford Escape use similar powertrain transmissions, as shown in Figure 4.3 as well as in Figure 1.23. It has an engine, two electric machines, and a planetary gear train in the transmission. The engine is connected to the carrier, electric motor MG2 is connected to the ring gear as well as the final drive, and the generator MG1 is connected to the sun gear. Hence, the speed and torque relationships are

$$\begin{aligned}
\omega_e &= \frac{N_s}{N_r+N_s}\omega_g + \frac{N_r}{N_r+N_s}\omega_r \\
T_r &= T_e\frac{N_r}{N_r+N_s} \\
T_g &= T_s = T_e\frac{N_s}{N_r+N_s} \\
T_{shaft} &= (T_m+T_r)^* i_1
\end{aligned} \tag{4.15}$$

where ω_e, ω_m, and ω_g are the speeds of the engine, the motor and the generator, respectively, ω_r is the ring gear speed, $\omega_m = \omega_r$; $\omega_s = \omega_g$; $i_1 = N_2/N_1$ is the gear final drive ratio, and N_1 and N_2 are the gear teeth number of the final drive.

Since there is no clutch, the planetary gear is always running whenever the vehicle is moving. It can be seen from the above equation and the diagram of the powertrain that the speed of the motor MG2 is directly proportional to the linear speed of the vehicle through the radius of the front tires and the final drive ratio. The ring gear speed and the motor speed are identical.

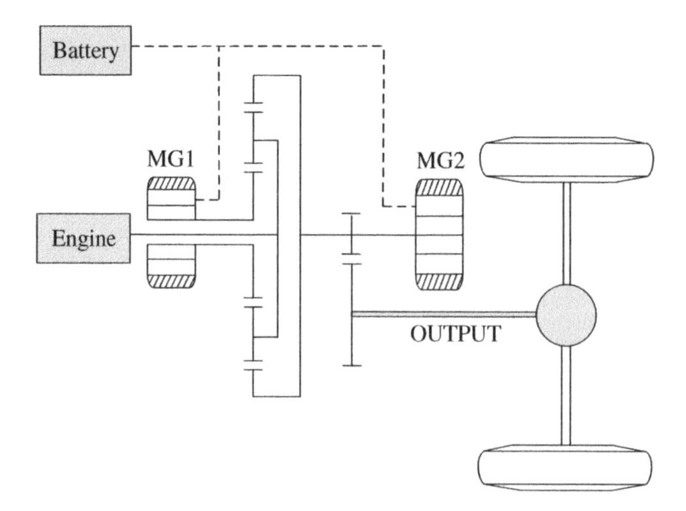

Figure 4.3 Toyota Prius transmission.

There are four different operating modes:

- **Mode 0:** Launch and backup – the motor is powered from the battery; the vehicle is driven by the motor only:

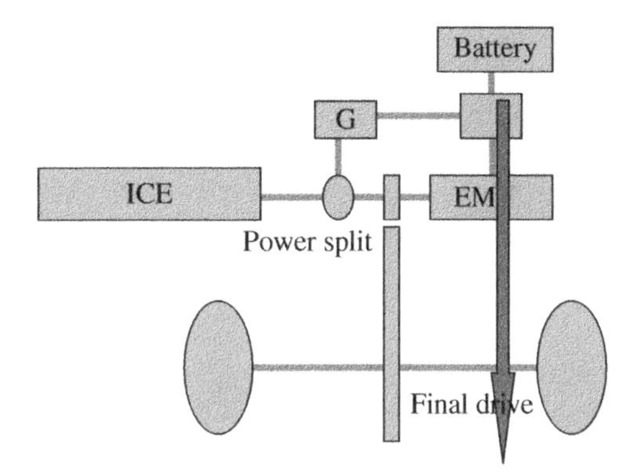

$$
\begin{aligned}
T_e &= 0 \\
T_g &= 0 \\
T_{shaft} &= i_1 T_m
\end{aligned}
\tag{4.16}
$$

- **Mode 1:** Cruising, or e-CVT mode 1:

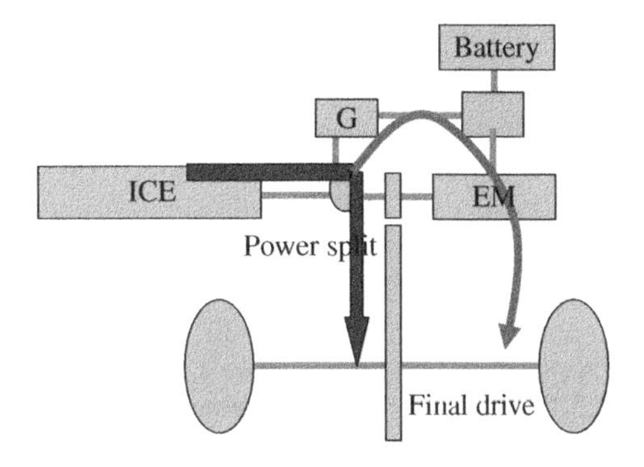

$$T_r = \frac{N_r}{N_r + N_s} T_e \tag{4.17}$$

$$T_g = \frac{N_s}{N_r + N_s} T_e \tag{4.18}$$

$$T_{out} = i_1 \left(T_r + T_m \right) \tag{4.19}$$

$$P_B = 0; \; P_m = \omega_m T_m = P_g = \omega_g T_s \tag{4.20}$$

- **Mode 2:** Sudden acceleration, e-CVT mode 2:

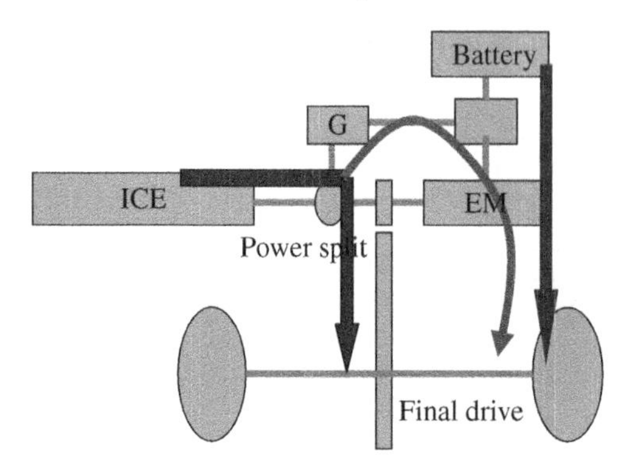

$$T_r = \frac{N_r}{N_r + N_s} T_e \tag{4.21}$$

$$T_g = \frac{N_s}{N_r + N_s} T_e \tag{4.22}$$

$$T_{out} = i_1 \left(T_r + T_m\right); \; P_m = \omega_m T_m = P_g + P_B \tag{4.23}$$

- **Mode 3:** Regenerative braking – MG2 is operating as a generator to produce electricity to charge the battery and at the same time to provide braking torque to the final drive. This operation is the reverse of launch and backup operation.

During normal operation (e-CVT or acceleration mode), the speed of the engine is controlled by the torque on the generator. Basically, the generator power is adjusted so that the engine turns at the desired speed. Hence, by adjusting the generator speed, the engine can operate at a relatively constant speed while the vehicle is driven at different speeds.

In the Prius, the engine is limited from 0 to 4000 rpm. The motor is limited from a small negative rpm for reverse and up to 6000 rpm (~103 mph or 165 km/h). The generator is limited to ±5500 rpm. The ring gear and sun gear have 78 and 30 teeth respectively. The four planetary gears each have 23 teeth. The final drive ratio is 3.93 and the wheel radius is 0.287 m. Hence, $\omega_e = 0.7222\omega_m + 0.2778\omega_g$.

The control strategy is as follows. For a given vehicle speed, and a desired output power determined by the drive cycle, or driver inputs, the desired operating point of the engine can be determined based on the maximum efficiency curve of the engine. From the vehicle speed and engine speed, the desired generator speed can then be calculated. The generator speed is regulated through the inverter by controlling the output power of the generator (either as generator or motor). Motor torque is determined by looking at the difference between the total vehicle torque demand and the engine torque that is delivered to the ring gear. The battery provides power to the motors along with the electricity generated by the engine.

Example 4.1 Consider a planetary gear train based transmission with an engine (carrier) providing 100 kW at 2000 rpm optimum operating point. The ring gear has 72 teeth and the sun gear has 30 teeth. The final drive ratio is 3.7865, and the wheel radius is 0.283 m. (1) For a vehicle speed of 45 mph or 20.6 m/s, the power demand under heavy acceleration at this speed is 120 kW. Find the speed and power for each component, assuming no losses. (2) For a vehicle speed of 70 mph, or 32.7 m/s, when cruising, the power demand is 70 kW. Calculate the speed and power of each component.

Solution

$$\omega_e = \frac{N_r}{N_r + N_s}\omega_r + \frac{N_s}{N_r + N_s}\omega_s = 0.706\omega_r + 0.294\omega_s$$

i) At 45 mph, the ring gear (same as motor speed) is calculated from the above to be 2632 rpm. Therefore, the sun gear (generator) speed needs to be 482 rpm in order to operate the engine at 2000 rpm:

$$T_c(\text{engine}) = P_{engine} / \omega_{engine(carrier)} = 477 \text{ Nm}$$

$$T_r(\text{ring gear}) = \frac{N_r}{N_r + N_s}T_c = 0.706 \times 477 = 337 \text{ Nm}$$

$$T_s(\text{generator}) = \frac{N_s}{N_r + N_s}T_c = 0.294 \times 477 = 140 \text{ Nm}$$

$$P_c(\text{engine}) = 100 \text{ kW}$$

$$P_r(\text{ring gear}) = T_r\ \omega_r = 337 \times 2 \times \pi \times 2632/60 = 92.9 \text{ kW}$$

$$P_s(\text{generator}) = T_s\ \omega_s = 140 \times 2 \times \pi \times 482/60 = 7.1 \text{ kW}$$

$$P_c(\text{engine}) = P_r + P_s$$

$$P_{vehicle} = 120 \text{ kW}$$

$$P_m(\text{motor}) = P_{vehicle} - P_r = 27.1 \text{ kW}$$

$$P_{bat} = P_m - P_s = 20 \text{ kW}$$

ii) At 70 mph, the ring gear (same as motor speed) is calculated from the above to be 4080 rpm. Therefore, the sun gear (generator) speed needs to be −2995 rpm in order to operate the engine at 2000 rpm:

$$T_c(\text{engine}) = P_{engine} / \omega_{engine(carrier)} = 477 \text{ Nm}$$

$$T_r(\text{ring gear}) = \frac{N_r}{N_r + N_s}T_c = 0.706 \times 477 = 337 \text{ Nm}$$

$$T_s(\text{generator}) = \frac{N_s}{N_r + N_s}T_c = 0.294 \times 477 = 140 \text{ Nm}$$

$$P_c(\text{engine}) = 100 \text{ kW}$$

$$P_r(\text{ring gear}) = T_r\ \omega_r = 337 \times 2 \times \pi \times 4080/60 = 144 \text{ kW}$$

$$P_s(\text{generator}) = T_s\ \omega_s = 140 \times 2 \times \pi \times (-2995)/60 = -44\ \text{kW}$$

$$P_{vehicle} = 70\ \text{kW}$$

$$P_m(\text{motor}) = P_{vehicle} - P_r = -74\ \text{kW}$$

$$P_{bat} = P_m - P_s = -30\ \text{kW}$$

4.3 GM Two-Mode Hybrid Transmission

The GM two-mode hybrid transmission was initially developed by GM (Alison) in 1996, and later advanced by GM, Chrysler, BMW, and Mercedes-Benz in a joint venture named Global Hybrid Cooperation in 2005.

The GM two-mode hybrid electric powertrain (or transmission) is shown in Figure 4.4 [1, 2]. This powertrain consists of two planetary gear sets P1 and P2, two electric machines MG1 and MG2, and three clutches C1, C2, and C3. The powertrain is capable of providing electric continuous variable transmission (eCVT) for both high-speed and low-speed operations, hence two-mode. The two-mode concept can be referred to and compared to the Toyota and Ford hybrid electric vehicle powertrain whose operation is limited to only one mode. In principle, two-mode operation can provide more flexibility for transmission control, increased drivability, and improved vehicle performance and fuel economy.

4.3.1 Operating Principle of the Two-Mode Powertrain

In the GM two-mode hybrid transmission, the engine is connected to the ring gear of planetary gear P1 through clutch C1. Electric machine MG1 is connected to the sun gear of P1. The carrier of P1 is connected to the final drive through the output shaft. MG2 is connected to the sun gear of planetary P2. The carrier of P2 is also connected

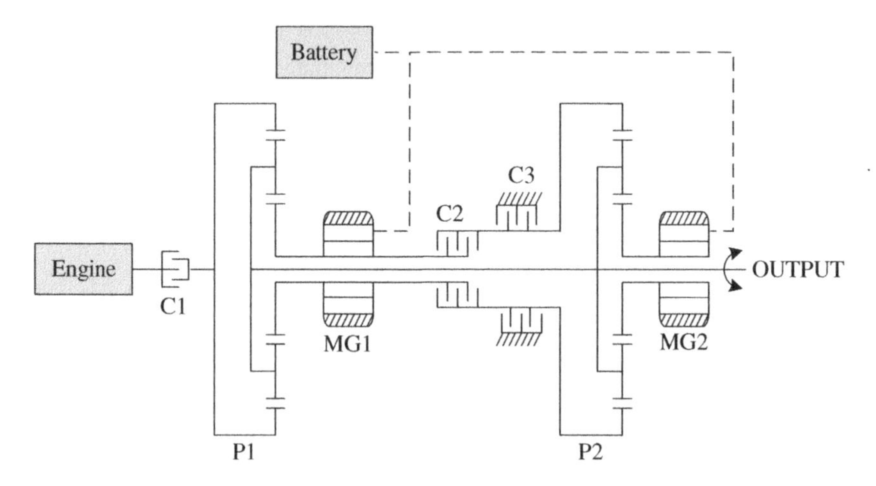

Figure 4.4 GM two-mode hybrid transmission.

to the output shaft. There is a dual-position clutch that connects either the ring gear of P2 to ground, or the ring gear of P2 to the shaft of MG1. Through control of C2 and C3, different operating modes can be realized. The engine in this system can be kept at the best speed and torque combinations in order to achieve the best fuel economy by controlling the input/output of the two electric machines. The engine may be stopped or may idle during vehicle launch and backup, as well as at low power demand. At cruising conditions, the engine efficiency is further enhanced by cylinder deactivation, also known as active fuel management (GM) or the multi-displacement system (Chrysler). Note that this discussion is generic and may not be the same as those implemented in a real vehicle by the automobile manufacturers.

In the following derivations, ω is the angular velocity, T is torque, N is the number of teeth of a gear, and P is power. Subscript s stands for sun gear, r for ring gear, c for planetary carrier, 1 for planetary gear set 1, 2 for planetary gear set 2, g for motor/generator 1, or MG1, m for motor/generator 2, or MG2, and *out* is for output or final drive.

4.3.2 Mode 0: Vehicle Launch and Backup

During vehicle launch and backup, the system is operating in motor-alone mode (Mode 0). C2 is open and C3 is engaged to ground the ring gear of P2. In this mode, there are two possibilities for engine operation, either off or idle at cranking speed (approximately 800 rpm) by adjusting MG1 speed. MG1 torque is not transmitted to the final drive. MG2 provides the needed torque to launch the vehicle forward or backward. Figure 4.5 shows the power flow during launch and backup. The speed/torque relationships are

$$\omega_{out} = \frac{N_{s2}}{N_{s2} + N_{r2}} \omega_m \tag{4.24}$$

$$T_{out} = \frac{N_{s2} + N_{r2}}{N_{s2}} T_m \tag{4.25}$$

In the final implementation, C1 was eliminated. Therefore, the engine is always connected to the ring gear of P1. Since the carrier of P1 is always connected to the final drive, MG1 needs to be controlled so that the engine is either at zero or at a certain speed:

$$\omega_g = \frac{N_{s1} + N_{r1}}{N_{s1}} \omega_{c1} - \frac{N_{r1}}{N_{s1}} \omega_e \tag{4.26}$$

If the engine is maintained at 800 rpm without fuel injection, there are still friction losses, but the engine can be ignited at any time without delay. If the engine is controlled to be at zero speed, then MG1 will need to be controlled so that the ring gear of P1 reaches 800 rpm before the engine can be started. In this case, the engine is controlled by cylinder deactivation.

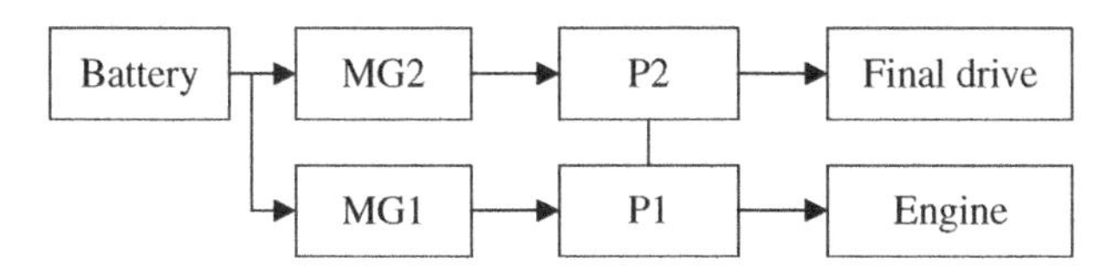

Figure 4.5 Power flow during launch and backup.

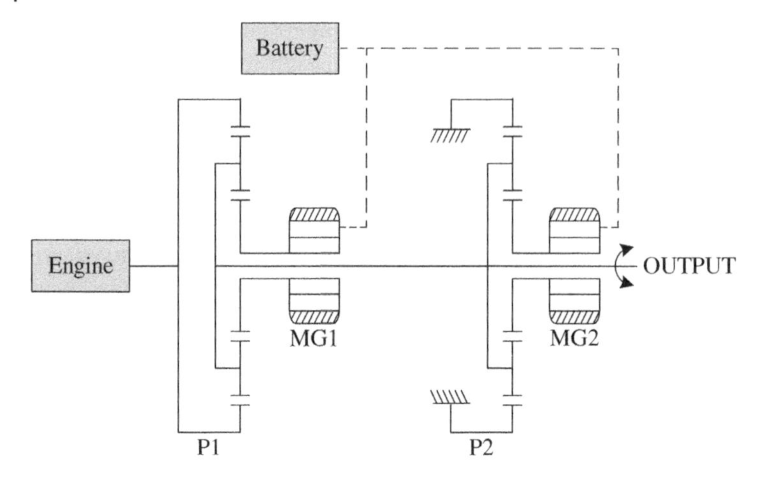

Figure 4.6 Low range.

4.3.3 Mode 1: Low Range

Mode 1 is also called the low-range or low-speed mode. In this mode, C1 is engaged, C2 is open, and C3 is engaged. The second planetary gear works as a speed reduction gear for MG2. Figure 4.6 illustrates the mechanical connections of the transmission. The engine may be controlled by partial cylinder deactivation to further save fuel and reduce emissions based on vehicle power demand. The torque and speed relationships during steady state operations can be expressed as

$$T_g = \frac{N_{s1}}{N_{r1}} T_e \tag{4.27}$$

$$\omega_{c1} = \frac{N_{r1}}{N_{s1} + N_{r1}} \omega_e + \frac{N_{s1}}{N_{s1} + N_{r1}} \omega_g \tag{4.28}$$

$$\omega_{out} = \omega_{c2} = \omega_{c1} = \frac{N_{s2}}{N_{s2} + N_{r2}} \omega_m \tag{4.29}$$

$$T_{out} = \frac{N_{r1} + N_{s1}}{N_{r1}} T_e + \frac{N_{s2} + N_{r2}}{N_{s2}} T_m \tag{4.30}$$

The different operations in Mode 1 can be described as follows:

1) **Engine alone (CVT 1):** MG2 is off (freewheel) and MG1 can be either in motoring mode or in generating mode. When MG1 is in motoring mode, P1 acts as a speed coupling mechanism to couple the speed of the engine and MG1. When MG1 is in generating mode, engine power is split between the final drive and MG1 with power generated by MG1 charging the battery. Since the battery can be quickly charged fully, this mode is generally brief.
2) **Combined mode (CVT 2):** MG2 is turned on to assist the driving. P2 acts as a torque coupling mechanism to add the torque of the engine (P1 carrier portion) and MG2. If needed, both MG1 and MG2 can work in motoring mode to maximize the driving torque.

3) **Power split mode (CVT 3):** MG2 is in generating mode to charge the battery. MG1 can be either motoring or generating.

4.3.4 Mode 2: High Range

Mode 2 is also called the high-range or high-speed mode. C1 is engaged, C2 is engaged, but C3 is open. In this mode, the sun gear of P1 is connected to the ring gear of P2 through MG1, that is, MG1, S1, and R2 will have the same speed. Figure 4.7 shows the mechanical connections of the transmission in Mode 2. In this operating mode, the engine is generally kept at a constant speed to achieve the best fuel economy. MG1 and MG2 are controlled in either motoring or generating mode depending on the vehicle speed and power demand. The torque and speed relationships during steady state operation are

$$T_g = \frac{N_{s1}}{N_{r1}} T_e + \frac{N_{r2}}{N_{s2}} T_m \tag{4.31}$$

$$\omega_{c1} = \frac{N_{r1}}{N_{s1} + N_{r1}} \omega_e + \frac{N_{s1}}{N_{s1} + N_{r1}} \omega_g \tag{4.32}$$

$$\omega_{c2} = \frac{N_{r2}}{N_{s2} + N_{r2}} \omega_g + \frac{N_{s2}}{N_{s2} + N_{r2}} \omega_m \tag{4.33}$$

$$\omega_{c2} = \omega_{c1} \tag{4.34}$$

$$\omega_{r2} = \omega_{s1} = \omega_g \tag{4.35}$$

$$T_{fd} = \frac{N_{r1} + N_{s1}}{N_{r1}} T_e + \frac{N_{s2} + N_{r2}}{N_{s2}} T_m \tag{4.36}$$

Similar to Mode 1, the engine may be controlled by partial cylinder deactivation to further save fuel and reduce emission, based on vehicle power demand.

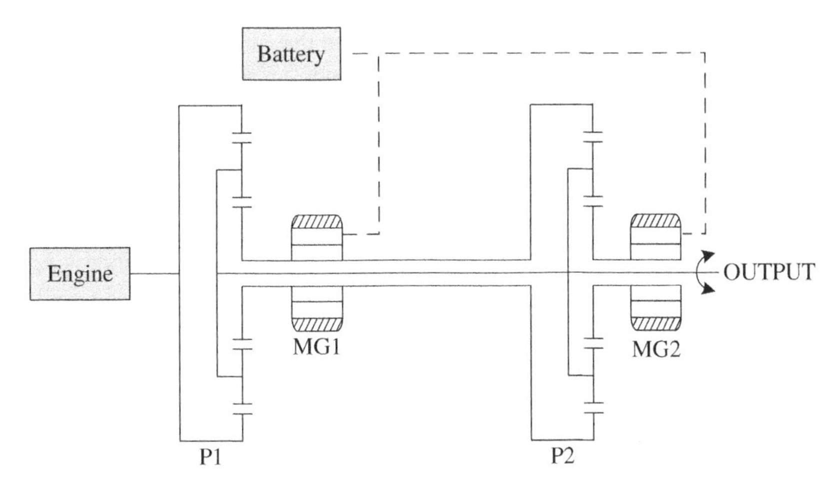

Figure 4.7 High range.

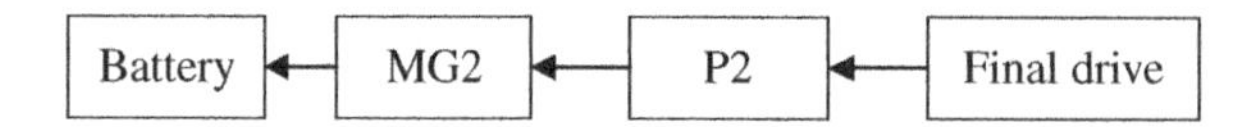

Figure 4.8 Power flow in regenerative braking.

4.3.5 Mode 3: Regenerative Braking

During regenerative braking, C1 is open, C2 is open, and C3 is engaged, to ground the ring gear of P2 (Mode 3). The engine and MG1 are off or freewheel. MG2 provides the needed braking torque for the vehicle and, at the same time, stores regenerative braking energy in the onboard battery. Figure 4.8 shows the power flow during regenerative braking. The speed/torque relationship is

$$\omega_m = \frac{N_{s2} + N_{r2}}{N_{s2}} \omega_{out} \tag{4.37}$$

$$T_m = \frac{N_{s2}}{N_{s2} + N_{r2}} T_{out} \tag{4.38}$$

Hydraulic/frictional braking may be controlled in coordination with regenerative braking to maximize the braking torque and/or maintain vehicle stability and prevent wheel locking. In this case, MG2 only provides a portion of the braking torque.

4.3.6 Transition between Modes 0, 1, 2, and 3

In general, transition is performed at a condition that can minimize mechanical disturbance to the overall vehicle system. The vehicle is generally launched by MG2 with the engine off (Mode 0). MG1 is turned on before transitioning to Mode 1 such that the engine speed reaches approximately 800 rpm. Transition from Mode 0 to Mode 1 is characterized by the engine turning on. This typically happens when the power demand reaches a certain level such that MG2 is no longer capable of providing the needed torque. The power demand is a combination of vehicle speed, acceleration demand, vehicle load, and road conditions.

Transition from Mode 1 to Mode 2 happens when the sun gear of P1 and the ring gear of P2 reach the same speed. In other words, since the ring gear of P1 is grounded (zero speed), transition from Mode 1 to Mode 2 will happen when the sun gear of P1 or MG1 reaches zero speed. Similarly, transition from Mode 2 to Mode 1 also happens when the speed of MG1 reaches zero.

Transition from Mode 1 to Mode 3, or Mode 2 to Mode 3, is triggered by a braking request from the driver (brake pedal is pressed).

Example 4.2 Both planetary gear sets have 30 teeth for the sun gear and 70 teeth for the ring gear. The engine is kept at 800 rpm in Mode 0, ramped up from 800 to 2000 rpm in Mode 1, and kept at 2000 rpm in Mode 2. The wheel radius is 0.28 m. Vehicle speed V is in kilometers per hour and ranges from −40 to 160 km/h. The final drive gear ratio (including axle) is 3.3.

Solution

The final drive speed is a function of vehicle speed V:

$$\omega_{out} = (V \times 1000/3600/0.28) \times 3.3 \times 60/2\pi = 31.2\text{V}(\text{rpm})$$

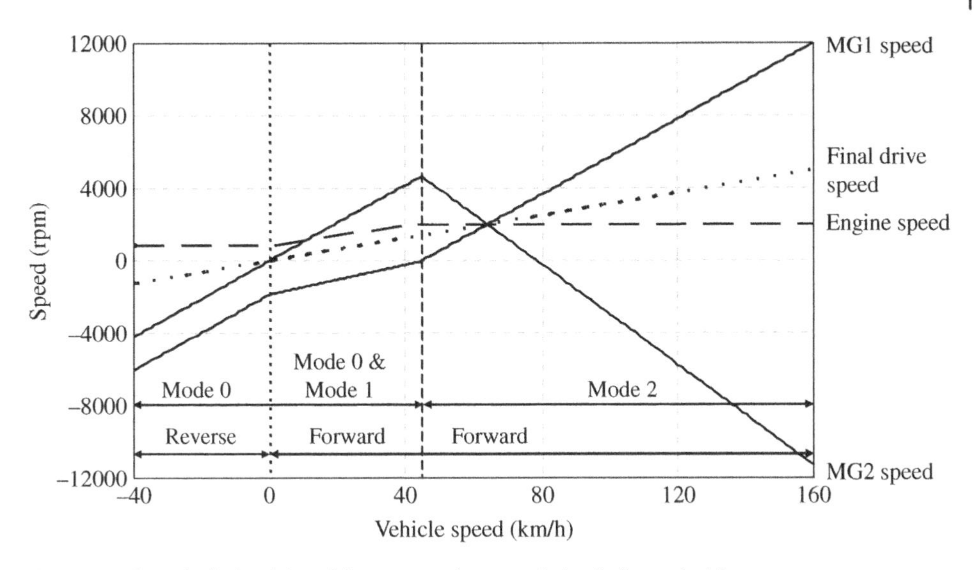

Figure 4.9 Speed relationships of the two-mode transmission in Example 4.2.

In Mode 0, the engine is kept at 800 rpm and the speed of the ring gear of P2 is zero. Therefore

$$\omega_m = \frac{N_{s2} + N_{r2}}{N_{s2}}\omega_{c2} = 3.33\omega_{out}$$

$$\omega_e = 800 \text{ rpm}$$

$$\omega_g = \frac{N_{s1} + N_{r1}}{N_{s1}}\omega_{c1} - \frac{N_{r1}}{N_{s1}}\omega_e = 3.33\omega_{out} - 2.33\omega_e$$

In Mode 1, the engine speed will ramp up from 800 to 2000 rpm. Note that the engine can be turned on or kept idling. Engine on/off is determined by vehicle power demand. The speed relationships are the same as in Mode 0.

When the speed of the sun gear of P1 reaches zero, the vehicle will shift from Mode 1 to Mode 2. In Mode 2, the engine speed is kept at 2000 rpm. The speed relationships are

$$\omega_e = 2000 \text{rpm}$$

$$\omega_g = \frac{N_{s1} + N_{r1}}{N_{s1}}\omega_{c1} - \frac{N_{r1}}{N_{s1}}\omega_e = 3.33\omega_{out} - 2.33\omega_e$$

$$\omega_m = \frac{N_{s2} + N_{r2}}{N_{s2}}\omega_{c2} - \frac{N_{r2}}{N_{s2}}\omega_{r2} = 3.33\omega_{out} - 2.33\omega_g$$

Figure 4.9 shows the speeds of the system: engine, MG1, MG2, and final drive.

Example 4.3 Planetary gear set 1 has 35 teeth for the sun gear and 65 teeth for the ring gear. Planetary gear set 2 has 30 teeth for the sun gear and 70 teeth for the ring gear. The engine is kept at 0 rpm in Mode 0, ramped up from 0 to 3000 rpm in Mode 1, and kept at 3000 rpm in Mode 2. The wheel radius is 0.28 m. Vehicle speed V is in kilometers per hour and ranges from −40 to 160 km/h. The final drive gear ratio (including axle) is 3.

Solution

The final drive speed is a function of vehicle speed V:

$$\omega_{out} = (V \times 1000/3600/0.28) \times 3.3 \times 60/2\pi = 31.2\text{V(rpm)}$$

In Mode 0, the engine is kept at 0 rpm and the speed of the ring gear of P2 is zero. Therefore

$$\omega_m = \frac{N_{s2} + N_{r2}}{N_{s2}} \omega_{c2} = 3.33\omega_{out}$$
$$\omega_g = \frac{N_{s1} + N_{r1}}{N_{s1}} \omega_{c1} = 2.86\omega_{out}$$

In Mode 1, the engine speed will ramp up from 0 to 3000 rpm. Note that the engine can be turned on or kept idling. Engine on/off is determined by vehicle power demand. The speed relationships are

$$\omega_m = \frac{N_{s2} + N_{r2}}{N_{s2}} \omega_{c2} = 3.33\omega_{out}$$
$$\omega_g = \frac{N_{s1} + N_{r1}}{N_{s1}} \omega_{c1} - \frac{N_{r1}}{N_{s1}} \omega_e = 2.86\omega_{out} - 1.86\omega_e$$

When the speed of the sun gear of P1 reaches zero, the vehicle will shift from Mode 1 to Mode 2. In Mode 2, the engine speed is kept at 3000 rpm. The speed relationships are

$$\omega_e = 3000\,\text{rpm}$$
$$\omega_m = \frac{N_{s2} + N_{r2}}{N_{s2}} \omega_{c2} - \frac{N_{r2}}{N_{s2}} \omega_{r2} = 3.33\omega_{out} - 2.33\omega_g$$
$$\omega_g = \frac{N_{s1} + N_{r1}}{N_{s1}} \omega_{c1} - \frac{N_{r1}}{N_{s1}} \omega_e = 2.86\omega_{out} - 1.86\omega_e$$

Figure 4.10 shows the speeds of the system: engine, MG1, MG2, and final drive.

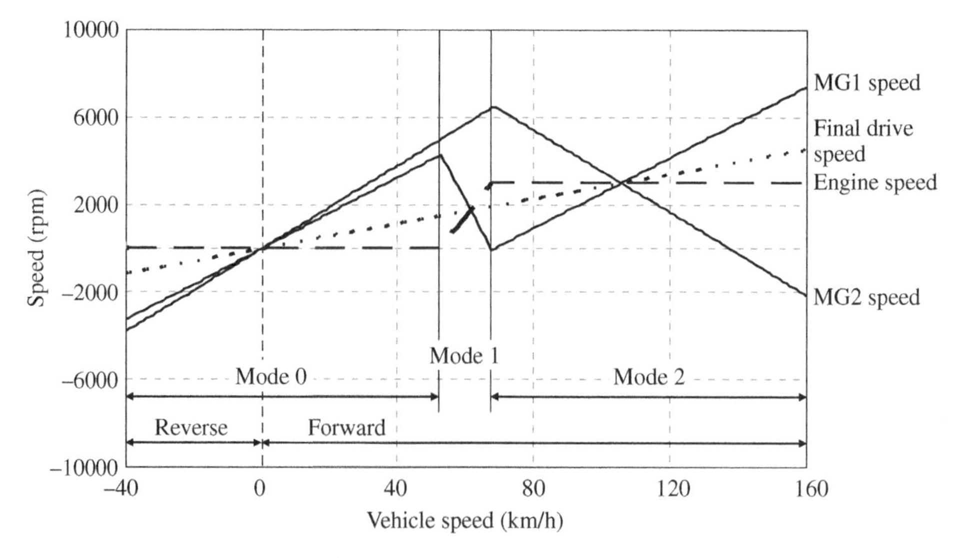

Figure 4.10 Speed relationships of the two-mode transmission in Example 4.3.

4.4 Dual-Clutch Hybrid Transmissions*

There are a few variants of automatic transmissions such as automated manual transmission (AMT), CVT, and dual-clutch transmission (DCT). Each of these technologies has its own penetration levels in different regions of the world (North America, Europe, or Asia). The advantages of DCT include high efficiency, low cost, and driving comfort. Conservative estimates peg DCT technology at around 10% of the global market by 2015. Table 4.1 compares the advantages and disadvantages of CVT, AMT, and DCT.

DCT technology is well suited for both high-torque diesel engines and high-revving gas engines. Some of the major drivers for DCT include

- flexible and software tunability
- gear ratio flexibility the same as that of manual lay shaft transmissions allowing greater compatibility with any engine characteristics.

4.4.1 Conventional DCT Technology

A typical DCT architecture has a lay shaft with synchronizers used for maximum efficiency. It also has launch clutches (either wet or dry) used with electronics, along with mechanical or hydraulic actuation systems to achieve the automatic shifting. Lay shaft transmissions yield an efficiency of 96% or better as compared to 85–87% efficiency of automatic transmissions [3, 4].

Figure 4.11 shows the diagram of a DCT-based transmission. It is a typical setup found in many of the latest vehicle models with a DCT. It consists of two coaxial shafts, each having the odd and even gears. It is tantamount to having two transmissions, hence the name.

In a DCT system, the two clutches are connected to two separate sets of gears. The odd gear set is connected to one of the clutches and the even gear set to the other clutch. It is necessary to preselect the gears to realize the benefits of the DCT system. Accordingly, the off-going clutch is released at the same time as the on-coming clutch is engaged. This gives an uninterrupted torque supply to the driveline during the shifting process. This preselection of gears can be implemented using either complicated controllers such as fuzzy logic or simple ones such as selections based on the next anticipated vehicle speed.

4.4.2 Gear Shift Schedule

Initially, when the vehicle starts, gear N1i is synchronized. Therefore, engine torque is transmitted to the final drive through gears N1i and N1m. Vehicle speed increases as

Table 4.1 Qualitative comparison of automatic and manual transmissions.

Aspects	Automatic transmission	Manual transmission	Desired transmission
Cost	Expensive	Lower	Low
Efficiency	Moderate	High	High
Ease of use	Easy	Hard	Easy
Comfort	Good	Poor	Good

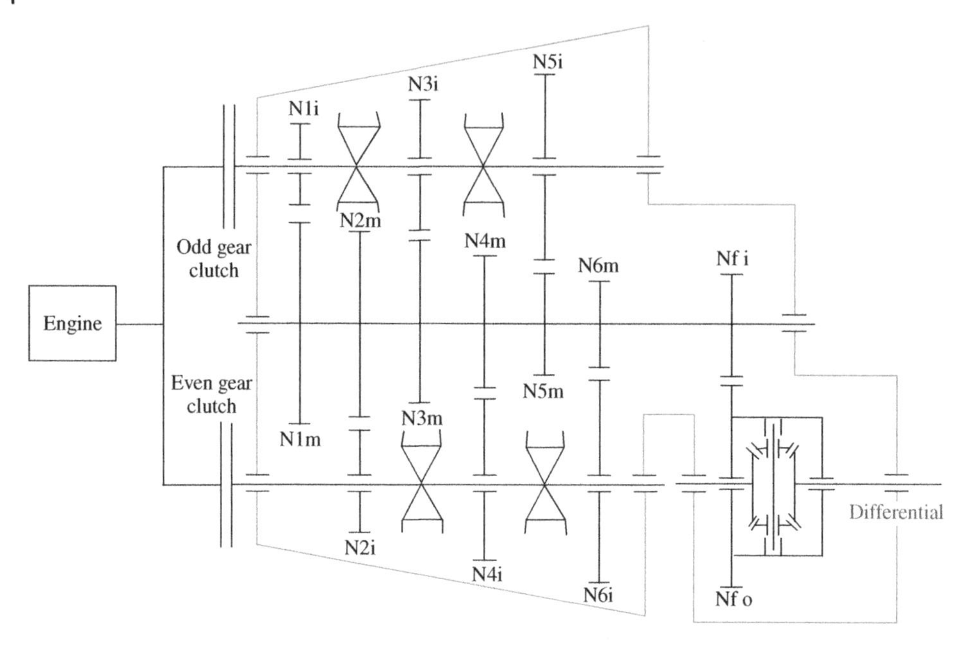

Figure 4.11 Dual-clutch transmission. Note that the reverse gear is omitted in the diagram.

the odd clutch engages. When vehicle speed reaches a certain threshold, gear N2i is synchronized. As the even clutch engages (the odd clutch disengages), engine torque is shifted from gear N1i to N2i. Hence engine torque is transmitted through gear N2i and N2m. As vehicle speed increases, N3i is synchronized. Then the odd clutch would engage and the even clutch would disengage. This process will continue until the vehicle speed becomes stable (from N3i to N4i, from N4i to N5i, and from N5i to N6i).

During downshift, the process is reversed. For example, assume initially that N4i is synchronized and the even clutch is engaged. During downshift, N3i is synchronized before the even clutch opens. When the even clutch disengages and the odd clutch engages, engine torque is transferred from N4i to N3i. Similarly, N2i would be synchronized before the even clutch engages.

Since all transitions in a DCT are managed by gear synchronizers and two clutches, there is no need for a torque converter in a DCT. The transitions (gear shifting and torque shifting) are very smooth. Control of the synchronizers and clutches, or shift controller, is computerized in the vehicle. The shift controller decides the upshifts or the downshifts of the transmission as per the gear shift schedule as shown from left to right in Figure 4.12. This controller intelligently preselects the higher or the lower gear depending on the current and desired vehicle velocity.

4.4.3 DCT-Based Hybrid Powertrain

The diagram for a DCT-based hybrid powertrain is shown in Figure 4.13 [5]. The transmission is a six-speed AMT. The hybrid powertrain consists of two motors with each coupled mechanically onto the two shafts using a standard gear reduction. Due to the presence of the motor/generator, the vehicle can be reversed without the reverse gear. The odd shaft houses gears 1, 3, and 5, and the even shaft houses gears 2, 4, and 6. The two motors can also be operated as generators as needed by the hybrid control strategy.

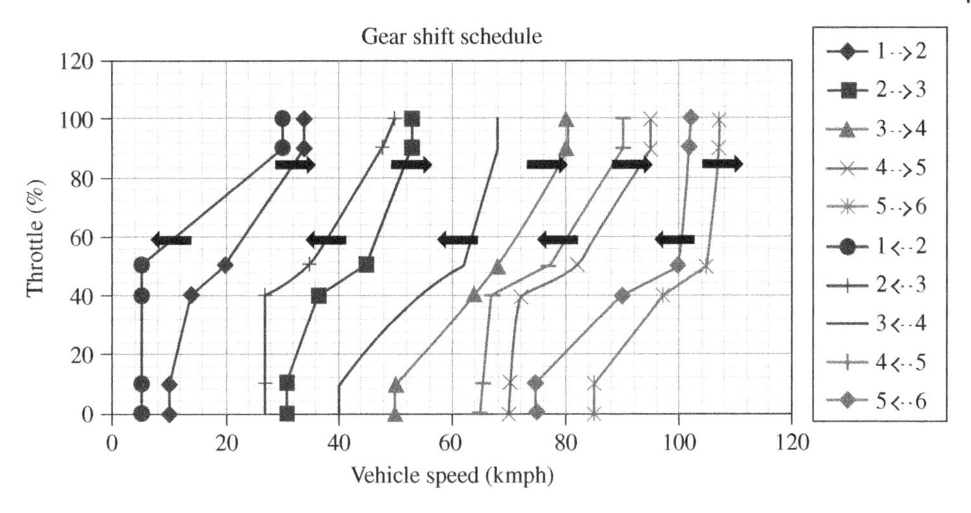

Figure 4.12 Gear shift schedule.

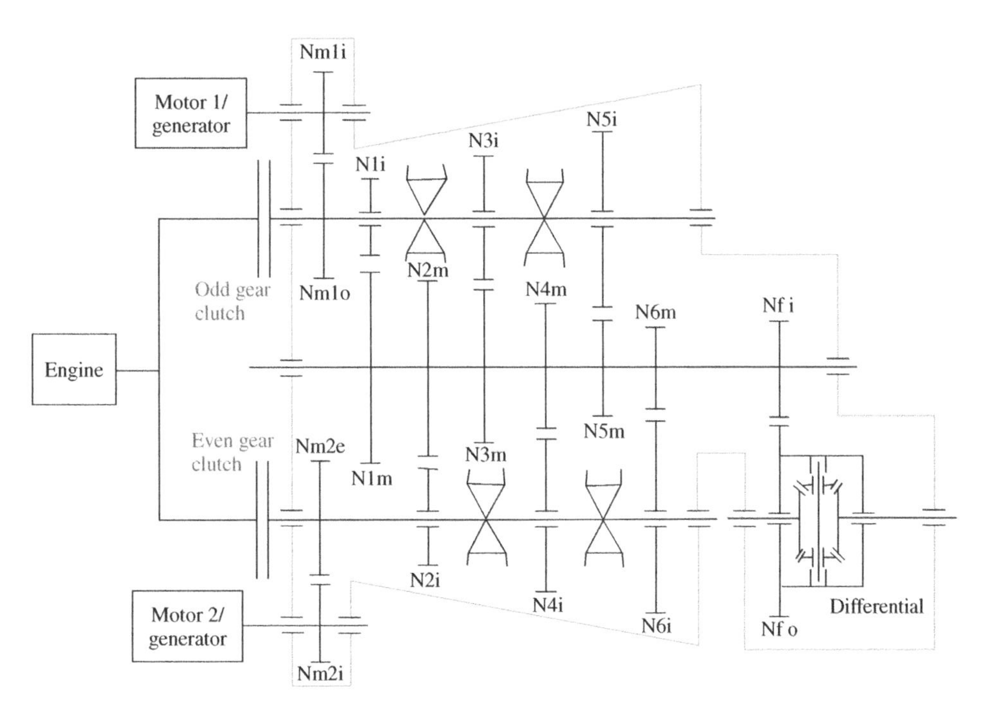

Figure 4.13 Hybrid powertrain based on dual-clutch transmissions. Reverse gear is not needed due to the fact that the motors can be used to back up the vehicle.

4.4.4 Operation of DCT-Based Hybrid Powertrain

The DCT-based hybrid powertrain shown in Figure 4.13 has seven operating modes when the vehicle is in motion and one additional operating mode for standstill charging.

4.4.4.1 Motor-Alone Mode

The vehicle is always launched in the motor-only mode unless the battery's state of charge (SOC) is below the minimum level. In this mode, the gears are selected according to the shift logic controller. The vehicle operates in this mode up to a maximum speed defined by the controller, provided the SOC is greater than the minimum SOC for the battery as per the system design. Since the engine does not operate in this mode, the dual clutches are disengaged to prevent any backlash to the engine. Either motor can be used for the launch and backup of the vehicle. The equations for this mode are

$$\omega_o = \frac{\omega_m}{i_g \; i_a \; i_m} \tag{4.39}$$

$$T_o = i_a \; i_g \; i_m \times T_m \tag{4.40}$$

4.4.4.2 Combined Mode

This mode is selected when a high torque is required for situations such as sudden acceleration or climbing a grade. This mode is also selected if the vehicle speed becomes more than the maximum speed defined by the controller in the motor-alone mode. Both the engine and the motor provide the propulsion power to the drive shaft. Depending on the vehicle speed, the transmission shift controller selects the appropriate clutch and the gears. The power flow is shown in Figure 4.14. The equations for this mode are

$$\omega_o = \frac{\omega_m}{i_g \; i_a \; i_m} = \frac{\omega_e}{i_g \; i_a} \tag{4.41}$$

$$T_o = i_g \; i_a \; i_m \; T_m + i_g \; i_a \; T_e \tag{4.42}$$

4.4.4.3 Engine-Alone Mode

This mode involves the engine as the only source of propulsion. The engine controller ensures that the engine transmits power to the lowest possible gear ratio such that the engine remains in the best efficiency window. The equations for this mode are

$$\omega_o = \frac{\omega_e}{i_g \; i_a} \tag{4.43}$$

$$T_o = i_a \; i_g \times T_e \tag{4.44}$$

4.4.4.4 Regenerative Braking Mode

The motor is coupled to the output shaft through gears, and it can function as a generator as well. It is used to recover the energy during braking to charge the battery. Depending on the current clutch that is used, the controller decides which motor is to

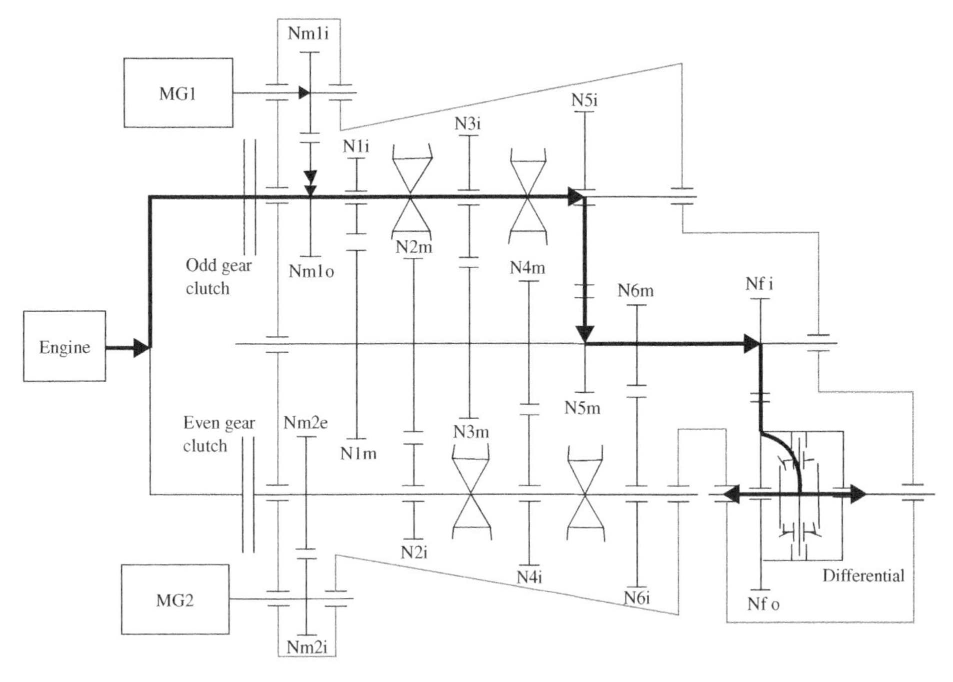

Figure 4.14 Power flow in the combined mode.

be operated in this mode. In case the motor torque is not sufficient to brake the vehicle, a conventional braking system is used to supplement the braking demand.

The equations for this mode are

$$\omega_{in} = \frac{\omega_m}{i_m \ i_g \ i_a} \tag{4.45}$$

$$T_{in} = -i_m \ i_a \ i_g \times T_m \tag{4.46}$$

4.4.4.5 Power Split Mode

This mode is used to charge the battery when the vehicle is in motion. The vehicle controller decides on this mode if the engine supplies more power than that required to drive the vehicle. The excess power is then used to charge the battery. The motor on the same lay shaft that drives the output shaft is selected to act as the generator to charge the battery. The motor controller selects the correct motor depending on the shaft that is transmitting the power to the final drive. The equations for this mode are

$$\omega_o = \frac{\omega_m}{i_a \ i_m \ i_g} = \frac{\omega_e}{i_a \ i_g} \tag{4.47}$$

$$T_o = i_a \ i_g \ T_e - T_m \tag{4.48}$$

4.4.4.6 Standstill Charge Mode

This mode can be used to crank-start the engine or charge the battery when the vehicle is in standstill position. The controller opts for this mode when the battery SOC is lower than the minimum SOC level permitted by the design. This is the only operating mode when the engine is cranked and the vehicle is in standstill position. Since the vehicle is not moving and no power is transmitted to the drive train, all the gears are disengaged for safety. The kinematic equations for this mode are

$$\omega_o = 0 \tag{4.49}$$

$$T_e = T_m \ i_m \tag{4.50}$$

$$\omega_e = \frac{\omega}{i_m} \tag{4.51}$$

4.4.4.7 Series Hybrid Mode

This mode offers a very interesting option for the DCT-based hybrid powertrain. The engine is operated in a region near its sweet spot (by adaptively changing the gear ratios) so that the torque generated from the engine is used by one of the motors to generate electricity. This electricity is then used by another motor on the other shaft to drive the vehicle. This therefore gives the option of having the DCT-based hybrid powertrain operating as a series hybrid. The kinematic equations for this mode are

$$\omega_o = \frac{\omega_m}{i_a \ i_m \ i_g} \tag{4.52}$$

$$T_o = T_m \ i_m \ i_g \ i_a \tag{4.53}$$

4.5 Hybrid Transmission Proposed by Zhang et al.*

An alternate hybrid transmission was proposed to use one electric motor, a planetary gear set, and four fixed gears to realize automated transmission and CVT for a parallel hybrid, as shown in Figure 4.15 [6]. The design is based on the concept of AMTs. It uses a combination of lay shaft gearing and planetary gearing for power transmission. The lay shaft gears on the input shaft and the motor shaft freewheel unless engaged by the shifter–synchronizer assemblies. The carrier of the planetary gear train is connected with the input shaft that picks up the engine torque. The sun is connected with the motor shaft if so engaged. One motor is used either as the driving assisting unit or as the generator in charging and regenerative braking operations. Mode switching and gear shifting are realized by shifters actuated by computer and controlled by step motors as in an AMT. The hybrid system has five operating modes for vehicle driving and one standstill mode for emergency or convenience operations. The six operating modes and the related kinematics are described in the following.

4.5.1 Motor-Alone Mode

The vehicle is always launched in the motor-alone mode. In this mode, the motor shaft is engaged by the shifter, and power is transmitted to the final drive by the motor gears.

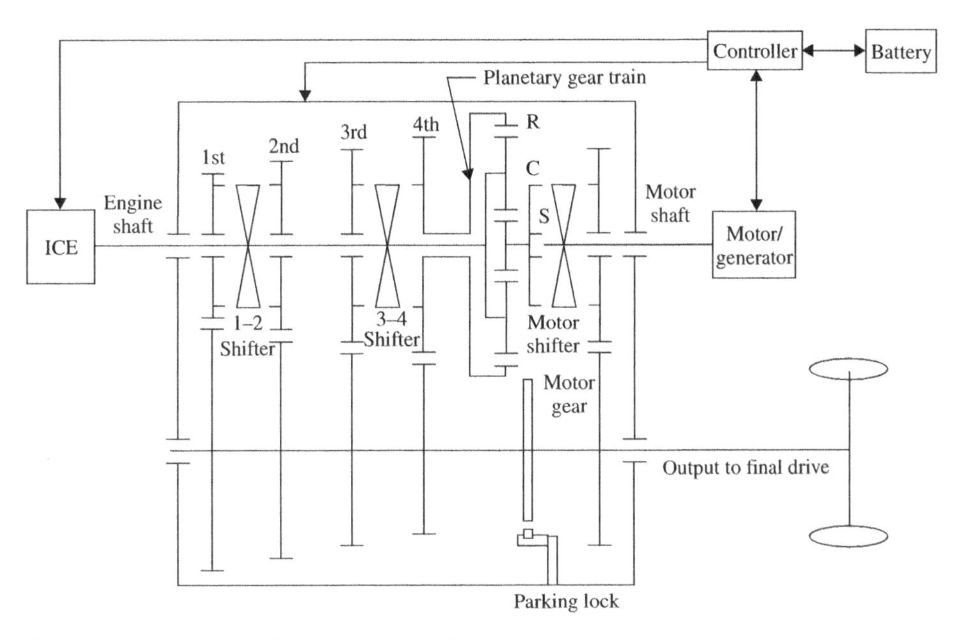

Figure 4.15 Hybrid transmission proposed by Zhang et al.

Vehicle backup is realized by reversing the motor rotation. All other gears freewheel in this mode. The operating parameters for the engine and the motor in this mode are related by

$$\omega_o = \frac{\omega_m}{i_m} \tag{4.54}$$

$$T_o = i_m \; T_m \tag{4.55}$$

where ω_o and ω_m are the angular velocities of the output shaft and the motor respectively, T_o and T_m are the output torque and the motor torque respectively, and i_m is the motor gear ratio.

4.5.2 Combined Power Mode

The combined mode is used when high power is required in situations such as accelerating and hill climbing. In this mode, the motor shaft and one of the lay shaft pairs are engaged. One of the four available gears as shown in Figure 4.15 can be selected according to the vehicle operating condition. The motor and the engine operate at speeds that are mechanically linked by the gear ratios to drive the vehicle jointly, with the operating parameters related as

$$\omega_o = \frac{\omega_e}{i_k} = \frac{\omega_m}{i_m} \tag{4.56}$$

$$T_o = i_k \; T_e + i_m \; T_m \tag{4.57}$$

where ω_e and T_e are the angular velocity and the torque of the engine respectively, and i_k (with $k = 1–4$) is the gear ratio of the lay shaft gear pairs.

4.5.3 Engine-Alone Mode

The engine-alone mode is the most efficient mode for highway cruising. In this mode, a lay shaft gear pair with low gear ratio is engaged to transmit the engine torque to the output shaft with the motor shaft in neutral. The vehicle runs like a conventional vehicle in this mode. The operating parameters are linked by the lay shaft gear ratio:

$$\omega_o = \frac{\omega_e}{i_k} \tag{4.58}$$

$$T_o = i_k \; T_e \tag{4.59}$$

4.5.4 Electric CVT Mode

The electric CVT mode provides two degrees of freedom for vehicle operation control that allows optimization of engine operation for best fuel economy. In this mode, the engine drives the vehicle and powers the generator for battery charging at the same time. The sun gear of the planetary gear train is coupled to the motor shaft by the shifter, and the fourth lay shaft gear is coupled to the ring gear. The operating parameters of the system are governed by the characteristics of the planetary gear train as

$$\omega_m = \frac{N_s + N_r}{N_s}\omega_e - \frac{N_r}{N_s} i_4 \; \omega_o \tag{4.60}$$

$$T_m = T_s = \frac{N_s}{N_s + N_r} T \tag{4.61}$$

$$T_o = T_r \; i_4 = \frac{N_r}{N_s + N_r} i_4 \; T_e \tag{4.62}$$

where N_r and N_s are the number of teeth of the ring gear and sun gear, respectively; T_r and T_s are the ring gear torque and sun gear torque, which are distributed from the engine torque at a constant proportion. The output torque T_o and the angular velocity ω_o are determined by the vehicle driving condition. The engine torque and the torque provided to the generator are determined by optimizing the engine efficiency. In the electric CVT mode, the engine speed ω_e is optimized at the point for the highest efficiency corresponding to the required torque. The generator speed ω_m is controlled such that the engine speed and torque are optimized.

4.5.5 Energy Recovery Mode

In the energy recovery mode, the motor is coupled to the output shaft through the motor gear pair by the shifter, and functions as a generator. The relations for the operating parameters are the same as in the motor-alone mode, with the power flow reversed.

4.5.6 Standstill Mode

In this mode, the motor is engaged by the shifter to the sun gear and the parking locker is applied to lock the output shaft (ring gear). The lay shaft gears freewheel. This mode can be used to crank-start the engine or use the engine to charge a low battery at standstill. It can also be used as a generator for household electricity or other conveniences if a bidirectional power converter is provided. The parameters are

$$\omega_e = \frac{N_s}{N_r + N_s}\omega_m \tag{4.63}$$

$$T_m = \frac{N_s}{N_r + N_s}T_e \tag{4.64}$$

4.6 Renault IVT Hybrid Transmission

In the Renault infinitely variable transmission (IVT) as shown in Figure 4.16, there are two electric motors MG1 and MG2, two planetary gears sets P1 and P2, but no clutches [7]. MG1 is connected to the sun gear of P1; MG2 is connected to the sun gear of P2; the engine is connected to the carrier of P2 as well as to the ring gear of P1; the carrier of P1 and the ring gear of P2 are coupled together and connected to the final drive. The system is capable of providing infinitely variable transmission by controlling the two electric motors to match the vehicle speed while optimizing operation of the engine.

Since there are no clutches, there is only one operating mode. The steady-state torque–speed relationships are

$$\omega_e = \omega_{c2} = \frac{N_{r2}}{N_{s2} + N_{r2}}\omega_o + \frac{N_{s2}}{N_{s2} + N_{r2}}\omega_{mg2} \tag{4.65}$$

$$\omega_o = \omega_{c1} = \frac{N_{r1}}{N_{s1} + N_{r1}}\omega_e + \frac{N_{s1}}{N_{s1} + N_{r1}}\omega_{mg1} \tag{4.66}$$

$$T_o = T_{r2} + T_{c1} = \frac{N_{r2}}{N_{s2} + N_{r2}}T_e + \frac{N_{r1} + N_{s1}}{N_{s1}}T_{mg1} \tag{4.67}$$

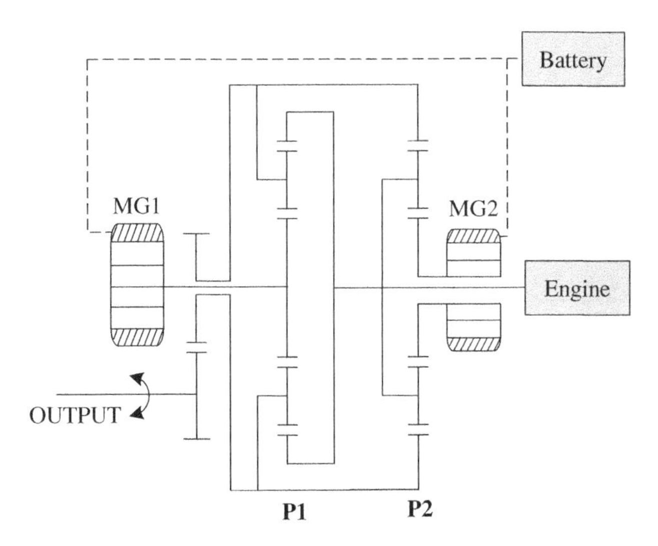

Figure 4.16 Renault two-mode transmission.

$$T_{mg2} = \frac{N_{s2}}{N_{s2} + N_{r2}} T_e \tag{4.68}$$

$$T_{mg1} = \frac{N_{r2}}{N_{r2}} T_e \tag{4.69}$$

4.7 Timken Two-Mode Hybrid Transmission

The Timken hybrid powertrain shown in Figure 4.17 is also a two-mode hybrid system [8]. The transmission contains two electric motors, MG1 and MG2, two planetary gears, P1 and P2, two clutches, C1 and C2, and two locks, B1 and B2.

The engine is connected to the ring gear of P1; MG1 is connected to the sun gear of P1 and via a clutch (C2) to the ring gear of P2; MG2 is connected to the sun gear of P2; the carrier of P2 is connected to the output shaft; the carrier of P1 is connected through C1 to the output shaft, or can be locked by B1. By controlling the two clutches and two locks, the system can operate in high range or low range, based on vehicle operating conditions.

4.7.1 Mode 0: Launch and Reverse

The vehicle can be launched by MG2. In this mode, B2 locks the ring gear of P2. MG2 torque is transferred through the sun gear of P2 to the carrier of P2. Since the sun gear of P1 is coupled to the ring gear of P2, the sun gear of P1 is also locked. In this case, the carrier of P1 needs to be locked by B1 so that the engine is stalled as well. The equations are

$$\omega_{mg2} = \omega_{s2} = \frac{N_{s2}}{N_{s2} + N_{r2}} \omega_o \tag{4.70}$$

$$T_o = \frac{N_{s2} + N_{r2}}{N_{s2}} T_{mg2} \tag{4.71}$$

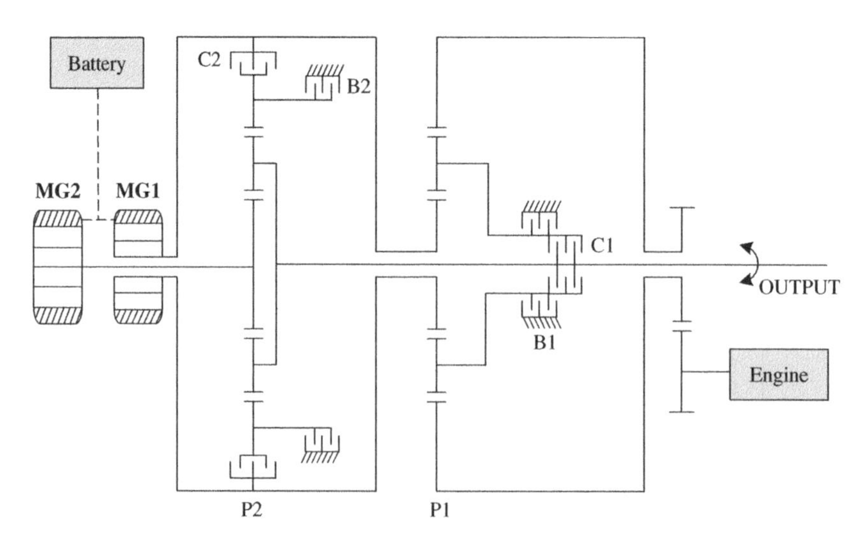

Figure 4.17 Timken two-mode transmission.

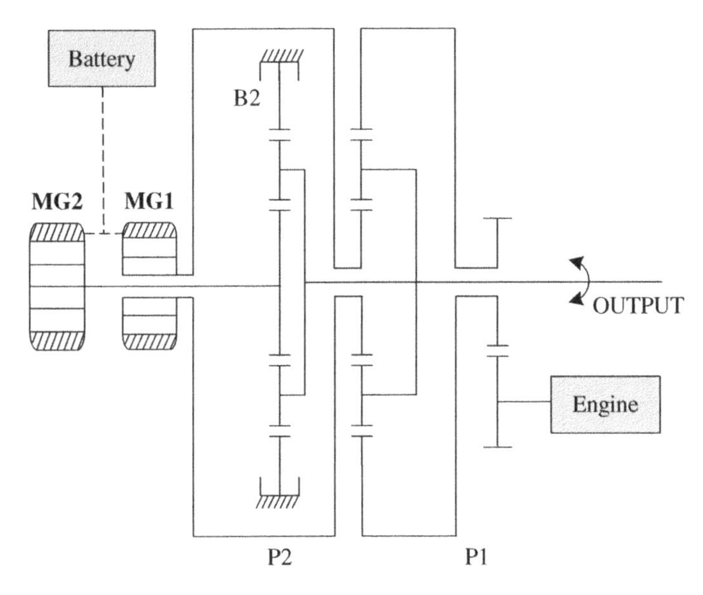

Figure 4.18 Low-speed mode of the Timken two-mode transmission.

4.7.2 Mode 1: Low-Speed Operation

In this mode, B2 locks the ring gear of P2; C1 engages the engine, as shown in Figure 4.18. The operation of this mode is exactly the same as the GM two-mode hybrid.

4.7.3 Mode 2: High-Speed Operation

In high-speed operation, C1 engages the carrier of P1; C2 engages MG1 (Figure 4.19). The sun gear of P1, the ring gear of R2, and MG1 will all have the same speed. This mode is also the same as the GM two-mode hybrid powertrain.

4.7.4 Mode 4: Series Operating Mode

The powertrain can also operate in series mode by locking the carrier of P1 and the ring gear of P2. In this mode (Figure 4.20), engine power is delivered to MG1 through the sun gear of P1 (with the carrier locked). The electricity generated by MG1 will be delivered to MG2, which drives the sun gear of P2, which in turn drives the carrier of P2 with the ring gear locked. The torque-speed equations are

$$\omega_o = \omega_{c2} = \frac{N_{s2}}{N_{s2} + N_{r2}} \omega_{mg2} \tag{4.72}$$

$$\omega_{mg1} = -\frac{N_{r1}}{N_{s1}} \omega_e \tag{4.73}$$

$$T_o = \frac{N_{r2} + N_{s2}}{N_{s2}} T_{mg2} \tag{4.74}$$

$$T_{mg1} = \frac{N_{s1}}{N_{r1}} T_e \tag{4.75}$$

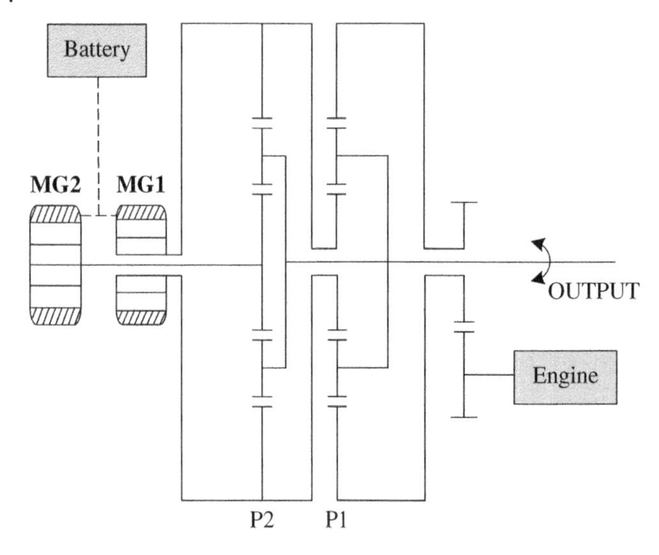

Figure 4.19 High-speed mode of the Timken two-mode transmission.

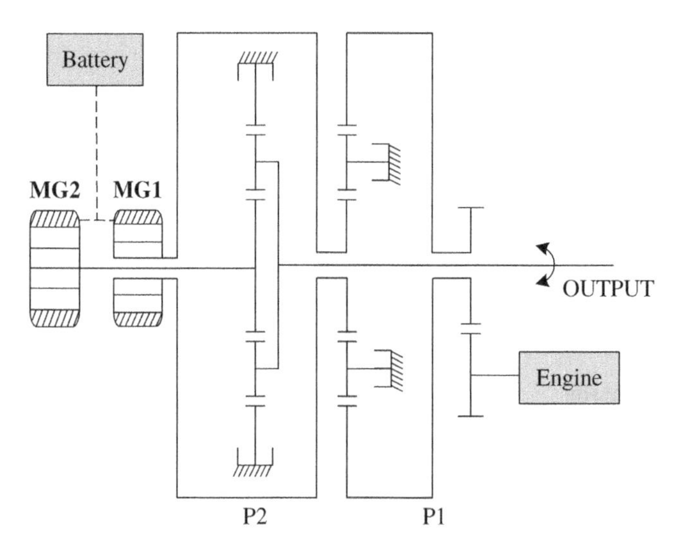

Figure 4.20 Series operating mode of the Timken two-mode transmission.

4.7.5 Mode Transition

Similar to any other hybrid powertrain, the transition between different modes needs to happen at the moment when mechanical disturbance to the system can be minimized. For example, with the carrier of P1 locked to B1, the engine can be started by MG1. In order to engage the carrier of P1 with the final drive shaft, first B1 needs to be released, then the speed of MG1 will be controlled such that C1 will accelerate to the same speed as the final drive shaft, and then C1 will engage the carrier of P1 to the final drive shaft. Similarly, in order to engage the ring gear of P2 to the sun gear of P1 (and MG1), the sun gear speed of P1 and MG1 is first brought down to zero and then C2 is engaged.

4.8 Tsai's Hybrid Transmission

In the hybrid system proposed by Tsai, as shown in Figure 4.21, the transmission includes one electric motor, two clutches, two planetary gear sets, and two locks [9, 10].

There are 14 different combinations of operating modes based on the different configurations of the clutches and locks, but there are only seven valid modes, as shown in Table 4.2.

The speed/torque relationships can be written in two conditions: C1 engaged or C2 engaged.

When C1 engages (B1 freewheel)

$$\omega_e = \frac{N_{s1}}{N_{s1}+N_{r1}}\omega_m + \frac{N_{r1}}{N_{s1}+N_{r1}}\omega_o \tag{4.76}$$

$$T_o = \frac{N_{r1}}{N_{s1}+N_{r1}}T \tag{4.77}$$

$$T_m = \frac{N_{s1}}{N_{s1}+N_{r1}}T_e \tag{4.78}$$

When C1 engages (B1 locked)

$$\omega_e = \frac{N_{s1}}{N_{s1}+N_{r1}}\omega_m + \frac{N_{r1}}{N_{s1}+N_{r1}}\omega_o \tag{4.79}$$

$$\omega_o = \frac{N_{s2}}{N_{s2}+N_{r2}}\omega_m \tag{4.80}$$

$$T_o = \frac{N_{r1}}{N_{s1}+N_{r1}}T_e + \frac{N_{s2}+N_{r2}}{N_{s2}}T_m \tag{4.81}$$

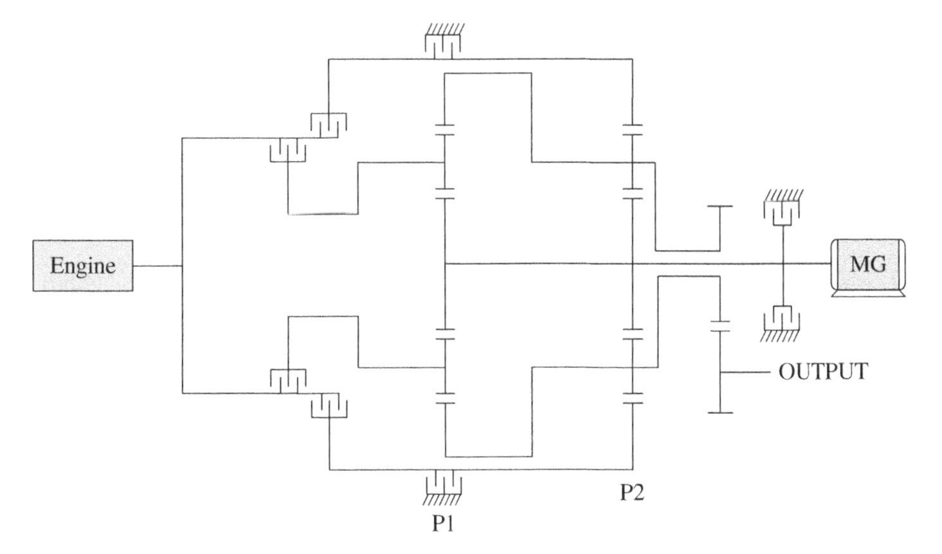

Figure 4.21 A multimode hybrid transmission proposed by Tsai et al.* [9]. *Source:* Tsai 2001. Reproduced with permission of ASME.

* The permission is for Figure 4.21 and Table 4.2.

Table 4.2 Different combinations of operating modes for Tsai's hybrid transmission.

C1	C2	B1	B2	State	Additional modes
0	0	0	0	None	—
0	0	0	1	None	—
0	0	1	0	Motor alone	Engine idle, generator/regen.
0	0	1	1	None	—
0	1	0	0	Engine + motor	Motor, generator/regen.
0	1	0	1	Engine alone	Motor stationary
0	1	1	0	None	—
0	1	1	1	None	—
1	0	0	0	Engine alone/CVT	Generator/charging
1	0	0	1	Engine + motor	Motor stationary
1	0	1	0	Engine + motor	Motor, generator/regen.
1	0	1	1	None	—
1	1	0	0	Engine + motor	Motor, generator/regen.
1	1	0	1	None	—
1	1	1	0	None	—
1	1	1	1	None	—

From [9]. © 2001 ASME. Reprinted, with permission, from *ASME Journal of Mechanical Design.*

When C2 engages

$$\omega_o = \frac{N_{s2}}{N_{s2}+N_{r2}}\omega_m + \frac{N_{r2}}{N_{s2}+N_{r2}}\omega_e \tag{4.82}$$

$$\frac{N_{s2}+N_{r2}}{N_{r2}}T_e = \frac{N_{s2}+N_{r2}}{N_{s2}}T_m \tag{4.83}$$

When C1 and C2 engage

$$\omega_o = \omega_m = \omega_e \tag{4.84}$$

$$T_o = T_m = T \tag{4.85}$$

4.9 Hybrid Transmission with Both Speed and Torque Coupling Mechanism

The hybrid configuration proposed by Ehsani et al. in [11] uses one electric motor, three clutches, and two locks to achieve both speed coupling and torque coupling functions (Figure 4.22) [11]:

- **Mode 0:** Vehicle launch and backup (motor-alone mode), and regenerative braking. C1 open, C2 closed, C3 open, L1 closed, and L2 open. Only the motor torque

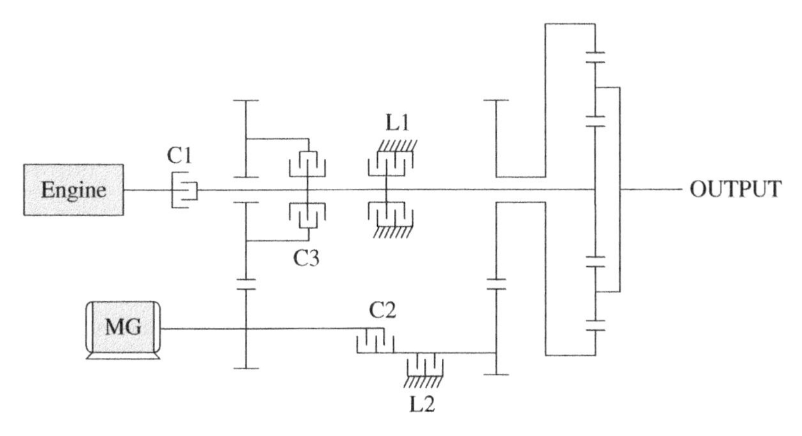

Figure 4.22 A hybrid transmission proposed in [11]. *Source:* Ehsani 2009. Reproduced with permission of Taylor & Francis.

is transmitted to the final drive. The sun gear of the planetary gear is locked. The torque/speed relationships are

$$\omega_{out} = \frac{N_r}{N_s + N_r}\omega_m \tag{4.86}$$

$$T_{out} = \frac{N_s + N_r}{N_r}T_m \tag{4.87}$$

- **Mode 1:** Engine-alone mode. C1 closed, C2 open, C3 open, L1 open, L2 closed. In this mode, the motor is off – only the engine is transferring torque to the final drive. The ring gear is locked. The torque/speed relationships are

$$\omega_{out} = \frac{N_s}{N_s + N_r}\omega_e \tag{4.88}$$

$$T_{out} = \frac{N_s + N_r}{N_s}T_e \tag{4.89}$$

- **Mode 2:** Low-speed mode. C1 closed, C2 open, C3 closed, L1 open, L2 closed. In this mode, the engine torque and the motor torque are added to provide the maximum drivetrain torque for acceleration needs. The ring gear is locked and the motor torque is added to the engine shaft. The torque/speed relationships are

$$\omega_{out} = \frac{N_s}{N_s + N_r}\omega_e \tag{4.90}$$

$$\omega_m = \frac{N_1}{N_2}\omega_e \tag{4.91}$$

$$T_{out} = \frac{N_s + N_r}{N_s}\left(T_e + \frac{N_1}{N_2}T_m\right) \tag{4.92}$$

- **Mode 3:** Combined and power split mode (CVT). C1 closed, C2 closed, C3 open, L1 open, L2 open. In this mode, the motor and the engine output are coupled to the planetary gear on the sun gear and ring gear, respectively. The output and input relationships are

$$\omega_{out} = \frac{N_s}{N_s + N_r}\omega_e + \frac{N_r}{N_s + N_r}\omega_m \tag{4.93}$$

$$T_{out} = \frac{N_s + N_r}{N_s}T_e = \frac{N_s + N_r}{N_r}T_m \tag{4.94}$$

The vehicle could be running between Mode 2 and Mode 3 during highway cruising.

Mode transition in this transmission is complicated. In order to reduce mechanical disturbance, the locks have to be engaged at zero speed and the clutches have to be engaged when the two sides of the gears have similar speeds.

4.10 Toyota Highlander and Lexus Hybrid, E-Four-Wheel Drive

The Toyota Highlander and Lexus hybrid vehicles feature an electric four-wheel drive, or e-four, with the front wheels driven by a planetary gear-based hybrid powertrain, and the rear wheels driven by an electric motor. The generalized schematics are shown in Figure 4.23.

In this scheme, the engine is connected to the carrier of the planetary gear set, the generator is connected to the sun gear, and the ring gear is connected to the final drive of the front axle. The total powertrain torque available during any driving condition is

$$T_{out} = \frac{N_r}{N_s + N_r}T_e + \frac{N_1}{N_2}T_{m1} + \frac{N_3}{N_4}T_{m2} \tag{4.95}$$

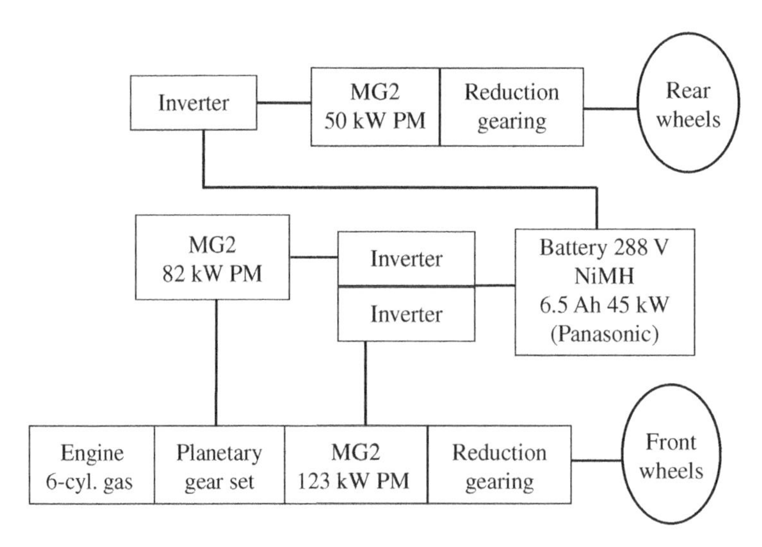

Figure 4.23 Schematics of electric four-wheel-drive hybrid system.

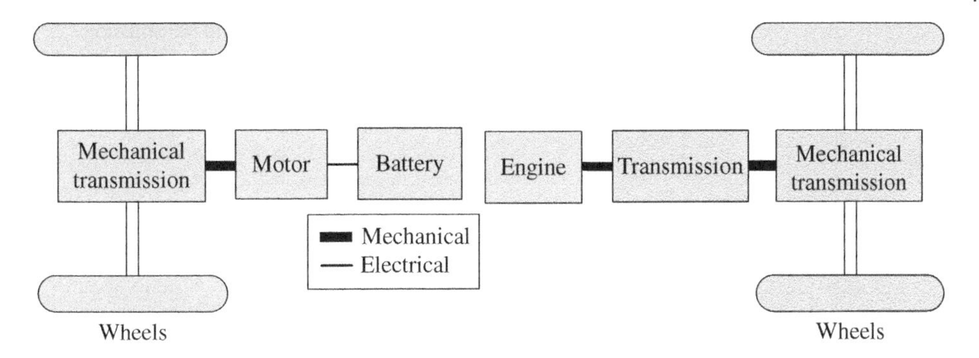

Figure 4.24 Hybrid powertrain with separate driving axles.

The generator torque is

$$T_g = \frac{N_s}{N_s + N_r} T_e \tag{4.96}$$

The power relationship, when neglecting losses, is

$$P_{out} = P_{m1} + P_{m2} + P_r \tag{4.97}$$

$$P_e = P_r + P_g \tag{4.98}$$

$$P_{m1} + P_{m2} = P_g + P_B \tag{4.99}$$

A simplified version of the e-four is shown in Figure 4.24, with the goal of reducing the overall system cost (http://reviews.cnet.com/suv/2006-toyota-highlander-hybrid/1707-10868_7-31352761.html). In comparison to the above configurations, this configuration has a significant cost advantage and is simple to fabricate and manufacture, because it does not involve any modification to the front axle design. However, this design does not allow the flexibility of engine speed control. Besides, the system is not very efficient during power split mode operation, because engine power needs to be transferred through the vehicle body to the rear axle and then to the electric motor. The total torque of the powertrain is

$$T_{out} = k_e \; T_e + k_m \; T_m \tag{4.100}$$

where k_e and k_m are the gear ratios of the engine transmission and motor transmission, respectively.

4.11 CAMRY Hybrid

In the Camry hybrid (Figure 4.25), there are two planetary gear sets. The engine, generator, and planetary gear set 1 are configured the same way as in the Prius, that is, the engine is coupled to the carrier, and the generator is coupled to the sun gear. However, the ring gear is connected to a counter gear, which is also connected to the ring gear of the motor speed reduction planetary gear. The motor is connected to the sun gear of planetary gear

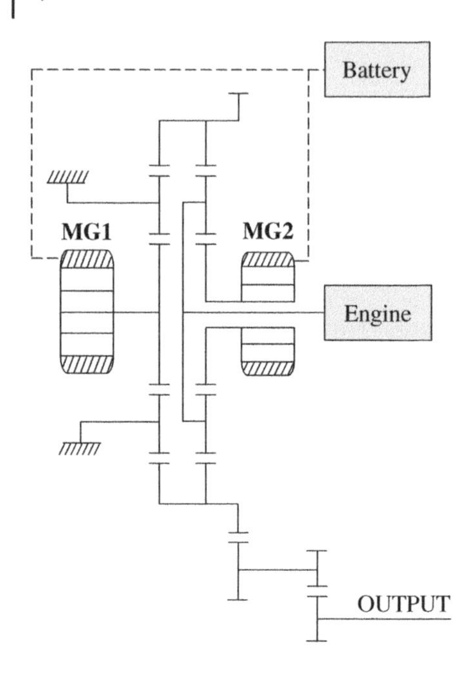

Figure 4.25 Toyota Camry hybrid transmission.

set 2, and the carrier of planetary gear set 2 is grounded. The multifunction gear connects the counter gear and the final drive.

The ring gear speed can be calculated from the vehicle speed:

$$\omega_r = \frac{N_2}{N_1}\frac{N_4}{N_3} i_{fd} V \tag{4.101}$$

The engine and generator satisfy

$$\omega_e = \frac{N_{r1}}{N_{r1} + N_{s1}} \omega_r + \frac{N_{s1}}{N_{r1} + N_{s1}} \omega_g \tag{4.102}$$

The motor and the ring gear satisfy

$$\omega_r = -\frac{N_{s2}}{N_{r2}} \omega_m \tag{4.103}$$

The torque at the final drive is

$$T_{out} = \left(\frac{N_{r2}}{N_{s2}} T_m + \frac{N_{r1}}{N_{r1} + N_{s1}} T_e \right) \frac{N_2}{N_1}\frac{N_4}{N_3} i_{fd} \tag{4.104}$$

4.12 Chevy Volt Powertrain

The Chevy Volt from GM (Figure 4.26) has been described as an extended range electric vehicle (EREV). The exact powertrain configuration is not yet public. However, a number of sources have suggested that the Volt employs two electric motors and a planetary gear

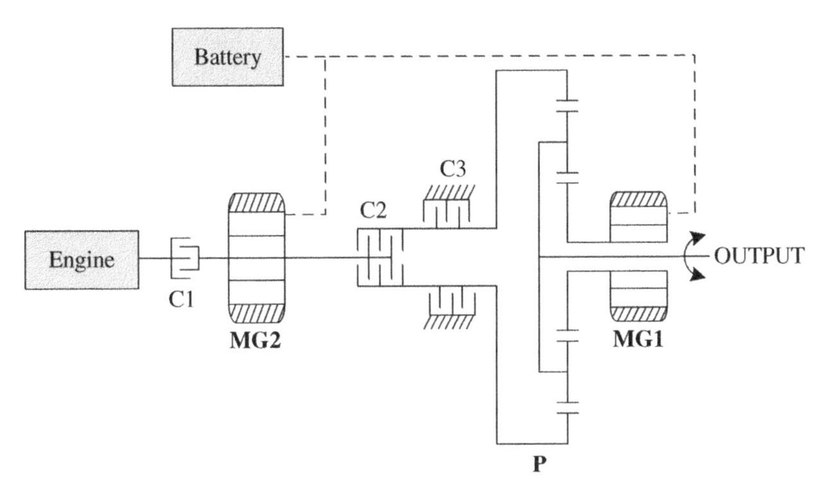

Figure 4.26 The Chevy Volt transmission.

set, along with the engine and three clutches [12, 13]. The Volt is equipped with a 16 kWh lithium-ion battery pack, a 125 kW induction motor, and a 1.4 liter four-cylinder engine. The initial driving range of 25–50 miles (40–80 km) can be achieved by using energy from the onboard battery, and additional driving range can be achieved by using gasoline.

In the Volt's transmission, MG1 is the main drive motor, which is connected to the sun gear of the planetary system. The ring gear can be grounded through clutch C3, or connected to MG2 by engaging clutch C2. The carrier is connected to the final drive. The engine can be connected to MG2 through clutch C1. MG1 can be connected either to the engine to become a generator through C1, or to the ring gear to become a motor through C2.

The operation of the system can be described as follows:

- **Mode 1:** Single-motor driving mode. In this mode, MG1 drives the sun gear with the ring gear locked by C3. The torque of the motor is transmitted through the carrier to the final drive. The engine and MG2 are idle. All driving power is provided by the battery. This mode is suitable for launch, backup, low-speed driving, and cruising. The equation is

$$T_{out} = \frac{N_s}{N_s + N_r} T_{MG1} \tag{4.105}$$

- **Mode 2:** Dual-motor driving mode. In this mode, MG2 is connected to the ring gear through C2, with C3 disengaged. The engine is idle with C1 open. Both motors receive power from the battery to drive the carrier, which delivers torque to the final drive. The equations are

$$\omega_{out} = \frac{N_s}{N_s + N_r} \omega_{MG1} + \frac{N_r}{N_s + N_r} \omega_{MG2} \tag{4.106}$$

$$T_{out} = \frac{N_s}{N_s + N_r} T_{MG1} = \frac{N_r}{N_s + N_r} T_{MG2} \tag{4.107}$$

Since the speed of the two motors is added, this can achieve a high cruising speed for the vehicle. Hence, the authors believe that this mode is used for highway cruising with the battery only.

- **Mode 3:** Extended driving range. In this mode, C1 is engaged, so the engine drives MG2 which is now a generator. C2 is open and C3 is engaged to lock the ring gear. Electricity generated by MG2 is delivered to MG1 through power inverters. Only MG1 is driving the final drive. Hence, the output torque expression is the same as in Mode 1:

$$T_{out} = \frac{N_s}{N_s + N_r} T_{MG1} \quad \text{and} \quad T_{MG2} = kT_e \tag{4.108}$$

where k is the gear ratio between the engine and MG2.

- **Mode 4:** With the engine turned on and C1 engaged, C2 is engaged as well (but C3 is open). Now, a portion of the engine torque can be transmitted to the ring gear to drive the carrier, which delivers the combined engine and MG1 torque to the final drive. Another portion of the engine torque is still used to drive MG2 to generate electricity. This mode is suitable for high speed and for heavy accelerations. Thus,

$$kT_e = T_{MG2} + T_r \tag{4.109}$$

$$\omega_{out} = \frac{N_s}{N_s + N_r} \omega_{MG1} + \frac{N_r}{N_s + N_r} \omega_{MG2} \tag{4.110}$$

- **Regenerative braking mode:** The engagement of the clutches is the same as in Mode 1. The only difference is that the wheels are now driving MG1 to generate electricity and at the same time generating the required braking torque to slow down the vehicle.

4.13 Non-Ideal Gears in the Planetary System

In the previous sections, the torque/speed relations were given under the assumption that the losses of transmission/gears are neglected, and the kinetic motion is also neglected. When gear losses are considered, the torque equations of a planetary gear set in steady state can be expressed as

$$T_r + \xi_r \ \omega_r = \frac{N_r}{N_r + N_s} \left(T_c - \xi_c \ \omega_c \right) \tag{4.111}$$

$$T_s + \xi_s \ \omega_s = \frac{N_s}{N_r + N_s} \left(T_c - \xi_c \ \omega_c \right) \tag{4.112}$$

where ξ_c, ξ_r, and ξ_s stand for the frictional loss coefficients of the carrier, ring gear, and sun gear, respectively. m – motor, r – ring gear, s – sun gear, c – carrier, g – generator, e – engine. ζ_m is the motor output gear.

When the gear losses are considered, the torque relationships of the Prius transmission are

$$T_r + \xi_r \ \omega_r = \frac{N_r}{N_r + N_s} \left(T_e - \xi_c \ \omega_c \right) \tag{4.113}$$

$$T_g + \xi_s\ \omega_s = \frac{N_s}{N_r + N_s}\left(T_e - \xi_c\ \omega_c\right) \tag{4.114}$$

$$T_o = T_r + \frac{N_{m1}}{N_{m2}}\left(T_m - \xi_m\ \omega_m\right) \tag{4.115}$$

where ξ is the friction coefficient of each gear, and subscript m represents motor, r, ring gear, s, sun gear, c, carrier, g, generator, e, engine. ξ_m is the friction coefficient of the motor output gear.

4.14 Dynamics of the Transmission

When the dynamics of the transmission is considered, there will also be transients in the transmission. For any given rotational system, the rotational dynamics can be written as

$$T_{in} = T_{out} + J\frac{d\omega}{dt} \tag{4.116}$$

In the following, the dynamics of the Toyota Prius planetary gear-based hybrid transmission is further analyzed. The analysis of other systems should be very similar.

The inertias of the final drive shaft and axle are transferred to the output shaft of the transmission [2]:

$$J_{fd} = \frac{1}{G_a^2}J_a + J_{fd\_sh} \tag{4.117}$$

where the subscripts are: a, axle, and sh, final drive shaft. G_a is the final drive gear ratio. The final drive and the ring gear are coupled directly to the motor shaft. Therefore, on the motor shaft, the total inertia is

$$J_{ma} = J_m + J_r + J_{fd} \tag{4.118}$$

The sun gear is coupled to the generator. Therefore, the total generator shaft inertia is

$$J_{gq} = \left(J_s + J_g\right) \tag{4.119}$$

The engine shaft inertia is the total of the engine crankshaft and the carrier:

$$J_e = J_{crank} + J_c \tag{4.120}$$

where J_{crank} is the crankshaft inertia. From generator to carrier,

$$J_{gc} = \frac{\left(N_r + N_s\right)N_r}{N_s^2}\left(J_g + J_s\right) \tag{4.121}$$

The generator shaft inertia can be transferred to the motor shaft and engine shaft. The equivalent inertias of the engine shaft and the motor shaft are

$$J_{eq} = J_e + \frac{\left(N_r + N_s\right)^2}{N_s^2}J_{gq} \tag{4.122}$$

$$J_{mq} = J_{ma} + \frac{N_r^2}{N_s^2}J_{gq} \tag{4.123}$$

On the generator shaft,

$$T_g = \frac{N_s}{N_s + N_r}\left(T_e - \xi_c\ \omega_c - J_{eq}\ \dot{\omega}_e - J_{gc}\ \dot{\omega}_m\right) - \xi_s\ \omega_s \tag{4.124}$$

On the output shaft,

$$\begin{aligned} T_o &= T_m - \xi_m\ \omega_m + \frac{N_r}{N_r + N_s}\left(T_e - \xi_c\ \omega_c\right) \\ &- \left(\frac{N_r}{N_r + N_s} J_{eq} + J_{gc}\right)\dot{\omega}_e - \left(\frac{N_r}{N_r + N_s} J_{gc} + J_{mq}\right)\dot{\omega}_m \end{aligned} \tag{4.125}$$

This torque will drive the final drive shaft at a certain speed. Due to slip of the wheels, $\lambda = (\omega r - V)/V$, there exists a traction force $F_{fd} = mg\mu(\lambda)$, where $\mu(\lambda)$ is the traction coefficient. This traction force is to overcome the resistive force of the vehicle during driving:

$$F_{fd} = mg\mu(\lambda) = mg\sin\alpha + \frac{1}{2}C_D\ A_F\rho V^2 + mg\left(C_0 + C_1\ V^2\right) + m\frac{dV}{dt} \tag{4.126}$$

4.15 Conclusions

It should be noted that most planetary gear-based HEVs, including the Toyota, Ford, and GM two-mode hybrids, do not include a separate dedicated starter for the engine. The engine is started by one of the motors/generators at an appropriate condition. Due to the fact that the engine usually starts at the time when the drive needs more power, such as on acceleration, there is usually a "jerk" or "hiccup" because a portion of the motor torque is diverted to start the engine. The battery has limited power capability.

Another issue that many drivers have experienced is the weakness of the 14V battery used to supply power to the vehicle's auxiliary power, such as for wipers, headlights, entertainment systems, power steering, and hydraulic compressor.

The authors feel that if a starter–alternator is added, the "jerk" during acceleration can be eliminated, because the engine may be started by the 14V onboard battery. Besides, this starter can also be used to charge the 14V battery when the engine is on, which may also ease the burden on the 14V battery.

In the case of the Toyota and Ford hybrid systems, an additional clutch between the engine and the planetary gear system could help smooth the acceleration.

Control of these powertrains is complicated. Often an advanced control algorithm is needed to manage the system. Fuzzy logic, dynamic programming, and wavelet transforms are popular in the power management of complex hybrid vehicles.

References

1 Holmes, A.G. and Schmidt, M.R. (2002) Hybrid electric powertrain including a two-mode electrically variable transmission. US Patent US6478705 B1, November 12, 2002.
2 Miller, J.M. (2006) Hybrid electric vehicle propulsion system architectures of the e-CVT type. *IEEE Transactions on Power Electronics*, 21 (3), 756–767.

3 Kulkarni, M., Shim, T., and Zhang, Y. (2007) Shift dynamics and control of dual-clutch transmissions. *Mechanism and Machine Theory*, 42 (2), 168–182.
4 Liu, Y., Qin, D., Jiang, H., and Zhang, Y. (2009) A systematic model for dynamics and control of dual clutch transmissions. *Journal of Mechanical Design*, 131, 061012.
5 Joshi, A., Shah, N.P., and Mi, C. (2009) Modeling and simulation of a dual clutch hybrid vehicle powertrain. 5th IEEE Vehicle Power and Propulsion Conference, September 7–11, Dearborn, MI.
6 Zhang, Y., Lin, H., Zhang, B., and Mi, C. (2006) Performance modeling of a multimode parallel hybrid powertrain. *Journal of Mechanical Design, Transactions of the ASME*, 128 (1), 79–80.
7 Villeneuve, A. (2004) Dual mode electric infinitely variable transmission. Proceedings of the SAE TOPTECH Meeting on Continuously Variable Transmission, March 8–11, Detroit, MI, pp. 1–11.
8 Ai, X., Mohr, T., and Anderson, S. (2004) An electro-mechanical infinitely variable speed transmission. Proceedings of the SAE Congress Expo, March 8–11, Detroit, MI.
9 Tsai, L.W., Schultz, G.A., and Higuchi, N. (2001) A novel parallel hybrid transmission. *Journal of Mechanical Design, Transactions of the ASME*, 123, 161–168.
10 Schultz, G.A., Tsai, L.W., Higuchi, N., and Tong, I.C. (2001) Development of a novel parallel hybrid transmission. SAE 2001 World Congress, March, Detroit, MI.
11 Ehsani, M., Gao, Y., and Emadi, A. (2009) *Modern Electric, Hybrid Electric, and Fuel Cell Vehicles: Fundamentals, Theory, and Design*, 2nd edn, Power Electronics and Applications Series, CRC Press, Boca Raton, FL.
12 Conlon, B.M., Savagian, P.J., Holmes, A.G., and Harpster, M.O. Jr. (2007) Output split electrically-variable transmission with electric propulsion using one or two motors. US Patent US2009/0082171 A1, filed September 10, 2007, and published March 26, 2009.
13 Amend, J.M. (2010) Charge up, Chevy Volt rises above sound, fury of introduction. Ward's AutoWorld, November.

5

Plug-In Hybrid Electric Vehicles

5.1 Introduction to PHEVs

Plug-in hybrid electric vehicles (PHEVs) have the potential to displace transportation fuel consumption by using grid electricity to drive the car. PHEVs can be driven initially using electric energy stored in the onboard battery, and an onboard gasoline engine can then extend the driving range. In the 1990s and early 2000s, pure electric cars were not successful, and one of the major reasons was the limited driving range available at that time. For example, the GM electric vehicle (EV) had a range of about 100 miles (160 km) and the Ford Ranger electric truck had a range of approximately 60 miles (96 km).

5.1.1 PHEVs and EREVs

PHEVs are sometimes called range-extended electric vehicles (ReEVs) or extended range electric vehicles (EREVs) in the sense that these vehicles always have onboard gasoline or diesel that can be used to drive the vehicle for an extended distance when the onboard battery energy is depleted. Furthermore, these vehicles can provide high fuel economy during the extended driving range due to the large battery pack that can accept more regenerative braking energy and provide more flexibility for engine optimization during the extended driving range.

However, EREVs, such as the GM Chevy Volt, must be equipped with a full-sized electric motor so that pure electric driving can be realized for all kinds of driving conditions. It is shown that, for some driving conditions, all-electric drive sometimes does not provide the most benefits, given the limited battery energy available.

For example, the powertrain motor of a PHEV is rated at 125 kW, and the battery pack has a capacity of 16 kWh. This means that at full powertrain power, the battery needs to supply a power eight times its nominal capacity, or 8 C. Not only is this high power requirement difficult to achieve, but it also results in inevitable heavy losses inside the battery pack, which makes the drive system very inefficient.

Example 5.1 A 400 V, 16 kWh lithium-ion battery pack has an internal impedance of approximately 0.5 Ω. The motor and inverter have a combined efficiency of 90% and rated output of 125 kW. The rated current is 347 A on the DC side of the inverter. At this current, the battery will drop 174 V internally. In other words, the battery terminal

Hybrid Electric Vehicles: Principles and Applications with Practical Perspectives,
Second Edition. Chris Mi and M. Abul Masrur.

voltage will be only 226 V at 347 A. This will result in a battery output power of 78 kW instead of the full 125 kW, and a loss inside the battery of 60.8 kW.

5.1.2 Blended PHEVs

Blended PHEVs have become more popular because of the reduced system cost (smaller electric motor, smaller battery pack, and lower battery power ratings), as well as the flexibility of optimizing fuel economy for different driving conditions. Compared to an EREV, a blended PHEV usually uses a parallel or complex configuration, in which both the engine and the motor can drive the wheels directly. Since the engine is available for propulsion at high power demand, the size of the electric motor and the power requirement for the battery pack can be much smaller than that of an EREV. Therefore, the cost of the vehicle is reduced.

Planetary gear-based hybrid vehicles, such as in the Toyota Prius and the GM two-mode hybrid, can be considered as parallel configurations since the electric motor (referred to as MG2 in Chapter 4) is in parallel with the engine output, while the generator (referred to as MG1 in Chapter 4) is used to realize the continuously variable transmission (CVT) and to optimize engine operation.

5.1.3 Why PHEV?

A survey [1, 2] showed that 78% of the US population drives an average of 40 miles (64 km) or less in their daily commuting. Figure 5.1 shows the distribution of daily miles driven versus percentage of population. Based on this survey, a PHEV with an electric range of 40 miles (or PHEV40) will satisfy the daily driving needs of 78% of the US

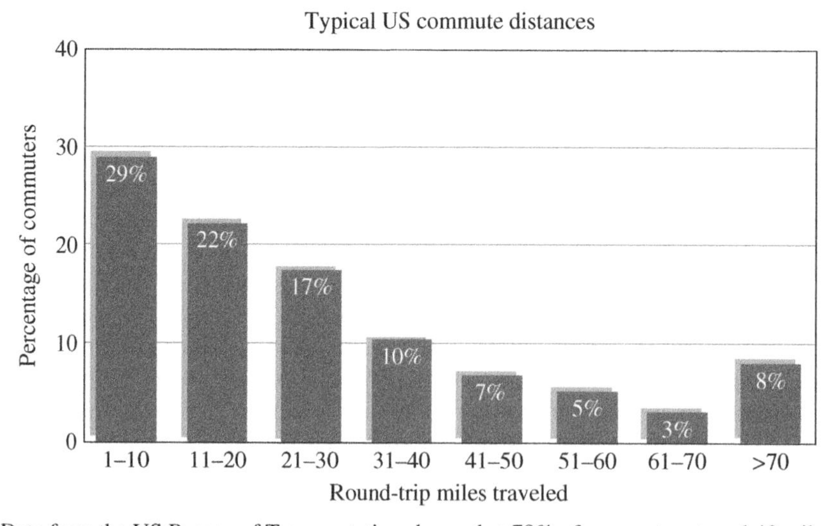

Data from the US Bureau of Transportation shows that 78% of commuters travel 40 miles or less each day – the expected battery-only range of PHEVs with routine overnight charging. For longer distances, the vehicles could run indefinitely in hybrid (gasoline/electric) mode.

Figure 5.1 Daily commuting distance versus population [1]. *Source:* Sanna 2005. Reproduced with permission of EPRI.

population while driving on electricity in their daily commuting. Furthermore, people owning a 40 mile electric range PHEV but driving less than 40 miles daily will not need to refuel gasoline if they charge their car at night on a daily basis.

PHEVs can produce significant environmental and economic benefits for society. The advantages of PHEVs can be evaluated by how much fuel is displaced, as well as by how much pollution, including greenhouse gas (GHG) emissions, can be reduced. The main purpose for developing PHEVs can be summarized as follows:

1) **Displacement of fossil fuel consumption in the transportation sector:** Since PHEV owners will not need to refuel gasoline or need less gasoline, a significant amount of fossil fuel can be saved. This will have a long-term impact on the economy, environment, and political arena.
2) **Reduction of emissions:** Due to the reduced use of gasoline, a significant amount of emissions can be reduced due to the large deployment of PHEVs. Centralized generation of electricity is much more efficient and produces much less emissions than gasoline-powered cars. The mitigation of emissions from urban (by cars) to remote areas (in power plants) where electricity is generated can also mitigate the heavy pollution in population-dense metropolitan areas. As more and more electricity in the future will come from renewable energy sources (which will be used by PHEVs), the emissions can be further reduced.
3) **Energy cost savings:** PHEVs use electricity for the initial driving range. Since electricity is cheaper than gasoline on an equivalent energy content basis, the cost per mile driven on electricity is cheaper than on gasoline. At the present time, gasoline costs about $3 per gallon in the United States, while electricity costs $0.12 per kWh. For a medium-size car, a gallon of gasoline can drive the car for about 30 miles, while a kilowatt hour of electricity can drive it 5 miles. This means that it will cost approximately $0.72 (or 6 kWh of electricity) to drive 30 miles while it costs $3 of gasoline to drive 30 miles. In other words, a driver can save $2.28 for 30 miles driven. However, PHEVs are generally more expensive than conventional cars due to the increased number of components, such as batteries and motors. The initial cost of the PHEV will take the owner some time to recover from fuel savings. At current fuel prices, it will take a number of years for the owner to recover the initial investment. As petroleum prices increase in the long run, the cost savings could be more advantageous. Also, as the production of PHEVs ramps up, their cost is expected to drop. Government incentives (tax rebates etc.) also help subsidize the initial cost of PHEVs. A differentiated electricity pricing structure could potentially help PHEVs. This potential structure will charge the consumer a lower rate if the PHEV is charged during off-peak times. Of course, the consumer still has the option to charge at any time.
4) **Maintenance cost savings:** PHEVs can generally save on maintenance costs. Due to the extensive use of regenerative braking, braking system maintenance and repair is less frequent, such as brake pad replacement and brake fluid change. Since the engine is not operating, or operating for much less time, there will be longer intervals for oil changes and other engine maintenance services.
5) **Backup power:** A PHEV can be used as a backup power source when a bidirectional charger is provided. A typical PHEV battery pack can provide a home or office with 3–10 kW of power for a few hours, and the onboard engine generator/motor can further extend the backup duration by using gasoline to generate electricity.

6) **End-of-life use of the battery:** Batteries that can no longer provide the desired performance in a PHEV can potentially be used for grid energy storage, which provides voltage regulation, system stability, and frequency regulation for a power grid. In particular, frequency regulation and stability become more and more important as more and more renewable energy generation is put on the power grid. These "retired" batteries, which may still have 30–50% of their original energy capacity, can provide this type of service.

5.1.4 Electricity for PHEV Use

Since PHEVs need to be charged from the power grid, the utility industry must be ready for the large deployment of PHEVs. Fortunately, the deployment of PHEVs will take place over a period of time, which will give the utility industry sufficient time to get ready.

PHEVs will mostly be charged at night. The electric power grid has capacity available in late evenings and at night when most of the PHEVs are charged. An Electric Power Research Institute (EPRI) study showed that with 1 million PHEVs plugged into the US grid, there is no need to build a single power plant in the United States [1]. With 10 million PHEVs on the road, the United States will only have to build three additional power plants (http://www.ornl.gov/info/ornlreview/v40_2_07/2007_plug-in_paper.pdf) [1–4].

On the other hand, electricity can be generated from a variety of sources, including nuclear and renewables such as hydroelectric, wind, photovoltaic, and ocean waves. Even coal-fired power plants will have advantages because there are more coal reserves than petroleum reserves. Using renewable energy for PHEVs will significantly reduce the consumption of petroleum-based fossil fuel. Finally, centralized generation will reduce the total emissions and move the source of emissions away from urban areas.

Figure 5.2 shows the current electricity generation portfolio in the United States [5]. Coal-fired power plants constitute the primary electricity generation in the world.

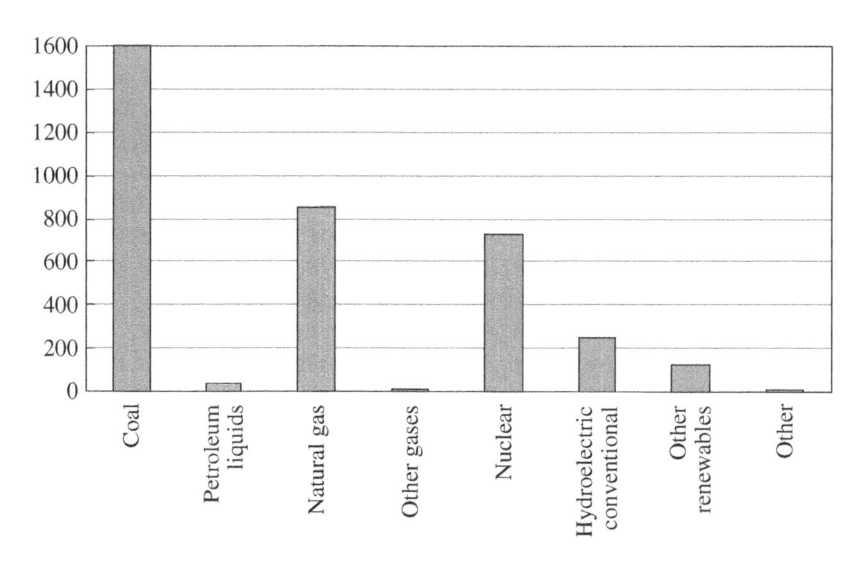

Figure 5.2 The US electricity generation portfolio [5]. *Source:* US Energy Information Administration.

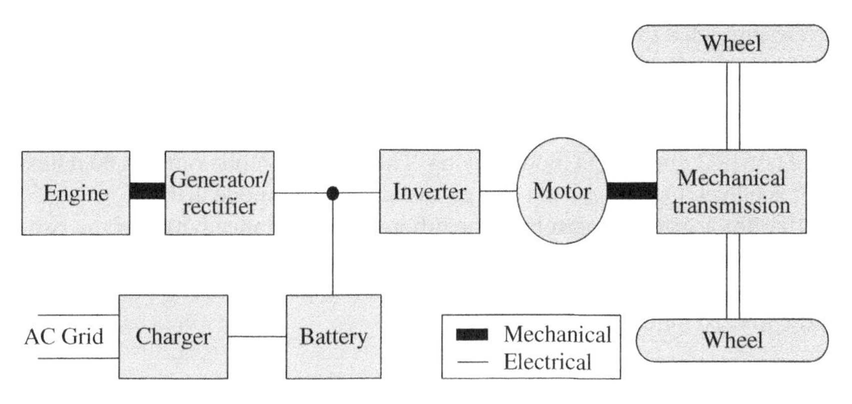

Figure 5.3 Series architecture of a PHEV.

However, it has been estimated that the world has coal reserves for about 400 years at the current rate of consumption while petroleum will last only 50 years. It can also be seen that there is potential for renewable energy to grow.

5.2 PHEV Architectures

Figure 5.3 shows the architecture of a series PHEV. In the series configuration, the gasoline engine output is connected to a generator. The electricity generated by the generator can be used to charge the battery or supply power to the powertrain motor. The electric motor is the only component driving the wheels. The motor can be an induction motor, a switched reluctance motor, or a permanent magnet motor. The motor can be mounted on the vehicle in the same way as in a conventional vehicle, without the need for transmission. In-wheel hub motors can also be chosen. In the series configuration, the motor is designed to provide the torque needed for the vehicle to drive in all conditions. The engine/generator can be designed to provide the average power demand.

Parallel and complex hybrids can also be designed as PHEVs. In parallel and complex configurations, the engine and the motor can both drive the wheels. Therefore, the motor size can be smaller than those in series configurations. In comparison to regular hybrid electric vehicles (HEVs), a parallel or complex PHEV will have a larger-sized battery pack, which provides a longer duration for extended electric drive. The engine is turned on whenever the vehicle's power demand is high.

5.3 Equivalent Electric Range of Blended PHEVs

For an EREV, the electric range can be easily calculated. For a blended PHEV, there may be no pure electric driving range available for some driving cycles. To find the equivalent electric range, it is useful to compare the fuel economy of a blended mode PHEV during charge-depletion (CD) mode to that of a comparable HEV.

Example 5.2 Assume that there are two vehicles, one regular HEV and one PHEV, both using the same powertrain architecture and components. However, the PHEV has a larger battery pack of 11 kWh while the HEV has a 2 kWh battery pack. Neglect the weight difference of the two vehicles. The HEV has a fuel economy of 40 MPG. The PHEV has a fuel economy of 60 MPG during CD mode driving. The total CD mode range is 60 miles for the PHEV. It achieves 40 MPG during charge-sustaining (CS) mode driving.

This means that for 1 gallon of gasoline, the difference in distance between the two vehicles is 20 miles. Therefore, the equivalent electric range of this PHEV is 20 miles. However, it must be pointed out that the 20 mile electric range is realized within a total 60 miles, not the first 20 miles, like the EREV.

Example 5.3 (Similar to Example 5.2): The HEV has a fuel economy of 40 MPG and the PHEV has a fuel economy of 120 MPG during CD mode driving and 40 MPG during CS mode driving. The total CD mode range of the PHEV is 45 miles.

For the first 45 miles, at 120 MPG, the PHEV consumes 0.375 gallons of gasoline. For this amount of gasoline, the HEV will run 15 miles. The difference between the two vehicles is 30 miles, which can be considered as the equivalent electric range of the PHEV.

5.4 Fuel Economy of PHEVs

The fuel economy of conventional vehicles is evaluated as fuel consumption (liters) per 100 km, or miles per gallon. In the United States, the Environmental Protection Agency sets the methods for fuel economy certification. There are usually two figures, one for city driving and one for highway driving. There is an additional fuel economy figure that evaluates the combined fuel economy by combining the 55% city and 45% highway MPG figures [6–8]:

$$FE_{MPG\_Combined} = \frac{1}{\dfrac{0.55}{FE_{city}} + \dfrac{0.45}{FE_{highway}}} \tag{5.1}$$

For pure EVs, the fuel economy is best described by electricity consumption for a certain range, for example, Wh/mile or kWh/100 km. For example, a typical passenger car consumes 120–250 Wh/mile. In order to compare the fuel efficiency of EVs with conventional gasoline or diesel vehicles, the energy content of gasoline is used to convert the figures. Since 1 gallon of gasoline contains 33.7 kWh energy (http://www.eere.doe.gov), the equivalent fuel economy of an EV can be expressed as

$$FE_{gas\_equivalent} = \frac{1}{\text{Wh/mile}} \times 33700 \tag{5.2}$$

Therefore, a passenger car that consumes 240 Wh/mile will have an equivalent gasoline mileage of 140 MPG from the energy point of view.

5.4.1 Well-to-Wheel Efficiency

The above fuel efficiencies are also called tank-to-wheel efficiencies. This does not reflect the losses during the refining and distribution. It is sometimes easier to compare the overall fuel efficiencies of conventional vehicles and EVs. For gasoline, this efficiency is

83%, which reflects a lumped efficiency from the refining and distribution of gasoline. For electricity generation, this efficiency is 30.3%, which reflects a lumped efficiency that includes electricity generation of 32.8% (assuming electricity is generated from gasoline) and distribution of electricity at 92.4%. Charge efficiency of the battery also needs to be reflected [9]. Thus,

$$FE_{EV\_well\_wheel} = \frac{1}{\text{Wh/mile}} \times 33700 \times \eta_{electricity} \tag{5.3}$$

$$FE_{ICEV\_well\_wheel} = FE_{mpg} \times \eta_{gasoline} \tag{5.4}$$

where $\eta_{electricity} = 30.3$ and $\eta_{gasoline} = 83\%$, and subscript *ICEV* stands for internal combustion engine vehicle.

Example 5.4 A car of 30 MPG will have a well-to-wheel fuel efficiency of 24 MPG, and an EV that consumes 240 Wh/mile will have a well-to-wheel efficiency of 42.5 MPG.

5.4.2 PHEV Fuel Economy

For PHEVs, it is usually confusing as to which figure should be used. Here, we discuss two different scenarios: all-electric capable PHEVs and blended PHEVs.

For all-electric capable PHEVs, it is useful to indicate the electric range, in miles or kilometers, and associated energy consumption during that range, in kWh/mile, and potentially gas equivalent MPG. Another set of figures is needed to show the MPG during CS mode driving. A suggested label is shown in Figure 5.4.

For blended PHEVs, since there is no pure electric driving range, it is useful to label the fuel economy in CD and CS mode separately as shown in Figure 5.5. It may be preferred to include the electric energy consumption during CD mode as well.

All electric range: 40 miles
240 Wh/mile
140 MPG gas equivalent

Fuel economy in CS mode
30 MPG city
33 MPG highway
32 MPG combined

Figure 5.4 Fuel economy labeling for all-electric-capable PHEV.

Fuel economy in CD mode
80 MPG City/75 MPG
Highway plus
Electric energy consumption
120 Wh/mile
CD mode range: 60 miles

Fuel economy in CS mode
30 MPG city
33 MPG highway
32 MPG combined
CS mode range: 360 miles

Figure 5.5 Fuel economy labeling for blended PHEV.

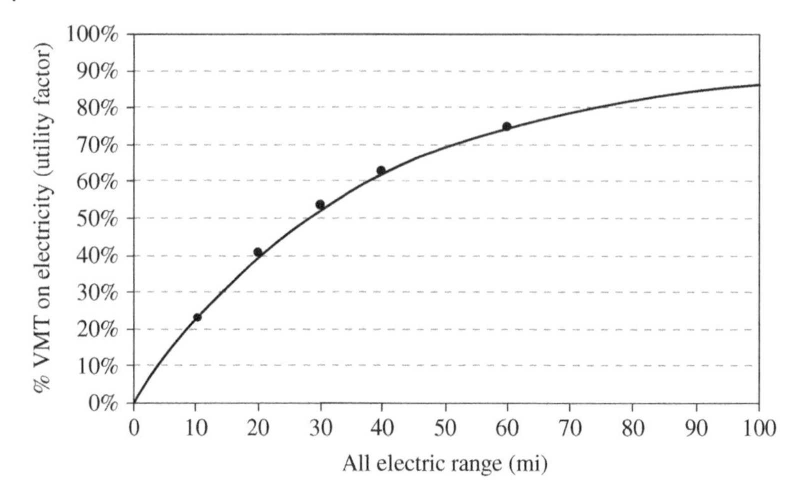

Figure 5.6 Definition of utility factor of PHEV (VMT, vehicle miles traveled).

5.4.3 Utility Factor

Another approach to fuel economy clarification is to use a utility factor. A utility factor is defined as the ratio of the CD range of a PHEV to the total distances driven in daily commuting by all the US population. For example, a CD range of 20 miles will result in a utility factor of 40% (Figure 5.6). Using the utility factor, the combined fuel economy can be expressed as [6].

$$FE_{Gas\_equivalent} = \frac{1}{\frac{UF}{FE_{CD}} + \frac{1-UF}{FE_{CS}}} \tag{5.5}$$

where UF is the utility factor, and FE_{CD} and FE_{CS} are the fuel economy during CD and CS operation of a PHEV, respectively.

Example 5.5 In CD mode, a blended PHEV consumes 10 kWh of electric energy from the onboard battery and 1.5 l of gasoline in its first 64 km. Beyond 64 km, the car enters CS mode, which consumes 6.25 l of gasoline per 100 km. The car has a total range of 724 km. Calculate the fuel economy of the car.

- **Scenario 1:** In CD mode, since the car consumes 1.5 l of gasoline in its first 64 km, its fuel consumption is

$$\begin{aligned} FC_{CD,\,Gas} &= 1.5/64 \times 100 \\ &= 2.34\ \text{l/100km or 100MPG, plus 156Wh/km electricity} \end{aligned}$$

- **Scenario 2:** If the total energy consumption (gasoline and electricity) during CD mode is converted to gasoline equivalent, then the total energy consumption is

$$FC_{CD,\,Total} = \frac{2.34\text{l}}{100\text{km}} + \frac{15600\text{Wh}}{100\text{km}} \bullet \frac{1}{33700 \times 1.609} = 2.63\text{l}/100\text{km or } 89\text{MPG}$$

- **Scenario 3:** Beyond 64 km, its fuel consumption is

$$FC_{CS} = 6.25 l/100\,\text{km or } 37.7\,\text{MPG}$$

- **Scenario 4:** The *UF* of this car is 0.62. The combined fuel economy by using the *UF* will be

$$FE_{Gas\_equivalent} = \frac{1}{\dfrac{UF}{FE_{CD}} + \dfrac{1-UF}{FE_{CS}}} = \frac{1}{\dfrac{0.62}{100} + \dfrac{1-0.62}{37.7}} = 61\,\text{MPG}$$

- **Scenario 5:** In this case, it is assumed that the car has a full tank of gasoline and a fully charged battery. It is then driven the full range of 724 km. If we consider the *UF* as the ratio of CD range versus total range, then the *UF* of this car is 64 km/724 km = 0.088. The overall fuel economy for the total range can be expressed as

$$FE_{overall} = \frac{452\,\text{miles}}{\dfrac{40\,\text{miles}}{FE_{CD}} + \dfrac{412\,\text{miles}}{FE_{CS}}} = 40\,\text{MPG, plus } 10\,\text{kWh electricity}$$

5.5 Power Management of PHEVs

A PHEV involves the operating conditions of both CD mode and CS mode. Typically, when the battery is fully charged, the vehicle is operated in CD mode, and when the battery state of charge (SOC) reaches a low threshold, it switches to CS mode. In CD mode, the vehicle will maximize the use of battery energy. In CS mode, the vehicle will use gasoline to power the vehicle while maintaining the battery SOC at the same level.

During CD mode operation, the goal of vehicle power management is to minimize the total energy consumption by distributing power between the battery and the gasoline engine/generator for a given driving scenario. In other words, the goal of power management in a PHEV is to minimize the fuel consumption for a given drive scenario.

For a series PHEV (or EREV), if the drive distance is less than the nominal electric drive range, then it is possible to operate the vehicle in all-electric mode, hence no fuel is consumed. If the drive distance is longer than the electric range, then there are three possible approaches for operating the vehicle:

1) Operate the vehicle in electric mode until the battery is depleted to a preset threshold, then run in CS mode.
2) Operate the vehicle in a blended mode with the engine turning on at high power demands, and deplete the battery to the preset threshold at the end of the total driving cycle.
3) Operate the vehicle in a blended mode with the engine turning on at high power demands but with an optimal battery discharge policy, so the battery will be depleted to the preset threshold before the end of the total driving cycle.

In these approaches, since the total drive distance is the same, the one that consumes the least fuel will be the best choice. Because the battery will exhibit a large

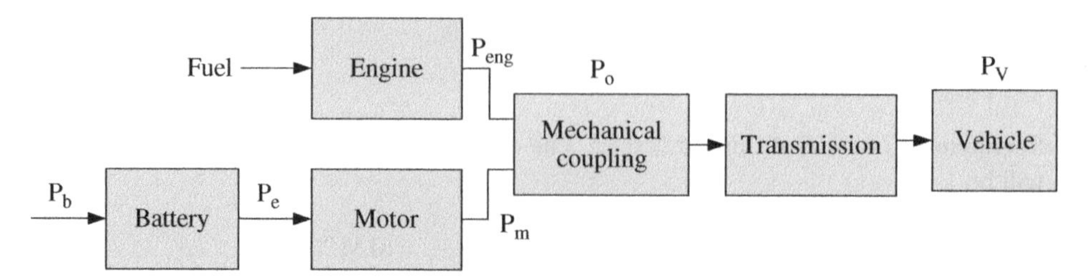

Figure 5.7 Idealized blended PHEV model for power management study.

power loss at high power output in comparison to its normal output power, it may be advantageous to operate the vehicle in blended mode. The optimization problem can be expressed as

$$\min\{\text{fuel consumption}\}$$

Subject to a given distance and drive cycle

For a blended PHEV, since there is no pure electric range available, the goal of the power management is to minimize fuel consumption for a given drive cycle and a given total available battery energy. This is strongly related to the characteristics of the power sources (battery and engine).

Figure 5.7 shows an idealized blended PHEV model for studying power management. In this model, the mechanical coupling and transmission losses are considered as part of the calculated vehicle power. The total vehicle power requested is satisfied by adding engine output and motor output:

$$P_o = P_m + P_{eng} \tag{5.6}$$

and

$$P_b = P_m + p_b + p_m \tag{5.7}$$

where lowercase p represents losses and uppercase P represents total power or output power.

The vehicle power demand can be calculated using the driving cycle profile. Figure 5.8 shows the distribution of vehicle power demand. Figure 5.9 is the normalized power demand of the vehicle, where $f(P_o)$ represents the total time that the vehicle spends at a certain power demand P_o.

The total battery energy consumed during the drive cycle is

$$E_b = \int_{P_{\min}}^{P_{\max}} P_b f(P_o)dP_o \tag{5.8}$$

The total fuel consumed is

$$F_e = \int_{P_{\min}}^{P_{\max}} f(P_{eng})f(P_o)dP_o \tag{5.9}$$

where $f(P_{eng})$ is the fuel consumption of the engine for a given power output of P_{eng}.

The optimization problem is to minimize the total fuel consumption F_e for a given battery energy E_b = constant.

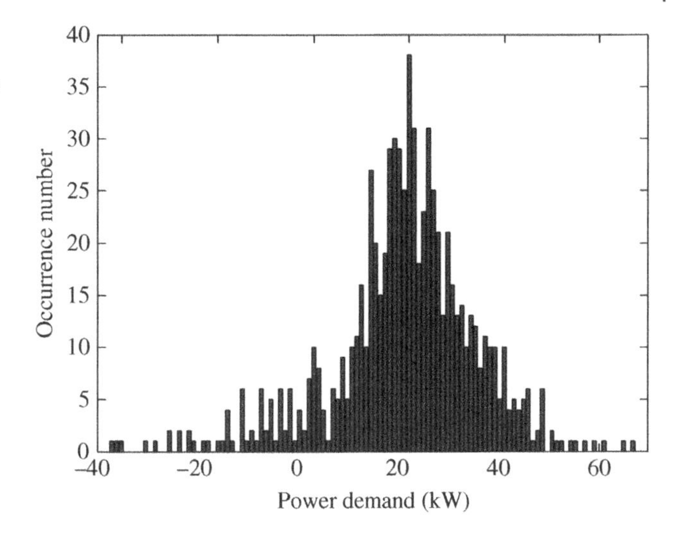

Figure 5.8 Power demand distribution in a passenger car under the US EPA's UDDS urban driving cycle. The horizontal axis is the power demand of the powertrain and the vertical axis is the occurrence of the power demand. The power demand is counted every second.

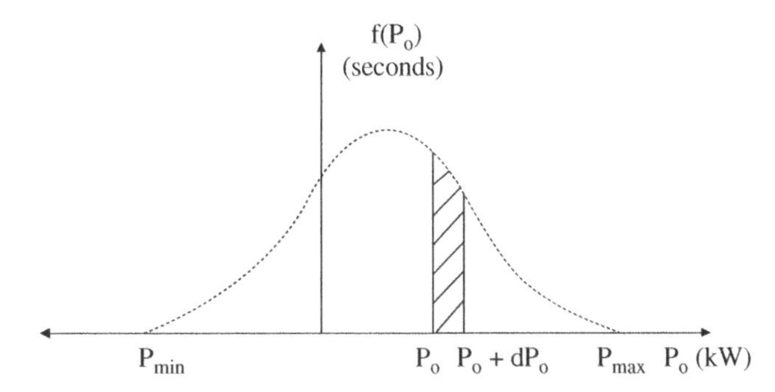

Figure 5.9 Normalized power demand distribution.

5.6 PHEV Design and Component Sizing

The main components of a PHEV are the powertrain motor and the battery pack. Using the vehicle resistive force and acceleration requirement, the driving motor can be sized. The total vehicle force is

$$F_{TR} = mg\sin\alpha + mgC_0 + mgC_1\ V^2 + \frac{1}{2}\rho C_D\ A_F\ V^2 + m\frac{dV}{dt} \tag{5.10}$$

where m is the vehicle mass, α is the road slope in radians, g is Earth's gravity, which is 9.8 m/s^2, C_0 and C_1 are rolling coefficients, ρ is the air density, C_D is the aerodynamic coefficient, A_F is frontal area in m^2, and V is vehicle speed. The total vehicle resistance is plotted in Figure 5.10.

At any given vehicle speed, the power at the wheels is

$$P = F_{TR}V = \left(mg\sin\alpha + mgC_0 + mgC_1\ V^2 + \frac{1}{2}\rho C_D\ A_F\ V^2 + m\frac{dV}{dt} \right) V \tag{5.11}$$

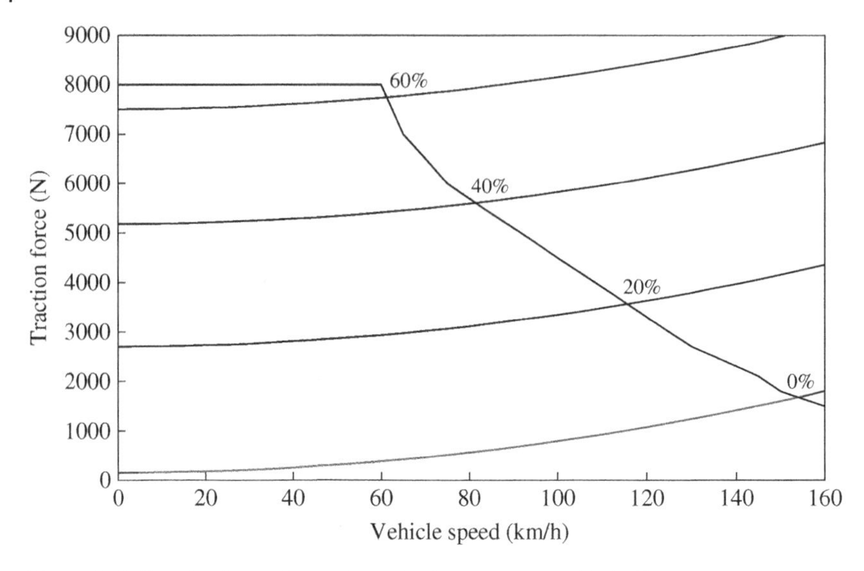

Figure 5.10 Traction force of a medium-sized passenger car. Motor Torque.

5.7 Component Sizing of EREVs

Assume the driveline efficiency is η_d. For a PHEV to be capable of electric driving, the powertrain motor needs to be sized such that the motor can provide sufficient torque for all driving conditions, that is,

$$P_m = P/\eta_d \tag{5.12}$$

The total acceleration time and gradability of the vehicle are other design criteria for traction motor sizing. In general, the traction motor of an EREV is very similar to the design of a series HEV. For example, the typical power rating of a passenger car is about 125 kW, and the typical traction motor for a middle-sized sports utility vehicle (SUV) is about 150 kW.

The engine/generator set of an EREV is designed to provide average power during the extended driving range. In addition to the driveline need, auxiliary power needs should also be considered.

For the traction battery, there are two very important parameters: energy capacity and power capacity. Since the EREV is designed to operate under all kinds of conditions without turning on the engine during CD mode, the battery power rating needs to match the traction motor rating. For example, if the traction motor is rated at 125 kW output with a system efficiency of 95% (inverter + motor), then the battery needs to have a power capacity of 131.6 kW.

Another parameter of the battery pack is the energy capacity. This can be calculated based on the total electric range to be designed. For example, if a vehicle is designed to have an electric range of 64 km, and the vehicle consumes an average of 150 W per km in city driving, then the battery needs to be able to provide 9.6 kWh of usable energy. If we allow the battery SOC only to go down to 30%, and consider the average efficiency of the battery to be 96%, then the nominal energy capacity of the battery will be 14.3 kWh.

Further, if we choose a battery pack that is composed of individual battery cells of 3.2 V, 40 Ah, then we will need 112 cells. Connecting all the cells in series will result in a nominal voltage of 358.4 V.

5.8 Component Sizing of Blended PHEVs

The design of the powertrain motor and engine of a blended PHEV is similar to the design of a parallel HEV. The only requirement is that the traction motor should be sized such that the vehicle can be driven in electric mode for the majority of city driving. For highway driving, the engine will provide power for the driving. In Figure 5.8, a motor rated at 50 kW will cover 95% of the city driving power demand.

The battery sizing is similar to the battery sizing of an EREV, that is, the equivalent electric distance required for the PHEV. Since the focus is on city driving, the battery is sized to satisfy city driving cycles.

For example, if a blended PHEV is equipped with a powertrain motor of 50 kW with an efficiency of 95%, then the battery needs to be capable of providing 52.6 kW of power. If the desired city driving distance in electric mode is 40 km, and the power consumption is 160 W per km, then the battery needs to have a usable capacity of 6.4 kWh. Again, if we assume that battery SOC can only go down to 30%, and that it has an average efficiency of 95%, then the battery nominal capacity will be 9.6 kWh. If we choose battery cells rated at 3.75 V, 32 Ah each, then the number of battery cells needed is 80. Connecting all cells in series will result in a nominal voltage of 300 V.

5.9 HEV to PHEV Conversions

Regular HEVs can be converted to PHEVs. Many independent conversions have been performed by various groups (such as Cal-Cars) and companies (such as A123 Hymotion and EEtrex) (http://www.eaa-phev.org/).

There are two potential ways to perform the conversion. One is to replace the original battery pack (usually an NiMH battery) with a larger battery (usually lithium-ion). Another way is to add an extra battery pack to the vehicle system.

Typical HEVs today are equipped with an NiMH battery pack of 1.2–2.2 kWh. A converted PHEV will usually have a battery pack of 7–16 kWh. This new or extra battery capacity will either make it possible to drive the vehicle in all-electric mode or significantly increase the fuel consumption of the car in the initial drive range.

5.9.1 Replacing the Existing Battery Pack

In this approach, the original battery pack is removed and a new battery pack is installed. In customized conversions of HEVs to PHEVs, the vehicle control is usually kept intact. In order to utilize the battery energy, usually the battery information is manipulated, or spoofed. One way of doing this is to replace the original battery electronic control unit (ECU) with a new battery ECU that duplicates all the battery

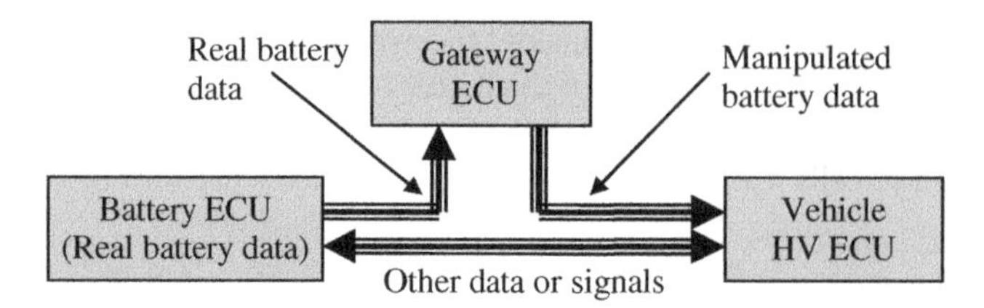

Figure 5.11 Gateway approach in the conversion of HEV to PHEV.

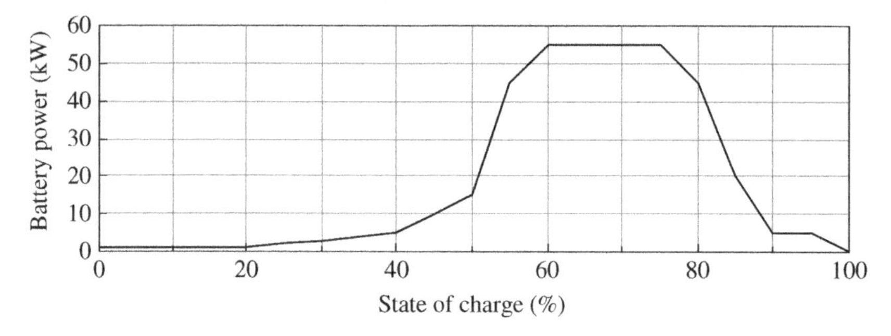

Figure 5.12 Typical battery power vs. SOC allowed by an HEV.

information. However, the real battery information is spoofed to a value such that the vehicle controller thinks that there is sufficient energy and power from the battery, so it uses more battery energy to drive the car.

Another approach for spoofing the battery data is to use a gateway, as shown in Figure 5.11. The real battery information is sent to a gateway, which then sends manipulated battery data to the vehicle controller. The gateway can be part of the new battery ECU, or a standalone ECU.

Most HEVs use high-voltage (HV) NiMH batteries (HV battery). The power and energy available from the battery are typically determined by the battery SOC and battery temperature. The battery ECU constantly monitors the HV battery conditions, including temperature, voltage, and amperage, and calculates the SOC. When the vehicle is in motion, the HV battery goes through repetitive charge and discharge cycles (discharged during acceleration and charged during regenerative braking). The battery could also be charged by the motor/generator using engine power as necessary. The ECU calculates the SOC and sends charge/discharge requests to the vehicle ECU to maintain the SOC at a median level. In the Prius, for example, the target SOC is 60% within a ±20% band. When the SOC drops below the controlled region (40%), the battery ECU sends a request to the vehicle ECU, which then sends a request to the engine ECU to increase engine power to charge the battery. If the SOC increases to above the controlled region (80%), then the battery ECU sends a request to the vehicle ECU, which then sends a request to the motor so that the battery power is discharged by using the motor to drive the vehicle (either alone or in combination with the engine).

Figure 5.12 shows a typical plot of battery power vs. the SOC of an HEV. The power available from the battery is closely related to the battery SOC. Only when the SOC is

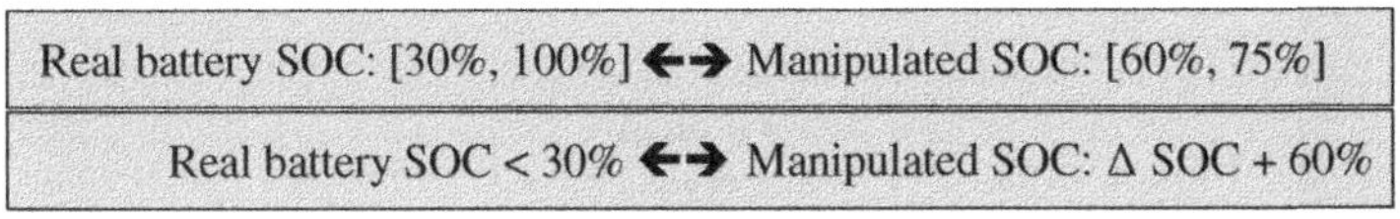

Real battery SOC: [30%, 100%] ←→ Manipulated SOC: [60%, 75%]
Real battery SOC < 30% ←→ Manipulated SOC: Δ SOC + 60%

Figure 5.13 Actual and manipulated battery SOC information for powertrain control purposes in a converted PHEV.

in a narrow range (55–80%) does the vehicle allow higher power from the battery. This approach is designed for battery life.

When a new and larger battery pack is added, the main goal is to use the battery energy as much as possible. The typical range is for the SOC to drop from 100% (fully charged) to 30% for a lithium-ion battery pack. Operating the battery below 30% will have a negative impact on battery health and life.

With no modification of the vehicle ECU, the SOC of the battery pack must be manipulated such that the vehicle will use more power from the battery. The mapping of the battery SOC for the limits shown in Figure 5.13 is as the follows. When the real battery SOC is above 30%, the real SOC is mapped to a range (60–75%) where the vehicle ECU will maximize the use of battery energy. When the real SOC drops below 30%, the calculation of the SOC is based on the original battery capacity as follows.

If $SOC_{real} > 0.3$, then

$$SOC_{real} = \frac{(Ah)_{new} - \int idt}{(Ah)_{new}} \tag{5.13}$$

$$SOC_{manipulated} = 60\% + 0.21\left(SOC_{real} - 0.3\right) \tag{5.14}$$

If $SOC_{real} < 0.3$, then

$$SOC_{manipulated} = \Delta SOC + 0.6 = \frac{\int idt}{(Ah)_{old}} + 0.6 \tag{5.15}$$

where $(Ah)_{old}$ is the capacity of the original pack, and $(Ah)_{new}$ is the capacity of the new pack.

5.9.2 Adding an Extra Battery Pack

The approach of A123 Systems' Hymotion (http://www.a123systems.com/hymotion/) is to add an extra battery pack to the vehicle as shown in Figure 5.14. The energy of the extra battery pack is slowly released to the original battery pack to be used by the vehicle. Other than controlling the new battery pack, there is no change to the original electrical and control systems.

There may be a need for a DC–DC converter between the new lithium-ion battery pack (Li) and the original NiMH battery pack. The DC–DC converter will take energy from the lithium-ion battery and charge the NiMH battery. In this way, the NiMH battery voltage is kept high. Since the battery ECU will reset the SOC of the NiMH battery pack based on the battery voltage, the energy that is charged to the NiMH battery is being used due to the detected high SOC.

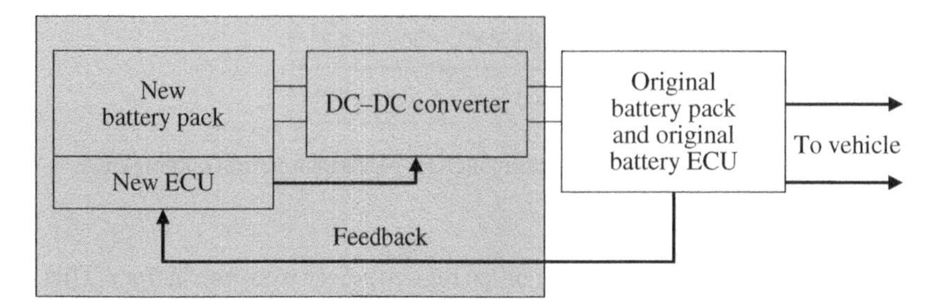

Figure 5.14 Adding an extra battery pack to the HEV.

5.9.3 Converting Conventional Vehicles to PHEVs

Some vehicles, such as pickup trucks, have abundant space available. Hybrid Electric Vehicles Technology Inc. and Raser Inc. have each converted a conventional gasoline-powered pickup truck to a PHEV (http://hevt.com/flyers/HEVT%20Ford%20F150%20Pickup%20Truck%20Plug-in%20Hybrid%20Electric%20Conversion.pdf, http://www.rasertech.com/). In the former approach, the company kept the existing front axle as gasoline engine driven, but modified the rear axle by adding an induction motor and an extra battery pack. The vehicle controller needs to be modified to take advantage of the battery energy.

5.10 Other Topics on PHEVs

5.10.1 End-of-Life Battery for Electric Power Grid Support

In general, battery energy capacity tends to fade over time and over discharge cycles. Typical battery energy capacity as a function of time is shown in Figure 5.15.

With 70% SOC depletion, a lithium-ion battery can typically last 3000–4000 charge cycles. This is approximately 10 years for a PHEV. After that time, the battery capacity may be only 50% of its initial capacity. While this is not satisfactory for the car owner due to the reduced electric driving range, the battery itself could be used for other purposes, such as for electric grid support. Since there is less space/weight constraint for power grid applications, these batteries can be used for grid energy storage for peak shaving, frequency regulation, and stability control. When more and more renewable energy is connected to the electric power grid, stability of the grid becomes extremely important due to the intermittent nature of renewable energy generation.

5.10.2 Cold Start Emissions Reduction in PHEVs

Emissions during cold-weather start of a vehicle have long been an issue. It has been shown (http://cfpub.epa.gov/ncer_abstracts/index.cfm/fuseaction/display.abstractDetail/abstract/1450/report/0) that the emissions from a vehicle during a cold-weather start are significantly more than those during normal weather starting. In particular, the emissions during the first two minutes can be as much as 80% of the total emissions during a standard driving cycle (1400 seconds). This is due to the fact that vehicle emissions are usually treated by a catalytic converter. A typical converter needs to heat up to function properly.

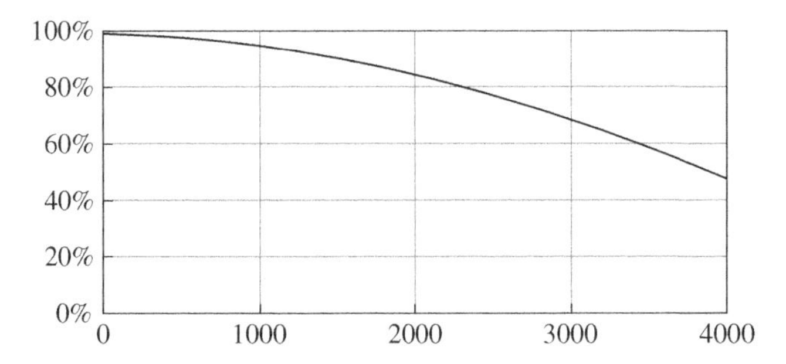

Figure 5.15 Typical battery capacity versus cycle life.

For conventional vehicles, a number of methods have been proposed to mitigate cold-weather start emissions, such as catalytic heaters and novel materials.

It is possible to reduce cold-start emissions in PHEVs by using the large onboard battery pack. The idea is to operate the vehicle in electric-only mode in the first few minutes and, at the same time, use electricity from the battery to warm up the catalytic converter to a certain temperature. Once the converter is normal and functional, the engine may be started as needed. However, almost all batteries do not work well under extremely cold weather conditions. Not only does the useful energy decrease and the internal impedance increase (efficiency drop), but there is also a negative impact on battery life if the discharge power is large. A reasonable compromise needs to be made in this regard.

5.10.3 Cold Weather/Hot Weather Performance Enhancement in PHEVs

The typical operating temperature range of a lithium-ion battery is from 0 to 50 °C. In extreme cold or hot weather conditions, the battery will exhibit large internal impedance and reduced power levels. In order to extend the driving distance and the battery life of a PHEV during cold weather conditions, additional measures may be necessary.

One method is to thermally insulate the battery pack so that no or very little heat exchange takes place between the battery cells and the ambient air, except at the cooling outlet. This thermal insulation will allow the battery to hold a steady temperature for an extended period of time during extreme cold or hot weather conditions. Another approach is to heat up the battery during cold weather by using an internal heater that consumes battery energy to warm up the battery pack. An additional method includes keeping the battery in trickle charge mode when plugged into the power grid.

5.10.4 PHEV Maintenance

One issue of PHEVs is that if the driver always drives less than the electric range and is always charging the battery, the engine may never start. This could cause the fuel to go stale and some mechanical parts including some in the engine to seize. Therefore, vehicles such as the Chevrolet Volt have been designed with an additional mode, namely, the maintenance mode [10]. This mode is activated either by the driver or automatically to operate the engine and the generator routinely to maintain the health of those components.

5.10.5 Safety of PHEVs

HEVs and PHEVs handle high voltages and a large amount of energy in the battery pack. A PHEV is safe under normal drive conditions but safety can be an issue during charging (rain or a wet plug connector), repair and service of the vehicle, as well as in an accident. Just like any HV system, the electrical system in an HEV or PHEV can be unsafe if mishandled. HV systems may cause electrical hazards if not handled properly, including electric shock, arcing, and blast [11].

The car's HV system is isolated from the ground and also isolated from the vehicle chassis. Therefore, electrical hazards can only arise when a person holds both the positive and the negative terminals of the HV system while repairing or servicing the car.

However, PHEVs must be plugged into an electrical outlet to charge the onboard battery. Advanced charging techniques, such as inductive and wireless charging, could potentially reduce the risk of electric shock during charging of the vehicle but will also reduce charge efficiency and increase system cost.

Charging the vehicle during rainy days with a contact charger can potentially cause leakage of current to the person handling the plug. Old and worn plugs and cables can cause current leakage. Therefore, extra caution needs to be taken when charging the vehicle. The leakage current can cause electric shocks, leading to muscle contraction, fibrillation, and tissue damage. A 10 mA current is enough to cause muscles to contract. A person may not be able to release the grip of their hand if it is exposed to sufficient electric current. If sufficient current goes through the chest, the person's chest muscle may paralyze and halt breathing.

More severe damage, such as fibrillation, can happen if the body's normal heartbeat is disrupted. The HV system, in particular the charger voltage from the electrical outlet, is enough to interrupt the brain current and cause the heart to stop beating. However, the current must follow a path through the body (hand to ground, hand to hand) in order for fibrillation to occur. Tissues can be damaged from prolonged exposure to electric shock due to heat concentration in the tissues.

The high-power high-voltage connectors may become short-circuited during an accident and cause sparks or overheating, which could lead to a fire or explosion. After-market PHEVs typically have the battery installed at the back of the vehicle. This may be an issue during a rear collision. The added battery pack also shifts the center of gravity of the vehicle, which may cause stability problems during braking. Mass-produced PHEVs will have been designed taking battery weight into consideration, and the batteries are not likely to be installed in the crash zone. Safety disconnection devices such as a service plug and fast fuse will provide additional safety measures.

Extra care should be taken when working on a PHEV during maintenance and repair. For example, insulated gloves are necessary in order to avoid electric shock when handling the HV system. Terminals of cables need to be insulated if they are disconnected from their original connection point. Tools and instruments should have insulated handles when dealing with the battery and other HV components in the PHEV. In fact, the handling and safety measures in the EV, HEV, and PHEV are the same as those used in home electrical repairs.

5.11 Vehicle-to-Grid Technology

Vehicle-to-grid, or V2G, is a concept referring to the capability of bidirectional power and energy exchange between the power grid and the vehicle battery (http://www.ornl.gov/info/ornlreview/v40_2_07/2007_plug-in_paper.pdf) [3, 4]. With a bidirectional charger, the vehicle could be used as a power backup for the home or office. It is also possible to use the PHEV battery to control the stability and regulate the frequency and voltage of the power grid, such as in a distributed power grid and with renewable energy generation.

PHEVs need to be charged from the electric power grid. During charging, the charger will generate inrush current and harmonics, and could cause the grid to malfunction if not coordinated properly.

In a broad sense, and in the foreseeable future, hundreds of thousands of PHEVs will be connected to the power grid as electric drive transportation prevails as our ultimate solution to becoming independent of fossil fuels. It is imperative to study the grid-to-vehicle (G2V) impact on power system operation and to consider various factors such as battery size, charging, PHEV distribution, and efficiency [12].

In order to optimize G2V it is important to educate consumers on the "smart grid" concept. They should be made aware of the fact that battery charging at night would improve utility generation efficiency, because at night-time the electricity is supplied by the base load generation units. Studies show that even with 50% penetration of PHEVs into the power system, no additional generating capacity and no new power plants would be required [12]. Although there are concerns with PHEVs straining the grid, PHEVs, if properly managed, could actually help to prevent brownouts, reduce the cost of electricity, and accommodate the integration of more renewable energy resources.

5.11.1 PHEV Battery Charging

There are three levels of charging for the PHEV, depending on the voltage: single-phase AC 120 V, single-phase AC 240 V, and three-phase AC 480 V. The different voltage levels will affect the charging time, ranging from hours to tens of minutes. In general, there are four types of charging algorithms for PHEV: constant voltage, constant current, constant voltage and current, and pulse charging. These different charging algorithms require corresponding controller designs for power electronics circuits. In this section, the pulse charging technique is studied in detail through simulation. Here PHEVs are connected to an example distribution system. The case of 10 PHEVs has been considered for simulations in which the PHEVs are connected to an IEEE 13-bus distribution system. Figure 5.16 shows this distribution system together with PHEVs. The system is relatively small and highly loaded. For this study, the PHEVs are connected between nodes 692 and 675 of the distribution system.

The PHEVs connected to the distribution system via a single-phase transformer are charged by the pulse charging technique. Initially the battery is assumed to have a 90% SOC. The battery is charged by the DC–DC converter with pulse current until it reaches a 95% SOC.

The AC–DC converter connected to the distribution system draws unity power factor, which shows that the PHEV is utility friendly. Figure 5.17 shows the unity power

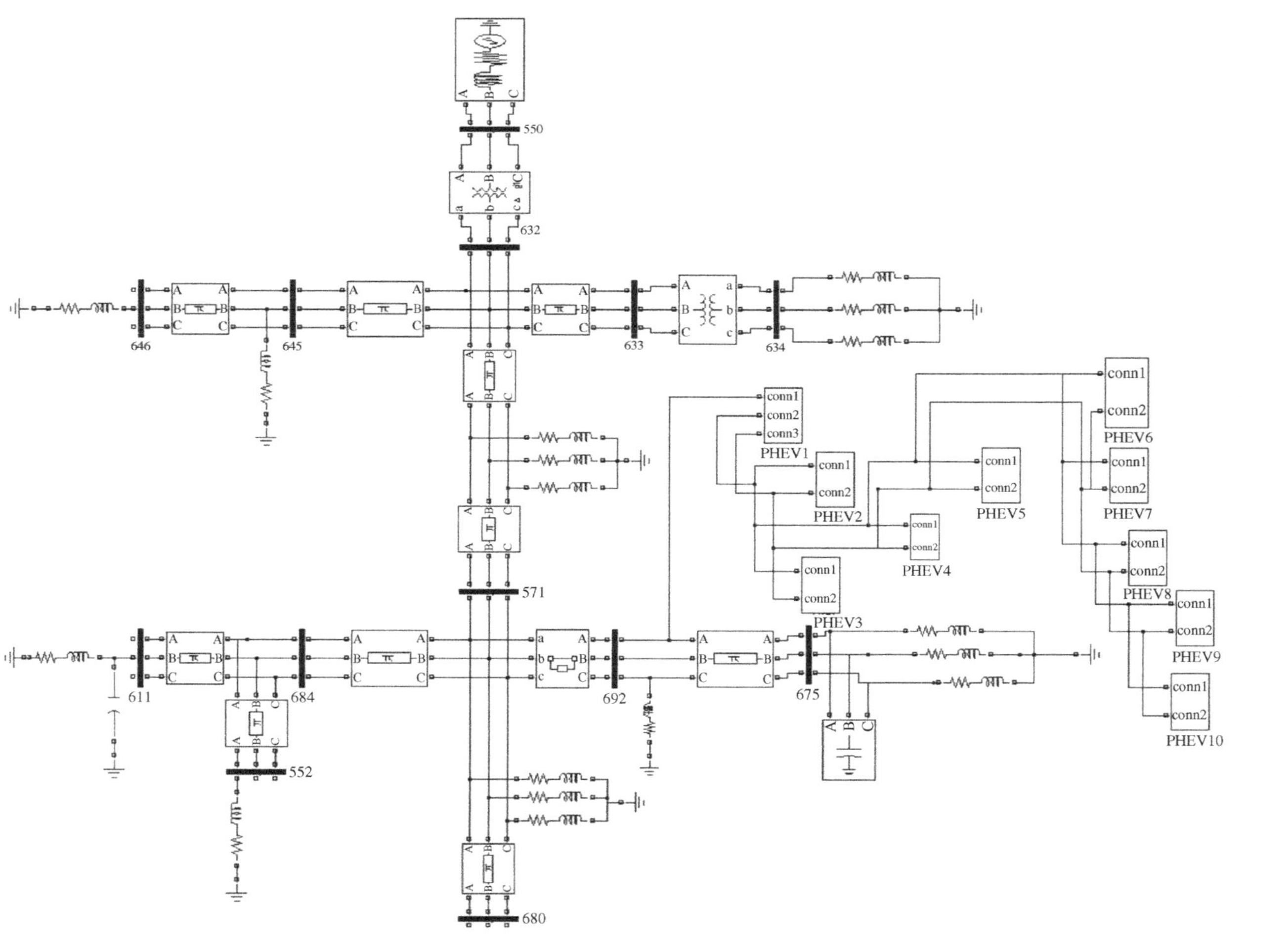

Figure 5.16 IEEE 13-bus distribution system with 10 PHEVs connected to phase A at node 692.

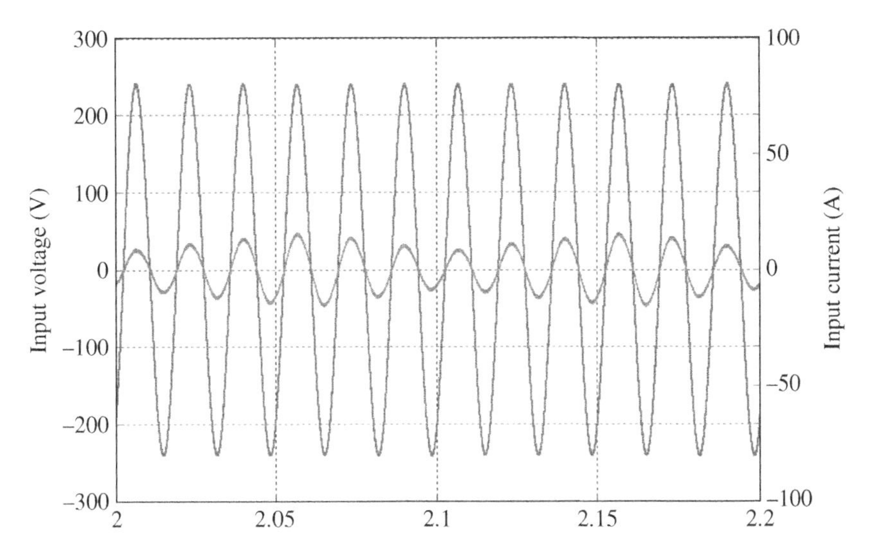

Figure 5.17 Unity power factor for input current and voltage during charging.

factor for input current and voltage at the secondary side of the single-phase transformer. The input voltage has a 240 V rms value. The reference current is generated by a phase-locked loop (PLL) block. The secondary side voltage of the transformer is the input to the PLL. This in turn generates a sine wave to form the reference input current for the unity power factor controller.

5.11.2 Impact of G2V

The IEEE 13-bus distribution system [13, 14] with 10 PHEVs has been simulated to study the impact of connecting many PHEVs to the power system/grid. The case to be discussed here is that of connecting 10 PHEVs to the grid at node 692 of the distribution system via a step-down transformer rated at 4.16 kV/240 V. The substation transformer rating is 5000 kVA with a primary side rated voltage at 115 kV and secondary side rated voltage at 4.16 kV. The types of loads connected to the distribution system are unbalanced spot load and distributed load. Figure 5.18 shows a one-line diagram of the distribution system.

The distribution system has initially no PHEVs connected to it. After 0.3 seconds, 10 PHEVs are connected for charging the onboard batteries. The study is carried out with the simulation results to see the impact of PHEV charging on the grid voltage, current, and active power. Figure 5.19 shows the voltage of phase A at node 692. From the figure, it can be clearly seen that phase A is heavily loaded when all the PHEVs are connected for charging; as a result the voltage in the line is reduced. The voltage drop is found to be 39%, which is not within permissible limits. When the voltage drops below a permissible value it has to be restored back to its original value. A capacitor bank can be used in cases where the voltage drop is below 20%. However, in this case the voltage drop is 39%, which is a very high value; local area generation will be needed to restore the voltage. At the generating end, automatic generation control (AGC) with a power system stabilizer (PSS) is usually used for maintaining voltage stability. Figure 5.20

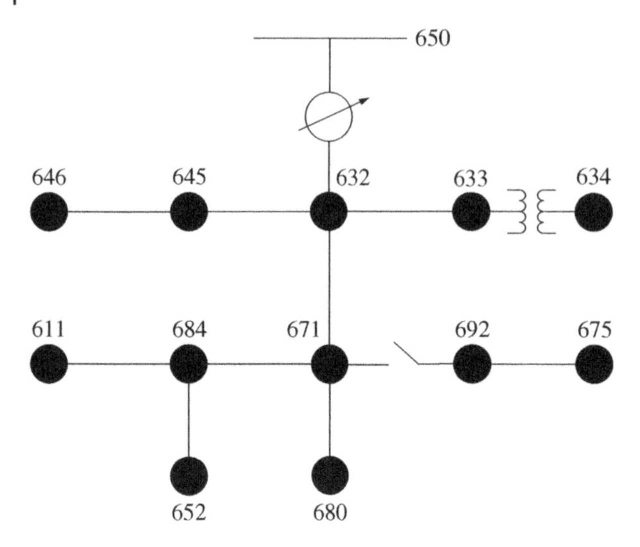

Figure 5.18 One-line diagram of IEEE 13-bus distribution system.

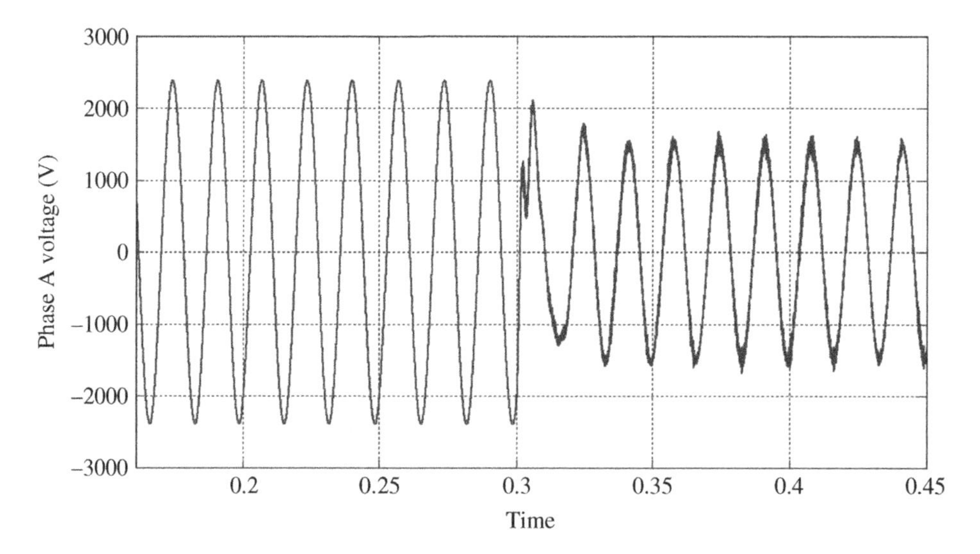

Figure 5.19 Phase A voltage with PHEV connected at $t = 0.3$ seconds.

shows the voltage waveform for sequential charging of PHEVs, where each PHEV is connected after 0.1 seconds. The voltage profile is greatly improved with the sequential charging method. In Figure 5.21, it can clearly be seen that after 0.3 seconds the value of the voltage is restored. Also, the total harmonic distortion (THD) of the grid side voltage was calculated after connecting the PHEVs. It was found to be 1.7%, which is well below the permissible value. Figure 5.22 shows a graph of the THD. Figure 5.23 shows the current of phase A at node 692. From the figure it can clearly be seen that the current increases after connecting a large number of PHEVs at $t = 0.3$ seconds. Figure 5.24 shows the current waveform for the sequential charging of PHEVs.

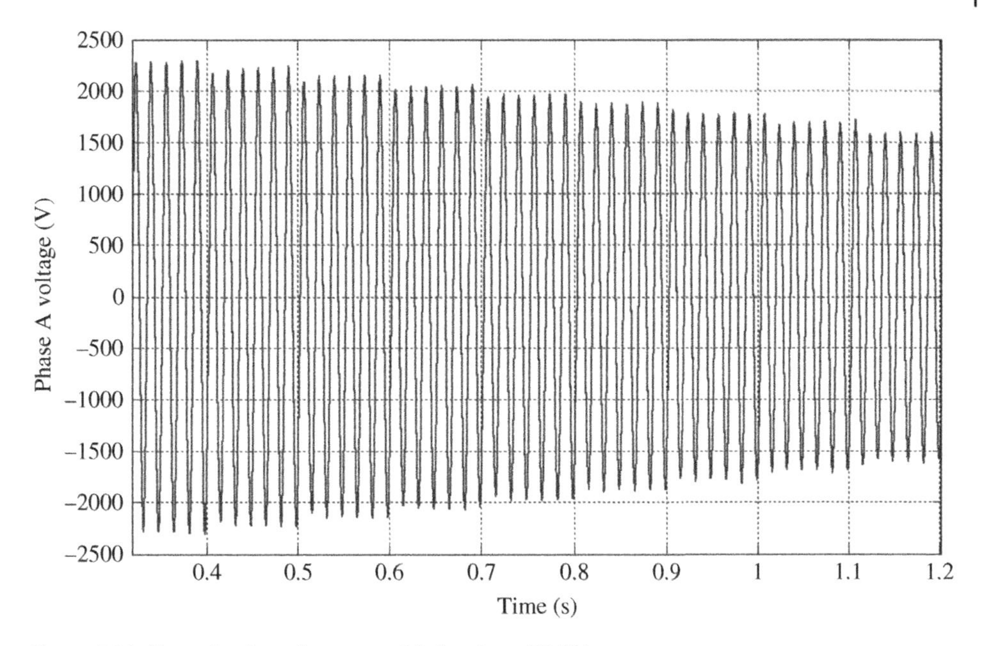

Figure 5.20 Phase A voltage for sequential charging of PHEVs.

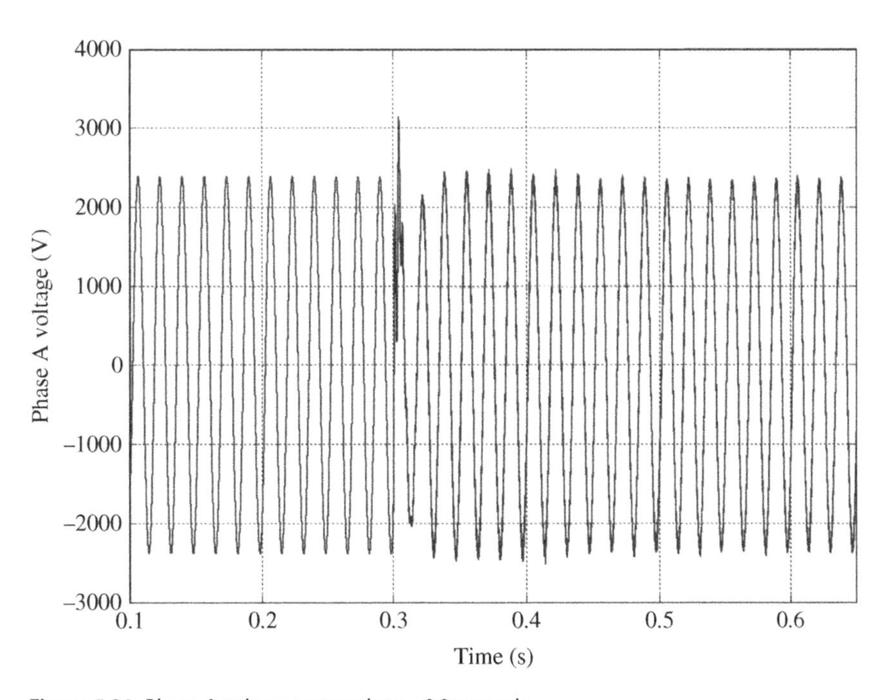

Figure 5.21 Phase A voltage restored at $t = 0.3$ seconds.

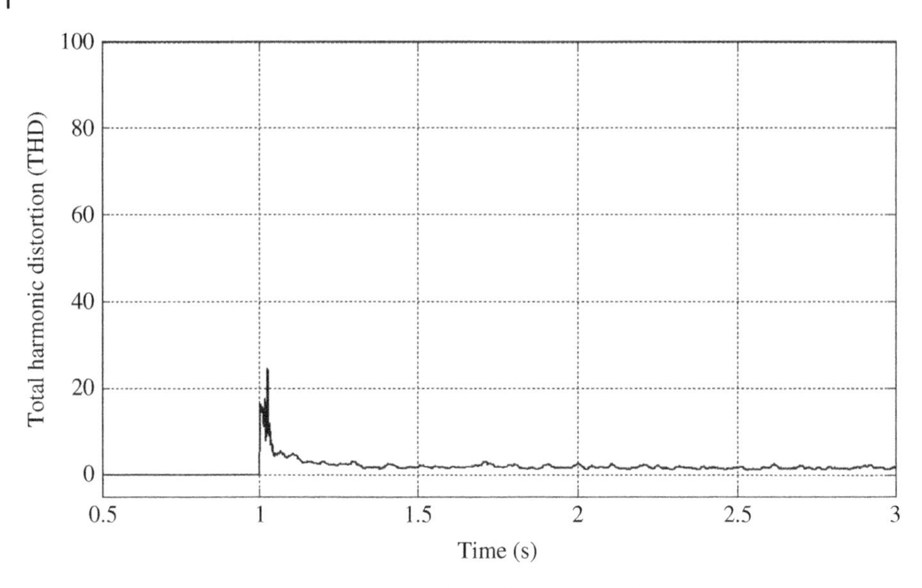

Figure 5.22 Total harmonic distortion (THD) of waveform input voltage.

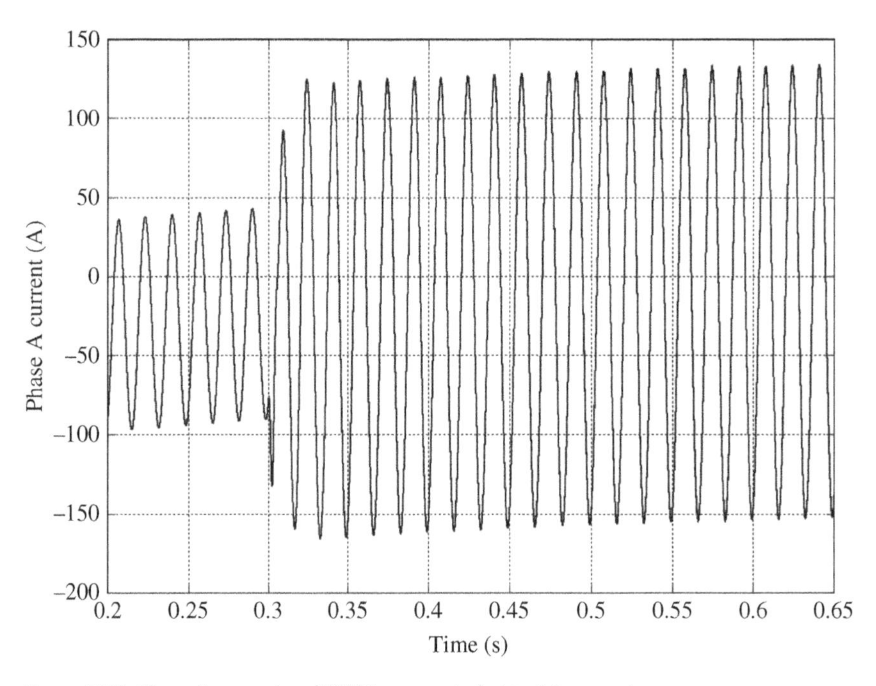

Figure 5.23 Phase A current and PHEVs connected at $t = 0.3$ seconds.

The current waveform of phase A shown in Figure 5.25 clearly depicts that the current is also restored because of the PSS. Figure 5.26 shows the average real power of phase A at node 692 with and without the PHEVs. It can be seen from the figure that the active power consumption increases when a large number of PHEVs are connected for charging at the same time.

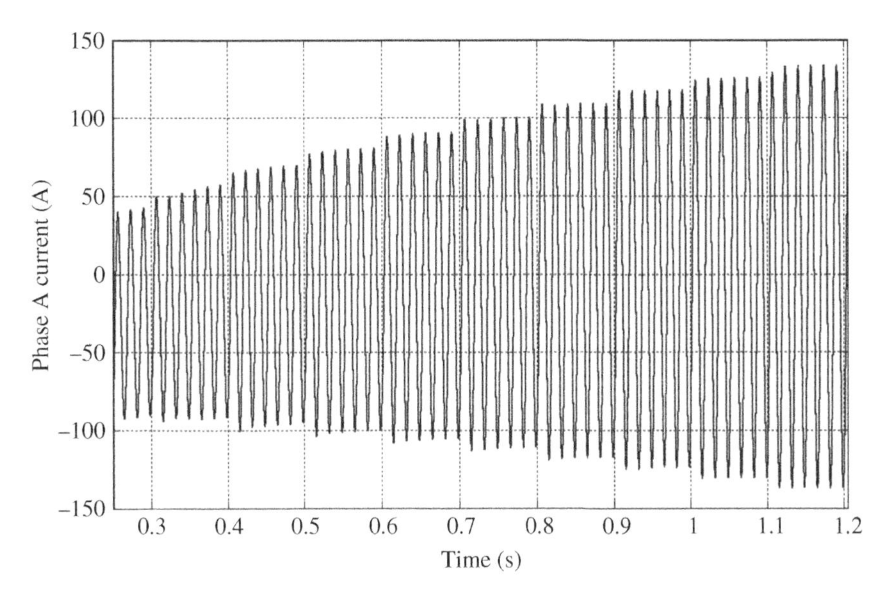

Figure 5.24 Phase A current for sequential charging of PHEVs.

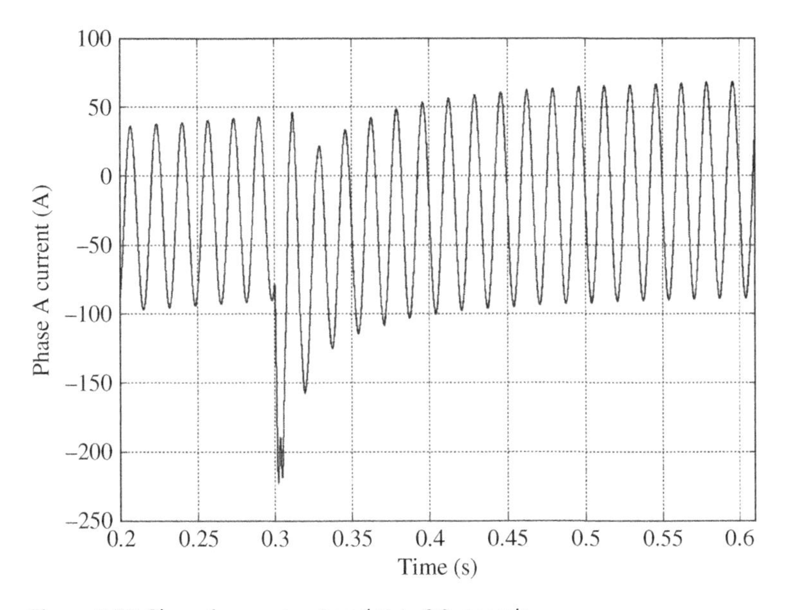

Figure 5.25 Phase A current restored at $t = 0.3$ seconds.

5.11.3 The Concept of V2G

The concept of V2G is that the energy stored within PHEV batteries can be utilized to send power back to the grid. The V2G technology supposes that if battery vehicles (BVs) or PHEVs become widespread, then they could supply peak load power with fast response when the vehicles are parked and are connected to charging stations [15, 16]. On average a car is driven for an hour per day and for the rest of the day it is parked.

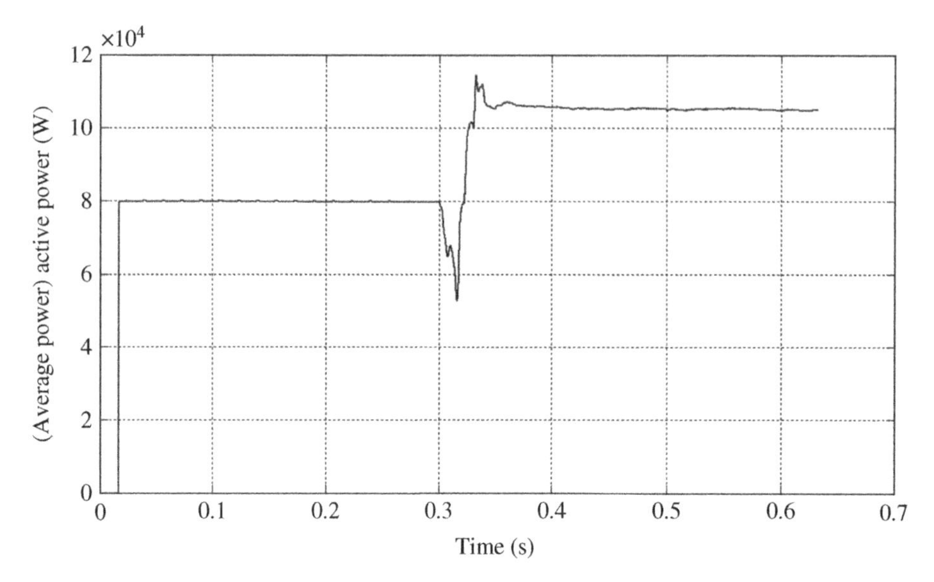

Figure 5.26 Phase A real power and PHEVs connected at $t = 0.3$ seconds.

Practically speaking, the amount of power drawn from the car can be around 10–20 kW. The V2G connection can provide a fast interface between the power system and the battery storage system of the PHEV. The system can produce both active and reactive power for the distribution system through proper control of the power electronics circuits, even though the batteries only store active power.

For EVs and PHEVs, the energy that can be used for V2G is limited by the onboard battery size. The unique aspect of power flow in PHEVs is that it is bidirectional, meaning that the vehicle can take power from the grid (during charging) and provide power (during discharge) to the grid. Apart from the concept of V2G, there is the concept of vehicle-to-home. The advantage of V2G is that it is parallel, which means that within a grid any car can be used to power any home by feeding its power back to the grid. On the other hand, vehicle-to-home is limited in the sense that a single vehicle can supply only a single home.

PHEVs can be treated as distributed energy resources via V2G and can provide voltage and frequency regulation, spinning reserves, and electrical demand side management [12, 16]. The V2G functions can be classified into two categories: local services and broad area services. Local services include supplying backup power for local houses or businesses, peak shaving, and voltage stabilization or power quality improvement. Broad area services include ancillary services for the grid. Ancillary services are power services by which the grid operators maintain reliable operation of the grid.

5.11.4 Advantages of V2G

There are a number of advantages of introducing V2G into the power system. Some of the advantages are stated below (http://www.udel.edu/V2G/docs/V2G-PUF-LetendKemp2002.pdf):

- **Improving security:** V2G inverters can respond quickly, to control the effects of any disturbance, as compared to the turbo-generator governor. This will help the power system to be more robust and reduce the vulnerability.

- **Improving reliability:** The advantage of locating the V2G system anywhere in the distribution system makes the backup supply available at a close distance even though it may not be installed at the consumer's location. This will have a major impact on consumer reliability as most interruptions are due to disturbances in the distribution networks.
- **Impact on generation:** By connecting a large number of PHEVs or V2G systems during daytime, the peak power can be curtailed during the daily peak load period. Also, during the light-load period PHEVs can be connected to charge the battery system, thus allowing the base load generators to operate efficiently without the need to carry large amounts of spinning reserve.
- **Environmental advantage:** Using PHEVs can reduce environmental pollution. They can promote the reduction of greenhouse gas emissions by indirectly using clean electricity as transportation fuel.

5.11.5 Case Studies of V2G

The distribution system [14], as discussed previously, has been considered for carrying out V2G simulations. The aim here is to study power system behavior due to V2G technology. Two scenarios, namely, PHEV for peak shaving and PHEV for reactive power compensation, are discussed and simulated by using the IEEE distribution model system. The PHEV is connected to phase A of the line between nodes 692 and 675 of the distribution system via a step-down transformer rated at 4.16 kV/240 V:

- **Case 1: V2G for peak shaving:** If the battery has enough charge, for example, with a 95% SOC, and if the PHEV is not in use, depending on the grid load condition, the PHEV can send some power back to the grid. In this section the simulation results for V2G connection are presented. The control strategy for sending the power back to the grid is shown in Figure 5.27. From the figure it can be seen that the reference power is divided by the battery voltage, which forms the reference current for the battery. This in turn is compared to the actual battery discharging current and an error signal is generated. The error is then processed through a PI (proportional integrator) controller that forms the duty cycle, which is then compared to the carrier wave for generating the pulses for switching the IGBT (insulated gate bipolar transistor) of the converter. Figure 5.28 shows the active power being sent back to the grid. When the car is plugged into the wall outlet, with the help of metering and the

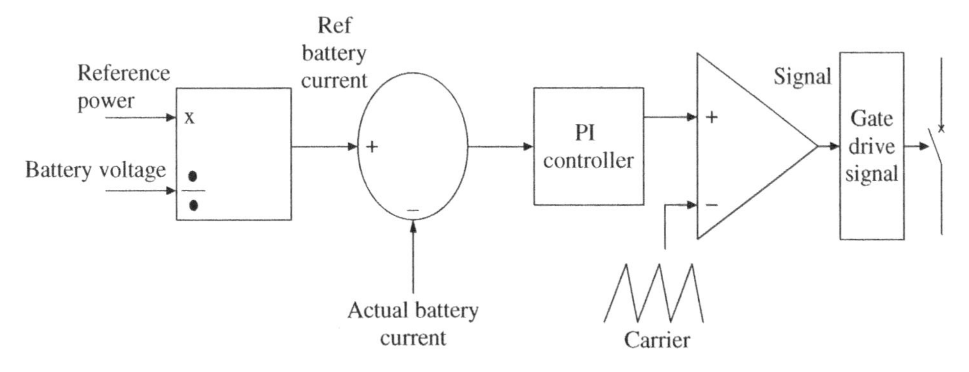

Figure 5.27 Control strategy for the battery system while sending the power back to the grid.

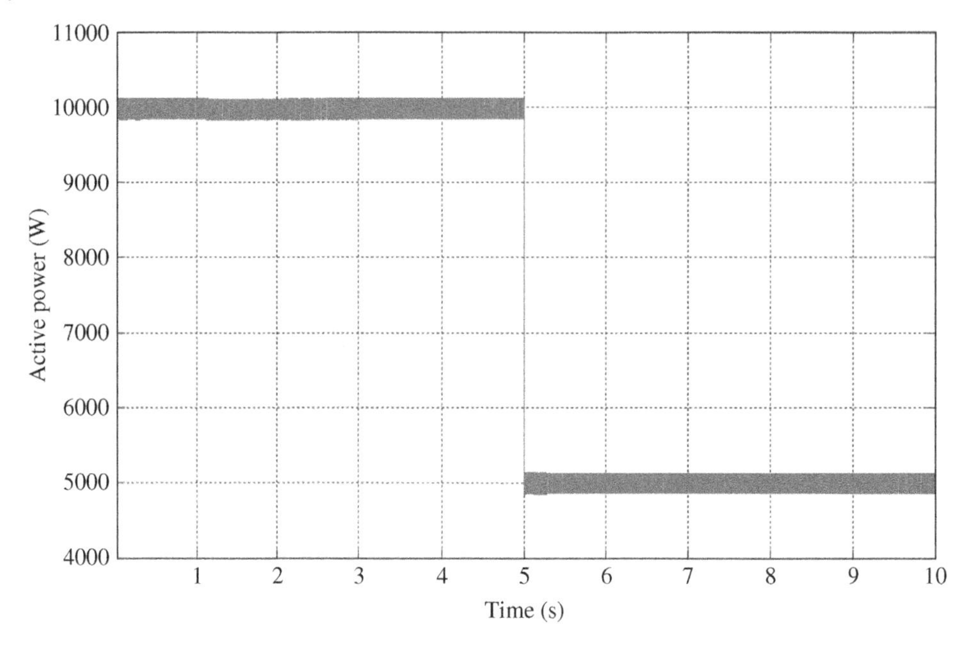

Figure 5.28 Real (active) power supplied by the battery.

communication system, the grid operator and consumer could interact and supply the available power to the grid. Such a case is depicted here, where initially the power being supplied by the PHEV is 10 kW; after time $t = 5$ seconds the power supplied is reduced to 5 kW. Figure 5.29 shows that the input current and voltage are 180° out of phase, meaning unity power factor when the power is delivered to the grid by the battery. This shows the utility-friendly nature of the charger. Also from the figure it can be seen that at time $t = 5$ seconds, when the power reduces from 10 to 5 kW the current at the grid side reduces correspondingly.

- **Case 2: Reactive power compensation:** A PHEV has the potential to act as a reactive power compensator for the power system. Capacitors are generally used for reactive power compensation in distribution networks. In Figures 5.30 and 5.31 the PHEV is shown as a reactive power compensator. Initially the PHEV is disconnected, and an inductive load is connected to the grid. From Figure 5.30 it can be seen that at the secondary side of the single-phase transformer at node 692, the current is lagging the voltage. After 0.5 seconds the PHEV charger is connected to the grid, and it can be seen in Figure 5.31 that the lagging current becomes in phase with the voltage. This demonstrates the reactive power compensation capability of the PHEV charger.

In this section, we briefly discussed the concept of G2V and V2G. Simulations were carried as an example to study the impact of connecting the PHEVs to the grid. From the results it can be seen that system voltage and current are greatly affected, hence local area generation with a PSS can be used to bring the system back to the normal state. V2G can partly spare the utility company from investing in creating spare capacity to meet electricity demand during peak hours. Some simulations are presented for cases where the PHEV is used for peak shaving and reactive power compensation.

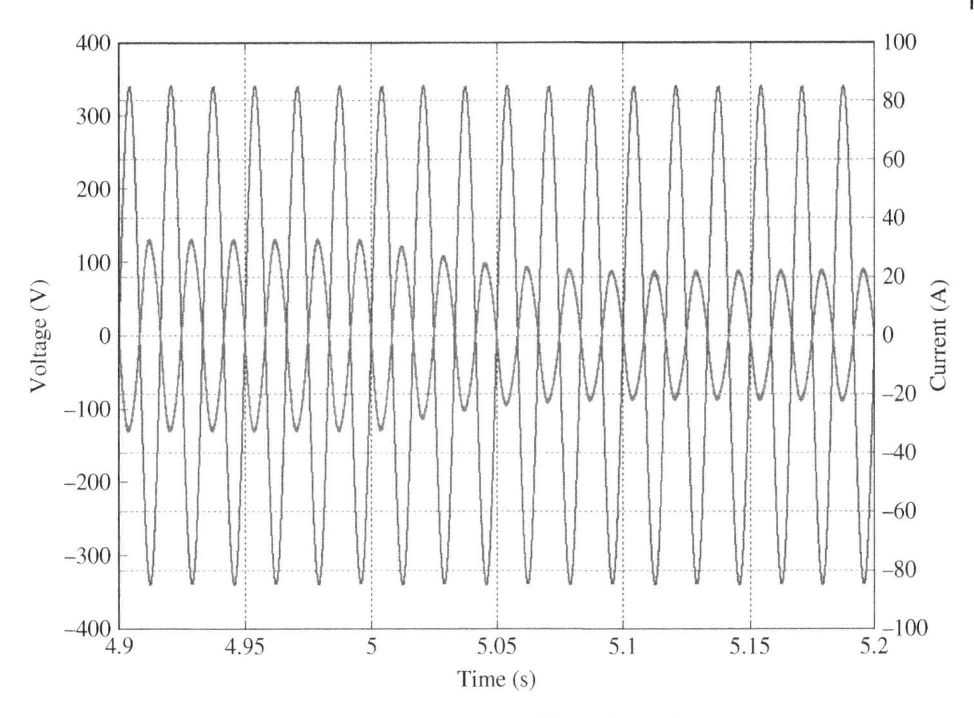

Figure 5.29 Input voltage and input current 180° out of phase during discharging.

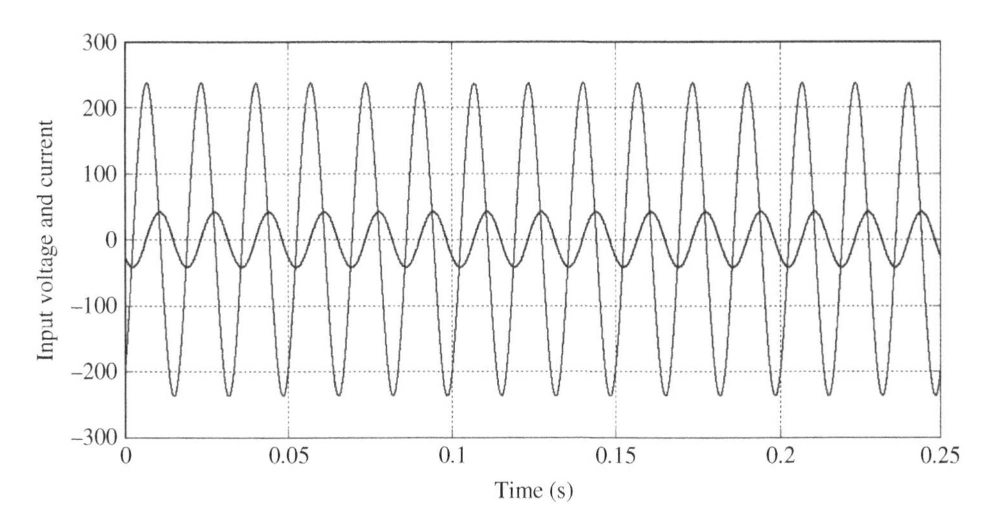

Figure 5.30 Current lagging voltage.

The bidirectional nature of the PHEV charger can prove to be beneficial during peak hours/periods. Also, the concept of reactive power compensation illustrates the fact that the PHEV can be used as an alternative to capacitors in distribution networks. On the other hand, a plug-in vehicle can also be designed to provide power for standby applications, through its V2G capability.

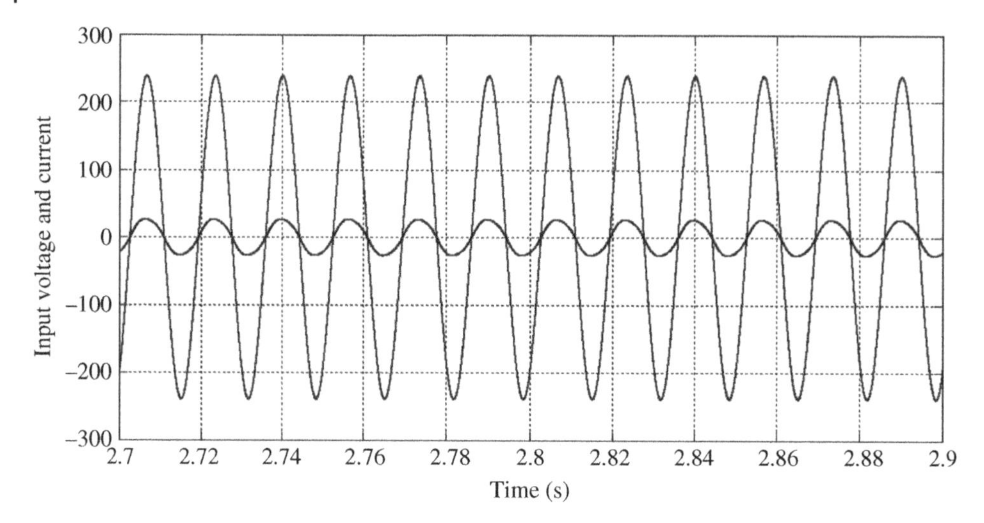

Figure 5.31 Reactive power compensation.

5.12 Conclusion

A PHEV equipped with a 10 kWh battery pack will require 8–10 hours of charging time at a regular 110 V/15 A outlet. Fast charging is only possible when higher power outlets are available, such as 110 V/50 A or 220 V/30 A, from which a 10 kWh battery back can be charged in less than two hours.

Finally, recycling of PHEV batteries has to be planned during the PHEV development phase, and be ready at the time of PHEV deployment. Continued or inappropriate use of batteries beyond their designed lifespan can lead to the release of toxic gases, injury, or fire. Inappropriate disposal of batteries can contribute to the contamination of soil, water, and air. The lithium and other metals in lithium-ion batteries can be extracted for other uses when the lithium-ion batteries are recycled.

References

1 Sanna, L. (2005) Driving the Solution – the Plug-in Hybrid Vehicle. http://mydocs.epri.com/docs/CorporateDocuments/EPRI_Journal/2005-Fall/1012885_PHEV.pdf (accessed February 2, 2011).
2 Douglas, J. (2008) Plug-in hybrids on the horizon – building a business case. *Journal of Electric Power Research Institute (EPRI)*, Spring.
3 Kempton, W. and Tomic, J. (2005) Vehicle-to-grid power implementation: from stabilizing the grid to supporting large-scale renewable energy. *Journal of Power Sources*, 144, 280–294.
4 Gage, T.B. (2003) *Development and Evaluation of a Plug-in HEV with Vehicle-to-Grid Power Flow*, AC Propulsion, Inc., San Dimas, CA, CARB Grant Number ICAT 01–2, December 17.

5 Electric Power Annual, http://www.eia.doe.gov/cneaf/electricity/epa/figes1.html (accessed February 2, 2011).
6 SAE International (2009) Surface Vehicle Information Report SAE J2841 – Utility Factor Definitions for Plug-in Hybrid Electric Vehicles Using 2001 U.S. DOT National Household Travel Survey Data, March.
7 SAE International (1999) Surface Vehicle Recommended Practice SAE J1711 – Recommended Practice for Measuring the Exhaust Emissions and Fuel Economy of Hybrid-electric Vehicles, March.
8 United States Environmental Protection Agency (2006) Fuel Economy Labeling of Motor Vehicles: Revisions to Improve Calculation of Fuel Economy Estimates, Final Rule, 40 CFR Parts 86 and 600, December 27, 2006, http://www.epa.gov/fedrgstr/EPA-AIR/2006/December/Day-27/a9749.pdf.
9 Williamson, S.S. and Emadi, A. (2005) Comparative assessment of hybrid electric and fuel cell vehicles based on comprehensive well-to-wheels efficiency analysis. *IEEE Transactions on Vehicular Technology*, 54 (3), 856–862.
10 Maintenance Mode of the Chevrolet Volt, http://gm-volt.com/2010/08/17/chevrolet-volt-maintenance-mode/(accessed February 2, 2011).
11 Ford Ranger EV User's Manual, http://www.eserviceinfo.com/download.php?fileid=18730 (accessed February 2, 2011).
12 Kramer, B., Chakraborty, S., and Kroposki, B. (2008) A review of plug-in vehicles and vehicle-to-grid capability. IEEE Industrial Electronics Conference IECON'08, November, Orlando, FL, pp. 2278–2283.
13 Kersting, W.H. (1991) Radial distribution test feeders. *IEEE Transactions on Power Systems*, 6 (3), 975–985.
14 Ma, W. and Jing, L. (1997) Distribution modeling and simulation based on MATLAB. 17th Conference of the Electric Power Supply Industry (CEPSi 2008), October, Macau SAR.
15 Haines, G., McGordon, A., and Jennings, P. (2009) The simulation of vehicle-to-home systems – using electric vehicle battery storage to smooth domestic electricity demand. Ecological Vehicles Renewable Energies, EVER'09, March, Monaco.
16 Kempton, W. and Tomić, J. (2005) Vehicle-to-grid power fundamentals: calculating capacity and net revenue. *Journal of Power Sources*, 144, 268–279.

6

Special Hybrid Vehicles

6.1 Hydraulic Hybrid Vehicles

Although this book is primarily dedicated to issues related to hybrid electric vehicles, it should be appreciated that non-electric hybrid vehicles are also viable and can sometimes be more beneficial than electric hybrid vehicles. Basically, there are a few reasons for going hybrid in the first place:

- A normal internal combustion engine (ICE) vehicle uses the engine over a wide speed range and hence the efficiency over this range is not the highest achievable efficiency for a particular engine.
- The maximum efficiency of ICE propulsion is very low, on the order of 30%.
- The electric energy storage system, including the battery and electric propulsion motor, has a high efficiency on the order of 80–90%.

Had it not been for the large size and cost, attributed to low energy storage per unit weight or volume capability of the battery, the pure electric vehicle probably would have replaced current vehicular technology. The next option, therefore, is the hybrid electric vehicle, where the ICE can be used to optimally charge the battery, and propulsion can be shared with the electrical method.

So, naturally, the question arises: is there anything else, other than ICE propulsion, available? The answer is that the hydraulic system is one such option. In a hydraulic system, the energy is stored in the form of a compressed fluid in a cylinder or by similar means. To pressurize the fluid needs energy, which comes from the ICE, to activate a hydraulic pump. While extracting the energy, we can use a hydraulic motor. In other words, the hydraulic pump is analogous to an electric generator, the hydraulic motor to an electric motor, and the pressurized fluid in the cylinder to a battery. Thus we see that the hydraulic system has a one-to-one equivalence to an electrical system. Although typically people assume the theoretical efficiency of the hydraulic pump, motor, and storage to be very high, around 90%, in reality they will be nearer to 70%, which is still much higher than that of an ICE. Hence all the items noted above to justify the use of a hydraulic system in a hybrid vehicle hold true.

To illustrate the hydraulic hybrid system, consider the diagrams in Figures 6.1 and 6.2 [1], giving a comparison of electric hybrid and hydraulic hybrid systems, side by side through direct analogy. In these figures the shaded areas show the subsystems

Hybrid Electric Vehicles: Principles and Applications with Practical Perspectives,
Second Edition. Chris Mi and M. Abul Masrur.

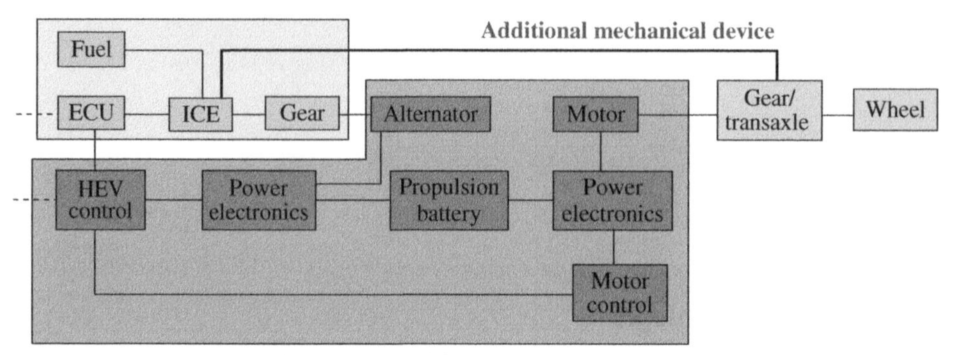

Figure 6.1 System-level diagram of HEV (ECU, engine control unit; ICE, internal combustion engine). *Source:* Masrur 2008. Reproduced with permission from IEEE.

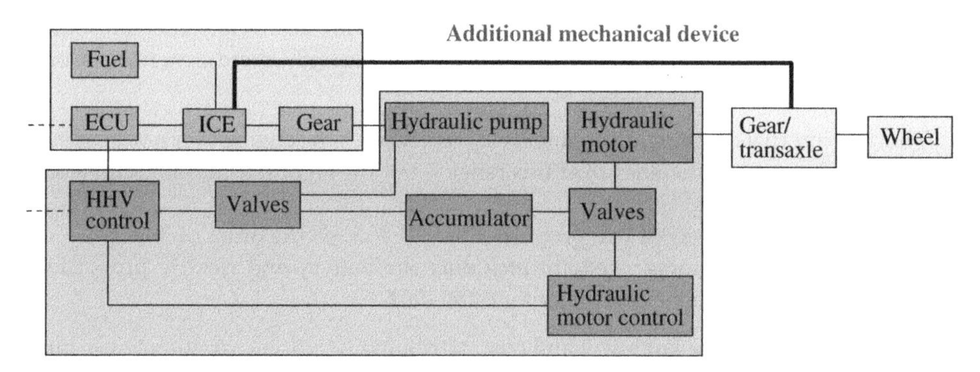

Figure 6.2 System-level diagram of HHV. *Source:* Masrur 2008. Reproduced with permission from IEEE.

specific to the regular ICE and the electric propulsion (Figure 6.1) and those specific to the hydraulic propulsion (Figure 6.2).

In the hydraulic hybrid vehicle (HHV) architecture shown in Figure 6.2, the alternator has been replaced by a hydraulic pump, the electric motor has been replaced by a hydraulic motor, the battery has been replaced by a hydraulic accumulator, the HEV controller has been replaced by a HHV controller, and the power electronics system has been replaced by the hydraulic valve system.

Figure 6.3 shows a full series hydraulic hybrid truck configuration [2]. The figure is consistent with the system-level diagram shown above. The accumulator includes a high-pressure (HP) accumulator containing some benign gas such as nitrogen. The pressure in this cylinder can be as high as 3000–5000 psi (21–35 MPa), whereas the low-pressure (LP) cylinder pressure can be very low, on the order of a few hundred psi. The ICE drives a pump, which takes the fluid from the LP cylinder side, pumps it to a very high pressure, and then delivers it to the HP cylinder side so that the mechanical energy can ultimately be stored in the form of HP gas. To drive the vehicle's wheels, the HP fluid from the HP cylinder side passes through a hydraulic motor drive assembly. The hydraulic motor takes in the HP fluid, converts it to mechanical power at the wheels, and when the fluid has passed through the hydraulic motor, its pressure drops and it is transferred to the LP cylinder.

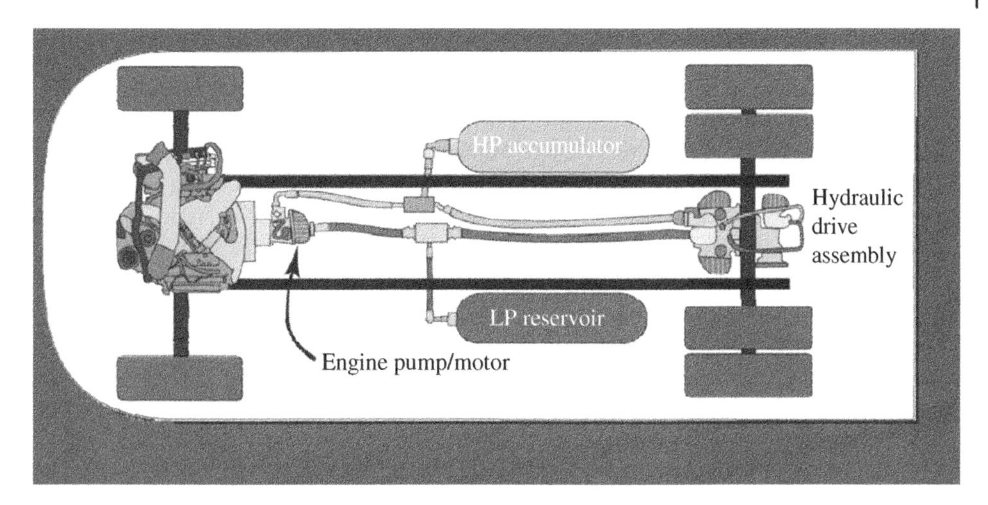

Figure 6.3 Physical architecture of HHV. *Source:* US EPA 2004.

Thus, the hydraulic circuit is completed. Note that the amount of energy storage in the hydraulic accumulator system is rather low in terms of Wh/kg. For example, the energy storage density in a hydraulic accumulator can be about 1.9 Wh/kg [2], whereas a battery can have an energy density of 30–240 Wh/kg. However, the power density of a hydraulic system can be 2500 W/kg, whereas the electrical system power density can be about 650 W/kg. It is therefore apparent that the hydraulic hybrid system is very suitable for a high-power and relatively low-energy system, particularly where short bursts of high-power acceleration and deceleration are involved.

As we see from the previous discussion, in a hybrid hydraulic system, the ICE can run in the optimum efficiency zone of its operating curve and deliver the energy through the hydraulic system to the wheels. A few scenarios can take place here. If the energy needed by the vehicle propulsion matches the energy from the ICE, the power generated will be used effectively in propulsion. Any excess energy will have to go to the accumulator for storage. The energy stored in the accumulator is rather low, hence this option cannot continue for long. If the HP accumulator cannot store any additional energy, then it will be necessary to change the engine operating point so as to match the propulsion need. However, even under this latter condition, it may be possible to stop the engine for a short while and use the energy from the hydraulic accumulator to propel the vehicle. The engine can be started again when the energy in the accumulator runs low. Thus it can be a stop and go situation.

One great benefit of having a hydraulic system for propulsion is that it can eliminate the need for a massive conventional transmission system. Also, hydraulic fluid can be more easily moved from one point to another without the need for an elaborate transmission gearbox, other mechanical linkages, and so on. This concept can be used in a regular non-hybrid vehicle as well, where the conventional transmission system is replaced by a hydraulic system. But to be able to drive an ICE at its most optimum point does require some sort of energy storage, which, in the case of a hydraulic hybrid, is an accumulator containing some gas. Without an energy storage, it is not possible to realize the optimum operating point of the engine, simply because the average propulsion needed in that case has to match the average power produced by the engine.

6.1.1 Regenerative Braking in HHVs

HHVs offer the benefit of regeneration when a vehicle is slowing down, and the ability to use the captured energy to accelerate again thereafter. As noted earlier, the specific energy of the hydraulic system, or Wh/kg, is relatively low compared to that of a battery. However, it can still be good enough for braking applications, since during braking the power is generally high, but the total energy involved is typically low. Hence hydraulic storage can be quite adequate for this purpose. The regeneration process is shown in Figure 6.4 [3].

In Figure 6.4, it can be seen that the efficiencies of the hydraulic pump and motor are both a little over 90%. The efficiency of the accumulator (HP and LP together) is about 98%. Hence the efficiency of the whole regeneration process is about 82%. In another situation, for a hybrid truck, the regenerative efficiency was shown to be 61%, which is still quite good.

One advantage of the hybrid hydraulic system is that the technology is very mature and has been around for many years. Components used in a hydraulic system, like the ones indicated in Figure 6.4, have a very high efficiency. Figures 6.5 and 6.6 show some typical hydraulic components used in an HHV [4].

Figure 6.5 is a bent-axis hydraulic motor from Bosch-Rexroth. The nominal gas pressure at the entrance can be over 5000 psi, the maximum speed can be 4500 rpm, and the torque at 5100 psi can be 371 lb-ft (503 Nm). This information is for a particular motor. The length of the unit, excluding the geared shaft, is about 5.15 in (13.1 cm). Hydraulic pumps also have similar appearance and characteristics.

Some accumulators are merely a cylinder with some valves to let the gas in or out with appropriate safety mechanisms. Other kinds of accumulators can be like the bladder type, in which a flexible bladder is submerged in a fluid or gas, and is pressurized to expand. It is the expansion of the bladder that pressurizes the gas in the accumulator.

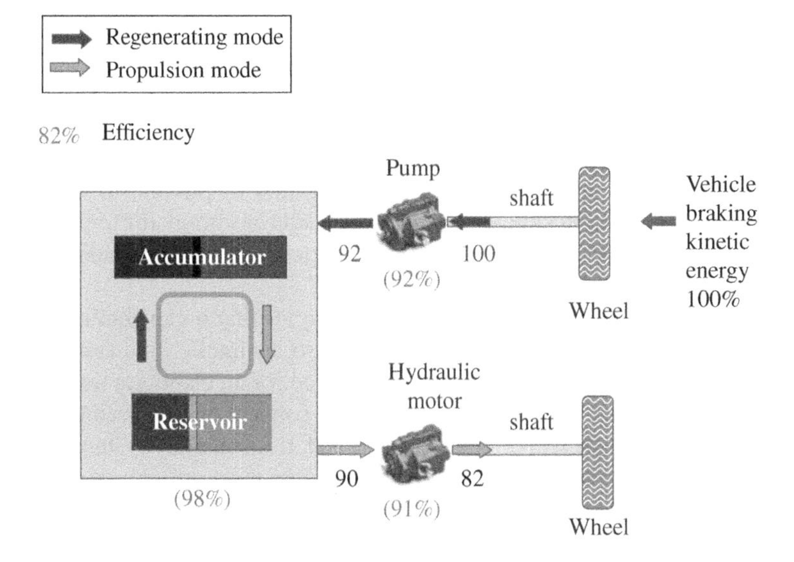

Figure 6.4 Regenerative braking efficiency distribution in HHV. *Source:* US EPA 2004.

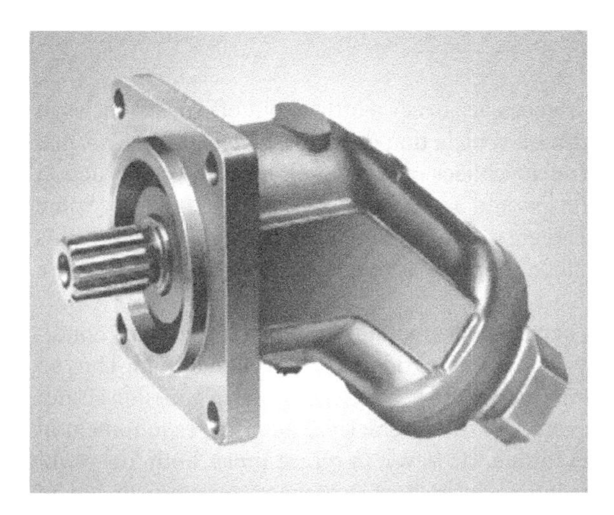

Figure 6.5 Bent-axis hydraulic motor. *Source:* Reproduced with Courtesy of Bosch-Rexroth.

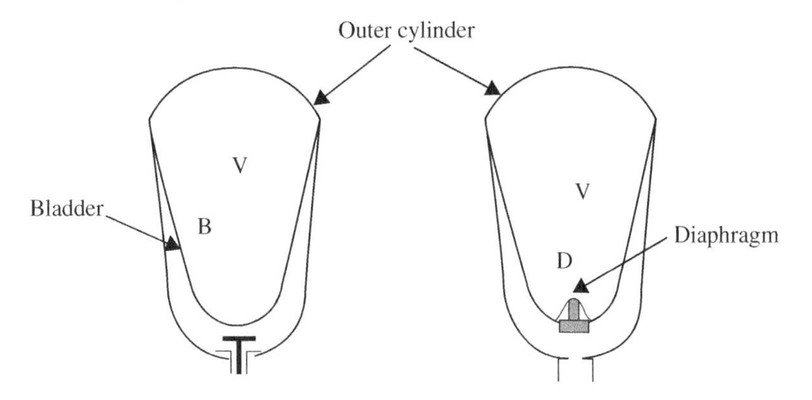

Figure 6.6 Bladder and diaphragm accumulators.

There are several working mechanisms for accumulators. In the hydro-pneumatic scheme a liquid pushes a piston, which then pushes a gas stored in the cylinder and compresses it, whereby energy is stored. The piston separates the liquid from the gas. Another type of accumulator is the diaphragm type.

Diagrams of some accumulators are shown in Figure 6.6, namely, diaphragm and bladder accumulators. The diagram on the right in Figure 6.6 shows a diaphragm accumulator. Here a diaphragm D is pushed by some fluid from the bottom, which then changes the volume V of the gas in the upper part of D, that is, between the diaphragm and the cylinder itself. The diagram on the left shows a bladder accumulator where the bladder B contains the gas (with volume V), which can be compressed. When the fluid enters the cylinder from the bottom, the bladder can expand or contract, thus energy can be stored in the gas when compressed and taken out from the gas when expanded. The energy is transferred through the fluid itself.

6.2 Off-Road HEVs

Before discussing off-road HEVs, it is necessary to say a little about off-road vehicles in general. An off-road vehicle is any ground vehicle that does not, in general, use normal roads for its operation. Examples of such vehicles include predominantly construction vehicles and equipment, mining vehicles, and agricultural vehicles like tractors. Some military vehicles also fall into this category. Off-road vehicles to be discussed in this section have quite different drive cycles and speed–torque demands compared to those of a regular automobile.

Construction and mining vehicles, in particular, operate either under stationary conditions or at relatively low speeds. Very often they also operate over rough terrain. Thus an extremely rugged system is needed for a successful vehicle. The issue of emissions is not of primary concern, due to the very nature of these vehicles. The prime requirement of these vehicles is very high power and torque. However, in recent times, both fuel economy and pollution control have come into focus to create a greener environment and, of course, for economic operation. Hence the possibilities of using hybrid technology are being considered.

A qualitative discussion on the torque–speed requirement for these vehicles and how to achieve it can be made at this point. Getting very high torque from a regular ICE can lead to a very large transmission system or gearbox, and so on. To alleviate these difficulties, in the past, hydraulic systems have been used. Hydraulic systems operate by continuously running an ICE, which is a relatively low-torque, high-speed system. This ICE is used to drive a hydraulic pump, which can drive a hydraulic motor. Using an incompressible fluid can achieve very high pressure, which can be transmitted to the drive end. However, it has been found that the hydraulic system, even though its technology is very mature, can cause certain problems. For example, in a complex plumbing system the pipes can, under rough terrain conditions, cause fluid to leak. This can lead to degraded performance and eventual failure. One disadvantage of hydraulic systems is that certain parasitic losses are unavoidable. Also, the ICE does not operate at its optimum efficiency point all the time due to the nature of the torque–speed demand of the load.

Hence, lately, various heavy off-road vehicle manufacturers, such as those in the mining and construction equipment industries, have been concentrating on transforming their system into an electric hybrid type of propulsion. The word "hybrid" here may have a slightly different connotation from the regular HEV in the automotive sector.

In very heavy vehicular applications like mining and construction, we noted earlier that a pure ICE vehicle can lead to a very large transmission system capable of providing the torque and power demands. This situation can be mitigated by using the ICE to drive a generator or alternator, which in general will generate a variable speed (hence variable frequency) and variable (amplitude) voltage system. This variable frequency, variable amplitude voltage is converted to a DC voltage of variable and ultimately constant value using a rectifier/regulator. The constant DC voltage is translated into a three-phase AC system whose amplitude, frequency, and phase can be controlled electronically through the use of an appropriate power electronics converter. To be more precise, instantaneous variable voltage as a function of time can be generated by using power electronics. This three-phase voltage can then drive an electric motor to handle the vehicular load. Note that this system can avoid the use of a battery. Although

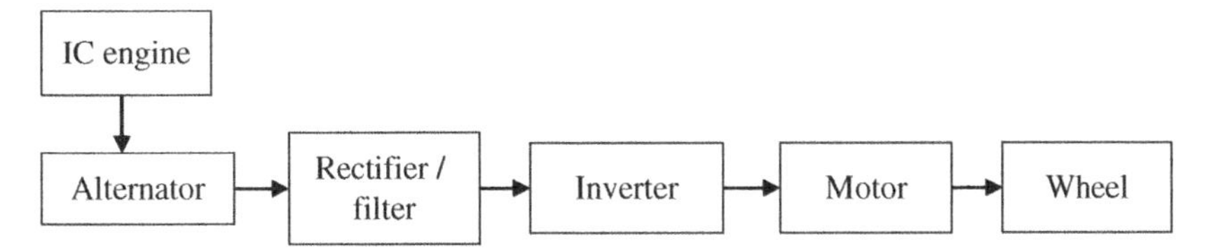

Figure 6.7 System-level architecture of a battery-less hybrid off-road vehicular system.

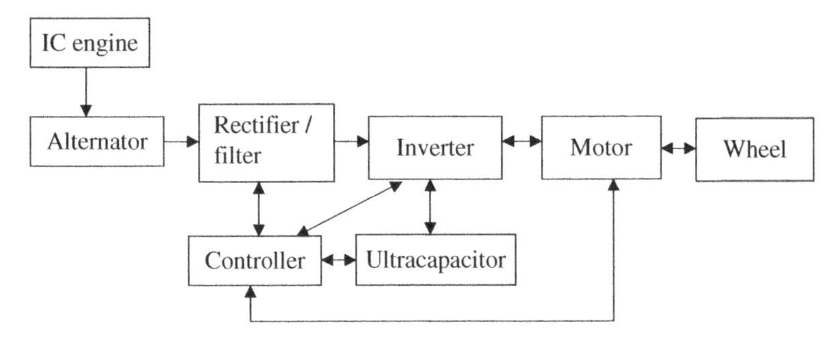

Figure 6.8 System-level architecture of a hybrid off-road vehicular system with ultracapacitor for storage.

the battery is a vital storage element in a regular HEV, in very large off-road vehicles the battery can require a very large and expensive system and hence it may be preferable to do without it. In other words, the system above is very similar if not identical to a diesel–electric locomotive system.

Various architectures of off-road vehicles are possible, two of which are shown in Figures 6.7 and 6.8. The previous paragraph described the architecture of the system in Figure 6.7.

Note that the propulsion motor in this application can be any motor, but for very heavy mining applications, induction motors are found to be more robust and suitable for the rough environment, rather than a permanent magnet motor. The motor is then used to drive the wheels of the vehicle. Generally, off-road vehicles greater than 150 t in weight have two axles, and two motors are used to drive them.

Although in the above architecture there is no battery, in principle a battery could be used for energy storage, but its size would be very large for such applications. In some applications an ultracapacitor bank can also be used. The ultracapacitor normally has a high specific power, that is, it can provide very high bursts of power input to a device, but the duration will be very short and hence the energy involved will be small. During the slowing down of a vehicle or some of its movable parts, the mechanical energy can be transformed into electrical energy and stored in the ultracapacitor. The stored energy can then be used to accelerate the vehicle at a later time. In this way, regeneration can be utilized to save energy, thus leading to better system efficiency and fuel economy. An architecture for such a system with an ultracapacitor is shown in Figure 6.8.

In the above architecture, if we want to drive the vehicle's wheels, the energy can come from the ultracapacitor or from the ICE, which drives the alternator, and then the rectifier creates direct current. The DC voltage from either the rectifier or the ultracapacitor can be translated to a proper AC voltage to drive the electric motor. The system

(a)

(b)

(c)

Liebherr T282 B

Figure 6.9 Typical mining vehicles: (a) Caterpillar; (b) Komatsu; (c) Liebherr. Courtesy: (a) Caterpillar; (b) Komatsu; (c) Wikimedia.

will have the necessary switches, which can choose between the ultracapacitor and the alternator/rectifier. When the vehicle slows down, the controller commands the ultracapacitor, inverter, and alternator/rectifier to coordinate properly, so that energy flows from the motor side (with the motor operating as a generator) into the ultracapacitor, through the inverter.

It is interesting to look at some typical sizes involved in the off-road vehicles indicated above. For example:

- A Caterpillar engine model Cat 3524B EUI has a gross power of 2648 kW (3550 hp) and a gross machine operating weight of 623,690 kg (1,375,000 lb) (http://catsays.blogspot.com/2005/01/caterpillar-797b-mining-truck.html).
- One Komatsu model has a gross horsepower of 2611 kW (3500 hp) and weighs 505,755 kg (1,115,000 lb).
- One Liebherr model has a gross horsepower of 2722 kW (3650 hp) at 1800 rpm and weighs 592 tonnes (652.5 t).

The vehicles indicated above are huge in size, as is obvious from the specifications above and the pictures shown in Figure 6.9.

Of the vehicles shown in Figure 6.9, the Caterpillar uses a conventional powertrain, whereas Komatsu and Liebherr use hybrid (HEV) powertrains. Caterpillar is also now moving toward an HEV platform.

From the Komatsu information sheet, the system has the following specifications:

Alternating current	
Alternator	GTA-39
Dual impeller in-line blower	453 m^3/min (16,000 cf/m)
Control	AC torque control system
Motorized wheels	GDY106 AC induction traction motors
Ratio	32.62 : 1
Speed (maximum)	64.5 km/h (40 mph)

*The authors most sincerely thank the Komatsu technical team who provided additional information on these items.

Although the exact details of the systems are normally proprietary information, some commonality can be noticed between them, and a reasonable idea of the sizes of the components can be obtained, based on the above information. For example, the traction motors in both the Komatsu and Liebherr models are induction motors. The alternators in the Komatsu vehicle (made by GE) have brushes, although brushless generators are also viable. The gear ratio reduction from motor to wheel is on the order of 28–42. The gross engine power of the vehicles is about 2.6–2.7 MW. The inverter used to drive the motor generally uses insulated bipolar gate transistors (IGBT) for switches. Assuming alternator and inverter efficiencies are on the order of 95% each, the motor power input is about $2.6 \times 0.95 \times 0.95 = 2.35$ MW. For the Komatsu 930E-4SE, the retarding power is 4 MW short term and 3 MW continuous. So it is reasonable to assume that an induction motor on the order of half of that can be used, assuming two motors for the system, with one at each axle. Hence each of these induction motors can be considered to be about 1.5 MW. The vehicle at the maximum speed amounts to a wheel revolution of about 84 rpm (assuming a tire diameter of about 4 m). With a gear ratio of about 32.62, the motor speed should be about 2700 rpm. Nominally, this leads to about 3000 rpm, which, at 50 Hz, leads to a four-pole induction motor. Such a motor is a common standard for such applications.

Regarding the rating of the motor for the mining truck, the nominal DC link voltage rating of these motors can be around 2600 V, leading to about 1600 V line-to-line rms input to the three-phase motor. For a 1.2 MW motor, this corresponds to about 541 A AC rms, assuming a power factor of about 0.8. In this case, the DC link current will be $1.2\,\text{MW}/2600 = 461$ A DC per motor. All these assume no additional loss in the system; the figures merely give an idea of the voltage and current values. Some systems use 3.3 kV IGBT technology [5] to drive these motors, which is consistent with the above ratings indicated for the motor.

6.2.1 Hybrid Excavators

A particular application of HEV in off-road vehicles involves a construction excavator. Although there are other construction vehicles such as wheel loaders, bulldozers, and industry utility vehicles, an excavator is considered to be more complicated in terms of its

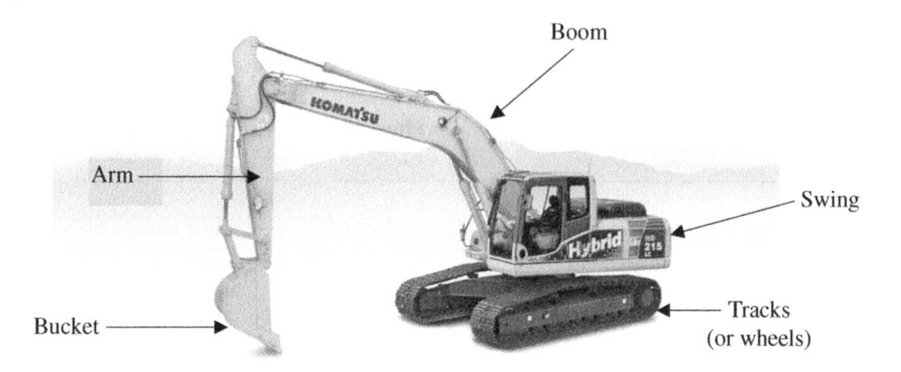

Figure 6.10 A typical excavator with parts labeled. *Source:* Reproduced with Courtesy of Komatsu.

architecture and implementation. Hence this particular machine is chosen as an example and will expose us to a more complex situation than other construction machinery.

In an excavator, the use of the HEV concept is not in the vehicle propulsion, but rather in the excavator arm movement. The excavator arms need very brief bursts of high power followed by low-power return and then slowdown of its speed to stop the arm. Such an application is very well catered for by using an ICE driving a generator to produce electricity to drive the excavator arm motor, which is an electric drive.

Let us first look at several possible architectures pertinent to an excavator.

Figure 6.10 shows a typical excavator with the various parts labeled. Although this particular excavator from Komatsu is a hybrid excavator, the physical look for a conventional hydraulic excavator is basically the same. The conventional hydraulic architecture is shown in Figure 6.11.

Here the engine drives a hydraulic pump and the fluid flow is controlled by a set of control valves. Fluid flow is channeled to various cylinders attached to different parts of the excavator – the boom, arm, and bucket – which are in one group in Figure 6.11. The second group involves a hydraulic motor that is also fluid driven and is used to physically move the excavator by turning the right and left wheels or tracks. The hydraulic motor torque can be passed through a gear before it goes to the wheels or tracks. The third group involves a hydraulic motor that is used to move the swing arm of the excavator.

As noted previously, the engine is running only one pump in the Figure 6.11, and the fluid goes to a group of valves to control the various moving members. It should also be noted that since every single moving member is not controlled by individual valves, there is always energy wastage in terms of fluid unnecessarily flowing through certain channels all the time. In other words, fluid energy can be wasted in the form of heat doing no significant work in a conventional excavator. To address such issues, the following possible hybrid architectures are implemented using electrical devices. It should be noted that the hybrid excavator still uses quite a number of hydraulic devices, since for construction machinery the amount of force or torque needed is sometimes very high. In principle, although such high forces can be generated electrically, the electric motor size and power electronics size may be prohibitive in that case. So, an electro-hydraulic system is considered the most optimum for this purpose.

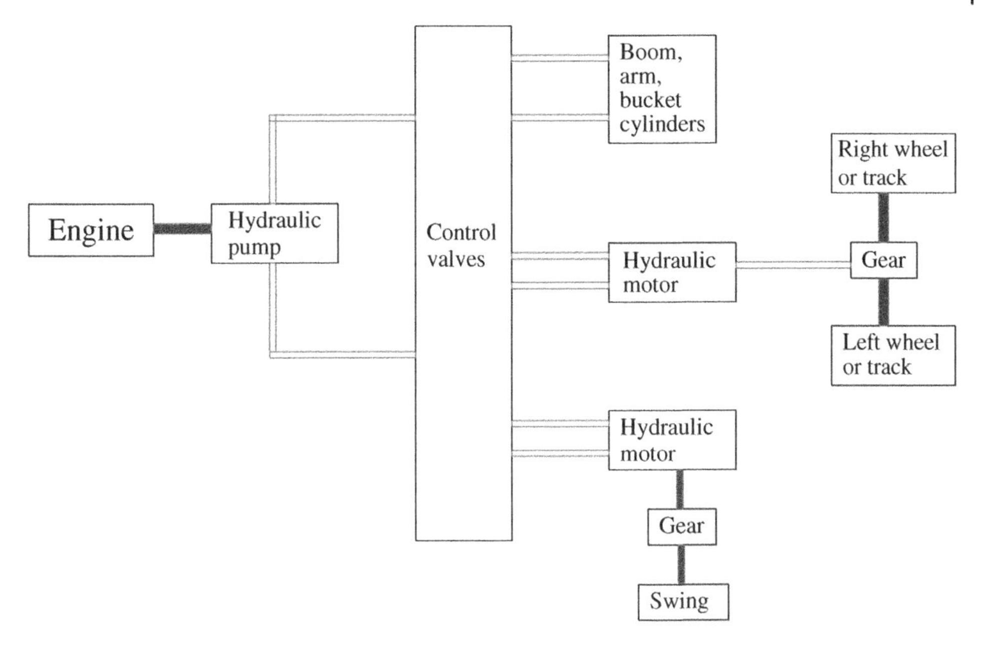

Figure 6.11 Architecture of a conventional off-road excavator.

The architecture is that of a series hybrid, shown in Figure 6.12. The operation principle of series architecture based excavator is as follows:

- The engine drives the generator and creates three-phase AC.
- Three-phase AC power is converted to DC.
- DC power goes to three separate inverters.
- DC power also goes to the ultracapacitor after voltage change through the DC–DC converter.
- For the boom cylinder – inverter power (three-phase AC) drives the electric motor; the electric motor drives the hydraulic pump; the hydraulic pump drives the boom cylinder, which is controlled by the control valve (C/V).
- For the arm and bucket cylinders, as above: inverter power (three-phase AC) drives the electric motor; the electric motor drives the hydraulic pump; the hydraulic pump drives the arm and bucket cylinders, which are controlled by a control valve (C/V) (separate from the boom).
- Swing operation is done completely by the electric motor.

It can be immediately noted that this architecture provides significant flexibility in controlling various components in the excavator by allowing for independent control of the motors that drive the hydraulic pumps and a completely electric operation of the swing system. However, this flexibility comes at the cost of several inverters and converters, and also several motors. Although costly due to its many additional components, this system allows for the best fuel economy and control flexibility which will compensate the initial investment.

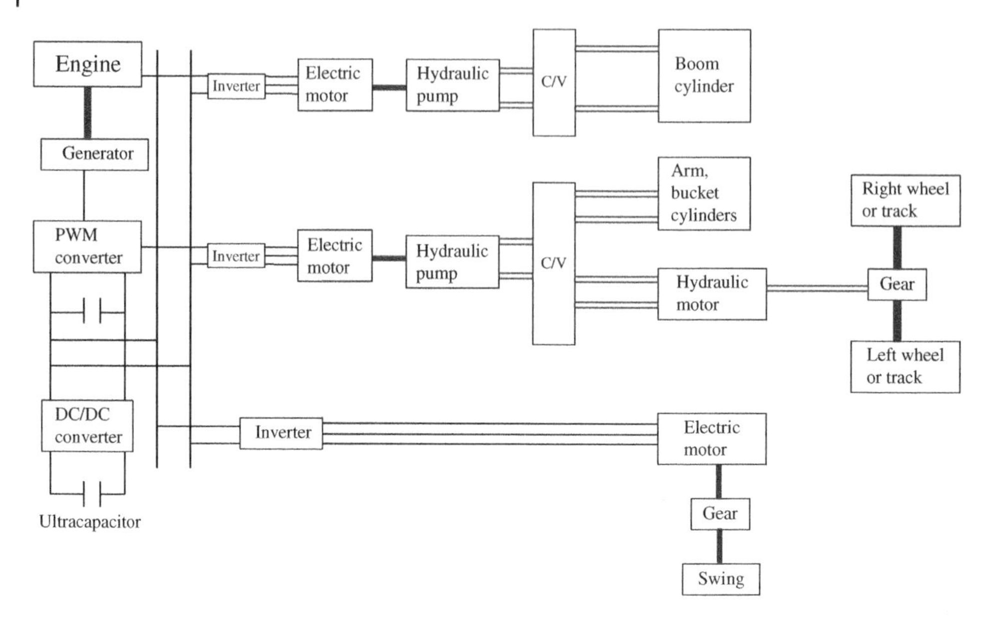

Figure 6.12 Architecture of a series hybrid excavator.

Figure 6.13 shows a parallel architecture. The operation principle of this architecture based excavator is as follows:

- The engine drives the generator and creates three-phase AC.
- Three-phase AC power is converted to DC.
- DC power goes to the ultracapacitor after voltage change through the DC–DC converter.
- The generator and the hydraulic pump are on the same shaft and are coupled.
- The speed of the generator depends on the speed of the hydraulic pump, dictated by the hydraulic load (all loads). If the engine has more available torque at a particular speed, and if the torque is higher than the hydraulic torque demand, then any additional torque can be used up by the generator, which is essentially used to charge the capacitor. On the other hand, if the hydraulic torque demand is higher, the generator can enter motor mode and provide some torque towards the hydraulic load, thus it is in a torque-assist mode. It is analogous to something like a mild hybrid passenger vehicle.
- It is not possible to recover any energy from the mechanical loads in this particular configuration.

Compared to the series architecture, this one needs fewer components, but provides less flexibility in controlling various components in the excavator. In a nutshell, this architecture is just like the conventional hydraulic architecture, except that there is a generator sandwiched between the engine and the hydraulic pump. The generator can be used to charge an ultracapacitor through a power electronic controller system and can also receive power from the ultracapacitor while running in the motor mode, adding any suddenly needed transient torque. In other words, this system can be equated more to a mild hybrid type of passenger electric vehicle with a little bit of power assist, and the main engine is designed for just catering for the average load demand. This can lead to a smaller engine and some additional fuel economy, though it is not as flexible as the series architecture.

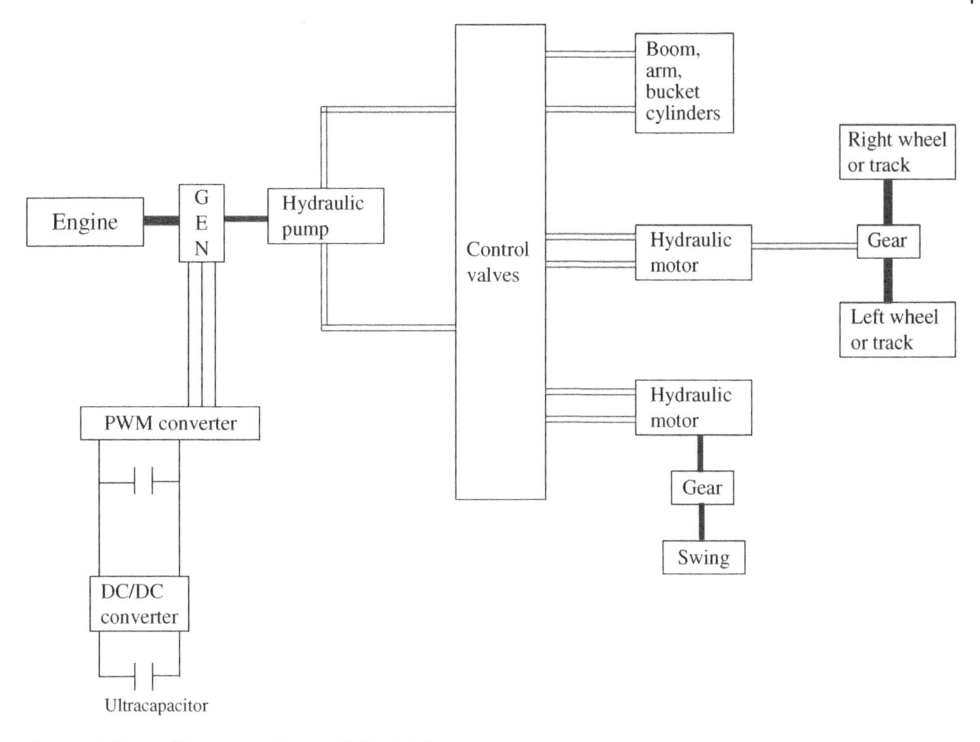

Figure 6.13 Architecture of a parallel hybrid excavator.

Figure 6.14 shows the compound hybrid architecture. The operation principle of this architecture based excavator is as follows:

- The engine drives the generator and creates three-phase AC.
- Three-phase AC power is converted to DC.
- DC power goes to the ultracapacitor after voltage change through the DC–DC converter.
- The generator and the hydraulic pump are on the same shaft and are coupled.
- The speed of the generator depends on the speed of the hydraulic pump, dictated by the hydraulic load (except the swing load). If the engine has more available torque at a particular speed, and if the torque is higher than the hydraulic torque demand, then any additional torque can be used up by the generator, which is essentially used to charge the capacitor. On the other hand, if the hydraulic torque demand (boom, arm, bucket) is higher, the generator can enter motor mode and provide some torque towards the hydraulic load, thus it is in a torque-assist mode. Hence this operation can be again compared to something like a mild hybrid vehicle.
- The swing load is completely controlled by the motor.
- It is possible to recover energy from the mechanical loads in this particular configuration, but only from the swing operation. In principle, it is also possible to recover energy from other moving members that are hydraulically operated, provided appropriate bidirectional control valves are used, but that can complicate the control, leading to additional cost.

This architecture is somewhat in between series and parallel. The upper part of the system is a conventional hydraulic system with power assist through the electrical

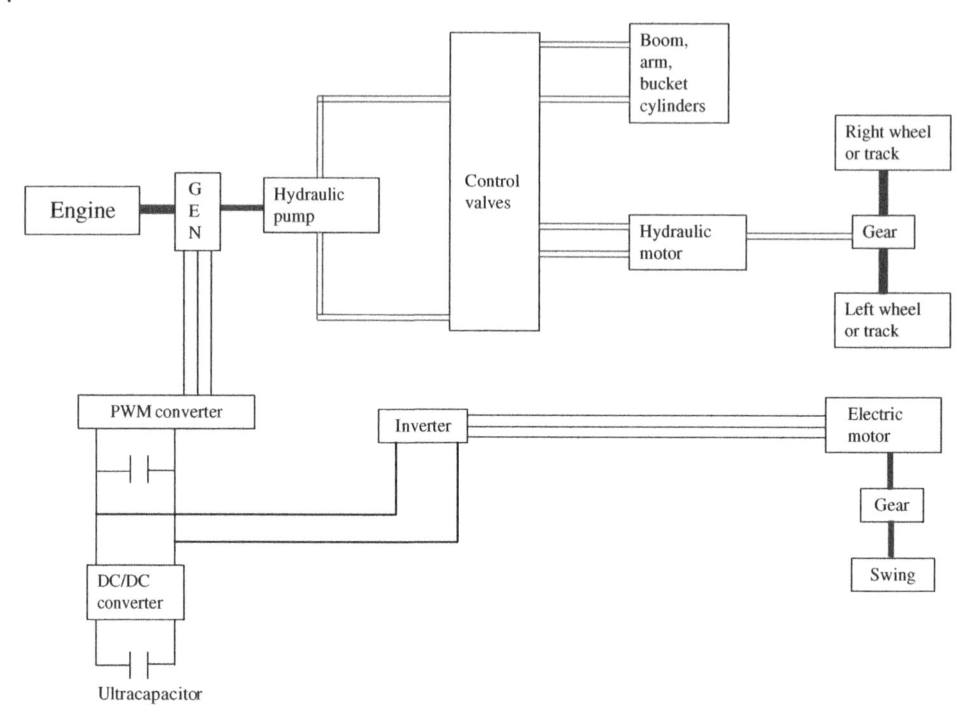

Figure 6.14 Architecture of a compound hybrid excavator.

generator, which can operate in the motor mode as well. This system also has the swing system, which is fully electrical. This system is a compromise between series and parallel systems with some additional components, but not as many as the series system. Hence quite a few commercial excavators use this architecture.

The architecture for the Komatsu excavator shown in Figure 6.15 is fundamentally the same as the compound architecture shown in Figure 6.14. It has been claimed that the Komatsu hybrid excavator leads to a fuel economy improvement of around 25% and that some specific users have achieved improvement of fuel economies as high as 41%. The pictures in Figure 6.15 were from the Komatsu web site (www.komatsu.com/CompanyInfo/press/2008051315113604588.html), which has since been updated by Komatsu, and more up-to-date information can be found at http://www.komatsu.eu/displayBrochure.ashx?id=82172. For the sake of illustration, some numerical values are provided here for this excavator: at 288 V DC, the DC link current will be about 70 A, the three-phase line-to-line rms voltage for the motor will be about $0.612 \times 288 = 176$ V, and the motor current will be about 82 A.

Even though ultracapacitors were shown in the architecture earlier, some manufacturers prefer to use a battery instead of an ultracapacitor for the purpose of storage and retrieval of energy. An example is the New Holland excavator, which uses a 36 hp diesel engine, 20 kW generator, and 288 V lithium battery. The choice of storage is highly dependent on the exact usage and duty cycle. If the vehicle is just a high-power excavator, then an ultracapacitor seems to be a suitable candidate. However, if in addition to the excavator application, there is any other auxiliary equipment needing energy that may not be immediately available from an ultracapacitor, a battery may be used.

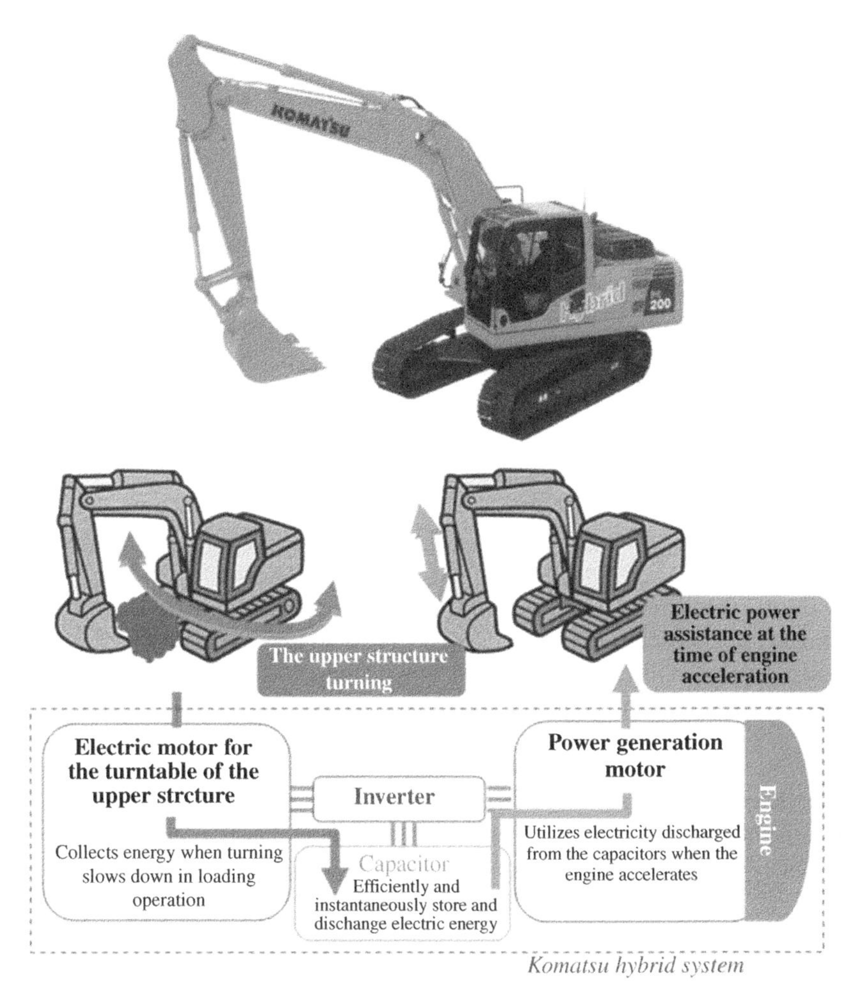

Figure 6.15 Komatsu hybrid excavator system. *Source:* Courtesy of Komatsu.

The duty cycle in terms of torque need, duration of the torque, auxiliary equipment used, cost, size, and similar factors leads to the decision process in terms of what will be best: ultracapacitor or battery. Hence on this matter there is no unique answer.

6.2.2 Hybrid Excavator Design Considerations

Hybrid technology as applied to regular vehicles remains the same when applied to construction machinery in general. However, load conditions in the two situations are very different. For example, heavy and light digging conditions in an excavator could be something like those depicted in Figure 6.16.

Based on these conditions and the architecture used, we have to choose the appropriate storage units for regeneration, e.g. during braking of the swing system.

Let us consider the parallel architecture as an example.

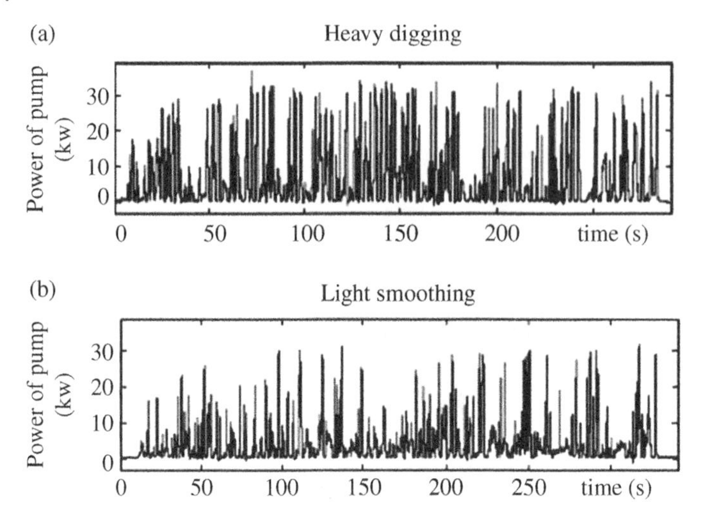

Figure 6.16 Hybrid excavator load demand. *Source:* Wang 2008. Reproduced with permission of IEEE.

As already noted, the parallel architecture is basically a conventional architecture, except for an electric motor/generator between the engine and the hydraulic pump, which acts as a power assist device during transient load demands. We have also seen in the architecture diagrams above that the energy storage device used to energize the power assist device can be a capacitor. This capacitor is charged by using the engine when it has any spare power available after driving the load. At times of sudden or transient excess load demand, the capacitor can deliver that demand through the electric machine, which becomes a motor. The benefits of this is that it avoids sizing the engine for the full transient expected load. In other words, the engine can be undersized to some extent, for example, in exchange for the generator, which will still be better when an electrical machine is used, compared to an oversized engine. This also allows the engine to run nearer its optimal speed and thus more efficiently.

A heavy digging load torque curve is shown in Figure 6.17 as an example.

Control strategy determination can be done as follows. As we know, the ICE provides power to the load and the electrical system. The job of the electric machine is to supply the deficit power to the load or the difference between the average power requirement and the peak transient.

$$P_{Elect} = P_{Load} - P_{ICE} \tag{6.1}$$

In the particular example in the reference [6]:

ICE power = 25 kW @ 2200 rpm and based on average load, where a turbocharged diesel engine was chosen.

$$\text{ICE torque} = 124.8\text{Nm}\,@\,1860\text{rpm} \tag{6.2}$$

Let us see the rationale behind the above figures. The maximum transients are about 30–40 kW (but closer to 30), so the deficit is mostly about 5 kW (once or twice we may encounter 15 kW).

So, it is reasonable to choose an electrical machine that is about 10 kW. Exceeding this limit by 50% once or twice may be acceptable. The torque rating chosen was 76 Nm @ 1860 rpm.

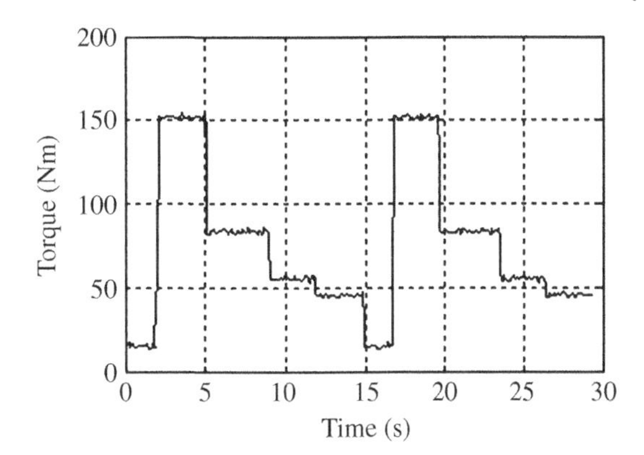

Figure 6.17 A torque demand curve. *Source:* Wang 2008. Reproduced with permission of IEEE.

Electrical machine speed range has to be chosen to match the ICE speed range, otherwise a gear box will be necessary.

The question is how was the engine speed chosen in the first place? That is based more or less on the load torque demand and we saw that torque demand from the load was about 150 Nm (Figure 6.17). Hence the total of the ICE + electrical machine torque is about 200 Nm, which should be quite adequate.

Next we have to understand how the speed of the ICE is chosen. Notice that the power demand of the load was about 30 kW. The torque was 150 Nm.

So, $\text{speed} = 30000/150 = 200\,\text{rad/sec} = 200 \times 60/(2\pi) = 1910\,\text{rpm}$, which is reasonably close to 1860 rpm. The important thing is that we need to know the torque demand and power demand of the load.

It should be noted that we can operate the load slowly or fast, with the same torque, but if it goes slowly, it will need less power, obviously. These are matters of choice and the designer has to decide (based on construction experience), how fast those loads will need to be handled.

Of course, we can use gears to amplify the torque and reduce the speed for the same power. Power will remain more or less invariant, but torque and speed could change, to match the power.

At this point we still have to decide a few more electrical design parameters which are now discussed.

We need to decide what voltage is good for the purpose, and on this there is no unique answer. In general, high voltage machines are physically smaller. Similarly, high speed machines are physically smaller. This means a smaller size for a given power capability.

The choice of motor voltage may have to do with whether we want a custom-built motor, or an off-the-shelf item based on what is currently available.

In this design, a 240 V machine was chosen. It does not have to be that way, though. If we go for a higher voltage, then the power electronics that drives the motor also has to be rated to the higher voltage. It should also be noted that a higher voltage needs better insulation and packaging than a lower voltage. Furthermore, a higher voltage means a smaller current and hence thinner windings. Similarly, a higher speed means lower torque and lower current. A high speed also implies a weaker magnetic field for a given voltage, meaning less magnetic material usage.

So, the ultimate choice is a compromise between several factors and this is a matter of experience and the availability of items, their cost, and so on.

If the speed of the motor chosen is different from the ICE and the load, then we will need appropriate gear, which will add to the cost.

A permanent magnet synchronous machine (PMSM), which is a relatively compact machine for a given power, was chosen in this case.

For energy storage design, we need to estimate the acceleration power that is to be met and the deceleration power that has to be absorbed. It is also good to know about a machine's lifetime digging cycle, i.e. how many times the various actuators are likely to move in a typical machine's life.

So, the process can start with the electric energy storage of the hybrid excavator, and can consist of the following possibilities [6]:

Battery only (BO)
Ultracapacitor only (UO)
Battery + ultracapacitor (BU)

As we should be aware, a battery is a more energy intensive device, whereas an ultracapacitor is a more power intensive device.

So, we can first compare the specific power and energy of each of these devices:

	Value	
Quantity	**NiMH battery**	**Ultracapacitor**
Specific power – W/kg	500	11,300
Specific energy – Wh/kg	35 @ 3C rate	3.2
Cycle life	1000	1,000,000

End of life can be defined as the number of allowable cycles with a depth of discharge defined by the manufacturer, after which the battery may not function well, i.e. not accept and retain charge for very long.

For an ultracapacitor, the cycle is defined based on the voltage excursion from nominal to half the nominal voltage.

Based on the power alone [6]:

$$W_{bat\_power} = (P_{ESU\_Max})/(\eta_{bat} \times S_{bat})$$

where the numerator term indicates the maximum power needed from the energy storage unit or ESU – battery in this case – and the first term of the denominator is the battery efficiency and the second term is the specific power of the battery.

If a conservative approach is taken, then the battery should be designed to handle the maximum average power of a heavy digging cycle, which is about half the peak power. The peak power was about 34 kW.

Hence:

$$W_{bat\_power} = (P_{ESU\_Max})/(\eta_{bat} \times S_{bat}) = (17\,\text{kW})/[(0.85 \times 0.5\,\text{kW/kg})] = 40\,\text{kg} \qquad (6.3)$$

Now, from the perspective of battery life time, i.e. the total number of charge discharge cycles with a given depth of discharge, we have

$$W_{bat\_life} = (T_{ESU} \times E_{cycle})/(t_{cycle} \times ED_{bat} \times k \times DOD \times n \times \eta_{bat}) \tag{6.4}$$

where,

T_{ESU} = battery life duration after which it needs replacement, assumed to be 4800 hours here.
E_{cycle} = digging energy each cycle = 15 Wh.
t_{cycle} = cycle time of digging = 15 sec
ED_{bat} = energy density of the battery = 35 Wh/kg
k = coefficient of energy that the battery has = 3 Wh per depth of discharge
DOD = depth of discharge = 0.8
n = total number of cycles in the life of the battery = 1000
η_{bat} = battery efficiency = 0.85

Hence the above equation becomes:

$$\begin{aligned} W_{bat\_life} &= (T_{ESU}) \times (E_{cycle})/(t_{cycle} \times ED_{bat} \times k \times DOD \times n \times \eta_{bat}) \\ &= \frac{(4800 \times 3600\text{ secs}) \times (15\text{ Wh/digging cycle})}{\begin{array}{l}(15\text{ sec/digging cycle}) \times (35\text{Wh/kg/discharge cycle}) \times 3 \times (0.8DOD) \\ \times (1000\text{ discharge cycle}) \times 0.85\end{array}} \\ &= 242\text{ kg} \end{aligned} \tag{6.5}$$

For the ultracapacitor:

$$W_{UC\_power} = (P_{ESU\_Max})/(PD_{UC}) = 17\text{kW}/(11.3\text{kW/kg}) = 1.6\text{kg} \tag{6.6}$$

From lifetime considerations:

$$W_{UC\_life} = (T_{ESU}) \times (E_{cycle})/(t_{cycle} \times k_1 \times ED_{UC} \times k_2 \times n_{UC} \times \eta_{UC}) \tag{6.7}$$

$$\begin{aligned} &= \frac{(4800 \times 3600\text{ secs}) \times (15\text{ Wh/digging cycle})}{\begin{array}{l}(15\text{ sec/digging cycle}) \times 0.75 \times (3.2\text{ Wh/kg/discharge cycle}) \\ \times (1000{,}000\text{ discharge cycle}) \times 0.8\end{array}} \\ &= 9.00\text{ kg, and } k_2 \text{ is assumed} = 1 \end{aligned} \tag{6.8}$$

From the energy perspective:

$$\begin{aligned} W_{UC\_energy} &= (E_{cycle})/(k_1 \times ED_{UC} \times \eta_{UC}) = 15\text{Wh}/(0.75 \times 0.8 \times 3.2\text{Wh/kg} \\ &= 7.82\text{kg}. \end{aligned} \tag{6.9}$$

So, it appears from consideration of power, energy, and lifetime together, that the weight of the battery above is 242 kg and the weight of the ultracap is 9 kg.

Hence UC-only appears to be better from the point of view of this application.

In regard to capacitor size, it is best to go to the heavy digging condition and see how fast the power of the load (as depicted by pump power in a conventional excavator) rises.

For this excavator, perhaps a 1 to 2 second rise is reasonable. If the power rises from 0 to 30 kW in 1 sec, let's see what capacitor size it needs.

Assume a voltage of 125 V as the nominal ultracap voltage and that it can go to 240 V while charging. In other words, we can assume that it was charged up to about 240 V and then was discharged down to 125 V in 15 sec.

So, $dV/dt = (240-125)/15 = 7.7$ V/sec

Average of 240 and 125 is = 182.5 V

For 15 kW average power, current = 15000/182.5 = 82.2 A

$C \times dV/dt = 82.2$, thus $C = 82.2/7.7 = 11$ farad.

The above calculation is from immediate power delivery requirements.

Thus far, the energy recovery was only with respect to the parallel architecture.

Here the capacitor was used only for the purpose of storing energy through the assist motor (or generator) or driving it using the power from the capacitor.

If the compound architecture is used, then we can recover the swing energy as well.

One particular motor used in a Kobelco [7] swing electric motor in an 8 ton class excavator is about 8 kW and the generator/motor (assist) is 10 kW.

So, it is quite reasonable to use a separate capacitor for storing the energy using a capacitor of similar size, i.e. about 11 F, as noted in the previous example.

Some people have tried to recover boom energy by using hybrid technology, which can be classified under the category of an extended compound machine, i.e. both the swing system and the boom system are electrically actuated, either directly (swing) or indirectly through the hydraulic devices. In that case, if we revisit the series architecture, we will see that the boom was driven by an electric motor, which drove a hydraulic pump.

So, if boom energy is to be recovered, we can potentially use a bidirectional hydraulic device, i.e. a hydraulic actuator at the end will become a pump, the hydraulic pump immediately in front of the electric motor will become a hydraulic motor, and the electric motor will become an electric generator, which will send energy to the capacitor. Figure 6.18 indicates the relative load demand of various systems within the excavator [8].

The areas under the curve in Figure 6.18 for various components give an idea of how much energy could potentially be captured for future use.

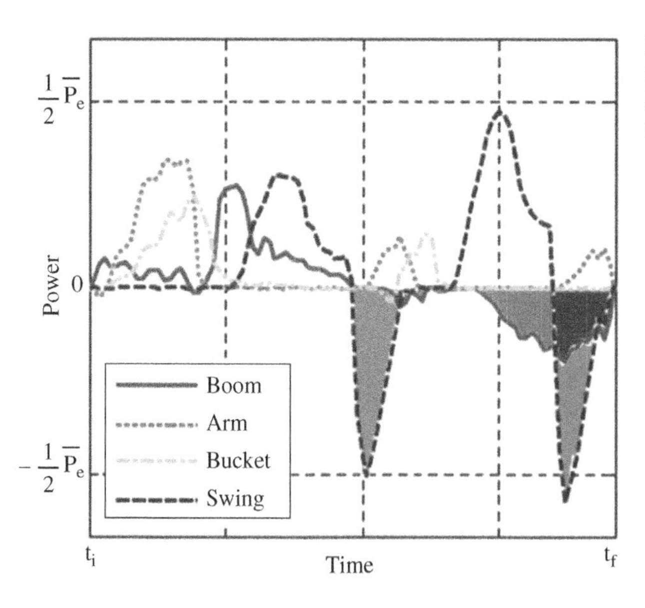

Figure 6.18 Excavator component power demand. *Source:* Yoo 2009. Reproduced with permission of IEEE.

6.3 Diesel HEVs

Fundamentally, a diesel HEV is nothing but a regular HEV, except that the ICE is a diesel engine instead of a gasoline engine. It is well known that a diesel engine is more efficient than a gasoline engine, hence by combining a diesel engine with an electric drive, we can get the best of both worlds. So, the question might arise: why should we not use a diesel hybrid and what are the motivations for not doing so? The following discussion will help shed some light on this issue.

The compression ratio of a diesel engine can be much higher than that of a gasoline engine – it can be around 15–20 compared to 9–12. The higher compression ratio means that the engine has to be heavier in construction to sustain the higher pressure. At low speeds, diesel engines are better in terms of fuel economy due to the absence of throttle valves, thus avoiding parasitic losses. Other benefits of diesel engines include lower greenhouse gas emissions such as CO_2. Low-speed characteristics of diesel engines are also better compared to those of a gasoline engine in terms of starting torque generation. The absence of an ignition system, and hence fewer components, leads to higher overall system reliability.

A diesel engine, being more efficient than its gasoline counterpart, obviously leads to smaller-sized engines, particularly at lower speeds. Hence for the same performance, and particularly during acceleration, it is more beneficial to use a diesel hybrid within the same packaging size. This is especially beneficial in a parallel hybrid configuration, where both the diesel and electric propulsion come into play to achieve performance in terms of power and acceleration. This performance issue is of no consequence in a series hybrid, since propulsion is done only by the electric motor.

In transit buses, diesel hybrid technology has benefits, due to the nature of the stop-start drive cycle and the need for acceleration at low speeds, which are very specifically offered by diesel hybrids. Hence diesel hybrids are ideal candidates for applications in buses and delivery trucks to name but a few. They are also suitable for very heavy duty vehicles like mining vehicles, locomotives, and so on, in a series hybrid configuration which can help eliminate the need for a very heavy transmission (gearbox, torque converter, etc.) system. In these applications (mining, locomotives) the engine is used to run a generator, which can be run at optimal speed. The electricity can then be coupled to a motor through some power electronics inverter.

If a comparison of diesel versus gasoline hybrid electrics is done on an equal basis, it can be briefly summarized as follows:

- A diesel engine operates more efficiently overall than a gasoline engine for a given size.
- A diesel engine has fewer components, so it needs less maintenance, compared to a gasoline engine.
- The higher the power need, the more important the size and cost become, so the diesel will have the edge compared to gasoline.

It therefore appears that there is a breakeven point in power requirement, above which diesel hybrids will prove more economical, in terms of both cost and fuel economy and overall life-cycle maintenance, compared to gasoline hybrids.

From this short overview of diesel engines, it can be concluded that the merits of diesel engines and electric propulsion, taken together, can lead to the best of both worlds, and to the most fuel-efficient vehicles, subject to the specific application needs noted above.

6.4 Electric or Hybrid Ships, Aircraft, and Locomotives

The success of HEVs in automotive applications has led certain other vehicular areas to consider this technology as well. These include ships and aircraft, which are non-ground vehicles. Diesel-electric locomotives have already been using this technology in a slightly different form, and we will discuss this as well.

6.4.1 Ships

Obviously, the need in a ship involves very large sizes of everything. It can ultimately be considered as an industrial power system, with both utility types of systems and a propulsion system. The overall power need in a ship could be anywhere from 1 MW to almost 100 MW. The focus of this book, however, will be on propulsion and not on the overall electric power system of a ship.

Historically, ships evolved in various stages, from steam or diesel propulsion to diesel-electric propulsion. Initially, the motors used in ship propulsion were DC motors. With the advent of power electronics, DC motors can now be replaced by robust induction motors with very good control, based on power electronics. The motors could also be field-excited synchronous motors. The benefits of using these motors are reliability and efficiency. In addition, recently, the technology of pod propulsion has become popular [9]. In this scheme, the propulsion motor is located separately in a pod, which is physically secured at a distance from the main body of the ship. The power electronics system is located within the main body of the ship and electrical wiring is run to the pod, which houses the motor. The size of the pod can be very large – something like 10–12 ft (3–4 m) in diameter. The propulsion motor shaft is connected to the propellers. Also, the pod is capable of being turned through a full 360° if needed. This eliminates the need to have a rudder in the ship. Let us now look at the architecture of the ship propulsion system [10] shown in Figure 6.19.

Figure 6.19 aims to give a complete overview of the possibilities in terms of generation, distribution, and propulsion in a ship. The choice of the particular architecture is dependent on the size, and this can significantly affect the cost. Not all of the components mentioned above will be suitable under all circumstances. If the ship generation system uses a diesel or gas turbine, then the overall ship system could be considered to be a hybrid system, whereas if there is no mechanical power system – for example, if a fuel cell is the only source of power – then the ship will be a completely electric ship. Even if the power is generated by nuclear energy – ultimately to get electricity – then it will need some kind of electric generator, which in turn will need to be turned by non-electrical means, and in this case the ship can be considered in the hybrid vehicle category. Whether it will be hybrid in terms of propulsion will depend on the exact propulsion means used, and what kind of devices are directly contributing toward propulsion.

So, a simpler subset of the above system could consist of a gas turbine as the prime mover, which drives a field-excited synchronous generator (or it could be a permanent magnet generator). The advantage of a field-excited generator is the ability to control the voltage by controlling the magnetic field, which can be done using semiconductors of relatively lower current rating. If a permanent magnet generator is used, the voltage control has to be at full power level at the stator terminal with much higher-rated semiconductors. The advantages of permanent magnet generators are higher efficiency and

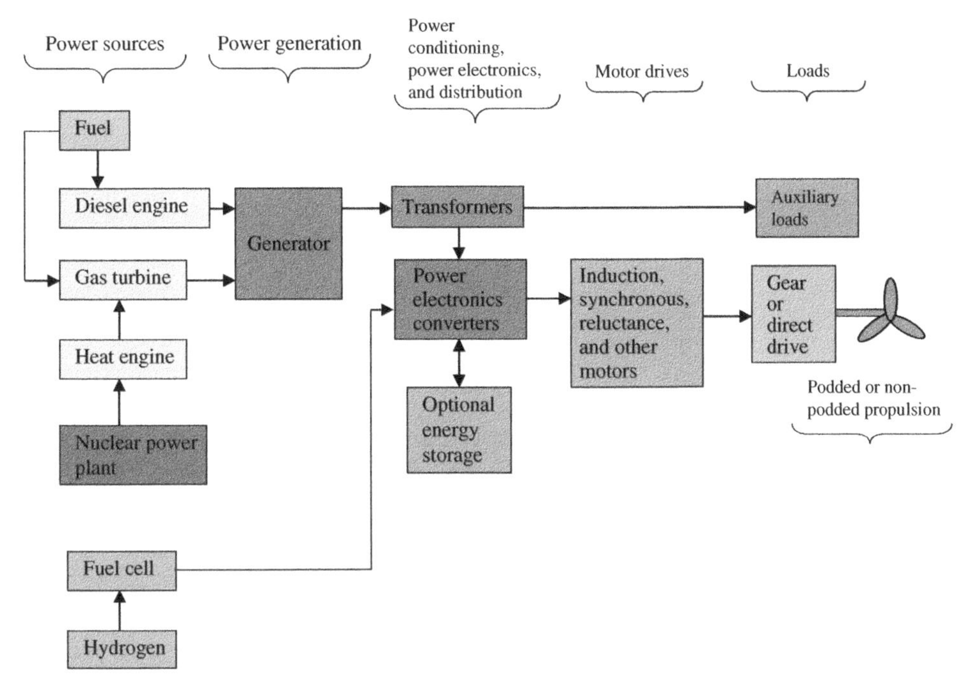

Figure 6.19 A generic architecture of a ship's electrical system. Paths shown by arrowheads entering a particular block merely imply multiple possibilities and do not necessarily indicate concurrent paths.

simplicity, with no need to use slip rings for the field. On the propulsion side, we can use synchronous or induction motors. Synchronous motors have to be doubly fed, whereas induction motors are fed only at the stators. The architecture of a possible system is shown in Figure 6.20.

The propulsion motor voltage may be rated at several hundred to several thousand volts, depending on the size and needs.

With the above architecture in mind, it will now be instructive to look at the pod propulsion we referred to earlier. As its name suggests, it merely involves the physical location of the propulsion motor. In other words, it is the items on the right of Figure 6.20 – the propulsion motor and propeller – which are housed inside the pod. The pod itself is outside the ship's main body structure (below the stern), but of course is secured to it through mechanical structures. The electric wiring runs from the main ship to the pod. So the items shown inside the dot-dashed lines in Figure 6.20 are in the pod. Figure 6.21 shows what the pod looks like.

The figure is reasonably self-explanatory, showing the various components. The slip ring unit has to be connected to the power system, which is located inside the main body of the ship. The man in the picture gives an idea of the size of the pod and its components.

There are only a handful of pod manufacturers in the world. The main ones are ABB, Rolls-Royce, and Schottel. ABB makes the Azipod [11] and Rolls-Royce makes the Mermaid pod. Schottel has a low-power pod known as Schottel Electric Propulsion (SEP) and there is a high-power version called Siemens Schottel Propulsion (SSP) in a

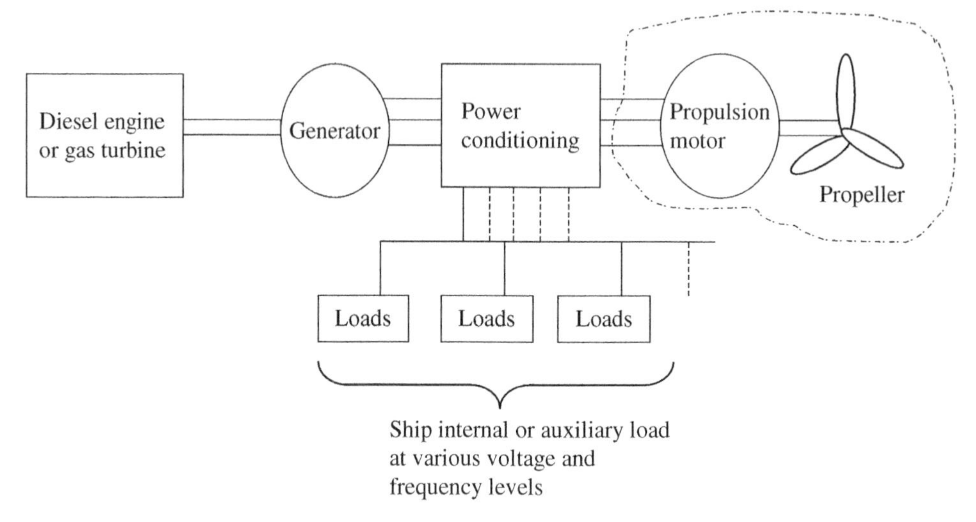

Figure 6.20 System architecture of a hybrid electric ship.

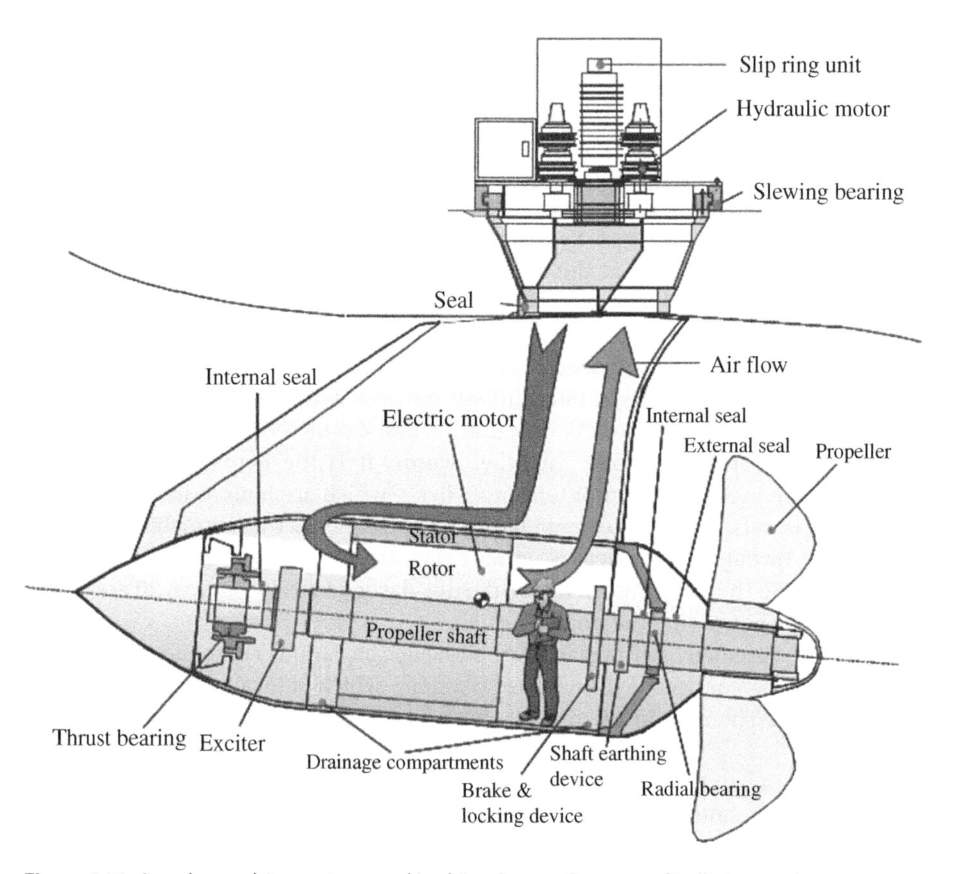

Figure 6.21 A pod propulsion system used in ships. *Source:* Courtesy of Rolls-Royce plc.

Figure 6.22 External view of actual pod propulsion systems in a ship: left, Azipod® by ABB Oy; right, Mermaid pod by Rolls-Royce. *Source:* Courtesy of Rolls-Royce plc.

joint venture with Siemens. The Azipod and the Mermaid are shown in Figure 6.22 to give an idea of their appearances.

6.4.2 Aircraft

Since this book is about electric and hybrid vehicles from a propulsion point of view, it might be instructive to consider an aircraft propulsion system and whether hybrid propulsion can be considered for this purpose or not. First, with the energy technology currently available to us, electric and hybrid propulsion systems cannot be considered feasible but we will give some explanation.

There are several fundamental issues that are different for aircraft compared to land or water-borne vehicles. First, why do we consider hybrid or electric technology? From our previous discussions in this book, the predominant considerations are: (1) fuel economy, (2) environmental friendliness, (3) size, (4) cost, (5) reliability, (6) weight of the mechanical transmission system in certain off-road applications, including locomotives, and (7) the drive cycle of the vehicle. The choice of hybrid or electric vehicle is dependent on the trade-off between these seven items. The drive cycle of a ground vehicle can fluctuate according to city or highway driving. Water vehicles, particularly ocean-going ships (including some smaller watersport and similar vehicles), and aircraft do not have a fluctuating drive cycle. Hence many of the considerations that are important in ground vehicles do not apply. Another consideration is available space. Ships have a lot of space available, whereas aircraft do not. Hence it is possible to place a high-power motor in a ship for propulsion, along with a large battery, which may be unrealistic in an aircraft.

Consider a Boeing 747 type of commercial aircraft, which can need around 90 MW of power during takeoff and about half that during cruising, depending on the speed. Regardless of the technology used, accommodating a motor with that kind of power or having a means (battery or energy source) to drive it is unrealistic – at least with the current technology of motors and energy sources. If that technology existed, we would certainly have quieter aircraft, and they would, generally, also be safer in the event of a crash, due to the absence of any combustible fuel that could result in fire. Just to give an idea of the specific energy and power required, it may be instructive to refer to Table 6.1, from the "Battery University" web site, with some modifications.

Table 6.1 Comparison of energy and power demands in different systems.

Specifications/ power and energy demand	Boeing 747 (jumbo jet)	*Queen Mary 2* or large ocean-going liner	Sports utility vehicle	Bicycle	Person on foot
Weight	369 t (fully loaded)	81,000 t	2.5 t	100 kg with person	80 kg (180 lb)
Cruising speed	900 km/h (560 mph)	52 km/h (32 mph)	100 km/h (62 mph)	20 km/h (12.5 mph)	5 km/h (3 mph)
Maximum power	77,000 kW (100,000 hp)	120,000 kW (160,000 hp)	200 kW (275 hp)	2000 W (professional)	2000 W
Power at cruising	65,000 kW (87–000 hp)	90,000 kW (120–000 hp)	130 kW (174 hp)	80 W (0.1 hp)	280 W (0.38 hp)
Number of passengers	450	3000	4	1	1
Power/passenger	140 kW	40 kW	50 kW	80 W	280 W
Energy/passenger per kilometer	580 kJ	2800 kJ	1800 kJ	14.4 kJ	200 kJ

Courtesy Battery University web site: http://batteryuniversity.com/parttwo-53.htm.

An interesting thing to notice is that the power/passenger is lowest for a bike and highest in a 747 jumbo jet. However, the energy/passenger per kilometer is somewhat different – it is lowest in a bike, but highest in a ship. Interestingly, an SUV needs more energy/person/kilometer than the jet. The fourth row is the power at cruising, leading to the sixth row upon dividing by the number of passengers (i.e., the fifth row). This shows that power/passenger is very high in a jet and a ship. The number of passengers indicated above for the 747 and *Queen Mary 2* are approximate, hence the sixth row is slightly different for those columns compared to those derived from fourth and fifth rows. But a ship has a lot more space available, hence an electric ship is a viable possibility, whereas a jet equivalent is not a viable option with the present technology. Thus, the only electric aircraft that we see today are some unmanned drones or very tiny propeller-driven planes.

With the above in mind, we can discuss the work done on electric aircraft and where things stand at this time. While the equivalent of a jet engine using electrical means is not currently possible, to alleviate the problem of energy requirement, small solar aircraft have been designed using solar panels. Some of these will now be discussed.

A very recent example of a solar panel aircraft is shown in Figure 6.23. This aircraft, called *Helios Prototype*, has been developed by NASA in the United States. It weighs 1600 lb (725 kg), has a wingspan of 247 ft (75.3 m), and a wing area of 1976 sq ft (184 m^2). The upper side of the wing carries the solar panels, which are very thin, like a sheet of paper. The solar power is fed into backup lithium–sulfur batteries so that the aircraft can fly in the absence of daylight.

Another aircraft is a hybrid, made by Falx Air Vehicles in the UK (Figure 6.24). It has a tilt rotor, uses a 100 hp combustion engine, a solar array, and an electric motor rated

Figure 6.23 NASA's *Helios Prototype* solar aircraft. *Source:* Wikimedia.

Figure 6.24 A hybrid electric solar aircraft by Falx Air. *Source:* Courtesy of NewsUSA.com.

at 240 hp peak power. The fuel consumption is claimed to be 10 l/h of flight. This is substantially lower than a regular helicopter, which consumes about 17 times more fuel.

It is obvious from the above that electric and hybrid aircraft are limited in size. The issue is in essence due to limitations of the energy storage mechanism and the extremely high power needed by large commercial aircraft during takeoff.

6.4.3 Locomotives

Locomotives have evolved over more than 200 years, the steam locomotive being around in the early 1800s. The power demand in these vehicles could be 3000–6000 hp on average, depending on the application. There are some exceptions where the size could be even bigger. Diesel locomotives began replacing steam ones starting about a decade before the mid-twentieth century. They are easier to maintain than steam locomotives and are more efficient (www.railway-technical.com/st-vs-de.shtml). Purely electric locomotives were introduced in 1894 (www.itdh.com/resource.aspx?ResourceID=GREAT21) by Kálmán Kandó, using a three-phase induction motor. However, this needed electrification of the railway track, to be successful in the long run. Then there was the gas turbine locomotive, where the gas turbine engine was used to run an electric generator, and the electricity was used to drive a propulsion motor. A gas turbine has the benefit of high specific power density. But the efficiency of gas turbines drops after a certain engine speed, so they become uneconomic in terms of fuel consumption. This is unlike a diesel engine, whose efficiency is flatter at a higher speed.

For these reasons, most locomotives now use diesel engines, but the propulsion system is implemented through an electric motor, as there are a couple of main advantages of the diesel-electric system. First, if the propulsion were purely mechanical, a rather large transmission system with a gearbox and other components, would be needed to create the necessary torque in the wheels. Second, particularly in short-haul trains, with frequent speed fluctuations, a diesel engine with a finite and large number of gears would need to be operated at speeds other than the most optimal in terms of efficiency. By using a diesel-electric system, we can remove both of these issues. In this system, the diesel engine can be run at the optimal speed, and a generator is run to produce electricity, which can be used to run a traction motor to drive the wheels. With the advent of power electronics, this system is very easy to realize, using reliable and efficient traction motors, which can be an induction motor or a synchronous motor – either permanent magnet or field excited. With this background then, let us look at the basic architecture of the diesel-electric system (Figure 6.25).

In general, the traction motors are placed on each axle of the locomotive to drive the wheel pairs. A couple of diesel-electric locomotives [12] are shown in Figure 6.26.

One particular locomotive example is indicated in Figure 6.27 [13], with some technical specifications. Its diesel engine is rated at 4000 hp, while the power at the wheels is 3350 hp. Six motors are used, one per axle – three axles at the front and three at the back. The motors are four-pole, squirrel cage, three-phase induction motors, maximum voltage 2030 V, with a 433 kW continuous power rating. So the total motor power is 2598 kW, or about 3500 hp. The maximum speed of the motor is 3220 rpm, and there is a gear ratio of 85:16 between the motor and the wheel. Note that the squirrel cage induction motor is a very reliable device for such applications. Although the specific power density can be somewhat smaller than in a permanent magnet motor, in a locomotive application where space may not be a premium, say in a small passenger car,

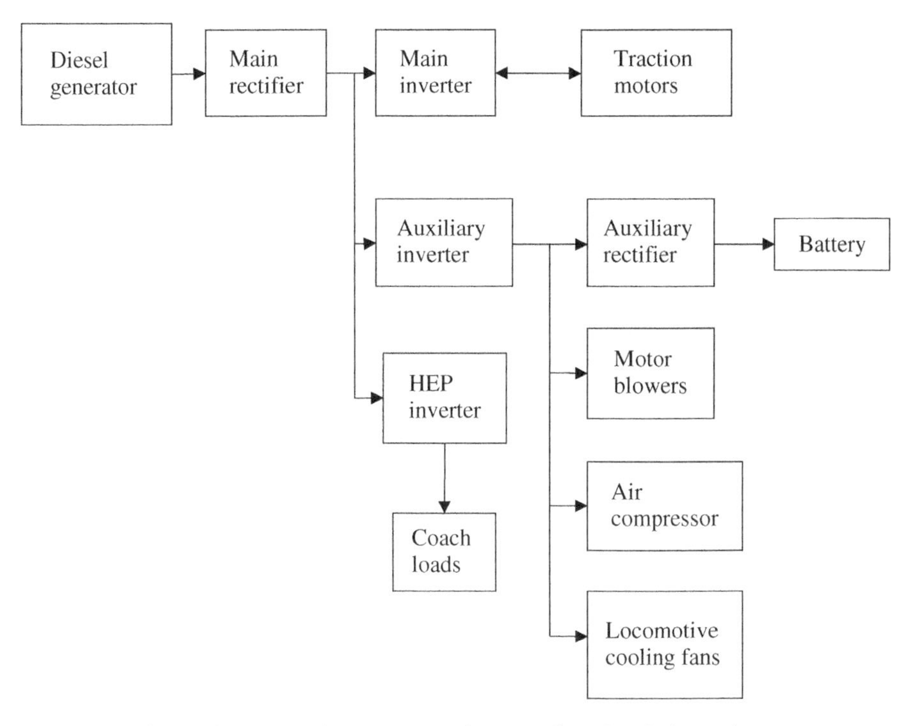

Figure 6.25 Electrical and propulsion system architecture for a diesel-electric locomotive.

Figure 6.26 Pictures of two diesel-electric locomotives by Siemens. Courtesy Siemens AG.

Figure 6.27 Picture of a large diesel-electric locomotive by Siemens. Courtesy Siemens AG.

this may be the ideal choice. In addition, induction motors are very resilient to temperature conditions, unlike permanent magnet motors.

It should be mentioned that in many applications there is no traction system battery in the locomotive (of course, the diesel engine needs a starting mechanism, which can be electric, hydraulic, or pneumatic; if it is electric, a small starter battery is needed). However, for capturing regenerative energy, it is necessary to have a storage battery, ultracapacitor, or a combination of the two, or even a flywheel storage unit, which can be used during regeneration. The regeneration can help improve fuel economy and is of more importance for short-haul trains but not so important for a long-haul train. The architecture of a locomotive system capable of regeneration is shown in Figure 6.28.

The system shown in this figure is essentially the same as Figure 6.25, except that it now has a propulsion battery and/or ultracapacitor, which can feed the propulsion motor. All the other principles remain the same. It is worth to mention that, nowadays, electrified rail transportation, especially high-speed trains, is more and more popular, which removes the need of a on-board diesel generator on a train. In these applications, the electric machines get power from its overhead cable.

6.5 Other Industrial Utility Application Vehicles

Industrial utility vehicles normally include industrial forklifts, airport tugs, golf carts, and vehicles used inside factories and airports for internal transportation. Some of them indeed qualify as off-road vehicles as well, but we will specifically reserve the name off-road vehicles for relatively heavy-duty vehicles such as mining vehicles, refuse trucks, and excavators. In a more specific way, in many instances, industrial utility vehicles are used within relatively confined spaces compared to off-road vehicles. Hence their propulsion systems sometimes use propane or other gas (including liquefied gas) to fuel their ICEs. The intent is to avoid exhaust-type pollution within a confined or closed space without adequate ventilation. However, to prevent noise pollution, in many cases propulsion is purely electrical.

So, the question arises: where do these vehicles appear within the scope of HEVs? First, let us note that the main motivation behind HEVs is fuel economy and overall size reduction of transmission gearboxes, if possible. Of these, fuel economy is more

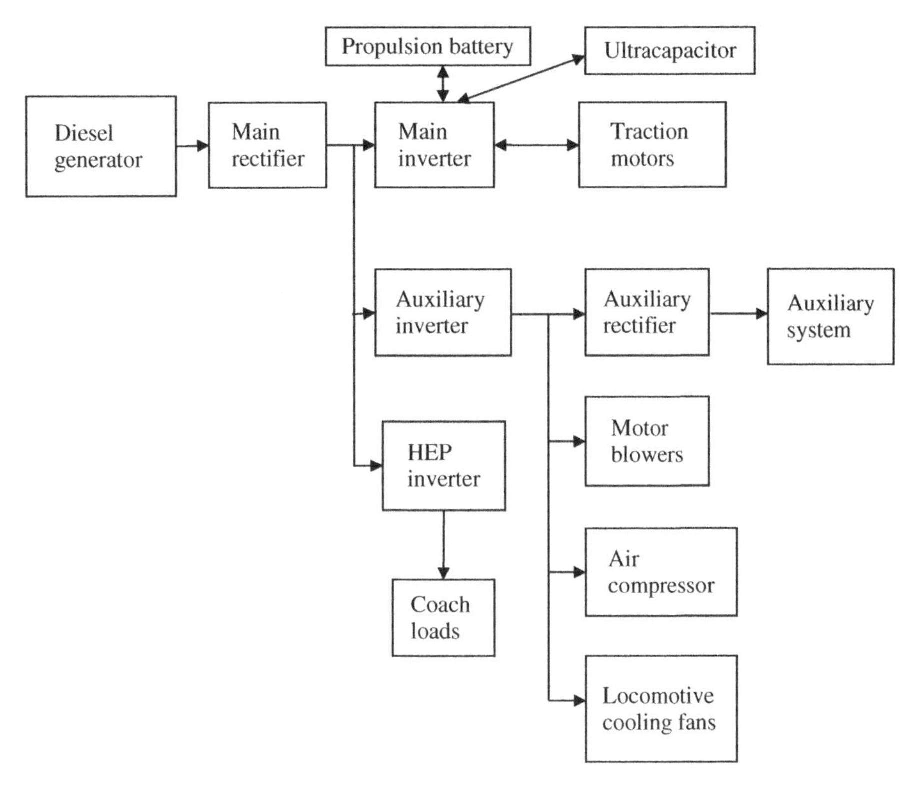

Figure 6.28 Electrical and propulsion system architecture for a diesel-electric locomotive including regenerative braking capability.

important in the case of passenger vehicles, whereas the issue of transmission size reduction plays an important role in off-road vehicles, locomotives, and so on. It is generally true that if the size of the gearbox and other components can be reduced, a concomitant fuel economy will take place. This is beside the fact that locomotives, for example, will also get better fuel economy by running the ICE at its most efficient point.

In an industrial utility vehicle, fuel economy is an issue, but noise reduction, pollution control, and reliability are more important. To that end, purely electric propulsion seems more attractive, which suggests a plug-in electric vehicle. This is generally suitable for indoor applications because, almost always, electricity will be available for charging the batteries of the vehicle for such applications. Therefore, multipropulsion HEVs are not particularly attractive for industrial utility vehicles that are operated within confined spaces and short drive cycles.

References

1 Masrur, M.A. (2008) Penalty for fuel economy – system level perspectives on the reliability of hybrid electric vehicles during normal and graceful degradation operation. *IEEE Systems Journal*, 2 (4), 476–483.

2 Kargul, J.J. (2007) Hydraulic hybrids – demonstration for port yard hostlers. EPA Presentation, July 11, 2007, http://www.epa.gov/otaq/presentations/diesel/hydraulic-hybrid-hostler.pdf (accessed February 2011).

3 US EPA (2004) Clean automotive technology – innovation that works. World's First Full Hydraulic Hybrid SUV, Presented at 2004 SAE World Congress, http://www.epa.gov/oms/technology/420f04019.pdf (accessed February 2011).
4 Rexroth-Bosch Group, Axial Piston Fixed Displacement Motor AA2FM (A2FM), Bosch product brochure.
5 Hierholzer, M., Bayerer, R., Porst, A., and Brunner, H. Improved Characteristics of 3.3kV IGBT Modules, http://www.scut-co.com/maindoc/techtrade/pdevice/eupec/documents/techsupport/ed_pcim_97.pdf (accessed February 2011)
6 Wang, D., Pan, S., Lin, S., and Guan, C. Design of Energy Storage Unit for Hybrid Excavator Power Management, *IEEE VPPC Conf., Sep* 2008.
7 SINM (Studio di Ingegneria Navale e Meccanica) (2004) Focus on Propulsion Pods. Report 060/2004.
8 Hebner, R.E. (2005) Electric ship power system – research at the University of Texas at Austin. IEEE Electric Ship Technology Symposium, July.
9 ABB, The World's First Azimuthing Electric Propulsion Drive, http://www04.abb.com/global/seitp/seitp202.nsf/0/589ea2a5cd61753ec12570c002ab1d1/$file/AzipodNew.pdf (accessed February 2011).
10 Siemens Product Information, Diesel-Electric Locomotives.
11 Siemens Technical Information, Diesel-Electric Locomotive SD70MAC.
12 Yoo, S., An, S., Park, C., and Kim, N. Design and Control of Hybrid Electric Power System for a Hydraulically Actuated Excavator, *SAE Paper #: 2009–0102927.*
13 Kagoshima, M. The Development of an 8 tonne Class Hybrid Hydraulic Excavator SK80H, *Kobelco Technology Review, no. 31, Jan 2013.*

Further Reading

Caterpillar, http://catsays.blogspot.com/2005/01/caterpillar-797b-mining-truck.html (accessed February 2011).
Comparing Battery Power, www.batteryuniversity.com/parttwo-53.htm (accessed February 2011).
Hungarians in the History of Transport, www.itdh.com/resource.aspx?ResourceID=GREAT21 (accessed February 2011).
Komatsu, http://www.komatsuamerica.com/default.aspx?p=equipment&f1=view&prdt_id=947 (accessed February 2011).
Komatsu. http://www.komatsu.com/CompanyInfo/press http://www.epa.gov/oms/standards/light-duty/udds.htm (accessed February 2011).
Wikimedia, http://commons.wikimedia.org/wiki/File:Liebherr_T282.jpg (accessed February 2011).
http://sections.asme.org/Fairfield/FEBRUARY%2023,%202005%20FAIRFIELD,%20DOLAN%20SCHOOL%20OF%20BUSINESS.htm
http://www.copyrightfreecontent.com/auto/prius-of-the-skies/(accessed February 2011).
http://commons.wikimedia.org/wiki/File:Helios_Prototype_on_Lakebed_-_GPN-2000-000198.jpg (accessed February 2011).

7

HEV Applications for Military Vehicles

7.1 Why HEVs Can Be Beneficial for Military Applications

There are several motivations, both direct and indirect, for HEVs for military applications. Military applications can include both direct vehicular applications, which are related to vehicular propulsion, and indirect applications in the sense of using electrically operated arms, or interfacing with the vehicular electrical system to create a microgrid to supply power to a military base etc.

One of the most important reasons for considering HEVs for military applications is the cost of fuel. Transporting fuel to the field through risky routes and over long distances can raise the cost of fuel significantly. The cost can rise from $1 in a regular civilian situation to $400 per gallon for carrying fuel to the battlefield (http://www.defensesystems.com/Articles/2010/03/11/Defense-IT-3-Greens.aspx), and if an airlift is needed, the cost can even rise to $1000 per gallon. In general, the cost to carry fuel to the field is about $100 per gallon (http://www.environmentalleader.com/2010/05/28/us-commanders-want-deployable-renewable-energy-generation). The bottom line is that we can assume that the cost of fuel will be several hundred dollars per gallon in military situations. So, even a small percentage saving in fuel can imply huge cost savings, to the tune of billions of dollars per year.

Other issues in military applications involve noise, which includes acoustic noise, electromagnetic interference (EMI), and also heat signatures in the form of infrared radiation, which can be detected elsewhere. A quiet HEV can help significantly towards achieving some of the above goals of noise elimination.

There are also some indirect benefits of HEVs. When the military move to a combat zone, there may be bases and other infrastructures that need stationary applications of electricity at utility-level voltages. With multiple HEVs, when properly connected and interfaced, it is possible to create utility-level voltages to run various items of stationary equipment. This can be achieved by using the HEV battery (and, if necessary, the ICEs in the HEV). Several HEVs can, in fact, form a microgrid with a reasonably robust source of utility power. Of course, when interfacing several HEVs to generate electricity, it will be necessary to do so properly through appropriate control and power electronics. This kind of utilization of vehicles for power generation can help significantly, in the sense that it can reduce the need for auxiliary power units, thus saving the cost and weight of these devices, and transporting them to the field.

Hybrid Electric Vehicles: Principles and Applications with Practical Perspectives,
Second Edition. Chris Mi and M. Abul Masrur.

Depending on the architecture used in the HEV – for example, if it has hub motors in each of the wheels, or even one motor per axle for propulsion, and uses a series HEV (SHEV) architecture – it will have the advantage of redundancy in case one of the motors fails. Then the vehicle can be made to run in a gracefully degradable mode, with a somewhat reduced performance, and moved to a safe zone as needed, prior to necessary repair and maintenance.

So, it can be seen that there are quite a few advantages in introducing HEVs for military applications.

7.2 Ground Vehicle Applications

7.2.1 Architecture - Series, Parallel, Complex

Military vehicles can have similar architectures to those of the regular commercial HEV vehicles mentioned earlier. A very informative research paper by Ucarol et al. [1] has made a comparison between ICEs, SHEVs, and parallel HEVs. There, based on simulated studies, from the point of view of weight, the ICE vehicle is noted to be the lightest, with the parallel HEV slightly heavier. However, the series hybrid vehicle was found to be quite a bit heavier. This analysis, although done for a commercial vehicle, is equally applicable to military vehicles.

If an ICE vehicle is considered as the baseline, we can compare other architectures. To match the performance of the baseline vehicle, it is obviously necessary to make the total size (in terms of power) of electric motor(s) in an SHEV the same as the ICE in the baseline vehicle. The authors of this book believe that it is really not necessary to do this, but rather that the motor should be sized so that over a given drive cycle the average energy going out of the generator is equal to the average energy consumed by the propulsion motor. The peak power can be handled by the battery, which should be charged within its bounds. A similar thought process also applies to the decision-making for selecting the size of the generator. The decision should be based on studies of different drive cycles, then presenting the worst case scenario. The key to this decision-making lies in the battery (or other energy storage devices) and whether they can provide the maximum power needed or not. Only if the battery is kept in a completely floating condition all the time, and the power from the ICE and generator is fed to the motor in parallel with the battery, will the generator be required to be equal in size to the maximum power demand. But the purpose of the battery, or any peak power source, is to address this eventuality of sudden peaks or higher demands; hence ICE size in SHEVs and the generator size could definitely be reduced.

In a parallel HEV, the size of the electric motor could even be less than half the size of that in an SHEV. Ucarol et al. [1] assign the balance of power – that is, any deficit between the load demand and the electric propulsion power – to the ICE in the parallel HEV. This mechanism to split the power assignment between the motor and the ICE in the parallel HEV assumes that the original ICE in the baseline vehicle was chosen based on the maximum power needed under the worst possible drive cycle scenario. There is also no generator in the PHEV in the study cited above. However, in some other architectures, like the Toyota Prius and GM two-mode hybrid, there is a generator that is separate from the main propulsion motor (which can also be used as a generator).

The reason for having this separate generator is highly dependent on the strategy for control. Under certain conditions of the battery's state of charge (SOC) and power demand for propulsion, it may be necessary to provide additional charging to the battery, which is where this generator comes into play.

A very important part of the SHEV or parallel HEV is the battery or any other storage device like the ultracapacitor. Ucarol et al. [1] use the battery capacity, that is, ampere-hours, as the metric for the battery. But in fact the ampere-hour itself is not sufficient for specifying a battery. It is also necessary to know the current or power rating of the battery, so that during the maximum power demand by the propulsion motor, the battery, in conjunction with the generator, is able to provide it. These are vital decision-making processes in creating the design specifications for an HEV.

The discussion above applies to both gasoline and diesel vehicles. Military vehicles are mostly diesel engine based (including JP-8 fuel). Improvements in diesel engine technology will help these vehicles, but hybridization will always allow even better fuel economy and other benefits regardless of the state of diesel engine technology.

The next question to be addressed is the choice of architecture in military vehicles for the purpose of hybridization, that is, whether it should be series, parallel, or complex. The question can be best answered depending on what the priority is – is it fuel economy, size or weight, or reliability? If fuel economy is the top priority, then the choice should be a parallel HEV. If size and light weight are of concern due to the requirement to carry the vehicle within an aircraft, and so on, then a parallel HEV will be better than an SHEV. In general, the SHEV will be to some extent heavier than the PHEV, which might affect fuel economy slightly. On the other hand, if performance of the vehicle is of utmost concern, that is, power output, it may be beneficial to use the SHEV architecture, which might help remove heavier transmissions and provide high power on demand very quickly due to the ability to control the motor much faster than an ICE. Reliability considerations might be better addressed in general by using a parallel HEV, because of the redundancy in propulsion by ICE and the motor. Reliability is not to be underestimated in a military vehicle. However, the parallel HEV is more complex in terms of control, and introduces complexity in its mechanical coupling. A thorough study of failure modes of these devices is needed before a decision can be conclusively made on the matter of reliability.

Another issue accompanying reliability in military vehicular systems is maintenance. It appears that maintenance of an SHEV is somewhat simpler for several reasons. Its control is simpler. Its mechanical linkage, unlike a parallel HEV, is also very simple. If there are hub motors in its propulsion, they can be quickly replaced in case of failure. The above discussion indicates the pros and cons of parallel and SHEVs, but overall reliability, and more specifically survivability needs, can probably be better served by a parallel HEV. Having said that, it should be noted that the decision could depend very much on the application and drive cycle. A logistic or support vehicle in the military (or non-combat type of vehicle) which travels longer distances, due to the drive cycle involved, probably will benefit more from using a parallel HEV. On the other hand, for a combat vehicle where power demand and quick response to such demand are most important, an SHEV will be more beneficial. Final decisions on such things are quite complex and systematic tradeoff studies by tabulating all the requirements of performance, fuel economy, reliability, and so on, should be done to determine the optimum decision.

7.2.2 Vehicles That Are of Most Benefit

There are many different types of military vehicles, as shown in Figure 7.1.

In the various types of vehicles shown in Figure 7.1, the drive cycles and needs are quite different. Some vehicles run for long distances, some are combat vehicles with extremely high power demand and performance needs, while others are unmanned robotic vehicles, where long-term mission and survivability can be more important than fuel economy. Some have weapons mounted, some do not. Based on these needs, the vehicular architectures could be significantly different.

Figure 7.1 Some military vehicles: (a) HMMWV (high-mobility multipurpose wheeled vehicle), (b) Stryker, (c) HEMTT (heavy expanded mobility tactical truck), (d) Bradley, (e) Abrams, (f) Fennek (European), (g) MRAP (mine-resistant ambush-protected), (h) Big dog robot, (i) Gladiator, and (j) Swords. Courtesy Wikimedia – US military public domain pictures.

In order to estimate the benefits of hybridization of military vehicles, some kind of metric should be established. A possible systematic way to do this analysis is suggested below, based on the itemized information noted.

Required information and questions to be answered:

- Number of vehicles currently deployed, N
- Cost of the vehicle initially, C
- Average fuel economy of the vehicle, i.e. miles per gallon, F
- Power rating of the vehicle, P
- Miles the vehicle travels in its life cycle, M
- Whether it is a combat vehicle or a support vehicle, on a scale of 1 (minimal combat and primarily support) to 10 (highly combat), K
- How the vehicle is transported to the field, on a scale of 1 (by itself), through ship transport (5), to air transport (10), T
- Ease of maintenance, on a scale of 1 (minimal) to 10 (extremely important), E
- How important its survivability is in terms of failure, on a scale of 1 (minimal) to 10 (extremely important), S
- Long-term sustainability in terms of mission without any refueling or other intervention from outside, on a scale of 1 (easily refueled, i.e. fuel is easily available to refuel it) to 10 (deployed and not able to be refueled externally, e.g. an unmanned robot), Q
- Additional cost of hybridization, H
- Loss or gain of reliability if hybridized, compared to the base vehicle on a scale of 0 to 1 (high gain = 1; low gain or loss = 0), R

Since these quantities are in different units, some are dimensionless, and their range can also be numerically very different, it is beneficial to normalize them in some way, to bring them within a comparable and manageable range. All these quantities can possibly be normalized except the number of vehicles, since this simply indicates the total benefit. In addition to using normalization, it is instructive to include a weighting factor for each of the criteria.

All these ideas are incorporated within Table 7.1 for ease of comprehension and with some placeholder numerical values for the purpose of illustration.

The above criteria provide a metric for a particular vehicle. This may not be a perfect metric but at least it allows a comparison of the vehicles against one another. For example, let us say that the benefit of hybridization of a single Abrams tank is much greater than the hybridization of a single HMMWV. But there may be many HMMWVs (in terms of numbers deployed in the field) compared to Abrams tanks. Hence it may be more important to introduce hybridization in a HMMWV, rather than in an Abrams, if there is an issue of choice. Similarly, other criteria proposed above are also important. The third column in the table, involving the weighting factor (i.e. how important this particular criterion is on a scale of 1–10), is critical. Normally all these items have to be initially settled after due discussions with various people who are experienced with the vehicles in terms of their applications and on matters related to the items in different rows of the table.

The concept of normalization and weighting factors introduced in this chapter has not appeared in the literature to the best of the authors' knowledge. Hence it may be an original process for hybridization in military vehicles. With more experience in the field, it may be necessary to revise this metric evaluation process and the entries in Table 7.1.

Table 7.1 Tradeoff mechanism in hybridization decision-making.

Criteria	Normalization factor	Weighting factor on a scale of 1 (less important) to 10 (most important)
N (number of vehicles)	1	1
C (initial cost per vehicle)	500,000	6
f (miles per gallon): use $1/f$	8	7
P (power rating)	300 kW	5
M (miles in life cycle)	m (to be defined based on the type of vehicle)	5
K (type of vehicle: combat, support, etc.)	5	5
T (transportability)	5	5
E (ease of maintenance)	5	7
S (survivability)	5	8
Q (long-term survivability without refueling)	5	6
H (cost for hybridization): use $1/H$	20,000	2
R (gain in reliability due to hybridization)	0.5	8

Total $T = [\{E_{11}\}/\{E_{12}\} \times \{E_{13}\}] \times [\Sigma\{E_{i1}\}/\{E_{i2}\} \times \{E_{i3}\}_{i=2,3\ldots12}]$, where E_{ij} = element in ith row, jth column in the table above, with $i = 1, 2\ldots12$ and $j = 1,2,3$.

Though the concept is introduced for military vehicles, the metric could be equally important for applications in regular commercial civilian vehicles. It could be useful in making decisions about moving to a hybrid platform instead of a regular non-hybrid one.

7.3 Non-Ground-Vehicle Military Applications

Military vehicles include non-automotive applications as well. This can include air- and water-borne vehicles, that is, aircraft, ships, and boats. In addition, some applications are not directly vehicular in nature. For example, a vehicle might use guns or other similar devices, some of which might be better operated electrically. An example could be an electromagnetic gun. Some of this equipment might need an ICE, or fuel cell, and so on, which could charge a battery or ultracapacitor. The equipment (gun or otherwise) itself could be directly activated by electrical power (from a battery or similar storage devices) which could later on be replenished by using an ICE or fuel cell.

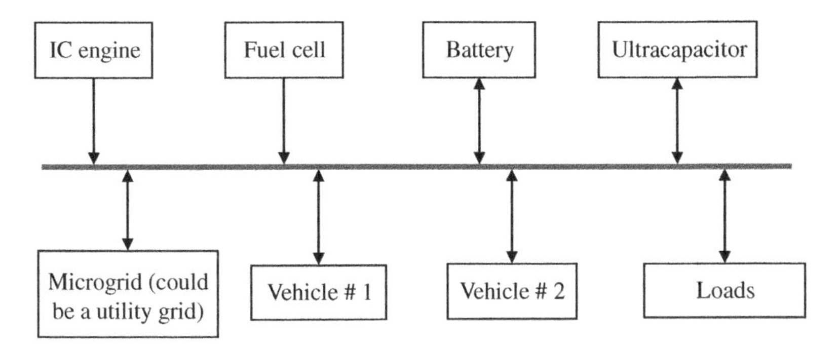

Figure 7.2 A generic hybrid power system architecture.

Hybridization, though not in the sense of a vehicle, can include matters like interfacing with a utility grid or a microgrid, where a number of vehicles could exchange power through a common bus. Here it is in principle possible to have an ICE-based generator, some fuel cells, and vehicles all properly synchronized and then exchanging power. A system-level diagram of this situation is shown in Figure 7.2.

Note that an HEV is merely a special case of the above hybrid power system at an internal vehicular level. Here (in Figure 7.2) the hybridization is extended to a higher system level. Vehicles 1 and 2 in the figure could be HEVs or EVs and could be exchanging power with the rest of the system. Similarly, if the vehicle was ICE based, it could be running the engine and generating power while it is stationary, and it could provide power to another vehicle that has a deficiency in power.

Example applications of non-propulsion types of ground vehicles are as follows.

7.3.1 Electromagnetic Launchers

These can use electromagnetic force to propel items at very high speed by accelerating them very quickly. In a regular gun the combustion of chemicals causes an explosion and creates the force of acceleration. Whereas the speed of combustion-based systems can be around 2 km/s for high-performance guns, electromagnetic guns can accelerate projectiles to around 6–7 km/s. This is quite spectacular, being three times more than combustion-based systems, and is about the speed needed to put satellites etc. into low earth orbit.

The principle of an electromagnetic launcher is shown in Figure 7.3 [2]. One of the enabling technologies related to this application is the availability of very high current and instantaneous power for a very short time. A simple power system cannot provide this, since such a high pulse power is needed. After each shot of high pulse power, it is necessary to recharge the system through some hybrid power system as noted earlier. If smaller-scale launchers were to be deployed in a mobile vehicle-based system, it would be almost mandatory to use a hybrid power system within the vehicle. This is an example which may not be a hybrid propulsion system in a direct vehicular sense, but it is definitely a hybrid power system.

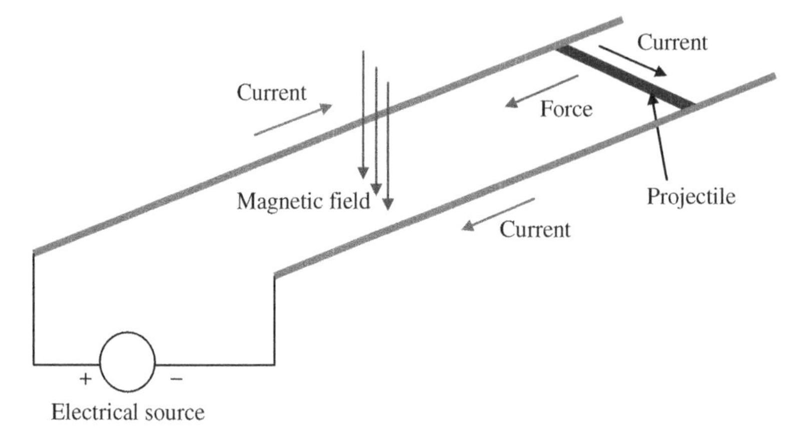

Figure 7.3 Basic principle of the electromagnetic launcher [2].

7.3.2 Hybrid-Powered Ships

In Chapter 6 we discussed electrically propelled ships extensively, and will not repeat that information here. The advantages of such systems were clearly shown in that chapter. In one particular effort by the US Navy, diesel can be directly reformed into hydrogen through a chemical process. This hydrogen can then be used in a fuel cell to generate electric power, which can be used to drive an electric motor. It has been indicated [3] that since ships travel at relatively low speed, peak power may not be needed, implying that an additional peak power source like a battery may not be necessary. The advantage of such a system is that an ICE is not needed and there will not be any pollution. Even if peak power is needed, it may be possible to use an ultracapacitor or some high-power battery to supplement the fuel cell, rather than have any ICE at all within the ship. Since in a ship there is much more space than in a car, this possibility of having some peak power source should not be ruled out. Further, whereas the gas turbine-based ship engine has about 16–18% efficiency, the fuel cell has 37–52% efficiency, which is significantly higher, thus leading to much better fuel economy. Of course, other benefits of a fuel cell-based system could be quiet operation and the ability to place a number of such fuel cell systems in different locations (strategically placed near the loads) since the connection to the propulsion motor loads will be through cables. This is unlike an ICE-based system, which is mechanically rigid in terms of location. This application, noted above, is a typical case of a hybrid power system in the sense that the diesel reformer system is an electrochemical system. This electrochemical system may need initial startup through some pilot electrical system of relatively small size. But once it is started, the fuel cell itself could maintain the diesel reformer system without the intervention of the startup system. The startup system could be a battery of relatively small size.

In a diesel reformer system, diesel, air, and water are used as the input. These components react and a gas mixture with around 30% hydrogen and other inert gases like carbon dioxide, steam, and nitrogen are obtained as the output. This mixture of gases is then fed to a catalytic burner to trigger the fuel cell system. The catalytic burner causes about 80% of the hydrogen to react with additional air, and produces heat and water as by-products. This is like a fuel cell with a conversion rate of about 80%. Next, another catalytic burner is used to make the remaining hydrogen react with additional air, and converts all the hydrogen to steam. In this particular system hydrogen is not stored as

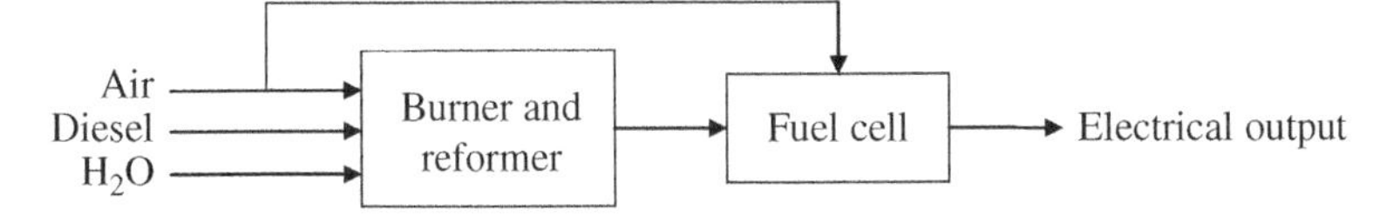

Figure 7.4 A diesel reformer for a fuel cell.

an intermediate process, and the hydrogen that is produced by the reforming process exists only for a very short time (about 15 seconds). This lack of hydrogen storage adds to the safety of the system, since hydrogen is a combustible gas. The system-level diagram for the reformation system is shown in Figure 7.4.

7.3.3 Aircraft Applications

As indicated in Chapter 6, due to the high power and thrust needed, the application of electric power for aircraft propulsion is very much limited at this time, other than in very light aircraft like certain unmanned drones and very small airplanes. For commercial aircraft it is reasonable to assume that the technology (including energy storage) is not yet available for a viable application in this field. Since details on this topic are already given in Chapter 6, they are not included here.

7.3.4 Dismounted Soldier Applications

A dismounted soldier implies a soldier who is not physically within a vehicle (i.e. a standalone individual) and does not have any direct connection for getting power or energy from a vehicle or otherwise. This implies that for whatever equipment is carried by the soldier, it is necessary to carry the power or energy sources on the person for energizing those devices. Questions might arise as to why it is even necessary to discuss the power issues for a dismounted soldier in a book on HEVs. The reason is that the architecture of the HEV has some similarity to the architecture of the dismounted soldier, since both are hybrid power systems in the first place. The concept of managing the power here also is similar to that of an HEV. In fact, even though the power level of the soldier's equipment is much smaller than in an HEV, the issue of managing it is conceptually the same. In fact, the issue of power management can be more important for dismounted soldiers who can be completely on their own, under various situations, and the difference between good and bad power management could very well imply the difference between life and death. Hence, even though the dismounted soldier power system may not be a propulsion system, the power system could be hybrid and use similar power management concepts which can be discussed well within the scope of this book.

A general architecture of the power system for the dismounted soldier is shown in Figure 7.5.

According to the UK Ministry of Defence [4], the equipment and supplies carried by a soldier in the field could be on the order of 45 kg. The environment in which the soldier operates is not very helpful in terms of temperature, which could be as high as 45°C, or as low as −20°C. The weight of the power sources carried by a soldier could be on the order of 4 kg or more, and the soldier may have to endure this condition, which could possibly last for 48 hours. In addition, soldiers could in the future carry cooling or heating jackets, if needed. The load demand of the dismounted soldier [4] could be as small as 1 W for some small devices or somewhere around 100–200 W for certain electronic equipment.

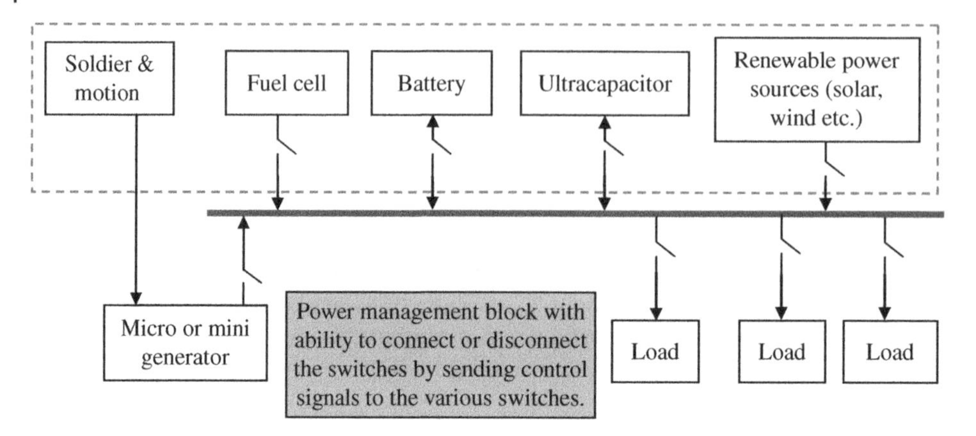

Figure 7.5 A generic hybrid power system for the dismounted soldier.

A dismounted soldier normally uses batteries of various sizes, from AA size to much larger ones which could weigh a few kilograms. These batteries could be either rechargeable or disposable. The energy requirement could be on the order of 65–200 Wh/kg [5, 6].

It is obvious from the above that the dismounted soldier has needs which can be quite demanding, even though the power levels might not compare to the power levels of a vehicle. The expectation is that, in the future, power sources which can be six to eight times lighter than existing batteries could be possible. The specific energy needed in that case could be even 600–800 Wh/kg.

It seems, therefore, that regardless of the available technology, it may be imperative for the dismounted soldier to have power sources that match the load demand in the best possible way. For example, some power sources may supply high-power but low-energy type of needs, whereas others could be the opposite. If only one type of source is used, then it will definitely lead to an oversizing of the power sources, leading to excessive weight. This is not acceptable for a dismounted soldier. Hence the architecture shown in Figure 7.5 could become important in considering the design of a power system for the dismounted soldier.

The architecture shown in Figure 7.5 implies that it would be ideal for the soldier if the whole system were placed in a single packaging, perhaps with various connecting jacks for different equipment, without requiring the soldier to connect individual sources. Such a single packaging could also include some controlled switches within the power system, as shown in Figure 7.5. Coordinating all these devices by proper switching is done by the power management block, which will monitor the voltage, SOC, and other information, and will accordingly manage the charging process. It will also give an indication to the soldier when the generator (which can be a hand cranked generator to charge a battery) needs to be cranked.

It may also be possible to use the walking, running, and other motions of the soldier to charge the batteries, without voluntary engagement of the soldier. This should be automatically done by using some mini or micro generators that can work on their own, based on the vibrational motion. Such machines come under the category of energy harvesting devices. The devices normally should work using inertial motion and periodic motion as in walking, even though the power amount may not be too high. A particular prototype was found to generate 7.3 W, which may still be reasonable [7] for certain loads. Normally, if such energy harvesting devices are carried by the soldier, there will not be any perceivable difference from the point of view of effort, even though the energy is ultimately coming from the soldier.

7.4 Ruggedness Issues

In military applications, devices are generally ruggedized. In a mechanical sense, this means that physical coverings or boxes containing any components are extremely robust compared to their commercial counterparts. Thus the devices are not easily damaged in physical encounters of any kind. In addition, it is very important that the electrical and other design aspects of the devices are attended to. This can be due to the fact that military applications can mean very large temperature extremes, both hot and cold. In general, high temperature is bad for items like power electronics and batteries. The devices have to be derated accordingly for military use. This means that they can become physically larger and more bulky, if performance has to be maintained under these extreme conditions. These considerations can in fact contribute to delays in HEV deployment in the field. It is expected that, in the future, with high-temperature and mechanically robust power electronics, the deployment of HEVs will be expedited by the military.

Basic ruggedness issues, therefore, pertain to mechanical vibration and temperature. Not only the power electronics, but also the regular electronic circuitry, are subject to these issues. In the power electronics area, it is expected that the use of silicon carbide semiconductor materials will significantly alleviate the temperature issues in terms of additional cooling needs. Even if the cooling system is enhanced for extreme temperature, that enhancement itself is also subject to adequate ruggedness questions.

For HEV applications using high-voltage and high-power battery packs, the cooling and heating issues are very important, otherwise HEV performance could be significantly compromised. Examples of ruggedized military batteries are shown in Figure 7.6 (www.mccdc.usmc.mil/OpsDiv/Integration/logistics_integration_files/Power%-20Systems.ppt).

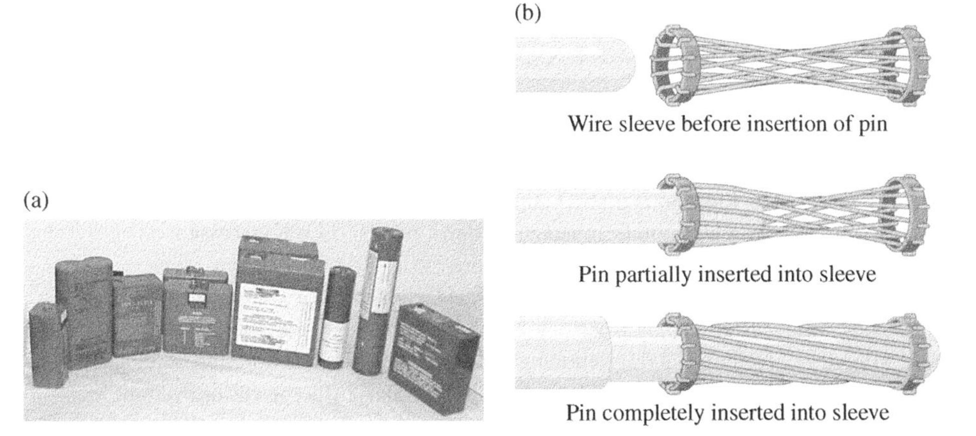

Figure 7.6 Examples of ruggedized battery (courtesy EaglePicher Technologies) and connectors (courtesy Hypertronics Technologies – Hypertac is a registered trademark).

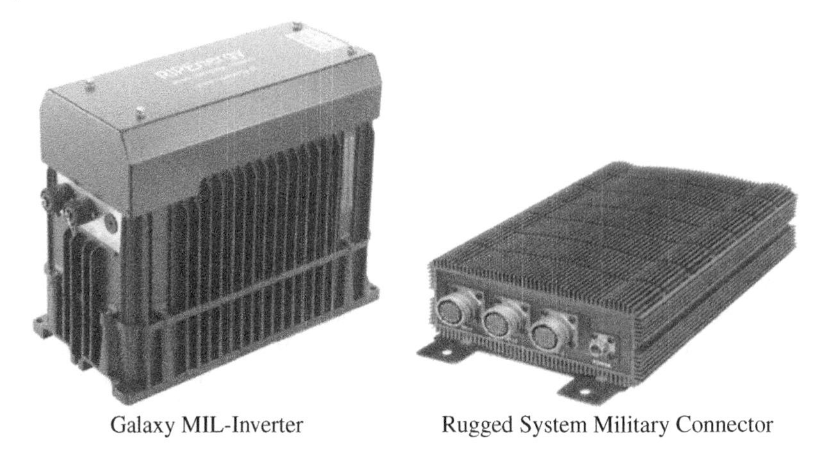

Figure 7.7 Examples of a ruggedized inverter (courtesy RIPEnergy AG) and connector (courtesy Adlink Technology).

Another important issue in connection with HEVs is the connector design. A reliable connector used in cable design is shown on the Hypertronics web site (www.hypertronics.com/en/Products.aspx). This example of such a connector, called the *Hypertronics' Hypertac®* connector, involves a wire basket socket contact system. When the pin is inserted into the sleeve or ring shown in Figure 7.6, the wires stretch around it, thus activating several contact paths. Thus, under harsh vibration and similar situations loss of contact can be eliminated or minimized. Pictures of a ruggedized military standard inverter and ruggedized interface connector are shown in Figure 7.7.

Several military standards are applicable in connection with ruggedization, which are applicable in design considerations of the power systems noted above. These include MIL-STD-461 (EMC), MIL-STD-901D, Grade A (shock), MIL-STD-167 (vibration), MIL-STD-810E (vibration), MIL-STD-1275 (ground–mobile), and MIL-STD1399 (interface standard for shipboard systems).

References

1 Ucarol, H., Kaypmaz, A., Tuncay, R., and Tur, O. (n.d.) A performance comparison study among conventional, series hybrid and parallel hybrid vehicles, www.emo.org.tr/ekler/c792a8279211dec_ek.pdf (accessed February 2011).
2 Fair, H. (2003) The Electromagnetic Launch Technology Revolution, Magnetics Magazine, http://www.magneticsmagazine.com/e-prints/UT.pdf (accessed February 2011).
3 Walsh, E. (2004) Hybrids on the High Seas: Fuel Cells for Future Ships, http://www.news.navy.mil/search/display.asp?story_id=12221 (accessed February 2011).
4 UK Ministry of Defence (2004) Reducing the Burden on the Dismounted Soldier Capability Vision Task 3 – The 'Energy-Efficient Soldier', http://www.innovateuk.org/_assets/pdf/competition-documents/briefs/energy_efficient_soldier.pdf (accessed February 2011).

5 Raadschelders, J. and Jansen, T. (2001) Energy sources for the future dismounted soldier: the total integration of the energy consumption within the soldier system. *Journal of Power Sources*, 96 (7), 160–166.
6 Atwater, T., Cygan, P.J., and Leung, F. (2000) Man portable power needs of the 21st century – I. Applications for the dismounted soldier. II. Enhanced capabilities through the use of hybrid power sources. *Journal of Power Sources*, 91, 27–36.
7 Priya, S. and Inman, D. (eds) (2008) *Energy Harvesting Technologies*, 1st edn, Springer.

Further Reading

Angel, J. (2009) Drone Takes Wing with Linux PCs Onboard, web article, http://www.linuxfordevices.com/c/a/News/Aurora-Excalibur-and-Parvus-DuraCOR-820 (accessed February 2011).

COTS Journal (2009) Rugged Stand-Alone Box System Roundup, http://www.cotsjournalonline.com/articles/view/101098htm (accessed February 2011).

Hypertronics, http://www.hypertronics.com/en/Products.aspx (accessed February 2011).

Marine Corps Electrical Power (2004) https://www.mccdc.usmc.mil/OpsDiv/Integration/logistics_integration_files/Power%20Systems.ppt (accessed July 2004).

Nova Electric, http://www.novaelectric.com/dc_ac_inverters/index.php (accessed February 2011).

The Green Optimistic, http://www.greenoptimistic.com/2008/04/30/HYBRID-ELECTRIC-AIRCRAFT-TO-TAKE-OFF-BY-2009 (accessed February 2011).

http://commons.wikimedia.org/wiki/File:Humvee_of_Doom.jpg (accessed February 2011).

http://commons.wikimedia.org/wiki/File:14_stryker.jpg (accessed February 2011).

http://commons.wikimedia.org/wiki/File:HEMTT.jpg (accessed February 2011).

http://commons.wikimedia.org/wiki/File:Two_M-3_Bradleys.jpg (accessed February 2011).

http://commons.wikimedia.org/wiki/File:M1A1_Abrams_Tank_in_Camp_Fallujah.JPEG (accessed February 2011).

http://commons.wikimedia.org/wiki/File:BundeswehrFennek.jpg (accessed February 2011).

http://commons.wikimedia.org/wiki/File:FBI_Mine_Resistant_Ambush_vehicle.jpg (accessed February 2011).

http://commons.wikimedia.org/wiki/File:Big_dog_military_robots.jpg (accessed February 2011).

http://commons.wikimedia.org/wiki/File:Gladiator_RSTA.jpg (accessed February 2011).

http://commons.wikimedia.org/wiki/File:SWORDS_robot.jpg (accessed February 2011).

8

Diagnostics, Prognostics, Reliability, EMC, and Other Topics Related to HEVs

8.1 Diagnostics and Prognostics in HEVs and EVs

Any vehicle, whether conventional, hybrid, or pure electric, ought to have some sort of diagnostics (find the cause of a problem that has "already happened" to the vehicle) and prognostics (find problems which "will or could possibly happen" in the future), given the present condition of the vehicle based on monitoring of various information within the vehicle. With this in mind, all modern automobiles have onboard diagnostic functions that can give a certain amount of diagnostics, but not prognostics. Diagnostics can be at several levels. One is at the vehicle level within the vehicle itself; this can inform the driver or basic level service personnel about what might have happened. The second level can be at somewhat deeper maintenance level when the vehicle can be dismantled at different subsystem levels in a repair shop, the problem pinpointed, and the item concerned replaced. The third level can be at the automotive dealer or eventually the vehicle manufacturer level to find out why a subsystem failed. Finally, the diagnostics can be at the subcomponent level or component manufacturer level when more microscopic analysis of the component can be initiated to find design or other flaws, if any. Up to the second level, that is, the maintenance level, the correctional step normally is replacement/repair of a part or subsystem, without necessarily trying to find the root cause of the failure. The third and final level can be analyzing the fundamental cause of a failure due to design or manufacturing flaws. Sometimes the fault may be not due to the design of a component or its material defect, rather it could be due to the wrong application or improper use of a device – which may not be due to the vehicle owner or driver, from the engineering design perspective. In that case, redesign of the system or subsystem will be needed on the part of the vehicle manufacturer. The aim of this chapter is not to delve into diagnostics or prognostics in a specific vehicle, or to be used as a repair or maintenance manual, rather it will discuss the general methodology for this, applicable to any vehicle, hybrid and electric vehicles included. It also includes a discussion of what can be incorporated in future vehicles, not only for diagnostics, but also in terms of prognostics.

8.1.1 Onboard Diagnostics

Since 1996, all vehicles have been required to have the second version of onboard diagnostics, that is, OBD II. Hybrid vehicles have OBD II as in a regular vehicle, the role of which in an HEV is the same as in a regular vehicle. As is well known, OBD II is a North

Hybrid Electric Vehicles: Principles and Applications with Practical Perspectives,
Second Edition. Chris Mi and M. Abul Masrur.

Figure 8.1 Picture of an OBD II scanner (picture taken in authors' lab).

American standard which deals with the engine control system, some parts of the chassis, body, and other devices, and the communication and control network diagnostics (controller area network or CAN). There are a few variants of the OBD II protocols corresponding to certain standards. In general, for example, GM vehicles use SAE 1850 VPW (variable pulse width modulation), Ford vehicles use SAE 1850 PWM (pulse width modulation), and Chrysler, all European, and Asian vehicles use ISO 9141 standards. The variations are reflected in the socket connector and the pin uses.

OBD II vehicles nowadays in general have a socket somewhere in the lower region in front of the driver. This may not be immediately visible while sitting, but one can normally see it by lowering the head to look above the pedal region around the plastic trim. This is where the technician or the owner can plug in the diagnostic tool or the scanner. The pins in the socket carry various signals from different sensors spread throughout the vehicle. There are various scanners available on the market. The ones made by the original equipment manufacturers (OEMs) can be quite costly, but aftermarket products are available at much lower cost. These scanners normally have display units on them. There are also some rather inexpensive devices (data loggers), which may not have a display unit like a fully fledged scanner, but can collect the same information as a regular scanner, and the information can then be displayed on a personal computer or laptop using the software provided by the data logger manufacturer. A typical scanner with accessories is shown in Figure 8.1. The price of such a scanner can be close to $100, but scanners are available whose price can be from a few hundred dollars to even around $2000, depending on features and capabilities.

It is not the intent here to discuss the details of various OBD II signals displayed on a scanner. That information is available in any of the scanner manuals. Basically there can be 300 or so readings coming from different sensors. They could indicate things such as ignition voltage or transmission shift points. As an example, the code P0032 could show up on the scanner. This can then be found to correspond to: Oxygen (A/F) sensor heater control circuit high. With a little more research, this code can then be related to perhaps a failed oxygen sensor, or a short in the sensor heating circuit, and so on.

There may be some other specific reasons connected to this code, which can be found from various manuals. In general, to save time, it might be better not to go further but just to replace the whole thing that could be involved with that code. This kind of code will be common to both an HEV and a regular vehicle. Similarly there are codes for other items related to the chassis, which can be common to any vehicle, not just HEVs. Some codes are generic OBD II items which can be read by any scanning tool and related to a corresponding cause.

Certain codes may be proprietary (in terms of how they are arrived at) but manufacturers' manuals will reveal the cause and what needs to be repaired or replaced, even though they may not release the methodology of how they come up with the cause, which may be proprietary.

For HEVs, the specifics could be as follows when it comes to diagnostics. The manufacturer's service manuals should normally reveal the details. This needs a knowledge of HEV technology to be able to correlate the diagnostic code against the symptoms. For example, in a Toyota HEV, the code P3005 will indicate that the "high-voltage fuse is blown", which can mean that the power cable or the fuse itself may need replacement. If after carrying out the replacement the code persists, then a deeper analysis might be called for. Similarly, P3006 will indicate that battery state of charge (SOC) levels are uneven in the battery modules. This may be cured by simply charging the battery for a while. But if the symptom persists, it might need a serious look at the cause. Unless a repair store technician is knowledgeable enough, the code itself may not reveal the true cause. That is why a very good training program for technicians to cater to the needs of HEVs is needed, since repair-related experience in this field is still evolving [1].

In general, Toyota and other OEMs will present detailed diagnostics charts with step-by-step procedures on how to proceed on a specific diagnostic code. As noted earlier, this will be available even though the details of how these methodologies are developed may be proprietary. The methodology may relate to one or more inner circuit-level diagnostics – and the mechanism of code generation itself might be a research issue. As an example, if the motor is not getting a proper voltage supply due to a failure in the inverter system, it may be necessary to find that out from external sensor readings of the voltages and currents. Although the exact method for doing this may be vehicle manufacturer specific, some idea can be obtained by following certain engineering principles. Some analyses may be simple electric circuit issues, while others might be quite complex.

In particular, the authors have done research [1, 2] on detecting if a specific inverter switch has become faulty or not, by monitoring the voltages and currents in the motor/inverter circuit. This kind of deeper diagnostic may be of importance because under certain circumstances it may be possible to reconfigure a partially failed system in real time (i.e. while the vehicle is running) by using software, and then run the vehicle in a gracefully degradable mode at a lower performance level. This can significantly contribute to reliability and safety. Of course, it will be necessary sooner or later to replace or repair the faulty component or system, to restore the full functionality of the vehicle.

Similar comments hold, for example, if a vibration is noticed in the vehicle. This could be due to many mechanical things, but it could also be related to a motor failure. Here OBD II codes and hybrid electric (HE) specific codes have to be analyzed together, and not in isolation, to come up with a reasonable diagnostic analysis. It is at this stage that an artificial intelligence method can come in handy, which can significantly reduce the burden of human guesswork or the personal experience of a technician.

In a purely electric vehicle, which does not have an ICE, OBD II codes related to emissions and the engine will be unnecessary. But a subset of the existing OBD II codes could still be used for matters related to the brakes, suspension, ABS, chassis, and so on. Everything related to the electric propulsion will be the same as discussed above in connection with HEVs.

8.1.2 Prognostics Issues

Prognosis involves the prediction of a problem before it happens, or it can also involve getting the health status of a system. In an HEV or EV, the prognostics can focus on basically the battery and power electronics, since these are weaker links in the system. The electric motor itself is important, but it is a more robust element than the battery and the power electronics. In addition, prognostics can be important for the steering and the braking system of the vehicle. These are safety-critical items or subsystems. On the contrary, there are less important items (in the sense of safety) – the air-conditioner, radio, and the like. Some of these are not HEV-specific items. Finally, a vital aspect of both HEVs and legacy ICE vehicles is the computational and networking elements. In fact, since a significant number of items in an HEV are controlled by computers/controllers and the CAN, a failure here can cause a lot of other elements to function incorrectly (even though those other elements could be in good condition), leading to a system shutdown, if not a dangerous accident. It is no less important to also know, if possible, the state of health of such computational elements and the communication network involved.

The health status of a system ("state of health" or SOH), be it battery, power electronics, or motor, needs to be compared against some benchmark or base system to get an idea of its health, even if it is functioning well. For that, we need to have various informational data available when the system is healthy, or rather when either it was new or certain components were repaired or replaced. Such data can be logged (i.e. stored in the database) from a healthy system. The alternative is a model-based system, where a simulation output is obtained based on ideal or new system parameters, and compared against the current system, supposedly not so new. Hence a prognostics system can have the generic structure shown in the block diagram of Figure 8.2.

In Figure 8.2, three methods of state of health estimation are indicated. A summary of the methodology is indicated in Table 8.1. In Figure 8.2, the three methods are also specifically demarcated by using braces with circled labels 1, 2, and 3.

Method 1 – This basically compares two input/output (I/O) sets, one obtained from the actual (physical) system at the present moment by direct measurements (with existing system parameters), and the other does the same thing but uses nominal system parameters and modeling and simulation (M&S) of the system. The nominal or new system parameters are obtained either from manufacturer's data or captured by measurements (along with any system identification techniques) when the system was new and stored in some database at that time. The I/O set using stored new (nominal) system parameters is obtained by M&S, since the real system will have aged, and hence parameters will have undergone changes in between. Once these two I/Os are compared, an average error between the two sets can be computed using some RMS error

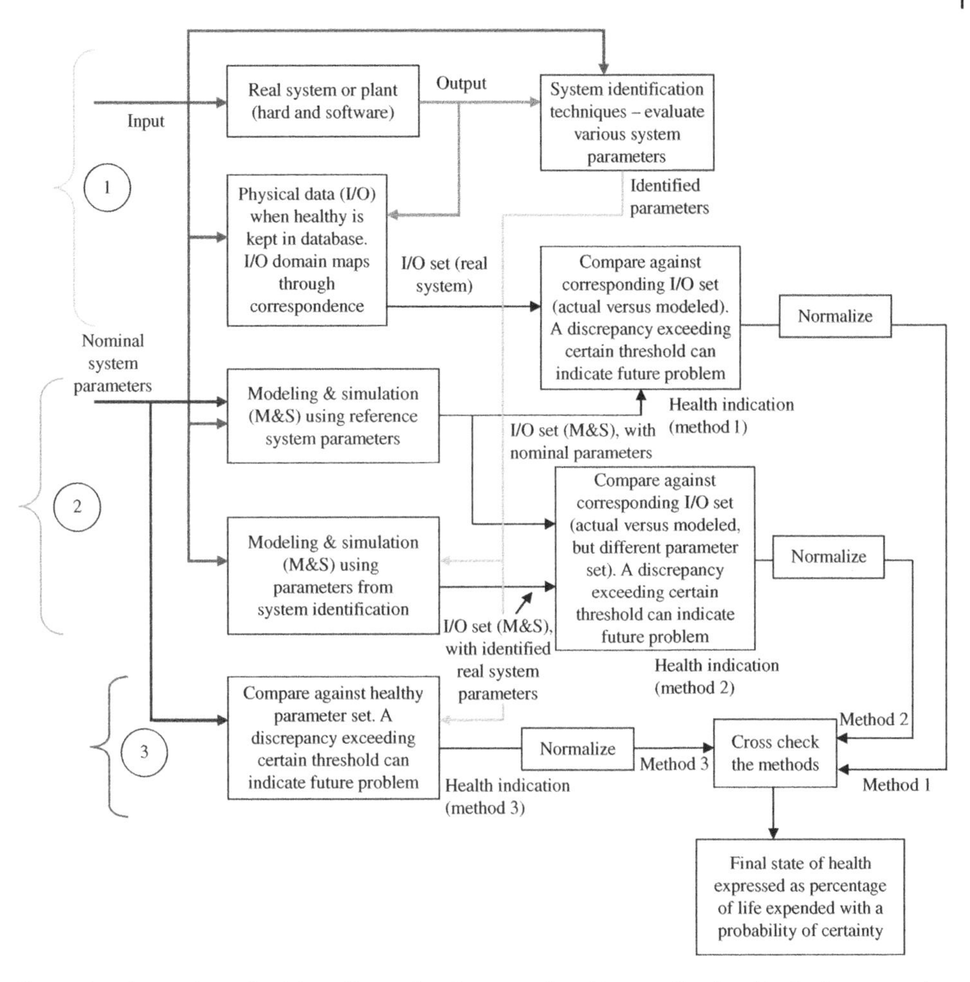

Figure 8.2 A generic methodology illustrating the state of health quantification (the basic concept is summarized in Table 8.1).

evaluation method or some other process. If the error exceeds some threshold, that can indicate a possible failure of the system in the near future.

Method 2 – This does the same thing as method 1, except that it is completely done by M&S, and in one case it uses the nominal system parameters (already stored) and in the other case it uses the existing system parameters, obtained from the present system I/O and using some system identification techniques. Method 2 allows the system to be run at operating points with overrated and stressed conditions, which cannot be done in a real system, since the real system can be damaged by running at such overrated operating points. Even though in Figure 8.2 the input lines to the M&S blocks are the same input which is going to the real system, it need not be so. In fact, the input to the M&S blocks could be artificially generated, which can be above the rated or allowable values, thus allowing the system to be exercised over a much wider I/O domain, beyond the rated values, which can sometimes be helpful to get additional information and ideas about the health of the system.

Table 8.1 Summary of the method indicated in Figure 8.2.

Input type	Parameter type used	Output types	Comparison between data types	RMS error or other type of comparison
Real input set	Nominal new system parameter values (NP)	M&S based system output with parameters when system was new (ONEW)	—	—
Real input set	Real present system parameter values (RP)	Real system output from measurement (OREAL)	ONEW vs OREAL (Health status, method 1) indicated by label in circle 1, Figure 8.2	Check threshold for health determination
Arbitrary input set under both nominal and also overrated operating conditions	Nominal system parameter values, based on new system (NP) conditions	M&S (X)	M&S (X) vs M&S (Y) (Health status method 2), indicated by label in circle 2, Figure 8.2	Check threshold for health determination
Arbitrary input set under both nominal and also overrated operating conditions	Present parameter values based on system identification, using real system input and real system output, for the present system at this time (SP)	M&S (Y)	NP vs SP, indicated by label in circle 3, Figure 8.2 (Health status method 3) – M&S not used, only system identification techniques are used	Check threshold for health determination

Method 3 – This method is not a direct comparison of the I/Os, rather it is a comparison between the nominal new parameter set and the existing parameter set obtained from the real system at the present moment, by using system identification techniques. In principle, this method involves comparing less data, since only the parameters are compared and not the I/Os. It is true that methods 1 and 2 involve I/Os that are affected by the system parameters, but some of the I/Os may be less affected (or even unaffected) directly by all the system parameters, depending on the nature of the system, while others may be more affected. The system parameter based method gives a slightly different perspective, and it may also be possible to incorporate some weighting factors on certain parameters which can affect the system more, compared to others, in terms of overall system health. It should also be noted that in method 3 it is not necessary to exercise the system by M&S, and only system identification technique based computations are needed to evaluate the parameters involved. Hence this method is less computationally intensive.

The bottom line is that in a complex system, multiple techniques to obtain the system diagnostics and health assessment allow some cross check between the techniques in terms of better decision-making about the system health. The actual comparison of the I/Os or the system parameters can be done by using various signal processing techniques, or artificial intelligence based techniques. It should also be noted that some normalization or scaling of the final data set may be needed in each of the above methods in order to bring the numerical values to the proper level prior to the cross checks.

To give some practical perspectives on using the concept in Figure 8.2, assume that we want to limit our focus only to the battery, power electronics, and motor system, which are HEV-specific items. In this case, the possible approach could be as follows.

For the battery, we need to monitor (experimentally in hardware) either voltage or current as input, and obtain current or voltage as output in response. In addition, temperature should be monitored as an input. In this immediate discussion, the terms input and output should be construed as input and output for the diagnostic process. They need not necessarily be actual system inputs or outputs in a normal electrical engineering sense, which drive the physical system.

For power electronics, we need to monitor the gate signals at each switch, the voltage across each switch, and the current through each switch. Also, we need to monitor the three-phase terminal voltages and currents coming out of the power electronics system. If placing a current sensor in each switch is too expensive, then it can be omitted, at the cost of losing some accuracy of the conclusion. In this system, diagnostic input can be considered to be the gate signal to each switch and also the voltage across each switch. The current through each switch or the current at the output terminals of the inverter can be considered to be the output of the system. Temperature too should be monitored as input. In this power electronics example, the voltages across the switches are to be monitored both when the switches are open and when they are closed.

For the motor system, terminal voltage and current can be considered as the input. The output can then be the speed and shaft torque. If torque monitoring hardware is considered too expensive or complicated, it is possible to monitor voltage, current, and speed, and infer torque from these measurements by using mathematical equations relating them.

Once all the above information is in place, we can use the generic process in Figure 8.2 to evaluate the state of health of the system described earlier and summarized in Table 8.1.

8.2 Reliability of HEVs

Recent events reported in the news media regarding faulty vehicles, including HEVs, relating to brake failure or sudden acceleration, have led to significant concerns about the safety and reliability of HEVs. While some of these concerns are genuinely important, it is important not to get too carried away with what is written in the media or make judgments based merely on that. Instead, the more scientific approach of studying reliability should be used to see the pros and cons of various aspects, and this will best serve the technical community and the users. In view of this, a reasonably detailed discussion of the subject will be presented here from a system-level perspective.

We all know that an HEV system is considered an important technology in the automotive industry these days. This is due to concerns about fuel economy, worldwide uncertainty of energy supplies, and pollution control. While discussing the subject, it seems that the focus in the technical community and the literature have been primarily on these issues, and also on control of the electric motor drives related to HEVs. In addition, people think in terms of cost premiums, that is, how long it will take to recover the extra cost of the vehicle (compared to a regular non-hybrid vehicle). Various figures have been indicated in the media and elsewhere in the technical community, suggesting that it can take anywhere from five to seven years to recover the extra cost of an HEV through any potential fuel savings. However, very little has been discussed on the issue of overall vehicular system reliability in HEVs. This issue is not trivial, and the overall acceptability of these vehicles in the long run will significantly depend on this, in addition to fuel economy and extra cost recovery. In this section, the aim is to show that, in HEVs, one of the penalties for fuel economy that has to be paid comes in terms of reliability. It is emphasized here that an HEV is not merely a collection of multiple propulsion sources and control systems to extract better fuel economy; rather it has a plethora of items in it. Overall reliable system-level functionality is no less important in making an HEV operate successfully, and be acceptable to the consumer in the long run, than the concern for fuel economy and cost reductions. Unfortunately, the literature on this topic is not available in the public domain to the best of the authors' knowledge. The three references [3–5] that are known to us and that deal with similar topics on the reliability of vehicular systems from a quantitative viewpoint, have been authored/co-authored by one of the authors of this book. The primary reason for this lack of published literature in this area of reliability is, we believe, that people have predominantly been involved until now only with the drive and control technology of hybrid vehicles, and matters related to fuel economy and emissions. A second reason is that hybrid vehicle technology is relatively new, and not much information about its reliability exists in the industry so far. Another important reason is that reliability data on components and subsystems takes a long time to monitor and collect, and even if it is collected in the industry, the data is normally retained as proprietary information.

The following discussion, prior to the section on software reliability, is extracted and/or modified as needed from Masrur [5].

We will consider here the architecture of a regular ICE-based vehicle, followed by series and parallel HEV architectures. The overall subsystem and component-level reliabilities are introduced by using some assumed figures for reliability, and then analyzing them. In addition, the concept of graceful degradation is introduced, and its implication from a quantitative point of view is discussed. The numerical values of reliability used in this section are merely to illustrate concepts; the exact reliability situation will depend on the system architecture and precise values of the reliability figures involved in the system under study. Of course, it should be recognized that finding accurate reliability figures for various components in a system requires prolonged efforts, sometimes using modeling and simulation studies, and also experimental tests; these issues are not within the scope of this discussion.

8.2.1 Analyzing the Reliability of HEV Architectures

For a system-level perspective in studying the reliability of HEVs, it is necessary to trace the individual reliability values of the subsystems and components noted in the architectures in Chapter 2 where a slightly different perspective was used. The architecture for parallel HEVs was also discussed briefly in Chapter 6. Some of these architectures are redrawn here.

For this discussion, the term reliability is defined as the probability that a component, subsystem, or system is functional, that is, performing its intended function, at the end of a particular time period, without any change or maintenance activities within

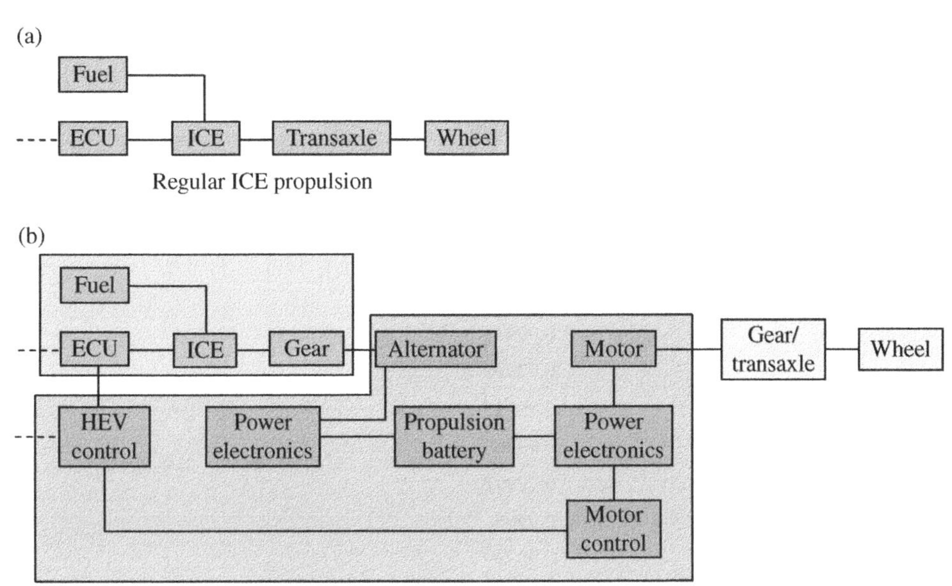

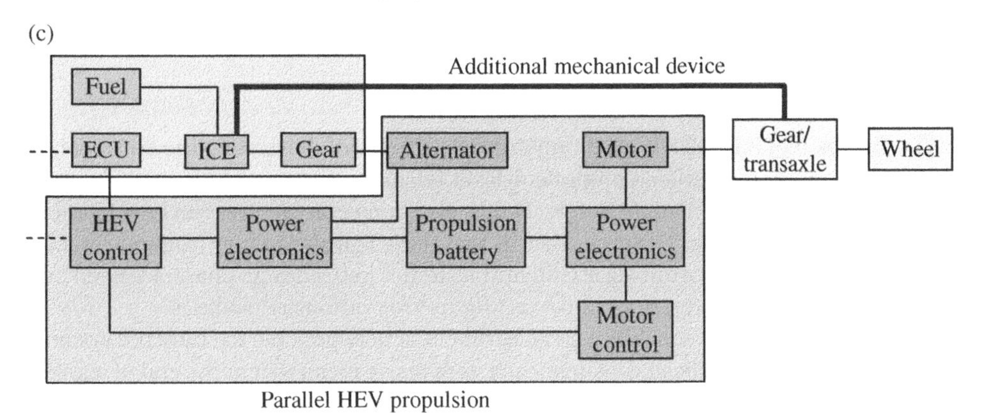

Figure 8.3 System-level block diagrams for: (a) regular ICE, (b) series HEV, and (c) parallel HEV architectures. From [5], © 2008 IEEE.

that time period. Thus, reliability is connected with both probability and time span. In addition we define the term availability where we take a hypothetical system with reliability equal to 1. Such a system will be said to be "fully" available.

We can study the overall reliability of each of these subsystems in Figure 8.3 by following the simple methodology below. Consider the various items (subsystems) in Figure 8.3, and let the reliability of each of those be as given in Table 8.2.

Note that each of the above items is constructed by using many constituent subsystems and components. However, we can use a single cumulative reliability figure for each of the items above; for example, for the motor an overall reliability of 0.99995 can be used, rather than delving into the individual constituent components within the motor [5].

Table 8.2 Assumed reliability numbers for subsystems shown in Figure 8.3.

Subsystem/component	Reliability
Fuel system	0.9999
ECU	0.99999
ICE	0.9999
Transaxle	0.99995
Wheel system	0.9999
Gear	0.99995
Alternator	0.99995
Motor	0.99995
Power electronics	0.99992
Propulsion battery	0.9999
Motor control	0.99999
HEV control	0.99999

Source: Masrur 2003. Reproduced with permission of IEEE.

The figures in Table 8.2 are used only for the purpose of illustrating the concepts in this section. As noted earlier, component-level reliability figures are generally kept as proprietary items by the manufacturers. Hence obtaining exact figures can be extremely difficult, if not impossible. The other issue is that these figures can vary from one manufacturer to another. Hence during architectural studies in the design phase, we need not be too concerned about trying to find exact figures for various reliabilities.

Using the definition of reliability given earlier, it is now possible to study the system. The numerical values indicated against each item above mean that at the end of a given (or chosen) period of usage time or mileage of the vehicle (e.g. 100,000 miles or 160,000 km), when the reliability is assigned a value of, say, 0.9999 for the ICE, it means that the chance of its failure is 1 in 10,000 (within that mileage or usage time, starting from when the item was newly installed). It is true that reliability is a function of time as the system ages. However, it will be assumed to be constant for the purpose of the discussion here.

It has been shown [5] that based on the above concepts, the reliability figures in this illustrative example in [5] are as follows:

1) For regular ICE propulsion,

$$R_{ICE} = 0.99964 \tag{8.1}$$

where R_{ICE} is the reliability (or the probability of being available) of the complete ICE vehicle system.

2) For series HEV propulsion,

$$R_{SH} = 0.999210280440917 \tag{8.2}$$

where R_{SH} (reliability of series hybrid architecture) is defined similarly as before.

3) For parallel propulsion,

$$R_{PH} = 0.999160319926895 \tag{8.3}$$

where R_{PH} (reliability of parallel hybrid architecture) is defined in an analogous manner to R_{SH}.

8.2.2 Reliability and Graceful Degradation

Consider the parallel HEV propulsion in Figure 8.3c, where demarcation between the ICE-based and the electric propulsion-based subsystems is shown using shaded areas. Using the terminologies EVP (for electric vehicle portion) and ICE (for the internal combustion engine portion of the propulsion system) we can write the following probabilities for various combinations of the EVP and ICE [5]:

a) Both ICE and EVP are good: $0.99974 \times 0.99962 = 0.99936$
b) ICE good and EVP bad: $0.99974 \times (1 - 0.99962) = 0.00037984$
c) ICE bad and EVP good: $(1 - 0.99974) \times 0.99962 = 0.00025987$
d) Both ICE and EVP bad: $(1 - 0.99974) \times (1 - 0.99962) = 0.0000000987763207$
e) Reliability of the wheel and final transaxle together $= 0.99985$.

Hence, the probability of having "some" amount of system functionality during partial failure conditions (i.e. under graceful degradation) is given by the sum of the items (a) through (c)

$$P_{GR} = (\mathrm{a} + \mathrm{b} + \mathrm{c}) \times (\mathrm{e}) = 0.999849906238495 \tag{8.4}$$

where P_{GR} is the reliability or probability of the system under graceful degradation, and (a), (b), (c), and (e) in Equation 8.4 are the reliability numbers corresponding to items (a), (b), (c), and (e) in the above list. Thus it can be seen that in a graceful degradation mode the system availability is higher than the situation where the partial availability or graceful degradation mode is not taken into account. However, it should be noted that under conditions (b) and (c), only partial functionality is obtained, not full functionality. Thus for items (b) and (c), we have to amend these items with some weighting factor to indicate that the functionality is degraded. This is explained in greater detail in [5].

Some additional issues pertaining to HEVs should also be considered. For example, if the EVP in a parallel HEV fails, the vehicle can still operate with the ICE, refilling the gas tank as needed, and keep running at a lower performance. If the ICE fails, the vehicle can still run with the EVP, but only as long as the battery lasts. Thereafter a plug-in operation is required, if there is provision for that, but otherwise, there is no option. Here the battery must not run below the level of allowable SOC, to save battery life. A similar analysis has been done [5] for the series vehicle as well, and the outcome can be summarized through the graph in Figure 8.4.

From this graph and from the earlier discussions it is apparent that the overall reliability numbers for a series HEV is quite a bit lower than for the parallel HEV. Overall system reliability of both the series and parallel HEV, without taking any graceful degradation into consideration, is of course lower than the regular ICE-based vehicle. This is due to the fact that the ICE has fewer components and hence initially it has the

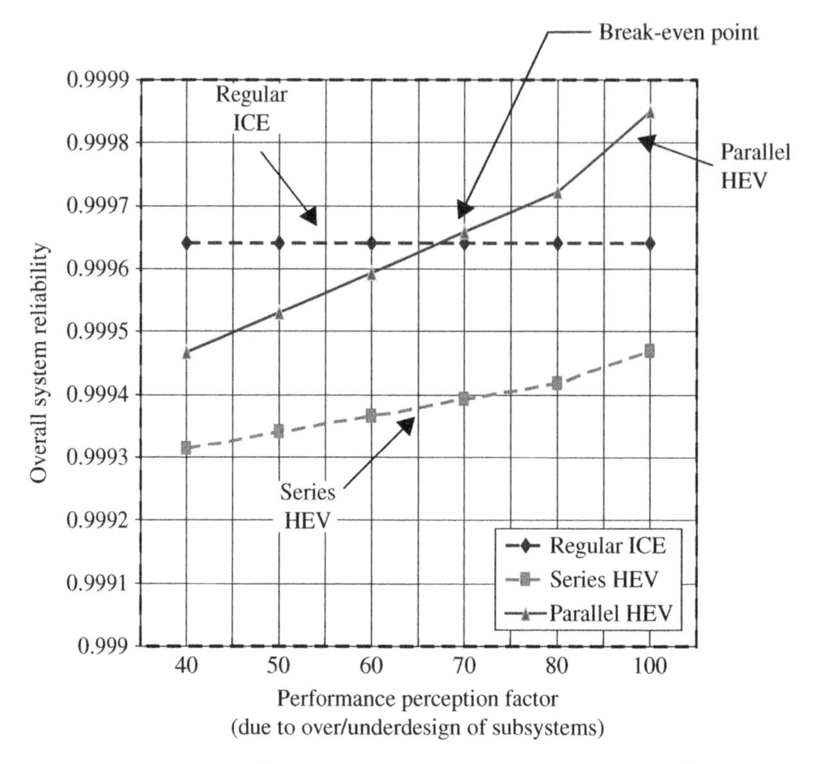

Figure 8.4 Comparison of system reliability vs. performance perception factor as a percentage for three different types of vehicles. From [5], © 2008 IEEE.

edge in terms of reliability. The basic idea here is that in a parallel HEV, in general, both the ICE and EVP will be designed at a somewhat lower rating. Hence during partial failure conditions of either the ICE or EVP, the performance of the vehicle is reduced. Thus the full performance functionality is not available during partial failure. For better performance during partial failure, it would be necessary to overdesign the ICE and the EVP to some extent. The definition of "performance perception factor" [5] during partial failure condition is essentially the ratio of the ICE or EVP rated power in an HEV, to the rated power of an ICE of a non-hybrid vehicle. In other words, say, the non-hybrid vehicle has 200 hp (150 kW) nominal. In an equivalent parallel HEV let us assume that the ICE and EVP are each sized to a lower value of, say, 80 hp for the EVP and 120 hp for the ICE. In that case, if the ICE of the HEV fails, it will only provide 80/200 of the nominal performance, that is, 40%. This will be called the perceived performance factor. For a higher performance, the size of the EVP has to be scaled up. If it is sized higher, obviously the system performance will be even better under graceful degradation. Hence its overall "quality" increases even under degradation. The horizontal axis (in Figure 8.4) basically says that if the EVP or ICE is overdesigned, then its reliability factor, due to better performance, will increase [5]. Thus, without overdesign, the non-hybrid ICE system is more reliable than an HEV, but with overdesign beyond a certain point the parallel HEV is more reliable, which includes the effect of performance along with reliability. But for a series HEV, due to the fact that it only has a single propulsion system, even with overdesign, its overall reliability cannot exceed that of a non-hybrid vehicle, since a non-hybrid vehicle has fewer components compared to a series HEV. The interested reader is referred to [5] for further elaboration and details.

8.2.3 Software Reliability Issues

With the proliferation of software in our daily lives and its usage in small electronically controlled devices, to automobiles, aircraft, spacecraft, ships, and so on, the importance of its reliable functioning cannot be overemphasized. In many cases software malfunctioning can merely cause some inconvenience. In a large number of other applications like automotive braking, steering, engine control, and aircraft stability control, or in certain medical and defense applications, malfunctioning software may result in serious accident, injury, and even loss of life. Some examples of catastrophic failures are presented in the article available at http://users.ece.cmu.edu/~koopman/des_s99/sw_reliability/. These include the tragedies with the Therac 25, a computer-controlled radiation therapy machine, caused by the software's inability to detect a race condition, and the case of the British destroyer *HMS Sheffield* which was sunk because the radar system identified an incoming missile as friendly. This website also indicates that sometimes small unnoticeable errors or drifts can cause failure. For example, a chopping error missed 95 ns in precision in every 10th of a second and, accumulated over 100 hours, made a Patriot missile fail to intercept a Scud missile, leading to the loss of 28 lives. Here is a case where errors, considered insignificant, but when accumulated, and apparently not corrected, caused failure. Recent events related to problems in the HEV and also regular vehicles, though not yet completely researched, may have both hardware and software issues and are yet to be seen.

Software and hardware reliability curves have been compared in the same website, through the diagrams in Figure 8.5. It can be seen that hardware passes through an initial burn-in phase, followed by a fairly stable rate, and finally physical wear out occurs and the failure rate rises. At that time hardware replacement may be needed. Unlike hardware, software initially goes through significant failure correction in its testing/debug phase. Thereafter, each upgrade leads to a sort of mini-replication of the initial phase. Eventually the software matures and the reliability factor settles down, but the software also tends to become obsolete as well at that point.

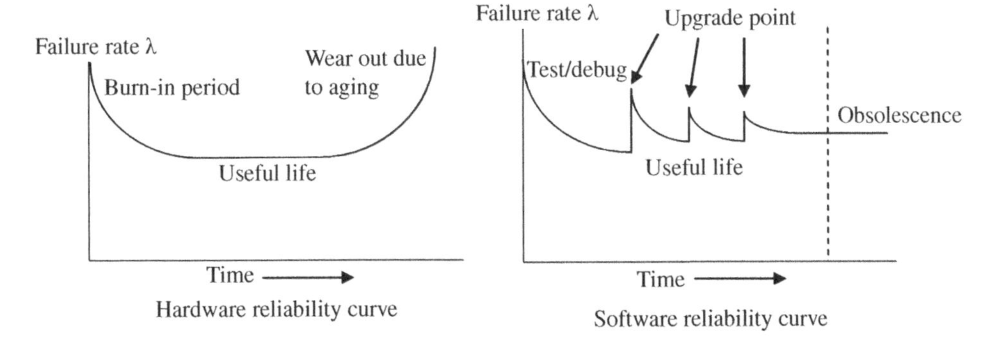

Figure 8.5 Hardware and software reliability curves.

Most people who use a computer operating system (OS) know that an older OS with upgrades works well. An example is Windows XP. But while it operates quite satisfactorily, one may think of new features, and hence obsolescence sets in. For Windows XP, this has led to Vista and Windows 7, and so on.

The bottom line is that there is still no definitive quantitative measure of software reliability and testing method that can assure software reliability under all conditions. Even though people have documented various models for reliability evaluation, in reality none are applicable to all situations.

Hence the question arises: how does all the above relate to the HEV? From the previous chapters it can be noted that the HEV is a complicated system, where various controllers are involved with algorithms that make a significant number of decisions. The controllers involved for successful HEV functionality also involve other controllers in the system, which may be legacy items upgraded for an HEV interface. Such examples include the body controller, engine controller, transmission controller, power electronics/motor drive controller, battery management controller, power management controller, to name a few, and their associated algorithms. All these involve significant amounts of software. Although testing can be done for some selected situations, there is really no guarantee that there does not exist a situation that can render the software ineffective. This means the situation creates a condition such that the output from the command leads to the following: (1) an output that is not supposed to be the desired response to an input as per design; (2) an algorithm that creates an unending loop with either the output remaining unchanged (which is not desirable), or the output flip-flops between a number of different output states. Neither (1) nor (2) is desirable, but as long as they do not lead to any catastrophically dangerous command to the HEV, they can at least be accepted as a system flaw requiring redesign or repair. If, however, a command is automatically generated, for example, the acceleration is continuously increasing, or if the brake command is not activating the brakes when the user is needing to stop the vehicle, then that can be catastrophic or even fatal.

Some of the reasons for software failure in HEVs are:

1) design flaw due to inability of the developer to account for all possible scenarios
2) improper specifications which can lead to indecision by the software due to lack of sufficient information to carry out the task, even though the software designer or developer is not at fault
3) lack of recognition of the need for a higher degree of precision in certain quantities, which can accumulate errors over time and operate with a wrong decision
4) lack of thorough testing of the software for all conceivable situations.

Several scenarios for catastrophic failure in the automotive area, and in particular in HEVs (e.g. sporadic noise), can cause a wrong input to the software in the engine or motor control. If the software does not have provision to correct the error, it can give the wrong signal to the engine or motor, leading to very undesirable behavior, perhaps in acceleration.

Another situation can be considered. Let us look at Figure 8.6 and assume that the driver pressed the brake, and regenerative action was in place, but the mechanical brake is not activated based on the brake algorithm. Let us also assume that a wheel of the car, which was previously on a smooth surface, momentarily runs over a pothole. At that moment, the tire loses contact with the wider road surface, the frictional torque ceases,

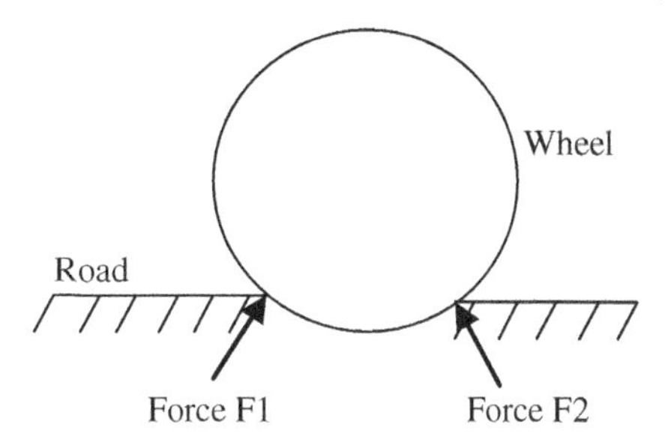

Figure 8.6 A wheel in a pothole.

and the edge of the pothole provides reaction forces F1 and F2, which will be more or less perpendicular to the tire surface. Even if there are any rotational forces (thus creating a torque) tangential to the tire, they will cancel for all practical purposes, due to the directions of forces F1 and F2 shown in Figure 8.6 (assuming the forces are the same, which is reasonable for the purposes of this discussion). At that very moment, the regenerative power from the brake will not be available, and if the mechanical brake does not activate, since it sees the other wheels still doing some regeneration, then that particular wheel of the car will accelerate forward even though the driver has pressed the brake. If the algorithm did not fully evaluate the situation based on the various wheel conditions and did not coordinate them properly, it could destabilize the vehicle or fail to slow it down properly. This is merely a hypothetical situation, since details of manufacturers' algorithms are in general proprietary and it is difficult to know exactly what will happen in this scenario. But recent issues such as this are coming to light, and although the problem could be due to sensor issues, or not having sufficient sensors, the software problem could still be one of the reasons.

Notwithstanding the fact that software problems can give rise to significant danger if the software malfunctions, the benefit of having software in general outweighs the drawbacks. Hence the only way to know the true quality of software is to deploy it over many years of service. Mere testing in a lab environment can never guarantee this attribute.

8.3 Electromagnetic Compatibility (EMC) Issues

Electromagnetic compatibility is an important issue these days, in view of the fact that HEVs and EVs use significant amounts of power electronics which use high-power, high-frequency switching. The EMC issue can be mitigated to a large extent by proper layout of the circuitry on various circuit boards and power distribution systems, in terms of both generation and propagation.

Electromagnetic interference (EMI) is generated by high-power circuits in the power converters and associated controllers. Several standards, such as EMC Directive 89/336/EEC, the Automotive Directive 95/54/EC, and IEC 61800–3, *EMC Requirements and Test Methods for Adjustable Speed Electrical Power Drive Systems*, are considered to be important. In addition, the CEI EN 55055 standard on radio disturbance characteristics for the protection of receivers onboard has also been considered [6]. The standard SAE J551 on surface vehicle EMC is considered more stringent compared to other applications, unlike automotive ones.

In an electric vehicle the inverter is located near the propulsion motor and housed in a metal box which shields noise emission [7]. However, the battery module is normally located at a distance, and unshielded power cables may be used to connect it to the power electronics. Hence the unshielded cables can lead to a significant amount of noise propagation. The studies by Chen and Xu [7] in an EV indicated that most noise energy is at frequencies below a few megahertz and this causes transients which can affect digital systems more than creating radiated noise to the environment. But the radiated noise energy level itself is also high, and not necessarily within limits.

In a very interesting presentation [8], the impact of EMI on HEVs and how it affects various items was studied. Although important for military applications, the information is equally valid for regular vehicles. However, the importance of EMI on communication systems is more vital for military as opposed to commercial vehicles. The standards relevant to the military in this context are MIL-STD-461E – *EMI Requirements, Design, and Test*, and MIL-STD-464 – *Vehicle EMC and Lightning Requirements*. The latter is more applicable to commercial aircraft. Other standards include RTCA-DO-160 – *EMI and LIT Requirements Including Test Methods*, and AC-20-136 – *FAA LIT Advisory Circular*. In addition, the FCC Rules and Regulations, Title 47, Part 15, Subpart B, are of importance. Several European standards were also indicated earlier.

Figure 8.7 shows the block diagram of an electric motor drive. A number of small capacitances, which can be ignored at low frequency, become important, especially when the switching frequency is high [9]. Here, the capacitors are shown between the individual housing elements to the ground or chassis/body of the vehicle. Various currents, in common and differential modes, could flow through these stray capacitances. They can also create problems in the shaft and bearings as follows.

Consider Figure 8.8, showing the bearing and shaft of a motor drive. The stray capacitance shown can give rise to a voltage across the grease layer shown shaded. This can

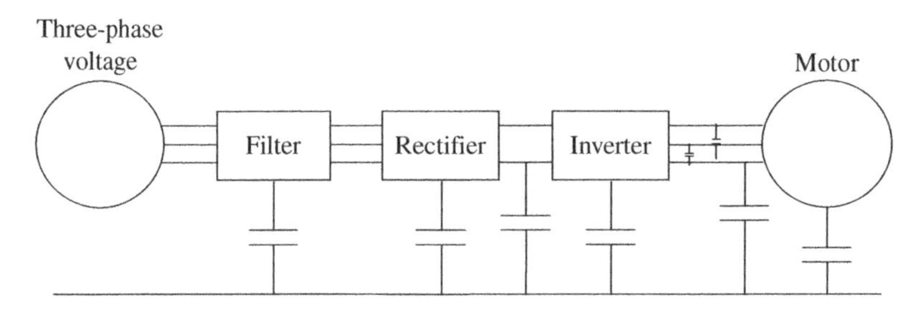

Figure 8.7 Power electronics architecture in an EV motor drive.

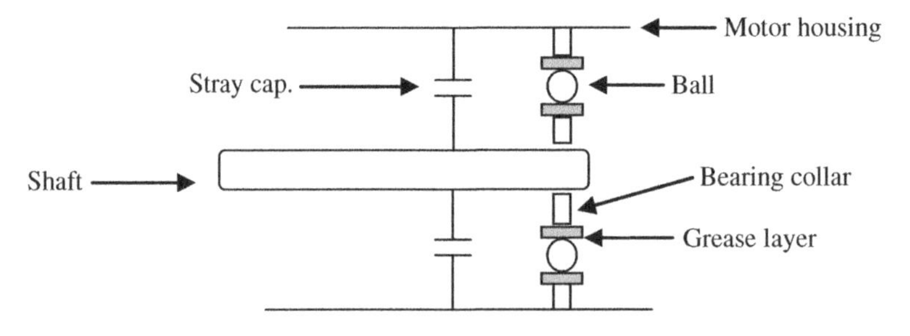

Figure 8.8 Completion of circuit through shaft, stray capacitance, and bearings.

potentially exceed the breakdown voltage of the grease, causing sparks in the bearing's steel material, thus leading to bearing damage and eventual failure. Of course, the other capacitances can also lead to additional stray currents and consequent radiation of noise.

Some of the mechanisms needed to reduce EMI effects include: (1) using very good conducting materials for the housing of power electronics and other equipment; (2) various joints in the housing should be tight and there should be no abrupt discontinuity in the housing structure, for example, an undesirable tiny gap or hole in the structure which can lead to high emission; (3) various cables should be shielded properly as far as practicable, and correct terminations should be used; (4) both magnetic and electric shielding should be used; (5) both common and differential mode filters should be used, as applicable; (6) screw spacing should be made correctly and a conducting gasket should be used as needed. Other techniques, including the use of laminated bus bars, high-frequency DC link capacitors, good snubbers, and separation of digital and high-power electronic circuits, have been suggested for the power electronics circuit. For the electrical wiring, benefits can be obtained by over-braiding high-power bundles. In addition, data buses should be properly shielded. In EV high-power wiring, even if it is not shielded, at least the data buses should be shielded to protect malfunctioning of these digital circuits due to EMI.

8.4 Noise Vibration Harshness (NVH), Electromechanical, and Other Issues

As noted earlier in this chapter, in connection with HEVs, normally people are more concerned with matters related to fuel economy and control mechanisms and algorithms, the battery, power electronics, and so on. But the fact that the HEV is subjected to frequent stopping and starting of the engine to conserve fuel, or even too much stopping or starting of the electric motor, deserves some consideration in terms of vibrations or stressing of the engine, motor, or mechanical members associated with the system.

Consider Figure 8.9 showing the configuration of a typical power split HEV. In this particular system, a power split device, that is, the planetary gear system, is used to connect the various items shown. This is typical for the Toyota HEV. In this system, since there are no clutches, the traction torque from the ICE and the electric motor is directly transmitted to the wheels. Hence, if there is any engine start/stop action, any accompanying vibration

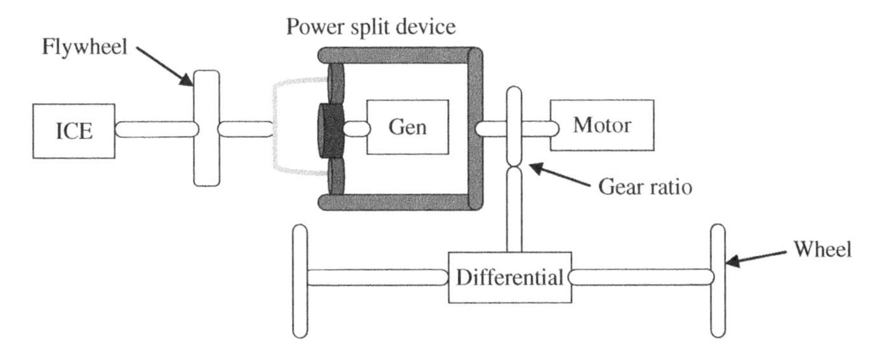

Figure 8.9 Configuration of a typical power split HEV.

is transmitted to the wheels through the various members in between, including the shafts and the connecting gears. Vibration from the engine can occur both before and after ignition [10]. When there is no ignition, there is compression and pumping pressure in the cylinders. After ignition, there will be a sudden torque change in the engine due to combustion. Note that these are initial changes due to a discontinuous process, that is, starting or stopping of the engine, which is separate from the regular pulsating torque in the engine under normal conditions. Ito et al. [10] have studied the vibration problem in an HEV engine in reasonable detail. The vibration was reduced by using two controllers, the first one controlling the ripple due to compression and pumping pressures in the engine cylinders, and the second controlling torsional vibration caused by rapid changes in the torque. The result was promising in terms of reducing the floor acceleration.

Ito et al. [10] present their experimental results according to the vibration control method used. A simplified qualitative comparison diagram is given in Figure 8.10 to indicate the difference. It is interesting to note that by using a new control methodology,

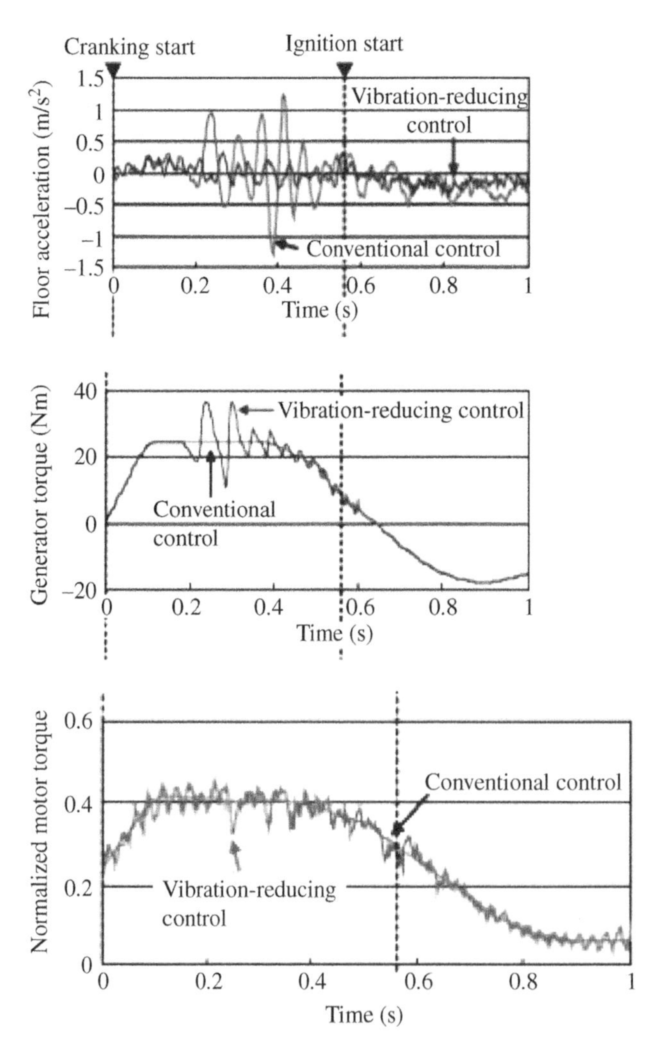

Figure 8.10 Qualitative comparison of reduction of vibration due to control action. Lines representing conventional control and lines showing the vibration reduction control proposed in [10] are indicated by arrows. Reprinted from *R&D Review of Toyota CRDL*, vol. 40, no. 2, p. 42.

the vibration in floor acceleration has been substantially reduced. However, it should also be noted that concurrently the torque generated at the motor and the generator has increased. This increase is essentially what nullifies the floor vibration. However, the above phenomenon leads to a reduction of mechanical torque, and indeed needs high-frequency control action through the motor and generator system. This may have some ramifications in terms of faster switching of the power electronics switches and higher-frequency currents and magnetic field in the motor, and it may actually lead to higher losses in those components. In other words, there is some penalty to pay in one place whenever some benefit is achieved in another. In this particular case, the reduction in mechanical vibration is essentially transferred into electrical form. These issues, related to high-frequency control and its ramifications, have not, to the best of the authors' knowledge, been studied.

In the patent [11], the issue of numerous starts and stops was addressed. It was indicated that vehicle designs are normally optimized for minimizing noise vibration harshness (NVH) for both idle speed and above, and particular for high-frequency conditions. But low-speed and low-frequency vibrations are normally difficult to manage. This patent [11] specifically indicates that at 0–800 rpm, the NVH issue and resonance are very problematic. Apparently this situation is caused by compression pulses in the engine and fluctuations in flywheel speed. In addition, the problem is important for gasoline engines, but more so for diesel engines. The former has a compression pressure around 120 psi (827 kPa), and the latter about 400 psi (2758 kPa). It appears that frequent start/stops every couple of minutes and shutdown of the engine for only a few seconds can be quite important in terms of NVH. The method suggested to improve the NVH issue includes opening the exhaust port (valve) during the shutdown sequence of the engine. The idea is to reduce the compression pressure in the cylinder, thus reducing the vibration. It was also indicated that the above process influences the change of rotational inertia in the engine which is prevented from transmission to the chassis.

The above issue of NVH is specific to the HEV which has multiple sources of power and involves some shutdown sequences. In a purely electric vehicle the issue is less likely to be of consequence due to the absence of an ICE and also that electric motor torque is much smoother compared to ICE torque.

8.5 End-of-Life Issues

In HEVs and EVs, the end-of-life issue is associated with the life of the various components and overall life cycle management. As far as the mechanical system is concerned, that is, the engine and transmission (power split based or other type of mechanical coupling), this is the same as in a regular ICE vehicle. So, the end of life can be considered to be around 100,000 miles (160,000 km) nominal, in terms of manufacturer warranty, and, depending on the quality, the life can easily be around 150–200,000 miles. However, important items to consider for HEVs are the battery, power electronics, and motor. The weakest link here is the battery, next is the power electronics, and then the motor. In an HEV, the battery is warranted for around 10 years, which is compatible with the warranted life of the vehicle. The cost of a new battery can be around $3000, depending on the specific vehicle size. An Oak Ridge National Laboratory (ORNL) report [12] says that the present cost of power electronics is about $200/kW. Attempts are underway to bring it down in the future to $7/kW. So, for a vehicle with a 60kW power electronics device, the cost was around $12,000 at the time of the report (2011) [12], and hybrid-specific item warranty is around 150,000 miles or 10 years. According to the ORNL study, the cost of a 75kW motor is a little over $800 to around $1500 depending on the size. The motor is considered the most robust entity within the hybrid-specific components. Although precise information is not available, the authors believe that the life expectancy of the motor itself is probably more than 10–12 years, that is, somewhere between 150,000 and 200,000 miles. The life of all the items which are hybrid specific (battery, motor, and power electronics) is significantly affected by ambient temperature and operational mode. This means that the variation between the continuous drive cycle demands and its peak affects the overall life cycle. This applies to the battery, power electronics, and motor. If there are very high peaks and valleys, with associated harmonics in the voltage and current, it will affect the overall life expectancy.

In connection with life cycle issues, it should be noted that some of the materials within a battery and motor could be recycled. For example, the battery casing could be directly reused, and some of the materials inside might be reprocessed. The same applies to the motor. The motor could most likely be reconditioned in terms of winding, and if it is a permanent magnet motor, the magnets could be replaced if they have lost some of their properties. With the power electronics, the silicon components themselves could be replaced, that is, the power electronics switch modules. Control electronics most likely could be retained. The replacement and reconditioning of the components will need some infrastructure, so that the operations of replacement and reconditioning can be done with least inconvenience to the users. All these processes are rather new and can only evolve with time.

References

1 Murphey, Y.L., Masrur, A., Chen, Z.H., and Zhang, B.F. (2006) Model-based fault diagnosis in electric drives using machine learning. *IEEE/ASME Transactions on Mechatronics*, 11 (3), 290–303.
2 Masrur, M.A., Chen, Z., and Murphey, Y. (2010) Intelligent diagnosis of open and short circuit faults in electric drive inverters for real-time applications. *IET Journal of Power Electronics*, 3 (2), 279–291.
3 Masrur, M.A., Shen, Z.J., and Richardson, P. (2004) Issues on load availability and reliability in vehicular multiplexed and non-multiplexed wiring harness systems. *Society of Automotive Engineers (SAE) Transactions, Journal of Commercial Vehicles*, 2003-01-1096, 31–39.
4 Masrur, A.M., Garg, V.K., Shen, J., and Richardson, P. (2003) Comparison of system availability in an electric vehicle with multiplexed and non-multiplexed wiring harness. IEEE Vehicular Technical Society Conference Proceedings, October, Orlando, FL, pp. 3277–3283.
5 Masrur, A.M. (2008) Penalty for fuel economy – system level perspectives on the reliability of hybrid electric vehicles during normal and graceful degradation operation. *IEEE Systems Journal*, 2 (4), 476–483.

6 Serrao, V., Lidozzi, A., Solero, L., and Di Napoli, A. (2007) EMI characterization and communication aspects for power electronics in hybrid vehicles. European Conference on Power Electronics and Applications, September, Aalborg, Denmark, pp. 1–10.
7 Chen, C. and Xu, X. (1998) Modeling the conducted EMI emission of an electric vehicle (EV) traction drive. IEEE International Symposium on Electromagnetic Compatibility, August, vol. 2, pp. 796–801.
8 Cortese, S. (2004) EMI in a Hybrid Electric World, www.dtic.mil/ndia/2004tactical/Cortese.ppt (accessed February 2011).
9 Zare, F. (2009) EMI in Modern AC Motor Drive Systems, https://ewh.ieee.org/soc/emcs/acstrial/newsletters/summer09/EMIinModernAC.pdf (accessed February 2011).
10 Ito, Y., Tomura, S., and Moriya, K. Vibration-reducing motor control for hybrid vehicles. Research Report. *R&D Review of Toyota, CRDL*, 40 (2).
11 Stone, K. (2009) Hybrid vehicle vibration reduction system and method. International Patent WO/2009/134695, May 2009, pp. 516–521.
12 ORNL Review (2000) Power Electronics: Energy Manager for Hybrid Electric Vehicles, http://www.ornl.gov/info/ornlreview/v33_3_00/power.htm (accessed February 2011).

Further Reading

Toyota Prius Diagnostics, http://www.aa1car.com/library/toyota_prius_diagnostics.htm (accessed February 2011).

9

Power Electronics in HEVs

9.1 Introduction

Power electronics is one of the enabling technologies propelling the shift from conventional gasoline/diesel engine-powered vehicles to electric, hybrid, and fuel cell vehicles. This chapter discusses the power electronics used in HEVs and PHEVs. However, the focus of the chapter is on the unique aspects of power electronics in HEVs and PHEVs.

To explain the types of power electronics circuits used in an HEV, we use the configuration of a typical series HEV powertrain as shown in Figure 9.1. In this configuration, the internal combustion engine (ICE) drives a three-phase permanent magnet synchronous generator, whose output is a three-phase voltage with variable frequency and variable voltage. This output needs to be rectified to DC.

The front wheels are driven by an induction motor which needs to be controlled by a voltage source inverter (VSI) or a current source inverter (CSI). An energy storage system is connected to the DC bus, between the generator/rectifier output and the inverter. However, there is a bidirectional DC–DC converter that manages the charge/discharge of the battery, as well as controlling the DC bus voltage.

In conventional vehicles, the air-conditioning (A/C) compressor is driven by the engine through a belt. In advanced HEVs, the engine is very often turned off during stop-and-go driving patterns. In order to have A/C while the engine is off, the compressor needs to be driven by an electric motor that runs from the hybrid battery. There may be an electrically driven hydraulic pressure pump for the vehicle's hydraulic systems, such as frictional brakes and power steering. The A/C motor and the compressor motor are typically brushless DC motors with an inverter.

In addition, auxiliary components, such as headlights, wipers, entertainment systems, and heat seats, run from the 14 V auxiliary battery. Most advanced hybrid vehicles no longer have an alternator, which means that the 14 V battery needs to be charged from the high-voltage (HV) battery. On the other hand, even if an alternator is present, when the engine is off, the 14 V battery can still be drained quickly without proper charge maintenance. Therefore, it is necessary to have a DC–DC converter to charge the 14 V battery from the HV battery. For a PHEV, there is also a battery charger installed on the vehicle or in the charging station.

Hybrid Electric Vehicles: Principles and Applications with Practical Perspectives, Second Edition. Chris Mi and M. Abul Masrur.

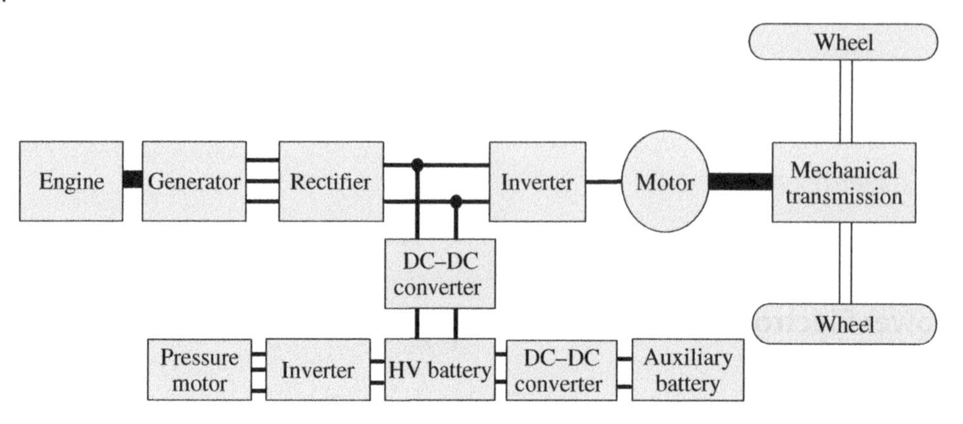

Figure 9.1 Power electronics converters used in a series HEV.

Even though there are commonalities between the power converters used in HEVs and the ones used in other industrial, commercial, or residential applications, there are some unique features specific to automotive applications. Examples include wider ambient operating temperature (−20 to 90°C), vibration and shock, electromagnetic compatibility, and thermal management.

Figure 9.2 shows the integrated main power electronics unit used to control the Toyota Highlander HEV (http://www.toyota.com). The power electronics unit consists of a bidirectional DC–DC converter that links the hybrid battery and the DC bus, and three motor drive circuits that control the front and rear motor/generators.

The issues to be addressed in the design of HEV power electronics circuits involve [1–23]:

- **Electrical design:** Includes main switching circuit design, controller circuitry design, switching frequency optimization, and loss calculations.
- **Control algorithm design:** Includes developing the control algorithm to achieve the desired voltage, current, and frequency at the output, and to realize bidirectional power flow as needed.
- **Magnetic design:** Includes the design of inductors, transformers, and other components such as capacitors needed for filtering, switching, and the gate driver units.
- **EMC design:** Includes understanding the electromagnetic interference (EMI) issues, analyzing switching transients, and circuit layout that minimizes parasitic inductance and capacitances.
- **Mechanical and thermal design:** Includes modeling of the loss of power devices and magnetic components; cooling system, heat sink, and enclosure design; and integration of the power electronics unit.

9.2 Principles of Power Electronics

In broad terms, power electronics is a discipline that studies power converters which process and control power flow by electronic means. It mainly involves the use and control of power semiconductor switches, such as power diodes, insulated gate bipolar transistors (IGBTs), and metal oxide field effect transistors (MOSFETs). Figure 9.3 illustrates the schematics of a power converter.

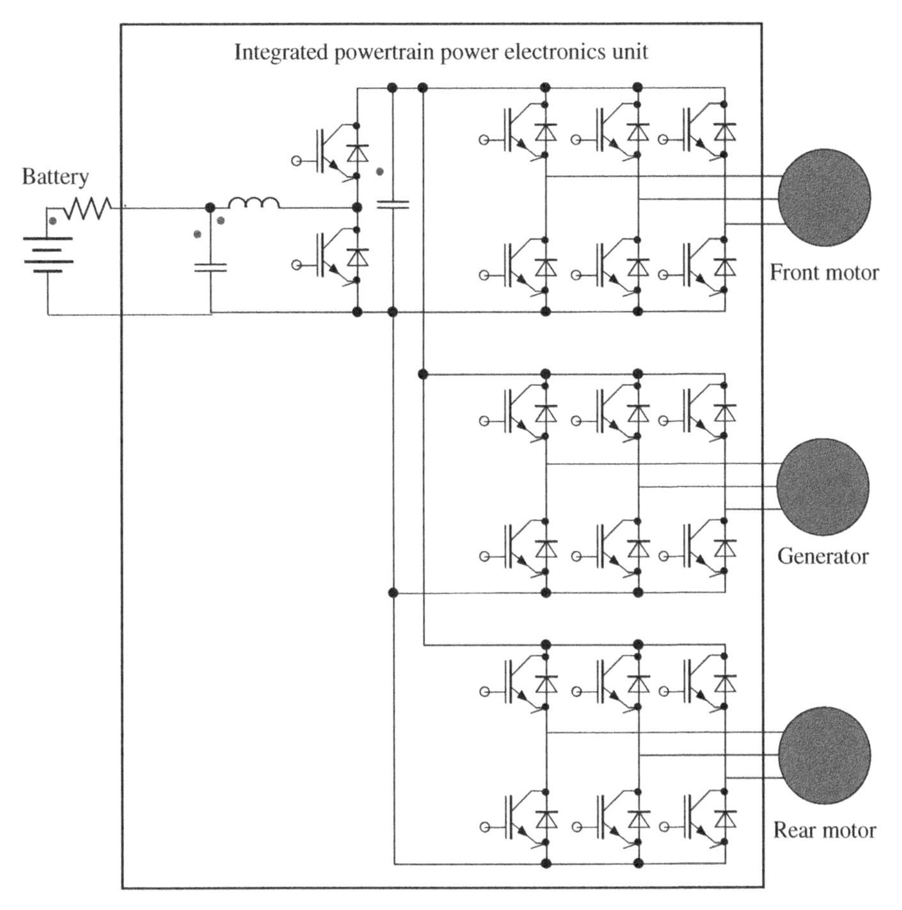

Figure 9.2 Integrated powertrain power electronics unit used to control the Toyota Highlander hybrid vehicle. Courtesy, CC Chan, The State of the Art of Electric, Hybrid, and Fuel Cell Vehicles, the IEEE Special Issue on Electric, Hybrid and Fuel Cell Vehicles, vol. 95, no. 4, pp. 704–718, 2007 © [2007] IEEE. Reprinted, with permission, from the Proceedings of the IEEE.

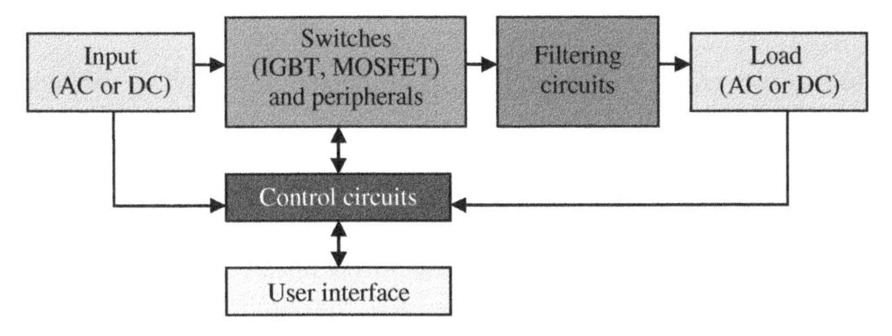

Figure 9.3 Schematics of a power converter.

As can be seen from Figure 9.3, a typical power converter will consist of four segments: switching and peripheral circuits, filtering circuits, control circuits and feedback, and an optional user interface:

- **Types of power converters:** Power converters are usually classified by their input and output. Since the input and output of a power converter can be either AC or DC, there can be four types of power converters:
  - DC–DC converter
  - DC–AC inverter
  - AC–DC rectifier
  - AC–AC cycloconverter.

  The first three types of power converters are all used in HEVs. The fourth type, the AC–AC cycloconverter, is only used in high-power AC–AC systems to control the voltage magnitude and frequency of large motors. However, AC–AC conversion involving an AC–DC circuit and a DC–AC circuit is not unusual. Depending on the powertrain configuration and the level of hybridization, an HEV can involve one or more power converters of different types.
- **Main circuit (switches, peripherals):** The main circuit consists of power semiconductor devices (switches and diodes) and peripheral circuits. The semiconductor switches are controlled to turn on and turn off at a frequency ranging from a few kilohertz to a few tens or hundreds of kilohertz for HEV applications. Depending on the voltage level of the systems, both MOSFETs and IGBTs are used in HEV power converters.
- **Filtering circuit:** Power electronics converters usually involve LC low-pass filters that will filter out the high-frequency components of the output voltage and let the low-frequency components or DC component pass to the load side.
- **Control and feedback circuit:** Control and feedback typically involve the use of microcontrollers and sensors. HEV powertrain applications usually involve feedback torque control. Current feedback is usually necessary.

As a well-developed discipline, power electronics has been very well covered in many textbooks. This chapter will only focus on the unique aspects of power electronics pertaining to HEV applications.

9.3 Rectifiers Used in HEVs

Rectifiers are used to convert an AC input to a DC output. Even though controlled rectifiers exist, they are rarely used in automotive applications. Therefore, we will only discuss uncontrolled passive rectifiers and their unique aspects in HEV applications.

9.3.1 Ideal Rectifier

Figure 9.4 shows a single-phase rectifier and a three-phase ideal rectifier operating from ideal voltage sources. With ideal diode characteristics, the output of a single-phase rectifier can be expressed as.

$$V_o = \frac{1}{T/2}\int_0^{T/2} \sqrt{2}V_i \sin(\omega t)dt = 0.9V_i \tag{9.1}$$

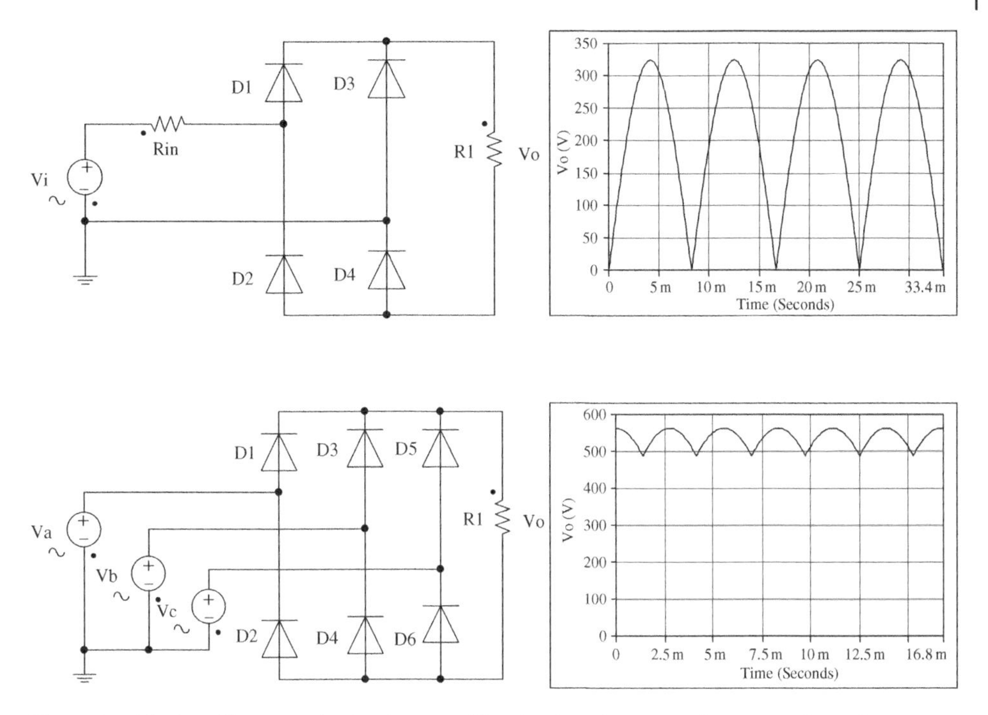

Figure 9.4 Ideal rectifiers: top, single-phase rectifier circuit and output voltage waveforms; bottom, three-phase rectifier circuit and output voltage waveforms.

where V_o is the output voltage, V_i is the rms value of the input voltage, T is the period of the input voltage, and ω is the angular frequency of the input.

The output of an ideal three-phase rectifier is

$$V_o = \frac{1}{\pi/3}\int_{-\pi/6}^{\pi/6}\left(V_a - V_b\right)dt = \frac{1}{\pi/3}\int_{-\pi/6}^{\pi/6}\sqrt{2}V_{LL}\ \cos(\omega t)dt = 1.35V_{LL} \tag{9.2}$$

where V_{LL} is the rms value of line-to-line voltage.

9.3.2 Practical Rectifier

In HEV applications, the input to a rectifier is usually the output of a synchronous generator (such as in a series HEV or complex HEV), or an alternator (in a belt–alternator–starter HEV). The circuit and output voltage waveforms of a practical HEV rectifier are shown in Figure 9.5. The generator impedance is in series with the voltage source, and the voltage drop of the diodes is also included. It can be seen that there is a significant amount of voltage drop in a practical rectifier when compared to an ideal rectifier. The voltage drop is caused by the impedance of the generator, which is generally not negligible, different from that of rectifiers connected to an infinite AC grid. Besides, there will be commutation loss due to the inductance of the generator. Therefore, the output voltage can be significantly different between no-load and loaded conditions. The difference is defined as voltage regulation.

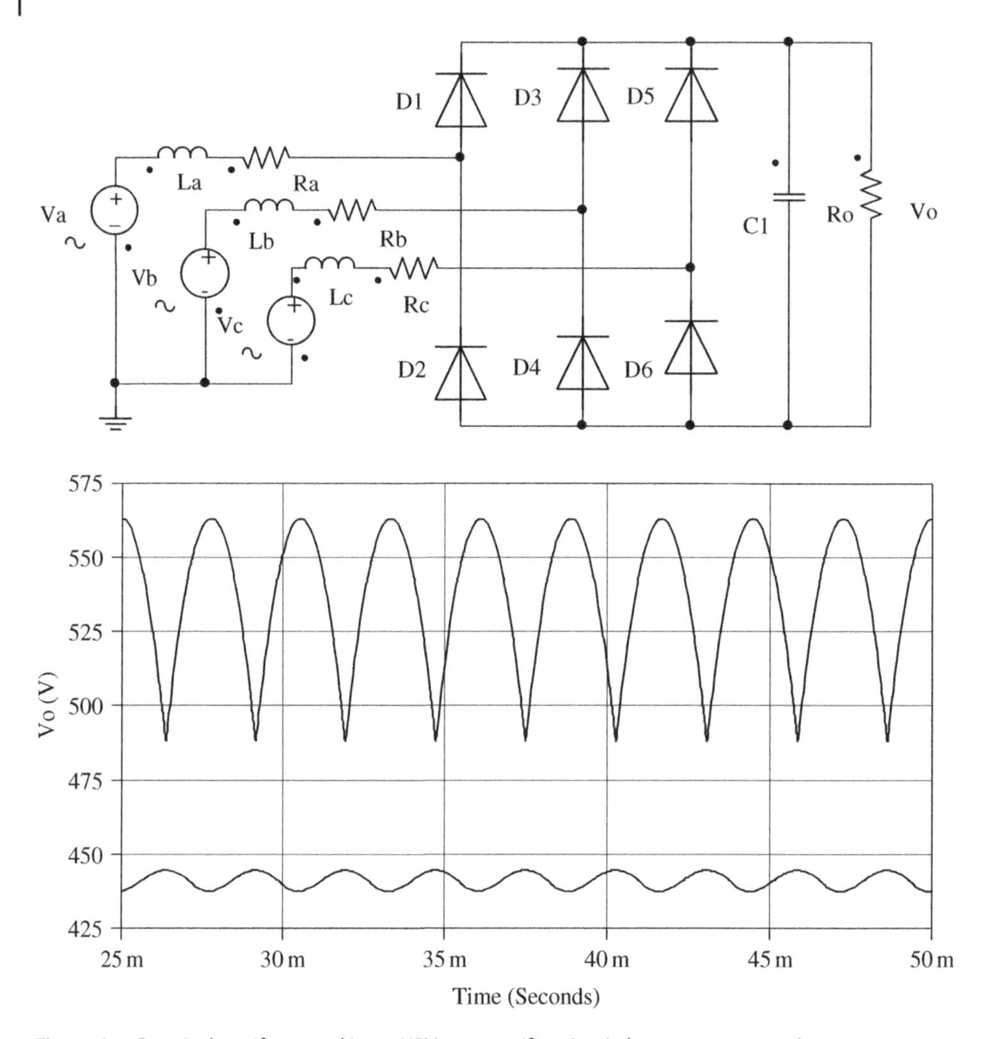

Figure 9.5 Practical rectifiers used in an HEV: top, rectifier circuit; bottom, output voltage in comparison to ideal rectifier (upper curve, ideal rectifier; lower curve, practical rectifier).

9.3.3 Single-Phase Rectifier

We will use a single-phase circuit to analyze the voltage regulation, voltage ripple, and commutation.

At no-load conditions, due to the existence of the output capacitor, the output voltage will equal the peak of the input voltage, that is,

$$V_o = \sqrt{2}V_a = 1.414V_a \tag{9.3}$$

When load current increases, the impedance of the generator, and the diodes, will each have a voltage drop on them. If the DC link capacitor is sufficiently large, then we can assume that the output voltage V_1 (DC link voltage) is constant. We further assume that diode voltage drop V_D is also a constant.

By solving $\sqrt{2}V_a \sin\omega t = 2V_D + V_1$, we get

$$\theta_1 = \omega t_0 = \arcsin\left(\frac{2V_D + V_1}{\sqrt{2}V_a}\right) \tag{9.4}$$

The analysis can be divided into two scenarios: discontinuous and continuous mode. In discontinuous mode, the AC side current is not continuous. The current starts from zero and builds up when $\omega t \geq \theta_1$; it reaches a maximum when

$$\sqrt{2}V_i \sin\omega t = 2V_D + V_o \quad (\omega t = \pi - \theta_1) \tag{9.5}$$

The current then drops to zero at θ_2, $\theta_2 < \pi + \theta_1$.

Continuous Mode: In continuous mode, the AC side current does not reach 0 at $\pi + \theta_1$. In other words, $\theta_2 > \pi + \theta_1$.

In continuous mode, the voltage equation is

$$V_i - L_a\frac{di}{dt} - R_a i - 2V_D = V_1, \quad i(t_0) = 0 \tag{9.6}$$

$$\frac{di}{dt} + \frac{R_a}{L_a}i = \frac{1}{L_a}\left(\sqrt{2}V_a \sin\omega t - 2V_D - V_1\right) \quad \text{when } \omega t \geq \theta_1 \tag{9.7}$$

Note that the AC input will not have current until the voltage exceeds the output voltage plus the diode drop. However, the current will continue to flow until it reaches zero at angle θ_2.

We further neglect the resistance. The above differential equation can be simplified and the following solution obtained:

$$i(t) = -\frac{\sqrt{2}V_a}{\omega L_a}\cos\omega t - \frac{2V_D + V_1}{\omega L_a}\omega t + C, \quad \theta_1 \leq \omega t \leq \theta_2 \tag{9.8}$$

Since $i(\theta_1) = 0$, from the above equation we get

$$C = \frac{\sqrt{2}V_a}{\omega L_a}\cos\theta_1 + \frac{2V_D + V_1}{\omega L_a}\theta_1 \tag{9.9}$$

Therefore,

$$i(t) = -\frac{\sqrt{2}V_a}{\omega L_a}\left(\cos\omega t - \cos\theta_1\right) - \frac{2V_D + V_1}{\omega L_a}\left(\omega t - \theta_1\right), \quad \theta_1 \leq \omega t \leq \theta_2 \tag{9.10}$$

To find θ_2,

$$\frac{1}{\omega L_a}\int_{\theta_1}^{\theta_2}\left(\sqrt{2}V_a \sin\omega t - 2V_D - V_1\right)d\omega t = 0 \tag{9.11}$$

$$\cos\theta_2 + \frac{2V_D + V_1}{\sqrt{2}V_a}\theta_2 = \cos\theta_1 + \frac{2V_D + V_1}{\sqrt{2}V_a}\theta_1 \tag{9.12}$$

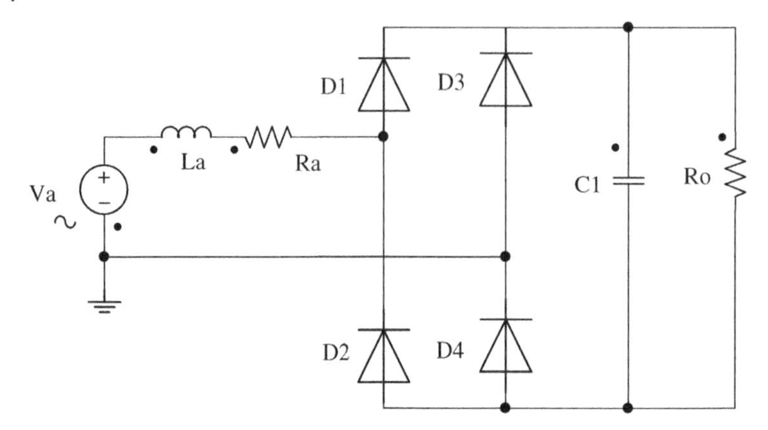

Figure 9.6 Voltage regulation, commutation of practical rectifier used in HEV.

The output power of the rectifier must be equal to the power consumed by the load:

$$P=\frac{1}{\pi}\int_{\theta_1}^{\theta_2}\sqrt{2}V_a\sin\omega t^{*}i(t)d(\omega t)=\frac{V_1^2}{R} \tag{9.13}$$

so the output voltage of the rectifier is

$$V_o=\sqrt{\frac{R}{\pi}\int_{\theta_1}^{\theta_2}\sqrt{2}V_a\sin\omega t^{*}i(t)d(\omega t)} \tag{9.14}$$

The above expression cannot be directly solved due to the fact that θ_1 and θ_2 are functions of V_o. But it can be seen that the rectifier output is closely related to the impedance of the generator.

The voltage regulation can then be calculated from Figure 9.6. That is,

$$\Delta V_o=\frac{V_1-V_o}{V_o}\times 100\% \tag{9.15}$$

Apparently, the voltage regulation is a function of the internal impedance of the generator, and the output power.

9.3.4 Voltage Ripple

The above derivation assumes that the output voltage is constant. However, it can be seen that the current from the AC input is discontinuous. This means that, during the portion of the cycle when there is no current from the AC side (e.g., from 0 to θ_1, and from θ_2 to π if $\theta_2<\pi$), the load current is supplied by the capacitor.

However, due to the nonlinearity of the current, when the load current is less than the AC side current, the capacitor still has to supply some current to the load. Therefore, we can assume that if the load current is constant, and also that the capacitor supplies current to the load 50% of the time, then the voltage ripple is

$$\Delta V_o=\frac{1}{2}\frac{1}{C}\frac{V_o}{R}\frac{\pi}{\omega} \tag{9.16}$$

Example

In Figure 9.6, $V_a = 220\,V$, 400 Hz, $L_a = 1\,mH$, $R_a = 0.05$ (which can be neglected in the analytical calculations), $C = 10\,mF$, and $V_D = 0.8\,V$. The DC output is equivalent to $10\,\Omega$. Find the output voltage and voltage ripple.

Solution

V_o can be solved using the above equations and is illustrated in Figure 9.7a. It is then solved in MATLAB (Figure 9.7b).

The solution is $V_o = 206\,V$. The calculated voltage ripple is 1.3 V.

Simulation: The same circuit is further simulated in Simplorer. The output voltage is 206.3 V, and the voltage ripple is 1 V.

Discontinuous mode: In discontinuous mode, the AC side current starts at $\omega t = \theta_1 > \theta_2$, and drops to 0 at $\omega t > \pi + \theta_1$:

$$\frac{di}{dt} + \frac{R_a}{L_a} i = \frac{1}{L_a}\left(\sqrt{2}V_a \sin\omega t - 2V_D - V_1\right), \quad \theta_2 \le \omega t \le \theta_2 + \pi \tag{9.17}$$

and

$$i(\pi - \theta_2) = i(\theta_2) = 0 \tag{9.18}$$

(a)

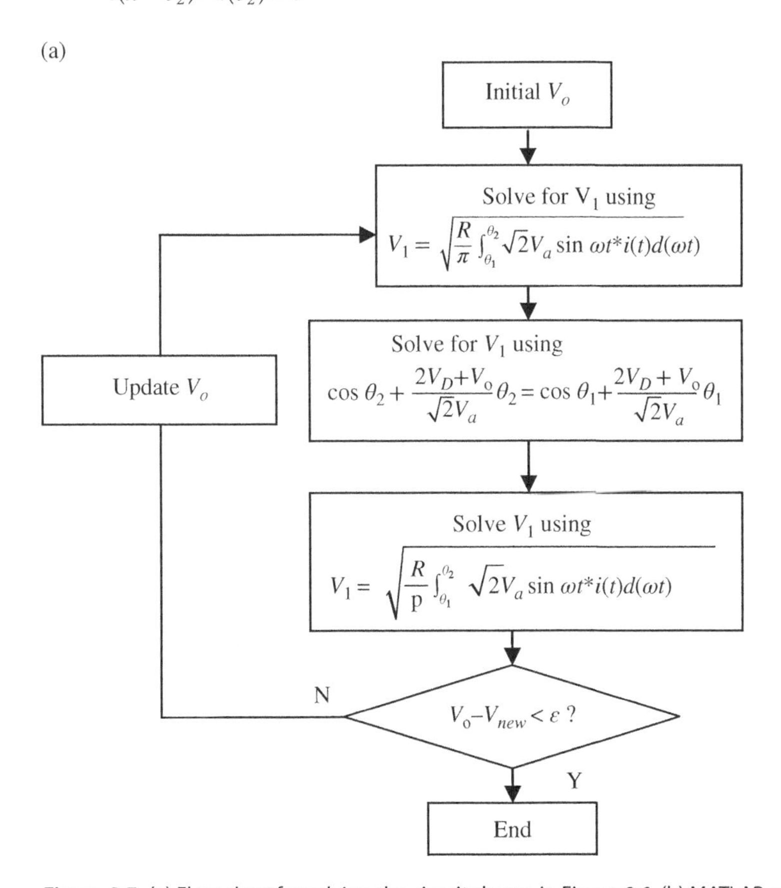

Figure 9.7 (a) Flow chart for solving the circuit shown in Figure 9.6. (b) MATLAB code corresponding to the flow chart in (a).

(b)

```
% Matlab Program to solve for Vo.
clear all
syms wt i v y
delta_error=0.01;
voltage_error=0.1;

VD=0.8;
Va=220*sqrt(2);
w=2*pi*400;
La=1e-3;
RLoad=10;
Ra=0.001;
RD=0.001;
R=RLoad+2*RD+Ra;
x=w*La;

%Initialization
Vo=Va; V1=0;

%Solve for Vo
while abs(V1-Vo)>voltage_error
    Vo=(Vo+V1)/2;
    c0=(2*VD+Vo)/Va;
    theta1=asin(c0);
    c1=cos(theta1)+ c0*theta1;

    %find theta2
    theta2 = pi - theta1;
    c2  = cos(theta2)+ c0*(theta2);
    while abs((c2-c1))>delta_error
        theta2=theta2+0.001;
        c2  = cos(theta2)+ c0*(theta2);
        if theta2 >= pi + theta1 break, end
    end

    i = -Va/x*(cos(wt)-cos(theta1)) -
(2*VD+Vo)/x*(wt-theta1);
    v = Va*sin(wt);
    y = simple(i*v);
    P1 = 1/pi*int(y, wt, theta1, theta2);
    P = abs(real(eval(P1)));
    V1 = real(sqrt(P*R))
    if V1>=Va V1=Va; end
end
Io=P/Vo;
Vo = Vo - (RD*2*Io + 2*VD)
DeltaV=1/C*Vo/RLoad*(theta1+pi-theta2)/w
```

Figure 9.7 (Continued)

When neglecting R_a,

$$i(t) = -\frac{\sqrt{2}V_a}{\omega L_a}(\cos\omega t - \cos\theta_2) - \frac{2V_D + V_1}{\omega L_a}(\omega t - \theta_2), \quad \theta_2 \le \omega t \le \pi + \theta_2 \tag{9.19}$$

To find θ_2, let

$$\frac{1}{\omega L_a}\int_{\theta_2}^{\pi+\theta_2}\left(\sqrt{2}V_a \sin\omega t - 2V_D - V_1\right)d\omega t = 0 \tag{9.20}$$

$$\theta_2 = \arccos\left[\frac{\pi(2V_D + V_1)}{2\sqrt{2}V_a}\right] \tag{9.21}$$

The boundary condition is $\theta_1 = \theta_2$. Therefore, the boundary condition occurs when

$$2V_D + V_1 = \frac{2\sqrt{2}V_a}{\sqrt{\pi^2 + 4}} \tag{9.22}$$

The above analysis is based on a single-phase generator. Since most generators and motors are three-phase, it is worth looking at three-phase circuits.

Again, if the output capacitor is sufficiently large, then we can assume that the output voltage is constant. The diodes only conduct 60° in each cycle. The voltage equation can be written as

$$V_{LL} - 2L_a\frac{di}{dt} - 2R_a i - 2V_D = V_1, \quad -\frac{\pi}{6} \le \omega t \le \frac{\pi}{6} \tag{9.23}$$

where V_{LL} is the line–line voltage, and V_1 is the DC link voltage with load. This equation can be solved using the same method as for the single-phase analysis.

If the generator is a three-phase salient-pole permanent magnet (PM) generator (such as an interior-type PM generator), then the circuit is even more complicated due to the fact that the generator has two equivalent inductances, direct-axis and quadrature-axis inductance.

9.4 Buck Converter Used in HEVs

9.4.1 Operating Principle

A buck converter will step down a higher voltage DC input to a lower voltage DC output. The typical application of a buck converter in an HEV is to step down the hybrid battery voltage (typically 200–400 V) to charge the auxiliary battery (14 V). The uniqueness is the large difference between the input and output voltage of the converter and the small duty ratio (3.5%) needed to control the switching. Figure 9.8 shows the main circuit of a buck converter. It consists of a switch, a freewheeling diode, and an LC filter.

The small duty ratio will make control and regulation very difficult. It also affects the design of the inductor, capacitor, current ripple, and voltage ripple. As a starting point

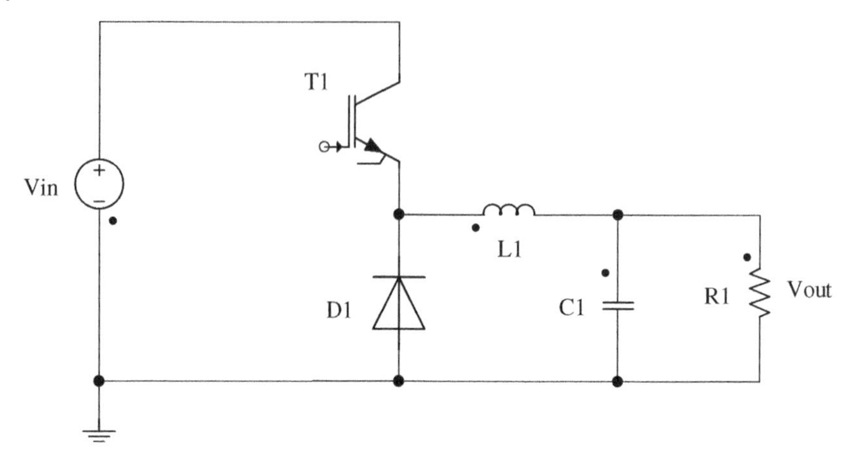

Figure 9.8 Buck converter circuit diagram.

for the analysis, we assume that the components are ideal, that is, the voltage drop is zero when turned on. We further assume that the output voltage is constant. When the switch is turned on, the voltage across the inductor is

$$V_L = V_d - V_o \tag{9.24}$$

When the switch is turned off, we assume the circuit is operating in continuous mode; then the voltage drop across the inductor is

$$V_L = -V_o \tag{9.25}$$

In steady state operations, the average voltage of the inductor must be zero. Therefore,

$$(V_d - V_o)DT_s = V_o(1-D)T_s \tag{9.26}$$

and

$$V_o = DV_d \tag{9.27}$$

The current ripple in the inductor is

$$\Delta I_L = \frac{1}{L}V_o(1-D)T_s \tag{9.28}$$

The voltage ripple of the output can be calculated. We assume the load current is constant; then all the current ripple will enter the capacitor,

$$\Delta V_o = \frac{1}{C}\frac{1}{2}\frac{\Delta I_L}{2}\frac{T_s}{2} = \frac{T_s^2}{8LC}V_o(1-D) \tag{9.29}$$

9.4.2 Nonlinear Model

The above analysis is based on assumptions that the output voltage is relatively constant and the parasitic parameters (resistance, inductance) can be neglected. Due to the fact that the ratio between the input and output voltages is very large, these assumptions may not hold true.

In order to accurately analyze the relationship and the influence of various parameters, we can use a detailed model to describe the system. In continuous mode, when the switch is turned on,

$$V_d = r_d\ i_L + r_s\ i_L + L\frac{di_L}{dt} + r_L\ i_L + V_o \tag{9.30}$$

$$i_L = i_c + i_o = C\frac{dV_o}{dt} + i_o = C\frac{dV_o}{dt} + \frac{V_o}{R} \tag{9.31}$$

where r_d, r_s, and r_L are the equivalent resistance of the diode, switch, and inductor, respectively, and i_o, i_c, and i_L are the current through the load resistance, capacitor, and inductor, respectively.

The above equations can be rewritten as

$$\begin{bmatrix} \frac{di_L}{dt} \\ \frac{dV_o}{dt} \end{bmatrix} = \begin{bmatrix} -\frac{r_d + r_s + r_L}{L} & -\frac{1}{L} \\ \frac{1}{C} & -\frac{1}{CR} \end{bmatrix} \begin{bmatrix} i_L \\ V_o \end{bmatrix} + \begin{bmatrix} \frac{1}{L} \\ 0 \end{bmatrix} V_d \tag{9.32}$$

When the switch is closed,

$$\begin{bmatrix} \frac{di_L}{dt} \\ \frac{dV_o}{dt} \end{bmatrix} = \begin{bmatrix} -\frac{r_D + r_L}{L} & -\frac{1}{L} \\ \frac{1}{C} & -\frac{1}{CR} \end{bmatrix} \begin{bmatrix} i_L \\ V_o \end{bmatrix} \tag{9.33}$$

These equations can be solved using numeric tools such as MATLAB.

9.5 Non-Isolated Bidirectional DC–DC Converter

The bidirectional DC–DC converter in an HEV is also sometimes called an energy management converter, or boost DC–DC converter. This DC–DC converter is a high-power converter that links the HV battery at a lower voltage with the HV DC bus. The typical voltage of a battery pack is designed at 300–400 V. The best operating voltage for a motor and inverter is around 600 V. Therefore, this converter can be used to match the voltages of the battery system and the motor system. Other functions of this DC–DC converter include optimizing the operation of the powertrain system, reducing ripple current in the battery, and maintaining DC link voltage, hence the high-power operation of the powertrain.

9.5.1 Operating Principle

The DC–DC converter provides bidirectional power transfer. The operating principle is show in Figure 9.9.

- **Buck operation:** In buck operation as shown in Figure 9.9b, the power is transferred from V_d to V_B. When T_1 is closed and T_2 is open, since $V_d > V_B$, $V_L = V_d - V_B$ and the inductor current I_L builds up. When T_1 is open, the inductor current I_L continues to flow through D_2. Thus, $V_L = V_B$.

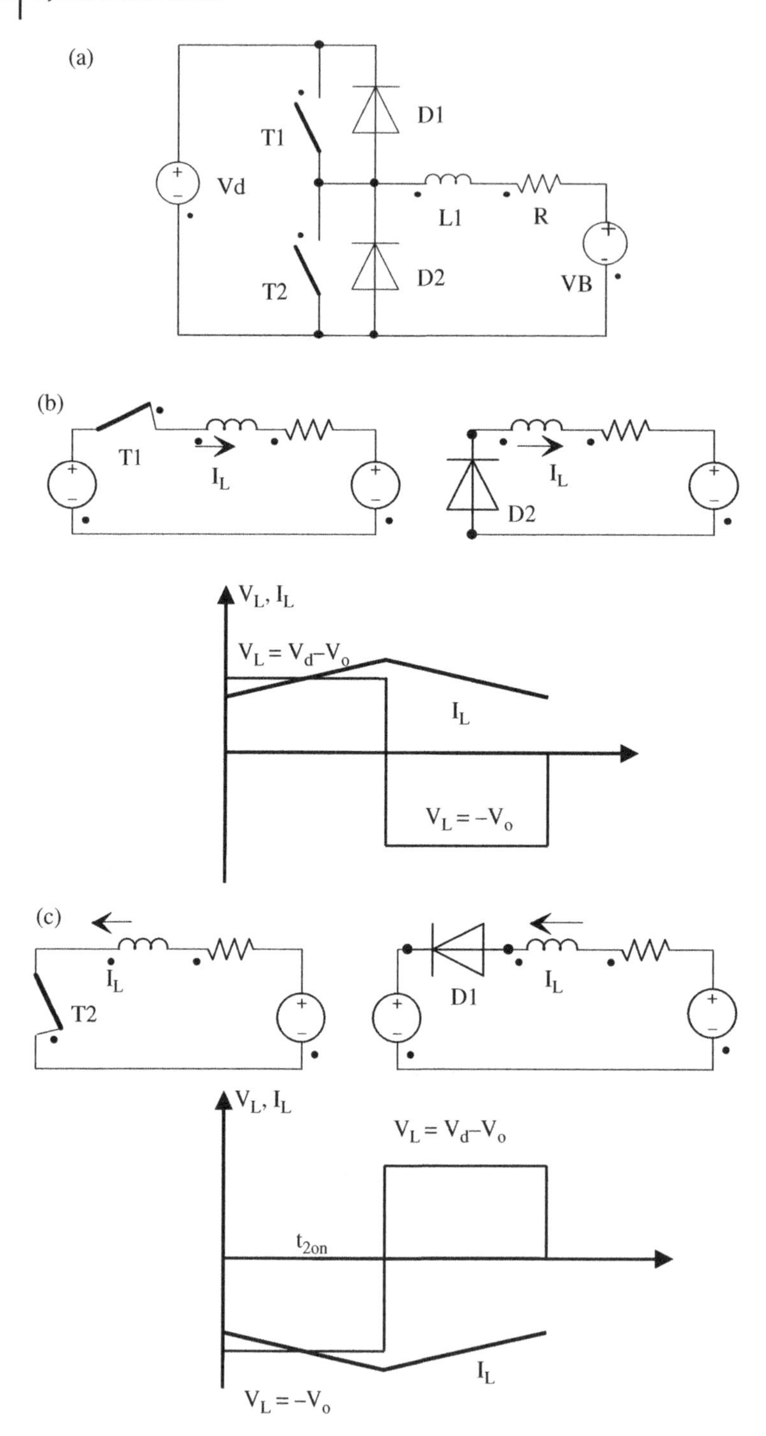

Figure 9.9 Operation of the bidirectional boost converter: (a) circuit topology; (b) inductor voltage and current waveform during buck operation; and (c) inductor voltage and current waveform during boost operation.

Assuming ideal components and a constant V_o, the inductor current over one cycle in steady state operation will remain the same, for example,

$$\int_0^{t_{1on}} (V_d - V_o)dt = \int_{t_{1on}}^{t_{1on}+t_{1off}} (-V_o)dt \tag{9.34}$$

$$V_o = \frac{t_{1on}}{T} V_d = D_1 \ V_d \tag{9.35}$$

where D_1 is the duty ratio defined as the percentage of on-time of switch T_1:

$$D_1 = \frac{t_{1on}}{T} \tag{9.36}$$

- **Boost operation:** In boost operation, the power is transferred from V_B to V_d. When T_2 is closed and T_1 is open, V_B and the inductor form a short circuit through switch T_2, as shown in Figure 9.9c, therefore $V_L = V_B$ and the inductor current I_L builds up. When T_2 is open, the inductor current continues to flow through D_1 to V_d, therefore $V_L = V_d - V_B$:

$$\int_0^{t_{2on}} V_o dt = \int_{t_{2on}}^{t_{2on}+t_{2off}} (V_d - V_o)dt \tag{9.37}$$

$$V_d = \frac{1}{1-D_2} V_o \tag{9.38}$$

where D_2 is the duty ratio defined as the percentage of on-time of switch T_2:

$$D_2 = \frac{t_{2on}}{T} \tag{9.39}$$

In the bidirectional boost converter control, since T_1 and T_2 cannot be switched on simultaneously, a practical control strategy is to turn T_2 off while T_1 is on and vice versa. In this case,

$$D_2 = 1 - D_1 \tag{9.40}$$

9.5.2 Maintaining Constant Torque Range and Power Capability

The above analysis neglected the internal impedance of the battery. In fact, the impedance is often not negligible. When an electric motor and inverter are directly connected to the battery without a bidirectional DC–DC converter as shown in Figure 9.10, as the current (power or torque) goes up, the battery terminal voltage starts to drop because of the voltage drop of the battery internal impedance. For example, a 16 kWh lithium-ion battery with iron phosphate chemistry will have an internal impedance of 0.5 Ω. If the powertrain inverter/motor is rated at 125 kW, 400 V, 90% efficiency, the rated current is 348 A at 400 V. The battery internal voltage drop is 174 V. This voltage drop will significantly affect the performance of the powertrain motors. In fact, in this example, the maximum power that can be delivered to the motor is only 78 kW. In addition, due to the available voltage at the input, the motor constant torque region is also affected. In the above example, the constant torque region is shortened by 43.5%.

Another factor is that battery voltage is related to battery state of charge (SOC). As the battery SOC drops, the battery voltage will also drop. Therefore, the available voltage at a motor/inverter terminal is also changed, which will make it difficult to maintain the constant torque range.

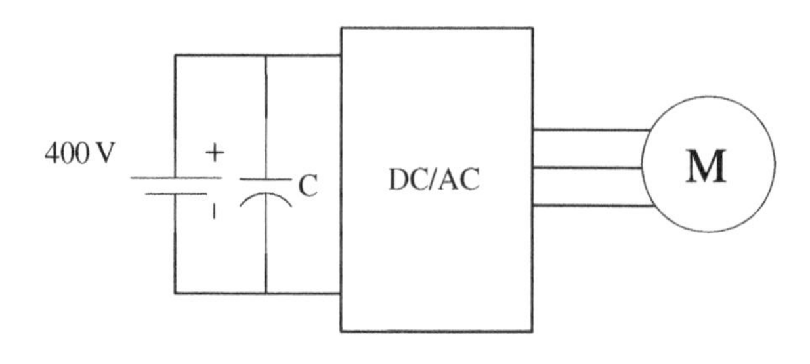

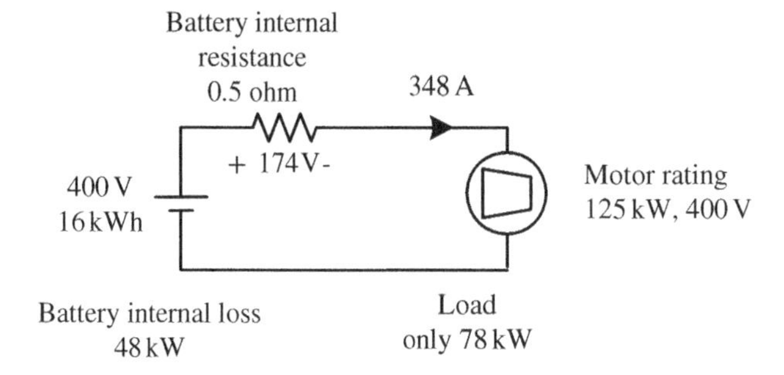

Figure 9.10 Powertrain motor directly connected to battery without the DC–DC converter: top, circuit topology; bottom, equivalent circuit of the circuit.

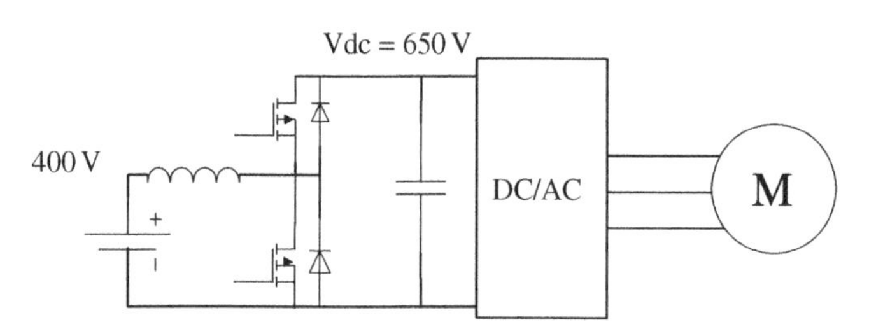

Figure 9.11 Powertrain motor connected to battery through a DC–DC converter.

When a DC–DC converter is inserted between the battery and inverter/motor as shown in Figure 9.11, the DC bus voltage before the inverter can be maintained as a constant. Therefore, the constant torque range will not be affected by the battery SOC or large power drawn by the inverter/motor.

The above analysis assumes that the battery system is designed to handle the large power dissipation during large power draw.

9.5.3 Reducing Current Ripple in the Battery

Due to the switching functions of the inverter used in the powertrain system, there are abundant high-frequency current harmonics on the DC side. The amount of current ripple that goes into/out of the battery depends on the switching methodology, switching frequency, and the capacitance on the DC bus. When there is no DC–DC converter in place, the amount of ripple current of the battery is determined by the DC bus capacitance C and the ratio of capacitor impedance to battery impedance. Without the capacitance, the battery current will be directly determined by the switching status of the DC–AC inverter, that is, the combination of the three-phase current of the motor, as shown in Figure 9.12. When there is a DC bus capacitor in parallel with the battery, the amount of current ripple flowing into/out of the battery is determined by the capacitance and parasitic impedance of the DC bus capacitor. For example,

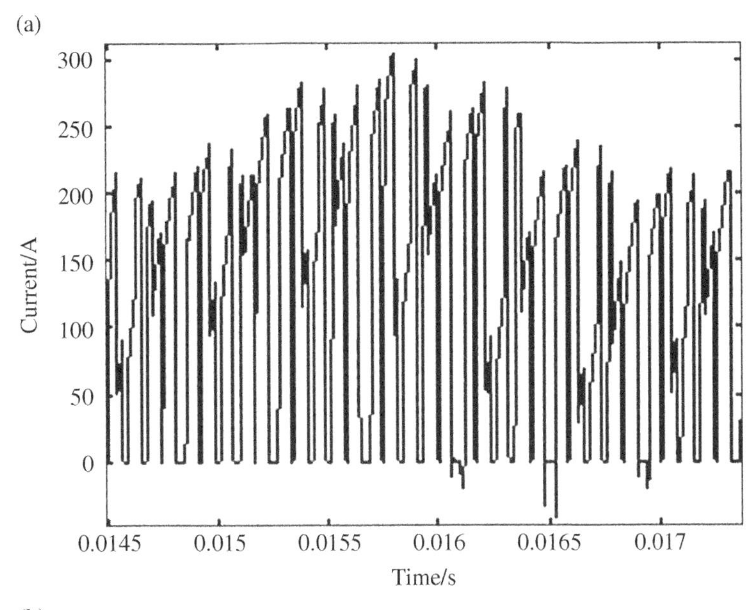

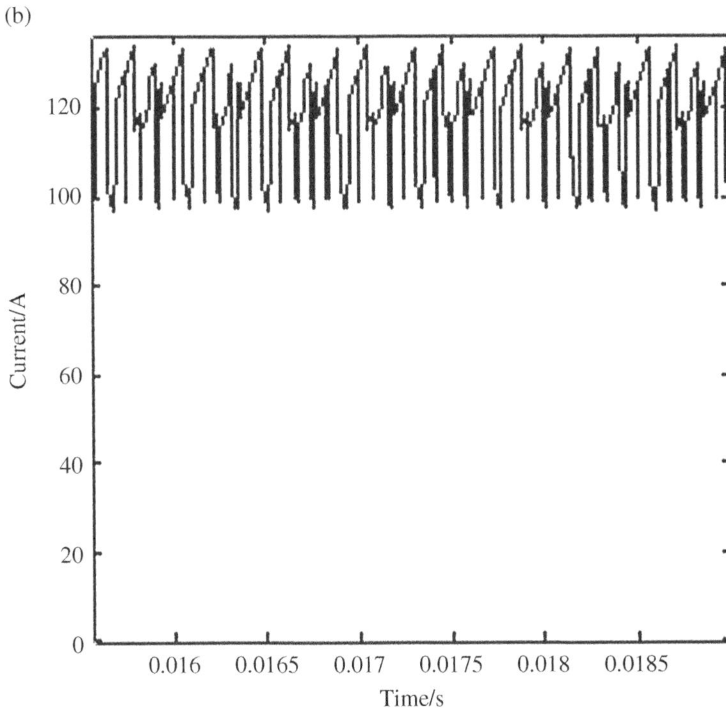

Figure 9.12 Battery current without a DC–DC converter: (a) DC bus current for no DC-bus capacitance; (b) DC bus current for $C = 1\,\text{mF}$, $R_c = 100\,\text{m}\Omega$; (c) DC bus current for $C = 10\,\text{mF}$, $R_c = 100\,\text{m}\Omega$.

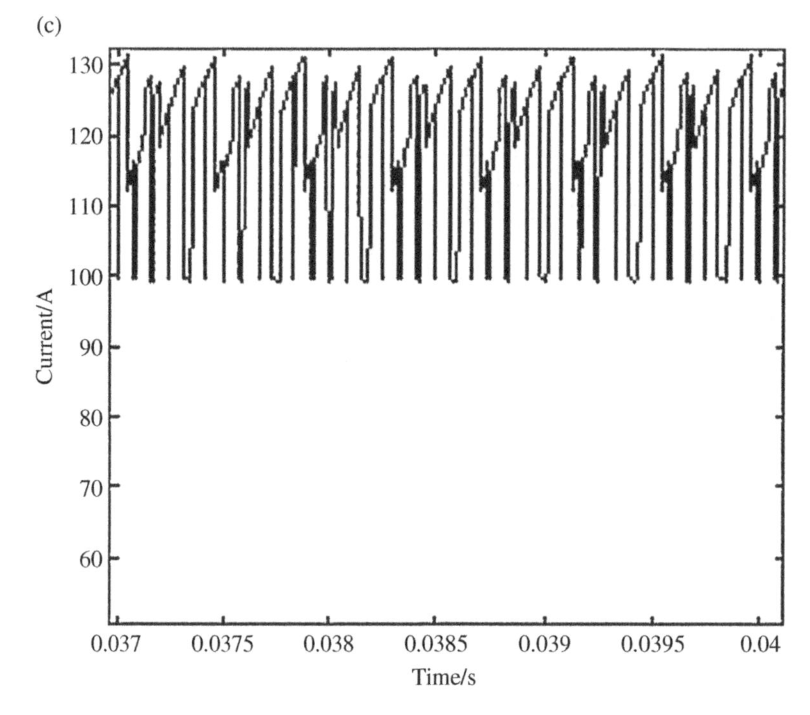

Figure 9.12 (Continued)

if $C = 10\,\text{mF}$, the capacitive impedance of the capacitor at switching frequency is only 2.65 mΩ, which is far less than the internal impedance of the battery. Ideally the high-frequency ripple will flow through the capacitor and the battery current is supposed to be constant.

However, the parasitic resistance of the capacitor is also not negligible. A high-quality 10 mF capacitor has 26 mΩ internal resistance and a second-class capacitor has 100 mΩ. The quality of the capacitor affects the current ripple of the battery. The lower the capacitor impedance, the lower the battery ripple, as shown in Figure 9.12. High-frequency ripple current is believed to be harmful to battery life.

When a DC–DC converter is added, the battery current can be maintained with a relatively small ripple, as shown in Figure 9.13.

9.5.4 Regenerative Braking

The regenerative braking of the two topologies – that is, with and without a DC–DC converter – will also be different. In the topology where there is no DC–DC converter, the DC bus voltage will fluctuate during transition from motoring to braking. For example, if the motor is initially motoring at 50 kW, and the battery internal voltage is 400 V with 0.5 Ω internal resistance, then the battery current is 155 A and the DC bus voltage is 322 V. If the motor is switched to braking at 50 kW, then the battery current is 110 A and the DC bus voltage is 455 V. This dramatic change of DC bus voltage will make motor control, such as vector control, very difficult.

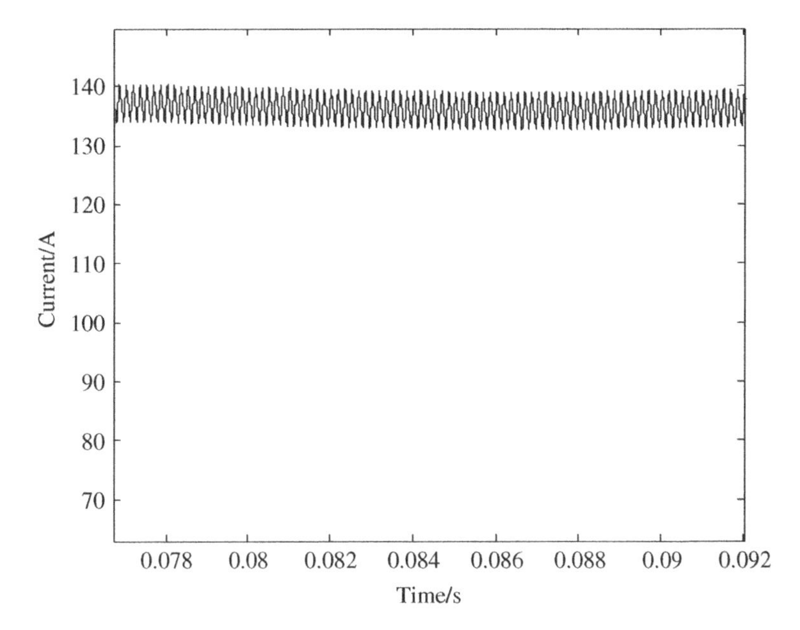

Figure 9.13 Battery current when a DC–DC converter is inserted between the inverter and the battery.

On the other hand, in a system that contains a DC–DC converter between the inverter/motor DC bus and the battery, the DC bus voltage can be maintained relatively constant. Hence, the transition between motoring and braking is easier to handle.

9.6 Voltage Source Inverter

Voltage source inverters (VSIs) are used in hybrid vehicles to control the electric motors and generators. Figure 9.14 shows the power electronic circuit arrangement of a VSI to control the induction motor, PM synchronous motors, and PM brushless motors. The switches are usually IGBTs for high-voltage high-power hybrid configurations or MOSFETs for low voltage designs.

The output of the VSI is controlled by means of pulse width-modulated (PWM) signals to produce sinusoidal waveforms. Certain harmonics exists in such a switching scheme. High switching frequency is used to move away the harmonics from the fundamental frequency.

9.7 Current Source Inverter

Figure 9.15 shows the circuit topology of a current source inverter (CSI). The CSI operates using the same principle as in a VSI, with the input as a current source. Three small commutating/filtering capacitors may be needed on the AC side.

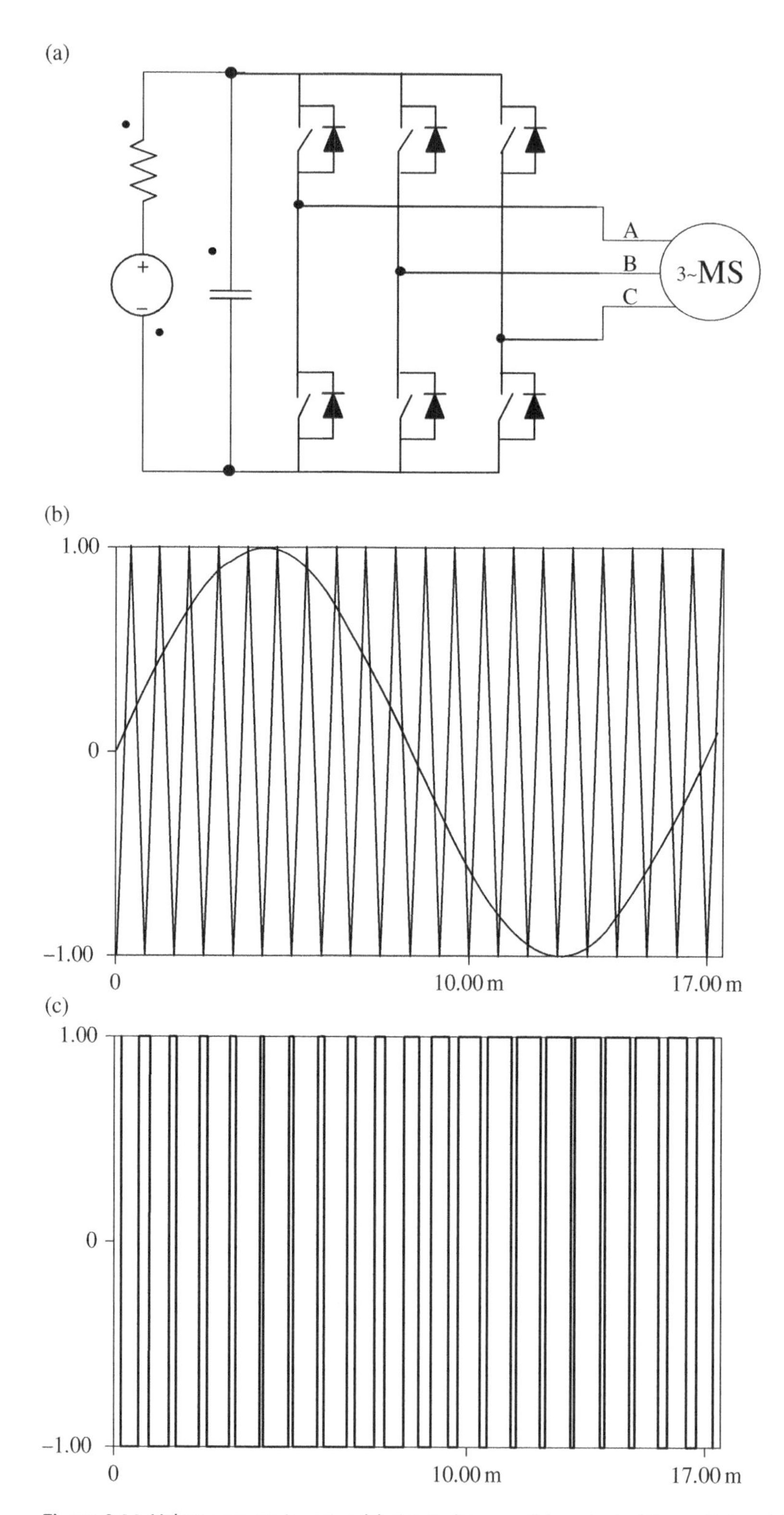

Figure 9.14 Voltage source inverter: (a) circuit diagram; (b) control of the switches; (c) gate control signal via PWM waveform.

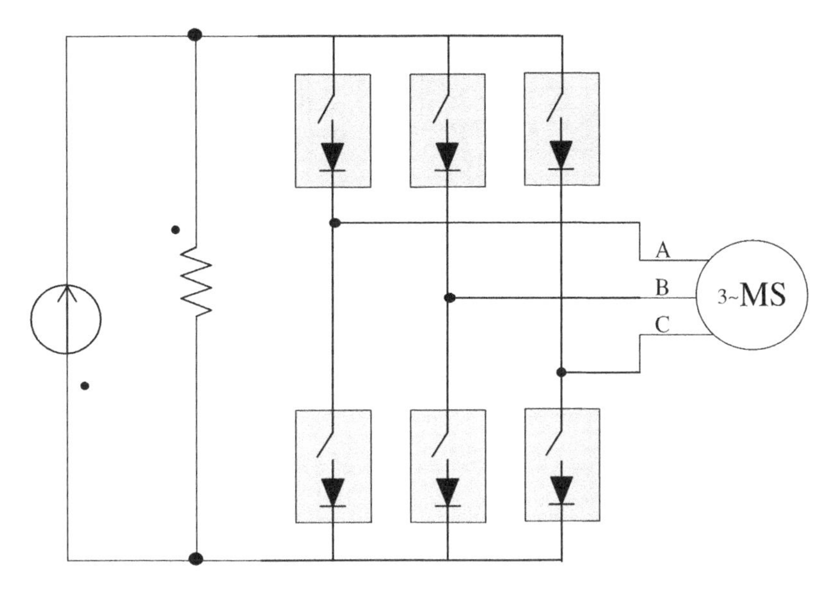

Figure 9.15 Current source inverter.

9.8 Isolated Bidirectional DC–DC Converter

In some applications, galvanic isolation between the battery and the load units is necessary and desirable [23]. Figure 9.16 shows a full bridge isolated bidirectional DC–DC converter.

In Figure 9.16, the primary bridge inverter switches at 20–50 kHz, with 50% duty ratio. The output of the primary is a square wave voltage which is applied to the primary winding of the isolation transformer. The secondary winding of the transformer will therefore have a square wave voltage. Without any control at the gating of the secondary bridge converter, the voltage of the secondary of the transformer is rectified through the four freewheeling diodes. The output voltage will fluctuate with load conditions and the primary voltage.

9.8.1 Basic Principle and Steady State Operations

Steady state operations of isolated bidirectional DC–DC converters have been studied in detail elsewhere [1, 3, 6]. In this section, we complement these studies by distinguishing the operating modes of isolated bidirectional DC–DC converters according to the phase shift angle, load conditions, and output voltage. In this analysis, the dead-band and switching dynamics will be neglected but will be analyzed later.

In the following analysis, the turns ratio of the transformer is n, the transformer primary voltage is V_1, and the switching frequency is f_s. For the convenience of analysis, we define T_s as one half of the switching period, that is, $T_s = 1/(2f_s)$. The duty cycle or phase shift is based on a half period, $D = t_{on}/T_s$. Therefore, DT_s is the phase shift between the two bridges. Further, I_{L_s} is the current of the leakage inductance of the secondary winding. The output voltage of the secondary bridge is V_2.

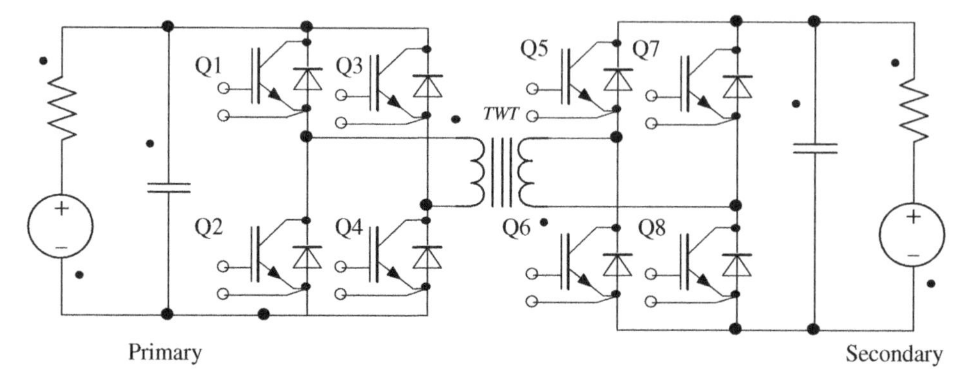

Figure 9.16 Isolated bidirectional DC–DC converter.

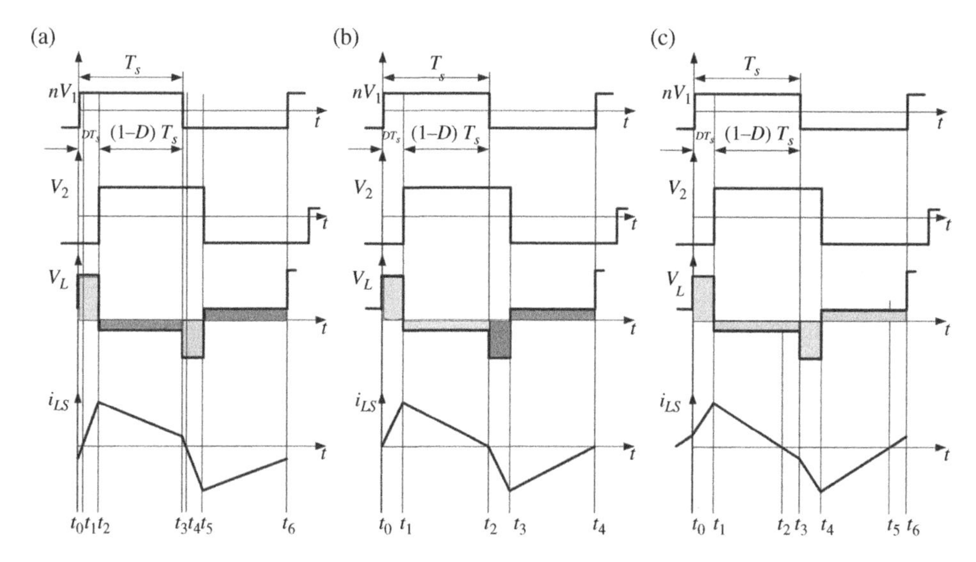

Figure 9.17 Typical voltage and current waveforms for $V_2 > nV_1$: (a) waveforms for $i(t_0) < 0$; (b) waveforms for boundary conditions $i(t_0) = 0$; (c) waveforms for $i(t_0) > 0$.

9.8.1.1 Heavy Load Conditions

There are a number of different operating modes based on the output current with a boundary condition, as illustrated in Figure 9.17. Under heavy load conditions, the inductor current increases from an initial negative value $i(t_0) < 0$ at the beginning of the switching cycle, and reaches a positive value $-i(t_0)$ at the end of the half switching cycle. Six different segments emerge in each switching cycle as shown in Figure 9.17a. In the following analysis, V_2 is assumed to be larger than nV_1.

- **Segment 0**, $[t_0, t_1]$: In this segment, Q_1 and Q_4 of the primary bridge are turned on. Therefore V_1 and nV_1 are positive. Q_6 and Q_7 of the secondary bridge are turned on. Due to the negative current in the inductor, D_6 and D_7 freewheel, and Q_6 and Q_7 do not conduct current. Thus, $V_{L_s} = nV_1 + V_2$. The inductor current increases linearly from a negative value. At t_1, the inductor current reaches 0.

- **Segment 1**, $[t_1, t_2]$: Switches Q_1 and Q_4 of the primary bridge, and Q_6 and Q_7 of the second bridge, are still turned on; $V_{L_s} = nV_1 + V_2$. The current continues to increase except that it becomes positive, that is, Q_6 and Q_7 conduct the current. The total current increment during interval $[t_0, t_2]$ (Segments 0 and 1) is

$$\Delta I_{L_s} = \frac{DT_s}{L_s}(V_2 + nV_1) \tag{9.41}$$

Hence,

$$i(t_2) = i(t_0) + \frac{DT_s}{L_s}(nV_1 + V_2) = I_{max} \tag{9.42}$$

- **Segment 2**, $[t_2, t_3]$: In this segment, switches Q_1 and Q_4 of the primary bridge continue to be turned on, but switches Q_6 and Q_7 are turned off, and switches Q_5 and Q_8 are turned on. Diodes D_5 and D_8 freewheel because the current is positive; $V_{L_s} = nV_1 - V_2 < 0$. The leakage inductor current increment during interval $[t_2, t_3]$ is

$$\Delta I_{L_s} = \frac{(1-D)T_s}{L_s}(nV_1 - V_2) \tag{9.43}$$

Hence,

$$i(t_3) = i(t_2) + \frac{(1-D)T_s}{L_s}(nV_1 - V_2) \tag{9.44}$$

A similar analysis could be done for the following three segments due to the symmetry of operation.

- **Segment 3**, $[t_3, t_4]$: Switches Q_2 and Q_3 of the primary bridge continue to be turned on, and switches Q_5 and Q_8 are turned on. The primary voltage of the transformer, hence the secondary voltage, is reversed and the current decreases from $i(t_3)$ to zero. D_5 and D_8 freewheel.
- **Segment 4**, $[t_4, t_5]$. Switches Q_2 and Q_3 of the primary bridge are turned on, and switches Q_5 and Q_8 are turned on. The current decreases linearly to the negative maximum. Switches Q_5 and Q_8 conduct current. Hence the current increment in L_s in Segments 3 and 4 is

$$\Delta I_{L_s} = -\frac{DT_s}{L_s}(nV_1 + V_2) \tag{9.45}$$

- **Segment 5**, $[t_5, t_6]$: Switches Q_5 and Q_8 are turned off, and D_6 and D_7 begin to freewheel. The current increment in L_s is

$$\Delta I_{L_s} = \frac{(1-D)T_s}{L_s}(V_2 - nV_1) \tag{9.46}$$

From the symmetry of the inductance current, $i(t_0) = -i(t_3)$. From Equations 9.41–9.46, the initial inductor current can be obtained:

$$i(t_0) = \frac{1}{4f_s L_s}[(1-2D)V_2 - nV_1] \tag{9.47}$$

The maximum current is

$$I_{max} = i(t_2) = \frac{1}{4f_s\ L_s}[-(1-2D)nV_1 + V_2] \tag{9.48}$$

The above analysis of operating modes is based on the assumption that $i(t_0) < 0$, that is, $(1-2D)V_2 < nV_1$. If $(1-2D)V_2 = nV_1$, or

$$V_2 = \frac{1}{1-2D}nV_1 \tag{9.49}$$

then $i(t_0) = 0$. This corresponds to the boundary condition as shown in Figure 9.17b which is very similar to the non-isolated boost converter in the steady state operation. Under this circumstance the inductor current increases from zero at the beginning of the switching cycle and drops to zero at T_s.

From Equation 9.49, it seems that V_2 will approach infinity when D reaches 0.5. But it can be seen from Equation 9.47 that when $D = 0.5$, $i(t_0) = -nV_1/4f_sL_s$, that is, $i(t_0)$ will never reach zero for $D = 0.5$. Therefore, the boundary can only be met when D is not equal to 0.5.

9.8.1.2 Light Load Condition

It can be seen from Equation 9.47 that when $(1-2D)V_2 > nV_1$, then $i(t_0) > 0$. This corresponds to light load conditions. The current and voltage waveforms are shown in Figure 9.17c, where the current increases from a positive value at the beginning of the cycle and drops to a negative value at the end of the half switching cycle.

9.8.1.3 Output Voltage

Equations 9.47–9.49 do not give the expression for the output voltage, except for the boundary condition. In order to derive the expression for the output voltage, let us start with the average current in the inductor. The average current of the leakage inductance in the half switching cycle can be derived from Figure 9.17a (note that $i(t_0) < 0$):

$$\bar{I} = \frac{1}{2T_s}[(I_{max} + i(t_0))DT_s + (I_{max} - i(t_0))(1-D)T_s] = \frac{1}{2f_s\ L_s}D(1-D)V_2 \tag{9.50}$$

The supplied power is

$$P_1 = nV_1\bar{I} = \frac{nV_1\ V_2}{2f_s\ L_s}D(1-D) \tag{9.51}$$

Assuming that the load has a fixed resistance, then the output power is

$$P_o = \frac{V_2^2}{R_L} \tag{9.52}$$

Neglecting the transformer and the switch losses, $P_1 = P_o$, then

$$V_2 = \frac{nV_1}{2f_s\ L_s}R_L D(1-D) \tag{9.53}$$

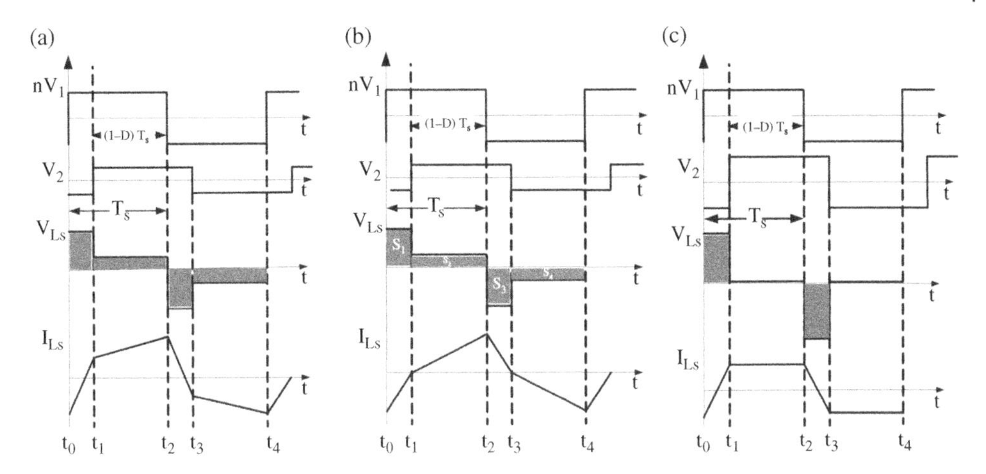

Figure 9.18 Typical voltage and current waveforms for $V_2 < nV_1$ and $V_2 = nV_1$: (a) common mode; (b) boundary mode; (c) extreme mode, $nV_1 = V_2$.

and

$$I_2 = \frac{nV_1}{2f_s\ L_s} D(1-D) \tag{9.54}$$

Equation 9.53 shows that, for a given switching frequency, leakage inductance, and input voltage, the output voltage is proportional to the load resistance and is a function of the duty ratio (phase shift angle). For a given load resistance, the output voltage varies with duty ratio, and reaches a maximum when $D = 0.5$. For a given duty ratio, the output voltage is directly proportional to the load resistance. Therefore, for a given phase shift angle, under heavy load conditions, or

$$R_L < \frac{2f_s\ L_s}{D(1-D)} \tag{9.55}$$

V_2 will be less than nV_1.

When V_2 drops to less than nV_1, the initial inductor current is confined to a negative value and the current waveforms are also different from the operations under the condition of $V_2 > nV_1$, as shown in Figure 9.18. The boundary condition occurs when $S_1 = S_2 = S_3 = S_4$ in Figure 9.18b, or $(nV_1 + V_2)DT_s = (nV_1 - V_2)(1 - D)T_s$. Therefore, at the boundary condition,

$$V_2 = nV_1(1-2D) \tag{9.56}$$

An extreme mode emerges when $nV_1 = V_2$ as shown in Figure 9.18c, where the inductor current will remain constant during the time interval $[DT_s, T_s]$.

It can also been seen from Equation 9.54 that the output current is proportional to the duty ratio (phase shift angle). This may be used to analyze conditions where the output voltage needs to be maintained as constant, and the current can be controlled through the phase shift angle.

9.8.1.4 Output Power

Substituting Equation 9.53 into Equation 9.51, the output power P_o can be obtained as

$$P_o = \left(\frac{nV_1}{2f_s\ L_s}\right)^2 D^2\ (1-D)^2\ R_L \tag{9.57}$$

From Equations 9.51 and 9.57, the output power always has maximum values when and only when $D = ½$, whether for a fixed resistance R_L or a constant output voltage V_2.

When the system is in the voltage closed-loop control mode, that is, V_2 = constant, then decreasing R_L will result in the output power increasing since the output power is inversely proportional to R_L as shown in Equation 9.52. To maintain V_2 as a constant, it can be seen from Equation 9.53 that D has to be adjusted accordingly: $D(1-D) \propto 1/R_L$ for a given V_1. For fixed V_1 and V_2, the output power reaches a maximum at $D = 0.5$ as shown in 9.51. At this condition, $R_L = R_C = 8f_sL_sV_2/(nV_1)$.

Further decrease of $R_L < R_C$ will result in the output voltage of the system collapsing and entering the open loop at $D = 0.5$, that is, V_2 cannot be maintained as a constant. Output voltage V_2 and output power P_o will decrease, and are both proportional to R_L as shown in Equations 9.53 and 9.57.

Consider Equations 9.47 and 9.57. When $D \in (0, 1/2]$, if V_1, V_2 are kept constant (closed-loop control), with the increase of D, the power will increase accompanying the decrease of the initial current $i(t_0)$. Hence the absolute value and the sign of $i(t_0)$ represent the load condition. The lower the inductor current $i(t_0)$, the higher the output power.

Under open-loop control, that is, for a given D, with the increase of load resistance, V_2 and P_o will increase according to Equations 9.53 and 9.57. At the same time, the initial current increases according to Equation 9.47 due to the increase of V_2. Hence the relations between the initial current and output power are very different in the open-loop and closed-loop operations of the converter.

In order to validate the theoretical analysis, a simulation model was set up with $V_1 = 200\,\text{V}$, $n = 2$, and $L_s = 120\,\mu\text{H}$. In Figure 9.19a,b, with the increase of power, the initial current decreases from a positive value to a negative value. In Figure 9.19c,d, the variation of initial current is the opposite. This result is in good agreement with the earlier analysis.

The maximum output power is also a function of L_s and f_s. Increasing the leakage inductance L_s will decrease the current impact to IGBTs but will reduce the capability of the maximum output power of the converter as shown in Equation 9.57. This will be discussed later in this section.

9.8.2 Voltage Ripple

The above switching modes are inherently nonlinear, which will cause voltage ripple on the output capacitor. The capacitor current is a combination of the inductor current I_L and the load current I_o at different operating modes. From Figure 9.17, the capacitor current can be written as

$$\begin{aligned} i_C &= -(I_o + i_L), \quad 0 \le t \le DT_s \\ i_C &= i_L - I_o, \quad DT_s \le t \le T_s \end{aligned} \tag{9.58}$$

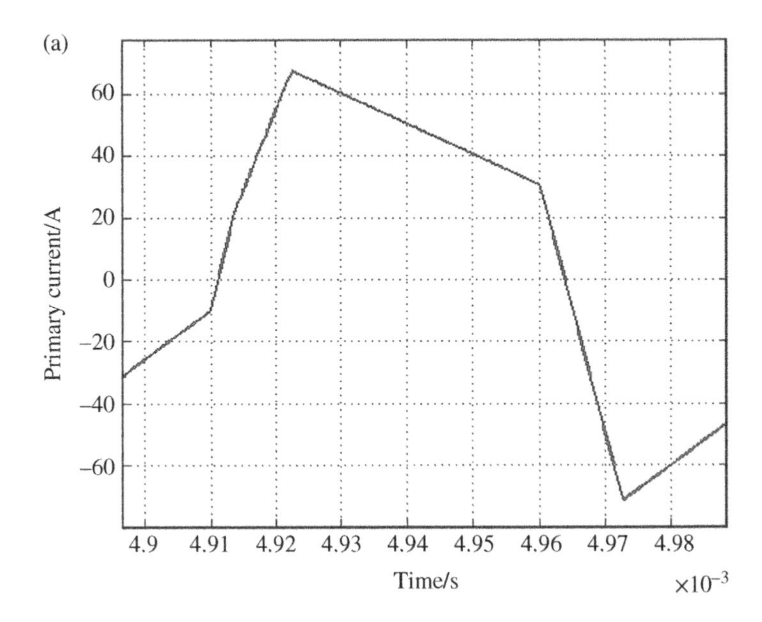

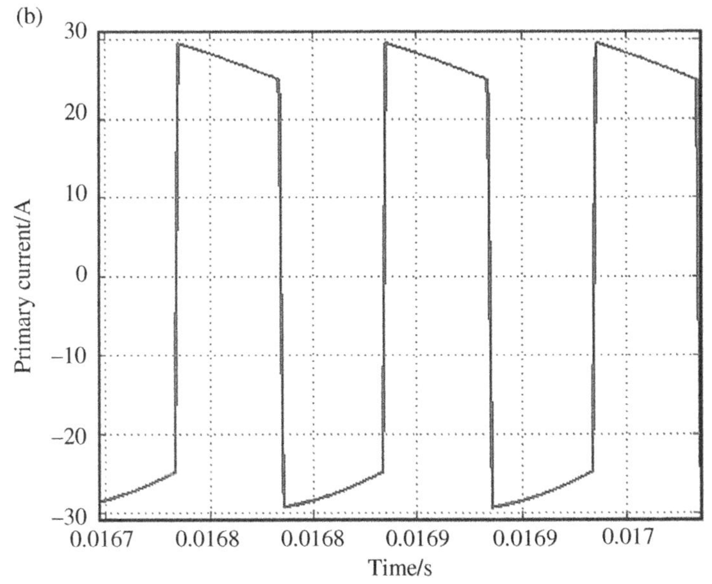

Figure 9.19 Simulated primary current at different voltage and power output: (a) primary current for $V_2 = 500\,V$, $R = 200\,\Omega$, $P = 1.8\,kW$; (b) primary current for $V_2 = 600\,V$, $R = 40\,\Omega$, $P = 9\,kW$; (c) primary current for $D = 1/8$, $R = 60\,\Omega$, $P = 9.2\,kW$; (d) primary current for $D = 1/8$, $R = 100\,\Omega$, $P = 13.6\,kW$.

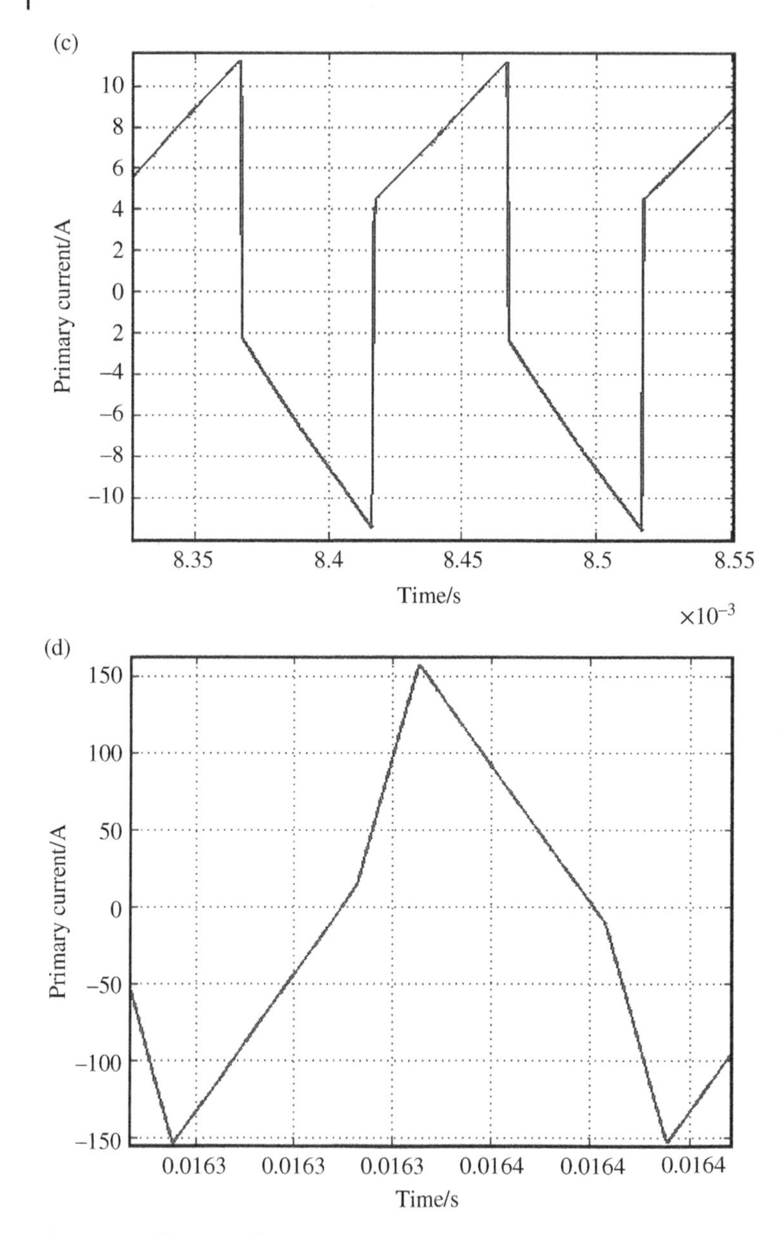

Figure 9.19 (Continued)

The capacitor current is shown in Figure 9.20, where at $t = t_1 + \Delta_2$, $I_{L_s} = I_o$, or $I_C(t_1 + \Delta_2) = 0$, that is,

$$I_{\max} + \frac{nV_1 - V_2}{L_s}\Delta_2 = I_o \tag{9.59}$$

Therefore

$$\Delta_2 = \frac{I_{\max} - I_o}{V_2 - nV_1} L_s \tag{9.60}$$

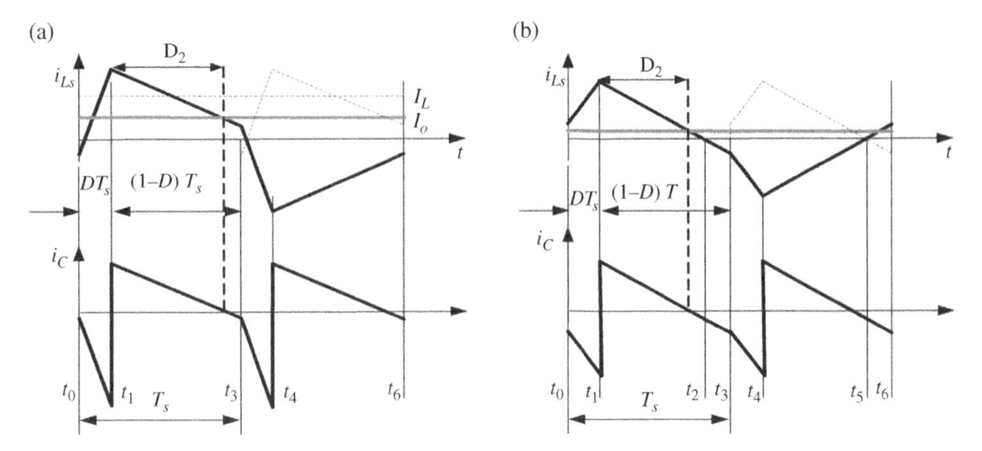

Figure 9.20 The current ripple in the capacitor: (a) current ripple for $i(t_0) < 0$; (b) current ripple for $i(t_0) > 0$.

Based on energy conservation, $P_1 = P_o$, or $nV_1\bar{I} = V_2 I_o$. Substituting Equation 9.51 for P_o, the output current can be obtained:

$$I_o = \frac{nV_1}{2f_s\ L_s} D(1-D) \tag{9.61}$$

The voltage ripple is

$$\Delta V_2 = \frac{1}{C}\int_{DT_s}^{DT_s+\Delta_2} (i_{L_s}(t) - I_o)\,dt = \frac{\Delta_2}{C}\frac{I_{\max} - I_o}{2} \tag{9.62}$$

Substituting Equations 9.59–9.61 into Equation 9.62, the voltage ripple can be derived as

$$\begin{aligned} \Delta V_2 &= \frac{[V_2 + (2D^2 - 1)nV_1]^2}{32 f_s^2\ L_s C(V_2 - nV_1)} = \frac{[D(1-D)R_L - 2(1-2D^2)f_s\ L_s]^2}{64 f_s^3\ L_s^2[D(1-D)R_L - 2f_s\ L_s]}\frac{nV_1}{C} \\ \Delta V_2\% &= \frac{\Delta V_2}{V_2} = \frac{[D(1-D)R_L - 2(1-2D^2)f_s\ L_s]^2}{32 f_s^2\ L_s C[D^2\ (1-D)^2\ R_L^2 - 2f_s\ L_s D(1-D)R_L]} \end{aligned} \tag{9.63}$$

Therefore the voltage ripple is a function of D, R, and nV_1. This expression is applicable to both open-loop control and close-loop control. The condition of Equation 9.63 is $V_2 > nV_1$, which implies $D(1-D)R_L > 2f_sL_s$. From Equation 9.63, it is easy to see that ΔV_2 is directly proportional to nV_1/C. In order to further study the influential factors, the voltage ripple under different conditions is shown in Figure 9.21.

From Figure 9.21a, the absolute voltage ripple decreases when R_L decreases under a given D, and the ripple reaches a maximum at $D = 0.5$. However, in Figure 9.21b, the relative ripple is an increasing function of D and only reaches a maximum when $D = 1$. All the curves cross at $D = 0.5$.

When $D < 0.5$, a high resistance will result in a high ripple under a fixed D. When $D > 0.5$ the absolute ripple decreases with D for a given load resistance, but the relative ripple increases with D.

From Figure 9.21c,d, increasing the switching frequency is beneficial for limiting both the absolute ripple and the relative ripple of the output voltage.

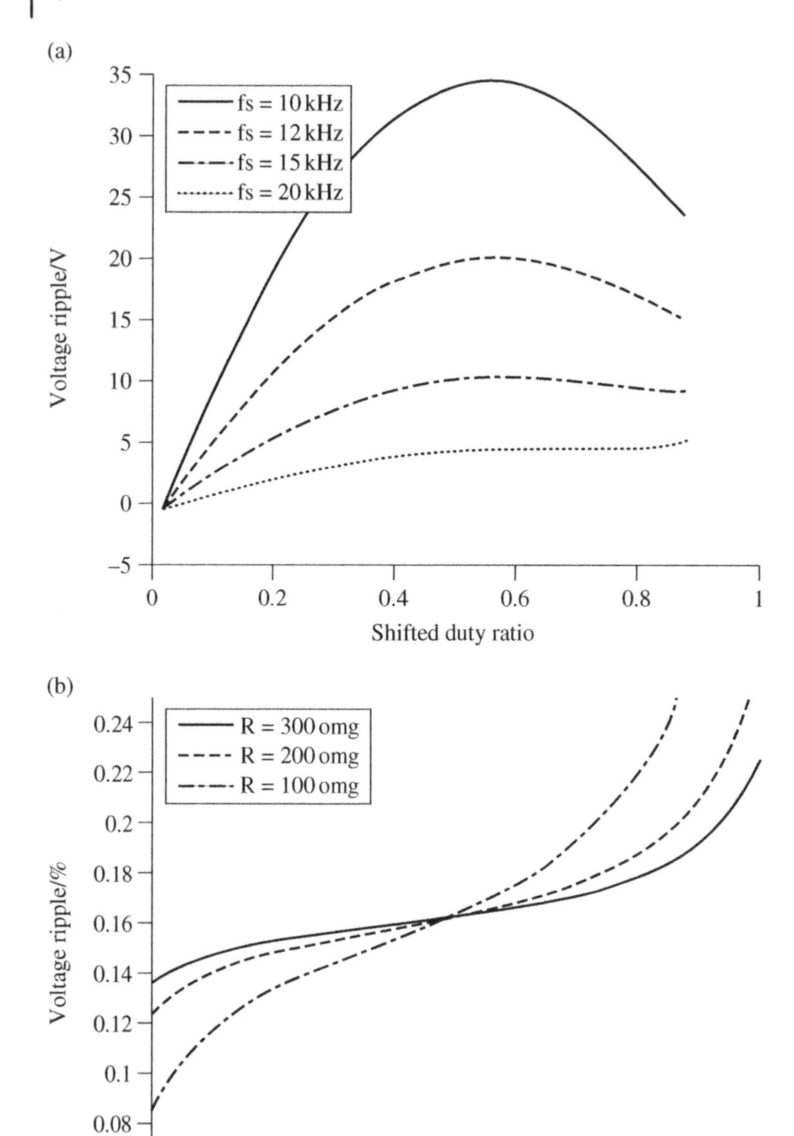

Figure 9.21 Voltage ripple under different operations simulation: (a) voltage ripple as a function of D and R_L; (b) voltage ripple (%) as a function of D and R_L; (c) Voltage ripple as a function of D and f_s; (d) Voltage ripple (%) as a function of D and f_s.

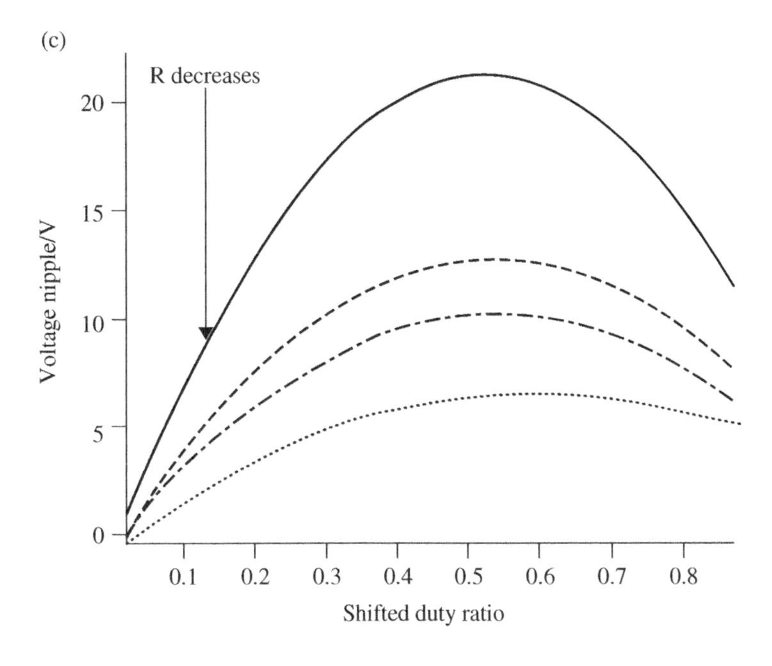

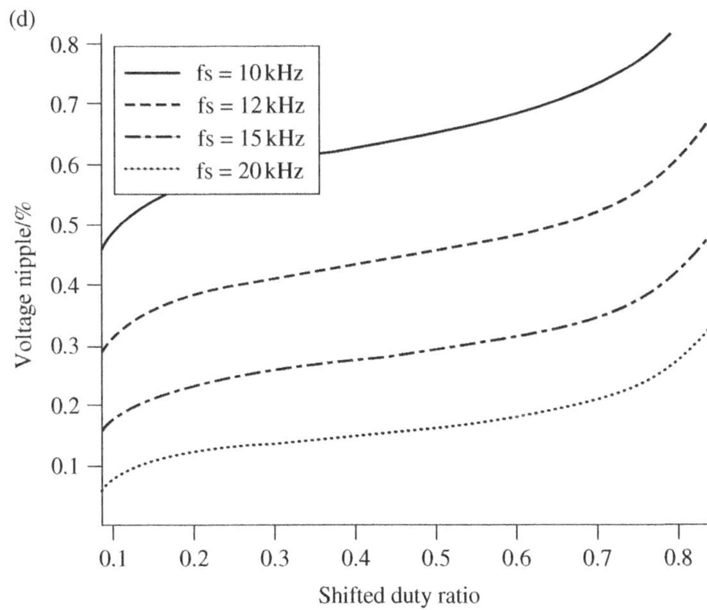

Figure 9.21 (Continued)

9.9 PWM Rectifier in HEVs

9.9.1 Rectifier Operation of Inverter

In braking mode, the motor is controlled to achieve regenerative braking. The VSI is operated as a PWM rectifier, as shown in Figure 9.22. The operating principle of the PWM rectifier, when the motor (operated as a generator) speed is below the base speed of the constant power region, is essentially the same as the boost operation of the isolated DC–DC converter in the previous section.

When the motor speed is above the base speed of the constant power region, the generator develops a back emf much higher than the DC link voltage, especially when a large constant power range is designed. In this case, the motor needs to be controlled in its field weakening region to reduce its winding terminal voltage. Neglecting the stator resistance and salience, the equivalent circuit of the control is illustrated in Figure 9.23, where E_o is

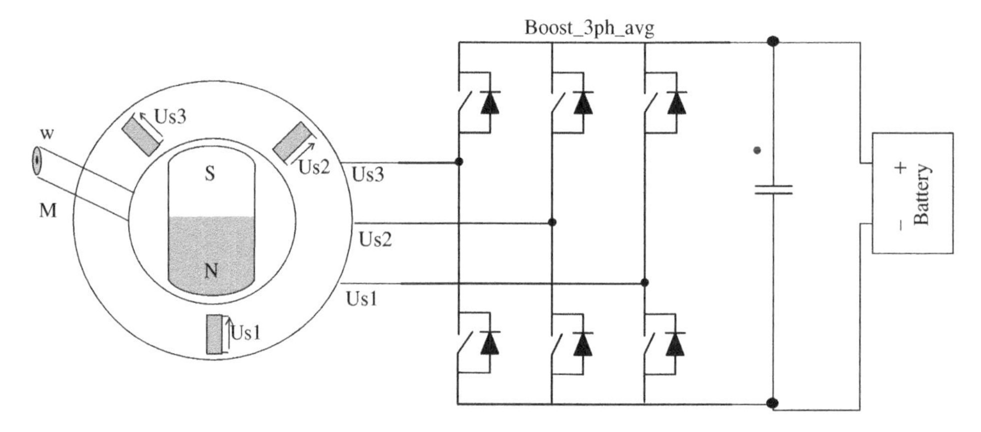

Figure 9.22 PWM rectifier.

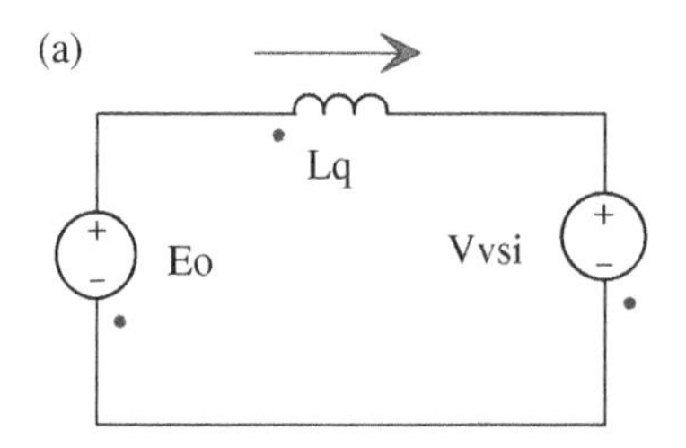

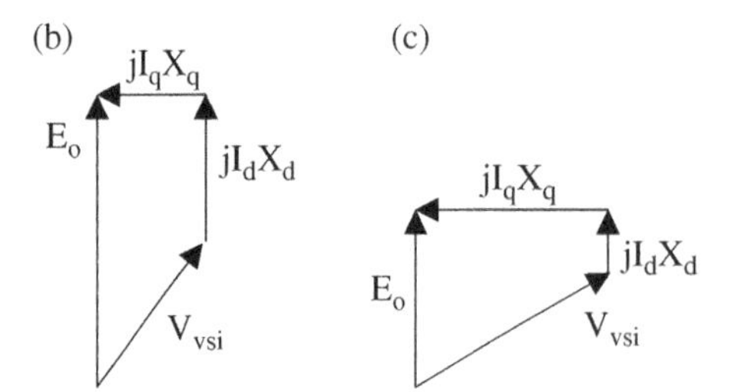

Figure 9.23 Regenerative braking in the constant power region: (a) equivalent circuit; (b) phasor diagram at high speed in the constant power region; and (c) phasor diagram at lower speed in the constant torque region.

the induced back emf of the PM motor, L_q is the synchronous inductance, and V_{vsi} is the equivalent sine wave output of the VSI:

$$V_{vsi} = E_o - jI_q X_q - jI_d X_d \tag{9.64}$$

The purpose of the field weakening control is to achieve a constant V_{vsi} for higher E_o, by applying a current opposite to the d-axis direction.

9.10 EV and PHEV Battery Chargers

The battery charging system is a critical part of EVs and PHEVs. The efficiency, charging speed, and cost of such chargers are crucial to the commercialization of PHEVs. There are many different topologies available for battery chargers. Due to safety requirements and standards, most battery chargers are required to be isolated from the AC electric grid (such as requirement of SAE J1772). In some cases, bidirectional power flow is necessary in order to realize vehicle-to-grid functions.

A survey conducted by the US Electric Power Research Institute (EPRI) revealed that 78% of the US population drive 40 miles (64 km) or less in their daily commuting. Therefore, PHEVs capable of a 40 mile electric driving range can significantly reduce gasoline usage in passenger cars. For a typical passenger vehicle (car and SUV), the average energy consumption is approximately 150–300 Wh per mile. To achieve a 40 mile electric range, a battery that contains usable energy of 6–12 kWh is appropriate. Lithium-ion batteries are considered as the only viable energy storage solution for PHEVs for the time being. To achieve safe and reliable operation while maintaining the cycle life and health of the batteries in a PHEV, the typical available energy from lithium-ion batteries is approximately 60–70% of the nominal capacity. Therefore, a 10–16 kWh lithium-ion battery pack is very typical in many PHEVs under development.

The battery onboard a PHEV needs to be charged from the grid through either an onboard or offboard charger. For private vehicles, these PHEVs will be charged at home or at a public charge station through either 110 or 220 V AC input. In order to make sure that the vehicle is ready for use by the following morning, the battery must be fully charged within a reasonable amount of time, typically 2–8 hours.

In the past few decades, various charging circuits have been developed targeting different applications, such as laptop computers, portable electronics, and uninterruptable power supplies (UPSs). Isolation is one of the basic requirements for all battery chargers for safety. The charging current is usually controlled by continuous feedback of critical battery parameters, such as battery voltage, SOC, and temperature. As far as the unidirectional charging system is concerned, chargers based on flyback and forward converters are typical examples for low-power applications. Both topologies need only one active switch. However, flyback and forward converters undergo HV spikes when the excessive energy stored in the leakage inductance of the isolation transformer is exhausted at the turn-off moment. Therefore, at higher power operations, an auxiliary snubber circuit is needed. Regardless of the limitations, flyback/forward topologies are being used in PHEV chargers.

Chargers based on half-bridge and full-bridge unidirectional DC–DC converters are favorable alternatives to chargers based on flyback and a forward converter. The magnetization of the isolation transformer in a half-bridge converter is bidirectional, therefore the demagnetizing circuit is eliminated. The leakage inductance of the transformer is a key parameter for energy transfer. The operation of a full-bridge DC–DC converter is similar to a half-bridge converter. The electrical stress of semiconductors in a half-bridge/full-bridge converter is significantly reduced. Soft switching is easy to implement in half-bridge/full-bridge converters. The disadvantage of these converters is the increased number of semiconductor switches.

Resonant converters can also be employed as battery chargers at extremely low cost. However, resonant chargers usually provide pulsed charge current, such as the case in inductive heating, other than constant current or constant voltage charging. The impact of pulsed current charging on battery capacity and life cycle is not fully understood at the present time.

Figure 9.24 shows the general architecture of a unidirectional PHEV battery charger. It consists of a front-end rectifier, a power factor correction (PFC) stage, and an isolated DC–DC stage.

9.10.1 Forward/Flyback Converters

Figures 9.25 and 9.26 show the circuit and operation of a forward and flyback converter respectively, where R is the battery internal resistance, E is the battery internal voltage, and V_o is the output voltage across the battery (including the battery internal voltage and voltage drop across the internal resistance). The operation of a forward converter is similar

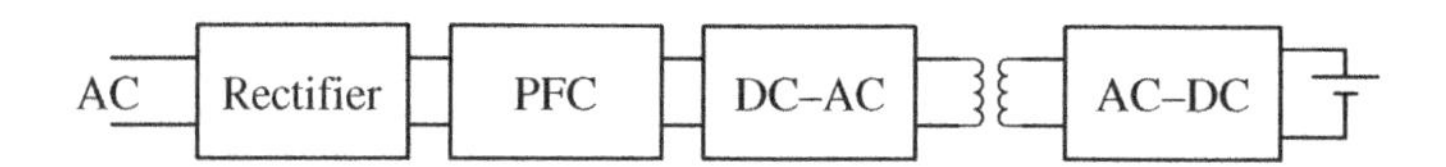

Figure 9.24 Basic PHEV charger architecture.

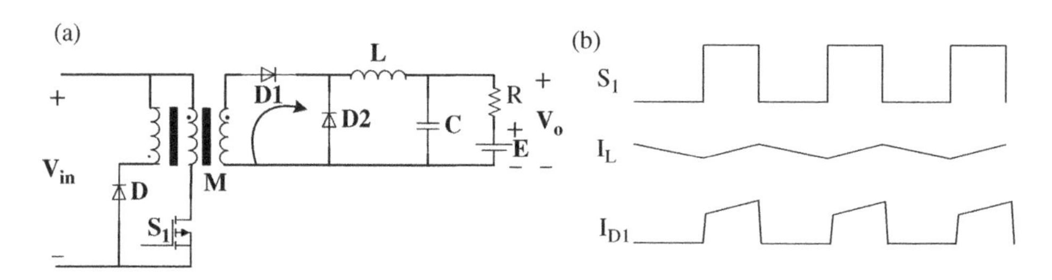

Figure 9.25 Forward converter: (a) circuit topology; (b) operation waveforms.

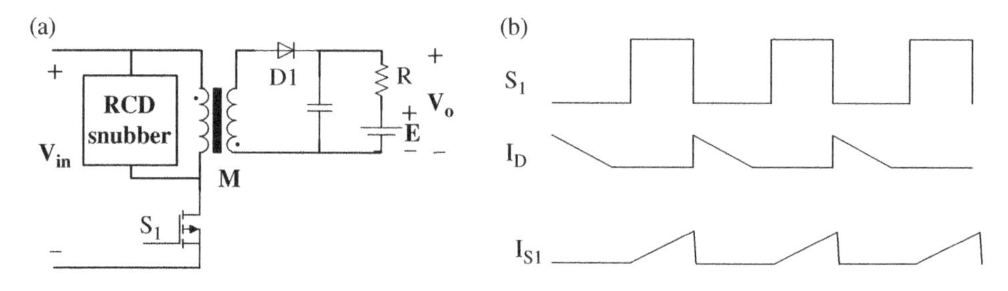

Figure 9.26 Flyback converters: (a) circuit topology; (b) operation waveforms.

to that of a buck converter. When switch S_1 in Figure 9.25a turns off, the leakage inductance of the transformer will exhaust all the excess energy through the switches. It therefore induces significant voltage spikes across the switches. The energy stored in the magnetizing inductance could be exhausted through the auxiliary winding shown in Figure 9.25a; however, the excess energy in the leakage inductance of the transformer could only be mitigated through a snubber circuit, which is also needed in a flyback converter.

In the flyback converter shown in Figure 9.26a, when D_1 conducts, the load voltage will be induced to the primary side. Therefore, in the off-state of S_1, the voltage across S_1 is $V_{in} + V_o/n$, where n is the turns ratio of the isolation transformer. This indicates that although it may not be necessary to have a filtering inductor in the flyback converter, the semiconductor switch will in fact undergo a higher voltage stress.

9.10.2 Half-Bridge DC–DC Converter

Figure 9.27 shows a half-bridge DC–DC converter. Switches S_1 and S_2 are switched with their phase shifted by 180°. The leakage inductance of the transformer serves as the component for energy transfer. If the parasitic inductance of the commutating loop is negligible, the voltage spike across the semiconductors will not be of concern.

9.10.3 Full-Bridge DC–DC Converter

Figure 9.28 shows the circuit topology and operation of a full-bridge DC–DC converter. Compared to the half-bridge converter where only half of the DC voltage is imposed on the primary side of the transformer in every switching cycle, the full-bridge converter utilizes the whole DC link voltage. Similar to a half-bridge converter, the leakage inductance of the transformer in a full-bridge converter does not contribute to any voltage spike across the switches. This leakage inductance should be designed appropriately for best performance [13].

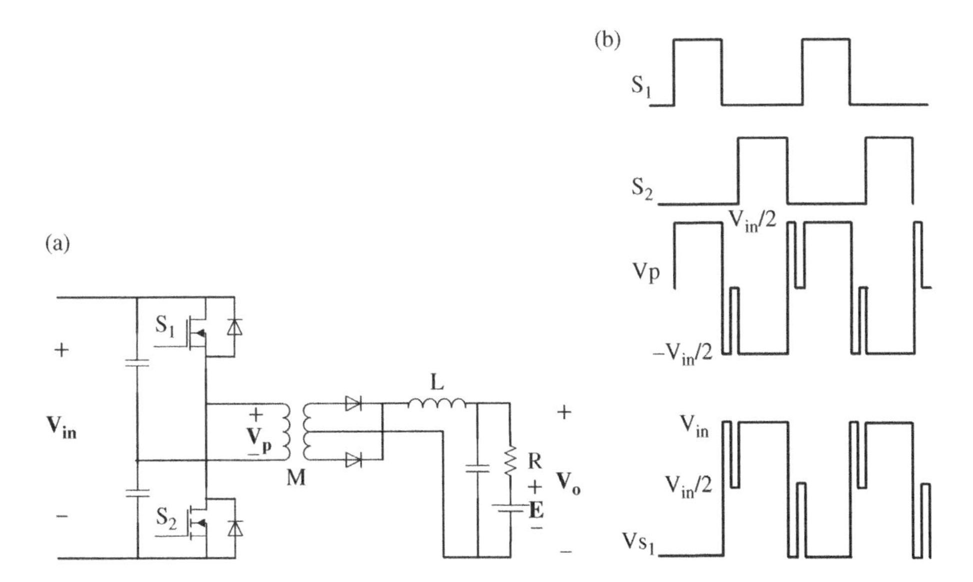

Figure 9.27 Half-bridge converter: (a) circuit topology; (b) operation waveforms.

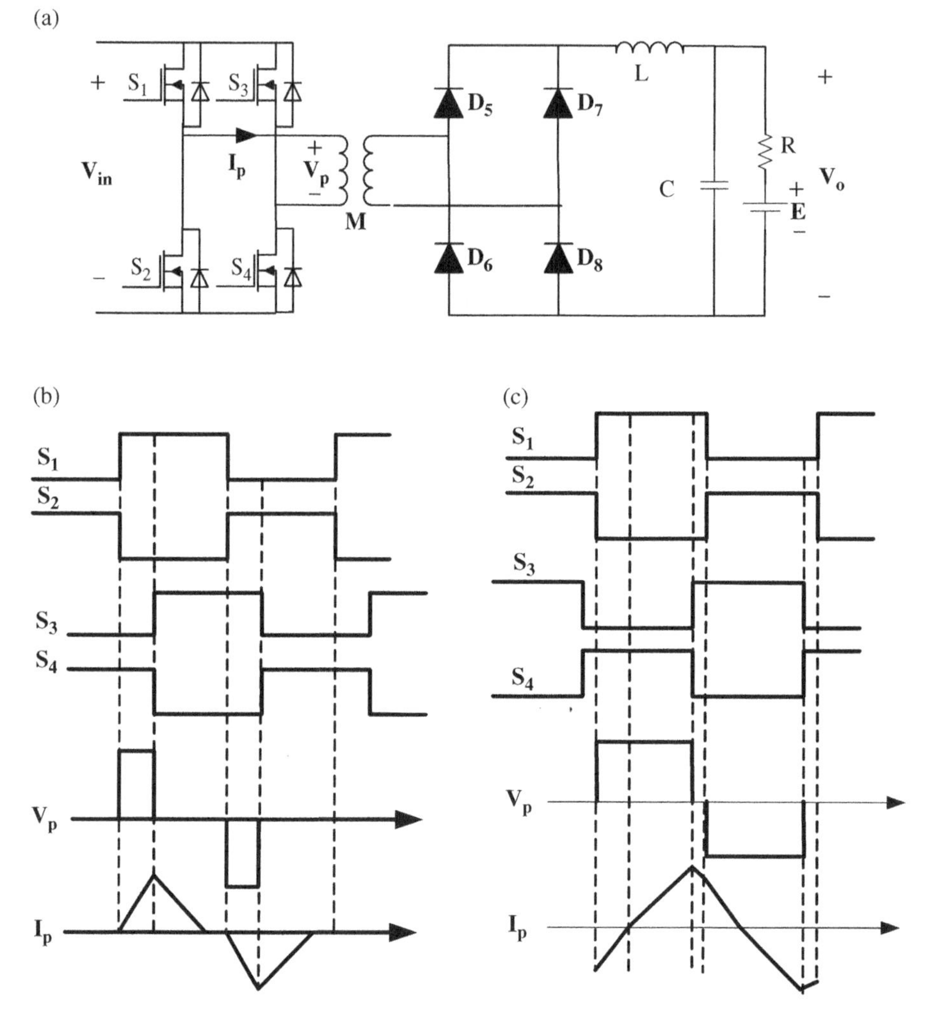

Figure 9.28 Full-bridge converter: (a) circuit topology; (b) discontinuous mode operation; (c) continuous mode operation.

9.10.4 Power Factor Correction Stage

A PFC stage is usually placed between the rectifier and the DC–DC stage to avoid harmonics pollution to the grid as well as to stabilize the DC link voltage. A typical PFC circuit, shown in Figure 9.29, consists of an inductor L, an active switch S, and a free-wheeling diode D. Bridgeless PFC circuits have also been used in battery chargers.

9.10.4.1 Decreasing Impact on the Grid

The AC grid side current with and without a PFC stage is shown in Figure 9.30, where the grid voltage is 110 V AC, and the output power of the charger is 5 kW.

It can be seen from Figure 9.30 that with a PFC, the input current is much better. The current is close to sinusoidal and the current peak is significantly decreased. The impact

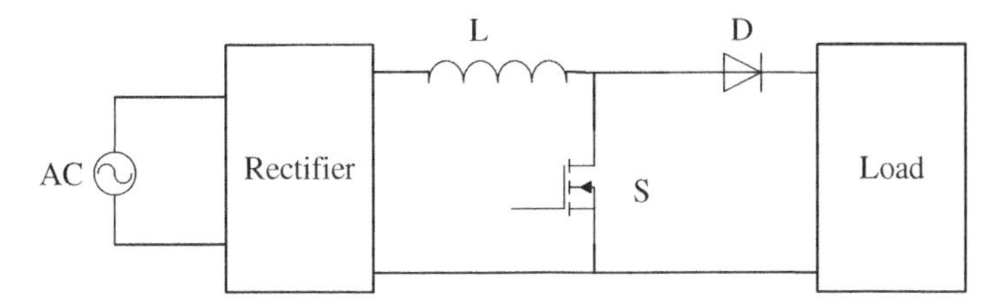

Figure 9.29 Power factor correction stage in a PHEV charger.

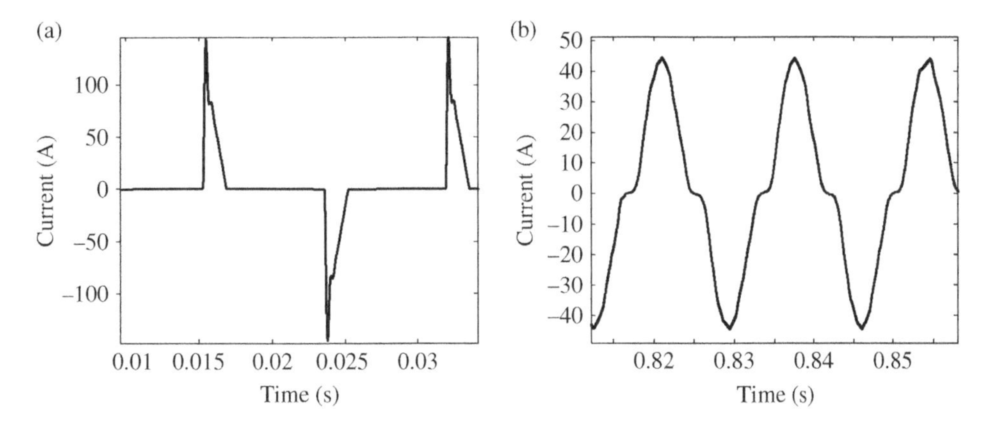

Figure 9.30 AC grid side current with and without PFC: (a) without PFC; (b) with PFC.

of harmonic currents on the grid is mitigated. Also the primary diode rectifier will undergo a smaller current stress. The power factor is close to 1.

9.10.4.2 Decreasing the Impact on the Switches

The PFC circuit also helps boost the DC link voltage to a higher level and stabilize the DC link voltage. Therefore, at the same output power, the current through the switches will be decreased to enhance safety and output capability. Figure 9.31a shows the switch current at different V_{in} with the same output power. When V_{in} is increased to 400 V DC, the switch current is significantly decreased. As long as the voltage across the switches does not exceed the breakdown value, a higher DC bus voltage will lead to higher power capability for the same devices used. Figure 9.31b shows the maximum charging current that the system can deliver. Increasing the DC link voltage will benefit the output capability. Here the maximum repetitive switched-off current of the semiconductor switch is set to 70 A.

9.10.5 Bidirectional Battery Chargers

It is possible to equip a PHEV with a bidirectional charger. With bidirectional power transfer capability, the energy stored in a PHEV battery can be sent back to the grid during peak demand hours for peak shaving of the AC electric grid, or to supply power to the home and office during a power outage.

The isolation can be achieved using high-frequency transformers in the DC–DC stage at high-frequency level as shown in Figure 9.32. The isolation can also be

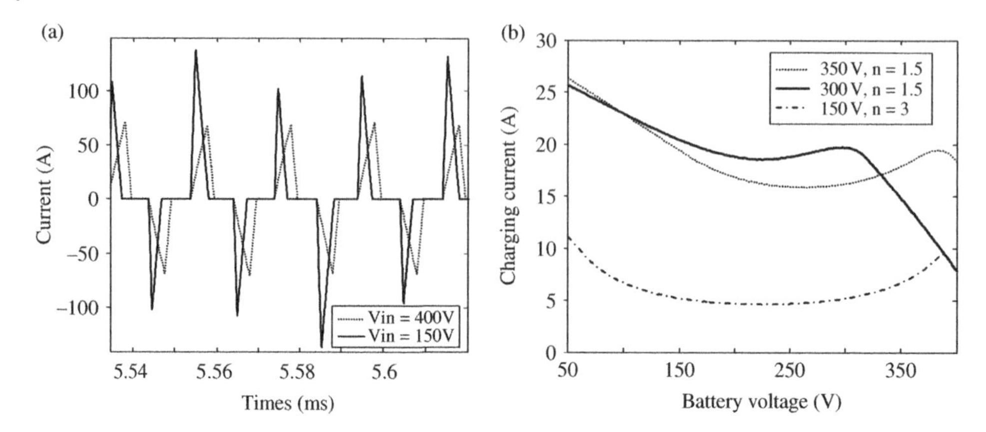

Figure 9.31 Comparison of the switch current without PFC: (a) comparison of switch current; (b) maximum charging current under different V_{in}.

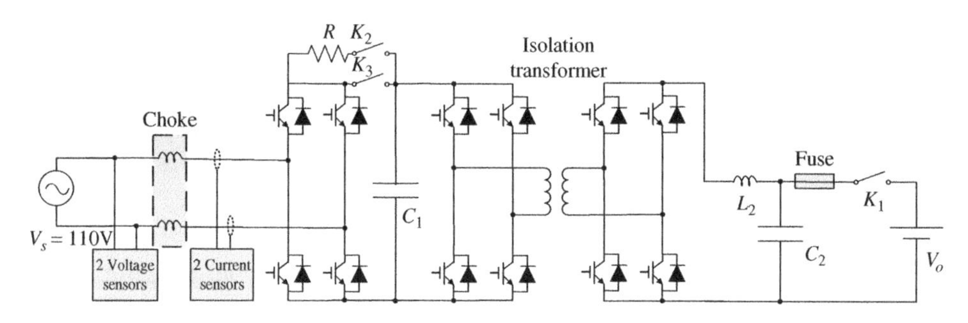

Figure 9.32 Isolation using a high-frequency transformer.

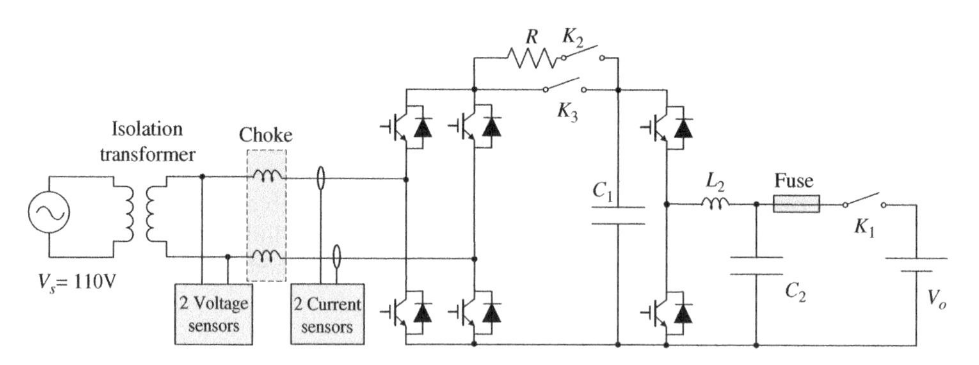

Figure 9.33 Isolation at the grid level with a line frequency transformer.

achieved using a transformer at the grid frequency level as shown in Figure 9.33. It is difficult to add a PFC stage to the bidirectional chargers but the AC side current can be controlled using the grid side inverter to limit the harmonics and improve power factor.

9.10.6 Other Charger Topologies

The above chargers are based on single-phase low-power inputs. In applications such as commercial transportation and recharge stations, it is necessary to charge the battery in a relatively short time. For example, for a 16 kWh pack, to charge the battery from 30 to 80% SOC in 5 minutes will require 16 kWh × (80% − 30%)/(5/60) = 96 kW. Assuming that the battery is able to accept this type of charging, the charging power will have to be supplied. This is typically done through a three-phase high-power input.

Excessive heat may be generated during high-power fast charging of EV and PHEV batteries. The high charge rate may also impact the long-term capacity or life of the battery.

With more and more renewable energy and distributed generation, DC grids are starting to emerge. Almost all renewable sources, including solar and wind, can be in the form of DC. If a PHEV is connected to a DC grid, then the battery can be charged directly from the DC source. Traditional chargers designed for AC input can be used to take DC input, but a specially designed DC charger will have no need for the rectifier and PFC stage.

9.10.7 Contactless Charging

The above chargers all need an electrical contact with the electric outlet. This hard-wired electrical connection has a few caveats. For example, if the cable is pulled out of the electric outlet (whether intentionally or unintentionally) when the battery is still being charged, then there could be a spark and potential damage or injury. Another example is that somebody (such as children) could get hurt if they happen to play with the cords etc. Charging the vehicle when it is raining could also be potentially dangerous. Wear and tear of the plug and cable could also be a source of danger.

Hence contactless charging has been considered an alternative. In 1995, the US SAE Electric Vehicle Charging System Group, Japanese manufacturers including Toyota, Nissan, Honda, and DENSO and IEC/ISO, together developed the inductive charging standard SAE J1773. In 1999, SAE J1773 was enacted and recommended as an international standard (http://www.toyota.com). GM already produced electric vehicles adopting electrically contactless charging, referred to as an *inductive charger* at that time, in GM's EV1 program in the 1990s. Since the contactless charging does not involve direct contact of electricity, it is a safe and convenient way to charge an EV battery. The task can even be completed by a child for 50 kW contactless charging.

Contactless charging involves an isolation transformer where the secondary winding and core can be removed from the primary. Figure 9.34 shows the topology used in inductive chargers.

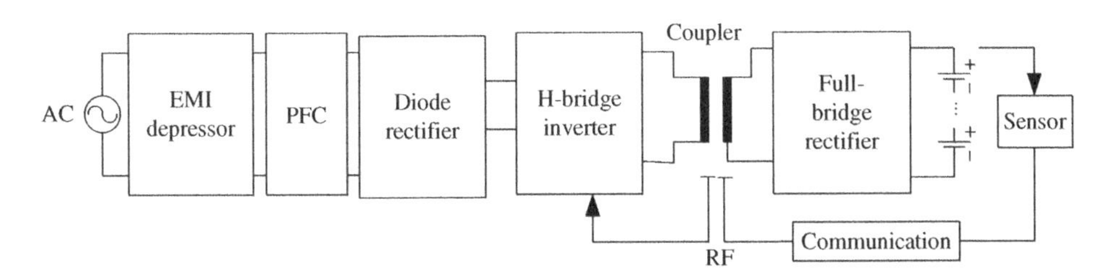

Figure 9.34 Isolated inductive charger.

9.10.8 Wireless Charging

In the past decade, the meaning of inductive chargers has evolved. For example, in SAE J1773 as well as the inductive charger used by GM in its EV1, an *inductive charger* is referred to as a contactless charger that may often be considered as a charger that does not have a direct electric contact. However, the pedestal may still have to be plugged in as shown in Figure 9.34. Today, an inductive charger is usually referred to a charger that uses the principle of a magnetic field as the medium and usually has a certain distance or gap between the transmitter and the receiver. Hence, it is more common to be called a wireless charger as discussed below, with more detail in Chapter 16.

Wireless charging involves the use of power and energy transfer at a much longer distance than the aforementioned contactless charging. Although the contactless charger discussed in the last section can eliminate the direct electric contact, it still needs a plug, cable, and physical connection of the inductive coupler. Wear and tear of the plug and cable could still cause danger as well.

Wireless charging could eliminate the cable and plug altogether. In this scenario, a driver can drive the car over a specially designed parking lot and the car battery is automatically charged without connecting any cable or plug, as shown in Figure 9.35. It provides the safest approach for EV battery charging. A brief introduction is provided in this chapter for the completeness of the contents. A more comprehensive discussion will be given in Chapter 16.

There have been a few different experiments carried out for wireless energy transfer. The most promising technology is using electromagnetic resonance as shown in Figure 9.36. In this setup, there are two antennas with one placed in the parking structure as the transmitter and one inside the car as the receiver. The two antennas are designed to resonate at the controlled frequency. The limitations are the level of power transfer, and efficiency due to the large air gap between the two antennas.

In the circuit shown in Figure 9.36, R_s is the internal impedance of the source and the resistance of the primary coil, R_o is the load resistance and the resistance of the secondary coil, L_l is the leakage inductance of the each coil, and L_m is the mutual inductance of the two coils. There are two ways to make the circuit resonant. If we design the circuit and select the frequency of the power supply V_1 such that

$$\omega L_l - \frac{1}{\omega C} = -2\omega L_m \tag{9.65}$$

the total equivalent impedance is

$$Z = R_s - \frac{j\omega L_m \ R_o}{R_o - j\omega L_m} \tag{9.66}$$

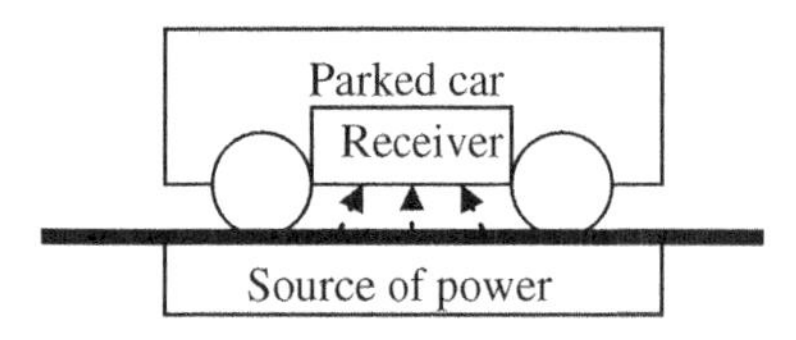

Figure 9.35 Wireless charging of a PHEV/EV on a parking floor.

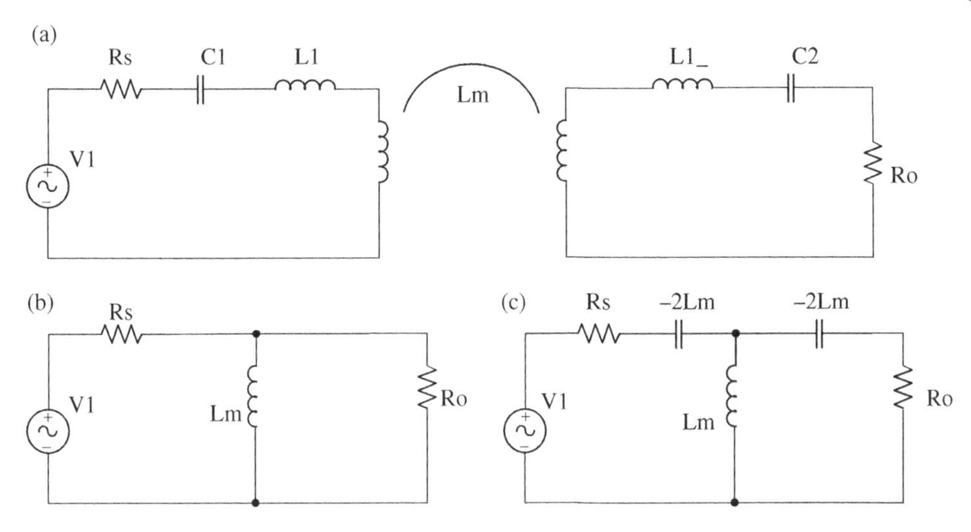

Figure 9.36 Circuits for electromagnetic resonance-based wireless charging: (a) circuit; (b) equivalent circuit at resonance frequency condition 1; (c) equivalent circuit at resonance frequency at condition 2.

If we design

$$\omega L_l - \frac{1}{\omega C} = 0 \tag{9.67}$$

and the total equivalent impedance is

$$Z = R_s + \frac{j\omega L_m \ R_o}{j\omega L_m + R_o} \tag{9.68}$$

then the circuit will be in resonance. However, since both the mutual inductance and leakage inductance change with distance between the two coils, the frequency will have to be tuned based on the distance in real-world applications. Figure 9.36b,c shows the equivalent circuit during resonance of the two conditions. Figure 9.37 shows the simulation results of the circuit. The first plot shows two resonant frequencies. The second plot shows that when the distance between the two coils increases, the two resonant frequencies gets closer.

9.11 Modeling and Simulation of HEV Power Electronics

Modeling and simulation play an important role in the design and development of power electronics circuits. The simulation of power electronics circuits in hybrid vehicle applications can be divided into two categories: device-level and system-level [11, 24–28].

9.11.1 Device-Level Simulation

Device-level simulation can reveal the details of the device behavior. To obtain detailed loss data, overvoltages, and other component stresses due to the non-ideal nature of

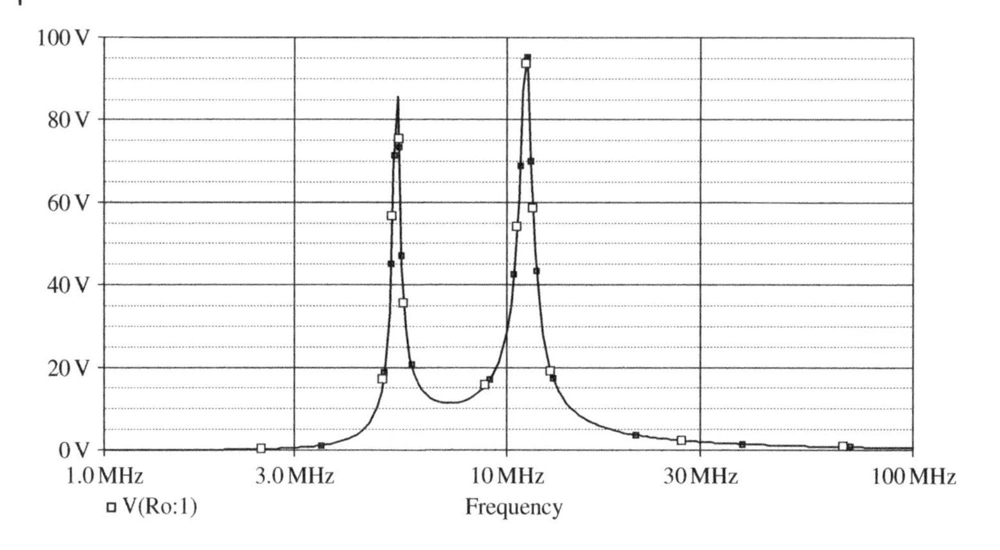

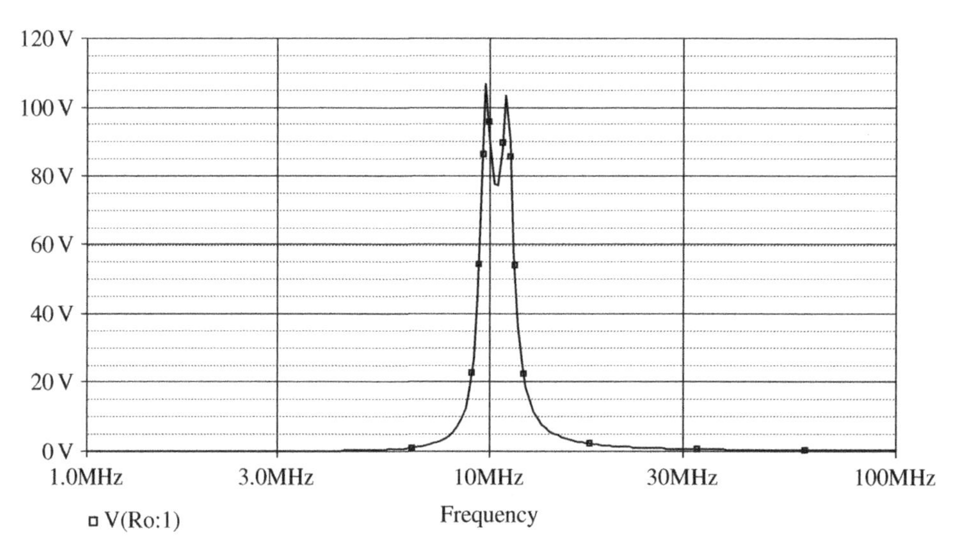

Figure 9.37 Resonance frequency of the wireless charging circuit: top, output voltage for shorter distance; bottom: output voltage for longer distance.

power electronics devices, and the stray inductance and capacitance of the circuitry, it is necessary to simulate a number of cycles of detailed switching pertaining to the worst case scenario.

9.11.2 System-Level Model

Detailed device-level simulation can take a significant amount of time due to the high switching frequency used in the power electronics circuits, whereas the mechanical constants of the vehicle system could be a few seconds or more. Therefore, device-level

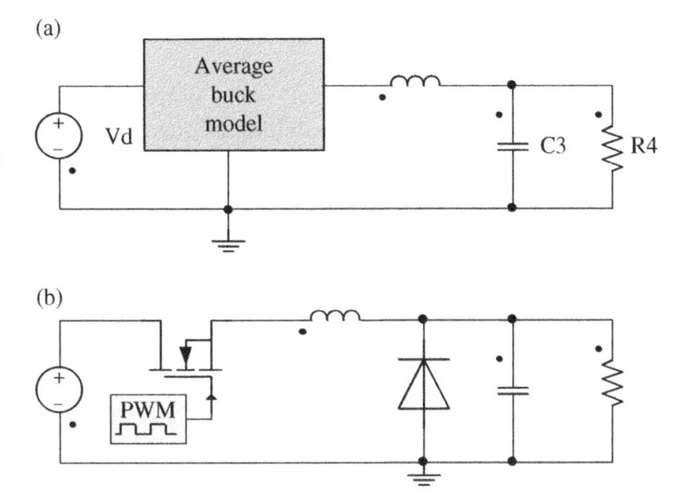

Figure 9.38 Device-level and system-level modeling of a buck converter: (a) system-level model only taking into account the nonlinear characteristics of the inductor; (b) device-level model involving detailed switching of the MOSFET.

simulation, while it can simulate the dynamic performance of the circuit, is not suitable for simulating the vehicle performance, such as gradability, acceleration, and fuel economy. On the other hand, the electronics circuits have very fast transients when compared to the vehicle dynamics.

In system-level simulations, the average model is generally used. For example, a buck converter can be represented in the simulation by an average model as shown in Figure 9.38a. A simulation of the system performance of one second only takes two seconds of simulation time, whereas in Figure 9.38b, which uses detailed device-level models, it takes about 20 minutes to obtain the system performance of one second.

9.12 Emerging Power Electronics Devices

The present silicon (Si) technology is reaching the material's theoretical limits and cannot meet all the requirements of hybrid vehicle applications in terms of compactness, light weight, high power density, high efficiency, and high reliability under harsh conditions. New semiconductor materials, such as silicon carbide (SiC), for power devices have the potential to eventually overtake Si power devices in hybrid vehicle powertrain applications [29–36].

SiC power devices potentially have much smaller switching and conduction losses and can operate at much higher temperature than comparable Si power devices. Hence, a SiC-based power converter will have a much higher efficiency than that of converters based on Si power devices if the same switching frequency is used. Alternatively, a higher switching frequency can be used to reduce the size of the magnetic components in a SiC-based power converter. In addition, because SiC power devices can be operated at much higher temperatures without much change in their electrical properties, ease of thermal management and high reliability can be achieved.

9.13 Circuit Packaging

Electromagnetic interference (EMI) is one of the most challenging problems in power electronics circuits. The high switching frequency and high current generate electromagnetic fields that will permeate the other components in the vehicle system and create large amounts of electrical noise. In order to minimize EMI, components must be carefully placed so that EMI is contained by shielding and will have minimal effect on the rest of the system. All paths must be kept as close as possible so that the generated fields will nullify one another. To minimize parasitics and aid in the EMI issue, the lengths of wires need to be kept as short as possible.

The control circuit needs to provide protection for overcurrent, short circuit, overvoltage, and undervoltage. The capability of detecting any fault signal and turning off the gate drive signals to the primary switches is a critical part of power electronic circuit design. Fast fuses need to be used in the circuit to protect the converter from being damaged by any other faults and used for safety.

9.14 Thermal Management of HEV Power Electronics

At power levels of 100 kW, even with an efficiency of 96–98%, the power loss of each power electronic unit is 2–4 kW. With two or three powertrain motors and associated power electronics circuits, together with the high-power bidirectional DC–DC converter, the heat generated in the hybrid vehicle system could be significant.

Significant advances in the thermal management of both power electronics and motors for HEV propulsion systems must be achieved to meet the automotive industry's goals of reduced weight, volume, and cost [36–39]. Through the optimization of existing technologies and the expansion of new pioneering cooling methods, hybrid powertrain components can achieve higher power densities, smaller volumes, and increased reliabilities. Investigations and advances in thermal issues can provide a viable path to bridging gaps still plaguing the successful achievement of automotive technical targets while simultaneously enhancing the ability to apply new technologies to automotive applications as they mature.

Thermal performance of a power module is measured by the maximum temperature rise in the die at a given power dissipation level with a fixed heat sink temperature. The lower the die temperature, the better the electrical performance. As the thermal resistance from the junction of the die to the heat sink is reduced, higher power densities can be achieved for the same temperature rise, or for the same power density a lower junction temperature can be attained. It is important to reduce thermal cycling or maintain low ambient temperature to improve the life and reliability of the die.

The main areas of concern in thermal management of power electronics are: operating temperature of IGBTs (it should be less than 125°C); contact resistance between various layers of a power module; low-thermal-conductivity thermal paste; heat flux limitations (ideally, faster IGBTs would have to reject heat at a rate of 250 W/cm^2); limitations on the inlet cooling fluid temperature (it is desirable to use the engine coolant at 105°C); the cost of the cooling system; and weight and volume.

The existing cooling technologies are depicted in Figure 9.39. It is shown that conventional cooling techniques such as forced convection and simple two-phase boiling

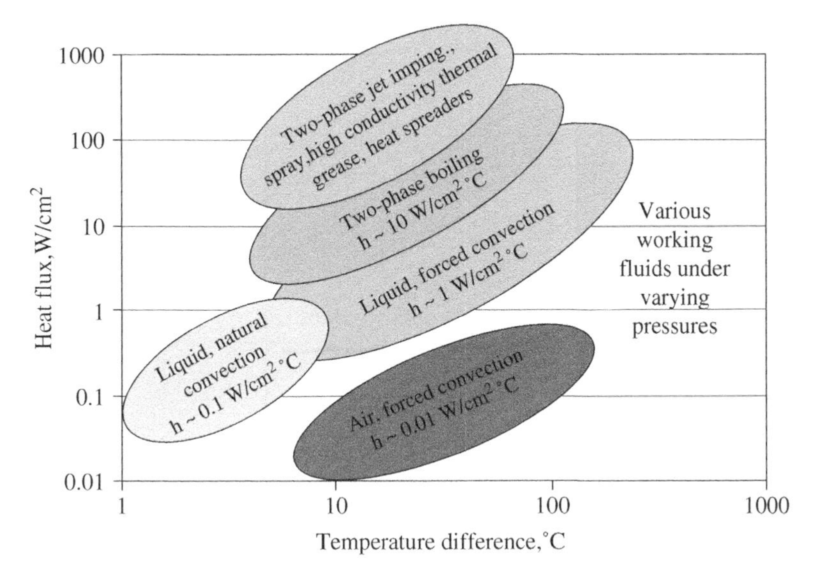

Figure 9.39 Existing cooling technologies.

techniques are not capable of removing high heat fluxes (in the range of 250 W/cm^2) at low temperature differences (20°C). However, this figure shows clearly that employment of enabling technologies such as spray cooling and jet impingement, along with some other innovative improvements will be able to meet the goals of the automotive industry.

Ideally, it would be more beneficial if IGBTs could be designed to operate at higher temperatures. The industry is pursuing various long-term research projects to evaluate and achieve that objective. However, to meet the immediate need of the automotive industry, existing IGBTs should be operated at temperatures below 125°C.

The existing power modules are constructed by bonding together the die, copper layers, substrate, and the base plate. The whole module is then mounted on a heat sink using thermal grease or a thermal pad. The existing thermal greases used by the industry that can stand high temperatures have very low conductivities on the order of 0.3–0.5 W/(m K). As a result of this low thermal conductivity, the thermal grease constitutes 30–40% of the total thermal resistance between the junction and the heat sink. Therefore it is crucial to reduce this resistance by increasing the thermal conductivity of the thermal grease. Figure 9.40 shows the impact of the conductivity of the thermal grease on the overall temperature difference between the junction and the heat sink. For a thermal conductivity of 0.5 W/(m K), the temperature difference is about 65°C. If the thermal conductivity of the thermal grease is doubled to 1.0 W/(m K), the maximum temperature difference can be reduced to 35°C.

To improve the power density of power modules, higher heat fluxes should be removed from the module because of increased switching frequencies. An important approach for removing higher heat fluxes from the IGBTs is to spread out their heat flux over a larger surface. The existing copper or aluminum base plates spread the heat 20–30%, reducing the maximum heat flux by a corresponding amount. Use of more effective heat spreaders such as highly conductive metal layers, mini heat pipes, and/or

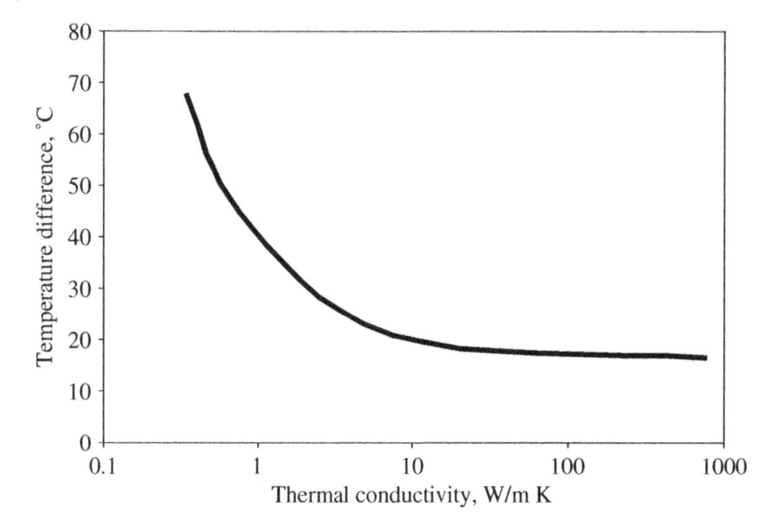

Figure 9.40 Impact of thermal interface material conductivity on temperature difference.

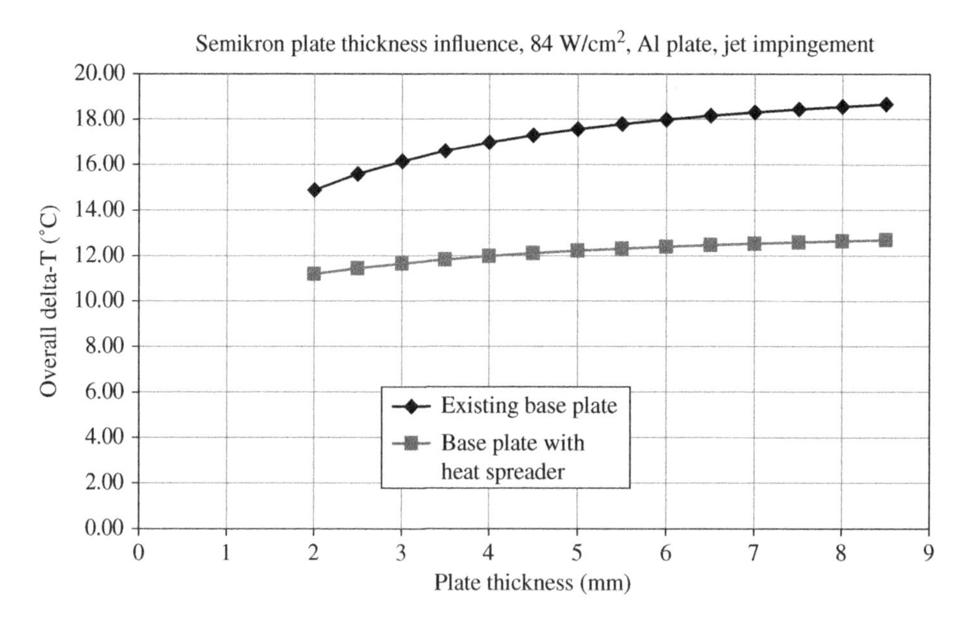

Figure 9.41 The impact of utilizing heat spreaders on junction temperature.

phase change materials inside the cold plate can spread out the high heat flux of the IGBTs over a larger surface area and reduce the maximum heat flux that needs to be removed by as much as 40%. This is a reduction of 50% in the maximum heat flux that needs to be removed, and this requirement can be easily met by one of the enabling cooling technologies such as jet impingement and spray cooling currently being considered. Figure 9.41 shows the effect of utilizing a heat spreader that results in an additional 4–6°C reduced temperature at the junction.

One approach to maintaining the power module's temperature at 125°C or below is to provide a separate cooling loop where the coolant can enter the heat sink at temperatures as low as 25°C, hence providing an adequate temperature difference to maintain the IGBTs' temperature at the desired value while removing heat fluxes as high as $250\,W/cm^2$.

9.15 Conclusions

This chapter presented an overview of the power electronics circuits for HEV applications with a focus on special circuit topologies, analysis, and thermal management. Novel power switching devices and power electronics systems have the potential to improve the overall performance of hybrid vehicles. Continued effort in power electronics circuit research and development will most likely be focused on innovative circuit topologies, optimal control, novel switching devices, and novel thermal management methods.

References

1 Bose, B.K. (2000) Energy, environment, and advances in power electronics. Proceedings of the International Symposium on Industrial Electronics, December, vol. 1, pp. TU1–T14.
2 Jahns, T.M. and Blasko, V. (2001) Recent advances in power electronics technology for industrial and traction machine drives. *Proceedings of the IEEE*, 89, 963–975.
3 Miller, J.M. (2003) Power electronics in hybrid electric vehicle applications. Proceedings of the 18th Applied Power Electronics Conference, February, vol. 1, pp. 23–29.
4 Ehsani, M., Rahman, K.M., and Toliyat, H.A. (1997) Propulsion system design of electric and hybrid vehicles. *IEEE Transactions on Industrial Electronics*, 44, 19–27.
5 Rahman, M.F. (2004) Power electronics and drive applications for the automotive industry, Proceedings of the Power Electronics Systems and Applications Conference, November, pp. 156–164.
6 Rajashekara, K. (2003) Power electronics applications in electric/hybrid vehicles. Proceedings of IECON'03, November, vol. 3, pp. 3029–3030.
7 Emadi, A., Williamson, S.S., and Khaligh, A. (2006) Power electronics intensive solutions for advanced electric, hybrid electric, and fuel cell vehicular power systems. *IEEE Transactions on Power Electronics*, 21, 567–577.
8 Hamilton, D.B. (1996) Electric propulsion power system – overview, *Power Electronics in Transportation*, pp. 21–28.
9 Vezzini, A. and Reichert, K. (1996) Power electronics layout in a hybrid electric or electric vehicle drive system. *Power Electronics in Transportation*, pp. 57–63.
10 Emadi, A., Ehsani, M., and Miller, J.M. (1999) Advanced silicon rich automotive electrical power systems. Proceedings of the Digital Avionic Systems Conference, October, vol. 2, 8.B.1-1–8.B.1-8.
11 Amrhein, M. and Krein, P.T. (2005) Dynamic simulation for analysis of hybrid electric vehicle system and subsystem interactions, including power electronics. *IEEE Transactions on Vehicular Technology*, 54, 825–836.

12 Namuduri, C.S. and Murty, B.V. (1998) High power density electric drive for an hybrid electric vehicle. *Proceedings of the 13th Applied Power Electronics Conference, February*, vol. 1, pp. 34–40.
13 Xingyi, Xu (1999) Automotive power electronics – opportunities and challenges [for electric vehicles]. International Conference on Electric Machines and Drives, May, pp. 260–262.
14 Ehsani, M., Rahman, K.M., Bellar, M.D., and Severinsky, A.J. (2001) Evaluation of soft switching for EV and HEV motor drives. *IEEE Transactions on Industrial Electronics*, 48 (1), 82–90.
15 Marei, M.I., Lambert, S., Pick, R. et al. (2005) DC/DC converters for fuel cell powered hybrid electric vehicle. Vehicle Power and Propulsion Conference, September, pp. 126–129.
16 Katsis, D.C. and Lee, F.C. (1996) A single switch buck converter for hybrid electric vehicle generators. *Power Electronics in Transportation*, pp. 117–124.
17 Chan, C.C., Chau, K.T., Jiang, J.Z., et al. (1996) Novel permanent magnet motor drives for electric vehicles. *IEEE Transactions on Industrial Electronics*, 43, 331–339.
18 Hayes, J.G. and Egan, M.G. (1999) A comparative study of phase-shift, frequency, and hybrid control of the series resonant converter supplying the electric vehicle inductive charging interface. 14th Applied Power Electronics Conference, March, vol. 1, 450–457.
19 Krein, P.T., Roethemeyer, T.G., White, R.A., and Masterson, B.R. (1994) Packaging and performance of an IGBT-based hybrid electric vehicle. *Power Electronics in Transportation*, pp. 47–52.
20 Tolbert, L.M., Peng, F.Z., and Habetler, T.G. (1998) Multilevel inverters for electric vehicle applications. *Power Electronics in Transportation*, pp. 79–84.
21 Di Napoli, A., Crescimbini, F., Rodo, S. et al. (2002) Multiple input DC-DC power converter for fuel-cell powered hybrid vehicles. 33rd Power Electronics Specialists Conference, June, pp. 1685–1690.
22 Solero, L., Lidozzi, A., and Pomilio, J.A. (2005) Design of multiple-input power converter for hybrid vehicles. *IEEE Transactions on Power Electronics*, 20, 1007–1016.
23 Gargies, S., Wu, H., and Mi, C. (2006) Isolated bidirectional DC/DC converter for hybrid electric vehicle applications. 6th Intelligent Vehicle Symposium, June.
24 Onoda, S. and Emadi, A. (2004) PSIM-based modeling of automotive power systems: conventional, electric, and hybrid electric vehicles. *IEEE Transactions on Vehicular Technology*, 53, 390–400.
25 Williamson, S., Lukic, M., and Emadi, A. (2006) Comprehensive drive train efficiency analysis of hybrid electric and fuel cell vehicles based on motor-controller efficiency modeling. *IEEE Transactions on Power Electronics*, 21, 730–740.
26 Filippa, M., Chunting, M., Shen, J., et al. (2005) Modeling of a hybrid electric vehicle powertrain test cell using bond graphs. *IEEE Transactions on Vehicular Technology*, 54, 837–845.
27 Mi, C., Hui, L., and Yi, Z. (2005) Iterative learning control of antilock braking of electric and hybrid vehicles. *IEEE Transactions on Vehicular Technology*, 54, 486–494.
28 Hak-Geun, J., Bong-Man, J., Soo-Bin, H., et al. (2000) Modeling and performance simulation of power systems in fuel cell vehicle. Proceedings of IPEMC 2000, August, 2, 671–675.

29 Johnson, R.W., Evans, J.L., Jacobsen, P., et al. (2004) The changing automotive environment: high-temperature electronics. *IEEE Transactions on Electronics Packaging Manufacturing*, 27, 164–176. See also *IEEE Transactions on Components, Packaging and Manufacturing Technology, Part C: Manufacturing*.
30 Ozpineci, B., Tolbert, L.M., Islam, S.K., et al. (2001) Effects of silicon carbide (SiC) power devices on HEV PWM inverter losses. 2, 1061–1066.
31 Kelley, R., Mazzola, M.S., and Bondarenko, V. (2006) A Scalable SiC Device for DC/DC Converters in Future Hybrid Electric Vehicles, p. 4.
32 Ozpineci, B., Tolbert, L.M., Islam, S.K., et al. (2002) Testing, characterization, and modeling of SiC diodes for transportation applications. 33rd Power Electronics Specialists Conference, June, pp. 1673–1678.
33 Dreike, P.L., Fleetwood, D.M., King, D.B., et al. (1994) An overview of high-temperature electronic device technologies and potential applications. *IEEE Transactions on Components, Packaging, and Manufacturing Technology, Part A*, 17, 594–609. [see also IEEE Transactions on Components, Hybrids, and Manufacturing Technology].
34 Ozpineci, B., Chinthavali, M.S., and Tolbert, L.M. (2005) A 55 kW three-phase automotive traction inverter with SiC Schottky diodes. Vehicle Power and Propulsion Conference, September, p. 6.
35 Ohashi, H. (2003) Power electronics innovation with next generation advanced power devices. International Telecommunications Energy Conference, October, pp. 9–13.
36 Traci, R.M., Acebal, R., and Mohler, T. (1999) Integrated thermal management of a hybrid electric vehicle. *IEEE Transactions on Magnetics*, 35, 479–483.
37 Alaoui, C. and Salameh, Z.M. (2005) A novel thermal management for electric and hybrid vehicles. *IEEE Transactions on Vehicular Technology*, 54, 468–476.
38 White, S.B., Gallego, N.C., Johnson, D.D., et al. (2004) Graphite foam for cooling of automotive power electronics. *Power Electronics in Transportation*, pp. 61–65.
39 Tatoh, N., Hirose, Y., Nagai, M., et al. (2000) Thermal management analysis of high-power electronic modules using Cu bonded AlN substrates. International Conference on Thermal and Thermomechanical Phenomena in Electronic Systems, May, vol. 2, pp. 297–302.

10

Electric Machines and Drives in HEVs

10.1 Introduction

Advances in electric machines, along with progress in power electronics, are the key enablers for electric, hybrid electric, and plug-in hybrid electric vehicles (HEVs). Induction machines, permanent magnet (PM) synchronous machines, PM brushless DC machines, and switched reluctance machines (SRMs) have all been considered in various types of vehicle powertrain applications [1–20]. Brushed DC motors, once popular for traction applications such as in streetcars, are no longer considered a proper choice due to the bulky construction, low efficiency, need for maintenance of the brush and commutator, high electromagnetic interference (EMI), low reliability, and limited speed range.

When using electric motors for powertrain applications, there are a few possible configurations. Today's electric motors, combined with inverters and associated controllers, have a wide speed range for constant torque operations, and an extended speed range for constant power operations, which make the design of the powertrain much easier. Depending on the configuration of the hybrid powertrain, the design and selection of electric motor drives can also be different. For example, for series hybrid vehicles, the powertrain motor needs to be able to provide the required torque and speed for all driving conditions. Hence, the size of the motor will be fairly large, usually rated at 100 kW or more for passenger cars. A PM motor or an induction motor is the preferred choice. For mild and micro hybrids, only a small-size motor of a few kilowatts is required. Therefore the motor can be a claw pole DC motor, or a switched reluctance motor.

Traditional automatic transmission or continuous variable transmission (CVT) used in conventional cars is no longer required in electric vehicles (EVs) and many HEVs. However, a fixed gear ratio speed reduction is often necessary. This is due to the fact that a high-speed motor has a smaller size and less weight than a low-speed machine. A two-speed automatic transmission may be beneficial in saving vehicle energy consumption.

Electric motors are extensively discussed in various textbooks and many technical publications. In this chapter, we will focus on a few unique aspects of electric motors that are specific to traction applications.

Hybrid Electric Vehicles: Principles and Applications with Practical Perspectives,
Second Edition. Chris Mi and M. Abul Masrur.

10.2 Induction Motor Drives

Induction motors are a popular choice for traction applications due to their robust construction, low cost, wide field weakening range, and high reliability. Especially for EVs, PHEVs, and HEVs that require a high-power motor, induction motors can provide more reliable operation than other types of electric motors [21–37]. However, when compared to PM motors, induction motors have a lower efficiency and less torque density.

One typical induction motor used for traction applications is the squirrel cage induction motor. An inverter is used to control the motor so that the desired torque can be delivered for a given driving condition at a certain speed. Advanced control methodologies, such as vector control, direct torque control, and field-oriented control, are popular in induction motor control for traction applications.

10.2.1 Principle of Induction Motors

The basic structure of an induction machine is shown in Figure 10.1. The two main parts of an induction motor are the stator (which houses the winding) and the rotor (which houses the squirrel cage). Both stator and rotor are made out of laminated silicon steel with thickness of 0.35, 0.5, or 0.65 mm. The laminated steel sheets are first punched with slots and are then stacked together to form the stator and rotor, respectively. Windings are put inside the stator slots while the rotor is cast in aluminum.

There are some additional components to make up the whole machine: the housing that encloses and supports the whole machine, the shaft that transfers torque, the bearing, an optional position sensor, and a cooling mechanism (such as a fan or liquid cooling tubes).

In Figure 10.1c, AX is phase a, BY is phase b, and CZ is phase c. The direction of the phase currents is for a particular moment $\omega t = 60$ electric degrees; "+" indicates positive and "–" indicates negative. It can be seen that conductor AZB forms one group and XCY forms another group. Together they create a magnetic field at 30° NW–SE. The direction of the field will change as the current changes over time.

The stator windings shown in Figure 10.1c are supplied with a three-phase AC sinusoidal current. Assume the amplitude of the currents is I_m amperes, and the angular frequency of the current is ω radians per second; then the three phase currents can be expressed as

$$\begin{aligned} i_a &= I_m \cos(\omega t) \\ i_b &= I_m \cos(\omega t - 2\pi/3) \\ i_c &= I_m \cos(\omega t - 4\pi/3) \end{aligned} \tag{10.1}$$

Since the currents of each of the three phases are functions of time, the direction of current as shown in Figure 10.1c will change with time. If we mark the direction of the current at any given time, we can see the magnetic field generated by the stator current with its peak changing position over time.

Mathematically, we can derive this magnetic field. Each of the three-phase currents will generate a magnetic field. Since the three windings are located 120° from each other

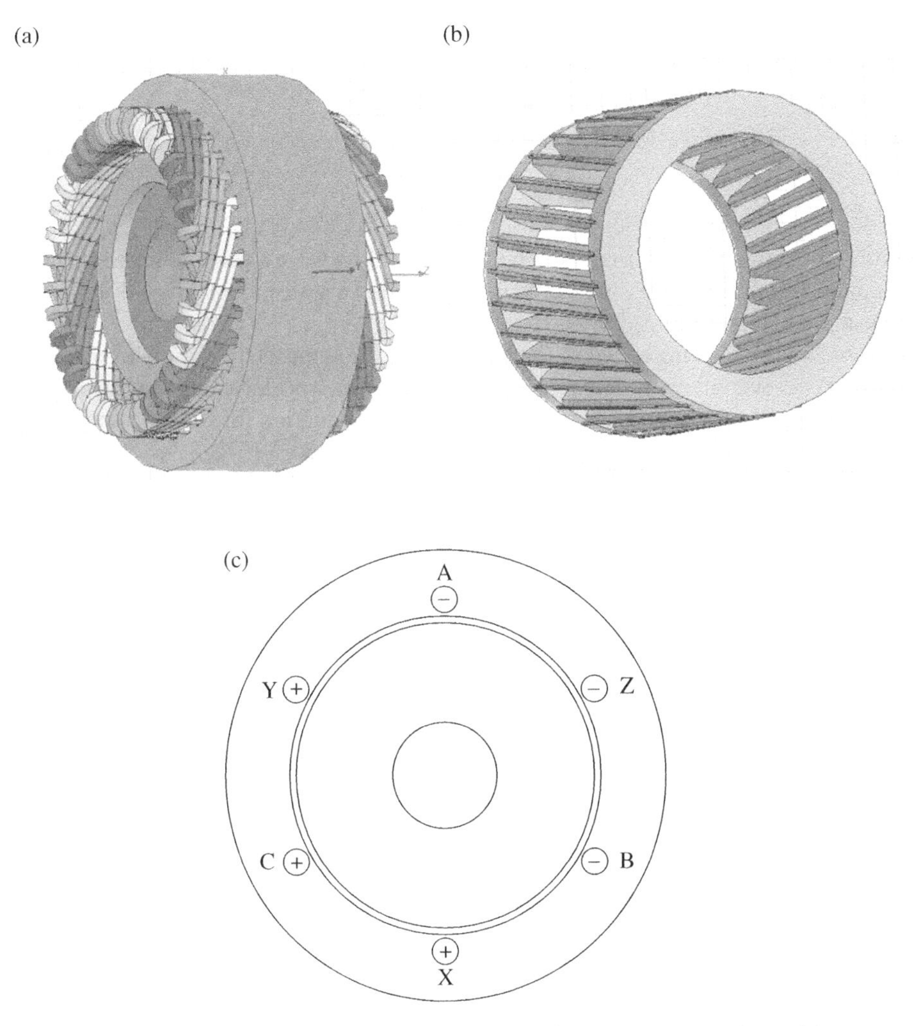

Figure 10.1 An induction motor: (a) rotor and stator assembly; (b) rotor squirrel cage; (c) cross-sectional view of an ideal induction motor with six conductors on the stator.

in space along the inside surface of the stator, the field generated by each phase can be written as follows, assuming the spatial magnetic field distribution in the air gap due to winding currents is sinusoidal by design:

$$\begin{aligned} B_a &= Ki_a(t)\cos(\theta) \\ B_b &= Ki_b(t)\cos(\theta - 120°) \\ B_c &= Ki_c(t)\cos(\theta - 240°) \end{aligned} \tag{10.2}$$

where K is a constant. Using Equations 10.1 and 10.2, considering that $\cos(\omega t)\cos(\theta) = [\cos(\omega t - \theta) + \cos(\omega t + \theta)]/2$ and $\cos(\omega t + \theta) + \cos(\omega t + \theta - 240°) + \cos(\omega t + \theta - 480°) = 0$, we get

$$\begin{aligned}B_{gap} &= Ki_a(\omega t)\cos(\theta) + Ki_b(\omega t - 120°)\cos(\theta - 120°) + Ki_c(\omega t - 240°)\cos(\theta - 240°)\\ &= \frac{3}{2}KI_m\cos(\omega t - \theta) + \frac{1}{2}KI_m[\cos(\omega t + \theta) + \cos(\omega t + \theta - 240°)\\ &\quad + \cos(\omega t + \theta - 480°)]\\ &= B_m\cos(\omega t - \theta)\end{aligned} \tag{10.3}$$

Equation 10.3 shows that the magnetic field is a traveling wave along the inner surface of the stator. In other words, the total magnetic field is a sinusoidal field with its peak rotating at angular speed ω rad/s.

Since $\omega = 2\pi f$, the rotating speed of the field will be the same as the supply frequency: f revolutions per second or $n_S = 60f$ revolutions per minute (rpm). Noting that the above derivation is based on one pair of poles, a more generic equation for the field speed (or synchronous speed) of an induction machine can be given as

$$n_S = \frac{60f}{p} \quad \text{and} \quad \omega_S = \frac{2\pi n_S}{60} = \frac{2\pi f}{p} = \frac{\omega}{p} \tag{10.4}$$

where p is the number of pairs of poles. Figure 10.2 shows the arrangement of a four-pole squirrel-cage induction motor with flux distribution.

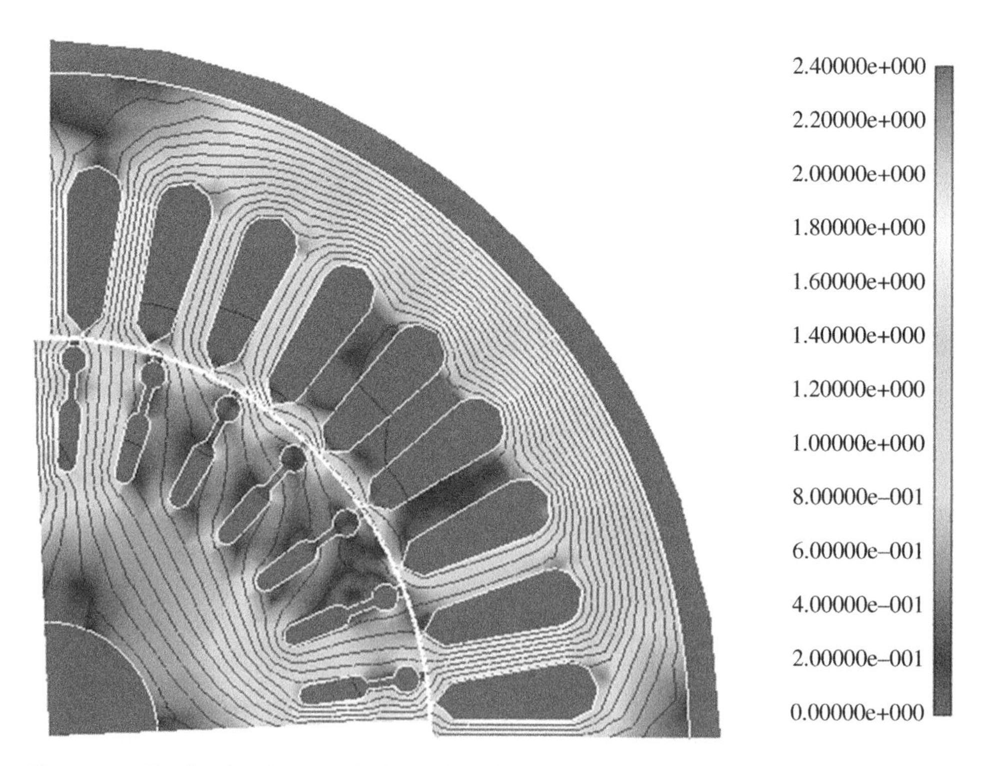

Figure 10.2 The flux distributions of a four-pole induction motor during transient finite element analysis.

Assuming initially that the rotor is stationary, an electromotive force (emf) will be induced inside the rotor bars of the squirrel cage. A current is therefore formed inside the rotor bars through the end rings. Similarly, since the field is rotating, this current will generate a force on the rotor bars (the rotor bar current is under the stator magnetic field). If the force (or torque) is sufficiently large, the rotor will start to rotate.

The maximum speed of the rotor will be less than the synchronous speed because, if the rotor reaches the synchronous speed, there will be no relative movement between the rotor bars and the stator field, hence no emf or force will be generated. The difference between the rotor speed and the synchronous speed is defined as slip s, that is, $s=(n_S-n_m)/n_S=(\omega_S-\omega_m)/\omega_S$, where n_m and ω_m are the rotor speeds in rpm and radians per second, respectively. Typical slips of induction motors are within 1–3%.

10.2.2 Equivalent Circuit of Induction Motor

We can represent the induction motor by two separate circuits, one for the stator and one for the rotor. Since the three phases are symmetrical, we only need to analyze one phase as shown in Figure 10.3. We use phasors for the analysis of the AC circuit. Here we have defined the direction of current flow using the transformer convention. It is worth noting that the rotor and the stator quantities will have different frequencies except when the rotor is stationary.

The voltage equation of the primary and the secondary circuit can be written as

$$\begin{aligned} V_S &= I_S\ R_S + j\omega L_S\ I_S + E_S \\ 0 &= I_R\ R_R + j\omega_R\ L_R\ I_R + E_R \end{aligned} \tag{10.5}$$

where V is the phase voltage, I the phase current, R the phase resistance, and L the leakage inductance of the winding. The subscripts S and R represent the stator and rotor respectively.

Since the field is rotating at synchronous speed ω_S and the rotor is rotating at speed ω_m, the speed of the magnetic field relative to the rotor bar is $\omega_S-\omega_m=s\omega_S=s\omega/p$, and $\omega_R=ps\omega_S=s\omega$ is the frequency of the rotor current.

If we multiply by k and divide by s for the both sides of the second equation in Equation 10.5, then we get

$$0=\left(k^2\frac{R_R}{s}\right)(I_R/k)+j\left(k^2\frac{\omega_R}{s}L_R\right)(I_R/k)+\frac{kE_R}{s} \tag{10.6}$$

The rotor has AC quantities at slipping frequency $\omega_R=s\omega$. By using the following, $R_R'=k^2\ R_R$, $X_R'=k^2\omega L_R=k^2\ X_R$, $I_R'=I_R/k$, $E_R'=kE_R/s$, we have

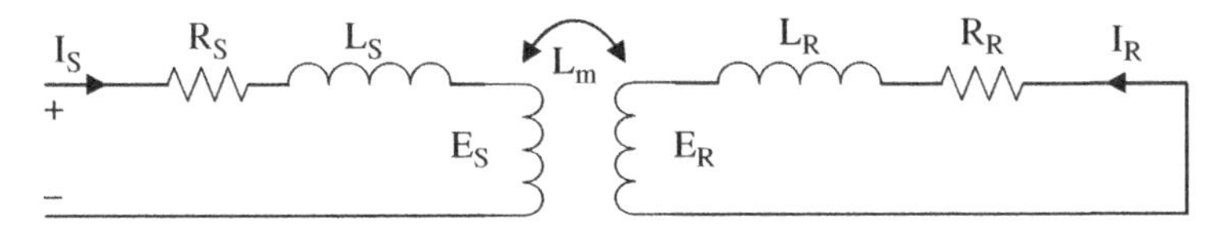

Figure 10.3 Stator and rotor circuits of an induction machine.

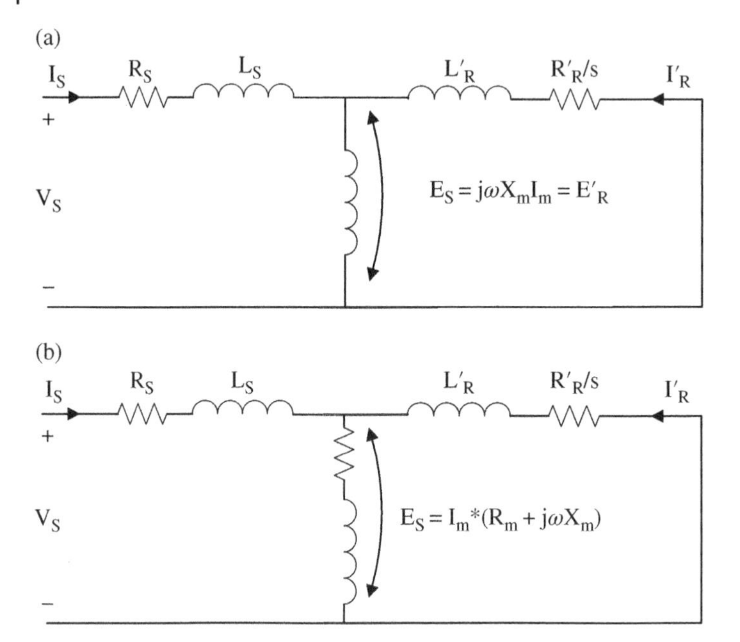

Figure 10.4 Modified equivalent circuit of an induction machine: (a), neglecting iron loss; (b), considering iron loss.

$$0 = \frac{R_R'}{s} I_R' + jX_R' \, I_R' + E_R' \tag{10.7}$$

We will choose k such that $E_S = E_R'$. We can then redraw the equivalent circuit of the induction motor as shown in Figure 10.4a. Here we neglected the magnetic loss in the stator core. If we include the magnetic loss, then the equivalent circuit can be illustrated as in Figure 10.4b.

In the above equivalent circuit, for a given voltage supply the current of the circuit can be written as

$$I_S = \frac{V_S}{R_S + j\omega L_S + (R_m + jX_m) \| (R_R'/s + jX_R')} \tag{10.8}$$

To simplify the analysis, we can neglect $R_m + j\omega X_m$. Under this assumption, the electromagnetic power transferred from the stator to the rotor is

$$P_{em} = mI_S^2 \frac{R_R'}{s} = \frac{mV_S^2}{(R_S + R_R'/s)^2 + (X_S + X_R')^2} \frac{R_R'}{s} \tag{10.9}$$

Noting that electromagnetic power or rotor power has two parts, namely, the loss of the rotor winding and power transferred to its shaft, the above equation can be rewritten as

$$P_{em} = \frac{mV_S^2}{(R_S + R_R'/s)^2 + (X_S + X_R')^2} \left[R_R' + \frac{(1-s)}{s} R' \right]_R \tag{10.10}$$

The first term represents the rotor copper loss and the second term the mechanical power on the shaft. The electromagnetic torque of the motor can be written as

$$\begin{aligned} T_{em} &= \frac{m}{\omega_S} \frac{V_S^2}{(R_S + R_R'/s)^2 + (X_S + X_R')^2} \frac{1}{s} R_R' \\ &= \frac{m}{\omega_m} \frac{V_S^2}{(R_S + R_R'/s)^2 + (X_S + X_R')^2} \frac{(1-s)}{s} R_R' \end{aligned} \tag{10.11}$$

We can plot torque T_{em} as a function of slip s from Equation 10.11 and obtain the torque–speed characteristics of an induction motor as shown in Figure 10.5.

10.2.3 Speed Control of Induction Machine

The speed of an induction motor, in rpm, can be expressed as

$$n = (1-s)n_S = (1-s)\frac{60f}{p} \tag{10.12}$$

Hence, we will have three approaches for changing the speed of an induction motor: change the number of poles, change the frequency, and change the slip:

1) **Change number of poles:** The stator winding is designed such that, by changing the winding configuration, the number of poles will change. For example, some induction motors are designed as 4/6, 6/8, or 4/8 pole capable. While changing the number of poles has been used in controlling the induction motor speed in the past,

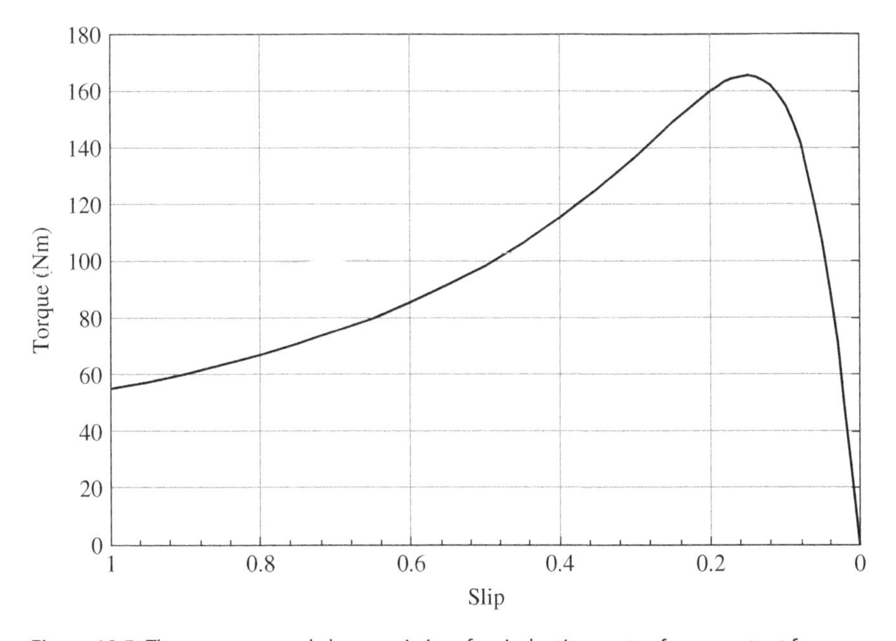

Figure 10.5 The torque–speed characteristics of an induction motor for a constant frequency and constant voltage supply.

it is used less and less today due to the complexity of the stator winding configuration and low efficiency.

2) **Change frequency of the supply voltage:** This is the most popular method for controlling induction motor speed in modern drive systems, including traction drives. This will be discussed in more detail in the next section.
3) **Change slip:** Since the electromagnetic torque of an induction motor is closely related to slip as shown in Equation 10.11, there are a few ways to change the slip to control induction motor speed:
   a) **Change the magnitude of the supply voltage:** As shown in Figure 10.6, as the voltage is changed, the speed of the motor is also changed. However, this method provides limited variable speed range since the torque is proportional to the square of voltage.
   b) **Change stator resistance or stator leakage inductance:** This can be done by connecting a resistor or inductor in series with the stator winding.
   c) **Change rotor resistance or rotor leakage inductance:** This is only applicable to wound-rotor induction motors.
   d) **Apply an external voltage to the rotor winding:** This voltage has the same frequency as the rotor back emf or rotor current. Modern, doubly fed wind power generators belong to this group. This method is only applicable to wound-rotor induction motors.

When an external resistance is in series with the stator or rotor winding, there is loss associated with this resistor. Hence the system efficiency is compromised. When an external inductor is in series with the stator or rotor, the power factor is compromised.

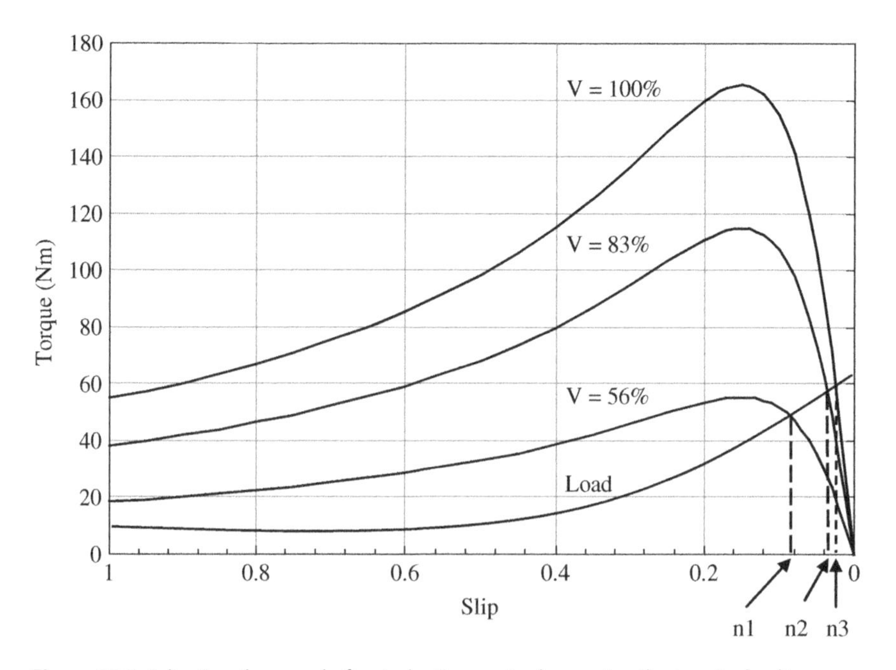

Figure 10.6 Adjusting the speed of an induction motor by varying the terminal voltage.

Hence, adding resistance or inductance is no longer a popular method in modern electric drive systems.

10.2.4 Variable Frequency, Variable Voltage Control of Induction Motors

Varying the frequency of power supply is by far the most effective and most popular method of adjusting the speed of an induction motor. If we neglect the stator resistance, leakage inductance, and the magnetic loss, the stator voltage equation can be written as

$$V_S = E_S = k_S \omega \phi = k_S 2\pi f \phi \tag{10.13}$$

where k_S is the machine constant and φ is the total flux. Hence, when changing the frequency, the stator voltage should also be changed proportionally in order to maintain a relatively constant flux so that the stator and rotor core do not get saturated, while the output torque can be maintained constant,

$$\frac{V_S}{f} = \text{constant} \tag{10.14}$$

When the frequency and voltage are adjusted, the torque–speed characteristics are as shown in Figure 10.7. Although the above expression is generally true, three observations can be made:

1) For low-frequency operations, the voltage drop across the stator resistance and inductance is no longer negligible, so the stator voltage has to be increased to compensate.

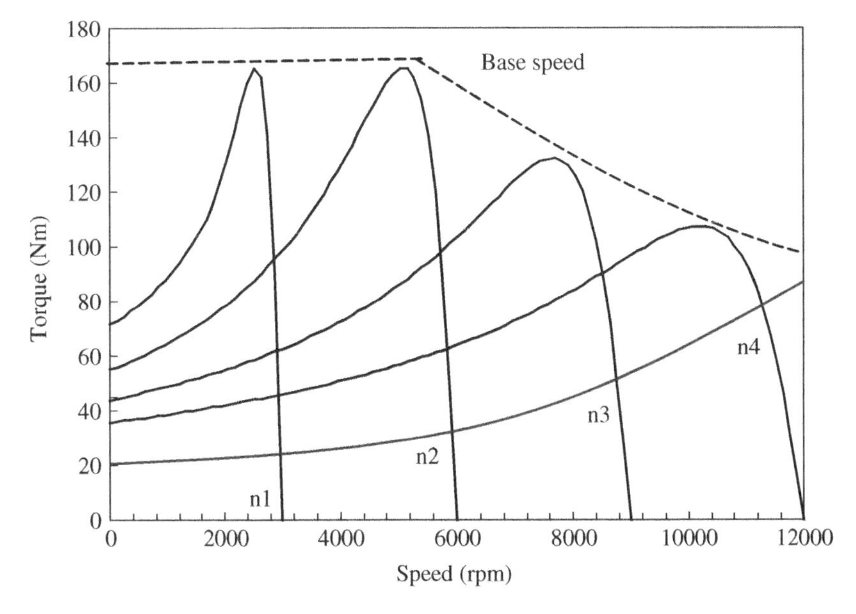

Figure 10.7 Adjusting induction motor speed using variable frequency supply. In this example, the rated speed is 6000 rpm, and the maximum speed is 12,000 rpm. The adjustable speed range $X = 2$.

2) The motor speed corresponding to the rated frequency and rated voltage is called the rated speed or base speed.
3) When the stator voltage reaches its rated supply (maximum), in order to further increase frequency (or speed), the stator flux must be reduced in order to satisfy Equation 10.13. This is called the flux weakening operation. The ratio of the maximum speed to the rated base speed of the motor is defined as the adjustable speed range, or X. Modern induction motors can achieve up to $X = 5$ adjustable speed range.

10.2.5 Efficiency and Losses of Induction Machine

The losses in an induction machine are shown in Figure 10.8. The losses include: (1) copper loss in the stator winding; (2) magnetic loss in the stator iron (or core loss or iron loss); (3) copper loss in the rotor winding; (4) windage loss due to the rotation of the rotor and frictional loss in the bearing; and (5) additional losses that cannot be accounted for by the above components, also called additional loss or stray load loss.

The power balance equations are

$$\begin{aligned} P_{em} &= P_1 - p_{cu1} - p_{iron} \\ P_{mec} &= P_{em} - p_{cu2} \\ P_2 &= P_{mec} - p_{fw} - p_{ad} \end{aligned} \tag{10.15}$$

P_1 is the input power from the voltage supply; P_{em} is the electromagnetic power transferred from the stator to the rotor; P_{mec} is the total mechanical power on the rotor shaft; P_2 is the output power to the load connected to the shaft; p_{cu1} is the copper loss of the stator winding; p_{cu2} is the copper loss of the rotor; p_{iron} is the iron loss of the stator core; p_{fw} is the frictional and windage loss; and p_{ad} is the stray load loss.

The efficiency can be expressed as

$$\eta = \frac{P_2}{P_1} = \frac{P_2}{P_2 + p_{cu1} + p_{iron} + p_{cu2} + p_{fw} + p_{ad}} \tag{10.16}$$

One aspect of traction motors for modern HEVs is high-speed operation. Traditionally, laminated silicon steel sheets were designed for use at low frequencies (50 or 60 Hz), and today's traction drives typically operate at about 6000–15,000 rpm. With four-pole designs, the operating frequency is 500 Hz. Some traction motors

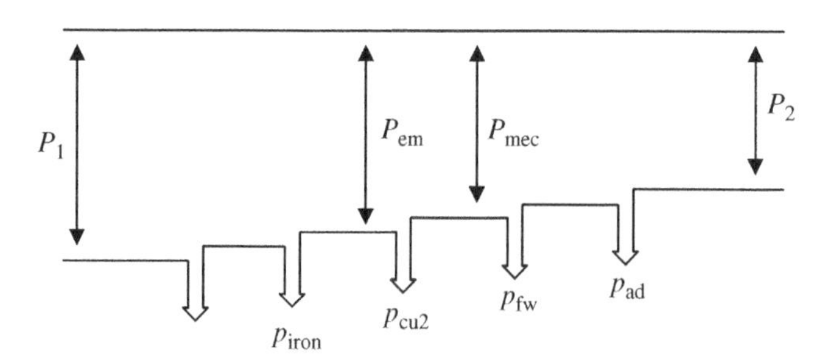

Figure 10.8 Losses in an induction motor.

operate at frequencies as high as 800–1200 Hz. Since eddy current loss and hysteresis loss are proportional to frequency or the square of frequency, the core loss will be significant at high frequencies. In order to keep the core loss within a reasonable range, the magnetic flux in the iron has to be relatively lower than that used in low-speed motors, and the thickness of the silicon steel sheets may have to be reduced as well.

The second aspect is that the inverter-operated induction motor will contain harmonics in its voltage and current. These harmonics will introduce additional losses in the winding and stator and rotor core. As is well known, the eddy current loss can be doubled in many induction motors due to the pulse-width-modulated (PWM) supply. These additional losses may cause excessive temperature rise which must be considered during the design and analysis of induction motors.

10.2.6 Additional Loss in Induction Motors Due to PWM Supply

To analyze the additional losses in an induction motor due to PWM operations, we first take a look at the PWM waveform. The general principle of bipolar PWM supply is shown in Figure 10.9, where the triangular waveform (carrier, V_{tri}) of switching frequency f_c is compared to the control signal ($V_{control}$) of fundamental frequency f_1. The intersections of the two waveforms determine the switching points of power devices. The ratio of switching frequency f_c to fundamental frequency f_1 is defined as the frequency modulation ratio, $m_f = f_c/f_1$. The ratio of the amplitude of the control waveform $V_{control}$ to that of the triangle waveform V_{tri} is defined as the amplitude modulation ratio, $m_a = V_{control}/V_{tri}$ [38, 39].

There are many different methods for determining the switching points, such as the natural sampling rule, regular sampling, and the selected harmonic elimination rule.

The symmetrical regular sampling rule is a basic method in sampling. The switch points can be determined from Figure 10.9 for this sampling method:

$$\alpha_k = \frac{\pi}{2m_f}\left\{2k-1+(-1)^k \; m_a \sin\left[(k+m)\frac{\pi}{m_f}\right]\right\} \quad k=1,2,\ldots,2m_f$$
$$\text{when } (-1)^k = 1, m=-1; \text{when } (-1)^k = -1, m=0 \tag{10.17}$$

where α_k is the switching point.

The output voltage of the inverter contains the fundamental voltage and other high-frequency harmonic voltages. The general Fourier series of the output voltage can be given by

$$v(t) = \sum_{\infty}^{n=1} v_n(t) = \sum_{\infty}^{n=1}(a_n \cos n\omega t + b_n \sin n\omega t) = \sum_{\infty}^{n=1} V_n \sin(n\omega t + \varphi_n) \tag{10.18}$$

where ω is the fundamental angular frequency, n is the order of harmonics, V_n is the amplitude of the nth harmonic, a_n and b_n are the Fourier coefficients; ϕ_n is the phase angle of the nth harmonic, and

$$a_n = \frac{V_{dc}}{n\pi}\sum_{2m_f}^{k=1}(-1)^k \sin(n\alpha_k) \tag{10.19}$$

$$b_n = \frac{V_{dc}}{n\pi}\sum_{2m_f}^{k=1}(-1)^k \cos(n\alpha_k) \tag{10.20}$$

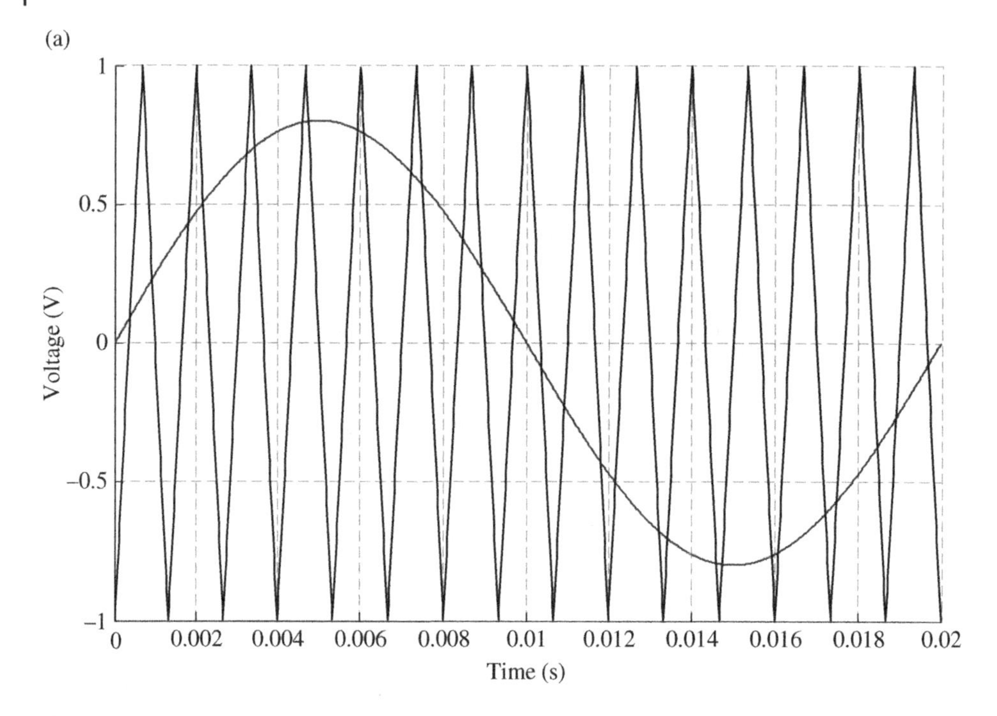

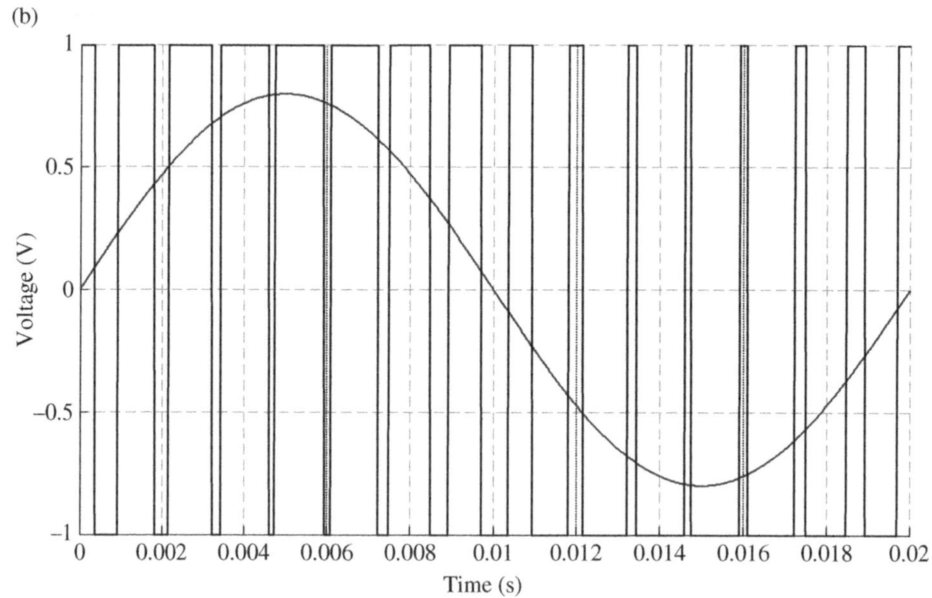

Figure 10.9 Principle of bipolar PWM supply ($m_a = 0.8$, $m_f = 15$, $f_1 = 50\,\text{Hz}$): (a) carrier waveform V_{tri} and control waveform $V_{control}$; (b) PWM output and its fundamental component.

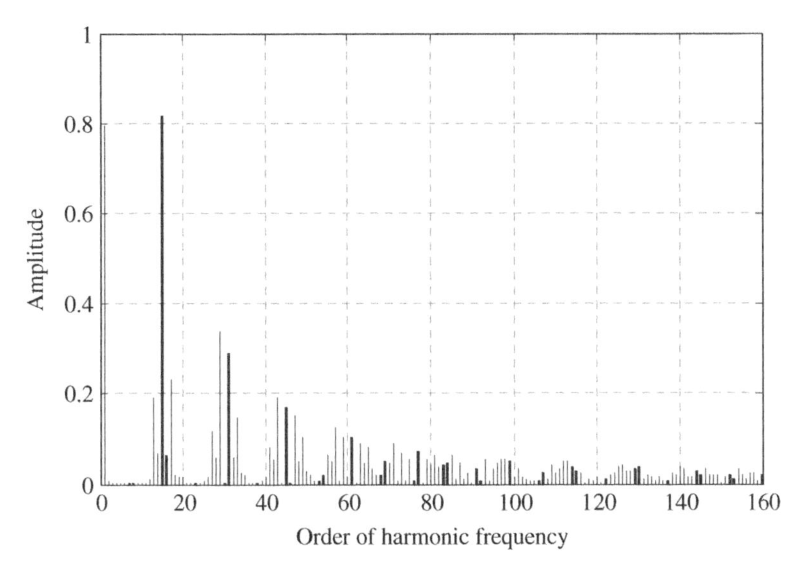

Figure 10.10 Harmonic frequency analysis of output voltage.

$$V_n = \sqrt{a_n^2 + b_n^2} = \frac{V_{dc}}{n\pi}\sqrt{\left[\sum_{k=1}^{2m_f}(-1)^k \cos(n\alpha_k)\right]^2 + \left[\sum_{k=1}^{2m_f}(-1)^{k+1}\sin(n\alpha_k)\right]^2} \tag{10.21}$$

$$\varphi_n = \arctan\left(\frac{b_n}{a_n}\right) \tag{10.22}$$

where V_{dc} is the DC link voltage.

According to Equation 10.21, the amplitude of the harmonics is dependent on the DC link voltage V_{dc}, the order of the harmonic waveform n, and the switching points α_k. These switching points are determined from the frequency modulation ratio m_f and amplitude modulation ratio m_a. Figure 10.10 shows the harmonic spectrum of the PWM output waveform for $m_f = 15$ and $m_a = 0.8$. The PWM output wave contains carrier frequency-related harmonics with modulation frequency-related sidebands in the form of $mf_c \pm nf_1$, where m and n are integers and $m + n$ is an odd integer.

The total iron loss density p_{iron} is commonly expressed in the following form for sinusoidal varying magnetic flux density B with angular frequency ω [40]:

$$p_{iron} = p_h + p_c = k_h\ B^{\beta}\omega + k_c\ B^2\ \omega^2 (\mathrm{W/m^3}) \tag{10.23}$$

where p_h and p_c are the hysteresis loss density and the classical eddy current loss density respectively, k_h and k_c are the hysteresis and eddy current constants respectively, and β is the Steinmetz constant – all of which depend on the lamination material. Because the heavy iron loss increase in the case of PWM supply is mainly induced by the eddy current loss increment, this section will focus on the analysis of the eddy current loss.

Equation 10.23 is only valid for flux densities under sinusoidal time-varying conditions. Under PWM supply, there are many high-frequency harmonic flux density components. With the assumption of a linear material, the flux density harmonic amplitude is proportional to the voltage harmonic amplitude. So the harmonic components of the PWM voltage output can be analyzed, and the eddy current loss of each harmonic component can be calculated. Although such an assumption is not in accordance with the magnetic property of the core material, considering that the objective here is to compare the iron losses influenced by PWM supply, the absolute error caused by the assumption does not affect the final results.

In order to give a clear view of the iron losses associated with the PWM parameters, a single phase was used. Under no-load condition (rotor winding open-circuit condition), the circuit equation is

$$v(t)=e(t)+i(t)R+L_{\sigma}\frac{di(t)}{dt} \tag{10.24}$$

where $v(t)$ is the applied voltage, $e(t)$ is the back emf, $i(t)$ is the current, L_{σ} is the leakage inductance of the stator winding, and R is the resistance of the stator winding.

When neglecting the voltage drop on winding resistor R and leakage inductance L_{σ},

$$v(t)\approx e(t)=N_1\frac{d\Phi(t)}{dt}=N_1A\frac{dB(t)}{dt} \tag{10.25}$$

where N_1 is the number of turns of the winding and A is the cross-sectional area of the core. Under the linear material assumption, the flux density is

$$B(t)=-\sum_{\infty}^{n=1}B_n\cos(n\omega t+\varphi_n) \tag{10.26}$$

where B_n is the flux density of the nth harmonic:

$$B_n=\frac{V_n}{N_1An\omega} \tag{10.27}$$

The eddy current loss can then be obtained by substituting Equation 10.27 into Equation 10.23:

$$\begin{aligned} p_c &= \sum_{\infty}^{n=1}p_n=\sum_{\infty}^{n=1}k_cB_n^2(n\omega)^2=\sum_{\infty}^{n=1}k_c\left(-\frac{V_n}{N_1An\omega}\right)^2(n\omega)^2 \\ &= \frac{k_c}{(N_1A)^2}\sum_{\infty}^{n=1}(V_n)^2 \end{aligned} \tag{10.28}$$

Combining Equations 10.18 and 10.26, the eddy current losses can be derived as

$$p_c=\frac{k_c}{(N_1A\pi)^2}\sum_{\infty}^{n=1}\left(\frac{V_{dc}}{n}\right)^2\left\{\left[\sum_{2R}^{k=1}(-1)^k\cos(n\alpha_k)\right]^2+\left[\sum_{2R}^{k=1}(-1)^k\sin(n\alpha_k)\right]^2\right\} \tag{10.29}$$

$$p_c=K\sum_{\infty}^{n=1}\left(\frac{V_{dc}}{n}\right)^2\left\{\left[\sum_{2R}^{k=1}(-1)^k\cos(n\alpha_k)\right]^2+\left[\sum_{2R}^{k=1}(-1)^k\sin(n\alpha_k)\right]^2\right\} \tag{10.30}$$

where

$$K = \frac{k_c}{\left(N_1 A\pi\right)^2}$$

From the above equations, it can be concluded that the eddy current loss is related to the amplitude of the fundamental and other high-order harmonics. Furthermore, it is related to the switching points α_k and DC link voltage V_{dc}. Switching point α_k is related to the sampling method of switching points. If the regular sampling method is adopted, α_k is determined by the frequency modulation ratio and the amplitude modulation ratio. Each harmonic component will contribute to the total eddy current loss. According to Figure 10.10, the fundamental voltage and the harmonics at the switching frequency contribute more to the eddy current loss.

Figure 10.11 shows the eddy current loss versus switching frequency for an induction motor, where the eddy current loss ratio is defined as the loss from PWM supply to the loss of sinusoidal supply. It can be seen that the eddy current loss is more than doubled with PWM supply. Figure 10.12 shows the iron loss of a 2 kW, 208 V induction motor when operated at 80 V with respect to modulation ratio (the DC bus voltage is adjusted to maintain the fundamental voltage at 80 V). It can be seen that the iron loss increases with the reduction of modulation ratio. The iron loss increases as the amplitude modulation ratio decreases.

An experiment was carried out on a 3 hp induction motor. The motor is a four-pole, 60 Hz, 208 V, Y-connected, three-phase induction motor. Figure 10.13 shows the cross-section of the machine. The machine has a double-layer, three-phase stator winding, with 36 slots in the stator and 24 slots in the rotor.

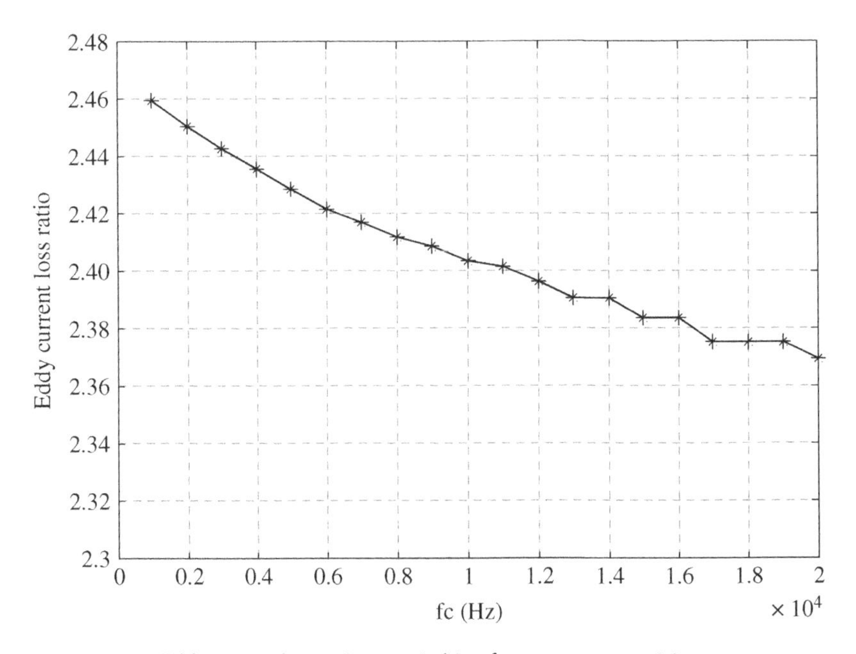

Figure 10.11 Eddy current loss ratio vs. switching frequency at $m_a = 0.9$.

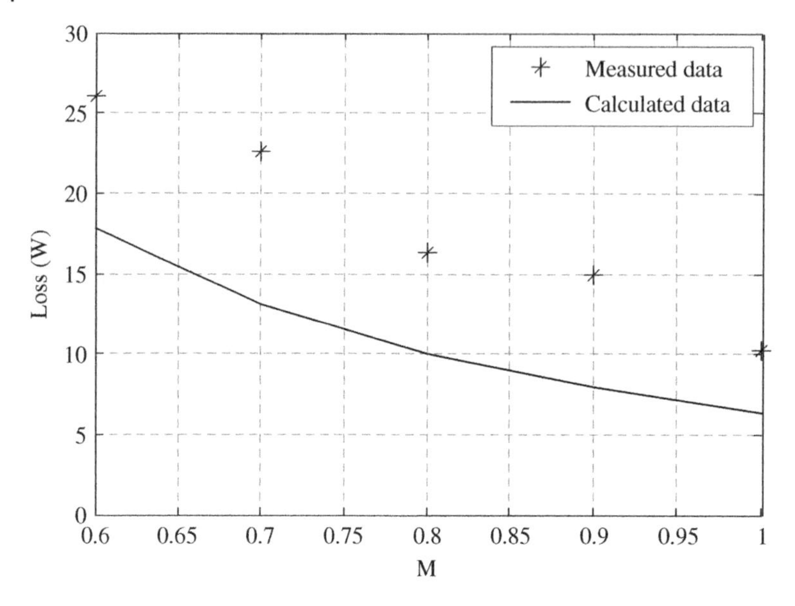

Figure 10.12 The effect of amplitude modulation ratio on PWM iron losses of induction motor for a 2 hp induction motor.

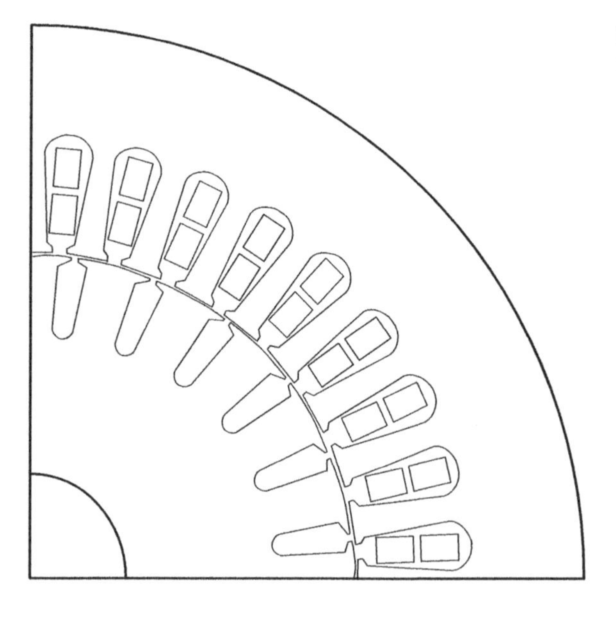

Figure 10.13 Experimental induction motor.

The experiment bench is shown in Figure 10.14, which includes the power supply, a unipolar PWM inverter, a power analyzer, an induction motor, a DC generator (as load), and a DC electronic load connected to the output of the DC generator. The solid line shows that the induction motor is operated by the sinusoidal supply directly and the dotted line shows that it is driven by the PWM inverter. The power analyzer is used to capture the line voltage and phase current.

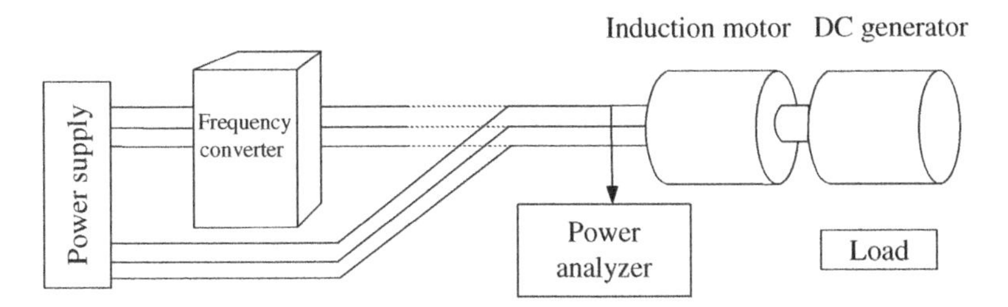

Figure 10.14 Experiment bench.

The input power of the induction motor can be obtained directly from the power analyzer. Both no-load and load tests were performed. Moreover, no-load tests with a fundamental frequency of 60 Hz and switching frequency ranging from 3.5 to 9 kHz were carried out to evaluate the influences of switching frequency on the additional losses.

To understand the mechanism of the additional losses and their impact on the temperature, a two-dimensional (2D) finite element model (FEM) was developed to compute the losses. A three-dimensional (3D) FEM model was set up to estimate the thermal profile within the motor.

The simulated iron losses are shown in Figures 10.15 and 10.16 for sinusoidal and PWM supply, respectively. Under no-load operation, the iron loss with PWM supply is 104 W, a 48.2 W increase over the iron loss when operated from the sinusoidal supply (55.8 W). However, when the motor is attached to the load, additional iron losses are induced by the harmonics in the PWM supply, with iron loss at 139.1 W or only 31.1 W more than the sinusoidal iron loss (108 W). All the losses from the simulation and experiments are listed in Tables 10.1 and 10.2 respectively.

The test results are consistent with the simulation results at fixed switching frequencies. In order to understand the impact of switching frequency on the additional losses, tests have been conducted using several different switching frequencies (3.5–9 kHz) for the PWM inverter. The results are shown in Figure 10.17. The total losses decrease proportionally with the increase of frequency, which is consistent with prior work [41].

By assigning all the losses into the 3D FEM model and incorporating the necessary boundary conditions based on the model of an actual motor, the temperature profile can be obtained. Figure 10.18 shows the temperature distribution in the stator with no load in sinusoidal supply. Tables 10.3 and 10.4 show the entire temperature of the motor, and the location of test points as shown in Figure 10.19 except point 1, which is in the center of the outer surface housing.

It can be seen from Tables 10.3 and 10.4 that the temperature rise of the induction motor is significantly higher when driven by the PWM inverter rather than by the sinusoidal power supply. In conclusion, for induction motor design it is important to consider the additional losses generated due to the harmonics in the PWM supply and possible excessive temperature rise inside the machine. Extra care needs to be taken in applications that experience a wide ambient temperature range, such as the ones used in HEVs.

(a)

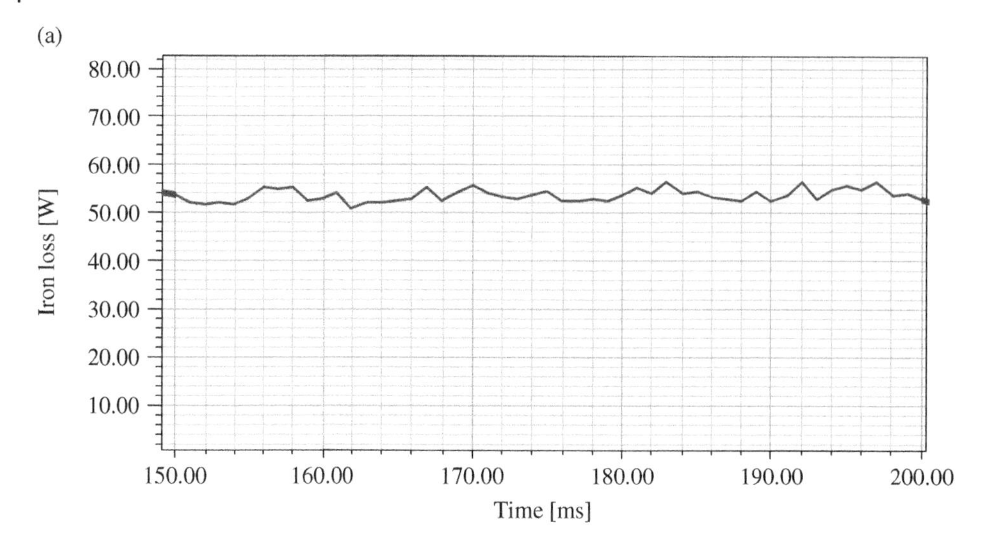

(b)

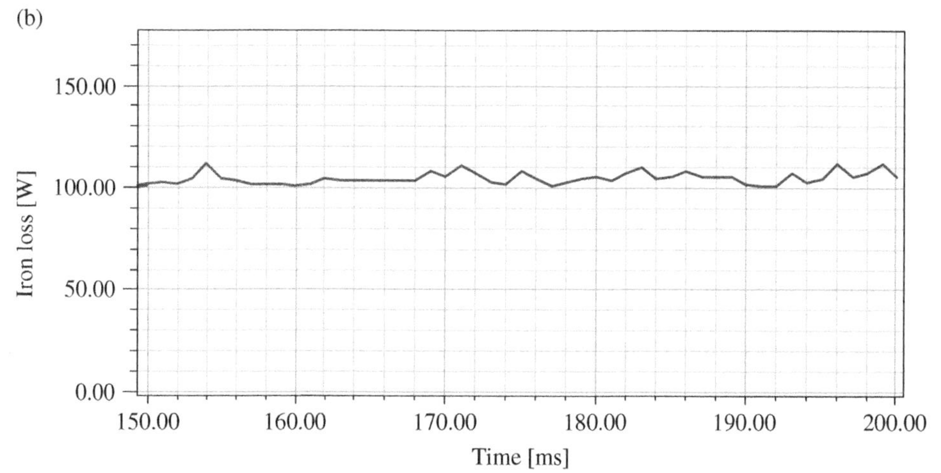

Figure 10.15 Iron losses from simulation for sinusoidal supply: (a) no load and (b) with load.

10.2.7 Field-Oriented Control of Induction Machine

With field-oriented control, an induction machine can perform somewhat like a DC machine. This section explains the theory and implementation of the field-oriented control of an induction machine [42].

When expressed in phasors, the voltage equation for a three-phase induction machine with three symmetrical stator windings is given as

$$V_S = R_S \ i_S + p\lambda_S \tag{10.31}$$

$$V_R = R_R \ i_R + p\lambda_R \tag{10.32}$$

where p is the differential operand d/dt, and V, I, and λ are phasors of voltage, current, and flux linkage respectively. Subscript S relates to stator quantities and R refers to rotor

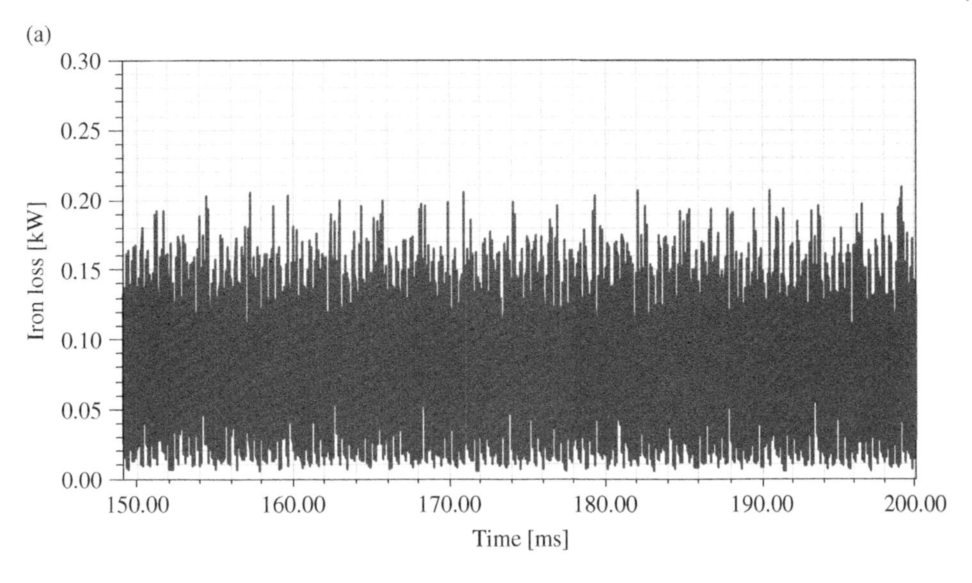

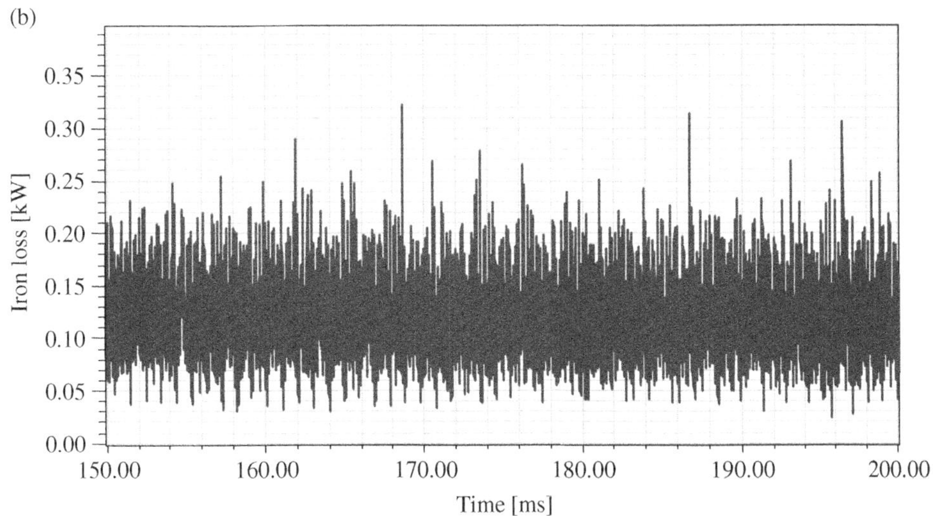

Figure 10.16 PWM losses from simulation: (a) no load and (b) with load.

Table 10.1 Losses from simulation.

	No load (W)		Load (W)	
	PWM	Sinusoidal	PWM	Sinusoidal
Iron loss	88.5	53.9	130.0	118.6
Copper loss	139.5	50.1	150.1	113.7

Table 10.2 Losses from experiments.

	No load (W)		Load (W)	
	PWM	Sinusoidal	PWM	Sinusoidal
Iron loss	104	55.8	139.1	108
Copper loss	133.7	58.8	148.9	124.5

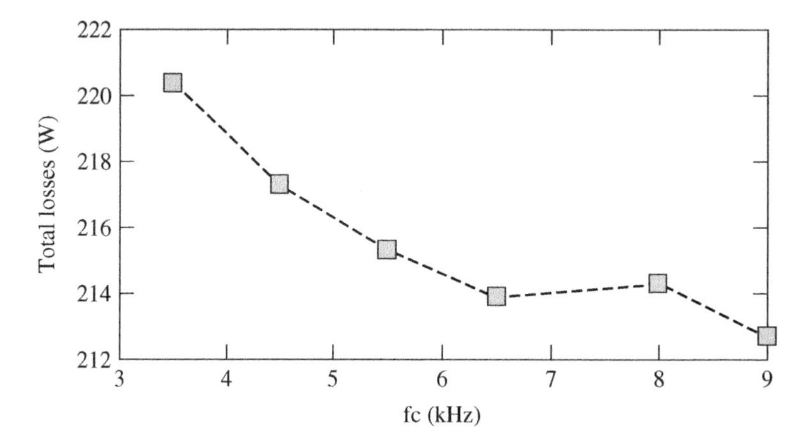

Figure 10.17 Loss with different switching frequencies.

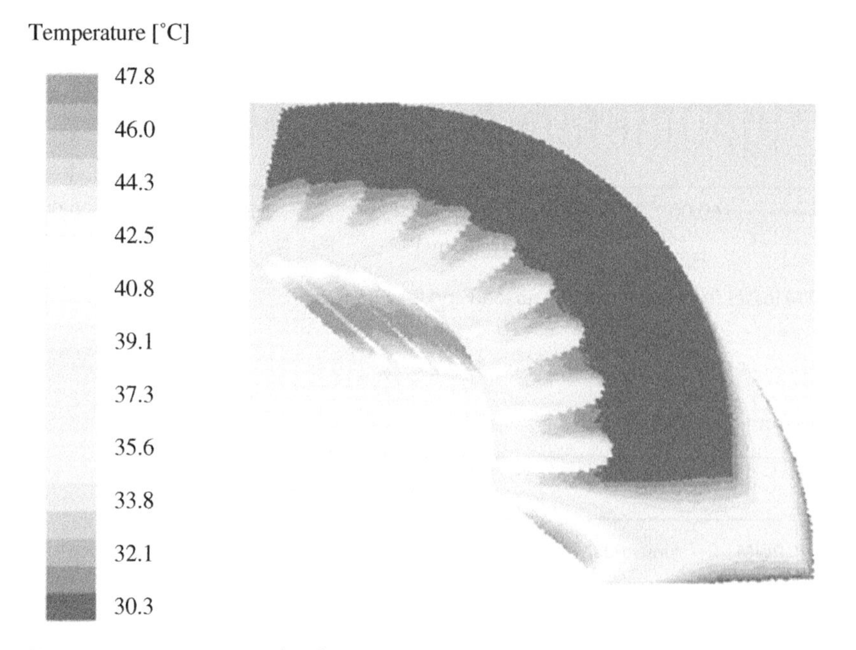

Figure 10.18 Temperature distribution in stator.

Table 10.3 Temperature profile with no load.

	Experiment (°C)		Simulation (°C)	
Point 1	31.4	40.9	30.3	37.3
Point 2	36.9	42.4	34.7	38.7
Point 3	38.2	47.1	35.6	45.6
Point 4	39.5	52.7	36.3	49.3
Point 5	40.2	51.1	43.3	47.3
Point 6	39.6	49.7	34.4	45.4
Point 7	50.8	55.7	53.6	53.6

Table 10.4 Temperature profile with load.

	Experiment (°C)		Simulation (°C)	
Point 1	34.2	41.8	32.7	37.7
Point 2	41.4	47.8	37.9	42.9
Point 3	47	52.8	44.6	50.6
Point 4	50.8	57.2	49.7	53.7
Point 5	60.5	63.9	55.5	56.3
Point 6	59.9	63.3	52.3	55.3
Point 7	67.6	71.7	63.6	66.7

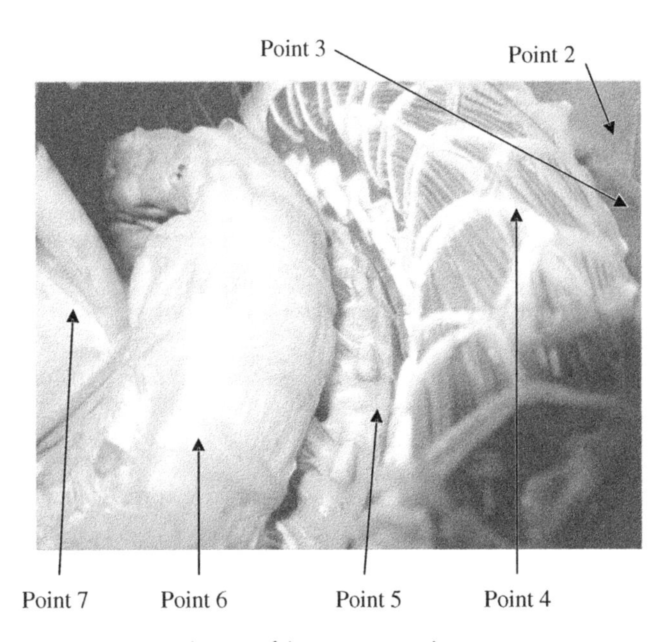

Figure 10.19 Inside view of the experimental motor.

quantities. Equations 10.31 and 10.32 are expressed in stator and rotor coordinates respectively. Therefore, stator frame S is stationary and rotor frame R is rotational (rotor quantities are at rotor frequency or slip frequency).

Suppose there is a frame B, and the angle between the stator and this frame B is δ, therefore the angle between the rotor and this frame is $(\delta - \theta)$. Multiplying Equation 10.31 by $e^{-j\delta}$ and Equation 10.32 by $e^{-j(\delta-\theta)}$, we get

$$\begin{aligned} V_S \cdot e^{-j\delta} &= R_S \; i_S \cdot e^{-j\delta} + p\lambda_S \cdot e^{-j\delta} \\ V_R \cdot e^{-j(\delta-\theta)} &= R_R \; i_R \cdot e^{-j(\delta-\theta)} + p\lambda_R \cdot e^{-j(\delta-\theta)} \end{aligned} \tag{10.33}$$

Let

$$\begin{aligned} V_S^{(B)} &= V_S \times e^{-j\delta}, \quad V_R^{(B)} = V_R \times e^{-j(\delta-\theta)} \\ i_S^{(B)} &= i_S \times e^{-j\delta}, \quad i_R^{(B)} = i_R \times e^{-j(\delta-\theta)} \\ \lambda_S^{(B)} &= \lambda_S \times e^{-j\delta}, \quad \lambda_R^{(B)} = \lambda_R \times e^{-j(\delta-\theta)} \end{aligned} \tag{10.34}$$

By employing the equation

$$p\left(\lambda_S^{(B)}\right) = p\left(\lambda_S \cdot e^{-j\delta}\right) = -j \cdot \lambda_S \left(p\delta\right) \cdot e^{-j\delta} + p\lambda_S \cdot e^{-j\delta} \tag{10.35}$$

or

$$p\lambda_S \cdot e^{-j\delta} = p\left(\lambda_S^{(B)}\right) + j \cdot \lambda_S \left(p\delta\right) \cdot e^{-j\delta} \tag{10.36}$$

$$p\lambda_R \cdot e^{-j(\delta-\theta)} = p\left(\lambda_R^{(B)}\right) + j \cdot \lambda_R \; e^{-j\delta} \, p\left(\delta - \theta\right) \tag{10.37}$$

Equations 10.31 and 10.32 can then be transferred to a general frame B, where all space phasors are expressed in frame B with the superscript (B) as shown below:

$$V_S^{(B)} = R_S \; i_S^{(B)} + p\lambda_S^{(B)} + j \cdot \lambda_S^{(B)} p\delta \tag{10.38}$$

$$V_R^{(B)} = R_R \; i_R^{(B)} + p\lambda_R^{(B)} + j \cdot \lambda_R^{(B)} p\left(\delta - \theta\right) \tag{10.39}$$

The superscript (B) will be omitted further in this section for convenience.

When expressed in phasors, the flux linkage can be expressed as

$$\lambda_S = \left(L_m + L_{1\sigma}\right) \cdot i_S + L_m \cdot i_R \tag{10.40}$$

$$\lambda_R = L_m \cdot i_S + \left(L_m + L_{2\sigma}\right) \cdot i_R \tag{10.41}$$

where L_m is the stator inductance and $L_{1\sigma}$ and $L_{2\sigma}$ are the stator and rotor leakage inductance respectively.

Note that although the phasors are in a different frame, the stator flux and rotor flux are rotating at the same speed.

For squirrel cage induction machines, the rotor current i_R is not accessible. Therefore, a fictitious rotor magnetizing current i_{mr} is defined such that the rotor flux can be expressed in terms of this fictitious rotor magnetizing current and stator inductance in the same way as in Equation 10.40:

$$\lambda_R = i_{mr} \cdot L_m \tag{10.42}$$

The rotor current can then be expressed as a function of magnetizing current and stator current from Equation 10.41:

$$i_R = \frac{i_{mr} - i_S}{1+\sigma} \tag{10.43}$$

where

$$\sigma = L_{2\sigma} / L_m \tag{10.44}$$

Substituting Equations 10.42 and 10.43 into Equation 10.41 and considering that V_R is normally set to 0 for squirrel cage induction motors, the rotor equation can be rewritten as

$$0 = i_{mr} - i_S + T_r \cdot p i_{mr} + j \cdot T_r \; i_{mr} \cdot p(\delta - \theta) \tag{10.45}$$

where T_r is the rotor time constant which can be expressed as

$$T_r = L_m(1+\sigma)/R_R \tag{10.46}$$

As stated above, the rotor magnetizing current is a fictitious current. The magnitude of this current can be observed through the following approach. If the rotor equation is written in the stator frame then $\delta = 0$, $p\theta$ is equal to the speed of the rotor ω, and Equation 10.45 has the following form:

$$0 = i_{mr} - i_S + T_r \cdot p i_{mr} - j \cdot T_r \; i_{mr} \cdot \omega \tag{10.47}$$

Since this equation is written in the stator frame, we can find the α and β components of phasors i_S and i_{mr}:

$$\begin{aligned} i_S &= i_{S\alpha} + j i_{S\beta} \\ i_{mr} &= i_{mr\alpha} + j i_{mr\beta} \end{aligned} \tag{10.48}$$

Therefore Equation 10.47 becomes

$$\begin{aligned} \frac{di_{mr\alpha}}{dt} &= \frac{1}{T_r}(i_{s\alpha} - i_{mr\alpha}) - i_{mr\beta} \cdot \omega \\ \frac{di_{mr\beta}}{dt} &= \frac{1}{T_r}(i_{s\beta} - i_{mr\beta}) + i_{mr\alpha} \cdot \omega \end{aligned} \tag{10.49}$$

Stator current can be easily transferred from the *abc* system to the $\alpha\beta$ system. Equation 10.49 can be implemented discretely in the time domain, therefore, $i_{mr\alpha}$ and $i_{mr\beta}$ can be observed. Once this has been done, i_{mr} and δ_r can finally be calculated:

$$i_{mr} = \sqrt{i_{mr\alpha}^2 + i_{mr\beta}^2}, \quad \cos(\delta_r) = i_{mr\alpha} / i_{mr}, \quad \sin(\delta_r) = i_{mr\beta} / i_{mr} \tag{10.50}$$

where δ_r is the angle between the fictitious current i_{mr} and the stator current $i_{S\alpha}$ as shown in Figure 10.20.

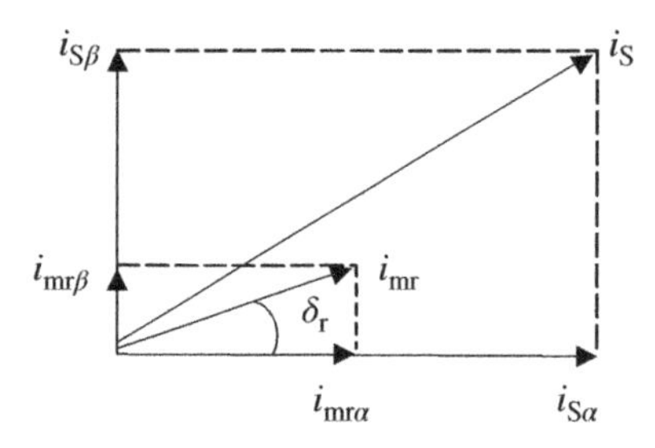

Figure 10.20 Stator and rotor current in α, β coordinates.

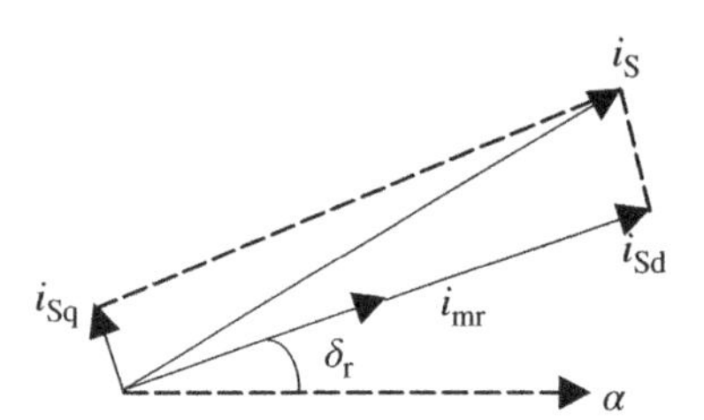

Figure 10.21 Stator current in *d, q* and α, β coordinates.

If the frame is chosen such that B is aligned with λ_R as shown in Figure 10.21, i_{mr} will only have real components. Therefore this rotor equation can then be decomposed into its direct and quadrature components as

$$\begin{aligned} i_{sd} &= i_{S\alpha}\cos\delta_r + i_{s\beta}\sin\delta_r \\ i_{sq} &= -i_{S\alpha}\sin\delta_r + i_{s\beta}\cos\delta_r \end{aligned} \tag{10.51}$$

From Equation 10.45, when i_s is decomposed to d,q components, the equation can be written as

$$i_{mr} - i_{sd} + T_r \cdot p i_{mr} = 0 \tag{10.52}$$

$$-i_{sq} + T_r \; i_{mr} \cdot p(\delta - \theta) = 0 \tag{10.53}$$

From Equation 10.52 it can be seen that i_{mr} is only related to i_{sd}. Therefore, i_{mr} can be controlled by controlling i_{sd}.

The torque in the machine is

$$T_q = \frac{3}{2} \cdot \frac{p}{2} \cdot (\lambda_S \times i_S) \tag{10.54}$$

which has to be balanced with the load and acceleration torque:

$$T_q = T_L + J \cdot p\omega \tag{10.55}$$

where T_q is the developed torque and T_L is the load torque; ω is the angular speed of the motor. If θ is the angle between the stator and the rotor, then $\omega = p\theta$.

It can also be proved that i_{sq} is directly related to motor torque as follows. By substituting i_S and i_R into the torque in Equation 10.54, the torque can be derived:

$$T_q = \frac{3}{2} \cdot \frac{p}{2} \cdot \frac{L_S}{1+\sigma} i_{mr} \; i_{sq} \tag{10.56}$$

Magnetizing current i_{mr} can be controlled by controlling the real component of stator current, and torque control is achieved by controlling the imaginary component of the stator current.

For ease of implementing control, we will introduce the per-unit system. A per-unit system is essentially a system of dimensionless parameters occurring in a set of wholly or partially dimensionless equations. This kind of system can extensively simplify the phenomena of problems. The parameters of the machines fall in a reasonable narrow numerical range when expressed in a per-unit system related to their ratings and therefore this is extremely useful in simulating machine systems and implementing the control of electrical machine by digital computers. Generally, rated power and frequency can be chosen respectively as the base values of power and frequency for normalization, whereas the peak values of rated phase current and phase voltage may be chosen respectively as the base values of current and voltage. Derived base values of impedance, inductance, and flux leakage are as follows (with subscript B indicating the variable as base value):

$$\begin{aligned} Z_B &= V_B / I_B \\ L_B &= Z_B / \omega_B \\ \lambda_B &= L_B \; I_B \end{aligned} \tag{10.57}$$

Normalized torque can be expressed as

$$T_{qB} = \frac{3}{2}\frac{p}{2}\lambda_B \; I_B = \frac{3}{2}\frac{p}{2} L_B \; I_B^2 \tag{10.58}$$

The torque equation can then be normalized. Dividing Equation 10.56 by Equation 10.58, we get

$$T_q^* = \frac{1}{1+\sigma}\frac{L_S}{L_B}\frac{i_{mr}}{I_B}\frac{i_{sq}}{I_B} = \frac{L_S^*}{1+\sigma} i_{mr}^* \; i_{sq}^* \tag{10.59}$$

Superscript * donates the normalized value. For convenience, superscript * will be omitted in further derivations. To implement the control strategy, a technique has to be developed to identify the magnitude of the magnetizing current i_{mr} and the angle δ_r.

There are two ways to implement the flux observer of Equation 10.49. One way is to take the Laplace transform of Equation 10.49 and apply a bilinear transformation to convert the Laplace transform to the z transform. The inverse z transform can be used to obtain $i_{mr\alpha}$ and $i_{mr\beta}$ in the discrete time domain. An alternative method is to discretize Equation 10.49 directly in the time domain. Assuming that the sample time is T_s, then the following equation can be obtained from Equation 10.49:

$$\begin{aligned} \frac{\left(i_{mr\alpha_i} - i_{mr\alpha_{i-1}}\right)}{T_s} &= \frac{1}{T_r}\left(\frac{i_{s\alpha_i} + i_{s\alpha_{i-1}}}{2} - \frac{i_{mr\alpha_i} + i_{mr\alpha_{i-1}}}{2}\right) - \frac{i_{mr\beta_i} + i_{mr\beta_{i-1}}}{2}\cdot\omega \\ \frac{\left(i_{mr\beta_i} - i_{mr\beta_{i-1}}\right)}{T_s} &= \frac{1}{T_r}\left(\frac{i_{s\beta_i} + i_{s\beta_{i-1}}}{2} - \frac{i_{mr\beta_i} + i_{mr\beta_{i-1}}}{2}\right) - \frac{i_{mr\alpha_i} + i_{mr\alpha_{i-1}}}{2}\cdot\omega \end{aligned} \tag{10.60}$$

Therefore $i_{mr\alpha}$ and $i_{mr\beta}$ can be derived from Equation 10.60:

$$\begin{aligned} i_{mr\alpha_i} &= \frac{1-\kappa}{1+\kappa} i_{mr\alpha_{i-1}} + \frac{\kappa}{1+\kappa}\left(i_{s\alpha_i} + i_{s\alpha_{i-1}}\right) - T_r \frac{\kappa}{1+\kappa}\left(i_{mr\beta_i} + i_{mr\beta_{i-1}}\right)\cdot\omega \\ i_{mr\beta_i} &= \frac{1-\kappa}{1+\kappa} i_{mr\beta_{i-1}} + \frac{\kappa}{1+\kappa}\left(i_{s\beta_i} + i_{s\beta_{i-1}}\right) + T_r \frac{\kappa}{1+\kappa}\left(i_{mr\alpha_i} + i_{mr\alpha_{i-1}}\right)\cdot\omega \end{aligned} \tag{10.61}$$

where κ is the ratio of sampling time to rotor constant,

$$\kappa = T_s/2T_r \tag{10.62}$$

The time variables can also be made dimensionless by multiplying ω_B to the both sides of the equations. Therefore both T_s and T_r are expressed in per-unit values in Equations 10.60 and 10.62. A block diagram of the flux observer is shown in Figure 10.22. The flux observer takes the phase currents and speed as input, and calculates i_{mr}, $\cos\alpha$, and $\sin\alpha$.

It has been shown in the previous sections that it is possible to control the magnetizing component and torque component of the stator current separately. A PI controller is one way to implement control. The numerical expression for a PI controller is

$$V_o = K_{PI}\left(T_{PI}\varepsilon + \int \varepsilon dt\right) \tag{10.63}$$

where V_o is the output of the PI controller and ε is the error signal of input V_i (here V_i can be the measured current or torque of the motor, and V_o can be the PWM signal). In order to get the time domain discrete expression, we differentiate Equation 10.63:

$$\frac{dV_o}{dt} = K_{PI}\left(T_{PI}\frac{d\varepsilon}{dt} + \varepsilon\right) \tag{10.64}$$

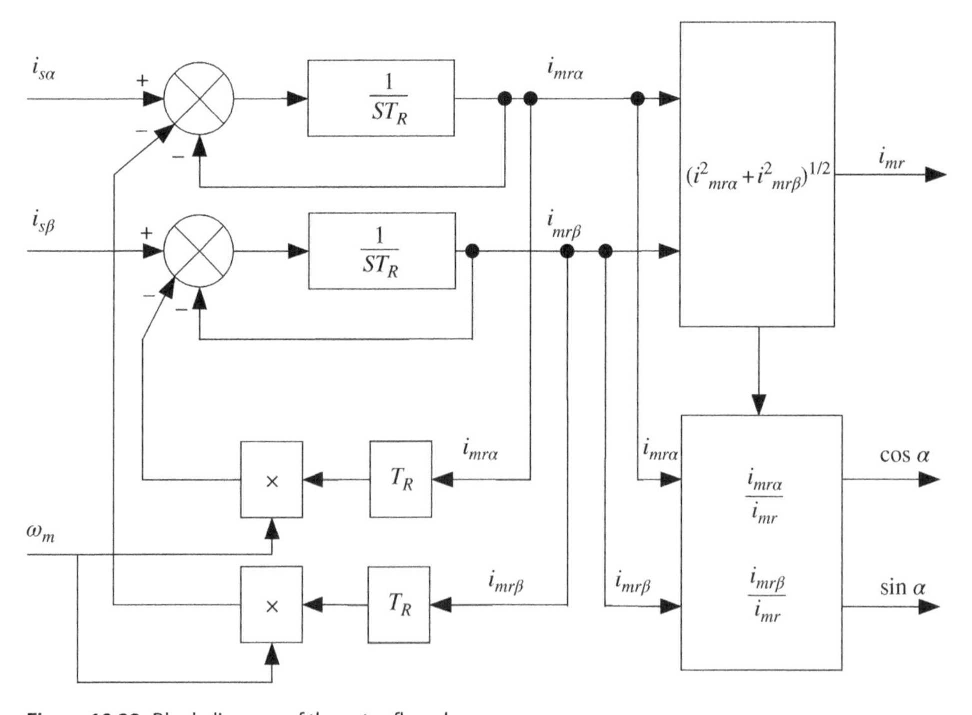

Figure 10.22 Block diagram of the rotor flux observer.

Further implementation is straightforward:

$$\frac{V_{oi} - V_{oi-1}}{T_S} = K_{PI}\ T_{PI}\frac{\varepsilon_i - \varepsilon_{i-1}}{T_S} + K_{PI}\frac{\varepsilon_i + \varepsilon_{i-1}}{2} \quad (10.65)$$

$$V_{oi} = V_{oi-1} + K_1\left(\varepsilon_i - K_2\ \varepsilon_{i-1}\right) \quad (10.66)$$

where K_1 and K_2 can be expressed as

$$K_1 = \left(1 + T_S/2T_r\right)K_{PI}\ T_{PI}$$
$$K_2 = \frac{\left(1 - T_S/2T_r\right)}{\left(1 + T_S/2T_r\right)} \quad (10.67)$$

For example, if the gain is chosen as $K_{PI} = 50$, $T_{PI} = 0.02$ seconds, sampling time $T_r = 0.02$ seconds, and $T_s = 0.67$ ms, then the constants K_1 and K_2 are $K_1 = 1.0168$, $K_2 = 0.9671$.

The purpose of field-oriented control is to control an induction machine in such a way that it behaves like a DC motor. A block diagram is shown in Figure 10.23 and the flow chart is shown in Figure 10.24. An incremental encoder is used to measure the speed of the motor. As shown in Equation 10.52, the magnetizing current does not change instantaneously with i_{sd} as it does in a DC motor. Rather, the magnetizing current lags a time constant T_r corresponding to the change of i_{sd}.

In this setup, the flux observer uses the speed signal of an incremental encoder and the current measurement through two external current sensors. Only the currents of two phases are needed to perform the coordinate transformations due to symmetry.

10.3 Permanent Magnet Motor Drives

PM motors are the most popular choices for EV and HEV powertrain applications due to their high efficiency, compact size, high torque at low speeds, and ease of control for regenerative braking [43–90]. The PM motor in an HEV powertrain is operated either as a motor during normal driving or as a generator during regenerative braking and power splitting, as required by the vehicle operations and control strategies. PM motors with higher power densities are also now increasingly the choice for aircraft, marine, naval, and space applications.

The most commercially used PM material in traction drive motors is neodymium–ferrite–boron (Nd–Fe–B). This material has a very low Curie temperature and high temperature sensitivity. It is often necessary to increase the size of magnets to avoid demagnetization at high temperatures and high currents. On the other hand, it is advantageous to use as little PM material as possible in order to reduce the cost without sacrificing the performance of the machine.

10.3.1 Basic Configuration of PM Motors

When PMs are used to generate the magnetic field in an electric machine, it becomes a PM motor. Both DC and AC motors can be made with PMs. Only PM synchronous

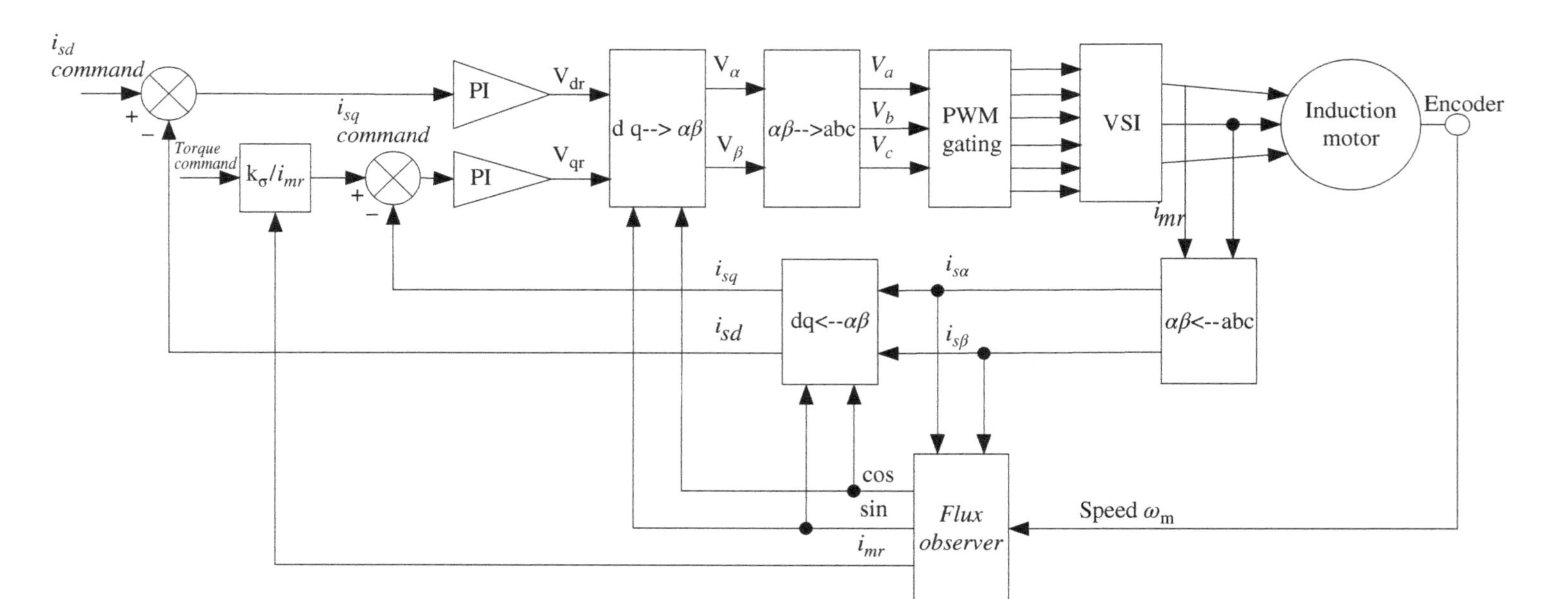

Figure 10.23 Field-oriented control of an induction machine.

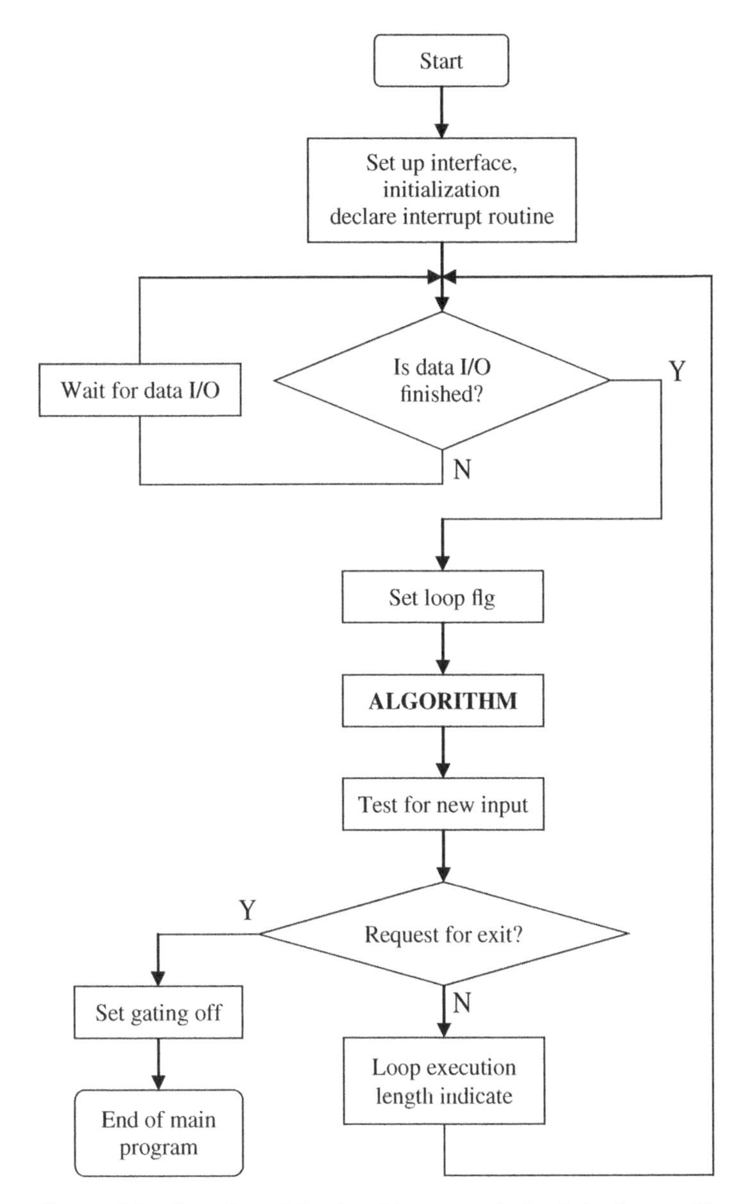

Figure 10.24 Flow chart of the closed-loop control of an induction machine.

motors and PM brushless DC motors are chosen for modern traction drives. We will primarily explain the operation of PM synchronous motors in this book.

A PM synchronous motor contains a rotor and a stator, with the stator similar to that of an induction motor, and the rotor contains the PMs. From the section on induction motors, we know that the three-phase winding, with three-phase symmetrical AC supply, will generate a rotating magnetic field. To generate a constant average torque, the rotor must follow the stator field and rotate at the same synchronous speed. This is also why these machines are called PM synchronous motors.

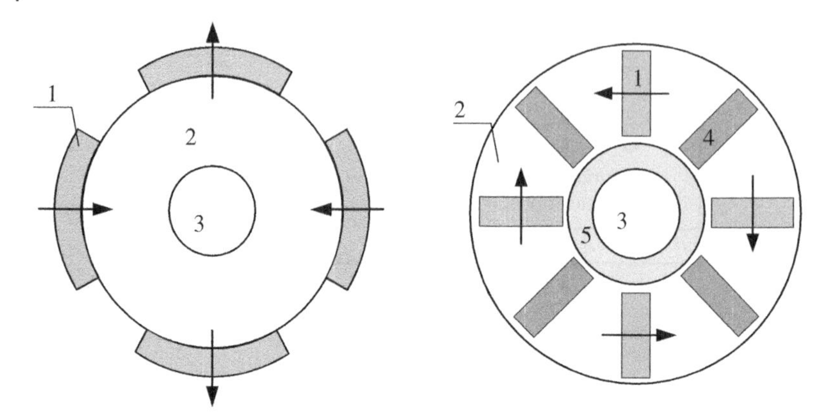

Figure 10.25 Surface-mounted magnets and interior magnets: left, SPM motor; right, IPM motor. 1 – magnet; 2 – iron core; 3 – shaft; 4 – non-magnet material; 5 – non-magnet material.

There are different ways to place the magnets on the rotor, as shown in Figure 10.25. If the magnets are glued on the surface of the rotor, it is called a surfaced-mounted PM motor (SPM). If the magnets are inserted inside the rotor in the pre-cut slots, then it is called an interior permanent magnet motor (IPM).

For an SPM motor, the rotor can be a solid piece of steel since the rotor iron core itself is not close to the air gap, hence the eddy current loss and hysteresis loss due to slot/tooth harmonics can be neglected. For the IPM motor, the rotor needs to be made out of laminated silicon steel since the tooth/slot harmonics will generate eddy current and hysteresis losses.

Due to the large air gap as well as the fact that the magnets have a permeability similar to that of air, SPM motors have similar direct-axis reactance x_d and quadrature-axis reactance x_q. On the other hand, IPM motors have different x_d and x_q. This difference will generate a so-called reluctance torque. It is worth pointing out that although there is a reluctance torque component, it does not necessarily mean that an IPM motor will have a higher torque rating than an SPM motor for the same size and same amount of magnetic material used. This is because, in IPM motors, in order to keep the integrity of the rotor laminations, there are so-called "magnetic bridges" that will have leakage magnetic flux. So for the same amount of magnet material used, an SPM motor will always have higher total flux. There are many different configurations for IPM motors as shown in Figure 10.26.

10.3.2 Basic Principle and Operation of PM Motors

The no-load magnetic field of PM machines is shown in Figure 10.27. When the rotor is driven by an external source (such as an engine), the rotating magnetic field will generate three-phase voltage in the three-phase windings. This is the generator mode operation of the PM machine.

When operated as a motor, the three-phase windings, similar to those of an induction motor, are supplied with either a trapezoidal form of current (brushless DC) or sinusoidal current (synchronous AC). These currents generate a magnetic field that is rotating at the same speed as the rotor, or synchronous speed. By adjusting the frequency of the stator current, the speed of the rotor or the synchronous speed can be adjusted accordingly.

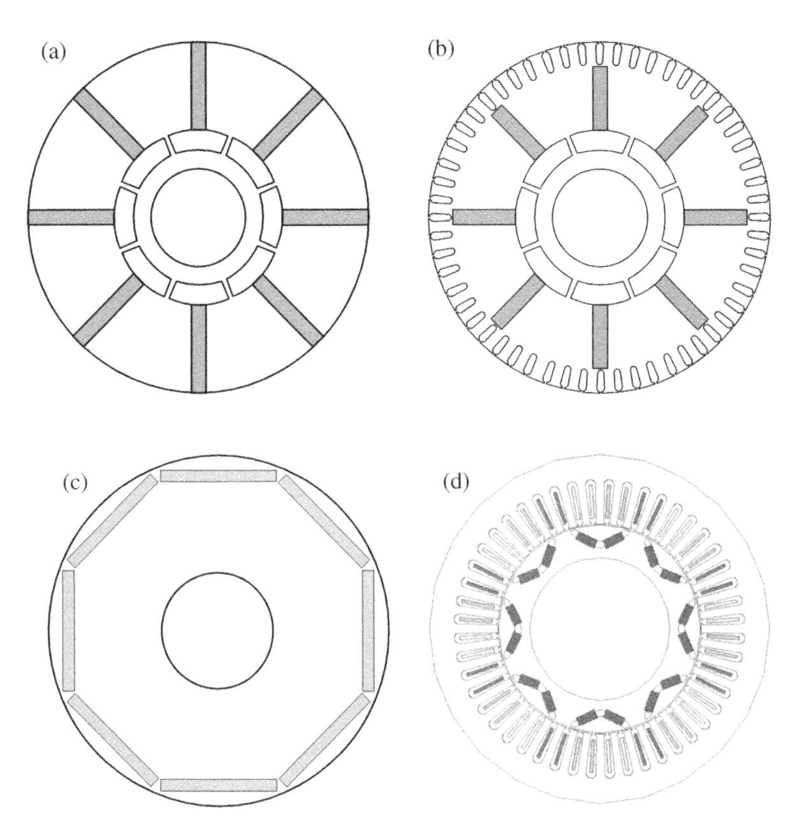

Figure 10.26 Four commonly used IPM rotor configurations: (a) circumferential-type magnets suitable for brushless DC or synchronous motor; (b) circumferential-type magnets for line-start synchronous motor; (c) rectangular slots IPM motor; (d) V-type slots IPM motor.

The torque is the attraction between the rotor magnetic field and the stator magnetic field in the circumferential direction. Hence, at no-load conditions, the rotor and the stator field are almost lined up. When the angle between the rotor field and the stator field reaches 90 electric degrees, the maximum torque is reached in SPM motors. For IPM motors, the maximum torque occurs at an angle slightly larger than 90° due to the existence of reluctant torque.

Figure 10.28 illustrates how a PM motor operates in different modes. The stator winding generates a rotating field that attracts the rotor magnets. If the two fields are lined up, the attraction between the two magnetic fields is in the radial direction, hence there is no electromagnetic torque. When the stator field is leading the rotor field, the stator will attract the rotor magnets. The machine then operates as a motor. When the stator field is lagging the rotor field, the machine becomes a generator.

At no load, the rotor magnetic field will generate a back emf E_o in the stator windings. When a voltage with the same frequency is applied to the stator windings, then a current will be generated and the voltage equation can be written as

$$V = E_o + IR + jIX \tag{10.68}$$

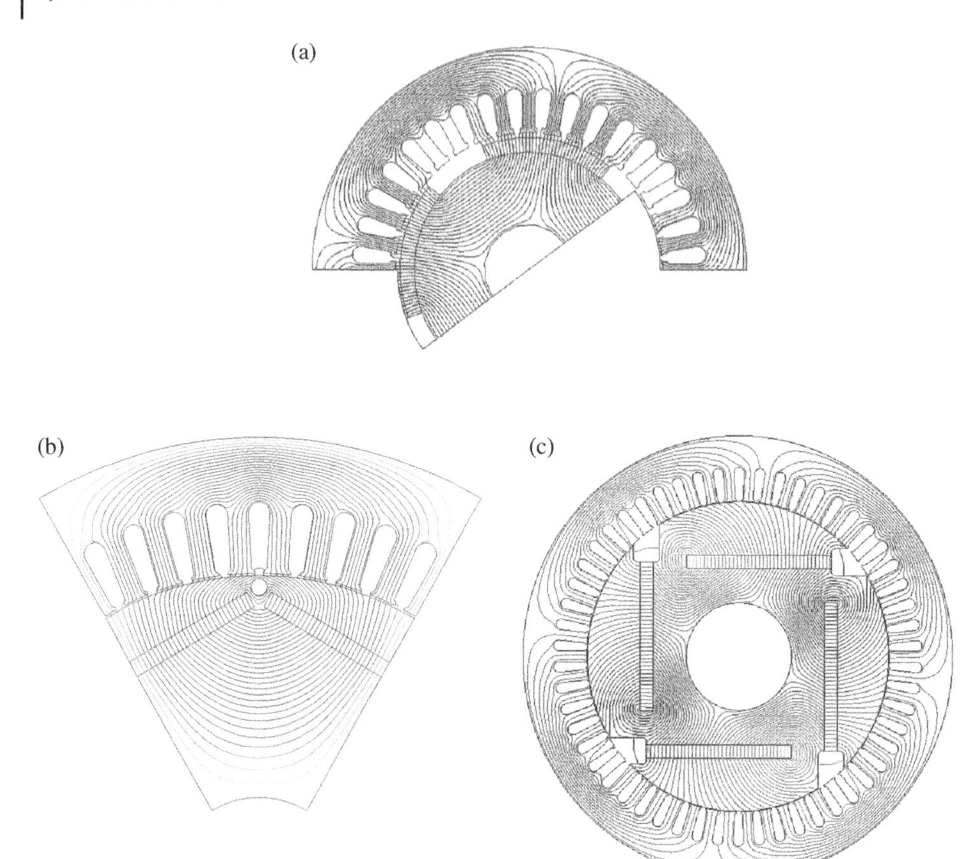

Figure 10.27 Magnetic field distribution of PM machines at no-load conditions (the stator current is zero): (a) a four-pole SPM motor; (b) an eight-pole symmetrical IPM motor; (c) a four-pole unsymmetrical IPM configuration.

where R is the stator resistance and X is the synchronous impedance. The phasor diagram is shown in Figure 10.29 when neglecting the stator resistance. From the diagram, the term jIX can be further decomposed into two components: jI_dX_d and jI_qx_q. In fact, in IPM motors, the d axis and q axis will have different reactances. By using Figure 10.29, Equation 10.68 can be rewritten for IPM motors as

$$V = E_o + IR + jI_d\ X_d + jI_q\ X_q \tag{10.69}$$

The real power can be calculated, since from Figure 10.29, $\phi = \delta + \theta$:

$$\begin{aligned} P_1 &= mIV\cos\varphi = mIE_o\cos\delta = mV\left(I\cos\delta\cos\theta - I\sin\delta\sin\theta\right) \\ &= mV\left(I_q\cos\theta - I_d\sin\theta\right) \end{aligned} \tag{10.70}$$

where ϕ is the power factor angle (the angle between the voltage and current), θ is the angle between the voltage and back emf, and δ is the inner power angle (the angle between the back emf and current). From Figure 10.29,

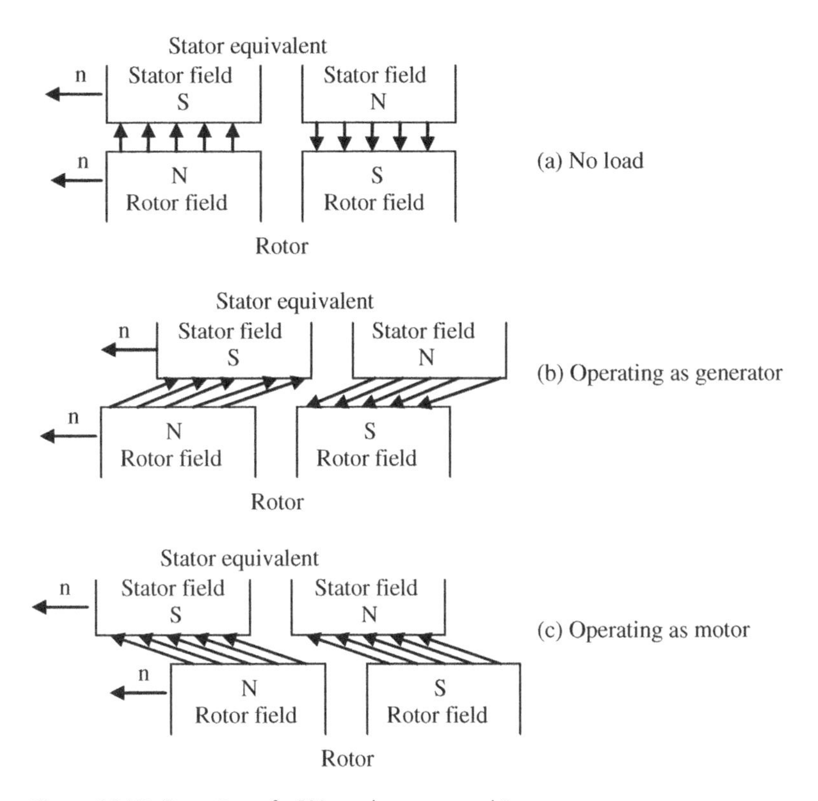

Figure 10.28 Operation of a PM synchronous machine.

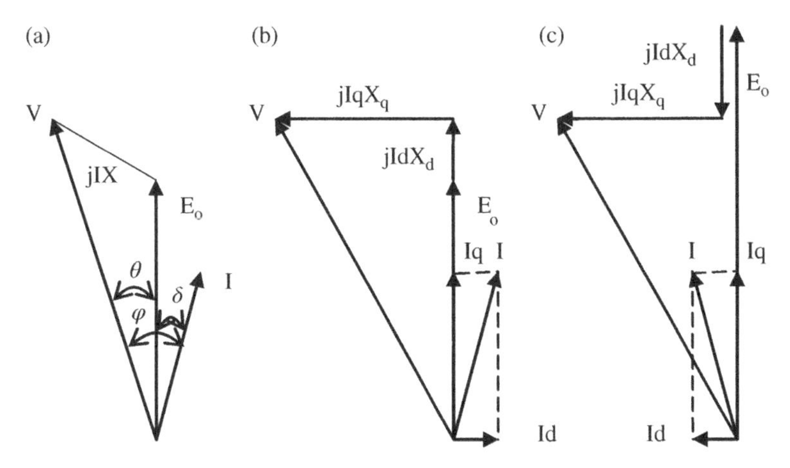

Figure 10.29 Phasor diagram of PM synchronous motors: (a) SPM; (b) IPM; (c) flux weakening mode of IPM.

$$\begin{aligned} I_q \ X_q &= V \sin\theta \\ I_d \ X_d &= V \cos\theta - E_o \end{aligned} \tag{10.71}$$

Therefore, the power of PM motors can be expressed as

$$P = \frac{mE_oV}{X_d}\sin\theta + \frac{mV^2}{2}\left(\frac{1}{X_q} - \frac{1}{X_d}\right)\sin(2\theta) \tag{10.72}$$

The torque can be derived by dividing Equation 10.72 by the rotor speed as shown in Figure 10.30, where the torque–speed characteristics of a typical PM motor are shown. For SPM motors, since $X_d = X_q$, the second term of Equation 10.72 is zero. For IPM motors, the q axis has less reluctance due to the existence of soft iron in its path, and the d axis has magnets in its path which has larger reluctance. Therefore, X_q is much larger than X_d.

On the other hand, from Equation 10.70, and neglecting losses, we can see that

$$\begin{aligned} T &= \frac{mIE_o\cos\delta}{\omega/p} = \frac{mIk\omega\varphi}{\omega/p}\cos\delta = mpkI\varphi\cos\delta \\ T_{\max} &= mpkI\varphi = \text{constant} \end{aligned} \tag{10.73}$$

Therefore, when inner power angle $\delta = 0$, for a given stator current, the torque of the motor reaches its maximum. In this condition, stator current is in phase with back emf E_o, and

$$\begin{aligned} V^2 &= E_o^2 + \left(I_q \ X_q\right)^2 = (k\omega\varphi)^2 + \left(I\omega L_q\right)^2 \\ \frac{V}{\omega} &= \sqrt{(k\varphi)^2 + \left(IL_q\right)^2} = \text{constant} \end{aligned} \tag{10.74}$$

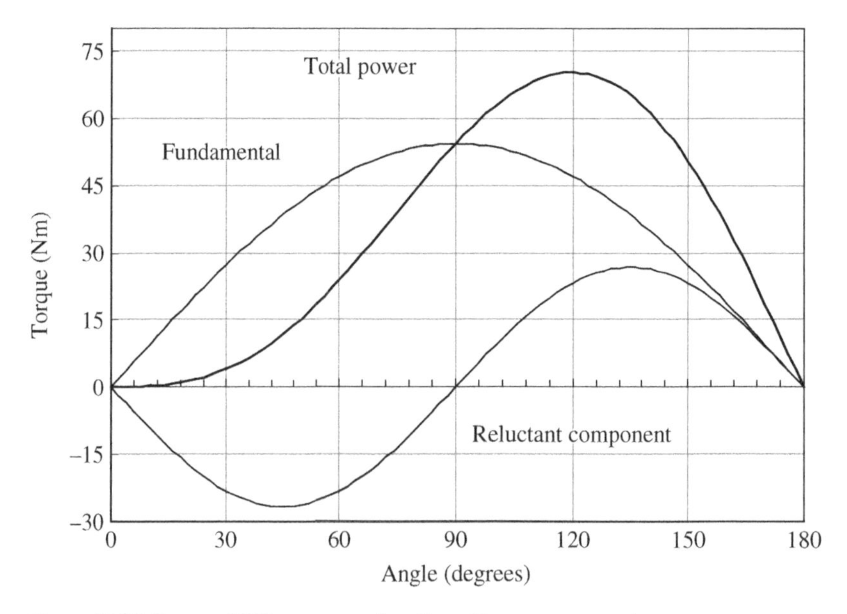

Figure 10.30 Power of IPM motor as a function of inner power angle.

Hence, the stator voltage must be proportional to frequency to satisfy Equation 10.74 and maintain maximum torque output at the same time. This operation is also called constant torque operation. It can also be seen from Equation 10.72 that for a given θ, the power is inversely proportional to frequency, since V, X_d, and E_o are all proportional to frequency ω. This is similar to the V/f control of induction motors.

When the stator voltage reaches its maximum, Equation 10.74 can no longer be maintained. As ω increases, V becomes constant, and a current in the d-axis direction must be supplied, as shown in Figure 10.29c. The relationship of voltage and frequency can be expressed as

$$\begin{aligned} V^2 &= \left(E_o - I_d\, X_d\right)^2 + \left(I_q\, X_q\right)^2 = \left(k\omega\varphi - I_d\omega L_d\right)^2 + \left(I_q\omega L_q\right)^2 \\ \frac{V}{\omega} &= \sqrt{\left(k\varphi - I_d\, L_d\right)^2 + \left(I_q\, L_q\right)^2} \end{aligned} \tag{10.75}$$

This operation is also called the flux weakening operation region because the d-axis current generates a magnetic flux in the opposite direction to the PM field. Note that due to constraints such as the current limit of the inverter, the q-axis current may have to be decreased from its rated value so that the total current from the inverter is kept the same. Additional losses at higher speeds may make it necessary to further reduce the torque output. It can also be seen from Equation 10.72 that for a given θ, the first term is constant since V is constant, and both X_d and E_o are proportional to frequency ω. In theory, the torque is inversely proportional to frequency in this operation, so the power is constant. Hence this mode is also referred to as the constant power operation range.

10.3.3 Magnetic Circuit Analysis of IPM Motors

Although today's motor design is usually aided by finite element analysis (FEA), the initial design stages are still realized through analytical methods. Air-gap flux is one of the most important parameters of PM motor designs and equivalent magnetic circuit analysis is used to calculate the air-gap flux in PM motors. For SPM motors, the equivalent magnetic circuit is straightforward. But for IPM motors, the PMs are buried inside the rotor laminations, with magnets inserted into the pre-punched slots. This arrangement protects the magnets from flying away from the rotor surface due to centrifugal force, fatigue, and aging of material during operation of the motor. Another advantage of IPM motors is that rectangular (cuboid) magnets can be used to simplify the manufacturing process and reduce the cost of manufacturing PM material. Flux concentration structures (such as magnets arranged in a V-shape) are often used to increase air-gap flux density in IPM motors [85].

Calculating air-gap flux in IPM motors is somewhat troublesome due to the existence of so-called magnetic bridges. When an integrated lamination is used for IPM motors, magnetic short circuits exist around the edges of the magnets. These magnetic bridges are designed to enhance the integrity of the rotor. The magnetic bridges introduce magnetic short circuits and complicate the design and analysis of IPM motors. On the other hand, there are also concerns about how to limit the leakage flux in these magnetic bridges while maintaining the mechanical strength of the rotor. The flux leakage and flux distribution in the magnetic bridges can be precisely obtained through numerical methods such as FEA. However, FEA can only be performed after the preliminary

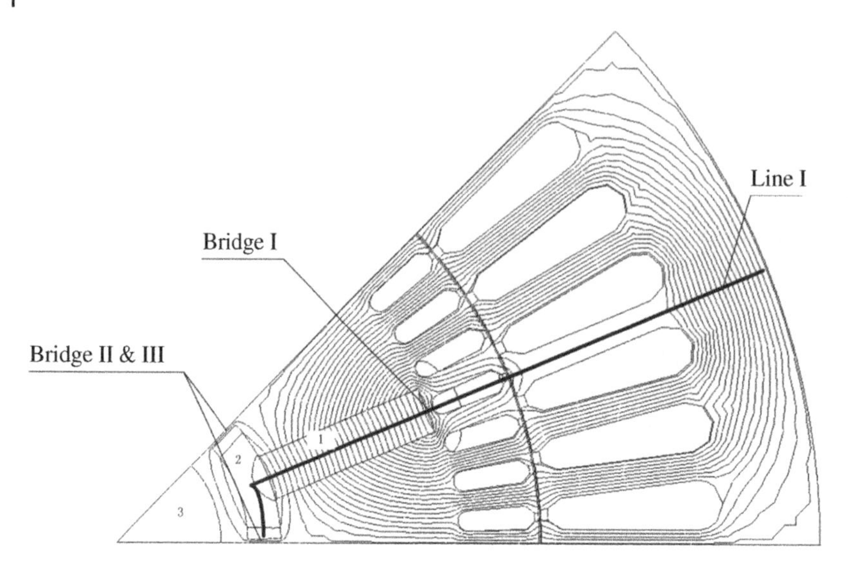

Figure 10.31 Flux distribution of an IPM line-start synchronous motor with circumferential-type magnets: 1, magnet; 2, non-magnetic material; 3, shaft.

dimensions of the motor have been determined. FEA is also cumbersome and time consuming in the early stages of PM motor design where numerous iterations are usually performed. Analytical calculation and analysis of all types of PM motors are essential in their early design stage.

This section discusses the analytical method for calculating the air-gap flux of IPM machines using an equivalent magnetic circuit model taking into account the assembly gap and saturation in the steel. Factors that affect the flux leakage in an IPM motor will also be discussed.

Figure 10.31 shows the configuration and no-load flux distribution of an eight-pole circumferential-type IPM line-start synchronous motor calculated using FEA. Integrated laminations are used to maintain integrity of the rotor. It contains three magnetic bridges in each pole: *bridge I* between the magnet and rotor slot; *bridges II* and *III* at the interpolar space between the magnet and the shaft. Figure 10.32 shows the flux density along *line I* of Figure 10.31. It can be seen from Figure 10.32 that the flux densities differ in the two bridges.

It will be shown later that the magnetic flux density in the magnetic bridges is related to the width and length of the magnetic bridge, rather than being constant. For the situation here, there is flux leakage in the rotor slot and the non-magnetic material between the magnet and the shaft. The flux leakage through the stator slot is negligible.

Modern rare earth permanent magnets (REPMs) have a straight demagnetization curve as shown in Figure 10.33. The low cost makes Nd–Fe–B REPMs ideal for motor applications. However, temperature effects, as shown in Figure 10.33, must be taken into consideration when designing a PM motor.

At room temperature (25 °C), the demagnetizing curve of a cuboid REPM can be represented by

$$\Phi_m = \Phi_r - F_m \cdot \Phi_r / F_c = \Phi_r - F_m / R_M \tag{10.76}$$

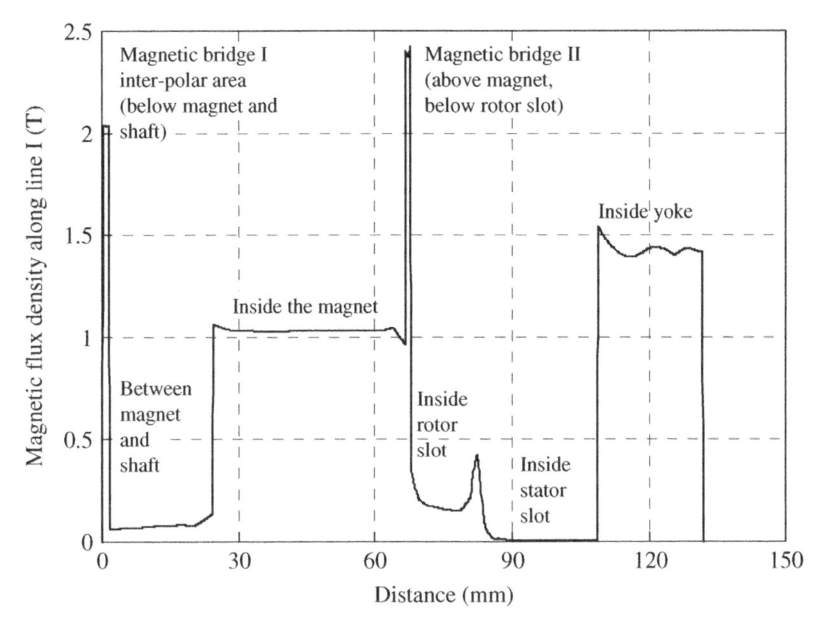

Figure 10.32 Flux density along *line I* of Figure 10.31.

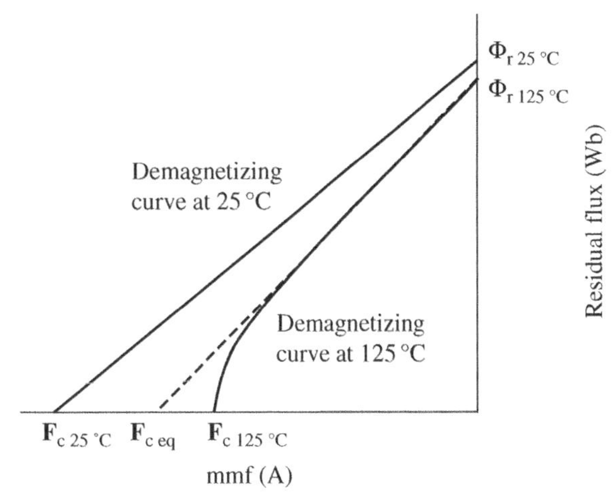

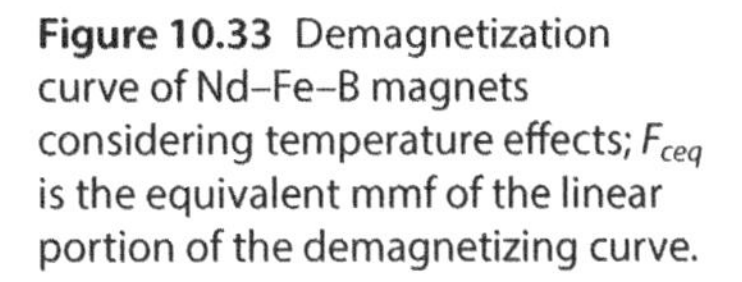

Figure 10.33 Demagnetization curve of Nd–Fe–B magnets considering temperature effects; F_{ceq} is the equivalent mmf of the linear portion of the demagnetizing curve.

where Φ_r and F_c are the residual flux and magnetomotive force (mmf) of each pole respectively, and R_M is the reluctance of the magnet, which is the reciprocal of magnet permeance λ_M:

$$R_M = 1/\lambda_M = F_c/\Phi_r \tag{10.77}$$

For parallel or circumferentially magnetized poles, as shown in Figure 10.26a,b,

$$\Phi_r = 2B_r\ A_m, \quad F_c = l_m\ H_c \tag{10.78}$$

while for series or radially magnetized poles, as shown in Figure 10.26c,d,

$$\Phi_r = B_r \ A_m, \quad F_c = 2l_m \ H_c \tag{10.79}$$

where B_r and H_c are the remanence and coercive force of the magnets respectively, l_m is the length of the magnet, and A_m is the cross-sectional area. Thus

$$A_m = b_m \ l_{fe} \tag{10.80}$$

where l_{fe} is the length of magnet along the shaft direction and usually equal to the rotor lamination stack length.

At the operating temperature, these parameters are replaced by their respective values at the operating temperature. It is possible that the demagnetizing curve may become nonlinear at the operating temperature. In this case, F_{ceq} should be used in place of F_c as shown in Figure 10.33. The Norton equivalent of a cuboid magnet is shown in Figure 10.34.

A generic, circumferentially magnetized IPM rotor configuration is shown in Figure 10.35. The equivalent magnetic circuit of this configuration is shown in Figure 10.36, where R_δ, R_{y1}, R_{y2}, R_{t1}, R_{t2}, $R\sigma$, R_1, R_2, R_S are the reluctances of the air gap, stator yoke, rotor yoke, stator teeth, rotor teeth, assembly gap between magnets and laminations, magnetic bridge I, magnetic bridges II and III (combined due to symmetry), and leakage through the rotor slots and the non-magnetic material respectively. End effects are neglected.

Fluxes passing through these bridges are leakage fluxes. These magnetic bridges are highly saturated as can be seen from Figure 10.32.

The magnetic circuit can be simplified as shown in Figure 10.36b, where λ_δ' is the total permeance combining the air gap, stator teeth, stator yoke, rotor teeth (if any), and rotor yoke; λ_1 is the permeance of magnetic bridge I; λ_2 is the permeance of magnetic bridges II and III; and λ_S is the total permeance leakage through the rotor slot and non-magnetic material. Let

$$\begin{aligned} \lambda_\delta' &= 1/(R_\delta + R_{t1} + R_{t2} + R_{y1} + R_{y2}) \\ \lambda_o &= \lambda_\delta' + \lambda_1 + \lambda_2 + \lambda_S \\ \lambda_{ex} &= \lambda_o || \lambda_\sigma \end{aligned} \tag{10.81}$$

where λ_{ex} is the total external permeance.

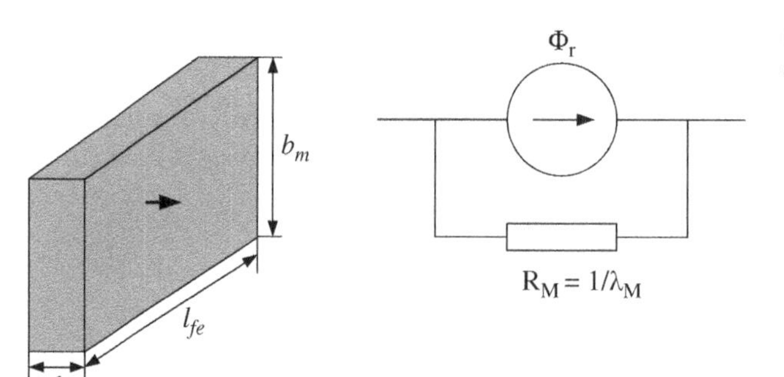

Figure 10.34 Norton equivalent of cuboid magnets.

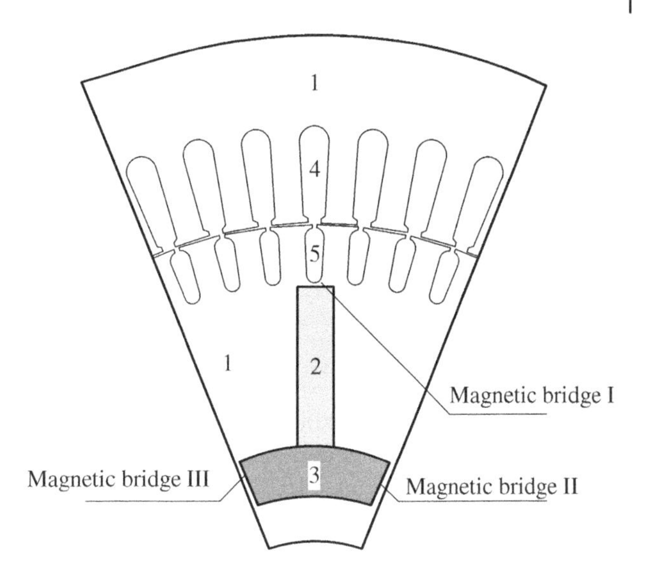

Figure 10.35 Arrangements of magnets of line-start IPM motors: 1, stator and rotor iron laminations; 2, permanent magnets; 3, non-magnetic material; 4, stator slots; 5, rotor slots. Magnetic bridges are part of the rotor laminations to maintain the integrity of rotor laminations.

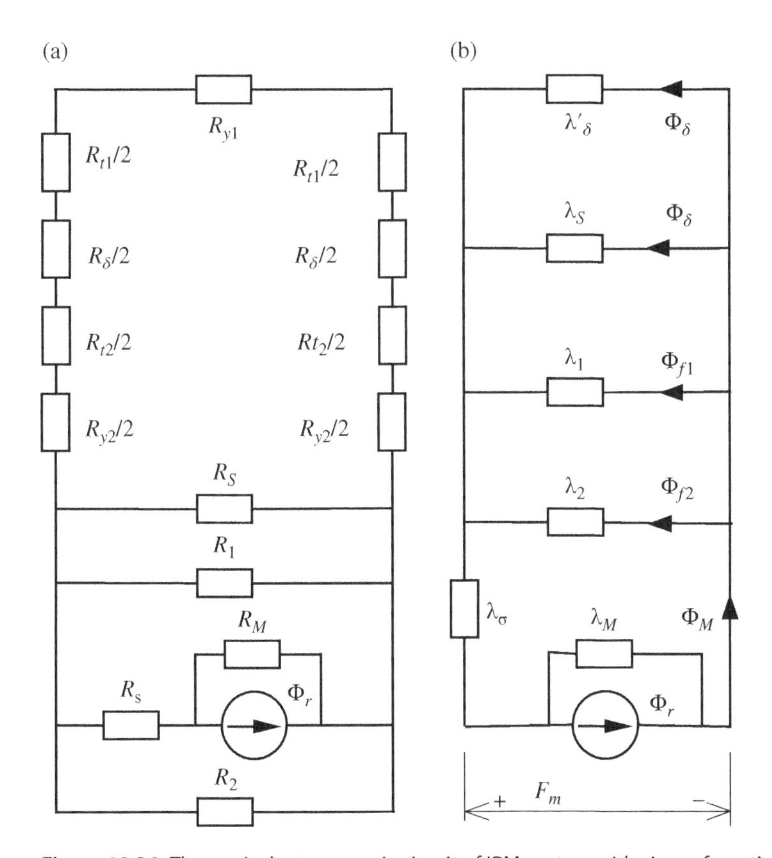

Figure 10.36 The equivalent magnetic circuit of IPM motors with circumferential magnets: (a) exact model; (b) simplified model.

Solving the magnetic circuit, the magnet operating point and all fluxes can be found:

$$F_m = \Phi_r / (\lambda_M + \lambda_{ex})$$
$$\Phi_m = \Phi_r \ \lambda_{ex} / (\lambda_M + \lambda_{ex}) \tag{10.82}$$

$$\Sigma F = F_m - \Phi_m / \lambda_\sigma$$
$$\Phi_\delta = \lambda_\delta' \Sigma F$$
$$\Phi_f = (\lambda_1 + \lambda_2) \Sigma F \tag{10.83}$$
$$\Phi_S = \lambda_S \Sigma F$$

$$\Phi_m = \Phi_\delta + \Phi_f + \Phi_S \tag{10.84}$$

where Φ_δ is the total air-gap flux, Φ_f is the total leakage flux in the magnetic bridges, and Φ_S is the total leakage flux in the rotor slot and non-magnetic material.

It can be seen from Figures 10.35 and 10.36 that the magnetic bridges should have the same mmf as that of the air-gap branch. Therefore, the magnetic field density of each of the bridges is

$$H_f = \Sigma F / l_f \tag{10.85}$$

where l_f is the length of the magnetic bridge along the flux path. It can be seen from Equation 10.85 that if the magnetic bridges have different paths for magnetic fluxes, then the field density will be different.

The flux density in each of the bridges can be found through table lookup or curve-fitting of the lamination material. The total flux leakage in each of the magnetic bridges can be expressed as

$$\Phi_f = B_f \ A_f \tag{10.86}$$

where B_f is the flux density in the magnetic bridges and A_f is the cross-sectional area of that magnetic bridge.

Although the above equations give an analytical expression for air-gap flux, it can hardly be solved due to saturation in the yoke, teeth, and magnetic bridges. Graphic analysis is an effective method for solving the operating point of the magnets.

10.3.3.1 Unsaturated Motor

If the yoke and teeth of the motor are not saturated but the magnetic bridges are highly saturated, and the assembly air gap is neglected, the air-gap flux can be calculated analytically.

Assuming that the flux density in the magnetic bridges is constant due to high saturation, then

$$\Phi_\delta = \frac{\lambda_\delta}{\lambda_\delta + \lambda_M + \lambda_S} (\Phi_r - \Phi_f) \tag{10.87}$$

where the air-gap permeance can be expressed as

$$\lambda_\delta = \frac{\alpha \tau l_{fe} \ \mu_0}{2 \delta k_\delta} \tag{10.88}$$

α is the effective pole width coefficient, τ is the pole pitch, k_δ is Carter's coefficient, δ is air-gap length, and μ_0 is the permeability of air, $\mu_0 = 4\pi \times 10^{-7}$ T m/A.

Carter's coefficient can be calculated for stator slots and rotor slots (if any) separately, that is,

$$k_\delta = k_{\delta 1}\ k_{\delta 2} \tag{10.89}$$

$$k_{\delta 1} \text{ or } k_{\delta 2} = t/(t - \gamma^* \delta) \tag{10.90}$$

$$\gamma = \frac{4}{\pi}\left[\frac{b_0}{2\delta}\arctan\left(\frac{b_0}{2\delta}\right) - \ln\left(\sqrt{1+\left(\frac{b_0}{2\delta}\right)^2}\right)\right] \tag{10.91}$$

where b_0 is the width of the slot opening and t is the slot pitch.

Leakage permeance λ_S can also be found for rotor slots and the non-magnetic material. The reluctance of the assembly air gap can be expressed as a function of σ, where σ is the average tolerance between the magnet and the lamination steels. The reluctance of the assembly air gap is

$$\lambda_\sigma = \frac{b_m\ l_{fe}\ \mu_0}{\sigma} \tag{10.92}$$

10.3.3.2 Saturated Motor

When the IPM motor is saturated, as is the case for most PM motor designs, graphic analysis can be used to solve for the magnetic operating point. Detailed steps of graphic analysis are given in Table 10.5 and Figure 10.37. The calculation procedure is detailed as follows:

1) Assume an air-gap flux density B_δ as in Table 10.5, row 1.
2) Calculate the flux density in the stator tooth (row 2); use table lookup or curve-fitting to find the magnetic field density H_t in the tooth (row 3); calculate the mmf of the tooth (row 4).
3) Repeat step 2 to calculate the mmf of the rotor tooth, stator yoke, and rotor yoke (rows 5–7).
4) Calculate the air-gap mmf (row 8).
5) Calculate the subtotal mmf by summing air-gap mmf and core mmf (row 9).
6) Calculate the magnetic field density of the magnetic bridges (rows 10 and 12); use lookup table or curve-fitting to find the magnetic flux density B_f in the magnetic bridges (rows 11 and 13).
7) Calculate the air-gap flux, leakage flux in the bridges, and the leakage flux through the rotor slots (rows 14–17).
8) Calculate the total flux (row 18) and total mmf (row 19).
9) Plot five curves in the second quadrant as shown in Figure 10.37 using Table 10.5. Find on the graph the crossing point A. This is the magnet operating point. Draw a horizontal line to intersect $\Phi_m \sim F_m$ at A′. Draw a vertical line down from point A′. The crossing point of this vertical line with curve $\Phi_\delta \sim F_m$ is the air-gap flux, with $\Phi_S \sim F_m$ the leakage flux through the rotor slot, and with $\Phi_f \sim F_m$ the leakage flux through the bridges.

Table 10.5 Magnetic circuit calculation of experimental motor.

1	B_δ	(Experimental motor)	0.8	0. 9
2	$B_t = t/b_t B_\delta$	$B_t = 2.02B_\delta$	1.6	1.8
3	$H_t(Lookup\_Table)$	H_t	4 250.0	13,280.0
4	$F_t = h_t H_t$	$F_t = 0.04H_t$	181.1	565.7
5	$B_y = \Phi_\delta/2h_y l_{fe}$	$B_y = 1.38B_\delta$	1.1	1.2
6	$H_y(Lookup\_Table)$	H_y	498.6	742.8
7	$F_y = l_y H_y$	$F_y = 0.12H_y$	59.3	88.4
8	$F_\delta = 1.6\delta k_\delta B/\mu_0$	$F_\delta = 566.8B_\delta$	907.1	1020.5
9	$\Sigma F = F_\delta + F_t + F_y$	$\Sigma F = F_\delta + F_t + F_y$	1147.5	1674.6
10	$H_{f1} = \Sigma F/l_{f1}$	$H_{f1} = \Sigma F/l_{f1}$	229,498.0	334,923.3
11	$B_{f1}(Lookup\_Table)$	B_{f1}	2.29	2.43
12	$H_{f2} = \Sigma F/l_{f2}$	$H_{f2} = \Sigma F/l_{f2}$	95,624.2	139,551.4
13	$B_{f2}(Lookup\_Table)$	B_{f2}	2.12	2.17
14	$\Phi_\delta = \alpha\tau l_{fe} B_\delta$	$\Phi_\delta = 14.8 \times 10^{-3} B_\delta$	0.011 263	0.012 671
15	$\Phi_{f1} = B_{f1} A_{f1}$	$\Phi_{f1} = 0.38 \times 10^{-3} B_{f1}$	0.001 743	0.001 848
16	$\Phi_{f2} = B_{f2} A_{f2}$	$\Phi_{f2} = 0.608 \times 10^{-3} B_{f2}$	0.001 287	0.001 322
17	$\Phi_s = \Sigma F \lambda_s$	$\Phi_s = 0.336 \times 10^{-6} F_m$	0.000 771	0.001 125
18	$\Phi_m = \Phi_\delta + \Phi_{f1} + \Phi_{f2} + \Phi_s$	Φ_m	0.015 063	0.016 966
19	$F_m = \Sigma F + \Phi_m/\lambda_\sigma$	F_m	1462.9	2029.9

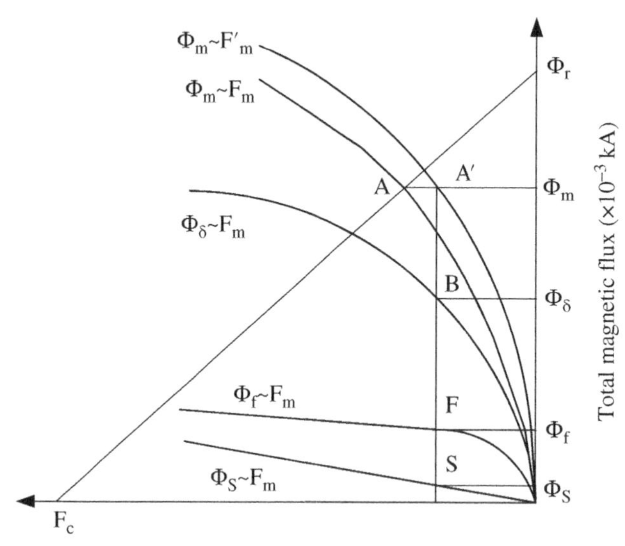

Figure 10.37 Graphic analysis of no-load IPM machine, where crossing point A is the operating point of the magnet, B represents the air-gap flux, F represents the leakage flux in the magnetic bridges, and S represents the leakage flux in rotor slots and the non-magnetic material. Note that the leakage flux in the magnetic bridges contributes a significant portion of the total flux supplied by the magnet as can be seen on the graph.

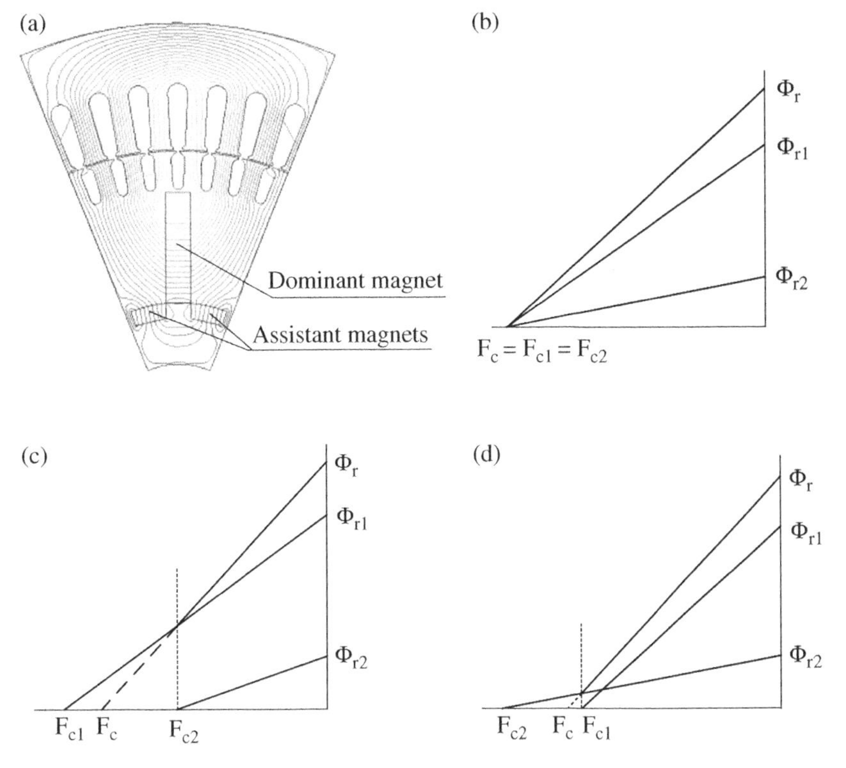

Figure 10.38 Flux concentration configurations: (a) configuration with the assistant magnets in series; (b) equivalent demagnetizing curve when the assistant magnets have the same mmf as that of the dominant magnet; (c) when the mmf of assistant magnets is more than that of the dominant magnet; (d) when the mmf of assistant magnets is less than that of the dominant magnet.

In Table 10.5, b_{t1} and h_{t1} are the width and height of the stator slots; b_{t2} and h_{t2} are the width and height of the rotor slots; l_{y1} and h_{y1} are the length and height of the stator yoke; l_{y2} and h_{y2} are the length and height of the rotor yoke; A_{f1} and A_{f1} are the cross-sectional areas of each magnetic bridge; and l_{f1} and l_{f2} are the lengths of each magnetic bridge, respectively.

10.3.3.3 Operation Under Load

The operating point of magnets at rated load can also be solved by shifting the air-gap curve $\Phi_\delta \sim F_m$ by F_{ad} to the left on the graph, where F_{ad} is the armature mmf.

10.3.3.4 Flux Concentration

Flux concentration configurations are often used in IPM machines to increase air-gap flux density, as shown in Figure 10.38a. In flux concentration configurations, the assistant magnets are usually designed to have the same mmf as the dominant magnet. However, due to dimensional and other constraints, the assistant magnets may have a different mmf.

There are three possibilities as shown in Figures 10.38b–d. The Norton equivalent of all magnets can still be expressed by (10.76), with Φ_r and F_c the equivalent residual flux and equivalent mmf, and R_M the equivalent reluctance. The equivalent Φ_r and F_c can be expressed as

$$\Phi_r = \Phi_{r1} + \Phi_{r2} \tag{10.93}$$

$$F_c = \frac{\Phi_r}{\Phi_{r1}\ F_{c2} + \Phi_{r2}\ F_{c1}} F_{c1}\ F_{c2} \tag{10.94}$$

where Φ_{r1} and F_{c1} are the residual flux and mmf of the dominant magnet, and Φ_{r2} and F_{c2} are the residual flux and mmf of the assistant magnet.

10.3.4 Sizing of Magnets in PM Motors

Sizing of magnets is one of the critical tasks of PM machine design. This section discusses the analytical methods to calculate the volume and size of magnets for PM motors. The proposed methods are validated by FEA and experiments [91].

In the following, the formulas will be derived based on a set of assumptions and then modified based on practical design considerations. The assumptions include the following:

- Magnetic pole salience can be neglected.
- Stator resistance is negligible.
- Saturation can be neglected.
- Air-gap flux is sinusoidally distributed.

Based on the above assumptions, and using the phasor diagram of a PM synchronous motor as shown in Figure 10.29, the input power of the PM synchronous motor can be written as

$$P_1 = mIV\cos\phi = mIE_0\cos\delta \tag{10.95}$$

where m is the number of phases, I and V are the phase voltage and phase current, E_o is the induced back emf per phase, ϕ is the power angle, that is, the angle between phasor I and phasor V, and δ is the inner power angle, that is, the angle between phasor I and phasor E_o.

The back emf of a PM synchronous machine with sinusoidal air-gap flux can be expressed as

$$E_0 = \sqrt{2}\pi K_w fW\Phi \tag{10.96}$$

where W is the number of turns per phase, Φ is the total air-gap flux per pole, and K_w is the winding factor.

The phase current can be expressed in terms of armature maximum direct-axis reactant mmf F_{adm}:

$$I = F_{adm}\frac{p}{0.9mWK_w\ K_{ad}\ K_m\sin\delta} \tag{10.97}$$

where K_{ad} is the d-axis armature reaction coefficient, K_m is the maximum possible armature current (per unit), and p is the number of poles.

Substituting Equations 10.97 and 10.98 into Equation 10.96, the input power can be expressed as

$$P_1 = \frac{\sqrt{2}\pi pf}{0.9K_{ad}\ K_m\tan\delta} F_{adm}\Phi \tag{10.98}$$

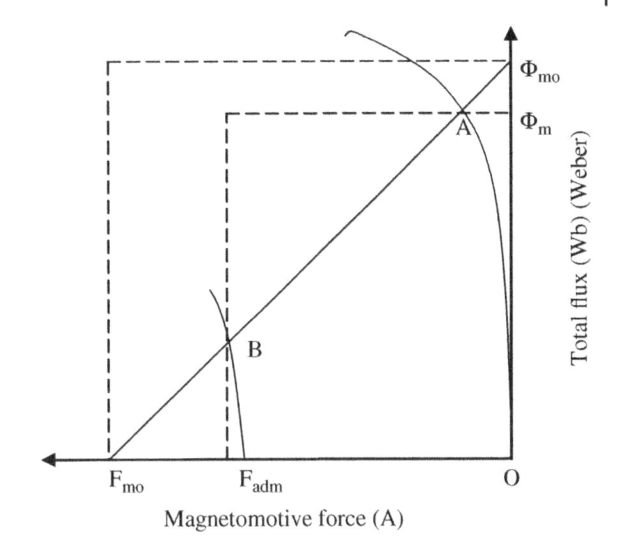

Figure 10.39 Illustration of magnet usage where A is the no-load operation point and B is the maximum reversal current point.

In this chapter, a new term – the magnet usage ratio ξ – is introduced and defined as follows:

$$\xi = \frac{F_{adm}\ \Phi_m}{F_{mo}\ \Phi_{mo}} \tag{10.99}$$

where Φ_{mo} is the total residual flux per pole, F_{mo} is the total mmf per pole, Φ_m is the total flux per pole at no-load condition, and F_{adm} is the maximum direct-axis reactant mmf of the motor.

The definition of this magnet usage ratio is illustrated in Figure 10.39, where point A is the magnet operation point at no load, and point B is the magnet operation point at maximum mmf. The air-gap flux per pole Φ can be expressed as a function of flux supplied by the magnet Φ_m and flux leakage coefficient σ_o:

$$\Phi = \Phi_m / \sigma_o \tag{10.100}$$

For series magnets, as shown in Figure 10.40a,

$$\left.\begin{aligned}\Phi_{mo} &= B_r\ S_m\\ F_{mo} &= 2H_c\ l_m\end{aligned}\right\} \tag{10.101}$$

For parallel magnets, as shown in Figure 10.40b,

$$\left.\begin{aligned}\Phi_{mo} &= 2B_r\ S_m\\ F_{mo} &= H_c\ l_m\end{aligned}\right\} \tag{10.102}$$

where l_m is the thickness of magnet per pole along the magnetizing direction, and S_m is the cross-sectional area of magnet under each pole.

Finally, the input power of the motor can be expressed as follows by substituting Equations 10.100–10.102 into Equation 10.99:

$$P_1 = \frac{\sqrt{2}\pi\xi f}{0.9K_{ad}\ K_m\ \sigma_0 \tan\delta} B_r\ H_C 2pS_m\ l_m \tag{10.103}$$

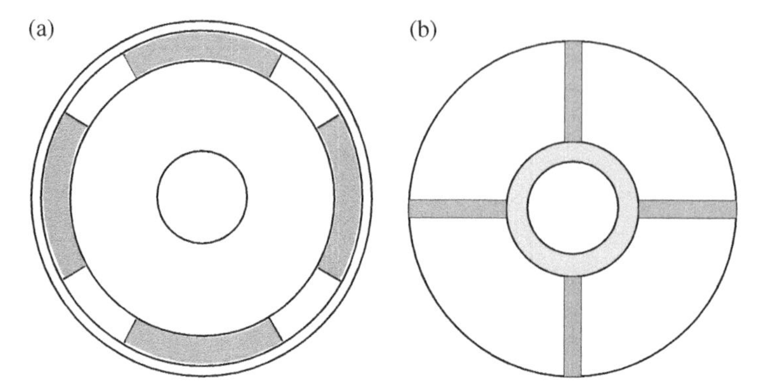

Figure 10.40 Configuration of series and parallel magnets: (a) surface mounted with sleeve rings; (b) parallel magnets.

Since the total magnet volume is

$$V_m = 2pS_m \; l_m \tag{10.104}$$

The magnet volume used in a PM synchronous motor can therefore be expressed as

$$V_m = \frac{0.2\sigma_0 \; K_m \; K_{ad}}{\xi} \tan\delta \frac{P_1}{fB_r \; H_C} = C_V \frac{P_1}{fB_r \; H_C} \tag{10.105}$$

where C_V is a coefficient

$$C_V = \frac{0.2\sigma_0 \; K_m \; K_{ad}}{\xi} \tan\delta \tag{10.106}$$

In order to use the above equations to determine the magnet volume needed for a PM motor, certain parameters of the motor need to be identified.

10.3.4.1 Input Power

At the design stage, the input power of a PM motor is given by

$$P_1 = \frac{P_N}{\eta \cos\phi} \tag{10.107}$$

where η is the target efficiency and $\cos\varphi$ is the target power factor of the motor.

10.3.4.2 Direct-Axis Armature Reaction Factor

Salience can be included in the direct-axis armature reaction factor. For a given magnet coverage α, K_{ad} is

$$K_{ad} = \frac{\alpha\pi + \sin\alpha\pi}{4\sin(\alpha\pi/2)} \tag{10.108}$$

10.3.4.3 Magnetic Usage Ratio and Flux Leakage Coefficient

The magnet usage ratio can be designed such that the demagnetization of magnets can be avoided. If one chooses 70–90% residual flux and 70–90% coercive force, then ζ is between 0.5 and 0.81.

Flux leakage for surface-mounted magnets is usually small, that is, σ_0 is approximately 1.0. For IPM motors, $\sigma_0 = 1.2$–1.5 and depends on the actual configuration of the motor.

10.3.4.4 Maximum Armature Current

Maximum armature current occurs during transient conditions, or during starting in the case of a line-start PM motor. During transient conditions, when the PM synchronous motor runs out of synchronization, the back emf and terminal voltage may run out of phase. Therefore, the maximum armature current always occurs when the terminal voltage is out of phase with the back emf,

$$I_{max} = \frac{E_o + V}{X_d} \tag{10.109}$$

where X_d is the direct-axis reactance of the motor.

A typical value of maximum current is 4–8 and must be verified during the design process.

10.3.4.5 Inner Power Angle

The power angle in Equation 10.107 refers to the rated operation point. In PM motor designs, this angle is usually kept around 25°–45°.

By substituting all the above coefficients into Equation 10.107, C_V can be determined. To a reasonable first approximation, C_V can be chosen to be 2. It should be adjusted during the design process.

Usually the length of magnet along the shaft direction is chosen to be the same as the rotor laminations stack length l_{fe}. The thickness of magnet along the magnetization direction is determined by the maximum armature current and operating temperature as shown in Figure 10.33.

For series magnets as, shown in Figure 10.40a, the magnet thickness is

$$l_m = K_A \frac{K_m \ F_{ad}}{H_{c,125°C}} \tag{10.110}$$

For parallel magnets, as shown in Figure 10.40b, the magnet thickness is

$$l_m = K_A \frac{K_m \ F_{ad}}{2H_{c,125°C}} \tag{10.111}$$

where K_A is a safety ratio, which can be chosen to be 1.1, therefore

$$F_{ad} = \frac{0.9mWK_w \ K_{ad} \sin\delta}{p} I \tag{10.112}$$

The width of rectangular magnets can be determined by

$$b_m = \frac{V_m}{2pl_m \ l_{fe}} \tag{10.113}$$

The radius of arc-shaped magnets can be determined by

$$R = \frac{V_m}{2\pi\alpha l_m \ l_{fe}} \tag{10.114}$$

Once the initial magnet volume and size have been determined, FEA can be used for further design analysis and optimization. The numerical calculation will help to identify whether the volume of magnets from the preliminary design is sufficient, insufficient, or excessive. Therefore, magnet volume can be further optimized during the numerical calculations.

In PM motor design and optimization, there are many conflicting design objectives. Multiobjective optimization is usually necessary in order to meet design criteria. In this chapter, the optimization objective is defined as the minimal usage of PM material while satisfying the performance requirements. The optimization problem is defined as

$$\min\{V_m(X)\} \tag{10.115}$$

subject to

$$f_i(X) \leq 0, \quad i = 1,2,\ldots,n \tag{10.116}$$

where X is a vector of the magnet width, thickness, and axial length:

$$X = \{b_m, l_m, l_{fe}\} \tag{10.117}$$

$f_i(X)$ is the motor performance requirements. Here, these requirements are defined as back emf, efficiency, maximum torque, and short-circuit current. These performances are calculated during the optimization process.

The optimization process starts with the preliminary design of the motor. The no-load magnetic field is first calculated using FEA to verify the back emf and short-circuit current. The load magnetic field is then calculated to confirm the maximum power/torque and efficiency. During each FEA, the magnet size is adjusted for given constraints.

The optimization implemented using FEA is shown in Figure 10.41. Saturation, salience, and air-gap flux waveform can also be verified during numerical calculations.

10.3.5 Eddy Current Losses in the Magnets of PM Machines

The eddy current loss in the magnets of PM motors in an HEV and PHEV is usually not taken into consideration in traditional motor design and analysis. However, due to the high conductivity of the rare earth magnet Nd–Fe–B, and slot/tooth harmonics, there is eddy current loss generated inside the magnets. This loss may not greatly affect the efficiency of the motor, but the temperature rise inside the magnets caused by this loss can lead to unpredictable deterioration of the magnets, such as degradation of performance and potential demagnetization. In addition, the output voltage of the PWM inverter contains abundant high-frequency harmonics, which induce excessive loss in the magnets. The excessive heat in PM motors induced by the eddy current loss combined with other losses can degrade the performance of the machine [92].

For sinusoidal supply, the eddy current loss in the magnets of an SPM motor is obtained from time-stepped FEA to be 102 W for a 50 kW PM motor. The eddy current loss in the magnet of an IPM was calculated to be 10 W for the same motor.

For PWM supply, the eddy current loss in the magnets of an SPM is 249 W for a 50 kW PM motor. The eddy current loss in the magnet of an IPMSM was 25 W for the same motor.

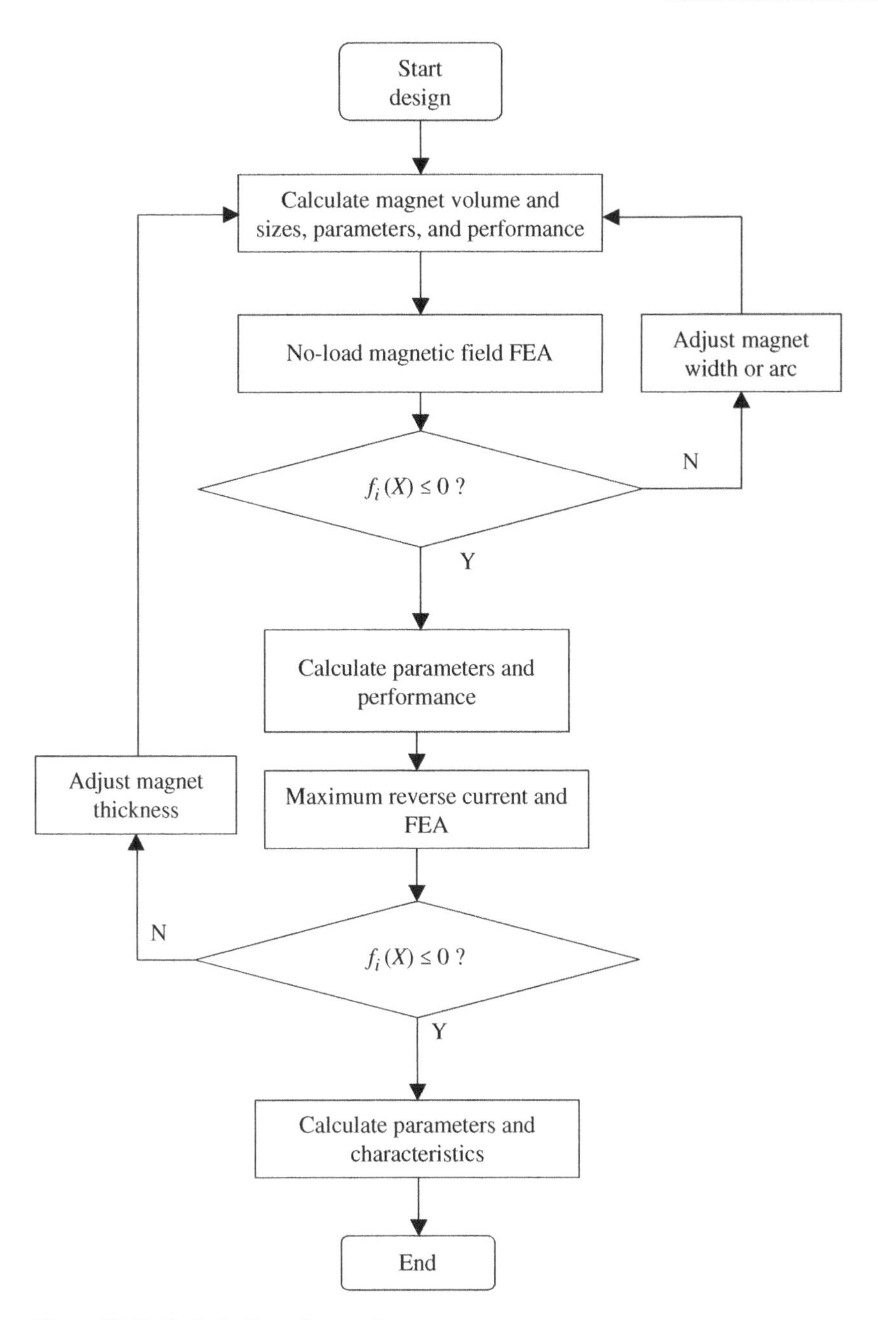

Figure 10.41 Optimization of magnet usage.

10.4 Switched Reluctance Motors

Both switched reluctance motors and synchronous reluctance motors have attracted attention in traction applications due to their simple structure, not needing squirrel cage or magnets on the rotor, very little loss on the rotor, and ease of control [93–114].

Although they have many advantages, PM motors and induction motors both have their own limitations. For example, PM motors face the possibility of demagnetization at extremely high temperature, limited speed range, and difficulty in protecting the powertrain during a fault condition. Induction motors have limited torque capabilities at low speeds, lower torque density and lower efficiency, noise due to stator/rotor slot combinations, and so on.

From the previous section we have seen that the torque of a synchronous motor has two terms, one related to E_o and X_d, which is induced by the rotor PM field, and one related to V, X_d, and X_q, which is induced by the difference in reactance of the d axis and q axis. In other words, even if the magnets are removed, an IPM motor can still generate torque with a sinusoidal supply due to the existence of salience of the rotor. This is called a synchronous reluctance motor. The stator and the rotor of a synchronous reluctance motor have the same number of poles.

Switched reluctance or synchronous reluctance motors do not use magnets or a squirrel cage. They simply use the difference in d-axis and q-axis reactance to produce reluctant torque. Therefore, they are similar to a synchronous motor without excitation and are therefore known as a switched reluctance motor. Hence only the second term of Equation 10.72 exists. The torque of a switched reluctance motor with sinusoidal supply is

$$T = \frac{mV^2}{2\omega_R}\left(\frac{1}{X_q} - \frac{1}{X_d}\right)\sin(2\theta) \tag{10.118}$$

In order to increase the torque of a switched reluctance motor, the q-axis and d-axis reactance is designed to have a large difference. A cross-section of a synchronous reluctance motor is shown in Figure 10.42.

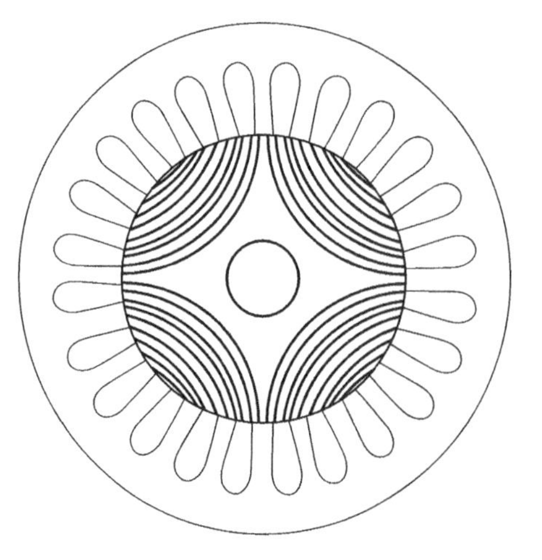

Figure 10.42 The synchronous reluctance motor [114].

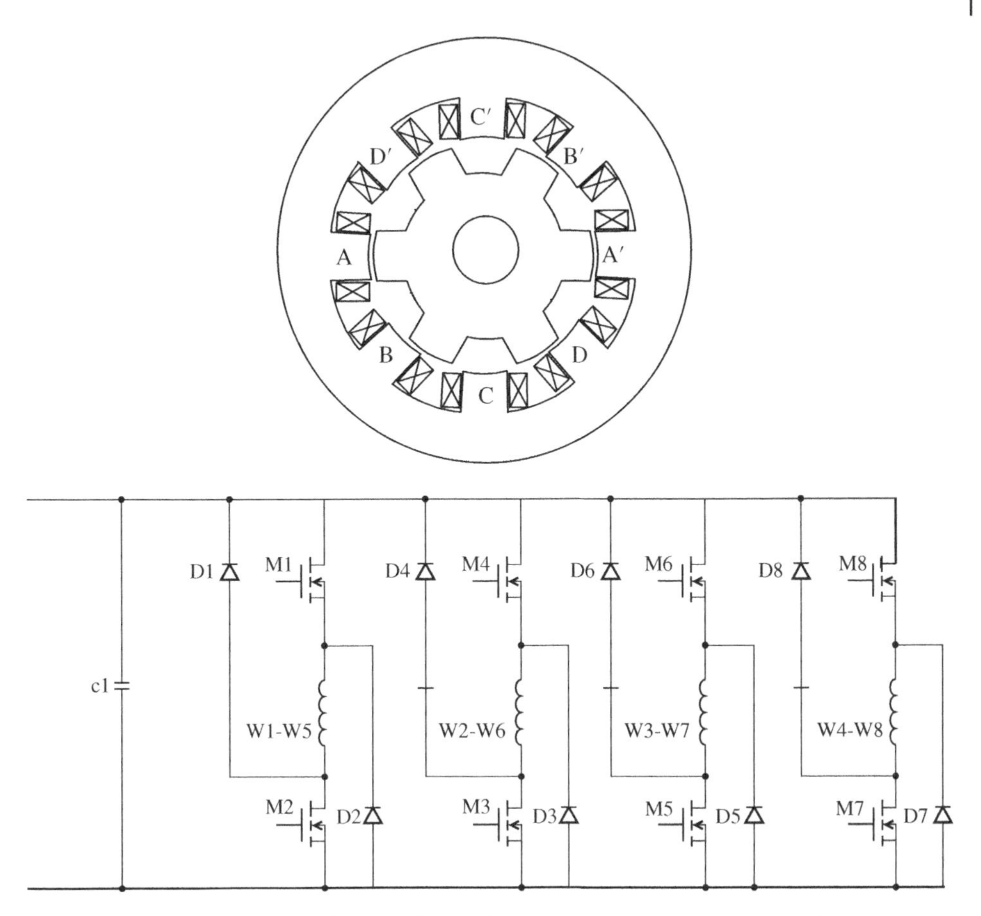

Figure 10.43 The cross-section of a 6/8-pole switched reluctance motor and its control circuit: top, cross-sectional area of the SRM; bottom, control circuit of the SRM.

Switched reluctance motors are similar to synchronous motors but will have different numbers of poles on the stator and the rotor. Figure 10.43 shows the cross-section of a switched reluctance motor and its control circuit.

10.5 Doubly Salient Permanent Magnet Machines

Doubly-salient permanent magnet (DSPM) machines, a new kind of inverter-fed electrical traction motor first proposed in the early 1990s [115, 116], are becoming more and more attractive because of their distinct features, such as high efficiency, high power density, and simple structure. Much progress has been made on the design of DSPM machines. For example, Liao et al. discussed the basic principles of DSPM machines in 1992, 1993, and 1995 [116–118]. Cheng and Chau and co-workers studied the steady-state characteristics and performance of DSPM machines using nonlinear varying-network magnetic circuit analysis [119–123].

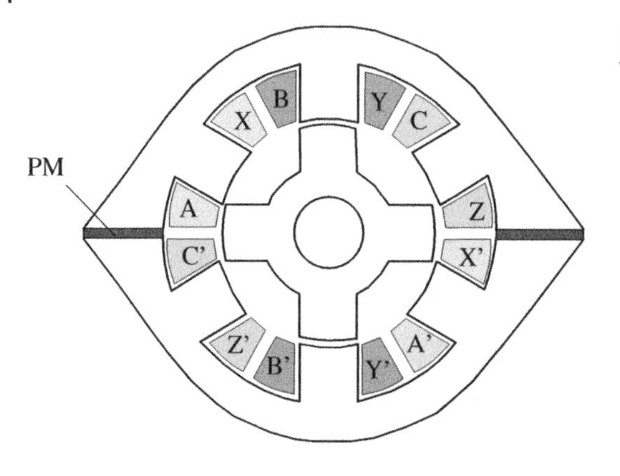

Figure 10.44 Typical DSPM geometry with 6/4 pole pairs.

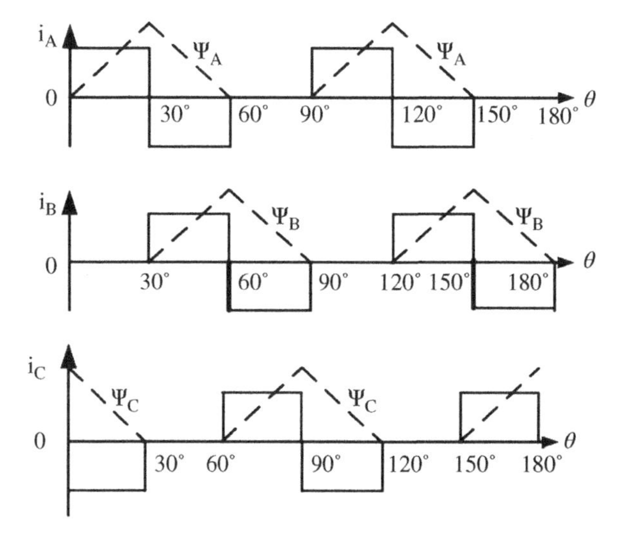

Figure 10.45 Flux linkage (dashed lines) and commutating mode (solid lines).

A DSPM machine resembles the structure of an SRM motor except that PMs are inserted in the stator. Therefore some common techniques used in the SRM motor are also adopted in the design and control of DSPM machines. For example, wider rotor pole arc, advanced shutoff angle control, and lagged firing angle control can all be used in the design and control of DSPM machines. However, due to the existence of PMs in the stator, the behavior of a DSPM machine is different from that of an SRM motor. Therefore, new design and control concepts need to be explored to optimize the performance of DSPM machines.

Figure 10.44 shows the typical geometry of a DSPM machine with 6/4 pole pairs, where AX, BY, CZ and A′X′, B′Y′, C′Z′ are the three-phase windings. It resembles the structure of an SRM motor, except that PMs are inserted in the stator. The three-phase flux linkages are shown in Figure 10.45. Because of the existence of PMs in the stator, PM flux plays a major role in the winding flux linkage. Therefore, dual-polarity control can be employed, as shown in Figure 10.45, to improve the power density.

In the design practices of SRM motors, in order to ensure the winding commutation and self-start capability at any rotor position and either rotating direction, there should

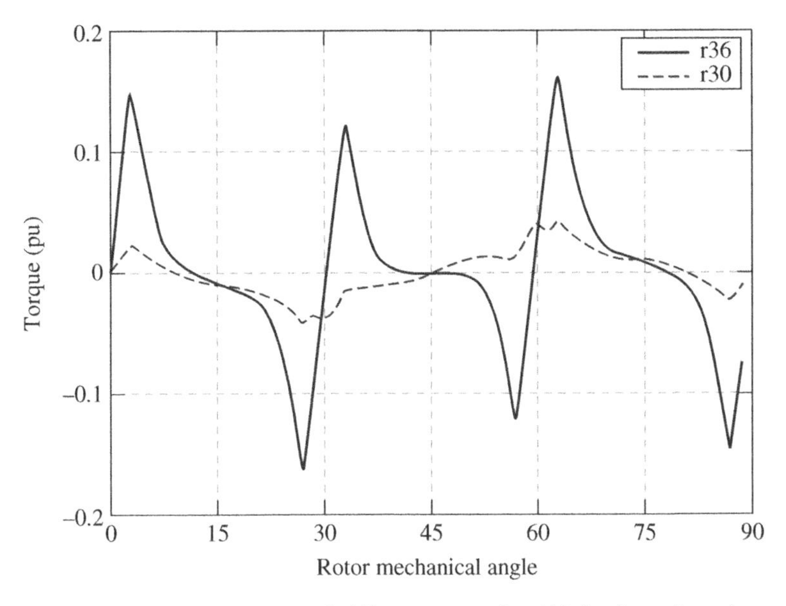

Figure 10.46 Cogging torque with different rotor pole width (unskewed rotor).

be a small overlap between the adjacent stator and rotor salient poles when the axis of the stator pole is aligned with that of the rotor pole. Therefore the width of the rotor pole is usually larger than that of the stator. This technology is also used in DSPM machines [119] because of the structural similarity. But due to the existence of PMs in the stator, cogging torque (due to the presence of permanent magnets and saliency of the slots or teeth in the laminations) exists in DSPM machines. According to the flux–mmf diagram of PM machines, cogging torque will reach its minimum if the resultant gap reluctance is uniform at any rotor position [124]. Therefore, for a 6/4 pole-paired DSPM machine, if the width of the rotor pole equals that of the stator and their width is one-half of the pitch, then cogging torque will reach its minimum value; if the rotor pole width is larger than the stator pole width, the cogging torque will increase significantly because the gap reluctance will not be uniform as the rotor position varies. Cogging torque is one of the most important issues of DSPM machines. The cogging torque obtained from FEA is shown in Figure 10.46 for the larger rotor pole width of 36° and uniform width of 30°. The cogging torque was calculated using 2D FEA. It can be seen from Figure 10.46 that the cogging torque is significantly less for the smaller rotor pole width. It can also be seen that the cogging torque has different peaks. This is caused by the different linkage flux from the two PMs. When the rotor poles are aligned with the stator poles that are adjacent to the magnets, there is less flux leakage. Therefore there is more cogging torque. When the rotor poles are aligned with the stator poles that are in the middle of the two magnets, more flux leakage is seen. Therefore there is less cogging torque.

In order to minimize the cogging torque of a DSPM machine, the width of the rotor pole is designed to be the same as that of the stator, both equal to one-half of the pole pitch,

$$\theta_s = \theta_r = \frac{\pi}{N_s} \tag{10.119}$$

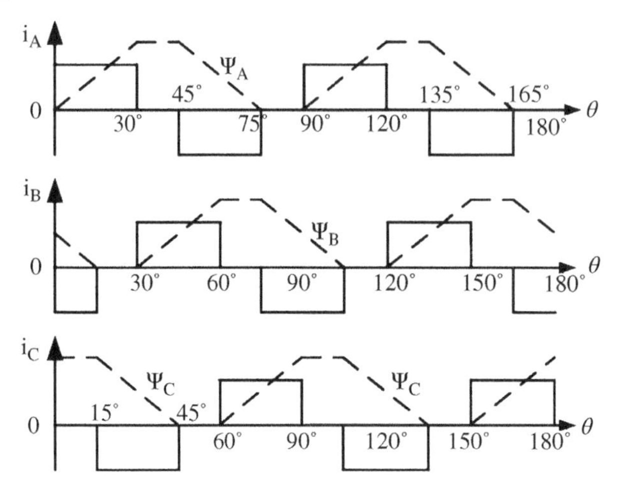

Figure 10.47 Flux linkage (dashed lines) and commutating mode (solid lines) of a skewed rotor DSPM machine.

For a 6/4 pole-paired DSPM machine, it is 30° mechanical degrees.

Second, a skewed rotor is used to ensure the capability of self-starting at any rotor position and either rotating direction. From the flux linkage curve, a skewed rotor can also lead to overlap between the adjacent stator and rotor salient poles, which provides the same effect as that of the larger rotor pole width used in SRM motors. In order to obtain the largest output capability, the skew angle is chosen to be one-half of the stator salient pole width,

$$\theta_{skew} = \frac{\theta_r}{2} = \frac{\pi}{2N_s} \tag{10.120}$$

For a 6/4 pole-paired DSPM machine, the skew angle is 15°.

The flux linkage of the skewed rotor DSPM machine is shown in Figure 10.47. It can be seen that the flux linkages of this machine are different from that of the non-skewed one shown in Figure 10.45; that is, the coverage of flux linkage is increased from 120° to 150°. As a result, the six-state commutating mode can be employed.

In the control of conventional DSPM machines, a three-state commutating mode was used, that is, +A − C, −A + B, and − B + C respectively, as shown in Figure 10.45. This simple commutating mode generally results in poor performance of the DSPM machine. It can be seen that the current commutating from positive to negative is in sequence. But in practice, because of the current continuity, the turn-off angle of the positive current should be advanced, or the freewheeling current will create a reverse torque. This will cause control complications and decrease output ability.

According to the same commutating principle as in conventional DSPM motors, that is, positive current conducts at the rising slope of flux linkage, while negative current conducts at the falling slope, a new commutating mode for the novel skewed rotor DSPM motor can be developed as shown by the solid line in Figure 10.47. It is a six-state commutating mode. The conducting sequences are + A − B, +A − C, +B − C, +B − A, +C − A, and + C − B respectively. Each state will be conducted for 60° electrical degrees continuously, and there is a 60° interval between the positive and negative current commutation. This commutating mode makes it possible to neglect the control of turn-off angle, therefore commutating performance can be improved and reliability enhanced.

10.6 Design and Sizing of Traction Motors

The most important task in the design of a traction drive is to calculate the size of the motor. Take a PM motor as an example. The power of the motor can be expressed as in Equation 10.70 and the back emf can be expressed as in Equation 10.96. Substituting Equation 10.96 into Equation 10.70, we have

$$P = mE_0 I\cos\delta = mI\cos\delta\sqrt{2}\pi K_w fW\Phi \tag{10.121}$$

The linear current density (or specific current density) along the inner stator surface is A (A/m) and the inner diameter of the stator is D. Then

$$2mWI = \pi DA \tag{10.122}$$

The total flux per pole can be expressed in terms of air-gap flux density B:

$$\Phi = \frac{\pi Dl}{2p}B\alpha \tag{10.123}$$

where l is the stack length of the stator, α is the pole enclosure as a percentage of the pole pitch, and B is the air-gap flux density. In PM motor control, we set the inner power angle δ to 0 to achieve maximum torque. Since $f = pn/60$,

$$P = m\frac{\pi DA}{2mW}\sqrt{2}\pi K_w \frac{pn}{60}W\frac{\pi Dl}{2p}B\alpha \tag{10.124}$$

Therefore

$$D^2 l = \frac{60}{\sqrt{2}\pi^3}\frac{4}{\alpha K_w}\frac{P}{AB}\frac{1}{n} \tag{10.125}$$

Since K_w, A, and B are in a relatively narrow range for all types of motors, this expression shows that the effective volume of the motor is proportional to power P and inversely proportional to speed n. Typical values for A are 100 kA/m for small air-cooled motors, and up to 400 kA/m for liquid-cooled motors. The typical level for B is around 0.4 T for small motors, and up to 1.2 T for high-density motors. Considering that $P = T2\pi n/60$, we have

$$D^2 l = k\frac{P}{n} \propto T \tag{10.126}$$

In other words, the size of an electric motor is proportional to its torque rating.

10.6.1 Selection of *A* and *B*

In the above equations for sizing traction motors, both A and B are experience-based selections. B shows how much the magnetic material (silicon steel) is utilized and A shows how much the electric material (copper or aluminum) is utilized. B is limited by magnetic losses in the teeth and yoke. A high B means less magnetic material but higher magnetic losses. A high A means less copper material but higher electric losses. Ambient temperature, operating frequency, and cooling method can impact the selection of A and B. Large motors will generally have larger A and B values.

10.6.2 Speed Rating of the Traction Motor

The rotor volume is inversely proportional to rotor speed. Hence, a higher speed rating means a smaller size. But a higher speed means a higher operating frequency, which results in more magnetic losses (eddy current and hysteresis losses). Smaller values of A and B may be necessary to limit the loss in high-speed motors.

For example, a four-pole 1500 rpm motor operates at 50 Hz, but a four-pole 12,000 rpm motor operates at 400 Hz. Since eddy current loss is proportional to f^2, and hysteresis loss is proportional to f^{α} $(1<\alpha<2)$, if the same magnetic flux density is chosen for the two motors, then the losses in the high-speed motor will be many times that of the low-speed motor even if the size of motor is much smaller. This is because loss increases 64 times, but size (D^2l) reduces by only a factor of 8.

10.6.3 Determination of the Inner Power

In Equation 10.125, the power used in the sizing is inner power (not shaft power). Therefore, for motors, that power is the shaft power divided by the efficiency and power factor, $P' = K_E P_N/(\eta_N \cos\phi_N)$, and for a generator, $P' = K_E P_N/\cos\phi_N$, where K_E is the ratio of back emf and rated supply voltage.

Since the efficiency η and power factor are not known until the design is finished, we have to start by making assumptions for these values and validate them at the end of the design.

10.7 Thermal Analysis and Modeling of Traction Motors

Thermal issues are important aspects in traction motor design and analysis. Traction motors exhibit high frictional and windage losses due to high-speed operation, high magnetic losses due to high-frequency operation, and additional losses caused by PWM harmonics. Thermal analysis is of particular importance for PM motors because the magnets can be demagnetized at high temperatures [125].

Thermal studies on electric motors often approach the subject using FEA. Time-stepping FEA can provide greater accuracy for the thermal distribution in a motor. However, it remains relatively time-consuming and does not provide as much insight as an analytical solution. Various analytical and numerical methods have been developed for the thermal analysis of various types of electric motors [126–138]. This section contains a discussion on how to calculate the temperature in a PM motor.

Thermal resistances in the lumped-parameter representation circumscribe the paths for heat transfer and are analogous to resistances in an electric circuit. The model thereby establishes the equivalent parameters between the thermal and electrical domains, as given in Table 10.6.

The analogy allows the development of a lumped-parameter thermal model for any electric motor. The thermal capacitances are usually neglected since only steady-state operation of the machine is of particular importance. Figure 10.48 shows the thermal equivalent circuit of an SPM. T_{in}, T_{stator}, T_{magnet}, T_{rotor}, T_w, and T_{shaft} respectively describe the temperatures of the inner surface of insulation between stator and casing, the inner surface of the stator core, the outer surface of the magnet, the outer surface of the rotor, the part of the winding within the stator core, and the part of the shaft under the rotor core. $T_{coolant}$ and T_{case} are assigned as boundary conditions.

Table 10.6 Analogy of thermal and electrical domains.

Electric circuit	Thermal circuit
Electric voltage u (V)	Temperature T (K)
Current I (A)	Heat loss Q (W)
Electrical resistance R (Ω)	Thermal resistance R (K/W)
Electrical conductivity σ (S/m)	Thermal conductivity k (W/(m K))

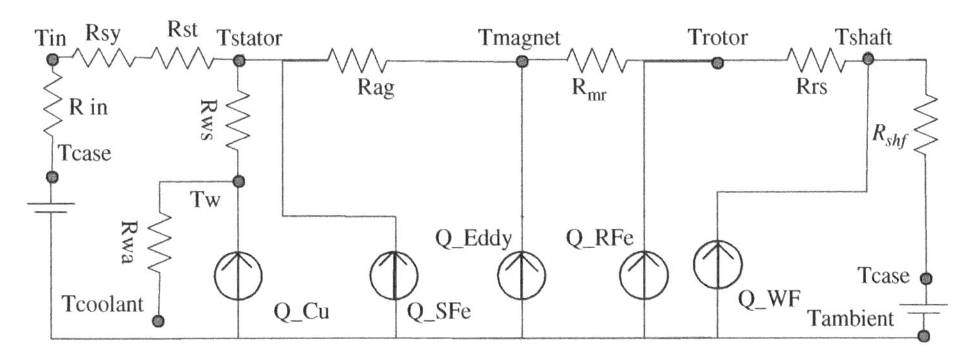

Figure 10.48 Equivalent circuit mode.

The individual thermal resistances are:

R_{ag}, convection thermal resistance of the air gap
R_{mr}, radial conduction thermal resistance of the pole
R_{rs}, radial conduction thermal resistance of the rotor core
R_{shf}, thermal resistance of the shaft
R_{sy}, radial conduction thermal resistance of the stator yoke
R_{st}, radial conduction thermal resistance of stator teeth
R_{ws}, conduction thermal resistance between the windings and stator
R_{in}, contact thermal resistance between the stator and housing.

10.7.1 The Thermal Resistance of the Air Gap, R_{ag}

Within the air gap the heat flow is greater than in the adjoining air as almost 99% heat emitted from the rotor surface is transferred directly across it to the stator.

Taylor [127] developed the dimensionless convection correlation method from testing two concentric cylinders rotating relative to each other, to consider the heat transfer across the air gap; the method was further modified by Gazley [128]. The thermal resistance can be defined in terms of the air-gap length l_g, a dimensionless Nusselt number N_{nu}, the thermal conductivity of motionless air K_{air}, and the average area of the air-gap cylindrical surface A_{ag}:

$$R_{ag} = \frac{l_g}{N_{nu}\ K_{air}\ A_{ag}} \tag{10.127}$$

$$N_{nu} = 2, \qquad N_{Ta} \le 41$$
$$N_{nu} = 0.212 N_{Ta}^{0.63} \; N_{\Pr}^{0.27}, \quad 41 < N_{Ta} \le 100$$
$$N_{nu} = 0.386 N_{Ta}^{0.5} \; N_{\Pr}^{0.27}, \quad N_{Ta} \le 41$$

The values of the Nusselt N_{nu}, Taylor N_{Ta}, and Prandtl N_{Pr} numbers for two rotating smooth cylinders were presented by Taylor. The flow in the air gap is laminar when $N_{Ta} \le 41$, whereas the flow through the vertex with enhanced heat transfer is in the range $41 < N_{Ta} \le 100$. If $N_{Ta} > 100$ there is fully turbulent flow in the air gap. Later, Gazley [128] modified the heat transfer with a 10% increase in the experimental results due to the slot effects.

The expression given in (10.127) is applicable to air-cooled electrical motors. In order to analyze liquid-cooled motors, the thermal conductivity of the air gap is assumed to have a constant value, and an equivalent air-gap resistance can be calculated considering the air gap as equal to a cylinder. Hsu et al. [139] showed that the thermal conductivity of the Toyota Prius traction motor is 10 W/(m °C) based on an oil and air convective mixture. In this case, the simpler expression for deriving the air-gap thermal resistance is

$$R_{ag2} = \frac{\ln\left(r_{is}/r_{magnet}\right)}{2\pi k_{ag} \; L_g} \tag{10.128}$$

where r_{magnet} is the outer magnet radius, r_{is} is the inner stator radius, and k_{ag} is the thermal conductivity of the air gap.

10.7.2 The Radial Conduction Thermal Resistance of the Rotor Core, R_{rs}

Figure 10.49 shows the rotor core as a cylinder made of laminations. The radial heat transfer is more pronounced than the axial heat transfer in the laminations, therefore the heat transfer coefficient is calculated for the radial direction:

$$R_{rs} = \frac{\ln\left(r_{rotor}/r_{shaft}\right)}{2\pi k_{rotor} \; L_s} \tag{10.129}$$

where r_{shaft} is the shaft radius, k_{rotor} is the thermal conductivity of rotor core, and L_s is axial length of the rotor core.

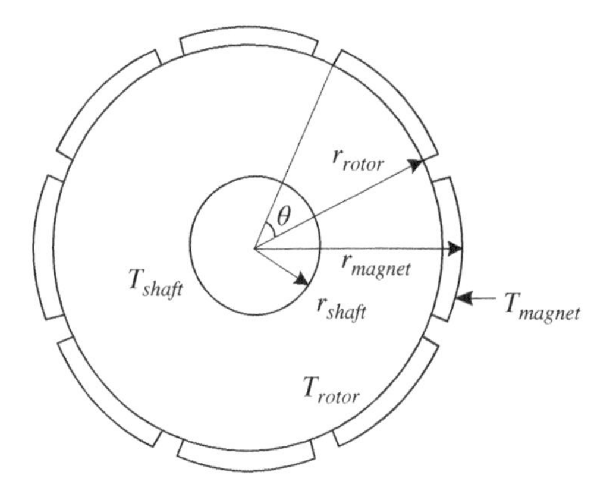

Figure 10.49 Radial dimensions of the rotor.

10.7.3 The Radial Conduction Thermal Resistance of the Poles, R_{mr}

The surface-mounted magnets distributed on the rotor core are assumed to be an equivalent cylinder with $n\theta$ radians:

$$R_{mr} = \frac{\ln\left(r_{magnet}/r_{rotor}\right)}{n\theta k_m\ L_s} \tag{10.130}$$

where r_{magnet} is the outer rotor radius, r_{rotor} is the rotor core radius, n is the number of poles, θ is the width of each pole in radians, k_m is the thermal conductivity of the magnet, and L_s is the axial length of the poles.

10.7.4 The Thermal Resistance of the Shaft, R_{shf}

The shaft is represented as a cylindrical rod with axial heat conduction and separated into three parts [126]: one that lies under the rotor core; a second that lies under the bearing; and a third that acts as a thermal connection between the mean temperatures of the previous two. As the bearings provide good thermal contact, it is sufficient to consider thermal contact that exists between the shaft and the thermal casing (Figure 10.50). The convection between the shaft and adjoining air is neglected as the heat transfer from the air to the shaft is negligible. Thus

$$R_{shf} = \left(R_a + R_b\right)/2 \tag{10.131}$$

where

$$R_a = \frac{1}{2\pi k_{shf}\ L_s} + \frac{L_{bs}}{2\pi k_{shf}\ \left(D_{shf}/2\right)^2}$$

$$R_b = \frac{1}{4\pi k_{shf}\ L_b} + \frac{L_{bs}}{2\pi k_{shf}\ \left(D_{shf}/2\right)^2}$$

k_{shf} is the thermal conductivity of the shaft, L_b is the thickness of the bearing, D_{shf} is the radius of the shaft, and L_{bs} is the distance from the bearing center to rotor mean.

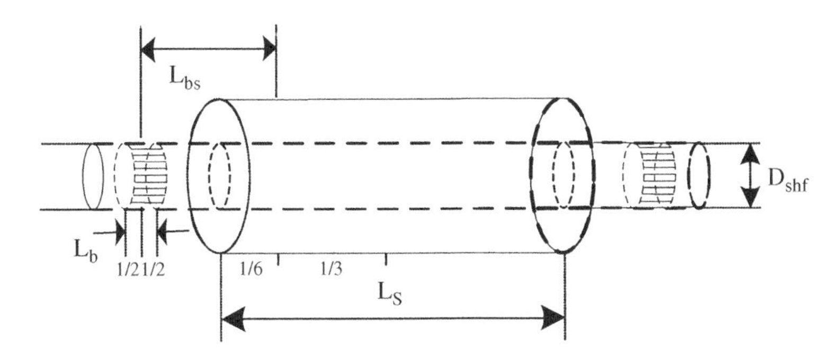

Figure 10.50 Axial dimensions of the shaft.

10.7.5 The Radial Conduction Thermal Resistance of Stator Teeth, R_{st}

As both the rotor and stator core consist of layers of laminations, only the thermal conductivity in the radial direction is considered (Figure 10.51). In order to calculate the thermal resistance of the stator precisely, the stator is modeled as two parts, one as the stator yoke and the other as the stator teeth. The equivalent cylinder with a reduction factor p is used to model the stator teeth. Thus

$$R_{st} = \frac{\ln\left(r_{ms}/r_{is}\right)}{2\pi k_{iro}\ L_s \rho} \tag{10.132}$$

where r_{is} is the inner stator radius, r_{ms} is the inner stator yoke radius, k_{iro} is the thermal conductivity of the stator, ρ is the percentage of the teeth section with respect to the total teeth plus all slots section.

10.7.6 The Radial Conduction Thermal Resistance of the Stator Yoke, R_{sy}

This is calculated from

$$R_{sy} = \frac{\ln\left(r_{os}/r_{ms}\right)}{2\pi k_{iro}\ L_s} \tag{10.133}$$

where r_{os} is the outer stator yoke radius.

10.7.7 The Conduction Thermal Resistance between the Windings and the Stator, R_{ws}

This is obtained from

$$R_{ws} = \frac{S_{slot} - S_{cu}}{l_s\ k_{cu,ir}\ A_{slot}} \tag{10.134}$$

where S_{slot} is the stator slot surface, S_{cu} is the copper section in the stator slot, l_s is stator slot perimeter, $k_{cu,ir}$ is the equivalent conductivity coefficient of the air and insulation material in the stator slot, evaluated by simulation, and A_{slot} is the interior slot surface.

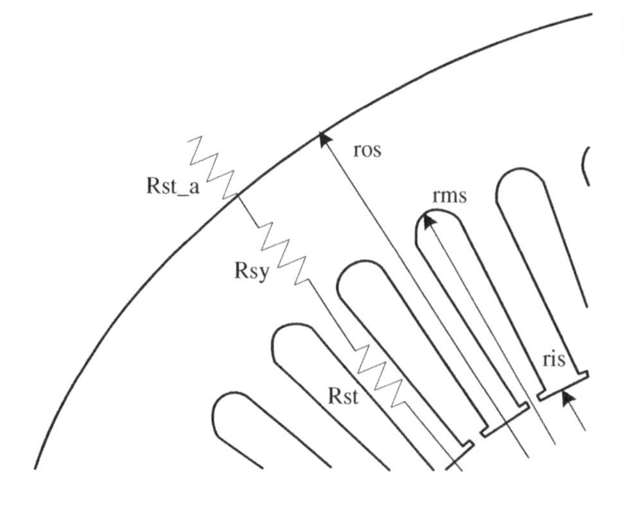

Figure 10.51 Radial dimensions of the stator.

10.7.8 Convective Thermal Resistance between Windings External to the Stator and Adjoining Air, R_{wa}

The convection coefficient between the end winding of the stator and the surrounding air is found from [126]

$$h_{wa} = 15.5(0.29v + 1) \tag{10.135}$$

where $v = r_{magnet}\omega\eta$ is the speed of inner air and v need be no more than 7.5 m/s; ω is the rotor angular velocity and η is fan efficiency.

The value of 50% was assumed for the fan efficiency, usually due to unavailable information on the radial air velocity. Sometimes, there is an oil and air mixture adjoining the winding, so an equivalent increase of fan efficiency is used.

The total surface of the winding external to the stator can be calculated as

$$S_{wa} = (L_c - L_s)2\pi r_{is} \tag{10.136}$$

where L_c is the external casing length.

Therefore from (10.135) and (10.137) the thermal resistance between the winding external to the stator and surrounding air can be calculated as

$$R_{wa} = \frac{1}{S_{wa}\ h_{wa}} \tag{10.137}$$

Assigning the losses and casing temperature (90 °C) to the thermal circuit, the temperature of individual components in the motor can be obtained after the calculation.

Consider the equivalent circuit model in Figure 10.48.

At node T_{in}

$$\frac{T_{stator} - T_{in}}{R_{sy} + R_{st}} - \frac{T_{in} - T_{case}}{R_{in}} = 0 \tag{10.138}$$

At node T_{stator}

$$\frac{T_{stator} - T_{in}}{R_{sy} + R_{st}} - \frac{T_w - T_{stator}}{R_l} - \frac{T_{magnet} - T_{stator}}{R_{ag}} = Q_{SFe} \tag{10.139}$$

At node T_w

$$\frac{T_w - T_{stator}}{R_{ws}} - \frac{T_w - T_{coolant}}{R_{wa}} = Q_{cu} \tag{10.140}$$

At node T_{magnet}

$$\frac{T_{magnet} - T_{stator}}{R_{ag}} - \frac{T_{rotor} - T_{magnet}}{R_m r} = Q_{Eddy} \tag{10.141}$$

At node T_{rotor}

$$\frac{T_{rotor} - T_{shaft}}{R_{rs}} + \frac{T_{rotor} - T_{magnet}}{R_{mr}} = Q_{RFe} + Q_{WF} \tag{10.142}$$

At node T_{shaft}

$$\frac{T_{rotor} - T_{shaft}}{R_{rs}} - \frac{T_{shaft} - T_{case}}{R_{shf}} = 0 \tag{10.143}$$

The above equations can be written in matrix form and solved by numerical methods:

$$\begin{bmatrix} -1/(R_{sy}+R_{st})-1/R_{in} & R_{sy}+R_{st}a_{12} & 0 & 0 & 0 & 0 \\ -1/(R_{sy}+R_{st}) & 1/R_{ag}+1/(R_{sy}+R_{st})+1/R_{ws} & -1/R_{ws} & -1/R_{ag} & 0 & 0 \\ 0 & -1/R_{ws} & 1/R_{ws} & 0 & 0 & 0 \\ 0 & -1/R_{ag} & 0 & 1/R_{mr}+1/R_{ag} & -1/R_{mr} & 0 \\ 0 & 0 & 0 & -1/R_{mr} & -1/R_{rs}+1/R_{mr} & -1/R_{rs} \\ 0 & 0 & 0 & 0 & 1/R_{rs} & -1/R_{rs}-1/R_{shf} \end{bmatrix} \cdot \begin{bmatrix} T_{in} \\ T_{stator} \\ T_w \\ T_{magnet} \\ T_{rotor} \\ T_{shaft} \end{bmatrix} = \begin{bmatrix} -90/R_{in} \\ Q_{SFe} \\ Q_{cu}+110/R_{wa} \\ Q_{Eddy} \\ Q_{RFe} \\ Q_{WF}+90/R_{shf} \end{bmatrix} \tag{10.144}$$

The temperature of a PM motor is solved with the temperature shown in Figure 10.52. The temperature of the different components is given in Tables 10.7 and 10.8 for a PM motor operated under pure sinusoidal supply and under PWM supply, respectively.

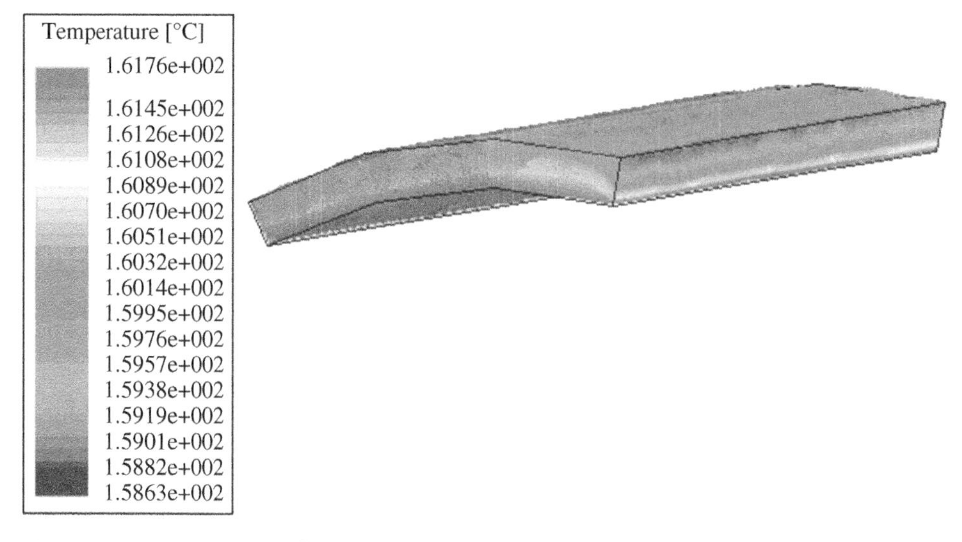

Figure 10.52 Temperature profiles in magnet.

Table 10.7 Temperature distribution under sinusoidal and PWM waveforms.

Name	Sinusoidal (°C)	PWM (°C)
Insulation	90–113.23	90–159.14
Stator	113.23–152.35	159.14–198.25
Winding	155.29	202.47
Magnets	154.88–155.47	200.92–201.54
Rotor	155.47–169.76	201.54–213.83
Shaft	90–169.76	90–213.83

Table 10.8 Temperature distribution within the motor fed by a sinusoidal waveform.

Name	Analytical method (°C)	Simulation (°C)
Insulation	90–113.23	90–119.45
Stator	113.23–152.35	120.21–156.73
Winding	155.29	162.63–180.39
Magnets	154.88–155.47	158.68–161.13
Rotor	155.47–169.76	156.34–162.90
Shaft	90–169.76	90–163.58

10.8 Conclusions

Electric motors and associated controllers are one of the key enabling technologies for electric, hybrid electric, and plug-in hybrid electric vehicles. Various types of electric motors and drive systems are available for the powertrain of electrified vehicles. Traction motors and drives experience very harsh environmental conditions, such as a wide temperature range (−30 to 60 °C), severe vibration and shock, high electromagnetic noise, size and weight constraints, and stringent safety and reliability requirements. As a result, there are many unique aspects in the design, development, analysis, manufacturing, and research of electric motors and drives for traction applications which are all important aspects but cannot all be covered in this chapter. Readers could consult the references below for further reading. For example, more in-depth studies about synchronous reluctance motor design and optimization can be found in [114] and studies of the uncontrolled generation in PM drive motors are covered in [140, 141].

References

1 Ehsani, M., Rahman, K.M., and Toliyat, H.A. (1997) Propulsion system design of electric and hybrid vehicles. *IEEE Transactions on Industrial Electronics*, 44, 19–27.

2 Muta, K., Yamazaki, M., and Tokieda, J. (2004) Development of new-generation hybrid system THS II – drastic improvement of power performance and fuel economy. SAE World Congress, March 8–11, paper no. 2004-01-0064.
3 Chan, C.C., Jiang, K.T., Xia, J.Z., et al. (1996) Novel permanent magnet motor drive for electric vehicles. *IEEE Transactions on Industrial Electronics*, 43 (2), 331–339.
4 Ehsani, M., Gao, Y., and Gay, S. (2003) Characterization of electric motor drives for traction applications. Industrial Electronics Society, IECON'03, November 2–6, vol. 1, pp. 891–896.
5 Rahman, Z., Butler, K.L., and Ehsani, M. (2000) Effect of extended-speed, constant-power operation of electric drives on the design and performance of EV propulsion system. SAE Future Car Congress, April, paper no. 2001–01-0699.
6 Rahman, K.M. and Ehsani, M. (1996) Performance analysis of electric motor drives for electric and hybrid electric vehicle application. *Power Electronics in Transportation*, pp. 49–56.
7 Rahman, K.M., Fahimi, B., Suresh, G., et al. (2000) Advantages of switched reluctance motor applications to EV and HEV: design and control issues. *IEEE Transactions on Industry Applications*, 36 (1), 111–121.
8 Honda, Y., Nakamura, T., Higaki, T., and Takeda, Y. (1997) Motor design considerations and test results of an interior permanent magnet synchronous motor for electric vehicles. Proceedings of the IEEE Industry Applications Society Annual Meeting, October 5–9, pp. 75–82.
9 Kamiya, M. (2005) Development of traction drive motors for the Toyota hybrid system. International Power Electronics Conference, April 4–8.
10 Miller, J.M., Gale, A.R., McCleer, P.J., et al. (1998) Starter/alternator for hybrid electric vehicle: comparison of induction and variable reluctance machines and drives. Proceedings of the IEEE 1998 Industry Applications Society Annual Meeting, October 12–15, pp. 513–523.
11 Wang, T., Zheng, P., and Cheng, S. (2005) Design characteristics of the induction motor used for hybrid electric vehicle. *IEEE Transactions on Magnetics*, 41 (1), 505–508.
12 Harson, A., Mellor, P.H., and Howe, D. (1995) Design considerations for induction machines for electric vehicle drives. Proceedings of the IEE International Conference on Electrical Machines and Drives, September 11–13, pp. 16–20.
13 Krishnan, R. (1996) Review of flux-weakening in high performance vector controlled induction motor drives. Proceedings of the IEEE International Symposium on Industrial Electronics, June 17–20, pp. 917–922.
14 Miller, T.J.E. (1993) *Switched Reluctance Motors and their Control*, Magna Physics Publishing and Clarendon Press, Oxford.
15 (a) Miller, T.J.E. (2001) *Electronic Control of Switched Reluctance Machines*, Reed, New York, pp. 227–245; (b) Williamson, S.S., Emadi, A., and Rajashekara, K. (2007) Comprehensive efficiency modeling of electric traction motor drives for hybrid electric vehicle propulsion applications. *IEEE Transactions on Vehicular Technology*, 56 (4), 1561–1572.
16 West, J.G.W. (1994) DC, induction, reluctance and PM motors for electric vehicles. *Power Engineering Journal*, 8 (2), 77–88.
17 Welchko, B.A. and Nagashima, J.M. (2003) The influence of topology selection on the design of EV/HEV propulsion systems. *Power Electronics Letters*, 1 (2), 36–40.

18 Zhu, Z.Q. and Howe, D. (2007) Electrical machines and drives for electric, hybrid, and fuel cell vehicles. *Proceedings of the IEEE*, 95 (4), 746–765.

19 Chan, C.C. (2002) The state of the art of electric and hybrid vehicles. *Proceedings of the IEEE*, 90 (2), 247–275.

20 Jahns, T.M. and Blasko, V. (2001) Recent advances in power electronics technology for industrial and traction machine drives. *Proceedings of the IEEE*, 89 (6), 963–975.

21 Akin, B., Ozturk, S.B., Toliyat, H.A., and Rayner, M. (2009) DSP-based sensorless electric motor fault-diagnosis tools for electric and hybrid electric vehicle powertrain applications. *IEEE Transactions on Vehicular Technology*, 58 (6), 2679–2688.

22 Kou, B., Li, L., Cheng, S., and Meng, F. (2005) Operating control of efficiently generating induction motor for driving hybrid electric vehicles. *IEEE Transactions on Magnetics*, 41 (1), 488–491.

23 Zeraoulia, M., Benbouzid, M.E.H., and Diallo, D. (2006) Electric motor drive selection issues for HEV propulsion systems: a comparative study. *IEEE Transactions on Vehicular Technology*, 55 (6), 1756–1764.

24 Wang, T., Zheng, P., Zhang, Q., and Cheng, S. (2005) Design characteristics of the induction motor used for hybrid electric vehicles. *IEEE Transactions on Magnetics*, 41 (1), 505–508.

25 Li, W., Cao, J., and Zhang, X. (2010) Electrothermal analysis of induction motor with compound cage rotor used for PHEV. *IEEE Transactions on Industrial Electronics*, 57 (2), 660–668.

26 Asano, K., Okada, S., and Iwamam, N. (1992) Vibration suppression of induction-motor-driven hybrid vehicle using wheel torque observer. *IEEE Transactions on Industry Applications*, 28 (2), 441–447.

27 Diallo, D., Benbouzid, M.E.H., and Makouf, A. (2004) A fault-tolerant control architecture for induction motor drives in automotive applications. *IEEE Transactions on Vehicular Technology*, 53 (6), 1847–1855.

28 Benbouzid, M.E.H., Diallo, D., and Zeraoulia, M. (2007) Advanced fault-tolerant control of induction-motor drives for EV/HEV traction applications: from conventional to modern and intelligent control techniques. *IEEE Transactions on Vehicular Technology*, 56 (2), 519–528.

29 Proca, A.B., Keyhani, A., and Miller, J.M. (2003) Sensorless sliding-mode control of induction motors using operating condition dependent models. *IEEE Transactions on Energy Conversion*, 18 (2), 205–212.

30 Salmasi, F.R., Najafabadi, T.A., and Maralani, P.J. (2010) An adaptive flux observer with online estimation of DC-link voltage and rotor resistance for VSI-based induction motors. *IEEE Transactions on Power Electronics*, 25 (5), 1310–1319.

31 Proca, A.B., Keyhani, A., and Miller, J. (2002) Sensorless sliding-mode control of induction motors using operating condition dependent models. *Power Engineering Review*, 22 (7), 50–50.

32 Sudhoff, S.D., Corzine, K.A., Glover, S.F., et al. (1998) DC link stabilized field oriented control of electric propulsion systems. *IEEE Transactions on Energy Conversion*, 13 (1), 27–33.

33 Khoucha, F., Lagoun, S.M., Marouani, K., et al. (2010) Hybrid cascaded H-bridge multilevel-inverter induction-motor-drive direct torque control for automotive applications. *IEEE Transactions on Industrial Electronics*, 57 (3), 892–899.

34 McCleer, P.J., Miller, J.M., Gale, A.R., et al. (2001) Nonlinear model and momentary performance capability of a cage rotor induction machine used as an automotive combined starter-alternator. *IEEE Transactions on Industry Applications*, 37 (3), 840–846.

35 Degner, M.W., Guerrero, J.M., and Briz, F. (2006) Slip-gain estimation in field-orientation-controlled induction machines using the system transient response. *IEEE Transactions on Industry Applications*, 42 (3), 702–711.

36 Kim, J., Jung, J., and Nam, K. (2004) Dual-inverter control strategy for high-speed operation of EV induction motors. *IEEE Transactions on Industrial Electronics*, 51 (2), 312–320.

37 Neacsu, D.O. and Rajashekara, K. (2001) Comparative analysis of torque-controlled IM drives with applications in electric and hybrid vehicles. *IEEE Transactions on Power Electronics*, 16 (2), 240–247.

38 Liu, R., Mi, C., and Gao, W. (2008) Modeling of iron losses of electrical machines and transformers fed by PWM inverters. *IEEE Transactions on Magnetics*, 44 (8), 2021–2028.

39 Ding, X. and Mi, C. (2011) Impact of inverter on losses and thermal characteristics of induction motors. *International Journal on Power Electronics*, in press.

40 Mi, C., Slemon, G.R., and Bonert, R. (2003) Modeling of iron losses of permanent magnet synchronous motors. *IEEE Transactions on Industry Applications*, 39 (3), 734–742.

41 Khluabwannarat, P., Thammarat, C., Tadsuan, S., and Bunjongjit, S. (2007) An analysis of iron loss supplied by sinusoidal, square wave, bipolar PWM inverter and unipolar PWM inverter. International Power Engineering Conference, IPEC 2007, December 3–6, pp. 1185–1190.

42 Mi, C., Shen, J., and Natarajan, N. (2002) Field-oriented control of induction motors. IEEE Workshop on Power Electronics in Transportation (WPET'02), September.

43 Parker, R.J. and Studders, R.J. (1962) *Permanent Magnets and their Application*, John Wiley & Sons, Inc., New York.

44 Merrill, F.W. (1955) Permanent magnet excited synchronous motor. *AIEE Transactions*, 74, 1754–1760.

45 Binns, K.J., Jabbar, M.A., and Parry, G.E. (1979) Choice of parameters in hybrid permanent magnet synchronous motor. *Proceedings of the IEE*, 126 (8), 741–744.

46 Honsinger, V.P. (1980) Performance of polyphase permanent magnet machines. *IEEE Transactions on Power Apparatus and Systems*, 99 (4), 1510–1518.

47 Rahman, M.A. and Little, T.A. (1984) Dynamic performance analysis of permanent magnet synchronous motors. *IEEE Transactions on Power Apparatus and Systems*, 103 (6), 1277–1282.

48 Rahman, M.A., Little, T.A., and Slemon, G.R. (1985) Analytical models for interior-type permanent magnet synchronous motors. *IEEE Transactions on Magnetics*, 21 (5), 1741–1743.

49 Zhou, P., Rahman, M.A., and Jabbar, M.A. (1994) Field and circuit analysis of permanent magnet machines. *IEEE Transactions on Magnetics*, 30 (4), 1350–1359.

50 Vaez, S., John, V.I., and Rahman, M.A. (1999) An on-line loss minimization controller for interior permanent magnet motor drives. *IEEE Transactions on Energy Conversion*, 14 (4), 1435–1440.

51 Uddin, M.N., Radwan, T.S., and Rahman, M.A. (2002) Performance of interior permanent magnet motor drive over wide speed range. *IEEE Transactions on Energy Conversion*, 17 (1), 79–84.

52 Rahman, M.A., Vilathgamuwa, M., Uddin, M.N., and Tseng, K.J. (2003) Non-linear control of interior permanent magnet synchronous motor. *IEEE Transactions on Industry Applications*, 39 (2), 408–416.
53 Jahns, T.M. (1984) Torque production in permanent magnet synchronous motor drives with rectangular current excitation. *IEEE Transactions on Industry Applications*, 20 (4), 803–813.
54 Morimoto, S., Sanada, M., and Takeda, Y. (1994) Wide-speed operation of interior permanent magnet synchronous motors with high performance current regulator. *IEEE Transactions on Industry Applications*, 30 (4), 920–926.
55 Jahns, T.M. (1987) Flux-weakening regime operation of an interior permanent-magnet synchronous motor drive. *IEEE Transactions on Industry Applications*, 23 (4), 681–689.
56 Morimoto, S., Sanada, M., and Takeda, Y. (1996) Inverter-driven synchronous motors for constant power. *IEEE Industry Applications Magazine*, pp. 18–24.
57 Zhu, Z.Q., Chen, Y.S., and Howe, D. (2000) Online optimal field weakening control of permanent magnet brushless ac drives. *IEEE Transactions on Industry Applications*, 36 (6), 1661–1668.
58 Zhu, Z.Q., Shen, J.X., and Howe, D. (2006) Flux-weakening characteristics of trapezoidal back-EMF machines in brushless DC and AC modes. Proceedings of the International Power Electronics and Motion Control, August 13–16, pp. 908–912.
59 Shi, Y.F., Zhu, Z.Q., and Howe, D. (2006) Torque-speed characteristics of interior-magnet machines in brushless AC and DC modes, with particular reference to their flux-weakening performance. Proceedings of the International Power Electronics and Motion Control, August 13–16, pp. 1847–1851.
60 Safi, S.K., Acarnley, P.P., and Jack, A.G. (1995) Analysis and simulation of the high-speed torque performance of brushless DC motor drives. *IEE Proceedings – Electric Power Applications*, 142 (3), 191–200.
61 Bose, B.K. (1988) A microcomputer-based control and simulation of an advanced IPM synchronous machine drive system for electric vehicle propulsion. *IEEE Transactions on Industrial Electronics*, 35 (4), 547–559.
62 Bose, B.K. (1988) A high-performance inverter-fed drive system of an interior permanent magnet synchronous machine. *IEEE Transactions on Industry Applications*, 24 (6), 987–997.
63 Lovelace, E.C., Jahns, T.M., Kirtley, J.L. Jr., and Lang, J.H. (1998) An interior PM starter/alternator for automotive applications. Proceedings of the International Conference on Electrical Machines, December, pp. 1802–1808.
64 Lipo, T.A. (1991) Synchronous reluctance machines – a viable alternative for AC drives? *Electric Machines and Power Systems*, 19, 659–671.
65 Soong, W.L., Staton, D.A., and Miller, T.J.E. (1995) Design of a new axially-laminated interior permanent magnet motor. *IEEE Transactions on Industry Applications*, 31 (2), 358–367.
66 Soong, W.L. and Ertugrul, N. (2002) Field-weakening performance of interior permanent-magnet motors. *IEEE Transactions on Industry Applications*, 38 (5), 1251–1258.
67 Chaaban, F.B., Birch, T.S., Howe, D., and Mellor, P.H. (1991) Topologies for a permanent magnet generator/speed sensor for the ABS on railway freight vehicles. Proceedings of the IEE International Conference on Electrical Machines and Drives, September 11–13, pp. 31–35.
68 Liao, Y., Liang, F., and Lipo, T.A. (1995) A novel permanent magnet machine with doubly salient structure. *IEEE Transactions on Industry Applications*, 3 (5), 1069–1078.

69 Chan, C.C., Jiang, J.Z., Chen, G.H., et al. (1994) A novel polyphase multipole square-wave permanent magnet motor drive for electric vehicles. *IEEE Transactions on Industry Applications*, 30 (5), 1258–1266.
70 Wang, J.B., Xia, Z.P., and Howe, D. (2005) Three-phase modular permanent magnet brushless machine for torque boosting on a downsized ICE vehicle. *IEEE Transactions on Vehicular Technology*, 54 (3), 809–816.
71 Russenschuck, S. (1990) Mathematical optimization techniques for the design of permanent magnet machines based on numerical field calculation. *IEEE Transactions on Magnetics*, 26 (2), 638–641.
72 Russenschuck, S. (1992) Application of Lagrange multiplier estimation to the design optimization of permanent magnet synchronous machines. *IEEE Transactions on Magnetics*, 28 (2), 1525–1528.
73 Rasmussen, K.F., Davies, J.H., Miller, T.J.E., et al. (2000) Analytical and numerical computation of air-gap magnetic fields in brushless motors with surface permanent magnets. *IEEE Transactions on Industry Applications*, 36 (6), 1547–1554.
74 Boules, N. (1990) Design optimization of permanent magnet DC motors. *IEEE Transactions on Industry Applications*, 26 (4), 786–792.
75 Proca, A.B., Keyhani, A., El-Antably, A., et al. (2003) Analytical model for permanent magnet motors with surface mounted magnets. *IEEE Transactions on Energy Conversion*, 18 (3), 386–391.
76 Pavlic, D., Garg, V.K., Repp, J.R., and Weiss, J.A. (1988) Finite element technique for calculating the magnet sizes and inductance of permanent magnet machines. *IEEE Transactions on Energy Conversion*, 3 (1), 116–122.
77 Miller, T.J.E., McGilp, M., and Wearing, A. (1999) Motor design optimization using SPEED CAD software – practical electromagnetic design synthesis. IEE Seminar, Ref. no. 1999/014, pp. 1–5.
78 ANSYS (2005) http://www.ansoft.com (accessed April 2005).
79 Kenjo, T. and Nagamori, S. (1985) *Permanent Magnet and Brushless DC Motors*, Clarendon Press, Oxford.
80 Miller, T.J.E. (1989) *Permanent Magnet and Reluctance Motor Drives*, Oxford Science Publications, Oxford.
81 Slemon, G.R. and Liu, X. (1992) Modeling and design optimization of permanent magnet motors. *Electric Machines and Power Systems*, 20, 71–92.
82 Bose, B.K. (1997) *Power Electronics and Variable Frequency Drives – Technology and Applications*, IEEE Press, Piscataway, NJ.
83 Balagurov, V.A., Galtieev, F.F., and Larionov, A.N. (1964) *Permanent Magnet Electrical Machines*, Energia, Moscow (in Russian, and translation in Chinese).
84 Gieras, J.F. and Wing, M. (2002) *Permanent Magnet Motor Technology: Design and Applications*, 2nd edn, Marcel Dekker, New York.
85 Mi, C., Filippa, M., Liu, W., and Ma, R. (2004) Analytical method for predicting the air-gap flux of interior-type permanent magnet machines. *IEEE Transactions on Magnetics*, 40 (1), 50–58.
86 Cho, D.H., Jung, H.K., and Sim, D.J. (1999) Multiobjective optimal design of interior permanent magnet synchronous motors considering improved core loss formula. *IEEE Transactions on Energy Conversion*, 14 (4), 1347–1352.
87 Borghi, C.A., Casadei, D., Cristofolini, A., et al. (1999) Application of multi objective minimization technique for reducing the torque ripple in permanent magnet motors. *IEEE Transactions on Magnetics*, 35 (5), 4238–4246.

88 Upadhyay, P.R., Rajagopal, K.R., and Singh, B.P. (2004) Effect of armature reaction on the performance of axial field permanent magnet brushless DC motor using FE method. *IEEE Transactions on Magnetics*, 40 (4), 2023–2025.

89 Li, Y., Zou, J., and Lu, Y. (2003) Optimum design of magnet shape in permanent magnet synchronous motors. *IEEE Transactions on Magnetics*, 39 (6), 3523–3526.

90 Fujishima, Y., Wakao, S., Kondo, M., and Terauchi, N. (2004) An optimal design of interior permanent magnet synchronous motor for the next generation commuter train. *IEEE Transactions on Applied Superconductivity*, 14 (2), 1902–1905.

91 Mi, C. (2006) Analytical design of permanent magnet traction drives. *IEEE Transactions on Magnetics*, 42 (7), 1861–1866.

92 Ding, X. and Mi, C. Modeling of eddy current loss and temperature of the magnets in permanent magnet machines. *Journal of Circuits, Systems, and Computers*, submitted.

93 Fahimi, B., Emadi, A., and Sepe, R. (2004) A switched reluctance machine-based starter/alternator for more-electric cars. *IEEE Transactions on Energy Conversion*, 19 (1), 116–124.

94 Rahman, K.M. and Schulz, S.E. (2002) Design of high-efficiency and high-torque-density switched reluctance motor for vehicle propulsion. *IEEE Transactions on Industry Applications*, 38 (6), 1500–1507.

95 Ramamurthy, S.S. and Balda, J.C. (2001) Sizing a switched reluctance, motor for electric vehicles. *IEEE Transactions on Industry Applications*, 37 (5), 1256–1263.

96 Mecrow, B.C. (1996) New winding configurations for doubly salient reluctance machines. *IEEE Transactions on Industry Applications*, 32 (6), 1348–1356.

97 Mecrow, B.C., Finch, J.W., El-Kharashi, E.A., and Jack, A.G. (2002) Switched reluctance motors with segmental rotors. *IEE Proceedings – Electric Power Applications*, 149 (4), 245–254.

98 Krishnamurthy, M., Edrington, C.S., Emadi, A., et al. (2006) Making the case for applications of switched reluctance motor technology in automotive products. *IEEE Transactions on Power Electronics*, 21 (3), 659–675.

99 Rahman, K.M., Fahimi, B., Suresh, G., et al. (2000) Advantages of switched reluctance motor applications to EV and HEV: design and control issues. *IEEE Transactions on Industry Applications*, 36 (1), 111–121.

100 Long, S.A., Schofield, N., Howe, D., et al. (2003) Design of a switched reluctance machine for extended speed operation. Proceedings of IEEE International Electric Machines and Drives Conference, June 1–4, pp. 235–240.

101 Schofield, N. and Long, S.A. (2005) Generator operation of a switched reluctance starter/generator at extended speeds. Proceedings of IEEE Conference on Vehicle Power and Propulsion, September 7–9, pp. 453–460.

102 Edrington, C.S., Krishnamurthy, M., and Fahimi, B. (2005) Bipolar switched reluctance machines: a novel solution for automotive applications. *IEEE Transactions on Vehicular Technology*, 54 (3), 795–808.

103 Dixon, S. and Fahimi, B. (2003) Enhancement of output electric power in switched reluctance generators. Proceedings of IEEE International Electric Machines and Drives Conference, June 1–4, pp. 849–856.

104 Cameron, D.H., Lang, J.H., and Umans, S.D. (1992) The origin and reduction of acoustic noise in doubly salient variable-reluctance motors. *IEEE Transactions on Industry Applications*, 26 (6), 1250–1255.

105 Colby, R.S., Mottier, F., and Miller, T.J.E. (1996) Vibration modes and acoustic noise in a four-phase switched reluctance motor. *IEEE Transactions on Industry Applications*, 32 (6), 1357–1364.

106 Long, S.A., Zhu, Z.Q., and Howe, D. (2001) Vibration behaviour of stators of switched-reluctance machines. *IEE Proceedings – Electric Power Applications*, 148 (3), 257–264.

107 Long, S.A., Zhu, Z.Q., and Howe, D. (2002) Influence of load on noise and vibration of voltage and current controlled switched reluctance machines. Proceedings of IEE International Conference on Power Electronics, Machines, and Drives, April 16–18, pp. 534–539.

108 Blaabjerg, F., Pedersen, J.K., Neilsen, P., et al. (1994) Investigation and reduction of acoustic noise from switched reluctance drives in current and voltage control. Proceedings of International Conference on Electrical Machines, December, pp. 589–594.

109 Wu, C.Y. and Pollock, C. (1993) Time domain analysis of vibration and acoustic noise in the switched reluctance drive. Proceedings of International Conference on Electrical Machines and Drives, October, pp. 558–563.

110 Gabsi, M., Camus, F., Loyau, T., and Barbry, J.L. (1999) Noise reduction of switched reluctance machine. Proceedings of IEEE International Electric Machines and Drives Conference, May 12–19, pp. 263–265.

111 Gabsi, M., Camus, F., and Besbes, M. (1999) Computation and measurement of magnetically induced vibrations of switched reluctance machine. *IEE Proceedings – Electric Power Applications*, 146 (5), 463–470.

112 Wu, C.Y. and Pollock, C. (1995) Analysis and reduction of acoustic noise in the switched reluctance drive. *IEEE Transactions on Industry Applications*, 31 (6), 91–98.

113 Long, S.A., Zhu, Z.Q., and Howe, D. (2005) Effectiveness of active noise and vibration cancellation for switched reluctance machines operating under alternative control strategies. *IEEE Transactions on Energy Conversion*, 20 (4), 792–801.

114 Matsuo, T. and Lipo, T.A. (2004) Rotor design optimization of synchronous reluctance machine. *IEEE Transactions on Energy Conversion*, 9 (2), 359–365.

115 Li, Y. and Mi, C. (2007) Doubly salient permanent magnet machines with skewed rotor and six-state communication control. *IEEE Transactions on Magnetics*, 43 (9), 3623–3629.

116 Liao, Y., Liang, F., and Lipo, T.A. (1992) A novel permanent magnet motor with doubly salient structure. Industry Applications Society Annual Meeting, October, pp. 308–314.

117 Liao, Y. and Lipo, T.A. (1993) Sizing and optimal design of doubly salient permanent magnet motors. Sixth International Conference on Electrical Machines and Drives, September, pp. 452–456.

118 Liao, Y., Liang, F., and Lipo, T.A. (1995) A novel permanent magnet motor with doubly salient structure. *IEEE Transactions on Industry Applications*, 31 (5), 1069–1078.

119 Cheng, M., Chau, K.T., and Chan, C.C. (2000) Nonlinear varying-network magnetic circuit analysis for DSPM motors. *IEEE Transactions on Magnetics*, 36 (1), 339–348.

120 Cheng, M., Chau, K.T., and Chan, C.C. (2001) Static characteristics of a new doubly salient permanent magnet motor. *IEEE Transactions on Energy Conversion*, 16 (1), 20–25.

121 Cheng, M., Chau, K.T., Chan, C.C., and Sun, Q. (2003) Control and operation of a new 8/6-pole doubly salient permanent-magnet motor drive. *IEEE Transactions on Industry Applications*, 39 (5), 1363–1371.

122 Cheng, M., Chau, K.T., and Chan, C.C. (2001) Design and analysis of a new doubly salient permanent magnet motor. *IEEE Transactions on Magnetics*, 37 (4), 3012–3020.

123 Chau, K.T., Sun, Q., Fan, Y., and Cheng, M. (2005) Torque ripple minimization of doubly salient permanent-magnet motors. *IEEE Transactions on Energy Conversion*, 20 (2), 352–358.

124 Deodhar, R.P., Staton, D.A., and Miller, T.J.E. (1996) Prediction of cogging torque using the flux-MMF diagram technique. *IEEE Transactions on Industry Applications*, 32 (6), 569–576.

125 Ding, X., Bhattacharyal, M., and Mi, C. (2010) Simplified thermal model of PM motors in hybrid vehicle applications taking into account eddy current loss in magnets. *Journal of Asia Electric Vehicles*, 8 (1), 1–7.

126 Mellor, P.H., Roberts, D., and Turner, D.R. (1991) Lumped parameter thermal model for electrical machines of TEFC design. *IEE Proceedings – Electric Power Applications*, 138 (5), 205–218.

127 Taylor, G.I. (1935) Distribution of velocity and temperature between concentric cylinders. *Proceedings of the Royal Society*, 159 (Pt A), 546–578.

128 Gazley, C. (1958) Heat transfer characteristics of rotational and axial flow between concentric cylinder. *Transactions of the American Society of Mechanical Engineers*, 80, 79–89.

129 Boglietti, A., Cavagnino, A., Lazzari, M., and Pastorelli, A. (2002) A simplified thermal model for variable speed self cooled industrial induction motor. Industry Applications Conference, 37th IAS Annual Meeting, October 13–17, vol. 2, pp. 723–730.

130 Tang, S.C., Keim, T.A., and Perreault, D.J. (2005) Thermal modeling of Lundell alternators. *IEEE Transactions on Energy Conversion*, 20 (1), 25–36.

131 Sooriyakumar, G., Perryman, R., and Dodds, S.J. (2007) Analytical thermal modelling for permanent magnet synchronous motors. Universities Power Engineering 42nd International Conference, UPEC 2007, September 4–6, pp. 192–196.

132 Staton, D., Boglietti, A., and Cavagnino, A. (2005) Solving the more difficult aspects of electric motor thermal analysis in small and medium size industrial induction motors. *IEEE Transactions on Energy Conversion*, 20 (3), 620–628.

133 Guo, Y.G., Zhu, J.G., and Wu, W. (2005) Thermal analysis of soft magnetic composite motors using a hybrid model with distributed heat sources. *IEEE Transactions on Magnetics*, 41 (6), 2124–2128.

134 Funieru, B. and Binder, A. (2008) Thermal design of a permanent magnet motor used for gearless railway traction. Industrial Electronics, IECON 2008, 34th Annual Conference, November 10–13, pp. 2061–2066.

135 Staton, D.A. and Cavagnino, A. (2008) Convection heat transfer and flow calculations suitable for electric machines thermal models. *IEEE Transactions on Industrial Electronics*, 55 (10), 3509–3516.

136 Cassat, A., Espanet, C., and Wavre, N. (2003) BLDC motor stator and rotor iron losses and thermal behavior based on lumped schemes and 3-D FEM analysis. *IEEE Transactions on Industry Applications*, 39 (5), 1314–1322.

137 Kim, W.-G., Lee, J.-I., Kim, K.-W., et al. (2006) The temperature rise characteristic analysis technique of the traction motor for EV application. Strategic Technology, 1st International Forum, October 18–20, pp. 443–446.

138 Chowdhury, S.K. (2005) A distributed parameter thermal model for induction motors. International Conference on Power Electronics and Drives Systems, PEDS 2005, November 28–December 1, vol. 1, pp. 739–744.

139 Hsu, J.S., Nelson, S.C., Jallouk, P.A., et al. (2005) Report on Toyota Prius Motor Thermal Management. http://www.ornl.gov/~webworks/cppr/y2001/rpt/122586.pdf (accessed February 2011).

140 Jahns, T.M. and Caliskan, V. (1999) Uncontrolled generator operation of interior PM synchronous machines following high-speed inverter shutdown. *IEEE Transactions on Industry Applications*, 35 (6), 1347–1357.

141 Liaw, C.Z., Soong, W.L., Welchko, B.A., and Ertugrul, N. (2005) Uncontrolled generation in interior permanent-magnet machines. *IEEE Transactions on Industry Applications*, 41 (4), 945–954.

11

Electric Energy Sources and Storage Devices

11.1 Introduction

Energizing an HEV or EV needs basically two types of devices. The first device is an energy source, or more appropriately, a source of power. Examples of these are an ICE, and a fuel cell. The ICE converts chemical energy to mechanical energy, but since this device was discussed earlier in this book, it will not be discussed here. This chapter will, instead focus on electrical sources. In this category is the fuel cell, which directly converts chemical energy into electrical energy. The other devices that do not convert power, but rather store energy, are devices like batteries, ultracapacitors or supercapacitors, and flywheels. Such devices are appropriate for EV, HEV, or PHEV (plug-in hybrid). Normally, in an HEV or PHEV, batteries or ultracapacitors can be charged by using the ICE, which can run a generator, and charge a battery or ultracapacitor. In case of an EV or PHEV, generally these will be charged from the utility power grid. Power sources noted above, or energy storage elements, can then drive an electric motor and provide propulsion. But energy storage systems offer the benefit of recuperating regenerative braking energy and store it for future use, normally to provide propulsion, thus leading to increased vehicle efficiency. In this chapter, modeling and simulation techniques will also be discussed to some extent, along with battery management and control, and we will discuss ultracapacitor applications for control of batteries and ultracapacitors.

The basic elements in a battery are cells, and several cells together constitute a module, and several modules constitute a battery pack. This is depicted in Figure 11.1. The cell is the smallest constituent element, and several cells make a module (normally a plastic box such as a 12 V automotive battery), and several modules connected in series and parallel make a complete battery pack, which can be perhaps a 240 V nominal pack.

In the HEV, due to the presence of the ICE, it is important to properly coordinate and control the HEV system in order to properly provide battery energy management, which implies charging, discharging, and charge equalization as needed.

The same also applies to ultracapacitors, and if both battery and ultracapacitor are present, it becomes important to coordinate these devices with the ICE, and also coordinate between the battery and the ultracapacitor. This may require the use of appropriate DC–DC converters, and the process may need to be bidirectional.

With a fuel cell, similar coordination and management is also necessary, except that a fuel cell can only give energy out, but cannot accept energy.

Hybrid Electric Vehicles: Principles and Applications with Practical Perspectives,
Second Edition. Chris Mi and M. Abul Masrur.

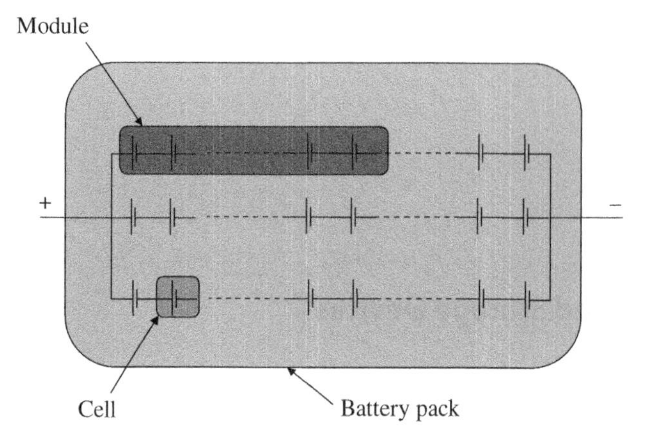

Figure 11.1 Relationship between cell, module, and battery pack.

A battery is generally considered to be an energy intensive device, whereas an ultracapacitor is considered to be more of a power intensive device. Loosely speaking, this implies that for a given size and weight, a battery is able to store much more energy compared to an ultracapacitor, and an ultracapacitor is able to provide a higher power compared to a battery. From an electrical circuit viewpoint, in terms of terminal characteristics, this can be attributed to a relatively low internal resistance of the ultracapacitor, compared to a battery. Hence, due to low time constant, the ultracapacitor has a faster time response compared to a comparable battery for delivering a sudden current demand. These properties are in fact due to more fundamental electrochemical properties of the two devices.

Although a battery is a highly nonlinear electrochemical device and no perfect model is available, a battery can be reasonably modeled from electrical circuit viewpoint in terms of its terminal characteristics. This involves its description through an equivalent circuit using internal resistance, capacitance, and diodes. In addition, the temperature dependence of a battery's behavior can be depicted by using equivalent circuit parameters which are dependent on temperature.

Ultracapacitors can also be similarly modeled, and sometimes inductive elements can be added as well, in addition to resistance and capacitance, to properly represent its terminal behavior. It should be noted that ultracapacitors use different technology (electric double-layer capacitor) from capacitors used in radio electronics and similar applications; they have a much higher energy density, which could be thousands of times greater than a regular electrolytic capacitor. Also, the ultracapacitor is a DC device but with bidirectional current capability which allows energy to be transferred to and from it, whereas electronic capacitors that are meant for filtering etc. are capable of high frequency applications with very little energy storage capacity. Ultracapacitor capacitance is normally measured in units of farads and could even be several hundred or more farads, unlike the communication engineering related electronic capacitors which are typically in terms of micro, nano, or pico farad units. Normally the state of charge of an ultracapacitor can vary over a wide range and it can even be drained to zero. In addition, a typical ultracapacitor can undergo around even a million charge discharge cycles, unlike a battery which can typically only go through a couple of thousand cycles.

11.2 Characterization of Batteries

11.2.1 Battery Capacity

Battery capacity implies the amount of electric charge that a battery is able to supply before it is fully discharged, but this statement needs to be qualified to some extent. Firstly, the capacity is dependent on the temperature of the battery, and secondly, the term "discharged" also needs to be clarified. When it is said that a battery is discharged, it means that the voltage has fallen to a specified value and also that it is not able to provide any further charge to a load. These will be further discussed here. The SI unit of battery capacity is the coulomb (C) which is an ampere-second. Another common unit for battery capacity is the ampere-hour (Ah), which is 3600 ampere-second = 3600 C. We can consider an example: a battery of 10 Ah implies that it can supply 1 A for 10 hours or 2 A for 5 hours etc. However, the amount of charge that a battery can accept while being charged, and can release during discharge, also depends on the rate at which the battery is either charged or discharged. In general, due to the chemical properties of the battery, there are some efficiency parameters that lead to higher efficiency when charging or discharging takes place at lower current. We will have the opportunity to discuss this further during the discussion on battery modeling.

One item of importance related to battery charge or discharge process is known as the C-rate. C-rate is that current which will charge a completely discharged battery to its rated ampere-hour (Ah) during charging, or discharge a fully charged battery (at its rated Ah) to its fully discharged condition. In other words, C-rate is numerically equal to the rated Ah capacity of a battery, but its unit is in amperes.

11.2.2 Energy Stored in a Battery

The stored energy in a battery depends on the terminal voltage and the amount of charge stored within it at a particular temperature. Since amp-hr represents the amount of charge, the energy will be the voltage multiplied by the charge with appropriate units. Although during the charging and discharging process the voltage changes, it may not change substantially, and hence it is acceptable to use the following equation for energy as long as the voltage is not varying substantially.

$$\text{Energy in watt} - \text{hr} = E = \text{amp} - \text{hr} \times \text{voltage} \tag{11.1}$$

11.2.3 State of Charge in Battery (SOC) and Measurement of SOC

Although state-of-charge or SOC, an important parameter related to battery, is used quite often, in reality its definition requires more elaboration. The simple definition of SOC is the amount of charge a battery has at a particular moment, compared to the rated charge or amp-hr of the battery, so, it is important to know the rated amp-hr. However, there are additional problems in defining SOC, since the amount of retrievable charge is dependent on the temperature and also on the age of the battery; also the amount of charge which was inserted during charging can be quite a bit more than what can be retrieved. So, these issues call for some discussion as follows.

Information known or otherwise available when a battery is first obtained is normally:

1) Battery type – lead acid, NiMH, Li-ion etc.
2) Model number, and manufacturer's characteristic curves to the extent possible.
3) It is assumed that the only things that can be physically measured are voltage *V*, current *I*, and temperature *T* (ambient temperature around the battery case outside, and not inside a battery module, for the type of batteries we will be using, so for the batteries we will be using, we will assume that no internal sensors exist inside the battery module itself).
4) It is possible to use battery "impedance" (the term will need to be defined more elaborately later), but it should be noted that battery impedance is a function of *V*, *I*, and *T*, and it is a derived quantity.
5) It is also assumed that the only things that can be controlled in a battery are the voltage across the two terminals or current through it, i.e. *V* or *I*, but not both independently. This can be done through power electronics, or by regulating the field of a generator used across the battery to charge it; or charging could be also done by some other device such as a fuel cell coupled with appropriate converter. It may be possible also to control the outside temperature, i.e. the battery ambient temperature in its immediate vicinity (and consequently the battery's inside temperature can be influenced) by using some external fan or other cooling method (perhaps using liquid coolant), in other words control the temperature *T*. Battery temperature can also be controlled to some extent by controlling its load – although this may not fully control the ambient temperature, but it may be able to do so to an extent. It (load control) can perhaps also help control the inside temperature of the battery to some extent.

During measurement of the battery variables in connection with the SOC, sampling time for *V* and *I* can be important only during sudden transients. Hence, in general, slow sampling of 100 msec or even 1 sec or more, may be quite adequate for many purposes. But the sampling can be made variable or adaptive, i.e. if a trend is seen that it (*V* or *I*) is not staying steady, then immediately the sampling time should be reduced, and vice versa. For temperature, which generally has a very long time constant – perhaps 10 secs or even 30 secs– sampling time may be quite adequate for most purposes.

11.2.3.1 SOC Determination

SOC can be measured based on two methods: Amp-hr balancing (the book keeping method) or the EMF based or direct method. But to be more accurate, it is desirable to use both of the above methods in combination and alternately, depending on the state of the battery.

11.2.3.2 Direct Measurement

In this method, which is not based on the cumulative Ah calculation, there is no historical data necessary. Here, we measure the open circuit voltage (OCV) and use a lookup table to correlate with the SOC, the table being based on the manufacturer's data made available under similar conditions. The other method involves measuring voltage *V* and current *I*, and then finding a *Z* (impedance), either through direct calculation or through matching with a model involving resistors and capacitors. Another method is to apply (or inject) a current step, and monitor the voltage time constant τ after the application of the step.

We can ask how, in a real-time environment, all these can be implemented when the system is delivering a load. One way is to apply just a very small percentage of the total load current through a current command to the battery, and monitor the voltage change (which can be quite small and require accurate measurement) and the time constant and record these. This won't significantly disturb the load and yet will provide real-time information. In addition to the above, we have to monitor the temperature T, since SOC is dependent on temperature. So, the end result of the above discussion can be expressed as: $SOC = f(V, I, T, Z, \tau)$, where

V = battery open circuit voltage
I = battery current
T = battery temperature
Z = battery impedance
τ = battery voltage relaxation time after application of a step current.

Some of these five quantities may be interdependent. This expression is a somewhat simplified version which may be adequate for power management activities, but in general it could be highly nonlinear, which may not be expressed as a direct functional form.

11.2.3.3 Amp-hr Based Measurement

In this method, the amp-hr is calculated by integration of I. But as noted earlier, this cannot calculate the actual retrievable amp-hr based on the amount while charging, or how much of the stored charge will come out while discharging. This is due to self-discharge and other losses within the battery, and it happens during both the charge and discharge process. In addition, these discharge and charge efficiencies depend on T. Another problem of amp-hr counting, which can be accurate in a simulation environment, will not hold in a real-time physical situation due to the drift of the integration of I over time. This drift amount can keep accumulating as time increases, leading to large inaccuracies. It has been indicated in various references that during such amp-hr bookkeeping it is necessary to calibrate the accumulated amp-hr value from time to time. The question is what to calibrate it against. If someone wants to calibrate it against "something else" as a reference, then that "other thing" obviously has to be found by some method other than coulomb counting. Sometimes that can be a voltage curve of the battery at different temperatures, and at different "SOC's", where "SOC" can be a quantity, which ***by "definition" is assigned a value of 100% at a certain voltage decided by the manufacturer.***

11.2.3.4 Some Better Methods

For the purpose of computing SOC, it is necessary to divide the battery states into the following states: Initial, Equilibrium, Transition, Charge, and Discharge – and the way to determine (after the initial state) which state the battery is in, is based on the current and the rate of change of voltage. The following figure and flow diagrams show the situation [1].

The diagrams in Figures 11.2 and 11.3 mean that we have to decide whether it is charging/discharging by looking at the sign of the current with respect to the voltage polarity, and also by noting whether its absolute value exceeds some tiny threshold or not. Similarly we have to look at dv/dt to decide if a transition between equilibrium to non-equilibrium state is taking place or not. Note that during both charging and discharging, dv/dt can

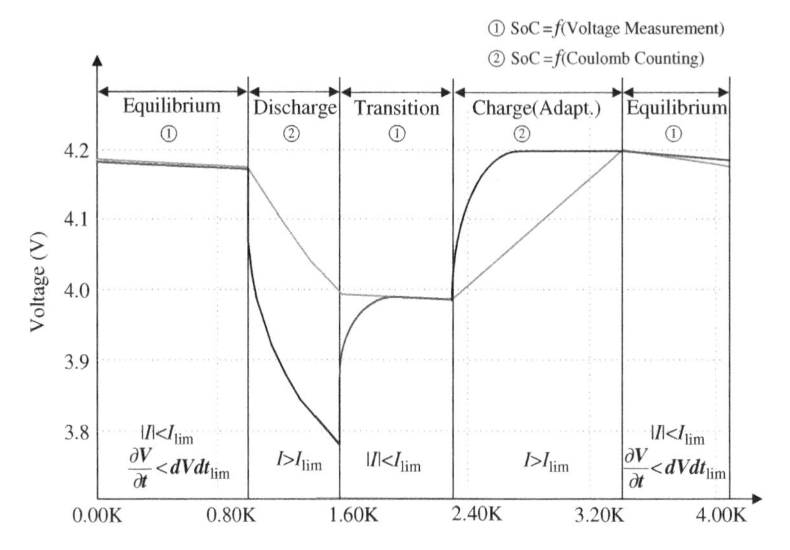

Figure 11.2 Possible states of a battery [1].

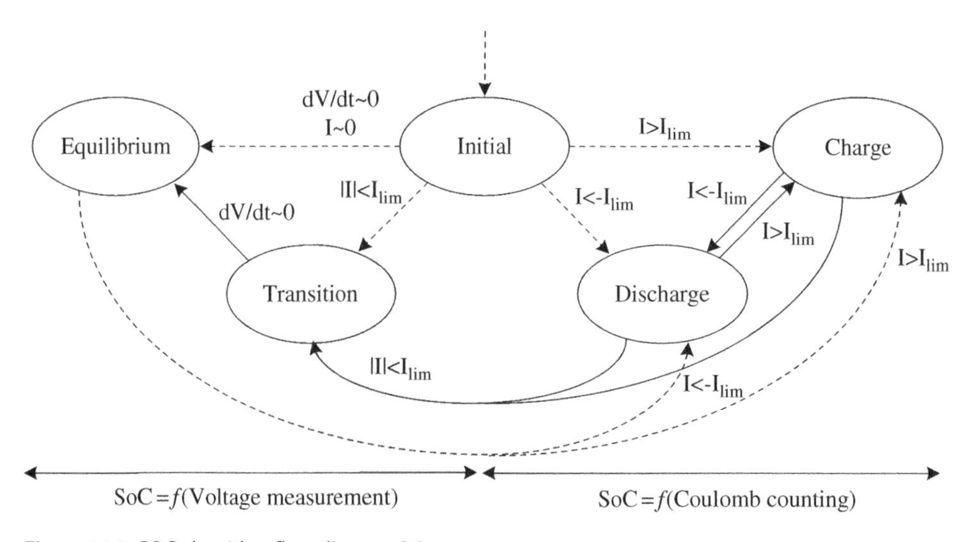

Figure 11.3 SOC algorithm flow diagram [1].

change. At the very beginning, when the process of computation is to begin, it is assumed that the battery, depending on whatever its condition, will have to transition to one of the four states noted above, regardless of where it was previously.

11.2.3.5 Initialization Process

The algorithm will begin by monitoring V, I, and T, and it should be monitored with a relatively small time step, to begin with. Several seconds should be allowed to elapse before using the data for evaluating the dV/dt, I, etc. As soon as the computation of these is complete, it will dictate which state we should transition to, based on the flow diagram.

If our state block is on the left-hand side of Figure 11.3, we should use SOC = *f*(*V*, *T*) and if it is on the right-hand side it should be SOC = f(accumulated amp-hr, *T*). These formulas can be enhanced later by making them a function of SOH as well.

So, the above discussion indicates that battery SOC is better evaluated as above, rather than using the simple method indicated by integrating the current over time.

11.2.4 Depth of Discharge (DOD) of a Battery

The depth of discharge (DOD) can be defined in terms of initial SOC when it was full, and the SOC at a particular time. In other words, it is the amount of depletion that the battery has undergone during the discharge process. The theoretical value of DOD is given by

$$\mathrm{DOD}(t) = \frac{Q_o - \int_{t}^{t_0} I_b(\tau)d\tau}{Q_o} \times 100\% \tag{11.2}$$

where Q_0 is the initial charge in the battery corresponding to SOC = 1.0 (full charge), and I_b is the battery current during discharge. In hybrid and electric vehicles, quite often (under normal circumstances) the amount of DOD can be on the order of 30% (corresponding to 70% SOC). Again, to get the correct depth of discharge it is necessary to find the SOC at a particular time, and compare it against the initial SOC at 100%. The accurate method of finding the SOC should be based on the previous section, otherwise the DOD is likely to be inaccurate.

11.2.5 Specific Power and Energy Density

Specific power represents the rated power of the battery for a given weight, e.g. kilogram of battery. Similar to this, there is volume related specific power defined by rated power of the battery for a given size or volume of the battery, e.g. per cubic meter. Similar to specific power, we can also define specific energy, i.e. energy stored (e.g. watt-hr) per kilogram, and the volume specific energy, i.e. energy stored per unit volume, e.g. per cubic meter. Sometimes the relationship between specific energy and specific power is represented by using a graph, known as a Ragone plot where the x-axis can be specific power and y-axis can be specific energy for a battery. The Ragone plot is generally used to compare various batteries or other types of energy devices (e.g. ultracapacitor), flywheel, various chemical sources and so on, in a single plot which shows where each of these devices are in terms of specific energy and power.

Table 11.1 shows the comparison of various devices in terms of specific energy.

11.2.6 Ampere-Hour (Charge and Discharge) Efficiency

Ampere-hour efficiency is the ratio between the electric charge given out during discharging a battery and the electric charge needed for the battery to return to the previous charge level. These quantities are not equal and could be on the order of 65–90% (which represents the efficiency). Efficiency depends on various things like battery chemistry, temperature,

Table 11.1 Specific energy of different energy sources.

Energy source	Specific energy (Wh/kg)
Gasoline	12,500
Natural gas	9350
Methanol	6200
Hydrogen	28,000
Coal	8200
Lead acid battery	35
Nickel metal hydride battery	50
Lithium-polymer battery	200
Lithium-ion battery	120
Sodium sulfur battery	150–300
Flywheel (steel)	12–30
Flywheel (carbon fiber)	30
Ultracapacitor	3.3

and rate of charge. Furthermore, it will be apparent during the discussion of battery modeling that the efficiency during charging of a battery (actual energy which was stored inside the battery versus actual energy input to charge the battery) is different from the discharging efficiency (actual energy output at the terminals during discharge versus actual energy which was available inside the battery) due to variation of equivalent internal resistances and chemical processes during charging and discharging. This also causes the charge and discharge efficiency to vary depending on whether the charging or discharging is done slowly (low current) or quickly (high current).

11.2.7 Number of Deep Cycles and Battery Life

EV/HEV batteries can undergo a few hundred to around a couple of thousand deep cycles, e.g. 80% DOD of the battery. Allowable DOD will in general depend on the battery chemistry and the usage pattern. The United States Advanced Battery Consortium (USABC) suggests a mid-term target of 600 deep cycles for EV batteries. This is a very important recommendation and affects battery life time in terms of deep-cycle number.

Example calculations on a particular battery The NiMH traction battery of the Toyota Prius 2004 model has the following specifications:

- 168 cells (28 modules)
- 201.6 V nominal voltage
- 6.5 Ah nominal capacity
- 28 hp (21 kW) output power
- 1300 W/kg specific power
- 46 Wh/kg specific energy.

Assume that the voltage is relatively constant. From the above, the energy rating of the battery is $201.6\,V \times 6.5\,Ah = 1310\,Wh = 1.31\,kWh$.

If the battery can be discharged at a maximum rate of 100 A, and only 40% can be discharged, the time for which the battery can be used when fully charged will be $\frac{(40\%)6.5\,Ah}{100\,A} = 0.026\,h = 93.6$ seconds.

If the battery can be charged at a maximum rate of 90 A, and the current SOC is 40%, the time it will take to charge the battery to 80% SOC will be $\frac{(80\% - 40\%)6.5\,Ah}{90\,A} = 0.0289$ hours $= 1.73$ minutes.

Assuming the battery has an internal resistance of 0.15 Ω, the efficiency at maximum charge rate will be $\eta = 1 - \frac{(90^2)(0.15)}{(90)(201.6)} = 93.6\%$.

If the battery has an internal resistance of 0.1 Ω, the efficiency at maximum discharge rate will be $\eta = 1 - \frac{(100^2)(0.1)}{(100)(201.6)} = 95\%$.

The voltage drop caused by this internal resistance at maximum charge/discharge will be as follows: At maximum charge, the voltage drop is $90 \times 0.15 = 13.5\,V$; at maximum discharge, the voltage drop is $100 \times 0.1 = 10\,V$.

The efficiency will change with maximum charge/discharge current, since it depends on the maximum charge/discharge current and internal resistance.

Assuming the leakage (self discharge) current is 20 mA, the number of days it will take for the battery to self-discharge from 80% SOC to 40% SOC will be $\frac{(80 - 40\%)6.5\,Ah}{20 \times 10^{-3}\,A} = 130$ hours $= 5.4$ days·

11.2.8 Some Practical Issues About Batteries and Battery Life

Let us revisit some of the issues related to SOC.

For the formula $SOC = f(V_{emf})$ it has been proposed that

$$V_{emf} = S \times A_S + V_{emf\_min} \tag{11.3}$$

where A_S is the slope of emf voltage with SOC S, and V_{emf_min} is the emf at zero SOC. So, how do we decide the voltage at which SOC should be considered to be 100%? Voltage can be measured easily, but when should we consider the SOC to be 100%? Also, given that, with aging, the open circuit voltage will be different even if the same amount of charge is given as input, a definition of SOC is needed, to provide a reference benchmark. That benchmark comes from the manufacturer specification, which may be defined as follows. Assume that a brand new battery which is charged as per manufacturer specification at a given current rate I_{CR} to a specified amp-hrs *Ahr*, arrives at a specified voltage V_{CR} (based on its design). We can consider that this voltage V_{CR} then corresponds to 100% SOC for that specific battery at the given temperature under which the charging was completed. The SOC, therefore, is based on the reference amp-hr rating *Ahr*, the rate I_{CR}, and the temperature, with respect to which other SOCs have to be compared. It is also necessary to have a set of curves from the manufacturer, taken

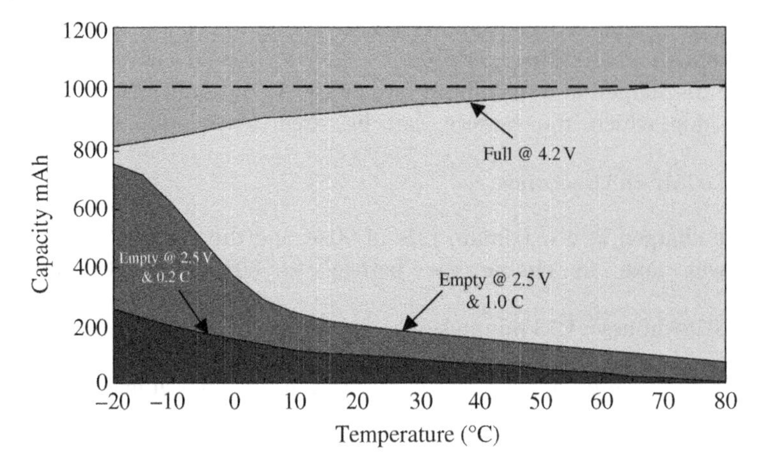

Figure 11.4 Battery capacity variation with temperature [3]. *Source:* Adapted from mpoweruk.com.

at different temperature and current rates, since the same battery can have a different SOC, by just charging it to say 100% at a certain temperature and then moving it to a different temperature.

Question: if a battery is charged at a certain temperature, and a certain amp-hr is inserted in it, and then the battery is moved to a colder temperature, we know we won't be able to retrieve all the amp-hrs we put into it. Where did the remaining amp-hr go upon changing the temperature? One explanation based on the circuit model is that the parameters of the battery will change due to temperature change and the energy is probably transformed into chemical energy in a form not directly retrievable without increasing the temperature again. Once the principles noted above are agreed upon, we can then just correlate a voltage and temperature, and map it to a corresponding SOC.

The above definition of SOC was based on a brand new battery. As the battery ages, its SOH (i.e. state of health) changes as well. So, the extractable output amp-hrs for a given SOC may be less. Hence we cannot really claim to have calculated the true SOC based on voltage alone.

For the purpose noted above, it may be very important to study curves as shown in Figure 11.4 [2, 3].

This gives an idea of how much capacity is left at a given temperature. Sometimes the manufacturers may not provide data in this way, but as shown in Figure 11.5 [4]. And the information may also be provided in the following equivalent form shown in Figure 11.6.

It is possible to translate Figure 11.6 into the one shown in Figure 11.4, which has temperature as x axis. Figure 11.4 tells us how much energy can be drawn from a battery at a given temperature. In other words, as noted previously, the battery might have been charged at one temperature and discharged at another, in which case the available energy may be different.

This issue may be quite important when survivability of a vehicle is concerned. Even though a battery might be thought to be dead initially, some temperature change could

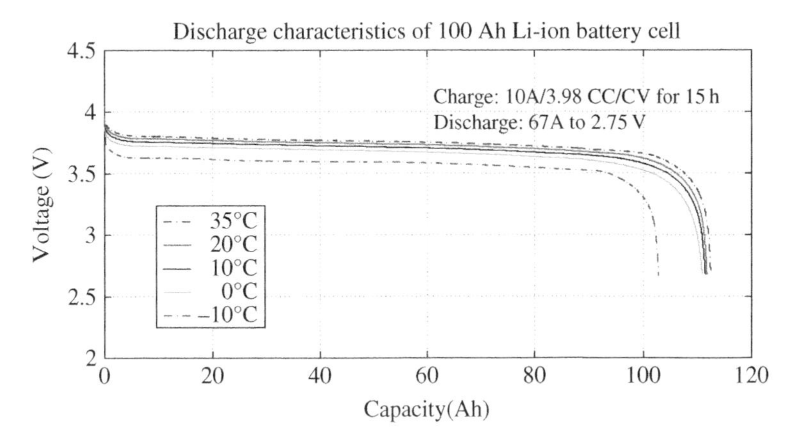

Figure 11.5 Battery voltage vs. amp-hr at different temperatures [2, 3].

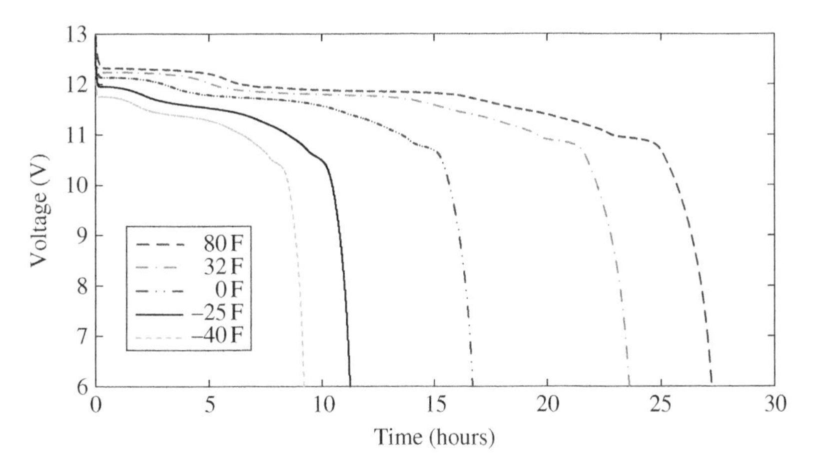

Figure 11.6 Battery discharge characteristics – terminal voltage at 5 A discharge rate (Exide battery specification sheet).

make it active again, even if for a brief period. This could perhaps be part of a battery management algorithm to indicate to the operator that some time should be given before trying to use the battery again, which could be vital – say, in a combat scenario.

It should be noted that Figures 11.4–11.6 are for a new battery. As the battery ages, the situation will change as noted in Figures 11.7 and 11.8 below [2, 3].

As the battery is cyclically passed through charge and discharge cycles of various amplitudes, the life of the battery will naturally be reduced as reflected in Figures 11.7 and 11.8. Since charge and discharge can happen at various amp-hr amplitudes, a count of amp-hr excursion and how many such cycles at a particular amp-hr took place, have to be considered. For this purpose, some techniques such as the "rainflow" counting method can be used. References exist on how to do such counting [5].

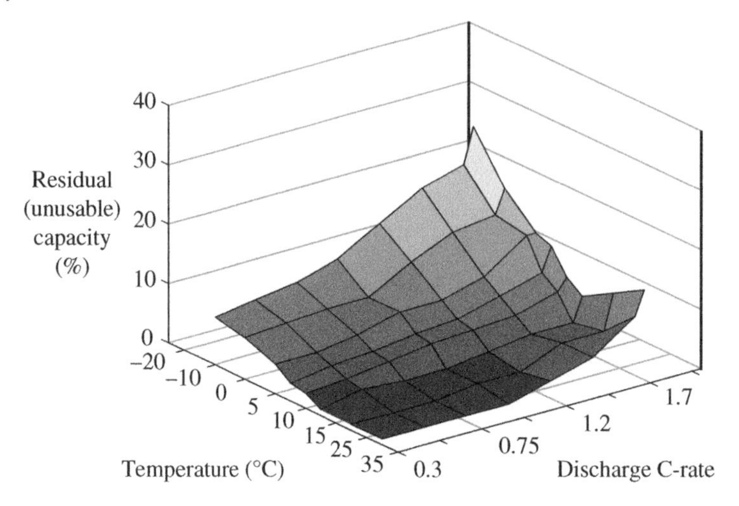

Figure 11.7 Battery capacity reduction at different temperature and discharge rates [2, 3].

Knowing the present amount of amp-hrs available after de-rating for temperature, cycles, and other health aspects, we can get a reasonable estimate of the energy available at a particular moment in the life of the battery usage. This amount, when compared with what was available in a fully charged new battery, will give the actual SOC of the battery at a particular temperature and at a particular moment in the battery's life history. To continue this discussion, it may be useful to define some terms as follows.

11.2.8.1 Acronyms and Definitions

OCV – open circuit voltage, i.e. voltage with no current flowing through the battery
SOCV – stabilized OCV, i.e. voltage at some rest time after opening the circuit
SOC – state of charge, i.e. actual charge/full charge capacity: AC/FCC (percentage or ratio)
FCC – maximum (or full) charge capacity in amp-hrs

FCC varies with age and hence it may need recalculation periodically as follows:

$$\mathrm{FCC} = [\Delta Q/\Delta(SOC)] \times 100 \times \eta \tag{11.4}$$

where ΔQ and $\Delta(SOC)$ are incremental change in charge and corresponding SOC respectively, and η is the efficiency of charging or discharging during which the incremental change in charge and SOC was measured.

FCC is frequently "defined" as the amount of amp-hrs at C/3 amp of charge rate. It was noted earlier that C is defined in terms of the rated amp-hr capacity of the battery and is *numerically equal* to that amp-hr, but its unit is in amps. This means that if the battery is discharged at FCC/3 amps, it should be able to deliver FCC amp-hrs, where FCC is the original new battery FCC. It is necessary that a manufacturer-defined voltage should also be associated with this definition, i.e. once the FCC amount of amp-hrs has been delivered, the voltage should be some predefined value. If this predefined voltage

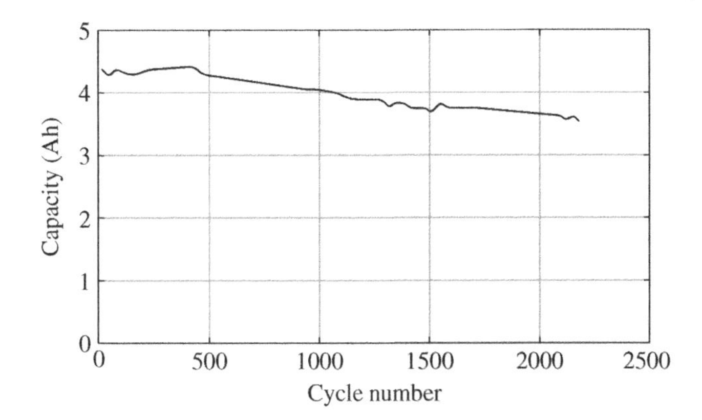

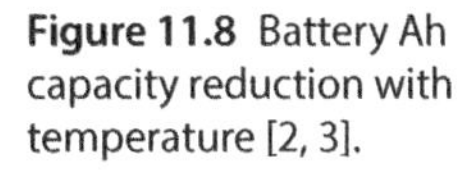
Figure 11.8 Battery Ah capacity reduction with temperature [2, 3].

is achieved before the FCC is delivered, then FCC value obviously has been reduced compared to new battery value.

$$\text{Actual capacity}: AC = FCC \times SOC \text{ (to be divided by 100 if SOC is expressed in \%)}$$

In the above, SOC is defined in terms of AC and FCC. Knowing the FCC offline, we can calculate SOC using AC/FCC. FCC, as we noted before, changes with age, temperature, and charging rate. These effects can be accounted through η, the battery efficiency at a particular moment. AC is the actual charge going in or out of the battery and can be measured. Hence, with age, SOC and FCC will change together, whereas AC will be an actual quantity measured. Both SOC and FCC are related to some voltage which is achieved by the battery upon charging and discharging. The 100% value of FCC and SOC is correlated with this voltage. It appears that one postulate that is used very often by people is that the voltage values (or the emf, to be more correct) corresponding to SOC at 100% or SOC at other percentages, will remain the same at corresponding SOC points, regardless of whether the battery was new or not and as per manufacturer specifications or if it ages, or if its operating environment changes.

Notice that, with age, AC, the actual charge capacity in an old battery, will be less (compared to a new battery) by the time the predefined voltage threshold has been achieved. Hence one can end up with 100% SOC if 100% SOC is defined to be the charge at which the voltage has achieved a predefined manufacturer value when the battery was new – yet the amount of charge or amp-hr that can be extracted from the battery will still be less than what it was when the battery was new. The reason is that now we are arriving at the same voltage with a much lower amount of charge input.

A matter of definition: if we agree that the FCC that we will use is the original new battery FCC (which we agree to keep constant by definition), then the AC will be lower, when the voltage threshold is reached, with an older battery. Hence at full voltage the SOC will be less than 100%. This seems to be a better indicator in terms of SOC, which will also indicate how much energy we can really get out of this battery at a particular time. On the other hand we can decide to use the other definition or equation for FCC, i.e. $FCC = [\Delta Q/\Delta(SOC)] \times 100 \times \eta$, but then FCC will change with age, AC will be whatever is needed to make the voltage = rated, yet the SOC can show 100%, but in reality that will not guarantee that we can extract the same amount of charge. In general, it seems that people like to rather have a variable value of FCC, hence we will reach FCC with a lower AC when the battery ages. Hence SOC 100% will not indicate the same amount of extractable energy for a new and an old battery. So, it is important to relate SOC with SOH, for it to be meaningful. In other words, FCC should be adjusted with age through the η parameter. Thus with this definition, the SOC is "state of charge" with respect to "maximum charge that the battery can now take" to bring the emf or voltage to a manufacturer defined threshold.

One important thing during SOC calculation is that a "reset" value of SOC has to be obtained from time to time while using the book-keeping method of charge integration to keep the numbers from drifting away.

SOCV after 15 minutes (or sometimes even more) of rest is reasonable to obtain a "reset" value of SOC for lead acid batteries. The relationship between SOCV and SOC is reasonably linear from 0 to close to about 100%, where the 100% should be based on a manufacturer specified voltage. We can start from a discharged battery at a certain voltage, specified by the manufacturer where it is considered to be at SOC = 0. We can first charge a battery at high current until some voltage is reached and then do a low current charge till the full charge is reached, as denoted by the manufacturer specified voltage. The amount of the total amp-hrs injected during this high and low current charging is the charge input AC. As soon as the final voltage is reached, we can say SOC = 100% and the corresponding amp-hr input can be considered to be the AC at SOC = 100% and here it will be numerically equal to FCC.

For an NiMH battery the relationship between SOC and SOCV is fairly constant (SOCV vs SOC is constant) over a wide range of SOC, and hence SOCV cannot be used as an estimate of SOC, since SOCV will not carry any new information. But we can still use the recharging technique indicated in the previous paragraph and use the manufacturer reference to determine the voltage when the manufacturer specified charge amount for 100% SOC has been inserted into the battery. But again, the voltage cannot be used as a measure to determine SOC.

One good way perhaps to do the above in real time is to initiate the charging process by controlling the alternator, when the load demand is low, and the battery is not needed for delivering load demand. The controlling of the alternator is intended to create a perturbation in the charging or discharging current and monitoring the corresponding perturbation in the voltage and the current. An Li-ion battery also has a linear relationship like lead acid and hence the previous method can be used. In Li-ion the SOCV is reached much faster than lead acid (can be as fast as 5 secs, compared to 15 minutes in lead acid) and hence SOC reset can be done more often, even in real time. Li-ion should not be trickle charged due to problems of explosion, safety etc.

Briefly then, we begin our SOC computation using the voltage method, as per the method indicated earlier in transition diagrams of Figures 11.2 and 11.3. We use both book-keeping and voltage method, depending on the state. Why then can we not use the voltage method all along, if we have to recalibrate using the voltage method after all? The reason is that the voltage method with SOCV needs resting. During stable condition it is ok, but the book-keeping method gives a good idea of the SOC during changing non-equilibrium conditions when charging or discharging is going on.

Due to the above, cumulative amp-hr error in Li-ion will be least, followed by lead acid, and the worst will be in NiMH, where the period will be longest before SOCV is reached.

A possible algorithm flow diagram for battery SOC computation accurately where model based adaptive method is used is given below in Figure 11.9.

So, the summary of the reset issue is that we need to know SOCV and then have to correlate it with the manufacturer information to map it to SOC estimate. Since the SOC correlation is dependent on temperature, we have to include the same as well, again based on manufacturer information. Similarly with age, we have to modify the FCC.

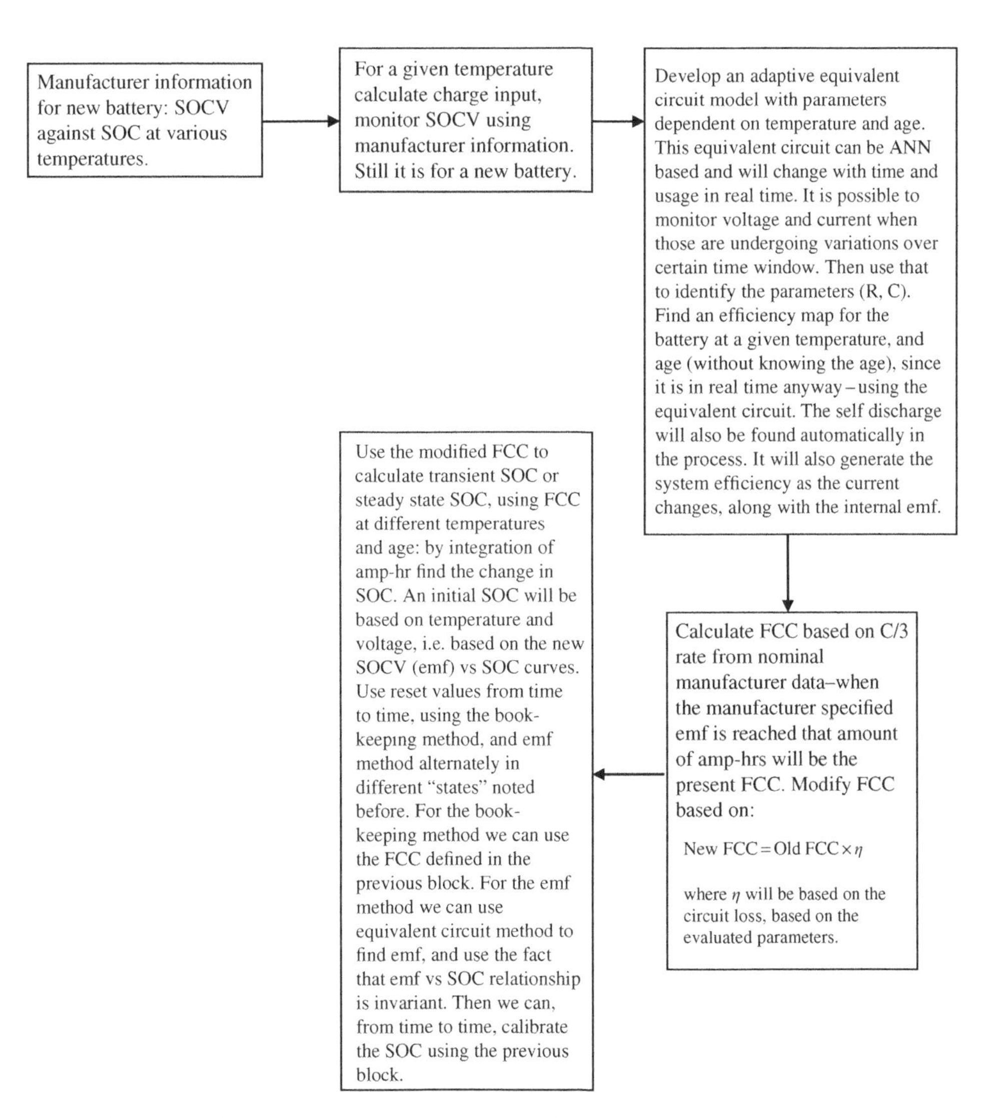

Figure 11.9 A possible algorithm for accurate battery SOC computation using model-based adaptive method.

So, we might need to revisit the circuit model shown before to evaluate the various parameters in that circuit from time to time, when the load demand is low and follow the block diagram in Figure 11.9 to evaluate those.

11.2.8.2 State of Health Issue in Batteries

Before discussing this, it is reasonable to refresh some terminology again.

Capacity – in amp-hrs

C-rate (*Cr*) – unit is amps. It is numerically equal to that value of current which will charge the battery to its rated value in 1 hour. In other words, it is numerically equal to amp-hrs.

SOH – One definition of SOH is: Q_{max} (at present age)/Q_{max} (new), where Q_{max} is the theoretical maximum amount of charge that can be drawn from the battery. We can interpret it as the amount of charge at 100% SOC that can be drawn now versus what could be drawn at 100% when it was new.

Notice that it is still 100% SOC in both cases, but that does not mean they have the same amount of charge.

SOH tells the overall battery conditions based on capacity, internal resistance, self discharge etc. One way to know about SOH is through counting its operational pattern at various temperature, since temperature has a major effect on the health of a battery.

Suppose a battery is rated by the manufacturer as having a life of X months if it is operated at a nominal temperature of T. But more likely than not, it will be operated at different temperatures. Hence the IEEE 450 standard on lead acid batteries indicates a method of calculating the life at a given point, using the following formula. It assumes that a thermal degradation chart is available from the manufacturer or perhaps through testing of the battery, which gives the following.

$$Lt_e = \frac{M}{\frac{mos@T_1}{\%Life} + \frac{mos@T_2}{\%Life} + ... + \frac{mos@T_n}{\%Life}} \tag{11.5}$$

where Lt_e = resultant % life, M = nominal life expectancy, $mos@T_1$ = months run @ T_1, and so on, and the term *%Life* is from the graph on the chart (Figure 11.10); the chart is defined by: x axis = temperature, y axis = % of life.

Obviously, then, as the temperature increases, the derated value of % of Life decreases.

When it comes to battery SOC, SOH determination, the issue is related to somehow identifying its parameters inside the battery, and how they change with temperature and other environmental parameters. These are parameters such as resistance, conductance, capacitance (impedance) which are used to model a battery as in the equivalent circuit shown previously. These parameters also vary with SOC, SOH, in addition to temperature.

The qualitative effect of temperature on battery health and performance can be stated as follows. In general, at cold temperature a battery loses its capacity but gets it back as the temperature rises. Low temperatures are not good for battery performance but they may help with life enhancement. Conversely, warmer temperatures are not good for life but will help performance. So, from an SOC point of view a warmer temperature can be considered better, whereas from an SOH point of view the colder temperature is better.

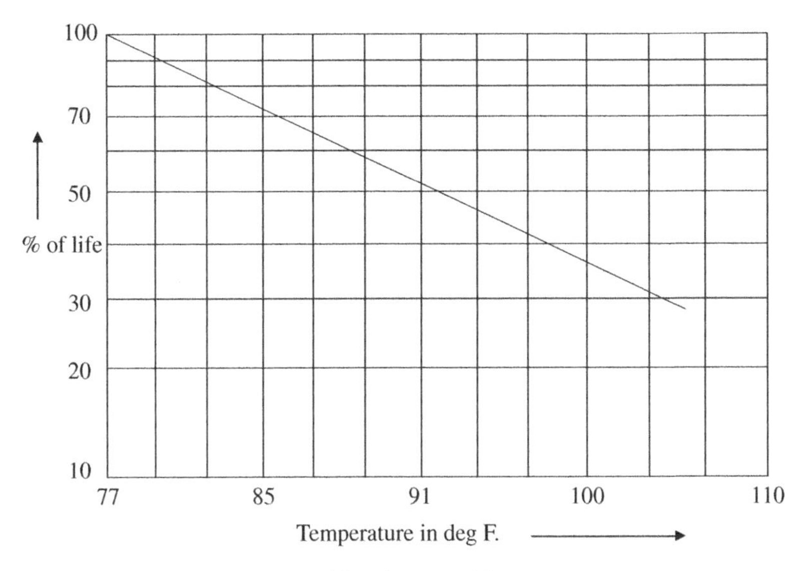

Figure 11.10 Battery percentage life reduction with temperature.

One issue that can arise with batteries is related to the using of the de-rating chart in Figure 11.10. In general, it is not always likely that the user will maintain a record of temperature and the period of time that the battery underwent at a given temperature. If this is indeed recorded, it can help in estimating the SOH. If this information is lacking, and suppose a battery is provided without any information about its history, then the following recourse might be adopted. It is assumed that, as a minimum, the battery type (lead acid, NiMH, Li-ion, etc.), the manufacturer chart such as voltage versus time, at different temperatures and at different C-rates are given.

11.2.8.3 Two-Pulse Load Method to Evaluate State of Health of a Battery [4, 6]

People have come up with different methods to find out SOH. One method is illustrated in Figure 11.11, which shows that a current pulse is injected to the battery and the corresponding voltage is observed. Only the delta-V from the second pulse is measured, since at the beginning, the first pulse may not be truly representative. This is because of the past history of the voltage trace, which can include an initial sudden dip in voltage during current initiation, known as the “coupe de fouet” (= “whip crack” in English).

The battery must be in an open condition for a minimum of 2 minutes to settle down. By applying the first pulse, stray data is avoided by the time of the second pulse, which is more stable in nature.

The initial difference between the first and second pulses is due to coupe de fouet. If the pulse test is continued for a long duration (say a few hours), the delta-V will increase when the battery is nearly empty, due to shortage of sulfuric acid and lead sulfate buildup on electrodes.

Delta-V (under the same current load and pulse time) increases, as the SOH and capacity decreases. This last item can be considered a measure of SOH.

Delta − V/Delta − I = R = internal resistance measure − as SOH decreases R increases.

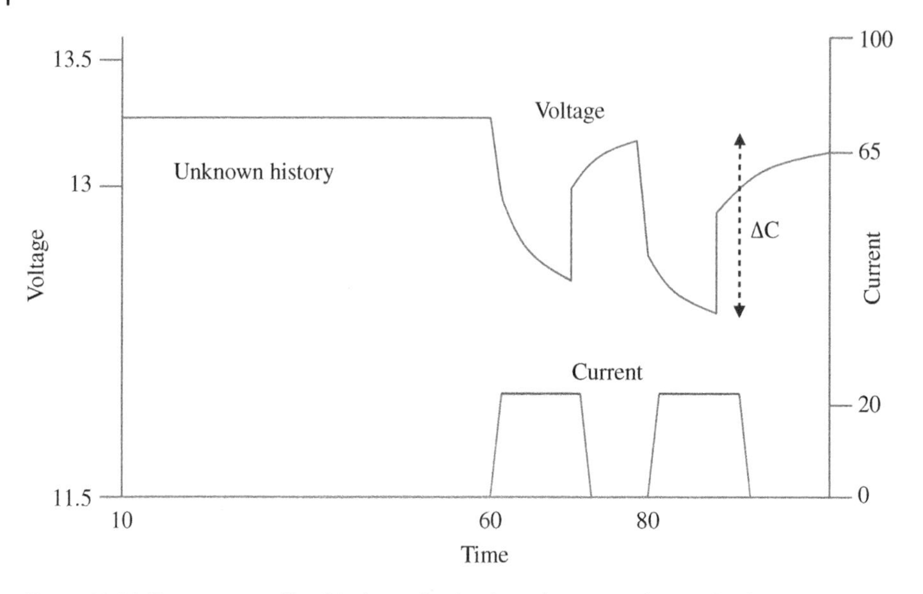

Figure 11.11 Battery state of health determination based on two-pulse method.

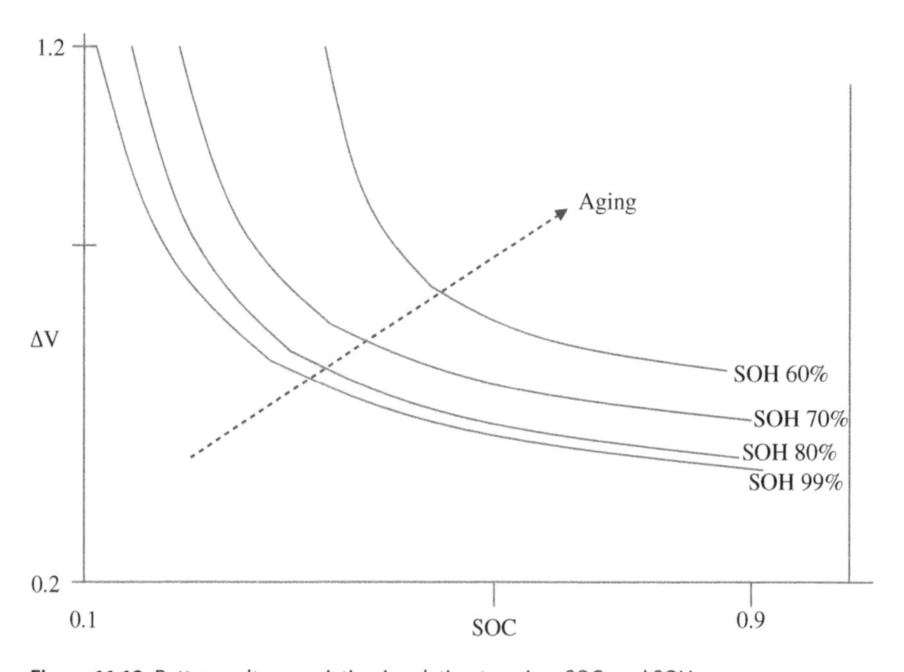

Figure 11.12 Battery voltage variation in relation to aging, SOC, and SOH.

SOH indicates the overall battery conditions based on capacity, internal resistance, self discharge, etc.

The correlation between aging and ΔV is indicated in Figure 11.12, and it can be used as a measure of SOH [4, 6].

Of course, it is necessary to have experimental results and archived information, which can help correlated aging or SOH by measuring ΔV (Figure 11.13).

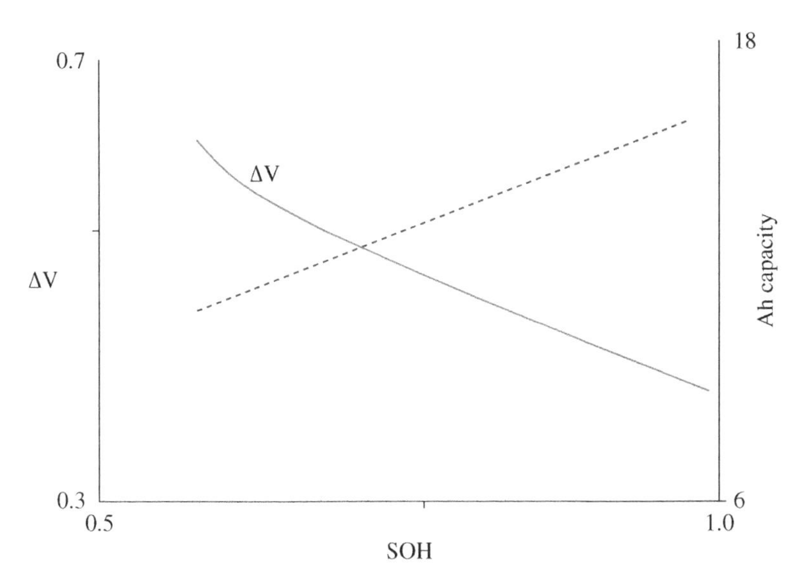

Figure 11.13 Battery voltage and amp-hr capacity variation in relation to SOH.

It is obvious, therefore, that SOC is a more short-term parameter whereas SOH is a long-term aspect. In addition, SOC may not be indicative of how much real energy can be extracted from a battery, as this is SOH-dependent [4, 6].

The above discussion, though related to a lead acid battery, is also helpful for other battery chemistries. Before continuing on this, it may be instructive to consider the following facts about different batteries.

- Lead acid battery (in terms of its model) is relatively more predictable
- Ni-MH apparently is less predictable
- Lithium-ion is apparently more predictable than NiMH

There are lots of research papers where people have tried to predict SOC, SOH etc. based on artificial intelligence methods such as neural networks and fuzzy logic algorithms. Under certain conditions they seem OK, but a battery is a highly nonlinear device, and there does not seem to be any perfect model yet.

From a user's perspective the goal is to get the maximum life and energy capacity out of a battery. To achieve this goal it is important to recognize the following good practices in connection with a battery [2, 3]:

1) Lead acid – during charging:
   Charge immediately after use
   Full discharge should be avoided for longer life
   Over-cycling (charge-discharge cycles) should be avoided
   Constant voltage charging is followed by floating condition should be applied
   Overheating shall be avoided
2) Lead acid – during discharging
   Avoid full cycle discharge
   80% depth of discharge is ok
   Recharge more often

3) Lithium-ion – during charging:
   Charge often
   Full discharge should be avoided for longer life
   Over-cycling (charge-discharge cycles) shall be avoided
   Constant voltage charging should **not** be followed by any trickle charge
   Fast charge possible
   Overheating should be avoided
4) Lithium-ion – during discharging:
   Same as lead acid
5) NiMH – during charging:
   Run it fully down every 3 months of usage
   Overheating should be avoided
   Constant current charge followed by trickle charge when full, should be applied
   Slow charge not recommended
6) NiMH – during discharging:
   Avoid too many full cycle (charge-discharge)
   Use 80% depth of discharge

Other factors in connection with all batteries are as follows:

1) Internal parameters in battery can be influenced by manufacturer.
2) External parameters can be influenced by operational pattern.
3) High temperature is harmful – hence the design of the enclosure or cabinet is important.
4) Thermal management is very important.
5) Deep discharge – long rest periods at low SOC is harmful.

11.2.8.4 Battery Management Implementation

To achieve the above objectives it is important to control the battery current, i.e. the charge-discharge mechanism, and control the environmental conditions (i.e. ambient temperature, humidity etc.) around it. This set of activities is normally referred to as battery management. Battery management implies that parameters external to the battery are to be influenced according to the different battery chemistries through some control mechanism. Basically this includes – data acquisition i.e. temperature, current, voltage, humidity etc. through various sensors. Both measured and calculated data can be used for battery management and control. The temperature control (thermal management) and current (or voltage) control at its two terminals are at the user's disposal.

Following facts should be noted in connection with battery management:

- Information and data acquisition is not costly.
- Power electronics and matrix switching with many modules can be expensive.
- Current sensors can be expensive compared to voltage sensors.
- Sometimes artificial intelligence techniques may be utilized, using only voltage sensor, and derived current based on AI algorithms.

On the other hand, based on voltage and current, and using AI, researchers have tried to come up with an estimation of internal parameter values and hence predict SOH.

The architecture for a generic battery management system (BMS) is shown in Figure 11.14.

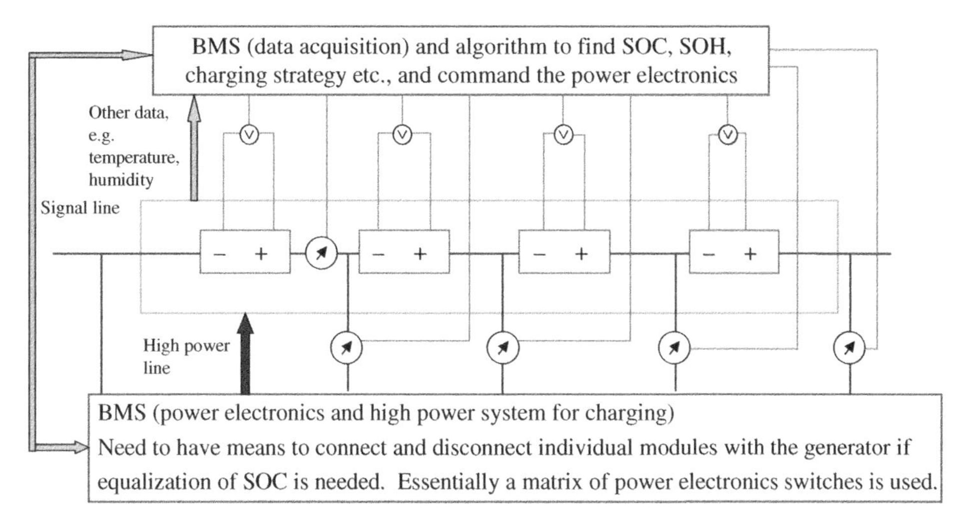

Figure 11.14 Battery management – system level architecture.

The data acquisition part involves collecting information from the battery and/or the power electronics interface through appropriate sensors, e.g. temperature, humidity, and other ambient condition information. The block shown in the bottom of Figure 11.14 involves power electronics, which deals with high currents. Ideally, it provides the interface between the battery on one side, and the loads and generator on the other. It should be possible to implement architectural reconfiguration of the battery modules such that the generator can selectively charge one or more modules, independent of the rest of the modules, and similarly it should also allow discharging of the batteries selectively, if necessary. This might call for the provision of a bleed resistor (in parallel with battery) which could dissipate the energy of the battery selected for discharge. The block at the bottom of Figure 11.14 will contain the generator, the bleed resistor, and connections to various loads. The decision to reconfigure, and how to reconfigure, has to come from the controller, i.e. the top block in Figure 11.14, which should direct the power electronics module accordingly.

11.2.8.5 What to Do with All the Above Information

Methods for determining battery SOC and SOH, and some good practices related to different types of batteries have been discussed here. Finding the correct SOC and SOH are important. The reason for this is that it gives a better indication of the battery condition and, depending on the needs of the vehicle, it helps to make optimum decisions as to whether to charge the battery or not, and whether to conserve the battery by lowering usage, to meet important survivability needs. The SOH is primarily very important from a prognostic point of view – deciding on the need to replace the battery before any unanticipated failure. This can also impact survivability.

As we have already observed, there are several commonalities in the operation of the different types of batteries, whether they are lead acid, NiMH, or lithium-ion. All of them are negatively impacted by high temperature or overheating, and by over-cycling, i.e. too many charge discharge cycles at full (specified maximum) depth of discharge. Therefore, this information about temperature can be used effectively as part of the battery management. For example, if the temperature is seen to be getting too close to some threshold, and if there is provision for either air or liquid cooling, then those cooling processes can be intensified by increasing the flow of air or liquid. Excessive cycling can be avoided by correct determination of the SOC, using methods indicated earlier, under different conditions such as charging, discharging, equilibrium, and transition. If it is seen that the discharge cycle is getting too close to the maximum depth of discharge, battery usage could be stopped, or it could be placed in charge cycle before it reaches the maximum depth. Depending on the load demand, it may be possible to dictate the battery usage and avoid going anywhere near the maximum depth of discharge. In addition, it has been suggested that life of lithium-ion batteries can be extended by reducing the charging threshold voltage. This means that before the battery has reached the maximum allowable voltage during charging, the charging should be cut off, somewhat ahead of that maximum. This more or less reduces the depth of discharge. For NiMH, as part of the good practice suggested earlier, it may be beneficial once in every three months to let it discharge fully, which can be monitored correctly through the proper evaluation of SOC. All the above requires accurate voltage, current, and temperature measurements using appropriate sensors. The battery management module then processes the information and commands the power electronics circuit to handle the charge or discharge mechanism.

Also, the same above information (related to SOC) can be used to better balance or equalize the SOC between various modules within the complete battery pack, by the following process for the purpose of charge equalization: (a) either charging the batteries collectively which can allow charging of the modules individually, or (b) collectively (followed by bleeding the charge of certain specific batteries which might have been charged at higher SOC compared to others during collective charging in series). The exact method will be very much dependent on the hardware involved. If every battery is to be individually controlled for equalization, it will most likely be necessary to have a mesh or matrix of power electronics switches connected to a generator or perhaps a fuel cell, which can be configured to charge the battery modules individually and selectively. This can lead to somewhat expensive power electronics, in addition to packaging space requirements.

SOH information, as noted earlier, will allow better maintenance, and allow replacement ahead of time. Although SOH is more for informational purposes, and truly speaking not much can be done if SOH for a particular battery module is indicated to be bad, still it can help in follow-up actions. At that point, the only thing that could be done is replacement. But if good practices are followed from the beginning, when the battery is new, then SOH will most likely be better at any particular point in time and can in general lengthen the overall life cycle of the battery. Even after following good practices, there is some likelihood of sudden failure of a system, and so accurate evaluation of SOH can at least help towards avoiding those events, or its continuous monitoring can point to some upcoming failure. These are very important things to know about beforehand, so that catastrophic survivability needs can be addressed.

In connection with SOH, it has been found that one method related to life cycle is based on de-rating on the basis of temperature only (based on IEEE 450 standards), i.e. how much the life is reduced due to temperature issues. Although the IEEE 450 is specifically for vented lead acid batteries, the general principle of de-rating will still hold for other batteries. It may need some amendment of the de-rating mechanism (i.e. the equation defining the life after de-rating for temperature), and particular de-rating curves from the manufacturers might be needed, regardless of whether it is lead acid, NiMH, or lithium-ion. The other method is based on cycle counting, i.e. how many cycles of charge discharge have been encountered. This includes both full and partial depth of discharge cycles and somehow including them in the counting mechanism through accumulation of the cycles. This cycle counting, of course, is intimately connected with correct SOC calculation. Hence incorrect or over-simplified SOC evaluation methods can lead to incorrect cycle counting. In the absence of adequate information on de-rating or cycle counting, it is also possible to use methods such as the two-pulse method already explained. Although no method can be perfect or precise when dealing with battery SOH and SOC, a combination of one or more methods – that is (a) life cycle de-rating based on temperature, (b) cycle counting, (c) two-pulse method – can together help make a reasonable estimate and cross check the situation on the life of a battery.

11.3 Comparison of Energy Storage Technologies

Comparison of several energy storage technologies is provided in Table 11.2 (Ref: US Council for Automotive Research LLC – http://www.uscar.org and USABC – US Advanced Battery Consortium, and http://www.energy.ca.gov). The advanced lead acid and Li-ion batteries are the most promising for application in HEVs. While battery and ultracapacitor technologies have their respective advantages and disadvantages, hybridization could result in better vehicle performance and longer battery life. The vehicle road load transients can be handled by ultracapacitors during acceleration and deceleration.

11.3.1 Lead Acid Battery

The lead acid battery is extensively used in the automotive industry, for both starting the engine and also for supplying auxiliary loads when the engine is not running, or for delivering sudden intermittent high power demands which cannot be met by the vehicle generator alone. The battery consists of two electrodes, one made of lead and the other made of lead dioxide. The electrolyte is sulfuric acid.

The chemical equation for a lead acid battery during discharge is

$$PbO_2 + Pb + 2H_2SO_4 \rightarrow 2PbSO_4 + 2H_2O \tag{11.6}$$

Table 11.2 Comparison of energy storage technologies suitable for HEVs.

Storage technology	Cycle life	Efficiency (%)	Specific power (W/kg)	Specific energy (Wh/kg)
Lead acid battery	500–800	50–92	180	30–40
Li-ion polymer battery	500–1000	80–90	>3000	130–200
NiMH battery	500–1000	66	250–1000	30–80
Ultracapacitor	1,000,000	90	1000–9000	0.5–30

The chemical equation for a lead acid battery during charge is

$$2PbSO_4 + 2H_2O \rightarrow PbO_2 + Pb + 2H_2SO_4 \tag{11.7}$$

During discharge, the electrolyte and the active material on the battery plates are used up and water and lead sulfate are produced. During charging, the reverse process takes place, and electrical energy is absorbed by the battery, water and lead sulfate are consumed, and electrolyte and the active material at the plates are produced.

11.3.2 Nickel Metal Hydride Battery

The NiMH battery is used in quite a number of hybrid vehicles, although other chemistries like lithium-ion are replacing it to some extent.

The overall reversible chemical reaction occurring in a NiMH cell is:

$$MH + NiOOH \Leftrightarrow M + Ni(OH)_2 \tag{11.8}$$

11.3.3 Lithium-Ion Battery

In Li-ion batteries, Li ions alternatively move into and out of host lattices during charging and discharging cycles. The Li-ion battery has anode and cathode plates like a lead acid battery, but they are made of lithium cobalt oxide (or other lithium composites) and carbon. These plates and the separator are immersed in a solvent which is most commonly ether [4]. This type of battery can be made with very high energy density. The overall reversible chemical reaction occurring in a Li-ion cell is

$$Li_xC + Li_{1-x}\ M_y\ O_z \Leftrightarrow C + LiM_y\ O_z \tag{11.9}$$

Li-ion batteries do not have the "memory effect" (like hysteresis at different maximum levels) that causes some other rechargeable batteries to lose their maximum charge level when repeatedly charged and discharged to capacities which can be different from the full capacity. Li-ion batteries impact the environment relatively less due to their composition. Unlike lead acid batteries, they have a much lower self-discharge rate, thus greatly increasing idle period capabilities. These batteries also have a higher power-to-volume ratio which is conducive to automotive applications where space is restricted [6].

It should be noted that lead acid batteries remains the preferred energy storage device in many applications for cost reasons and also due to the fact that Li-ion batteries have certain safety concerns requiring special attention. Overcharging or overdischarging these batteries can severely damage the plates inside the case. Overcharging can also cause gassing of the electrolyte and buildup of pressure in the case, which can lead to an explosion. These issues are not trivial and have actually caused problems in various applications, including aerospace. The reduction in life due to the above issues is much greater in lithium batteries than in lead acid batteries.

11.4 Ultracapacitors

Ultracapacitors are fundamentally different from batteries due to the fact that the energy is stored in the electric field within the battery, and the materials do not change chemically (as they do in a battery) during charging and discharging. Ultracapacitors

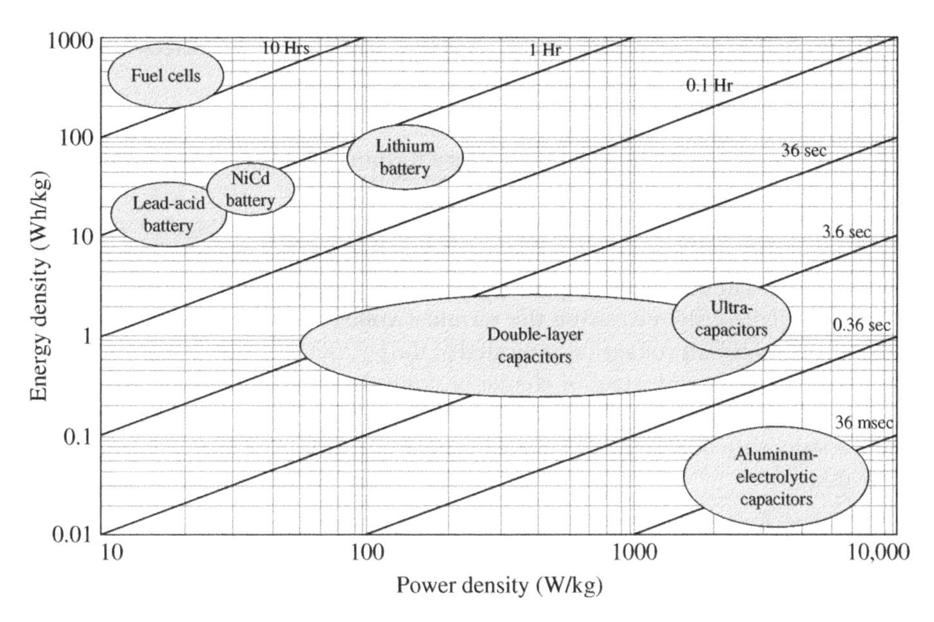

Figure 11.15 Comparison of power density and energy density for ESS in HEVs. *Source*: Courtesy of US Defense Logistics Agency.

have a very long shelf life, with much lower maintenance requirements, enhanced performance at low temperature, and environmental friendliness. However, ultracapacitors are still considered more costly than batteries and they have relatively low energy density [7–9]. The ultracapacitor's SOC is easier to estimate than that of a battery since the energy is proportional to the square of the voltage and the behavior is more predictable. Also, ultracapacitors can be charged to a specific value and, due to their shelf life and charging mechanism, they can hold that charge with very little self discharge. In addition, unlike a battery, repeated charge discharge cycles in an ultracapacitor are not detrimental to its lifespan.

Ultracapacitors allow rapid charging and discharging. This is especially useful for faster and more efficient regenerative energy recovery in HEVs as well as for rapid charging of PHEVs. Ultracapacitors have a long cycle life (on the order of a million cycles), with little degradation over hundreds of thousands of discharge/charge cycles. In comparison, rechargeable batteries last for only a few hundred deep cycles.

As noted in Table 11.2, the ultracapacitor's energy density is much lower than that of an electrochemical battery (3–5 Wh/kg for an ultracapacitor compared to 30–40 Wh/kg for a lead acid battery, and 120 Wh/kg or more for a Li-ion battery), and its volumetric energy density is only about 1/1000th of that of gasoline. The energy stored is proportional to voltage squared. Effective storage and retrieval of energy requires complex electronic control and balancing circuits involving power electronics switches. The internal resistance of ultracapacitors is very low, resulting in high efficiency (95% or more).

A comparison of the power density and energy density of different energy storage systems (ESSs) is illustrated in Figure 11.15 (Ragone plot or chart).

11.5 Electric Circuit Model for Batteries and Ultracapacitors

11.5.1 Battery Modeling

A simple equivalent circuit model of the battery terminal characteristics is shown in Figure 11.16.

It consists of an ideal battery with open-circuit voltage E_v and a series equivalent internal resistance R_i. The battery terminal voltage is V_t. E_v can be obtained experimentally from the open-circuit voltage measurement, and R_i can be evaluated by connecting a load and simultaneously measuring the terminal voltage and current. It should be noted that the open circuit voltage (as indicated in the bottom left curve in Figure 11.16) is a quantity that varies with state of charge or equivalently state of discharge (SoD). A fully charged battery has SoD = 0. Similarly the bottom right curve in Figure 11.16 shows the relationship between the terminal voltage and SoD. The terminal voltage V_t can be written as

$$V_t = E_v - I_b R_i \tag{11.10}$$

The model indicated above is very simple and works reasonably well under most normal circumstances. However, it is not a dynamic model and hence cannot respond well to transient events. Also, the internal resistance shown in the above model is a quantity that is dependent on the state of charge. Furthermore, since the battery is a chemical device with different chemical reactions during charging and discharging, the resistance during charging is different from that when discharging. If all the above are to be taken into account, a more sophisticated model as shown in Figure 11.17 can be used.

This is a bidirectional battery model and its parameters need to be captured. Since it is a fourth order system (four storage elements in independent locations) with nonlinearities i.e. diodes, nonlinear resistors, V_{emf} etc., its equivalent mapped parameters, dependent on physical parameters, can be estimated through an artificial neural network (ANN) type of systems. By monitoring a sufficient number of samples of physical variables of the battery within a sliding window, it is possible to estimate these equivalent parameters. However, it is necessary that the system should not be in steady

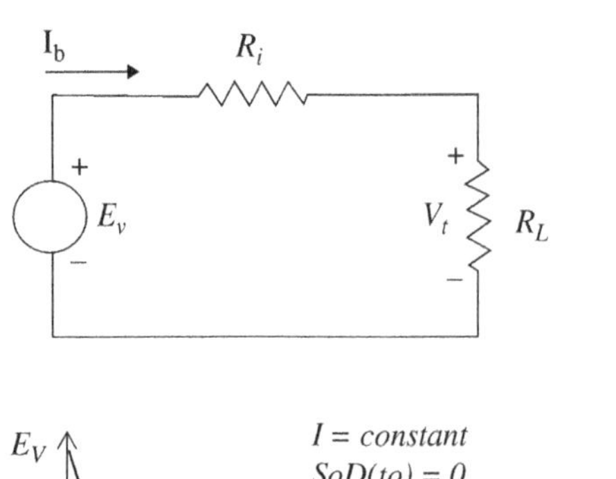

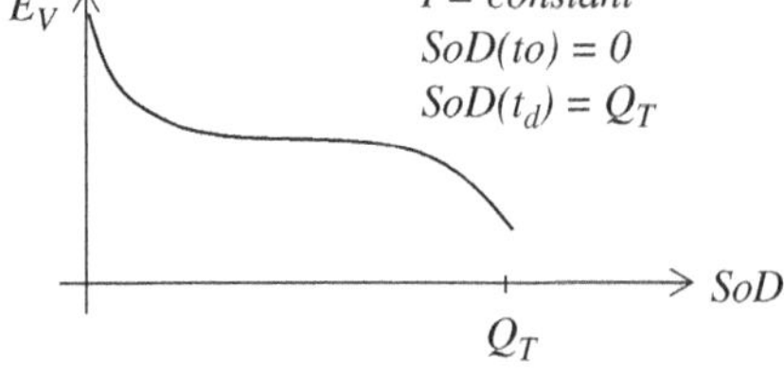

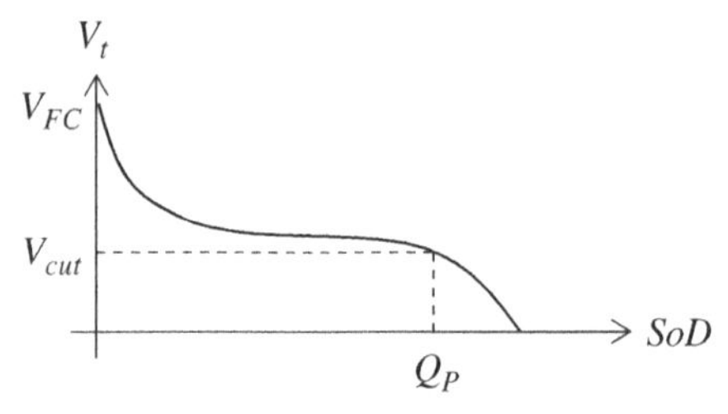

Figure 11.16 Battery equivalent circuit model.

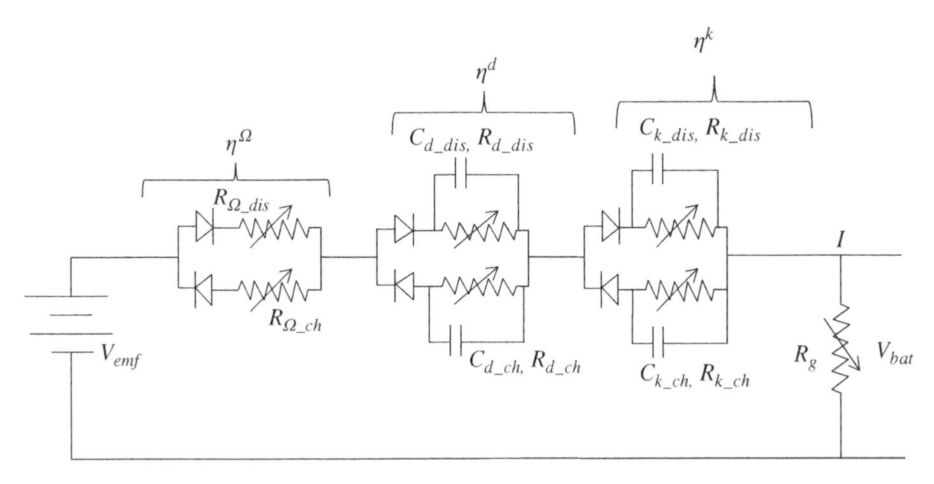

Figure 11.17 A detailed battery model.

state to estimate these parameters, especially the capacitors. Since these nonlinear elements are also temperature dependent, that fact can also be incorporated in the training process. This ANN training process can therefore require the availability of a lot of manufacturer data. The other option, i.e. an offline experiment performed by the user, obtaining data for batteries, is very tedious and labor intensive. To estimate the parameters *while delivering load, and on a real-time basis,* it may be necessary to periodically disturb the load current by a small percentage of the actual value at a particular moment, and monitor the corresponding change in voltage, or vice versa.

An advantage of using the above model is that it can take into account self discharge of the battery, represented by the parallel resistance R_g at the right-hand side of Figure 11.17, which is important. It can also take into account the effect of transients.

11.5.2 Electric Circuit Models for Ultracapacitors

In a capacitor, the electrical energy is stored in the electric field between the positive and negative plates (or electrode surfaces). The energy is stored in polarized liquid later at the interface between ionically conducting electrolyte and electrodes as shown in Figure 11.18.

This is unlike a battery where the energy is stored in the chemical reaction process. In ultracapacitors, special carbon-based electrodes are made to provide an extremely large internal active surface area. The energy stored in a capacitor are given by

$$E = \tfrac{1}{2}CV^2 \tag{11.11}$$

where C is the capacitance in farads, V is the voltage across the capacitor, E is the energy of the capacitor.

For ultracapacitors, the SOC can be computed with high accuracy:

$$\text{SOC} = \frac{C(V - V_{\min})^2}{C(V_{\max} - V_{\min})^2} = \frac{(V - V_{\min})^2}{(V_{\max} - V_{\min})^2} \tag{11.12}$$

where $V_{\max}$ and $V_{\min}$ are maximum and minimum allowable voltages for the ultracapacitor.

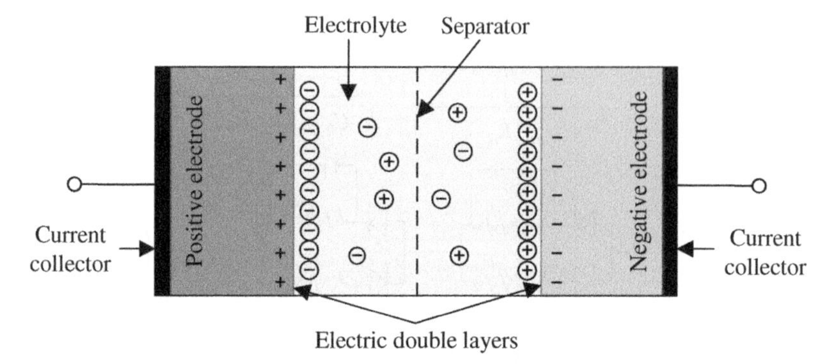

Figure 11.18 Structure of an ultracapacitor.

The capacitance C of a capacitor is given by

$$C = \varepsilon A / d \tag{11.13}$$

where ε is the permittivity of the dielectric medium, A is the plate area, and d is the distance between the plates.

The equivalent circuit of an ultracapacitor can be represented as follows in Figure 11.19.

In this equivalent circuit the major components are the series resistance R_s, capacitance C, parallel leakage resistance R_L, and the variables are the capacitor voltage V_C, terminal current i, capacitor internal current i_C, leakage current i_L, and the terminal voltage is V_t.

The dynamic equations for the above equivalent circuit are

$$V_t = V_C - Ri \tag{11.14}$$

$$C dV_C / dt = -i_C = -i_L - i \tag{11.15}$$

$$i_L = V_C / R_L \tag{11.16}$$

For the purpose of implementation in simulation the above equivalent circuit can be represented as shown in Figure 11.20.

Ultracapacitors can store much higher energy than electronic capacitors used in communication engineering and similar low power applications. However, the energy stored in an ultracapacitor is much lower compared to a battery of comparable size.

A typical ultracapacitor generally comes in a cell which is normally 2.5 volts. A lot of such cells may need to be connected in series to be able to work at higher voltages. Similarly by connecting multiple parallel sets of ultracapacitors, current capability can be increased. The situation is identical to that shown in Figure 11.1 for creating a battery bank out of smaller entities. Exactly the same applies in the case of ultracapacitor as well.

A typical discharge curve of an ultracapacitor is shown in Figure 11.21.

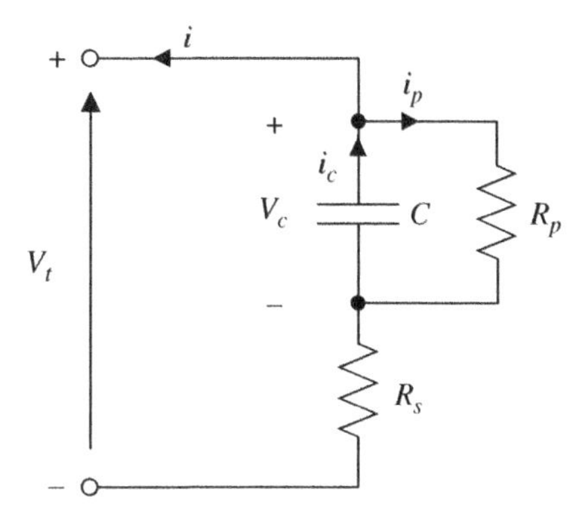

Figure 11.19 Equivalent circuit of an ultracapacitor.

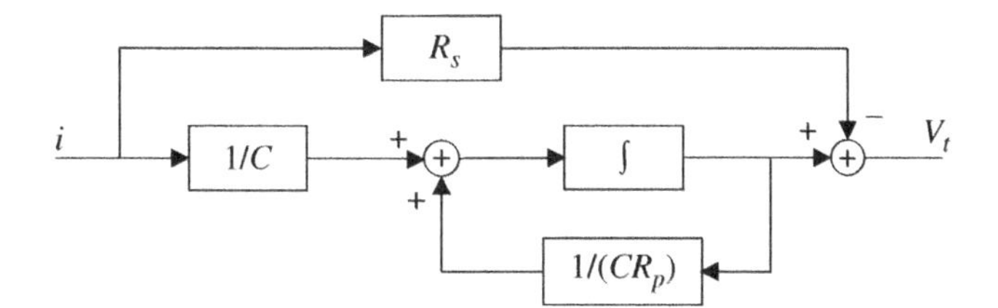

Figure 11.20 Simulation representation of equivalent circuit in Figure 11.19.

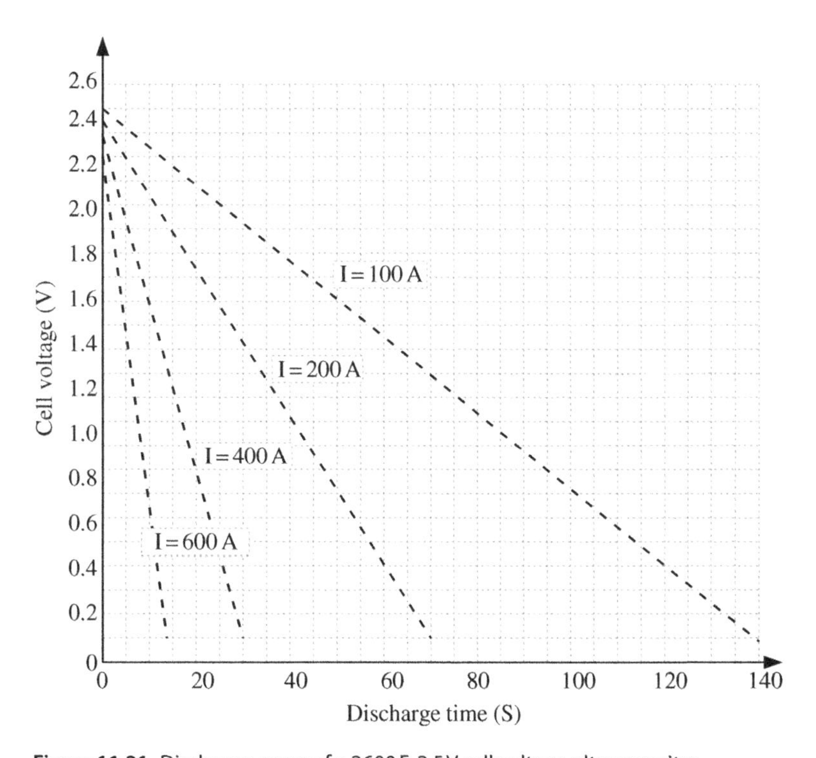

Figure 11.21 Discharge curve of a 2600 F, 2.5 V cell voltage ultracapacitor.

11.6 Flywheel Energy Storage System

Flywheels are drawing some interest these days in hybrid vehicle design, particularly for larger passenger transit vehicles. There are a number of reasons for this. In particular, it allows treatment of the requirements on specific power and specific energy separately, implying that power intensive sources can be kept separate from energy intensive ones, while considering which ones are to be used to meet the load demand characteristics. As can be seen from Equation 11.17, the amount of energy stored in a flywheel is dependent on the flywheel's speed of rotation and its moment of inertia.

$$E = \frac{1}{2} J\omega^2 \tag{11.17}$$

where J is the moment of inertia and ω the rotational speed. The above equation clearly indicates that there is a squared relation between energy stored in the flywheel and its rotational speed. Several companies have developed practical flywheel storage systems [10–12].

Increasing the speed of rotation produces improved specific energy, but increases the potential safety hazard, and also the cost, since special bearings and high-strength composite materials may become important for the purpose of constructing the system. The flywheel can also mitigate the problem of limited cycle life suffered by other sources because the life cycle of a flywheel is practically unlimited or at least longer than vehicle life [10–12].

A flywheel system diagram is shown Figure 11.22.

Some basic concepts related to flywheels may be considered at this point. A flywheel system can involve the transfer of one form of energy into another, e.g. the kinetic

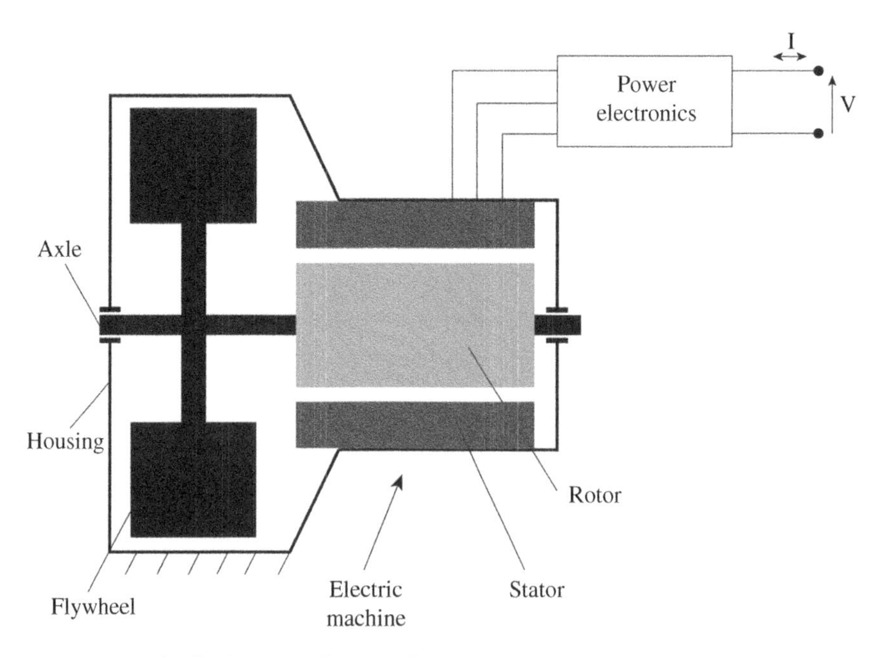

Figure 11.22 Flywheel system schematic diagram.

energy (mechanical energy) of the flywheel which could somehow be converted to electrical and vice versa, or the mechanical energy of the vehicle could be transferred to the flywheel, i.e. transfer of mechanical energy from one entity to another (between mechanical systems). A common method of transferring energy from one mechanical entity to another is through an electrical interface, which is what Figure 11.22 shows. If no electrical system is used, then it will call for a relatively more cumbersome system with gears, clutches etc. for the transfer of energy from one mechanical system to another. The method shown in Figure 11.22 works as follows. If energy of a rotating flywheel is to be transferred to the vehicle wheel, the simplest way is to connect the flywheel to the electrical machine. The flywheel and the electrical machine can be kept mechanically separate by using a clutch while its rotational energy is not consumed. To minimize friction, the flywheel itself can be enclosed in a very low air pressure (vacuum) chamber and it could also use magnetic bearing in certain cases so that bearing friction is minimized. Once the flywheel is engaged with the electrical machine, it can run as an electric generator. If the generator is already running by being coupled to an ICE, then the flywheel torque can add to the ICE torque to produce additional power by the generator. Thereafter the generator power can, as usual, be processed through a power electronics system and fed to a propulsion motor. While capturing mechanical energy from the wheel and storing it in the flywheel, the process can be reversed.

A system level diagram illustrating the interface concept between a flywheel, ICE, and vehicle propulsion system is shown in Figure 11.23.

In Figure 11.23, if the rotational energy of the flywheel is to be moved to the vehicle wheel, clutch 3 has to be engaged and also clutch 2. Clutch 1 may or may not be engaged, depending on whether the ICE is to be used or not to generate the required power needed for vehicle propulsion. Electrical machine 1 can run as a generator if power is flowing from left to right. Thereafter, power electronics will change the power to the proper voltage, frequency etc., and drive electrical machine 2 (run as a motor) for vehicle wheel propulsion. On the other hand, if energy from the vehicle wheel is to be captured (e.g. during regenerative braking) then electrical machine 2 can run as a generator, and a power electronics box can translate it into appropriate form and make electrical machine 1 run as a motor. Only clutches 2 and 3 have to be engaged to couple electrical machine 1 to the flywheel. Between the ICE, the clutches, the flywheel, and electrical machine 1 there can be appropriate gears of a suitable ratio. Similarly, between electrical machine 2 and the vehicle wheel there can be appropriate gears of a suitable ratio. For all the above to be

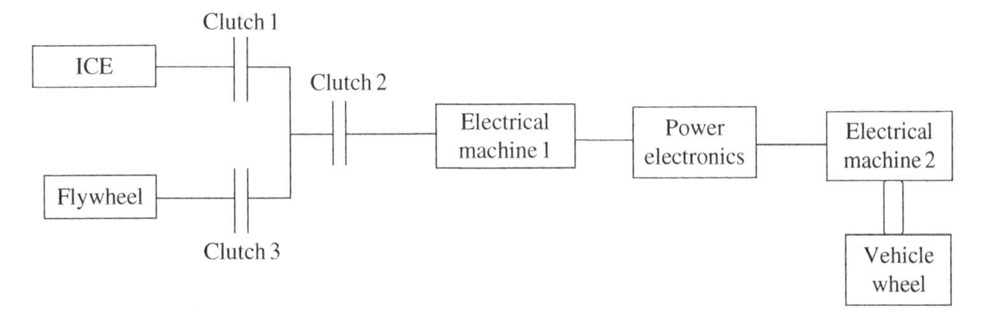

Figure 11.23 Flywheel system interfacing diagram.

implemented, it is necessary to have appropriate power management controller, and also power electronics and electrical machine control system to run the devices properly.

One final note regarding the flywheel is that, when it is rotating, it can cause a gyroscopic effect if the vehicle tries to turn quickly. This problem can be generally remedied by having two rotational masses with the same inertia but moving in opposite directions. This can be easily implemented by placing the two rotating members in a single housing with a gear in between. Sometimes there are also other mechanisms such as gimbals to handle the gyroscopic effect.

11.7 Fuel Cell Based Hybrid Vehicular Systems

11.7.1 Introduction to Fuel Cells

Fuel cells can be used as alternative power sources for electric and hybrid vehicular systems. It can help to completely replace the ICE, thus making it a completely electric vehicle. It can also be used in an HEV environment. Normally even a pure electric vehicle, using a fuel cell will need a battery to initiate the fuel cell activation process. In a fuel cell electric vehicle it is possible to eliminate the alternator as a power generating source. In place of a battery it is possible to have an ultracapacitor and even though the ultracapacitor could be eliminated, it is recommended to retain it under normal circumstances.

A system level architecture of a fuel cell vehicle is shown below in Figure 11.24.

11.7.1.1 Types of Fuel Cells

There are different types (or classes) of fuel cells. This classification is shown in Figure 11.25.

In addition to the above classification, fuel cells can also be divided based on their chemistries. This is shown in Figure 11.26.

11.7.2 System Level Applications

It should be noted that a fuel cell system is not just a stack of cells, but also contains several components to make up the package:

- stacked cells
- fuel tank (hydrogen)
- if a methanol or other fuel is used to generate the hydrogen, it will also need: vaporizer, reformer, gas purification process, etc.
- various pumps (motor)
- compressor if needed (motor)
- cooling fan (motor)
- heating system, as needed

The peripherals of a fuel cell are also important. In this regard, in order to operate a fuel cell it also needs:

- the cells be at a minimum temperature to initiate the cell chemical reaction
- a battery or some auxiliary means of initiating the process
- a means of keeping the cell voltages very accurately close, otherwise the life of the system can be significantly reduced because the fuel cells are very sensitive.

Typical characteristics of a fuel cell are depicted in Figure 11.27.

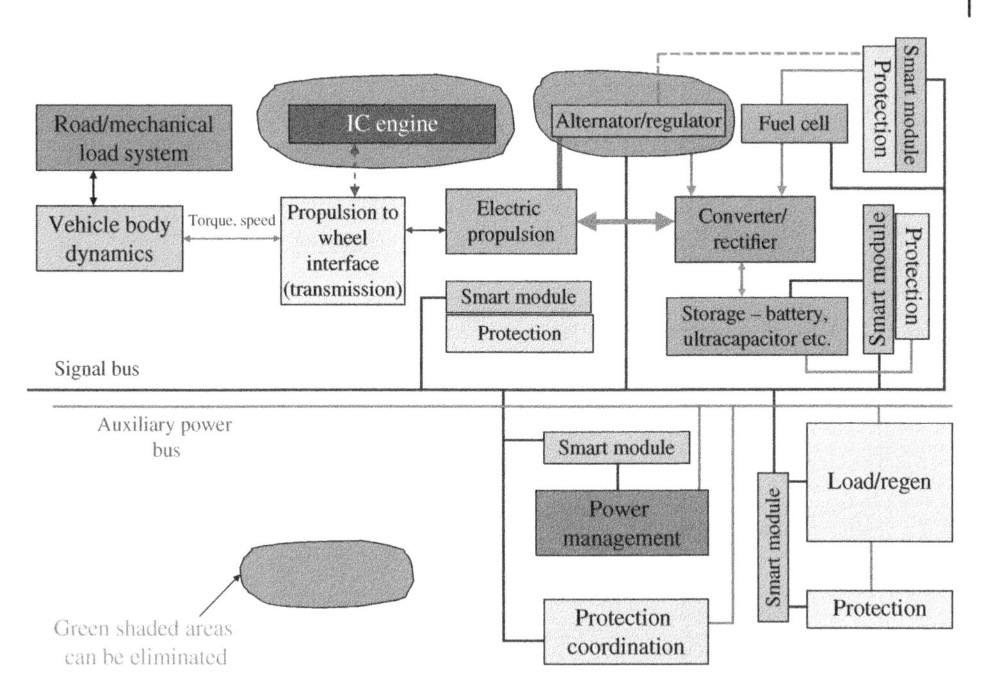

Figure 11.24 Fuel cell vehicle system level architecture.

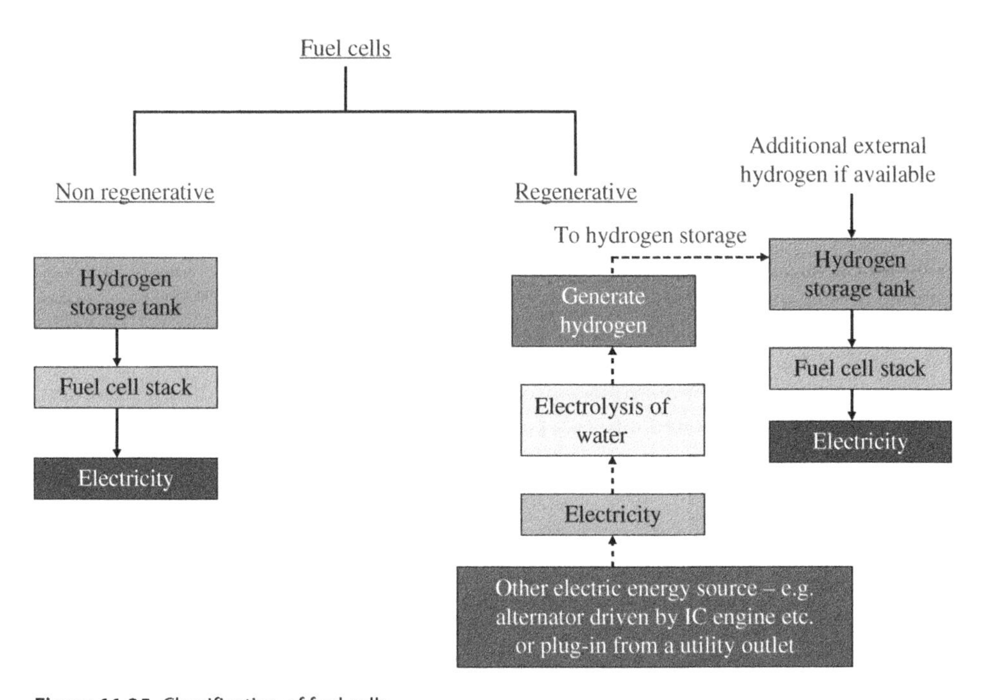

Figure 11.25 Classification of fuel cells.

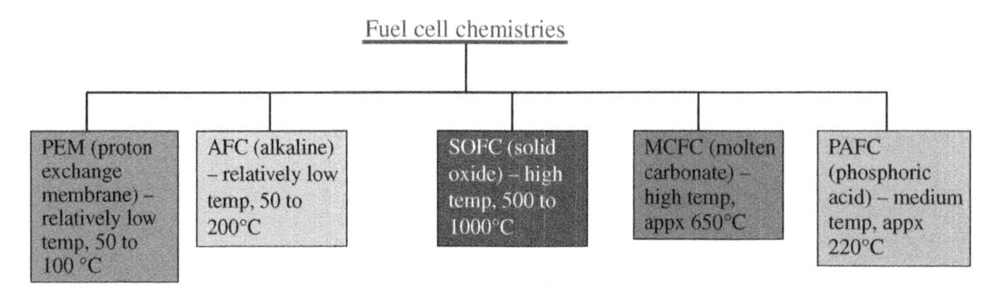

Figure 11.26 Fuel cells based on chemistries.

Typical curves for *V-I*, and η:

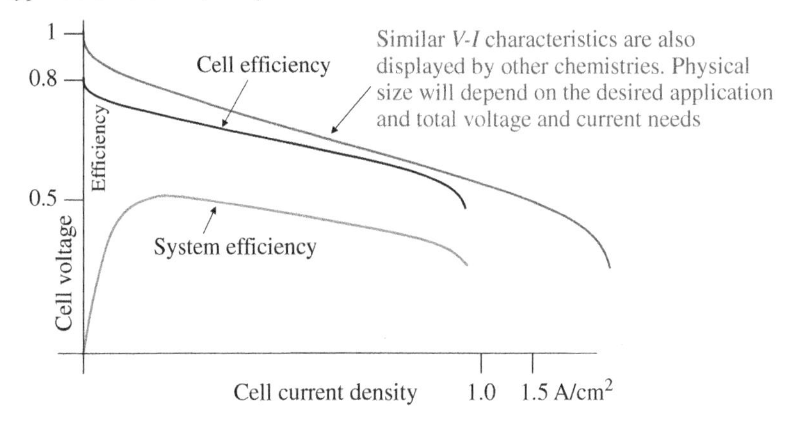

Figure 11.27 Fuel cells characteristic curves.

11.7.3 Fuel Cell Modeling

A simple equivalent circuit model of a fuel cell can be as shown in Figure 11.28.

This circuit can represent the relationship between cell current and output terminal voltage where the various parameters – i.e. resistance, capacitance – and internal emf are based on the chemical process within the cell. The model in Figure 11.28 is in fact a dynamic model with equations as shown below.

In this equivalent circuit model, E_{cell} represents the internal potential of the fuel cell. The output of fuel cell voltage V_d is E_{cell} with three types of voltage drop subtracted: activation voltage drop, ohmic voltage drop, and concentration voltage drop. Activation voltage drop can be separated into two parts, one affected by internal temperature and the other caused by the equivalent resistance of activation. Capacitor C is the equivalent capacitor due to the double-layer charging effect. The formula for calculating these parts is as follows [13, 14]:

$$V_d = E_{cell} - V_{act1} - V_{act2} - V_{conc} - R_{ohmic} I \tag{11.18}$$

where

$$E_{cell} = E_{0,cell} + \frac{RT}{2F} \ln\left[P_{H_2}^{*} \left(P_{O_2}^{*} \right)^{0.5} \right] - k_E (T - 298)$$

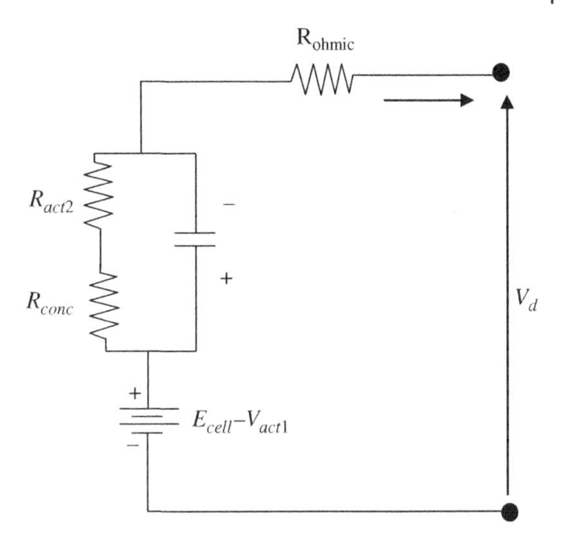

Figure 11.28 Fuel cell electric equivalent circuit model.

and $E_{0,cell}$ is the standard reference potential at standard state (298 K and 1 atm pressure), R is the gas constant, and F is the Faraday constant. P_{H_2} and P_{O_2} represent the partial pressure of H_2 and O_2 activated inside the fuel cell. The asterisk * represents the effective value of the parameter. k_E is an empirical constant in volts per kelvin and T is the actual temperature when the fuel cell is operating. Thus

$$V_{act1} = \eta_0 + a(T - 298) \tag{11.19}$$

where η_0 is the temperature-invariant part of V_{act1} and a is an empirical constant in volts per kelvin. Similarly,

$$V_{act2} = R_{act2} I \tag{11.20}$$

$$V_{conc} = R_{conc} I \tag{11.21}$$

$$V_C = \left(I - C\frac{dV_C}{dt} \right)(R_{act} + R_{conc}) \tag{11.22}$$

At steady state, the static characteristics of the fuel cell are as depicted in the following equation, in which the relationship between the output voltage and current of the fuel cell can be approximately considered as linear by ignoring V_{act2} and V_{conc}, due to the small values of the two parameters at steady state:

$$V_d = E_{cell} - V_{act1} - R_{ohm} I \tag{11.23}$$

At steady state, E_{cell}, V_{act1}, and R_{ohm} can be considered as constants, so the above equation can be modified approximately with the following expression:

$$V_d = K_1 + K_2 I \tag{11.24}$$

where

$$K_1 = E_{cell} - V_{act1} \quad \text{and} \quad K_2 = -R_{ohm}$$

For reference, typical values of the parameters of the fuel cell model are given in Table 11.3.

Table 11.3 Parameters of a typical fuel cell.

$E_{0,cell}$ (V)	R (J/(mol K))	F(C/mol)	P_{H_2} (Pa)	P_{O_2} (Pa)
58.9	8.3143	96 487	1.5	1.0
k_E (V/K)	η_0 (V)	a (V/K)	R_{ohm} (Ω)	T (K)
0.00085	20.145	−0.1373	0.2793	307.7

11.8 Summary and Discussion

This chapter has covered discussion of various important energy storage devices – batteries, ultracapacitors, flywheels – and the power source of fuel cells. An architectural and system level viewpoint has been taken in the discussions since these are important from a practical user standpoint. Practical methodologies to compute battery state of charge (SOC) and state of health (SOH) have been discussed. Battery management system has also been described. Combinations of battery and ultracapacitor, and possibly other storages or sources such as flywheel and fuel cell can help optimize the system architecture to meet a particular load demand. However, while using multiple sources, weight and size – i.e. packaging – can become significant issues, along with the total cost of these devices, and hence can preclude their use. These are practical matters, and proper tradeoff studies are needed to make a choice, and there may not be a unique solution to the issue. Although fuel cell technologies are promising for future clean vehicles, the technology still needs further improvement for portable vehicular applications. So, the idea here is that when deciding about the choice of these components, a holistic view has to be taken before developing the architecture.

References

1 Steen, H. v., (2006), Real-time Evaluation System for State-of-Charge Estimation, doctoral dissertation, University of Twente, Netherlands.
2 http://batteryUniversity.com
3 http://www.mpoweruk.com
4 M. Coleman, Lee, C., Zhu, C., and Hurley, W., "State-of-Charge Determination From EMF Voltage Estimation: Using Impedance, Terminal Voltage, and Current for Lead-Acid and Lithium-Ion Batteries", IEEE Trans. on Ind. Electronics, vol. 54, no. 5, Oct 2007.
5 ASTM E 1049–85 (2005) Rainflow Counting Method, 1987.
6 M. Coleman, Lee, C., Zhu, C., and Hurley, W., "State-of-Health Determination: Two Pulse Load Test for a VRLA Battery. From EMF Voltage Estimation: Using Impedance, Terminal Voltage, and Current for Lead-Acid and Lithium-Ion Batteries", IEEE Trans. on Ind. Electronics, vol. 54, no. 5, Oct 2007.

7 Gregory, B., Ultracapacitor Sizing and Packaging for Cost Effective, IOXUS white paper, http://www.ioxus.com/wp-content/uploads/2013/08/Cost-Effective-Micro-Hybrids-130711.pdf
8 Maxwell Technologies, (2006) Fuel Cells and Ultracapacitors – A Proven Value Proposition Versus Incumbent Technologies, San Diego.
9 Maxwell Technologies, (2011) Ultracapacitors Help P21 to Provide Fuel Cell-Based Backup Power for Telecoms, San Diego.
10 Http://Www.Compositesworld.Com/Blog/Post/Composite-Flywheels-Finally-Picking-Up-Speed.
11 Chan, C.C. and Chau, K.T. (2001) *Modern Electric Vehicle Technology*, Oxford University Press, Oxford.
12 Jefferson, C.M. and Barnard, R.H. (2002) *Hybrid Vehicle Propulsion*, WIT Press, Boston, MA.
13 Wang, C., Nehrir, M.H., and Gao, H. (2006) Control of PEM fuel cell distributed generation systems. IEEE Transactions on Energy Conversion, 21 (2), 586–595.
14 Wang, C. and Nehrir, M.N. (2003) A dynamic model for PEM fuel cells using electrical circuit. Proceedings of 35th North American Power Symposium, October, Rolla, MO, pp. 30–35.

Further Reading

1 V. Pop, H. Berveld, D. Danilov, and P. Regtien, "Battery Management Systems – Accurate State-of-Charge Indication for Battery-Powered Applications", Springer, 2008
2 H. Berveld, W. Kruijt, P. Notten, Battery Management Systems – Design by Modelling, Kluwer Academic Publishers, 2002.
3 P. Sinclair, R. Duke, and S. Round, "An Adaptive Battery Monitoring System for an Electric Vehicle".
4 S. Lee, J. Kim, J. Lee, and B. Cho, "The State and Parameter Estimation of a Li-Ion Battery Using a New OCV-SOC Concept", 2007 IEEE.
5 F. Huet, "A review of impedance measurements for determination of the state-of-charge or state-of-health of secondary batteries", *Journal of Power Sources*, 70, 1998, pg. 59–69.
6 C. Zhu, M. Coleman, and W. Hurley, State of Charge Determination in a Lead-Acid Battery: Combination EMF Estimation and Ah-balance Approach. 2004 IEEE.
7 A. Hande and T. Stuart, "A Selective Boost Equalizer for Series Connected NiMH Battery Packs", 2004 IEEE.
8 S. Sato and A. Kawamura, "A New Estimation Method of State of Charge using Terminal Voltage and Internal Resistance for Lead Acid Battery", 2002 IEEE.
9 G. Corey, "Battery Charging in Float vs. Cycling Environments", Sandia National Labs., SAND2000-1001C.
10 N. Kutkut and D. Divan, "Charge Equalization for Series Connected Battery Strings", IEEE Trans. on Ind. Appl. vol. 31, no. 3, May/June 1995.
11 C. Motloch, et. al., "Implications of NiMH Hysteresis on HEV Battery Testing and Performance", INEEL/CON-02-00052, 19th Int. EV Symp., Aug. 2002.
12 A. Jossen, V. Spaeth, H. Doering, and J. Garche, "Reliable battery operation – a challenge for the battery management system", JPS 84, 1999, pg. 283–86.

13 Rainbow Power Company Ltd. Literature, "Can I increase the life of my battery", July 2004.
14 Electropaedia – Battery & Energy Technologies.
15 S. McCluer, "Wanted: Real world battery life prediction".
16 Maxim Semiconductor Application Note 3958, "Battery fuel gauges: Accurately measuring charge level".
17 V. Marano, S. Onori, Y. Guezennec, and G. Rizzoni, "Lithium-ion Batteries Life Estimation for Plug-in Hybrid Electric Vehicles", 2009 IEEE.
18 http://www.compositesworld.com/blog/post/composite-flywheels-finally-picking-up-speed.

12

Battery Modeling

12.1 Introduction

Chapter 11 contained quite a significant amount of discussion on energy sources, including batteries, ultracapacitors, and flywheels. It also gave details on items related to battery and other energy systems, state of charge, state of health, life cycle and other details with particular reference to lead acid battery. In that chapter, battery and other energy device modeling were also covered to some extent, so this chapter will not deal too much with those items and will directly focus on hybrid and electric vehicle batteries, particularly the nickel metal hydride and lithium based batteries especially with reference to modeling. Although the previous chapter pertained primarily to lead acid batteries, the methodologies and philosophies will remain more or less the same in other battery chemistries covered here.

In this chapter the main focus will be on modeling of batteries. Battery modeling is useful: (1) for calculation of SOC, (2) for calculation of SOH, (3) as part of an overall vehicle model and to study the steady state and dynamic system behavior of the hybrid or electric vehicle as affected by the presence of the battery. This means obtaining the performance characteristics of the overall system with the battery model incorporated.

As is well known, a battery is a highly nonlinear device which displays hysteresis – in other words, not just its present situation or state of the battery, but also how it came to that state or operating point is also of importance. Although mathematical models have been developed for various battery chemistries, they do not necessarily predict the behavior correctly under all situations. Also, a model based on elaborate details of battery chemistry can be quite complicated and can represent a significant computational burden, if it is to be implemented with accuracy. Hence, relatively simpler electrical circuit or impedance based models to represent battery terminal characteristics are used quite often. In addition, sometimes a simple empirical model based on experimental studies of the battery can be used. Therefore there is no single battery model that can predict battery behavior accurately all the time and under all circumstances.

Hybrid Electric Vehicles: Principles and Applications with Practical Perspectives,
Second Edition. Chris Mi and M. Abul Masrur.

12.2 Modeling of Nickel Metal Hydride (NiMH) Battery

12.2.1 Chemistry of an NiMH Battery

In Chapter 11, it was indicated that an NiMH battery can be described by the following overall reversible chemical reaction:

$$MH + NiOOH \Leftrightarrow M + Ni(OH)_2 \tag{12.1}$$

This reaction can be split into two separate reactions, one for the positive electrode and the other for the negative electrode:

Positive electrode:

$$Ni(OH)_2 + OH^- \Leftrightarrow NiOOH + H_2O + e^- \tag{12.2}$$

Negative electrode:

$$M + H_2O + e^- \Leftrightarrow MH + OH^- \tag{12.3}$$

Adding these separate equations in (12.2) and (12.3) immediately gets us back to (12.1).

In the above, M stands for metal, which, in the case of the negative electrode, is an intermetallic compound. This metallic compound is a mixture of certain rare earth materials including some of the following: lanthanum, cerium, neodymium, praseodymium, nickel, cobalt, manganese, and/or aluminum. These compounds basically form some kind of a mixture of metal hydride compounds.

Even though it is possible to go into details of the chemical reaction indicated in (12.1)–(12.3), and use the various design parameters such as size of the battery plates, amount of chemical involved, and so on, to find out the amount of chemical energy and hence SOC within the battery, the process can become very cumbersome. Hence, for the purpose of HEV and EV studies, it is simpler to focus primarily on two-terminal electrical characteristics model here. The same circuit model indicated in Chapter 11 and shown in Figure 12.1, can serve as the basic generic model. The model can be utilized in NiMH batteries, just like the lead acid ones.

The circuit shown in Figure 12.1 can be considered as a modified version of the Thévenin equivalent, to represent the two-terminal behavior of the battery. However, it should be noted that the Thévenin equivalent circuit holds only under the assumption that the superposition theorem holds true, which applies in the case of linear systems. In a battery, the linearity does not hold over its complete operating domain in terms of temperature, SOC, SOH etc., implying that the above resistances and capacitances should be variable quantities, depending on various conditions, particularly temperature, and the current flowing at a particular moment, including the historical path followed in arriving at a particular state. Hence the circuit shown in Figure 12.1 can be considered to be a quasi-linear situation, assuming that the variation is not too much at a particular moment. In addition, since an NiMH battery also displays hysteresis behavior, this can sometimes be incorporated in the above model by including a hysteresis element as shown within the V_{emf}. The variability included in V_{emf} implies that the open circuit voltage is not a constant and depends on various factors including the fact that its change during charging does not follow the same path as during discharge, thus displaying hysteresis.

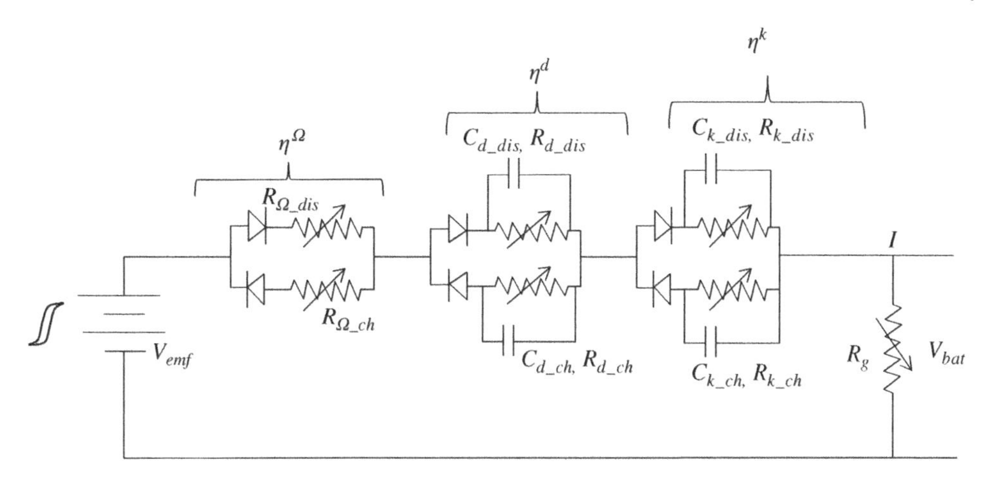

Figure 12.1 A detailed battery model applicable to NiMH including hysteresis.

Sometimes, the circuit in Figure 12.1 can be simplified by retaining only one series branch and the discharge resistance which is across the terminals on the right-hand side. How much detail will be retained in the circuit will depend on the needs. If multiple branches are retained, the corresponding parameters will have to be evaluated. The evaluation of parameters can be done by experimental studies with sufficient data at various operating points. It is also possible to model the battery using artificial neural network (ANN) systems, which is also a quasi-steady state situation. There have been studies on nonlinear dynamic ANN systems as well in the literature, which can be utilized for battery modeling in order to get better fidelity. If a circuit model is used, evaluation of parameters like capacitors will need data capture under transient conditions. Since these nonlinear elements are also temperature dependent, this can be incorporated during the parameter evaluation, or during the training process of an ANN. The ANN training process will also require the availability of as much manufacturer provided data as possible. It was indicated in the previous chapter that to estimate the parameters in a running system while delivering load, and on a real-time basis, it may be necessary to periodically disturb the load current by a tiny percentage of the actual value at a particular moment, and monitor the corresponding change in voltage, or vice versa. This can help evaluate the parameters which contribute to transient behavior, i.e. capacitors.

Although the circuit of Figure 12.1 may be useful for obtaining the two-terminal electrical behavior of the battery, it will not directly provide SOC and SOH information. SOC and SOH is information based on historical data, so they can be evaluated by dynamical storage of information. Although methods involving measurement of open circuit terminal voltage, current, and temperature can be utilized and compared against known curves for the battery to evaluate SOC and SOH, these are not really point functions. Hence to be accurate, it is necessary to have historical data for the purpose, and use the methodology indicated in Chapter 11.

If this bidirectional battery model represented by Figure 12.1 is used, then its parameters will have to be captured for both directions. Since this is a fourth order system (four storage elements at independent locations) with nonlinearities i.e. diodes, nonlinear resistors, V_{emf} etc., its equivalent parameters, dependent on physical parameters, can be estimated by using ANN systems. By monitoring sufficient samples of physical variables of the battery within a sliding window, it is possible to estimate these equivalent parameters. However, it is necessary that the system should not be in a steady state in order to estimate these parameters, such as the capacitances. Since these nonlinear elements are also temperature dependent, this can be incorporated in the training process. The other option – performing an offline experiment, and obtaining data for batteries to create a lookup table over a wide range of temperature, SOC, SOH etc. – will be very tedious and labor intensive, and may need significant computational resources.

12.3 Modeling of Lithium-Ion (Li-Ion) Battery

12.3.1 Chemistry in Li-Ion Battery

In Chapter 11, it was indicated that an Li-ion battery can be described by the following overall reversible chemical reaction:

$$\mathrm{Li}_x\mathrm{C} + \mathrm{Li}_{1-x}\,\mathrm{M}_y\,\mathrm{O}_z \Leftrightarrow \mathrm{C} + \mathrm{LiM}_y\,\mathrm{O}_z \tag{12.4}$$

This reaction can be split into two separate reactions, one for the positive electrode and the other for the negative electrode:

During discharge:

$$\text{Cathode}: \mathrm{Li}_{1-x}\mathrm{CoO}_2 + \mathrm{xLi}^+ + \mathrm{xe}^- = \mathrm{LiCoO}_2 \tag{12.5}$$

$$\text{Anode}: \mathrm{Li}_x\mathrm{C}_6 - \mathrm{xe}^- = 6\mathrm{C} + \mathrm{xLi}^+ \tag{12.6}$$

This reaction is reversible during charging.

Adding the above two equations, we get during discharge the following:

$$\mathrm{Li}_{1-x}\mathrm{CoO}_2 + \mathrm{Li}_x\mathrm{C}_6 \Leftrightarrow \mathrm{LiCoO}_2 + 6\mathrm{C} \tag{12.7}$$

Once again, the same statements indicated for NiMH batteries relating to the path of chemical reaction to find out SOC are true for Li-ion. There are various models that have been reported in the literature for various purposes. Some of these focus on SOC, and some focus on SOH. For our purpose, we will focus on an equivalent circuit model that mimics the behavior of the battery when the model is connected to the rest of the system. For this we can use the reference paper [1], in which a circuit with a series resistance and a resistance in parallel with a capacitor have been used. This model can be actually easily be derived by using the same generic model template in Figure 12.1 (as also used in Chapter 11).

As noted in Chapter 11, Li-ion batteries do not have the "memory effect" (like hysteresis at different maximum levels) that causes other rechargeable batteries to lose their maximum charge level when repeatedly charged and discharged to capacities other than the full capacity. Hence a hysteresis element in the circuit model can be omitted here.

12.4 Parameter Estimation for Battery Models

The discussion in the previous sections has indicated that it is possible to use a generic template like Figure 12.1 for the battery models, regardless of the chemistry. For NiMH though, a memory element can be included. The degree of sophistication of the model can depend on the number of elements included in the circuit models, i.e. resistance, capacitances, etc. It is also necessary to include the self-discharge resistance for the sake of completeness. Regardless of everything, the battery model development eventually depends on the estimation of the resistance and capacitance values in the circuit model and validating those against experimental results. Once that is done, the model can then be used in an overall system model environment for studying the performance of the system. SOC and SOH estimation then becomes part of the model variable collection while using the circuit model in Figure 12.1 in a physical system model environment, and use those variables for estimating the SOC and SOH as indicated in Chapter 11.

For the purpose of parameter estimation, basically the same circuit in Figure 12.1 with some reorganization and simplification can be used [1]. This is shown in Figure 12.2.

The method used to find E and R_0 is based on simple circuit theory. Initially it is assumed that all the capacitor voltages are zero. This assumption is true if the battery was rested for long enough time with load open. This implies that even if the capacitors had any charge, this would have been discharged through the corresponding resistors. Thereafter two measurements are needed. The first is with load open. The monitored voltage will then be equal to E or $V(0^-)$. Similarly the next experiment is done with the load, by placing some known constant voltage source ($V0^+$) across the terminals, and noting the value of the current (I). Since the initial capacitor voltages are zero, R_0 will be then given by:

$$R_0 = (V(0^-) - V(0^+))/I \tag{12.8}$$

In addition to the above, it is also possible to have a self-discharge resistance R_d across the load (and in parallel to the load) which indicates continuous discharge current (I_d)

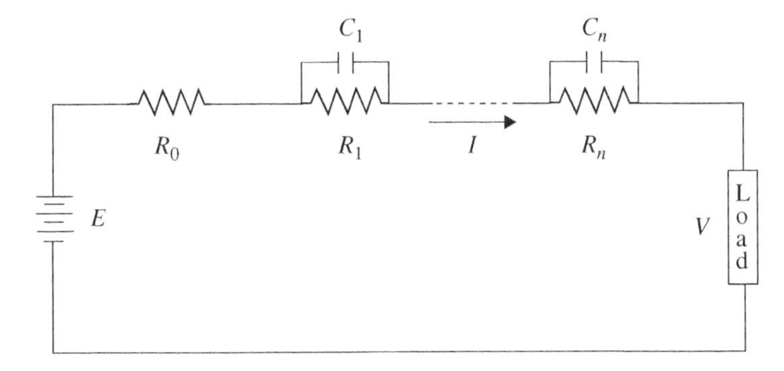

Figure 12.2 Generic battery model.

from the battery even when the load is open. In this case the total current measured at the terminals can be used to compute R_d and R_0 as follows:

$$R_0 = (V(0^-) - V(0^+))/(I_t - I_d) \tag{12.9}$$

where I_t is the current flowing through the terminals.

$$I_d = V(0^+)/R_d \tag{12.10}$$

$$I_t = \text{terminal current} = I - I_d \tag{12.11}$$

From Equations (12.9–12.11) we see that it is not possible to find the four variables R_0, I, I_d, R_d, based on the measured values of $V(0^-), V(0^+)$, and I_t. This situation can be resolved by doing one additional test – as a possible option, the momentary short circuit test across the battery and measuring the short circuit current, which can immediately give the value of $R_0 = V(0^-)/I_{sc}$, where I_{sc} is the measured value of the short circuit current through the terminals. Knowing R_0, we can immediately find I_d, and hence R_d can be evaluated from the above equations.

The remaining parameters – R_1, R_2, ... R_n, and C_1, C_2, ... C_n – can be determined by collecting the voltage V and current I data at the terminals, over a sufficient time duration at various intervals. The reference paper [1] suggests that by controlled constant current through the terminals, while collecting data, the method can be mathematically simplified. This is discussed below to some extent.

The simpler method to implement the above can be used by having a constant current load at the two output terminals. In that case, the terminal voltage $v(t)$ can be expressed as [1]

$$v(t) = E - R_0 I - \sum_{k=1}^{n} R_k I(1 - \exp[-t/(R_k C_k)]) \tag{12.12}$$

The basic idea here is that a constant current I is injected through the output terminals and $v(t)$ across the terminals is measured at different time steps. This leads to a set of equations with parameter R_k and C_k with k running from 1 to n. The value of n, which determines the degree of the equation and hence the complexity, is a matter of choice, based on the accuracy desired from the model. One reference paper [2] suggests that either too small, or too large a degree for n, can lead to problems, where the data obtained from the model cannot be well validated by experiment. According to that reference paper, $n = 3$ seems to be a reasonably good choice, and higher values can lead to more complexity and difficulty in optimizing the parameters for the best fit. Incidentally, Hu et al. [1] used a third-order model (which is consistent with [2]) and proceeded as follows.

$$v(t) = E - R_0 I - \sum_{k=1}^{3} R_k I(1 - \exp[-t/(R_k C_k)]) \tag{12.13}$$

To find the six parameters for R_k, C_k, with $k = 1$ to 3, we can choose six equally spaced time instants for convenience (although they do not have to be equally spaced) and form a set of equations as follows.

$t_k = kT$, where k = 1 to 6 for the equally spaced time intervals.

Define:

$$d_j = \exp[-T/R_j C_j], j = 1 \text{ to } 3 \tag{12.14}$$

$$v(kT) = E - R_0 I - \sum_{j=1}^{3} R_j I(1 - d_j^{\,k}), \text{ where } d_j^{\,k} = (d_j)^k, \text{ with } k = 1 \text{ to } 6. \tag{12.15}$$

In the above, it is assumed that E and R_0 have already been found by previously described open and short circuit methods in the Thévenin circuit.

Hence, in Equation (12.15) there will be six values of $v(kT)$ by measurements, leading to six separate equations. Although the six equations are nonlinear with higher powers, in principle those can be solved for the three pairs of unknown quantities, i.e. R_j, C_j, the resistances and capacitances in the model. Although such equations may appear impossible to solve by any analytical method, reference [1] has indicated that by making some changes in variables and reorganization of the equation set, a closed form solution can be found. This is apparently possible for $j = 1$ to 3. However, if a model order of higher than three is chosen, a numerical method has to be resorted to. The details of this numerical method have also been indicated in [1].

The method shown above is actually generic enough and can be adopted for use in batteries with different chemistries, e.g. lead acid, Li-ion, Li-polymer, NiMH, and also, in fact, for fuel cells.

Note that, in the above discussion, it is assumed that the operating conditions (SOC, SOH, temperature, load, history of charge/discharge) over a short period of time were not changing. However, if the model is to be valid and used over wider range of those working conditions, then the model will become very complex, and the parameter set will need to be evaluated under different SOC, SOH, temperature etc., and in that case, it may be necessary to create a multidimensional lookup table to use the model. This can necessitate significant computational resources.

12.5 Example Case of Using Battery Model in an EV System

In this section, a study [3] is described which was co-authored by one of the co-authors (C. Mi) of this book.

Figure 12.3 shows a typical EV system. The study involves the characteristics of battery packs with parallel-connected lithium-ion battery cells. Here, the problem of the influence of cell inconsistency in parallel-connected cells was studied, using a group of different degraded lithium-ion battery cells to build various battery packs. An equivalent circuit model was used to simulate the operation of the parallel-connected packs. The experimental results and simulation indicate that, with different degraded cells in parallel, there could be capacity loss and a large difference in discharge current, which may cause further accelerated degradation and more serious inconsistency problem.

The equivalent circuit model is given in Figure 12.4 which is the same equivalent circuit shown earlier in this chapter (Figures 12.1 and 12.2), but with much reduced order.

In Figure 12.4, V_{oc} is the open circuit voltage (OCV), which depends on the state of charge (SOC) of the battery, as shown in Figure 12.3. The parameters of the model are R_t, which represents the ohmic resistance. R_p and C_p arise due to the polarization effect of the battery and as noted earlier in this chapter, where C_p represents the transient effect under a dynamic operation. These internal parameters very much depend on other variables such as SOC, SOH, and temperature. If these conditions are assumed to

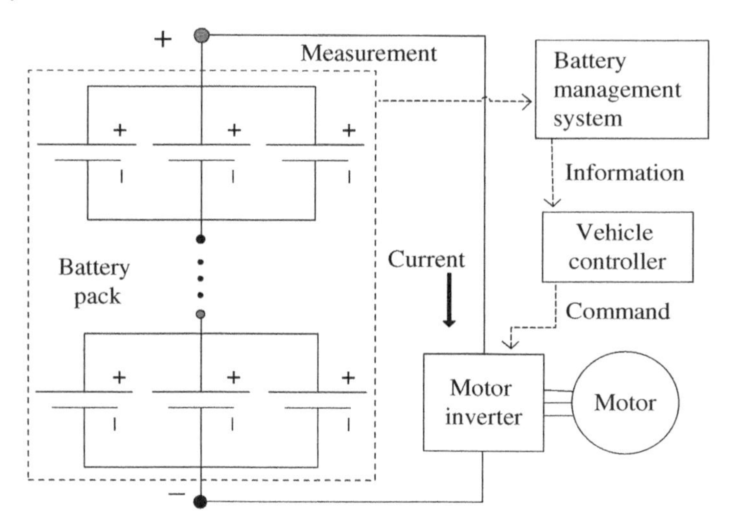

Figure 12.3 Typical battery powered electric vehicle drive system.

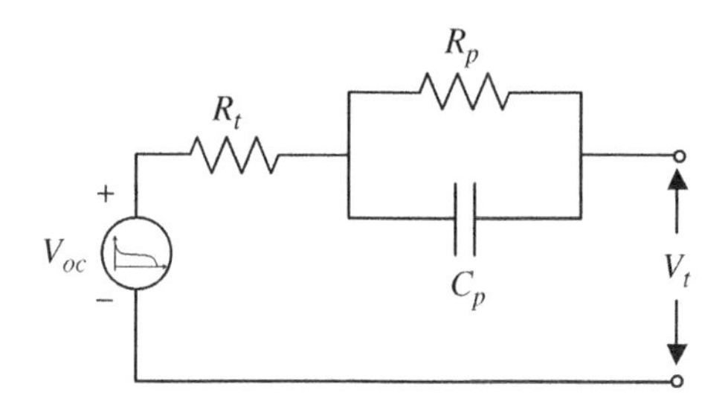

Figure 12.4 First-order equivalent circuit model of the battery.

be known at a particular operating point, i.e. under a quasi-static condition, then the mathematical procedure indicated in the previous section could be used to estimate those parameters.

Figure 12.5 shows the relationship between OCV and SOC. This relationship can be obtained by measurement and can be used at the left-hand side in the model shown in Figure 12.4 for V_{oc}.

Once the parameters have been evaluated, the whole battery package can then be modeled by simply combining a number of individual battery models shown in Figure 12.4 with proper series-parallel connection, thus creating the complete battery pack.

The model thus created in the reference paper is given in Figure 12.6.

The model has two RC network in parallel, along with the load represented by a current source. The currents i_{L1} and i_{L2} in Figure 12.6 depend on the internal parameters R_{t1} and R_{t2}, and polarization capacitances C_{p1} and C_{p2}, resistances R_{p1} and R_{p2}, and the open circuit voltage OCV. As indicated earlier in this chapter, these parameters are functions of SOC, SOH, temperature, etc. Assuming a quasi-static situation, these parameters were obtained and some examples are shown below in Figures 12.7–12.9.

In the above figures, the various curves correspond to the various cells in the package, although Figure 12.6 shows only two cells for the purpose of illustration. Figure 12.10

Figure 12.5 OCV versus SOC.

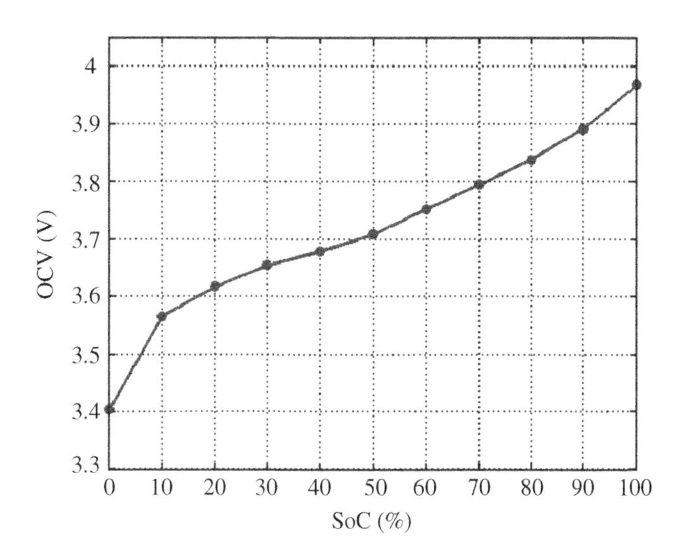

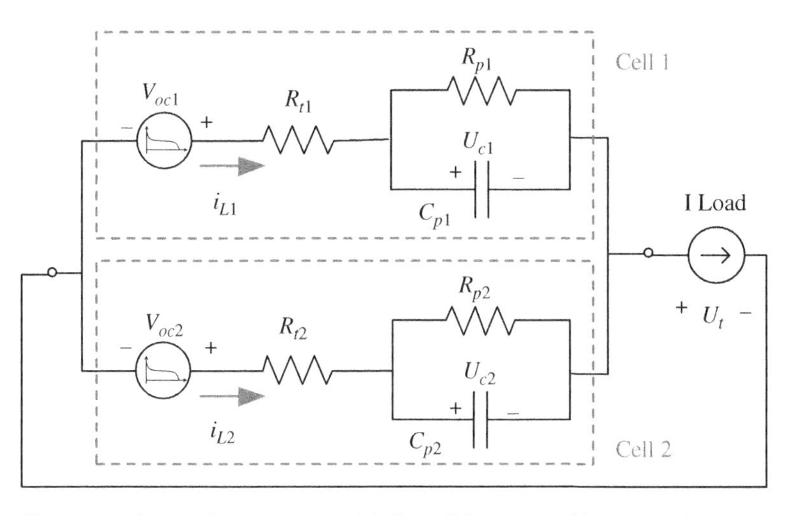

Figure 12.6 Lumped parameter model of parallel-connected battery pack.

Figure 12.7 Ohmic resistance R_t identification result.

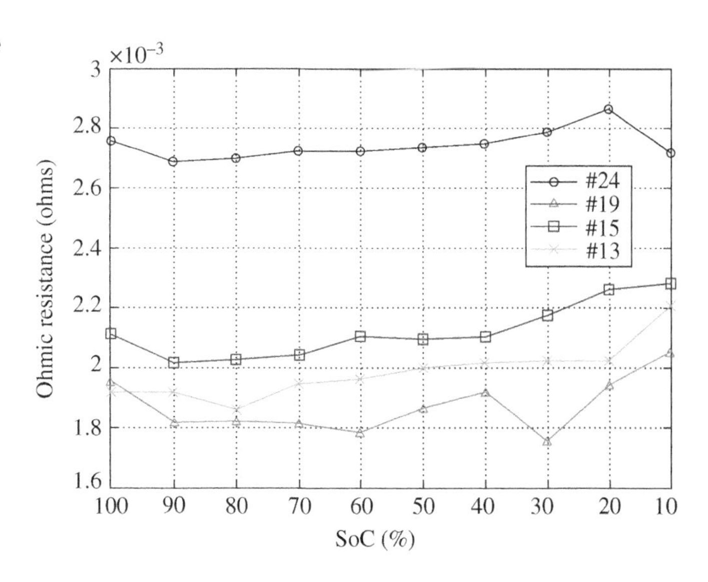

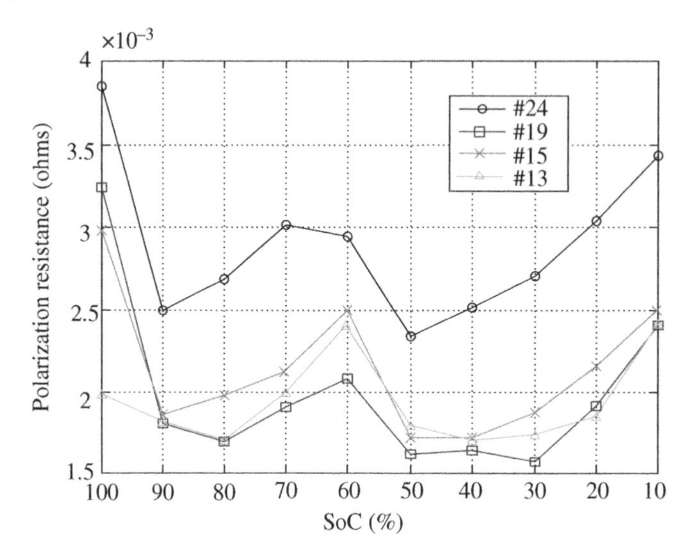

Figure 12.8 Ohmic resistance R_p identification result.

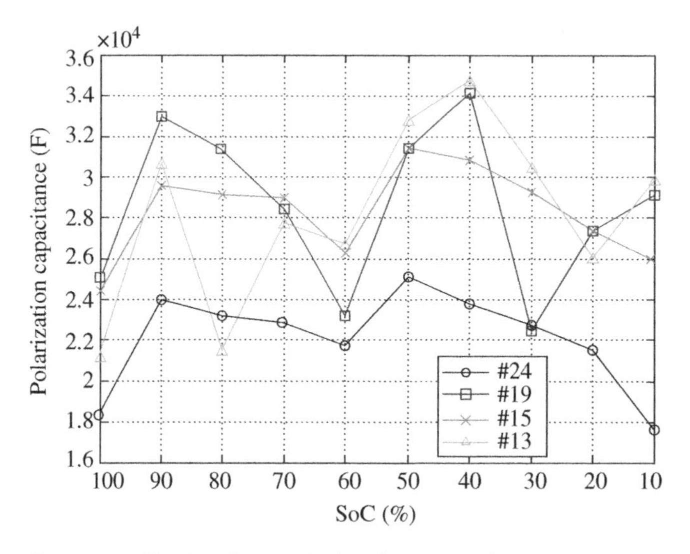

Figure 12.9 Ohmic resistance C_p identification result.

shows the MATLAB-Simulink implementation diagram of the battery model with two cells, utilizing the parameters estimated by using mathematical equations and experimental data.

Findings from the experimental results leading to parameter estimation (Figures 12.7–12.9) indicated that cell inconsistency has strong influence on ohmic resistance and polarization capacitance, and is also influenced by temperature, SOC and SOH of the battery. Consequently, these items influence the model parameters and hence the outcome of the model itself when used in an overall system model environment.

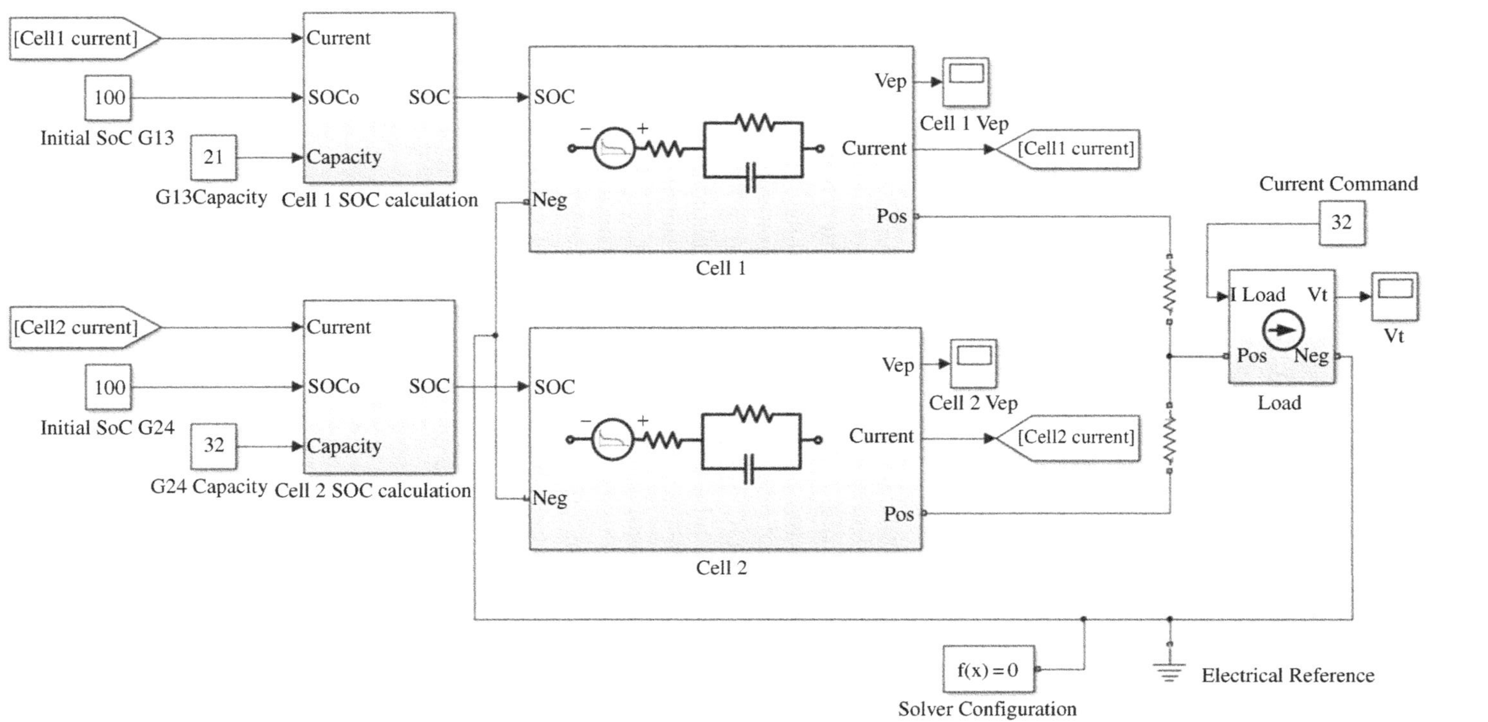

Figure 12.10 MATLAB Simulink implementation of the model.

12.6 Summary and Observations on Modeling and Simulation for Batteries

Based on what has been discussed so far in this chapter, the following observations can be made.

Firstly, there are several types of models that are applicable to batteries. These can be broadly classified as: (a) Electrochemical model, based on complex nonlinear algebraic and differential equations. These are difficult to implement due to the complexity of the chemical system, if accurate representation is needed. (b) Computational or artificial intelligence type of models e.g. using artificial neural networks, which are generally based on learning over wide operating domains, although learning systems for transient processes also exist. These can be used to find SOC. (c) Analytical models, which are simplified electrochemical system models, e.g. using Peukert's equation, kinetic battery model (KiBaM), or diffusion model. These can be used to track SOC. (d) Electric circuit model based on impedance. Some of these can be used to track SOC. This type of model is suitable for integration within an overall system-level simulation. In addition to the above, there is the stochastic model, which describes the battery as a Markov process with probabilities described in terms of parameters that are dependent on the physical characteristics of the chemical cell process. It has been indicated in references [4] that such models give good qualitative description of the behavior of a lithium-ion battery under pulsed discharge. The model does not handle load profiles with varying discharge currents and various battery nonlinearities.

The type of model which is to be used in a particular situation will depend upon the application and computational resources available. For example some models are well suited for providing better SOC. Some models are better in terms of handling short-term transient effects, while others may be more suitable for long-term effects like aging. In general, the electric circuit models are more suitable for situations where simulations at the bigger system level have to be integrated with the battery model.

Sometimes it may be better to use hybrid models [4]. This implies that certain items within the battery may be computed by using one type of model, and then certain other items can be modeled by another method. An example of such a situation is indicated by Figure 12.10.

In Figure 12.11, the SOC can be modeled by an accurate coulomb counting method alternated with battery emf based computation indicated in Chapter 11. This SOC can

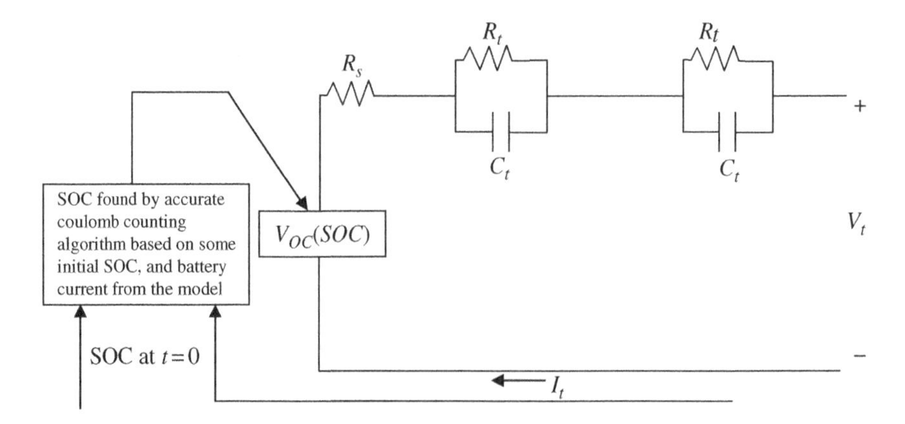

Figure 12.11 Hybrid battery model.

then be utilized as input to the emf of the electric circuit model. Such hybrid models can lead to better accuracy of the overall model results.

Hence it can be concluded that no single model can be best for every application and judgment has to be based on the needs, computational resource available, and accuracy demanded.

References

1 T. Hu, B. Zanchi, and J. Zhao, "Determining battery parameters by simple algebraic method", 2011 American Control Conference, San Francisco, CA, USA, June 29–July 01, 2011.
2 R. Jackey, M. Saginaw, P. Sanghvi, J. Gazzarri, T. Huria, and M. Ceraolo, "Battery Model Parameter Estimation Using a Layered Technique: An Example Using a Lithium Iron Phosphate Cell", SAE Paper # 2013-01-1547.
3 X. Gong, R. Xiong, and C. Mi, "A Data-Driven Bias Correction Method Based Lithium-ion Battery Modeling Approach for Electric Vehicles Application", 2014 IEEE Transportation Electrification Conference and Expo (ITEC).
4 T. Kim, "A Hybrid Battery Model Capable of Capturing Dynamic Circuit Characteristics and Nonlinear Capacity Effects", MSc Thesis, University of Nebraska, 2012.

Further Reading

1 V. Pop, H. Berveld, D. Danilov, and P. Regtien, "Battery Management Systems – Accurate State-of-Charge Indication for Battery-Powered Applications", Springer, 2008.
2 S. Lee, J. Kim, J. Lee, and B. Cho, "The State and Parameter Estimation of a Li-Ion Battery Using a New OCV-SOC Concept", 2007 IEEE.
3 F. Huet, "A review of impedance measurements for determination of the state-of-charge or state-of-health of secondary batteries", Journal of Power Sources, 70, 1998, pp. 59–69.
4 C. Zhu, M. Coleman, and W. Hurley, State of Charge Determination in a Lead-Acid Battery: Combination EMF Estimation and Ah-balance Approach. 2004 IEEE.
5 C. Motloch, et. al., "Implications of NiMH Hysteresis on HEV Battery Testing and Performance", INEEL/CON-02-00052, 19th Int. EV Symp., Aug. 2002.
6 Electropaedia – Battery & Energy Technologies.
7 H. van Steen, (2006), Real-time Evaluation System for State-of-Charge Estimation, doctoral dissertation, University of Twente, Netherlands.
8 http://batteryUniversity.com
9 http://www.mpoweruk.com
10 M. Coleman, C. Lee, C. Zhu, W. Hurley, "State-of-Charge Determination From EMF Voltage Estimation: Using Impedance, Terminal Voltage, and Current for Lead-Acid and Lithium-Ion Batteries", IEEE Trans. on Ind. Electronics, vol. 54, no. 5, Oct 2007.
11 C.C. Chan and K.T. Chau (2001) *Modern Electric Vehicle Technology*, Oxford University Press, Oxford.
12 C.M. Jefferson and R.H. Barnard (2002) *Hybrid Vehicle Propulsion*, WIT Press, Boston, MA.

13

EV and PHEV Battery Charger Design[1]

In this chapter, efficiency oriented design considerations are discussed in detail on the operation mode analysis of the LLC converter including the characteristics of charging profiles. The mode boundaries and distribution are obtained from the precise time domain model. The operation modes featuring both-side soft-switching capability are identified to design the operating trace of the charging process. Then the design constraints for achieving soft-switching with the load varying from zero up to the maximum are discussed. Finally, a charging trajectory design methodology is proposed and validated through experiments on a prototype converting 390 V from the DC power source to the battery emulator in the range of 250–450 V at 6.6 kW with a peak efficiency of 97.96%.

13.1 Introduction

In today's PHEVs and EVs, an onboard charger is installed to charge the high power lithium-ion battery pack through the utility power [1–6]. According to a thorough survey, the most common EV/PHEV charger architecture consists of a boost type AC–DC converter for active power factor correction (PFC) and an isolated DC–DC converter as the second stage, as shown in Figure 13.1(a) [7–10]. The characteristic of this type of charger is mainly dependent on the DC–DC stage, since the output voltage and current are regulated in this stage. Therefore, an efficient and compact isolated DC–DC converter is one of the most important components for EV and PHEV battery chargers. The LLC resonant converter with soft-switching capability for a wide operating range is considered to be a favorable topology to achieve both high efficiency and high power density [11].

A typical schematic of a full-bridge LLC resonant DC–DC converter used in EV/PHEV charger applications is shown in Figure 13.1(b). The resonant tank consists of three reactive components: L_r and C_r in series, and L_m in parallel with the primary of an n:1 ideal transformer. C_r denotes the resonant capacitor, L_m is the magnetizing inductance, and L_r

1 Reprint, copyright IEEE. J. Deng; C.C. Mi; R. Ma; S. Li, "Design of LLC Resonant Converters Based on Operation-Mode Analysis for Level Two PHEV Battery Chargers," in IEEE/ASME Transactions on Mechatronics, vol. 20, no. 4, pp. 1595–1606, Aug. 2015. doi: 10.1109/TMECH.2014.2349791.

Hybrid Electric Vehicles: Principles and Applications with Practical Perspectives,
Second Edition. Chris Mi and M. Abul Masrur.

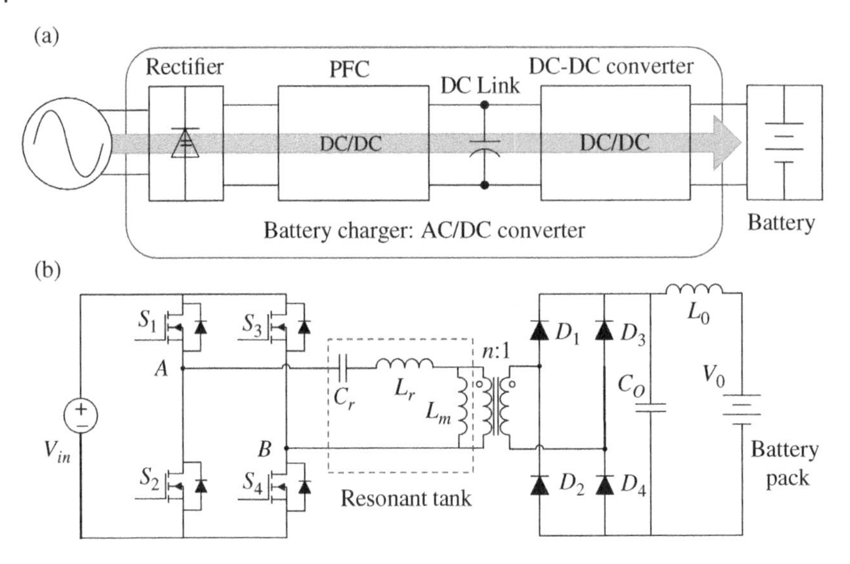

Figure 13.1 Full-bridge LLC resonant converter. (a) Typical EV/PHEV Charger system. (b) LLC DC–DC converter stage for the EV/PHE Charger.

is the leakage inductance reflected in the primary side. The LLC converter modifies the gain characteristics of a series resonant converter (SRC) by utilizing the transformer magnetizing inductance to form multiple resonant stages. It greatly improves the light load efficiency and allows the boost mode operation. However, its multiple resonant stages and various operation modes make it difficult to design [12].

Many design methodologies have been proposed for this converter in the past few decades. Exact analysis of LLC resonant converters [13] ensures accuracy, but cannot be easily used to get a handy design procedure due to the complexity of the model. First harmonic approximation (FHA) analysis [14] gives quite accurate results for operating points at and above the resonance frequency of the resonant tank [15], and has been widely used in constant output voltage applications where the LLC converter is designed to work at resonance at nominal condition. Designing a wide output range LLC resonant converter based on the FHA method is investigated in [16]. The expanded range is mainly designed at frequencies above the resonant frequency. The secondary rectifiers' zero current switching (ZCS) operation appears only in light-load condition in this region, which causes additional diode reverse losses compared to in the region below resonance [17]. The accuracy of the FHA is barely satisfactory for optimal design in the below-resonance region. Optimal design methods are developed based on operation mode analysis in [18, 19], and these approaches can give fairly good design results, but call for utilizing sophisticated calculation tools. A simple yet accurate design oriented model and step-by-step design procedure that ensures most of the merits of LLC converters is presented in [20]. As an extension of [20], another design methodology for optimizing efficiency by minimizing the reactive energy is presented in [21], but it involves solving nonlinear equations. Recently, an efficiency-oriented and straightforward design flow without a recursive loop was developed in [22], based on the operation mode analysis method presented in [17].

In the literature mentioned, the load is usually assumed to be a pure constant resistor and the output is usually fixed. The wide voltage gain range is normally required to resist the input variation or to meet the holdup time [11,23]. However, the design requirements of an LLC converter for a high voltage lithium-ion battery charger can be distinguished from those of the aforementioned applications. First of all, the nonlinear load i–v characteristics related to the charging profile exist in the design of a resonant converter for battery charger applications. Second, not only does the voltage ripple coming from the front-end stage need to be resisted, but the output voltage also varies significantly throughout the whole charging process. In addition, the LLC converter should be able to handle a wide adjustable regulated output voltage range, even when the load current varies. Third, the charge process for a lithium-ion battery usually contains several stages, and the output voltage and the load power change significantly during the whole charging process. It may go through different combinations of no load, full load, and light load conditions according to the control of the battery management system. Hence, high efficiency should be maintained in different load conditions. As a result, it is inappropriate to pick just one load condition out of the whole charging process to be the nominal condition to be targeted, which is normally done in the resistive load applications. The whole operating trajectory has to be taken into account for an optimal design.

In this chapter, the time domain model of the LLC converter is introduced first. The operation mode characteristics and the mode distribution are discussed and summarized in Section 13.2. The operation modes that promise both-side soft-switching are identified to be targeted. Based on the charging profile, the whole charging process is projected to the operation mode distribution domain as operating trajectories in Section 13.3. The key parameters and constraints that lead to the desirable charging trajectory are investigated in Section 13.4. Then the design procedure is proposed in Section 13.5. Experimental results are presented in Section 13.6, and the conclusions are drawn in Section 13.7.

13.2 Main Features of the LLC Resonant Charger

13.2.1 Analysis in the Time Domain

Because the switching action exists in both the input switches and the output rectifier bridge, the LLC resonant converter represents a nonlinear and time-variant system, which makes the analysis complicated. However, by dividing the circuit operation into different subintervals, the converter can be described between transitions [21]. Under a symmetrical condition of the resonant converter, the steady state can be characterized by a half-period of operation. During the positive half switching cycle in which S_1, S_2 are turned ON, the converter can be described by three equivalent circuits as shown in Figure 13.2. The capital letters O, P, N are used to denote the three different subintervals, which are characterized by the voltage polarity (off, positive, and negative) across the magnetizing inductor L_m [17]. The parameter n is the transformer turns ratio.

The subinterval O happens when the output is blocked by the diode rectifier, which forces the magnetizing inductor L_m to participate in the series tank resonance. Following

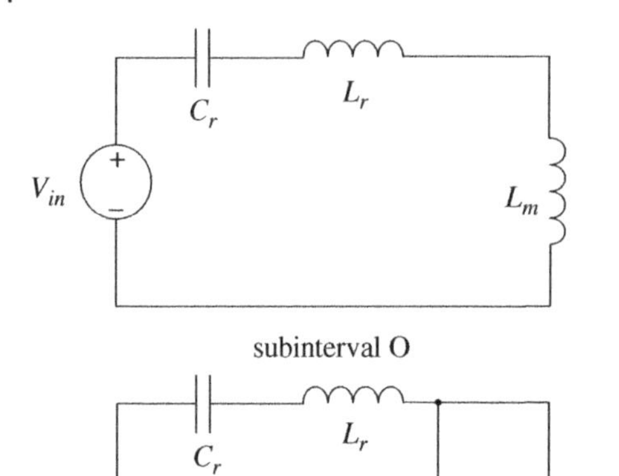

Figure 13.2 Intervals of the LLC converter.

Table 13.1 Abbreviations and normalizations.

Circuit variable	Symbol	Normalized variable
Resonant frequency	$\omega_0 = 1/\sqrt{L_r C_r} = 2\pi f_0$	–
Characteristic impedance	$Z_0 = \sqrt{L_r / C_r}$	–
Inductance ratio	$l = L_r / L_m$	–
Voltage gain	$M = nV_o / V_{in}$	–
Time	t	$\theta = \omega_0 t$
Second resonant frequency	$\omega_1 = 1/\sqrt{(L_r + L_m)C_r}$	$k = \omega_1 / \omega_0$
Switching frequency	$f_s = 1/T_s$	$f_n = f_s / f_0$
Half period	$1/2f_s$	$\gamma = \omega_0 / 2f_s = \pi / F_n$
Resonant capacitor voltage	$v_{Cr}(t)$	$m_{Cr}(\theta) = v_{Cr}(t)/V_{in}$
Series resonant inductor current	$i_{Lr}(t)$	$j_{Lr}(\theta) = i_{Lr}(t)Z_0 / V_{in}$
Magnetizing inductor current	$i_{Lm}(t)$	$j_{Lm}(\theta) = i_{Lm}(t)Z_0 / V_{in}$
Magnetizing inductor voltage	$v_{Lm}(t)$	$m_{Lm}(\theta) = v_{Lm}(t)/V_{in}$
Reflected output voltage	nV_o	M

the basic principles of circuit theory together with the normalizations in Table 13.1, the normalized equations describing the resonant states are [13, 19, 24]

$$m_{Cr,O}(\theta) = \left[m_{Cr,O}(0) - 1\right]\cos(k\theta) + \frac{1}{k} j_{Lr,O}(0)\sin(k\theta) + 1 \tag{13.1}$$

$$j_{Lr,O}(\theta) = k\left[1 - m_{Cr,O}(0)\right]\sin(k\theta) + j_{Lr,O}(0)\cos(k\theta) \tag{13.2}$$

$$j_{Lm,O}(\theta) = j_{Lr,O}(\theta) \tag{13.3}$$

$$m_{Lm,O}(\theta) = \frac{1}{1+l}\left[1 - m_{Cr,O}(\theta)\right] \tag{13.4}$$

with unknown starting values $m_{Cr,O}(0)$ and $j_{Lr0,O}(0)$. Note that in charger applications, the input voltage ripple is negligible compared to the wide output voltage range. So the input voltage should be chosen as the base voltage in the normalization rather than the widely changed output, which is different in resistive load applications.

For the subintervals P and N, the rectifier bridge will conduct. The voltage across the magnetizing inductor is positive clamped and negative clamped respectively, $m_{Lm}(\theta) = \pm M$. Therefore, the normalized waveforms for these subintervals can be identified as

$$\begin{aligned} m_{Cr,P/N}(\theta) = (1 \mp M)(1 - \cos\theta) + m_{Cr,P/N}(0)\cos\theta \\ + j_{Lr,P/N}(0)\sin\theta \end{aligned} \tag{13.5}$$

$$\begin{aligned} j_{Lr,P/N}(\theta) = \left[1 \mp M - m_{Cr,P/N}(0)\right]\sin\theta \\ + j_{Lr,P/N}(0)\cos\theta \end{aligned} \tag{13.6}$$

$$j_{Lm,P/N}(\theta) = j_{Lm,P/N}(0) \pm Ml\theta \tag{13.7}$$

$$m_{Lm,P/N}(\theta) = \pm M \tag{13.8}$$

where $m_{Cr,P}(0)$, $j_{Lr,P}(0)$, and $j_{Lm,P}(0)$ are used to denote the unknown initial conditions of subinterval P, while $m_{Cr,N}(0)$, $j_{Lr,N}(0)$, and $j_{Lm,N}(0)$ are for the subinterval N.

13.2.2 Operation Modes and Distribution Analysis

According to analysis and simulation, a total of nine operation modes can be found by combining the intervals in different sequences. The operation modes can be named by the appearance order in which the subintervals occur in a half-period. For instance, PO mode means that the subinterval P is $\theta \in [0,\alpha)$ and is followed by subinterval O for the rest of the half cycle: $\theta \in [\alpha,\gamma)$. All the operation modes and their characteristics are summarized in Table 13.2. It can be seen that the main power interval P occurs in every half-period except for the O mode under zero-load condition. So O mode is regarded as the cutoff mode since no power is delivered to the output [13]. There are six major operation modes (PN, PON, PO, OPO, NP, and NOP) that can be observed above the second resonant frequency when the operating frequency and load condition vary. Meanwhile, there are another two special modes that happen only under the specific conditions, which are P mode at resonant frequency and OP mode at the boundary between NOP and OPO mode [17].

Each operation mode is constrained by several conditions. First, as the capacitor voltage and inductor current should be continuous, their values should be the same at the joints of adjacent subintervals. So it is intuitive that the continuity constraint conditions should be met. Second, by symmetry, the end values of the capacitor voltage and inductor current in the last subinterval should be opposite to their initial values during the first subinterval in steady state. Third, the magnetizing current and resonant tank

Table 13.2 Operation modes of LLC resonant converter.

	$F_n < 1$					$F_n=1$		$F_n > 1$				
θ	Mode PN	Mode PON	Mode PO	Mode OPO	Mode O	Mode P	Mode O	Mode NP	Mode NOP	Mode OP	Mode OPO	Mode O
$[0, \alpha)$	P	P	P	O	O	P	O	N	N	O	O	O
$[\alpha, \beta)$	N	O	O	P	$(\alpha=\beta)$	$(\alpha=\beta)$	$(\alpha=\beta)$	P	O	P	P	$(\alpha=\beta)$
$[\beta, \gamma]$	$(\beta=\gamma)$	N	$(\beta=\gamma)$	O	$(\beta=\gamma)$	$(\beta=\gamma)$	$(\beta=\gamma)$	$(\beta=\gamma)$	P	$(\beta=\gamma)$	O	$(\beta=\gamma)$
Primary switches	ZVS/ ZCS	ZVS/ ZCS	ZVS	ZVS	ZVS	ZVS	ZVS	ZVS	ZVS	ZVS	ZVS	ZVS
Secondary rectifier	Hard switching	ZCS	ZCS	ZCS	OFF	ZCS	OFF	Hard switching	Hard switching	ZCS	ZCS	OFF
Conduction continuity	CCM	DCM	DCM	DCM	OFF	CCM	OFF	CCM	DCM	DCM	DCM	OFF
Voltage gain	Buck and boost mode		Boost ($M>1$) mode		Cutoff mode	Unity ($M=1$) mode	Boost mode		Buck ($M<1$) mode			Cutoff mode
Load condition			Heavy → Light			Independent	Light			Heavy → Light		
Unknowns at given F_n, l and P_n						$[j_{Lr}(0), m_{Cr}(0), \alpha, \beta, M]$						

Table 13.3 General constraint conditions of operation modes.

Constraints	Expressions
Continuity conditions	$m_{Cr,X}(\theta_{X,end}) = m_{Cr,Y}(0)$ $j_{Lr,X}(\theta_{X,end}) = j_{Lr,Y}(0)$ $j_{Lm,X}(\theta_{X,end}) = j_{Lm,Y}(0)$
Symmetry conditions	$m_{Cr,B}(0) + m_{Cr,E}(\gamma) = 0$ $j_{Lr,B}(0) + j_{Lr,E}(\gamma) = 0$ $j_{Lm,B}(0) + j_{Lm,E}(\gamma) = 0$
Inductor current condition	$j_{Lr,P/N}(\theta_{P/N,end}) - j_{Lm,P/N}(\theta_{P/N,end}) = 0$

current are the same at the end of subinterval P or N when entering the next subinterval. In addition, the voltage across the magnetizing inductor should at least resonate to the output voltage at the end of subinterval O if it is followed by subinterval P or N [15]. If X, Y denote the two different adjacent subintervals, and B, E indicate the start subinterval and the end subinterval respectively in a half cycle, the aforementioned constraint conditions can be listed as in Table 13.3.

Since the load is disconnected from the resonant tank in subinterval O, the energy can be delivered only in subintervals P and N. Therefore, the normalized output power is derived as [13]

$$P_n = \frac{nV_o\overline{I_o}}{(V_{in})^2/Z_0} = M\overline{j_o} = \frac{M}{\gamma}\int_{\theta_{P,0}}^{\theta_{P,end}}\left[j_{Lr,P}(\theta) - j_{Lm,P}(\theta)\right]d\theta + \frac{M}{\gamma}\int_{\theta_{N,0}}^{\theta_{N,end}}\left[j_{Lm,N}(\theta) - j_{Lr,N}(\theta)\right]d\theta \tag{13.9}$$

Generally, the continuity conditions are used with Equation (13.1–13.8) to get the intermediate values $x(\alpha)$ and $x(\beta)$ from the preceding subintervals. The normalized phase angles α and β are used to denote the switching moment of adjacent subintervals. Then they are substituted into the following subinterval as its initial values. Lastly, the final values $x(\gamma)$ at the half-period of an operation mode can be expressed by the corresponding initial values $x(0)$ of this mode. By applying the symmetry and inductor current conditions at AC given P_n defined as (13.9), knowing f_n and l, a sufficient number of equations can be found to solve for the unknown variables as listed in Table 13.1. However, the analytical solutions can only be found under cutoff mode and continuous conduction modes (CCMs), which include O, P, PN, and NP. The discontinuous conduction modes (DCMs), which have been summarized in Table 13.1, involve nonlinear equation solving. Hence, a numerical-based computing tool is required. To avoid the complexity of solving all the voltage gain characteristic curves, the boundaries of different operation modes can be done first to provide essential insight into the gain characteristic with parameters and load variation. The boundary curves are actually mode edges, which can be seen as a critical case of one of the six major modes or of a particular boundary mode [17]. Since a boundary condition can be applied as an extra constraint, the mode equations can be solved without knowing P_n. For instance, the boundary of PO and PON modes is the solution of PO mode satisfying the condition

Table 13.4 Boundary conditions of operation modes.

Boundaries	**Solve as**	**Extra constraints**
PN/PON	PN	$m_{Lm,N}(\alpha) = -M$
PON/PO	PO	$m_{Lm,O}(\gamma) = -M$
PO/OPO	PO	$m_{Lm,O}(0) = M$
OPO/NOP	OP	$m_{Lm,O}(\alpha) = M$
NOP/NP	NP	$m_{Lm,N}(\alpha) = M$
O	O	$m_{Lm,O}(\gamma/2) = M$
P	P	$M = 1$
Peak gain of PON	PON	$j_{Lr,P}(0) = j_{Lr,N}(\gamma) = 0$
Peak gain of PN	PN	$j_{Lr,P}(0) = j_{Lr,N}(\gamma) = 0$

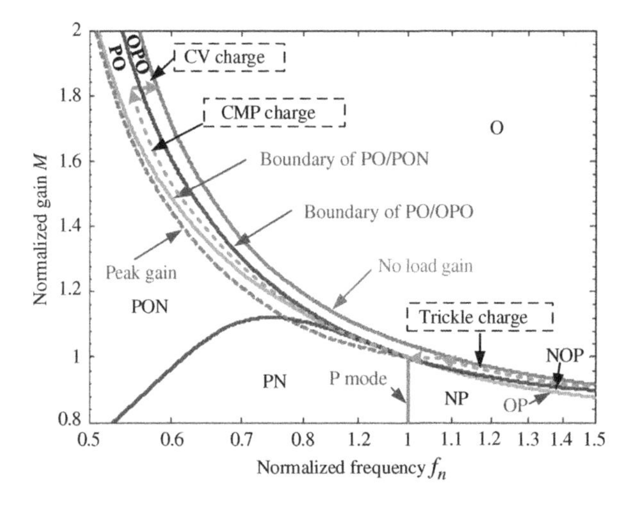

Figure 13.3 Gain-frequency mode boundaries of LLC resonant converter with $l = 0.2$.

that the value of m_{Lm} at the end of subinterval O is $-M$. Subinterval N is unavoidable before the half-cycle ends if this lower limit is exceeded. All the boundary conditions have been summarized in Table 13.4. The distribution of the operation modes is plotted by solving these boundary mode equations using MATLAB function fsolve(x).

As shown in Figure 13.3, the distribution of the operation modes in a range of switching frequency and gain is mainly determined by parameter l. A shrinking effect – which means all the boundary curves are distributed in a smaller frequency range with higher gain values – can be observed by increasing the value of l. Notably, all the boundaries of active modes converge to unity gain at $f_n = 1$, which confirms the load independent characteristic at resonance. In addition, the peak gain occurs in PN and PON modes, and the curves converge to the peak of the PN edge at the corresponding frequency.

Based on the above analysis, the LLC resonant converter can be described and solved precisely, as long as the operation conditions are known. The characteristics and distribution of the operation mode provide important design guidelines for different applications.

13.3 Design Considerations for an LLC Converter for a PHEV Battery Charger

The charge rate of a battery charger should be controlled according to the charging profile and the battery condition. As shown in Figure 13.4, there are usually three stages in a typical charging profile of a high voltage battery pack. A trickle charge stage with a constant current of 10% rated value (I_0) is performed first when the battery is deeply depleted. Bulk charge follows after the voltage has risen above the trickle charge threshold (U_0). Different constant current (in the range of 20% to 100% rated current) charge stages may exist in the bulk charge stage, limited by the power of the charger. Subsequently, the constant voltage stage is applied when the battery voltage reaches a certain value. However, the charging actions may be modified as a response to the battery condition variation at any time. Also, various types of charging profiles are requested for different battery packs, but all the profiles should be limited to the maximum output power of the converter. So maintaining the maximum power output during the whole charge process (at minimal input voltage), which is referred to as a constant maximum power (CMP) charging profile in this chapter, is reasonable to be seen as the overall charging profile, though it is not practical, due to safety and long cycle life considerations [25].

In this case, the LLC converter actually starts with a light load operation at the lowest gain. Then it goes through the full load operation within a wide gain range. The transition from full load back to light load operation occurs gradually at the highest gain. Besides, no-load operation may happen at any stage, dependent on the battery status.

Soft-switching is the most desirable advantage of resonant converters because it reduces switching loss and EMI. LLC topology is capable of realizing zero voltage switching (ZVS) for input inverting choppers and zero current switching (ZCS) for output rectifiers at the same time [26], which minimizes the switching losses of the MOSFETs and diodes. Therefore, from a designer's point of view, the design target is to achieve soft-switching under all operating conditions, as well as no-load operation

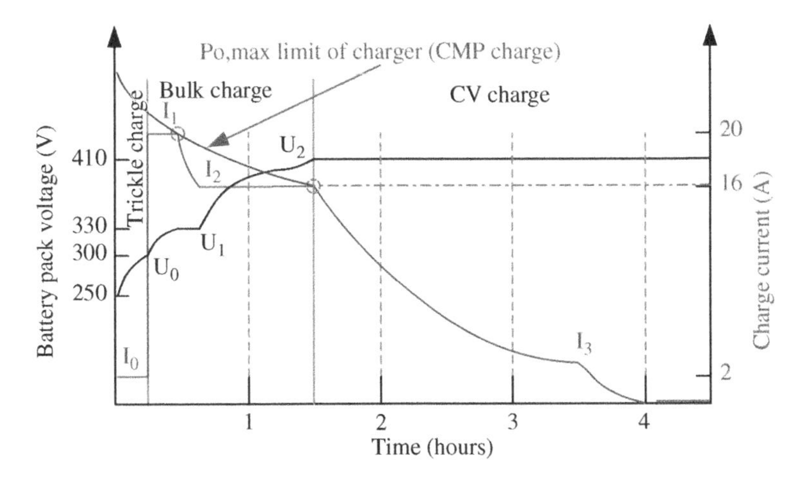

Figure 13.4 Charging profile of a 410 V lithium-ion battery pack.

ability. It has been summarized in Table 13.2 that there are several operation modes that feature primary ZVS and secondary ZCS capability: OPO, OP, P, and PO modes. In particularly, PON mode also possesses both-side soft-switching capability when the operation point is located in the zone formed by the peak gain curve and the PON/PO boundary shown in Figure 13.3. Among these candidates, OPO mode can be operated as both buck mode and boost mode under light load condition, making it suitable for the trickle charge stage and the CV charge stage. As for the CMP charge stage, a larger charging current is needed when the battery voltage is low. Hence, the equivalent load is relatively heavier at the beginning of the CMP charge. As a result, P mode, which features a load independent characteristic, is suitable for the start point of the CMP charge stage. So far, PO mode and PON mode are left to be targeted for the remaining part of the CMP charge. The peak gain points in PON mode form the critical boundary of realizing primary ZVS operation. So the peak voltage gain in a specific design is usually chosen as a design point since it indicates the converter's voltage regulation capability at the lowest permissible operating frequency. However, the most important feature of peak gain mode is that the switching current equals zero as summarized in Table 13.4. In practice, the switching current should be slightly larger than zero for reliable ZVS of the MOSFETs. In consequence, peak gain points in PON mode are not likely to be the best choice to operate at the required peak output voltage condition, let alone the complexity to be solved. In comparison, the entire PO mode is within the both-side soft-switching region. The gain in PO mode increases monotonically with the decrease of operating frequency, which ensures control stability in the controller design [18]. Besides, when the inductance ratio is larger, the energy stored in the magnetizing inductor is lower due to the reduced circulating tank current. This part of the energy will be released faster during the N subinterval of the PON mode, which causes the PO/PON boundary to get closer to the peak gain curve and finally to intersect at a certain low frequency. Therefore, PO mode is the most preferable boost operation mode. Meanwhile, the PO/PON boundary can be regarded as a more practical gain limitation curve than the actual peak gain curve to ensure soft-switching operation. Last but not least, O mode exists in the whole frequency and gain range at no-load condition as summarized in Table 13.2 and shown in Figure 13.3, which ensures that the charging can be shut down at any battery voltage due to charging completion or a fault detection.

Based on the above analysis, a possible operation trajectory that promises soft-switching under all charging stages can be plotted in Figure 13.3 (dotted line with arrow). It can be seen that the trajectory starts with O mode at the highest frequency waiting for the charging command. It goes through OPO mode above resonance during the trickle charge stage. The CMP charge stage is initiated from P mode to take advantage of its load independent property. By properly designing parameters, the whole CMP charge stage can be restricted within the PO mode region. In the CV charge stage, the converter maintains the highest required voltage gain while the equivalent load decreases gradually and ends with O mode eventually.

In addition, to further confirm the possibility of the proposed operation trajectory, the boundaries and distribution of the normalized load power of operation modes are plotted in Figure 13.5. The corresponding preferable operation trajectory in Figure 13.3 is also plotted in power–frequency and power–gain mode distribution coordinates respectively. As can be seen, the whole CMP charge can be limited in the PO mode

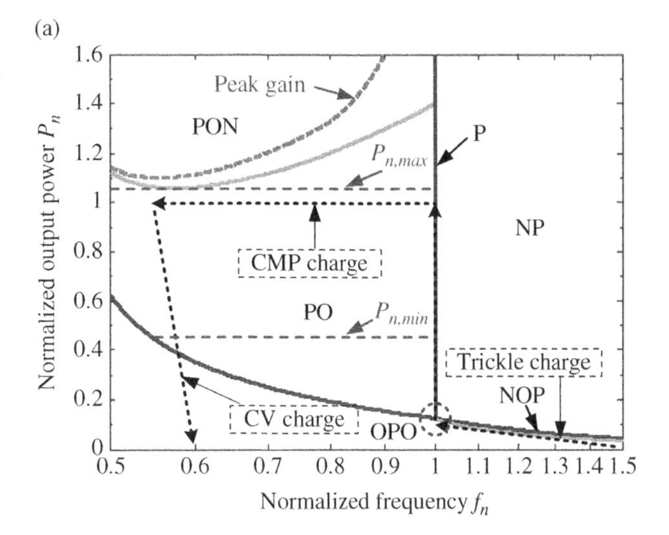

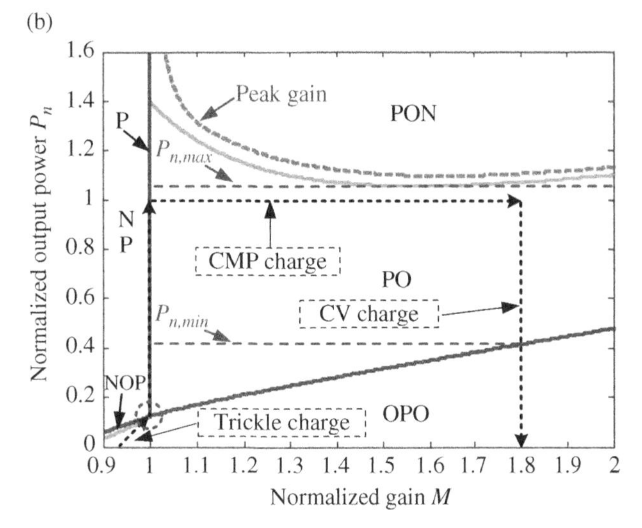

Figure 13.5 LLC mode boundaries and distribution with $l = 0.2$: (a) Power–frequency distribution and the peak gain limit (dashed line). (b) Power–gain distribution.

region as long as the normalized rated output power $P_{n,full}$ is designed to be between $P_{n.max}$ and $P_{n,min}$ as indicated in Figure 13.5, for the required gain within the allowable frequency range. However, in order to locate the trickle charge trajectory within the OPO zone, the upper limit may be lower than $P_{n.max}$. For instance, as mentioned, a constant 10% rated charge current is usually carried out in the trickle charge stage, which means the normalized output power during this stage is about 1/10 of the rated output power. Therefore, the normalized power at the crossing point of P mode and the OPO boundary, which is highlighted by a dashed circle in Figure 13.5, can be regarded as the maximum value of $0.1P_{n,full}$. The actual upper limit is shown as the possible CMP charge trajectory in Figure 13.5.

In the CV charge stage, the output power decreases gradually while the gain remains the same, so the trajectory of the CV charge can also be drawn in Figure 13.5.

13.4 Charging Trajectory Design

13.4.1 Key Design Parameters

To design the LLC converter operating along the trajectory that offers the whole range with soft-switching capability during the charging process, three key elements should be considered thoroughly: inductance ratio l, transformer turns ratio n, and characteristic impedance Z_0.

First of all, the unity gain operation is designed under the heaviest load condition when the CMP charge is applied to the battery pack at its lowest voltage. Therefore, the transformer turns ratio can be calculated from

$$n = V_{in,nom} / V_{out,\min} \tag{13.10}$$

After determining n, the minimum gain is given as

$$M_{\min,0} = nV_{out,\min} / V_{in,\max} = V_{in,nom} / V_{in,\max} \tag{13.11}$$

In general, the minimum gain occurs when the battery is at its lowest voltage and waiting for the charging command from the battery management system. The switching frequency should be adjusted to its maximum value to step down the output for the no-load operation when the maximum input voltage is applied to the converter. In this case, the converter operates in O mode and the relationship between the variables can be solved analytically [27]. The inductance ratio is

$$l = \left(\frac{1}{M_{\min,0}} - 1 \right) \frac{8 f_{n,\max}^2}{8 f_{n,\max}^2 - \pi^2} \tag{13.12}$$

The mode distribution is determined once the inductance ratio is chosen. According to (13.11), the required minimum gain depends on the ripple of input voltage, which is actually the output ripple of the front-end power factor correction (PFC) stage in charger applications. As a result, the maximum operating frequency $f_{n,max}$ becomes the most important parameter predefined by the designer. Typically, the output voltage ripple of the boost type PFC is ±5% to ±10%. In this case, the relationship between l and $f_{n,max}$ can be plotted as shown in Figure 13.6. Also shown in Figure 13.6, a smaller expected operating frequency range gives rise to a higher inductance ratio. However, it has been found in [22] that the resonant tank RMS current tends to be lower as l decreases at the same normalized load power p_n, which reduces conduction losses. Further, this effect is notably weakened when l is lower than 0.25. On the other hand, the available voltage gain in the same specific frequency range decreases with a smaller l. Therefore, the objective of the optimal design is to reduce conduction loss while maintaining the required gain within the frequency limit.

As already mentioned, the PO/PON boundary can be regarded as a conservative design boundary for the primary ZVS realization of MOSFET-based inverters in some circumstances, especially when the PO/PON boundary is close to the peak gain boundary. Additionally, an l value lower than 0.25 is preferable for reducing conduction losses, which makes the two boundaries close enough without losing any useful region according to the aforementioned operation mode analysis method (as can be seen in Figure 13.3). So the gain of the PO/PON boundary at the lowest operating frequency is reasonable to be the gain limitation. The gain limited by l when $f_{n,min} = 0.5$ can be drawn

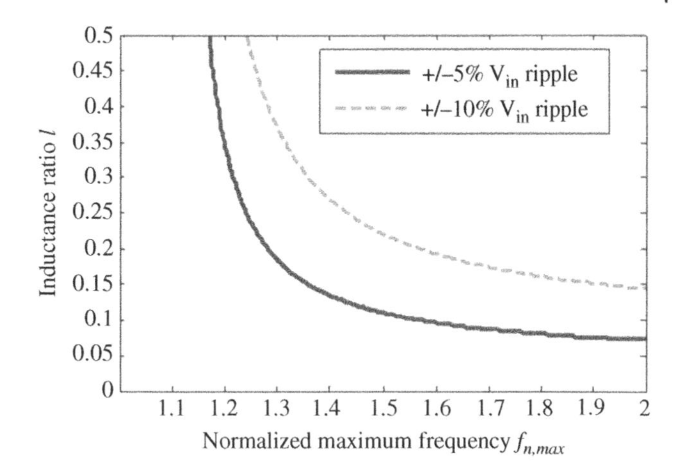

Figure 13.6 The relationship between the inductance ratio and the normalized maximum frequency under no-load condition.

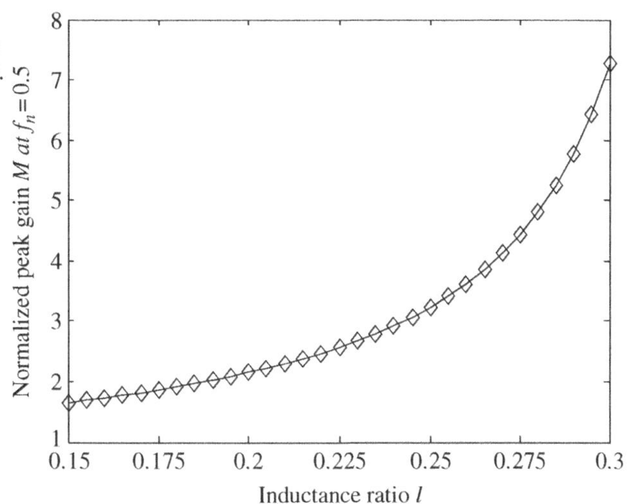

Figure 13.7 The relationship between the peak gain limitation and inductance ratio at $f_{n,min}=0.5$.

as in Figure 13.7. According to the above analysis, l can be chosen based on the required maximum gain and the input voltage ripple using Figure 13.6 and Figure 13.7.

The next step is to select a suitable value for the characteristic impedance Z_0 once l is determined. According to the definition of normalized rated load power:

$$P_{n,full} = \frac{P_o}{(V_{in,\min})^2} Z_0 \tag{13.13}$$

The normalized output power is proportional to Z_0 since the input voltage is fixed by the front-end stage. So the locus of CMP charge actually depends on the characteristic impedance. Assigning different values to $P_{n,full}$, the variation of the gain–frequency curve and the CMP charge trajectory can be shown as in Figure 13.8. It can be seen that the gain curves are pushed towards the PO/PON boundary by increasing Z_0. The CMP trajectories are also lifted towards the PO/PON boundary with a larger Z_0.

Obviously, a small Z_0 makes the LLC converter over-qualified because the operation trace will be far away from the boundary. Besides, a larger Z_0 value is preferable for

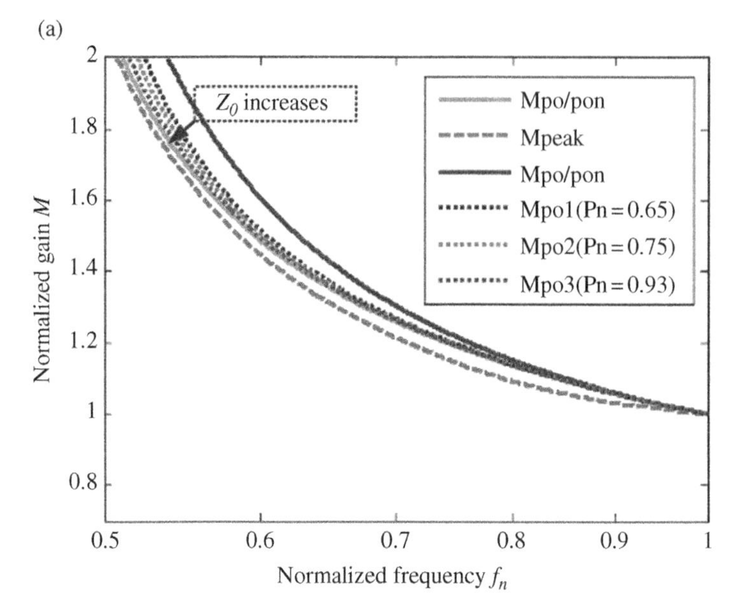

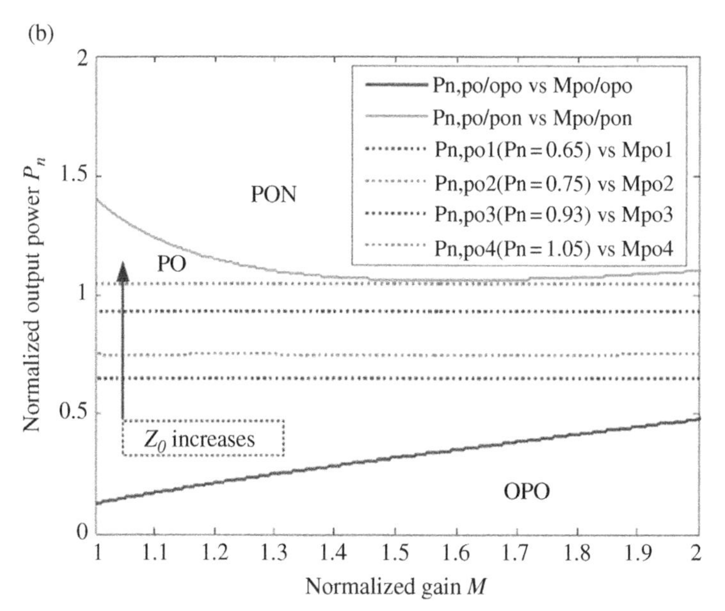

Figure 13.8 (a) Normalized gain *M* curves (dotted lines) for various designed P_n values in the PO region with $l = 0.2$. (b) The CMP charge trajectories (dotted lines) for various designed P_n values in the PO region with $l = 0.2$.

reducing conduction losses. In theory, the bottom of the PO/PON boundary $P_{n,PO/PON,min}$ in Figure 13.8 (b) can be targeted as the rated normalized power $P_{n,full}$ to calculate the corresponding Z_0 value, since it ensures that the whole CMP charging process is located in the PO operation region. This condition can be written as

$$Z_{0,PON} \le \frac{P_{n,PO/PON,\min}}{P_o / V_{in,\min}^{2}} \tag{13.14}$$

However, three more constraints should be taken into account when designing Z_0.

13.4.2 Design Constraints

The trickle charge occurs only when the battery is deeply depleted, which is normally supposed to be avoided for the sake of battery health [26]. However, it is still worth having the trickle charging region benefit from the soft-switching operation, because the constraints can be easily combined into the whole design procedure. In the aforementioned, the normalized output power at the end of the trickle charge is about $0.1P_{n,full}$. This value should not be larger than the value of the OPO/NOP boundary at the resonant frequency to restrict the trace within the OPO region as shown in Figure 13.5. Consequently, the constraint condition can be expressed as

$$Z_{0,TCK} \leq \frac{P_{n,NOP/OPO}\big|_{f_n=1}}{0.1P_o/V_{in,\min}^{2}} \tag{13.15}$$

Besides, the switching current, which is also the initial resonant inductor current $j_{Lr,X}$,(0) in each cycle, should be evaluated in different operation modes. Generally speaking, the worst-case situation for primary ZVS operation occurs when the output voltage is regulated at its minimum value, and the maximum input voltage is applied to the converter under the no-load condition. In this case, the switching frequency is usually regulated to its maximum value and the converter operates in cutoff mode. At this time, the switching current should be considered as its smallest value globally to reduce no-load conduction loss. On the other hand, this current has to be large enough to discharge the MOSFETs' junction capacitors within the dead time for soft-switching realization. The required minimum switching current $I_{sw,min}$ is calculated according to the parasitic parameters of the selected MOSFET and the dead time. So the third constraint of Z_0 is

$$Z_{0,OSW} \leq \left| j_{Lr0,O}\left(M_{\min}, f_{n,\max}\right)\right| \frac{V_{in,\min}}{I_{SW,\min}} \tag{13.16}$$

So far, there are three constraints applied to the selection of Z_0 in order to confine the LLC converter to operating in the desirable modes and ensure soft-switching capability. Based on (13.14)–(13.16), the upper limit of the characteristic impedance can be expressed as

$$Z_{0,\max} = \min\left\{Z_{0,PON}, Z_{0,TCK}, Z_{0,OSW}\right\} \tag{13.17}$$

Finally, the switching current variation is analyzed to verify that the primary ZVS operation has been promised under a certain designed Z_0. The variation of switching current in cutoff mode above the resonance should be solved first, since the switching current at the maximum frequency (or at the lowest gain) is set to be the lowest value that ensures that the parasitic capacitors of the MOSFETs can be fully discharged. The switching current variation of the PO/PON and PO/OPO boundaries can also be solved easily to draw the outline of the PO region. The variation of normalized switching current j_{Lr0} under different designed $P_{n,full}$ values along the CMP charge trajectory can be shown in Figure 13.9 when $l = 0.2$. In Figure 13.9, the normalized switching current at the minimum gain ($M_{min} = 0.9$) is −0.2, where negative means that the tank impedance is inductive. Hence, the absolute value of the normalized switching current during CMP charge should be larger than 0.2 to guarantee ZVS operation. Apparently,

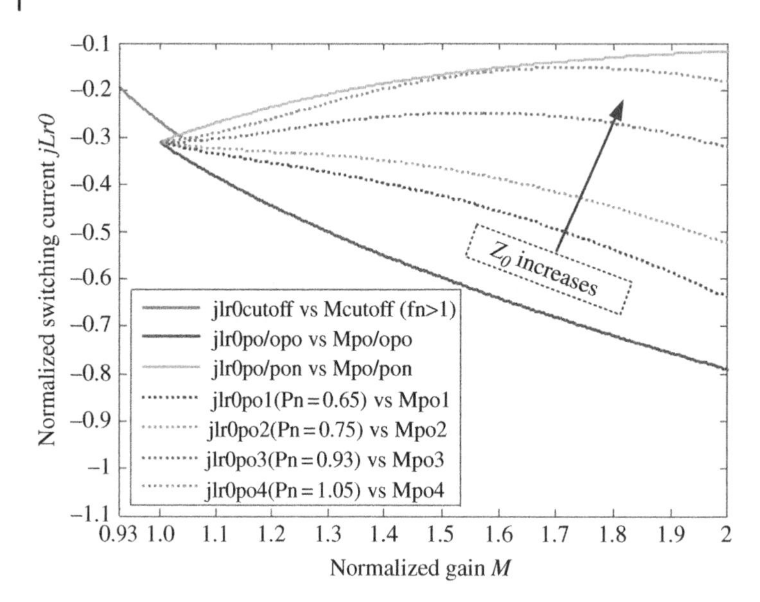

Figure 13.9 Normalized switching current curve (dotted lines) for various designed P_n values in the PO region and cutoff mode above resonance with l=0.2.

a larger Z_0 results in a smaller switching current in the PO region. However, the normalized switching current does not always vary monotonically with the voltage gain, which makes it impossible to get a handy analytical constraint expression for Z_0 design as before. As shown in Figure 13.9, the switching current increases with the gain when the designed normalized load power is 0.65–0.75. But the monotonicity is lost in the range of 0.9–1.05. The absolute value of the normalized switching current first decreases and then goes up, as the gain increases. The goal is to have ZVS operation when the output power is kept constant under different output voltage (gain). The criterion is that the switching current should never be smaller than the lower limit that has been set in the cutoff mode at minimum gain, which means 0.2 in Figure 13.9. Although the design $P_{n,full}$=1.05 guarantees the PO operation, as can be seen in Figure 13.8(b), it does not ensure that the switching current is large enough to keep the ZVS characteristic, because in this curve, the absolute value of the normalized switching current is lower than 0.2 when the gain is higher than 1.4. That is to say, the maximum characteristic impedance calculated by (13.17) may not qualify for use as the final design value.

As a result, one more step is necessary to draw the conclusion. The corresponding maximum normalized load power $P_{n,\max} = (P_o / V_{in}^2) Z_{0,\max}$ calculated by (13.13) and (13.17) should be assigned to the PO mode equations to solve for the normalized switching current curve. If the plotted curve is located below the line $j = j_{Lr0,O}(M_{\min}, f_{n,\max})$, $Z_{0,max}$ is eligible to be used. Otherwise, the Z_0 value needs to be reduced further to meet the criterion. With the accepted Z_0, the resonant circuit parameters are calculated as

$$L_r = \frac{Z_0}{2\pi f_0}, C_r = \frac{1}{2\pi f_0 Z_0}, L_m = \frac{L_r}{l} \tag{13.18}$$

13.5 Design Procedures

The proposed design procedure for the PHEV battery charger application is illustrated in Figure 13.10. This design method focuses on restricting the whole charging trajectory within the preferable operation modes that offer the converter both-side soft-switching capability. Efforts to minimize the circulating energy have been made by two steps of crucial importance during the design process. First, the inductance ratio l is minimized according to the given required gain range. Second, minimization is further accomplished by searching the allowable maximum characteristic impedance Z_0 limited by several design constraints that ensure reliable ZVS operation. These two optimization steps have induced two recursive loops in the whole design flow. The accuracy and existence of a solution can be promised by the precise time domain model. In particular, for EV charger applications, the input ripple is fixed by the front-end stage (±5% to ±10%) and the output range depends on the specifications of the battery pack (typically 250–500 V). It has been found that an l value around 0.2 is able to cover this gain range within a reasonable frequency range ($f_n = 0.5 \sim 1.6$) according to Figure 13.6 and Figure 13.7. The recursion number can be reduced greatly by starting from an experimental value. As for the second recursive loop, it is found that the switching current of the cutoff mode at $f_{n,max}$, which relies on the parasitic capacitance of the selected MOSFET, plays the most important role in designing Z_0. An appropriate estimation is helpful in reducing the number of recursions.

13.6 Experimental Results

To demonstrate the proposed charging trajectory design method, a prototype of a full-bridge LLC resonant converter for a Level 2 charger based on the specifications given in Table 13.5 was built. The semiconductor device and the key circuit components of the converter are listed in Table 13.6.

Following the proposed procedure, the turns ratio is calculated by (13.10) as 1.56. The inductance ratio is solved as $l = 0.1984$ from (13.12). The mode boundary and distribution solver mentioned in Section 13.2 is used to solve for the limitation of Z_0 set by (13.14)–(13.16). The calculated results are listed as follows: $P_{n,PO/PON,min} = 1.05$, $P_{n,NOP/OPO}$ $(f_n = 1) = 0.1254$, $|j_{Lr0,O}$ $(M_{min} = 0.9512,\ f_{n,max} = 1.29)$ $| = 0.2185$. The corresponding constraints are $Z_{0,PON} = 21.78$, $Z_{0,TCK} = 26.01$, $Z_{0,OSW} = 22.45$. It is to be noted that the minimum switching current $I_{SW,min}$ in (16) is set to 3.6 A, based on the parasitic parameters of the selected MOSFET. Therefore, the upper limit of Z_0 is chosen as 21.78 according to (13.17). So the upper limit is actually set by the PO/PON boundary $P_{n,PO/PON,min} = 1.05$ in this design example.

To verify the necessity of the second recursive loop of the proposed design procedure, the above calculated Z_0 is temporarily accepted as the final design value. The resonant tank parameters can be calculated by (13.18) as follows: $L_r = 22.36\,\mu H$, $C_r = 47.14\,nF$, and $L_m = 112.7\,\mu H$. A resonant tank has been built and tested based on the above parameters at full load conditions. The experimental waveforms are shown in Figure 13.11. In this figure, $v_{AB}(t)$ denotes the voltage applied on the resonant tank. The voltage across the resonant capacitor is shown as $v_{Cr}(t)$, $i_{Lr}(t)$, and $i_{Lm}(t)$ are used

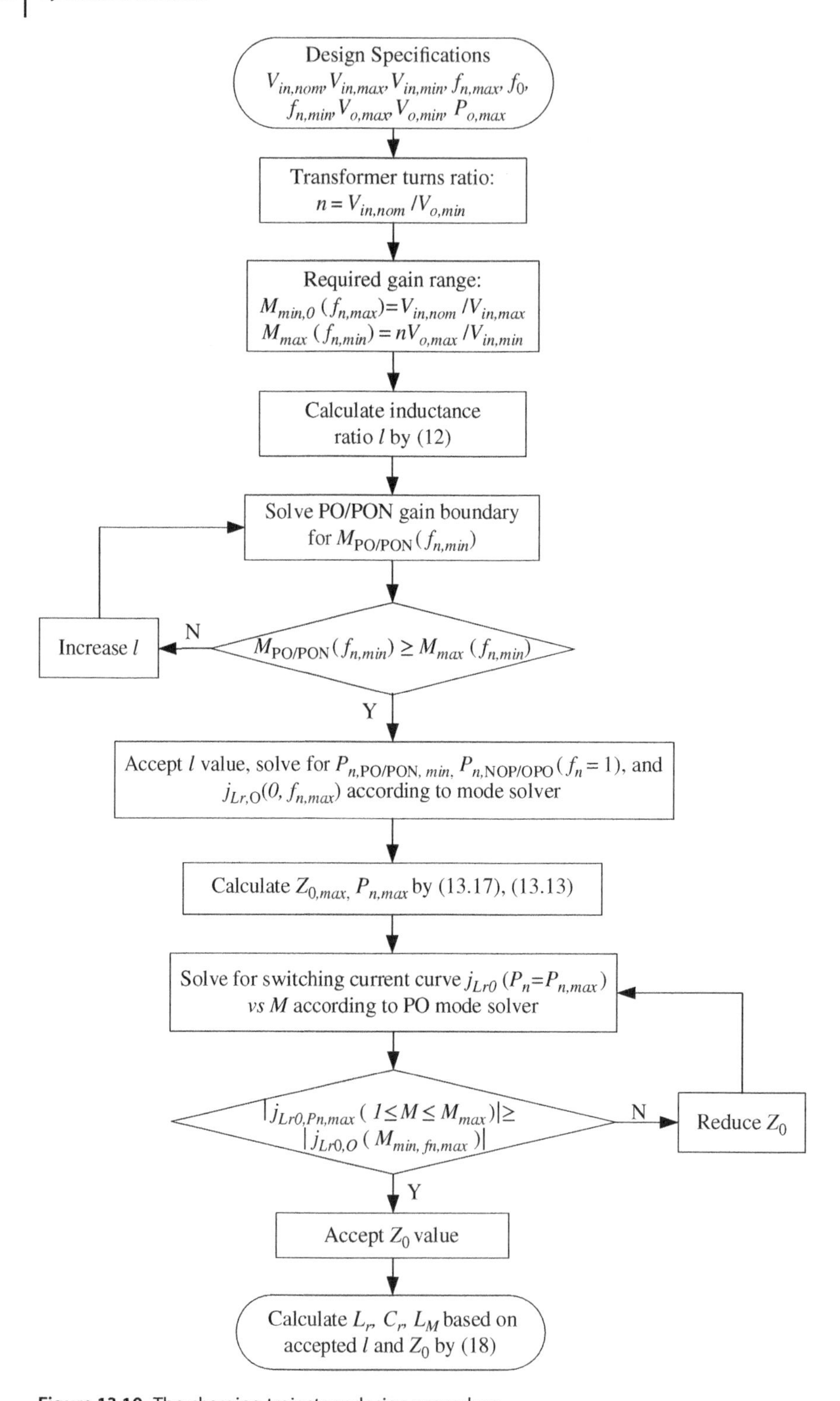

Figure 13.10 The charging trajectory design procedure.

Table 13.5 Design specification for the LLC resonant converter.

Parameter	Designator	Value
Input voltage range	$V_{in,\min} \sim V_{in,\max}$	370–410 V
Input voltage nominal	$V_{in,nom}$	390 V
Output voltage range	$V_{out,\min} \sim V_{out,\max}$	250–450 V
Maximum output power	$P_{out,\max}$	6.6 kW
Resonant frequency	f_0	155 kHz
Operating frequency range	$f_{s\min} \sim f_{s\max}$	85–200 kHz

Table 13.6 Components used in the prototype converter.

Component	Manufacturer	Part #
MOSFET	Infineon Technologies	IPW60R041C6
Diode rectifiers	Fairchild Semiconductor	FFH60UP60S
Resonant film capacitors	EPCOS	MKP 20×3.3 nF MKP 1×2.2 nF
Magnetic ferrite core of transformer	TDK	PC40 EC90×90×30
Litz wire	HM Wire International	AWG38×1000
Output film capacitors	Vishay	MMKP383 6×1μF

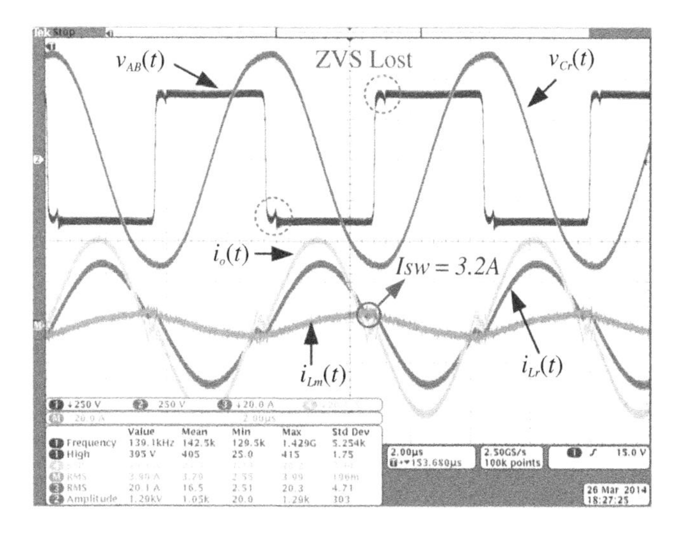

Figure 13.11 Experimental waveforms of the LLC converter prototype with the parameters skipping the last step of the procedure at $Vin = 390$ V, $Vo = 280$ V, $Po = 6.6$ kW.

to indicate the primary resonant tank current and the transformer magnetizing current respectively. The secondary current through the rectifier is indicated as $i_o(t)$. As predicted in Figure 13.9, though the converter operates in the preferable PO mode, the switching current $I_{sw} = 3.2$ A (lower than the designed cutoff switching current 3.6 A) is not large enough to discharge the parasitic capacitors of the MOSFETs to ensure ZVS realization. An oscillation can be observed when the dead-time ends and gate drive signals are applied to the corresponding switches, which induces switching losses, and this should be avoided. Hence, the value of the characteristic impedance needs to be reduced further. The variation of the normalized switching current shown in Figure 13.9 is used as a design reference, since the designed l is very close to 0.2. It can be seen that the switching current varies monotonically when the normalized output power is lower than 0.75, so this normalized output power is chosen as the rated output power for reliable ZVS operation. The resonant tank parameters are recalculated as follows: $L_r = 15.97\,\mu$H, $C_r = 66$ nF, and $L_m = 80.51\,\mu$H. The actual measured values are given for comparison: $n_a = 1.58$, $l_a = 0.198$, $L_{r,a} = 15.3\,\mu$H, $C_{r,a} = 68.2$ nF, and $L_{m,a} = 77.3\,\mu$H. Note that the magnetic integration is adopted to downsize the resonant tank, which makes it difficult to adjust the parameters to fit the designed values exactly.

The operating trajectory during the whole charging procedure is verified first. Four special operating points are picked to confirm the accomplishment of the design target. The experimental waveforms of these operations are given in Figure 13.12, in which the definitions of the symbols are the same as those in Figure 13.11. First, the no-load operation ($i_o(t) = 0$) at the lowest output voltage (250 V) is assured when the switching frequency is regulated to its maximum value (201 kHz) by Figure 13.12(a). Second, the OPO mode operation can be observed in Figure 13.12(b) when the output current is set around $0.1I_{o,full}$ at 250 V output, which stands for the end of the trickle charge stage. The oscillation of $i_o(t)$ is caused by the parasitic capacitor of the output rectifier when they are cut off from the resonant tank during the O subintervals of OPO mode. Third, the maximum output power is delivered at 250 V output voltage while P mode operation is promised as shown in Figure 13.12(c). So the heaviest load condition is designed in the load independence mode. Last, the typical PO mode operation waveforms can be seen in Figure 13.12(d) when the output voltage is regulated to the maximum value 450 V while the maximum output power is transferred, which indicates that the whole CMP charge is restricted within the PO region. In addition, it can be concluded from Figure 13.12 that both-side soft-switching is realized, since the charging trajectory is going along the desirable trace in the preferable region.

The performance of the CMP charge is then tested. An electronic load is used to simulate the load characteristic of the battery pack in the experiments. The efficiency of the converter as a function of output voltage for CMP charge is shown in Figure 13.13(a). It can be seen that the efficiency is higher than 97.2% during the whole CMP charge stage and hits its peak of 97.96% at 410 V output. The efficiency curves of the CV charge stage at 250 V and 410 V output are also given in Figures 13.13(b) and 13.13(c) for the light load performance evaluation. It can be seen that the conversion efficiency maintains above 90% from 10% load to full load. The peak efficiency of 97.96% is exhibited at an output current of 11 A at 250 V (2.75 kW).

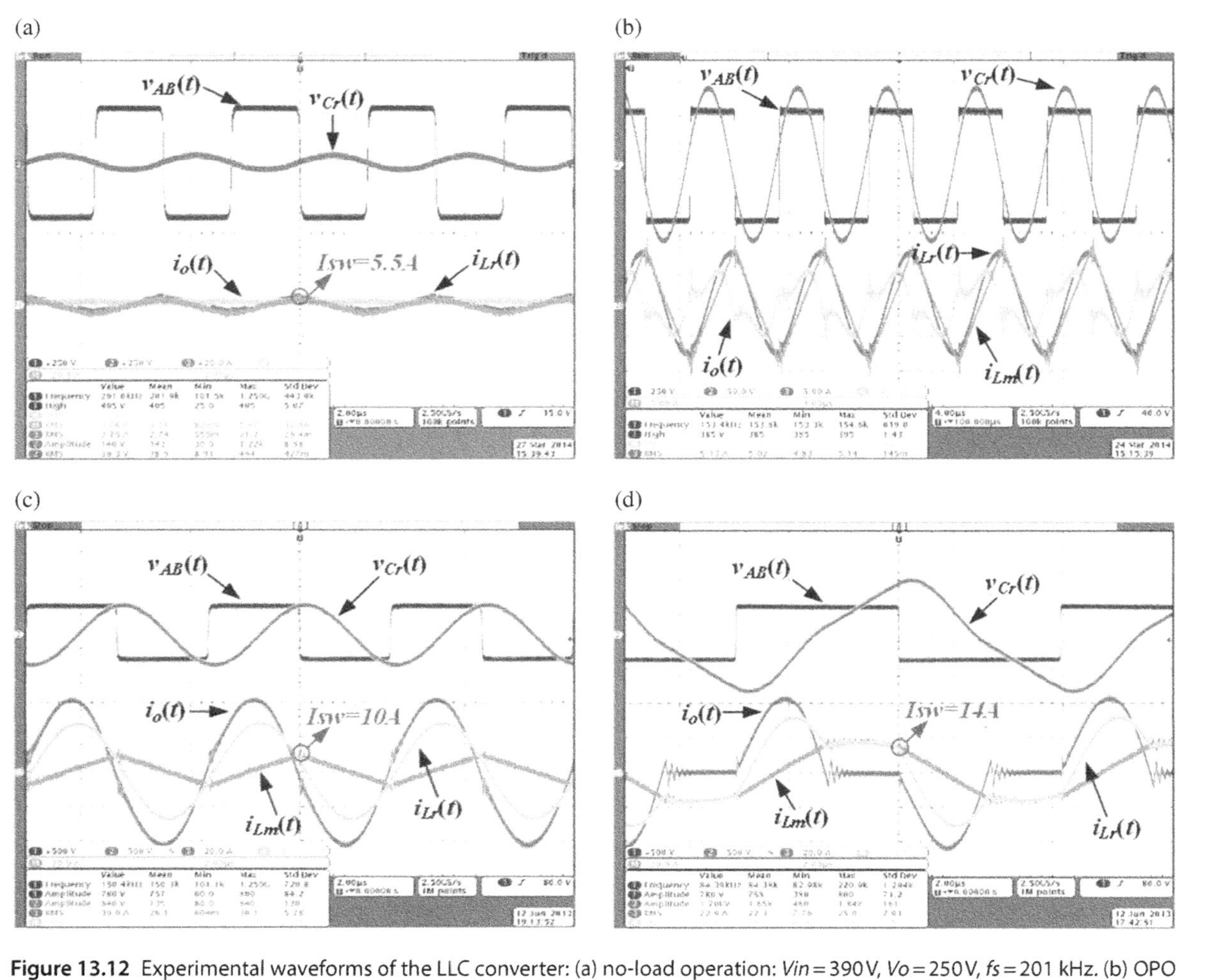

Figure 13.12 Experimental waveforms of the LLC converter: (a) no-load operation: $Vin = 390\,V$, $Vo = 250\,V$, $fs = 201$ kHz. (b) OPO mode operation in trickle charge: $Vin = 390\,V$, $Vo = 250\,V$, $Io = 2$ A, $fs = 153.4$ kHz. (c) P mode operation at the beginning of CMP charge: $Vin = 390$ V, $Vo = 250\,V$, $Po = 6.6$ kW, $fs = 150.4$ kHz. (d) PO mode operation at the end of CMP charge: $Vin = 390$ V, $Vo = 450$ V, $Po = 6.6$kW, $fs = 84.19$ kHz.

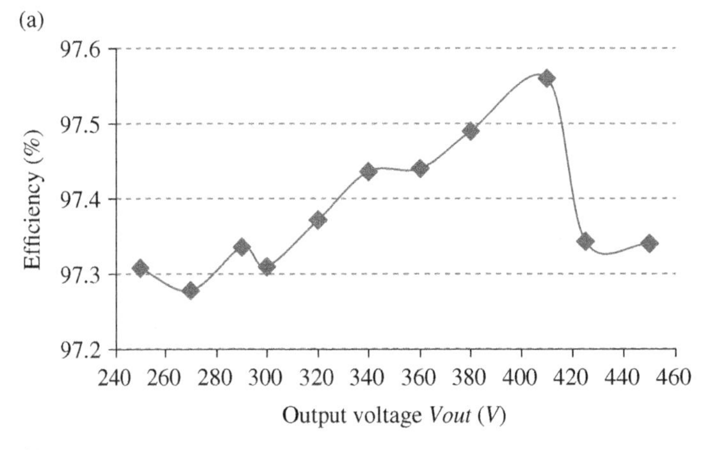

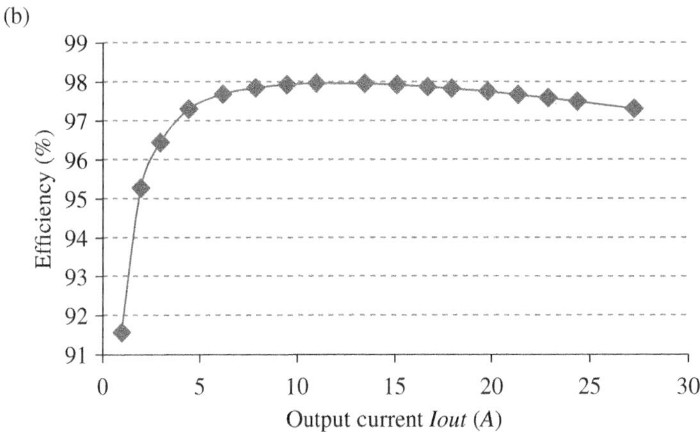

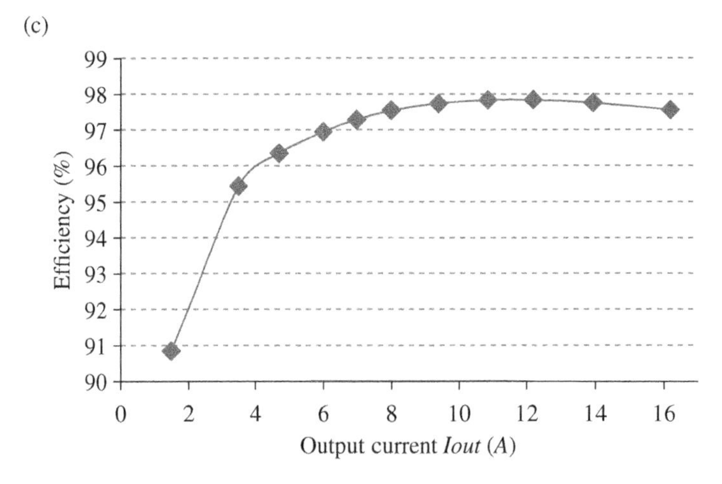

Figure 13.13 The LLC converter prototype performance: (a) Measured efficiency vs. output voltage at $P_o = 6.6\,kW$ and $V_{in} = 390\,V$. (b) Measured efficiency vs output current at $V_o = 250\,V$ and $V_{in} = 390\,V$. (c) Measured efficiency vs output current at $V_o = 410\,V$ and $V_{in} = 390\,V$.

13.7 Conclusions

The LLC resonant converter applied in EV/PHEV battery charger systems has been analyzed in this chapter, and the design methodology is presented. Different from the resistive load applications, which come with a single nominal condition, full load operation is required in a very wide output voltage range for battery chargers, so it is a trajectory design target rather than a single point design target. The mode boundaries and distribution of the LLC converter are discussed for the purpose of mapping the operation trace to the preferable region. The key parameters that affect the designed operating trajectory are identified. Finally, all the discussion has led to a design procedure that ensures soft-switching under all operating conditions. A 6.6 kW, 390 V DC input and 250–450V output LLC converter is built using the proposed method, which achieves 97.96% peak efficiency.

References

1 C. Mi, M. A. Masrur, and D. W. Gao, *Hybrid Electric Vehicles: Principles and Applications with Practical Perspectives*: Wiley, 2011.
2 X. Zhang and C. Mi, *Vehicle Power Management: Modeling, Control and Optimization*: Springer, 2011.
3 R. Sabzehgar and M. Moallem, "A Boost-Type Power Converter for Energy-Regenerative Damping," *Mechatronics, IEEE/ASME Transactions on*, vol. 18, pp. 725–732, 2013.
4 A. Dardanelli, M. Tanelli, B. Picasso, S.M. Savaresi, O. di Tanna, and M.D. Santucci, "A Smartphone-in-the-Loop Active State-of-Charge Manager for Electric Vehicles," *Mechatronics, IEEE/ASME Transactions on*, vol. 17, pp. 454–463, 2012.
5 M. Yilmaz and P.T. Krein, "Review of Battery Charger Topologies, Charging Power Levels, and Infrastructure for Plug-In Electric and Hybrid Vehicles," *Power Electronics, IEEE Transactions on*, vol. 28, pp. 2151–2169, 2013.
6 L. Siqi, D. Junjun, and C.C. Mi, "Single-Stage Resonant Battery Charger With Inherent Power Factor Correction for Electric Vehicles," *Vehicular Technology, IEEE Transactions on*, vol. 62, pp. 4336–4344, 2013.
7 F. Musavi, M. Craciun, M. Edington, W. Eberle, and W. G. Dunford, "Practical design considerations for a LLC multi-resonant DC–DC converter in battery charging applications," in *Applied Power Electronics Conference and Exposition (APEC), 2012 Twenty-Seventh Annual IEEE*, 2012, pp. 2596–2602.
8 K. Keun-Wan, K. Dong-Hee, W. Dong-Gyun, and L. Byoung-Kuk, "Topology comparison for 6.6 kW On board charger: Performance, efficiency, and selection guideline," in *Vehicle Power and Propulsion Conference (VPPC), 2012 IEEE*, 2012, pp. 1520–1524.
9 A. Khaligh and S. Dusmez, "Comprehensive Topological Analysis of Conductive and Inductive Charging Solutions for Plug-In Electric Vehicles," *Vehicular Technology, IEEE Transactions on*, vol. 61, pp. 3475–3489, 2012.
10 L. Jun-Young and C. Hyung-Jun, "6.6-kW Onboard Charger Design Using DCM PFC Converter With Harmonic Modulation Technique and Two-Stage DC/DC Converter," *Industrial Electronics, IEEE Transactions on*, vol. 61, pp. 1243–1252, 2014.
11 L. Bing, L. Wenduo, L. Yan, F. C. Lee, and J.D. van Wyk, "Optimal design methodology for LLC resonant converter," in *Applied Power Electronics Conference and Exposition, 2006. APEC '06. Twenty-First Annual IEEE*, 2006.

12 B. Yang, R. Chen, and F. C. Lee, "Integrated magnetic for LLC resonant converter," in *Applied Power Electronics Conference and Exposition, 2002. APEC 2002. Seventeenth Annual IEEE*, 2002, pp. 346–351 vol. 1.
13 J. F. Lazar and R. Martinelli, "Steady-state analysis of the LLC series resonant converter," in *Applied Power Electronics Conference and Exposition, 2001. APEC 2001. Sixteenth Annual IEEE*, 2001, pp. 728–735 vol. 2.
14 T. Duerbaum, "First harmonic approximation including design constraints," in *Telecommunications Energy Conference, 1998. INTELEC. Twentieth International*, 1998, pp. 321–328.
15 S. De Simone, C. Adragna, C. Spini, and G. Gattavari, "Design-oriented steady-state analysis of LLC resonant converters based on FHA," in *Power Electronics, Electrical Drives, Automation and Motion, 2006. SPEEDAM 2006. International Symposium on*, 2006, pp. 200–207.
16 R. Beiranvand, B. Rashidian, M.R. Zolghadri, and S.M.H. Alavi, "A Design Procedure for Optimizing the LLC Resonant Converter as a Wide Output Range Voltage Source," *Power Electronics, IEEE Transactions on*, vol. 27, pp. 3749–3763, 2012.
17 F. Xiang, H. Haibing, Z.J. Shen, and I. Batarseh, "Operation Mode Analysis and Peak Gain Approximation of the LLC Resonant Converter," *Power Electronics, IEEE Transactions on*, vol. 27, pp. 1985–1995, 2012.
18 F. Xiang, H. Haibing, J. Shen, and I. Batarseh, "An optimal design of the LLC resonant converter based on peak gain estimation," in *Applied Power Electronics Conference and Exposition (APEC), 2012 Twenty-Seventh Annual IEEE*, 2012, pp. 1286–1291.
19 R. Yu, G.K.Y. Ho, B.M.H. Pong, B.W.K. Ling, and J. Lam, "Computer-Aided Design and Optimization of High-Efficiency LLC Series Resonant Converter," *Power Electronics, IEEE Transactions on*, vol. 27, pp. 3243–3256, 2012.
20 C. Adragna, S. De Simone, and C. Spini, "A design methodology for LLC resonant converters based on inspection of resonant tank currents," in *Applied Power Electronics Conference and Exposition, 2008. APEC 2008. Twenty-Third Annual IEEE*, 2008, pp. 1361–1367.
21 C. Adragna, S. De Simone, and C. Spini, "Designing LLC resonant converters for optimum efficiency," in *Power Electronics and Applications, 2009. EPE '09. 13th European Conference on*, 2009, pp. 1–10.
22 F. Xiang, H. Haibing, F. Chen, U. Somani, E. Auadisian, J. Shen, et al., "Efficiency-Oriented Optimal Design of the LLC Resonant Converter Based on Peak Gain Placement," *Power Electronics, IEEE Transactions on*, vol. 28, pp. 2285–2296, 2013.
23 J. Mahdavi, M.R. Nasiri, A. Agah, and A. Emadi, "Application of neural networks and State-space averaging to DC/DC PWM converters in sliding-mode operation," *Mechatronics, IEEE/ASME Transactions on*, vol. 10, pp. 60–67, 2005.
24 A. Pawellek, C. Oeder, J. Stahl, and T. Duerbaum, "The resonant LLC vs. LCC converter – comparing two optimized prototypes," in *Energy Conversion Congress and Exposition (ECCE), 2011 IEEE*, 2011, pp. 2229–2235.
25 S.S. Zhang, "The effect of the charging protocol on the cycle life of a Li-ion battery," *Journal of Power Sources*, vol. 161, pp. 1385–1391, 10/27/2006.
26 N.K. Ure, G. Chowdhary, T. Toksoz, J.P. How, M.A. Vavrina, and J. Vian, "An Automated Battery Management System to Enable Persistent Missions With Multiple Aerial Vehicles," *Mechatronics, IEEE/ASME Transactions on*, vol. PP, pp. 1–12, 2014.
27 J. Deng, S. Li, S. Hu, C.C. Mi, and R. Ma, "Design Methodology of LLC Resonant Converters for Electric Vehicle Battery Chargers," *Vehicular Technology, IEEE Transactions on*, vol. 63, pp. 1581–1592, 2014.

14

Modeling and Simulation of Electric and Hybrid Vehicles*

14.1 Introduction

Compared to conventional vehicles, there are more electrical components used in electric, hybrid, and fuel cell vehicles, such as electric machines, power electronics, electronic continuously variable transmissions (CVTs), and embedded powertrain controllers [1, 2]. Advanced energy storage devices and energy converters, such as Li-ion batteries, ultracapacitors (UCs), and fuel cells are introduced in the next-generation powertrains. In addition to these electrification components or subsystems, conventional internal combustion engines (ICEs), mechanical systems, and hydraulic systems may still be present. The dynamic interactions between the various components and the multidisciplinary nature make it difficult to analyze a newly designed hybrid electric vehicle (HEV). Each of the design parameters must be carefully chosen for better fuel economy, enhanced safety, optimum drivability, and a competitive dynamic performance – all at a price acceptable to the consumer market. Prototyping and testing each design combination is cumbersome, expensive, and time-consuming. Modeling and simulation are indispensable for concept evaluation, prototyping, and analysis of HEVs. This is particularly true when novel hybrid powertrain configurations and controllers are developed.

Furthermore, the complexity of new powertrain designs and their dependence on embedded software are a cause for concern among automotive research and development engineers. This results in increasing difficulty in predicting interactions between various vehicle components and systems. In such situations, a modeling environment that can model not only components but also embedded software, such as the electronic throttle controller (ETC) software, is needed. Effective diagnosis also presents a challenge. Modeling can play an important role in the diagnostics of the operating components. For example, running an embedded fuel cell model and comparing the actual fuel cell operating variables to those obtained from the model can help fault diagnosis of fuel cells.

A face-off with modeling and simulation tools in the electronics industry has demonstrated that similar tools in the automotive domain still lack the power, sophistication, and automation required by and available to the electronics designers [3]. Advances in electronic design tools have validated Moore's law (as applied to the complexity of integrated circuits) and have helped achieve amazing standards in computing power, while simultaneously decreasing costs. For designers of automotive systems to duplicate and manage similar levels of complexity, design tools that automate the low-level details of the design process need to be developed [3, 4].

Depending on the level of detail at which each component is modeled, the vehicle model may be steady-state, quasi-steady, or dynamic [5–15]. For example, an ADVISOR (ADvanced VehIcle SimulatOR) [5, 6] model can be categorized as a steady-state model, a powertrain system analysis toolkit (PSAT) [7] model as quasi-steady, and PSIM [8] and virtual test bed (VTB) [9] models as dynamic. On the other hand, depending on the direction of calculation, vehicle models can be classified as forward-looking or backward-facing models [5]. In forward-looking models, vehicle speed is controlled to follow a driving cycle during the analysis of fuel economy, thus facilitating controller development.

The main advantage of employing a steady-state model or quasi-steady model is fast computation, while the disadvantage is inaccuracy for dynamic simulation. By contrast, physics-based models can facilitate high-fidelity dynamic simulations for the vehicle system at different timescales. This kind of dynamic model should be useful for developing an effective powertrain control strategy [10]. The models are tied closely to the underlying physics through a link such as a lumped-coefficient differential equation or some digital equivalent model.

This chapter addresses different modeling and simulation methods for electric and hybrid vehicles. The chapter is organized as follows: Section 14.2 reviews the fundamentals of vehicle system modeling. Sections 14.3 and 14.4 provide an overview of the existing vehicle modeling tools ADVISOR and PSAT, with application examples, using ADVISOR to study a hybrid battery–ultracapacitor energy storage system, and using PSAT to optimize a parallel powertrain design. Section 14.5 looks at physics-based dynamic modeling, introducing the resistive companion form (RCF) of modeling method with examples of a DC machine, a DC–DC boost power converter, and vehicle dynamics including a wheel slip model. Section 14.6 looks at bond graphs and other modeling tools such as PSIM, Simplorer, V-ELPH, Saber, and Modelica for hybrid powertrain modeling. Section 14.7 addresses the issue and mitigation methods of numerical oscillations for dynamic simulation involving power electronics. Finally, conclusions are given in Section 14.8.

14.2 Fundamentals of Vehicle System Modeling

It is important to define the common terms used in modeling. The following definitions are based on the text by Dr Peter Fritzson of Linköping University in Sweden [16] and are related to HEV modeling:

- **System:** The object or objects we wish to study. In the context of this chapter, the system will be an electric vehicle or HEV.
- **Experiment:** The act of obtaining information from a controllable and observable system by intelligently varying system inputs and observing system outputs.
- **Model:** A surrogate for a real system upon which "experiments" can be conducted to gain insight about the real system. The types of experiments that can be validly applied to a given model are typically limited. Thus, different models are typically

required for the same target system to conduct all of the experiments we wish to conduct. Although there are various types of models (e.g. scale models used in wind tunnels), in this chapter we will mainly discuss physics-based mathematical models.
- **Simulation:** An experiment performed on a model.
- **Modeling:** The act of creating a model that sufficiently represents a target system for the purpose of simulating that model with specific predetermined experiments.
- **Simulator:** A computer program capable of performing a simulation. These programs often include functionality for the construction of models and can often be used in conjunction with advanced statistical engines to run trade (sensitivity) studies, design of experiments, Monte Carlo routines, and other routines for robust design.

Vehicle system modeling is conducted over various areas of interest to answer vastly different questions (i.e. different experiments). Traditional areas include modeling for the analysis of noise vibration harshness (NVH); modeling of vehicle performance (e.g. acceleration, gradability, and maximum cruising speed); modeling for the prediction, evaluation, and optimization of fuel economy; modeling for safety, stability, and crashworthiness; modeling of vehicle controls; modeling for structural integrity; modeling to facilitate component testing and validation; modeling for preliminary conceptual design and design exploration; modeling for cost and packaging; and modeling for the prediction of emissions.

There are various types of mathematical models and simulators available to perform vehicle system simulations. For example, some simulators can be used to construct models that use macro statistics from duty cycles and cycle-averaged efficiencies of components for near-instantaneous prediction of fuel consumption and performance, whereas other simulators perform detailed sub-second transient simulations for more detailed experiments. There is also a typical tradeoff in the vehicle modeling between the amount of engineering assumptions the modeler has to make and the amount of time required to set up and construct a model. A simple high-level model can estimate fuel consumption using the engineer's knowledge of "typical" cycle-averaged component efficiencies. A more detailed model would actually simulate each of the components over time and mathematically determine cycle-averaged efficiencies. In addition to the assumption/specificity tradeoff, there is also a tradeoff between model detail and run time. In general, the more detailed the results, the longer the total time for model setup, simulation, and interpretation of the results.

Detailed vehicle system models typically contain a mix of empirical data, engineering assumptions, and physics-based algorithms. Good simulators provide a large variety of vehicle components along with data sets to populate those components. The components can then be connected together as the user desires to create a working vehicle powertrain, body, and chassis. Connections between components mathematically transmit effort and flow (e.g. torque and speed or voltage and current) during a simulation.

Depending upon the desired degree of detail, there are various models available such as steady-state spreadsheet models, transient power–flow models, and transient effort–flow models (effort–flow refers to the combinations of torque–angular speed, voltage–current, force–linear speed, etc.).

The transient vehicle system models can be divided into two categories based on the direction of calculation. Models that start with the tractive effort required at the wheels and "work backward" toward the engine are called *backward-facing models*. Models that start from the engine and work in transmitted and reflected torque are called forward-facing models. So-called non-causal models allow for forward or backward operation depending on the experiment being performed. Backward-facing models are typically much faster than forward-facing models in terms of simulation time. Forward-facing models better represent real system setup and are preferred where controls development and hardware-in-the-loop (HIL) will be employed. Forward models must typically use some kind of "driver model" such as a proportional–integral–derivative (PID) controller to match a target duty cycle. Some "hybrid" models include both concepts.

In addition, vehicle system models may interact with any number of more detailed models such as structural analysis models, vibrational models, and thermal models.

Driven by the need for fast simulation times, complex components such as engines and motors are typically simulated using "lookup maps" of energy consumption versus shaft torque and angular speed. Once the average torque and angular shaft speed for a given time step are determined, an interpolation on empirical data is performed to determine the component's energy consumption rate.

There have been extensive studies in the modeling and simulation of hybrid and electric vehicles [4–15]. Modeling tools such as ADVISOR, Autonomie, and PSAT are available in the public domain, and are discussed in more detail in the following sections.

14.3 HEV Modeling Using ADVISOR

ADVISOR is a modeling and simulation tool developed by the US National Renewable Energy Laboratory (NREL) [5, 6]. It can be used for the analysis of performance, fuel economy, and emissions of conventional, electric, hybrid electric, and fuel cell vehicles. The backbone of the ADVISOR model is the Simulink block diagram shown in Figure 14.1, for a parallel HEV as an example. Each subsystem (block) of the block diagram has a MATLAB file (m-file) associated with it, which defines the parameters of that particular subsystem. The user can alter both the model inside the block and the m-files associated with the block to suit the modeling needs. For example, the user may need a more precise model for the electric motor subsystem. A different model can replace the existing model as long as the inputs and the outputs are the same. On the other hand, the user may leave the model intact and only change the m-file associated with the block diagram. This is equivalent to choosing a different maker of the same component (e.g. choosing a 12 Ah battery manufactured by Hawker Genesis instead of a 6 Ah battery manufactured by Caterpillar). ADVISOR provides modeling flexibility for a user.

ADVISOR models fit empirical data obtained from the component testing to simulate a particular subsystem. In general, the efficiency and limiting performances define the operation of each component. For example, the ICE is modeled using an efficiency map that is obtained via experiments. The efficiency map of a Geo 1.0 l (43 kW) engine is shown in Figure 14.2. Various contour lines on this diagram indicate particular efficiency values labeled beside the corresponding contour. The maximum torque curve is also shown in this map. The engine cannot perform beyond this maximum torque constraint. Maximum torque change is another constraint on the engine subsystem. In other words, the model considers the inertia of the component in the simulation.

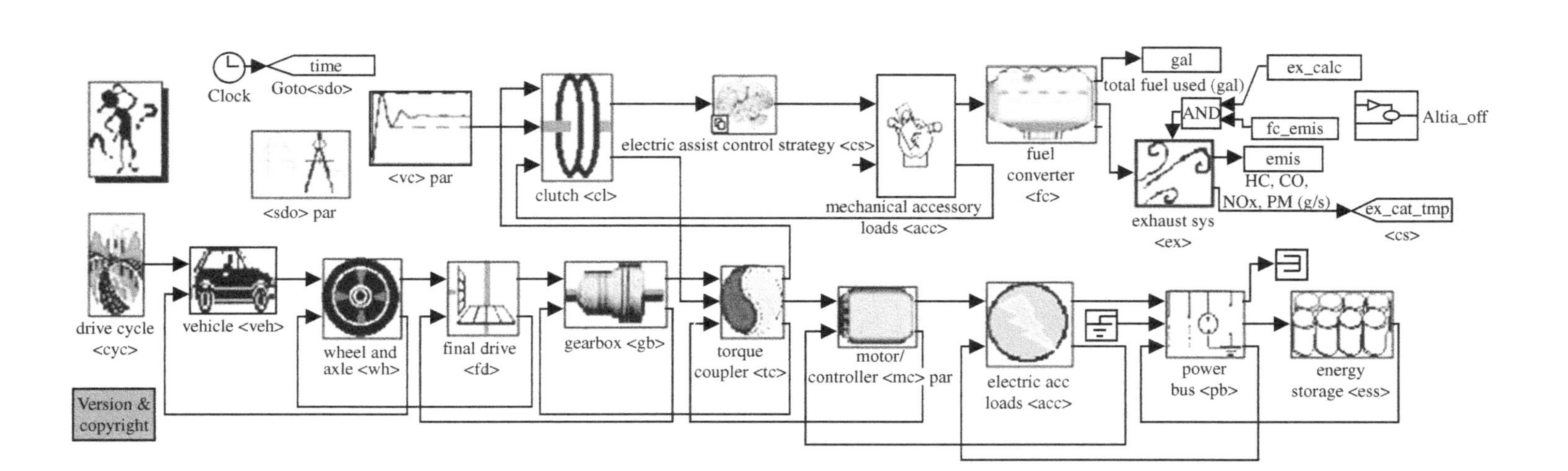

Figure 14.1 Block diagram of the parallel HEV in ADVISOR [6]. *Source:* Markel 2002.

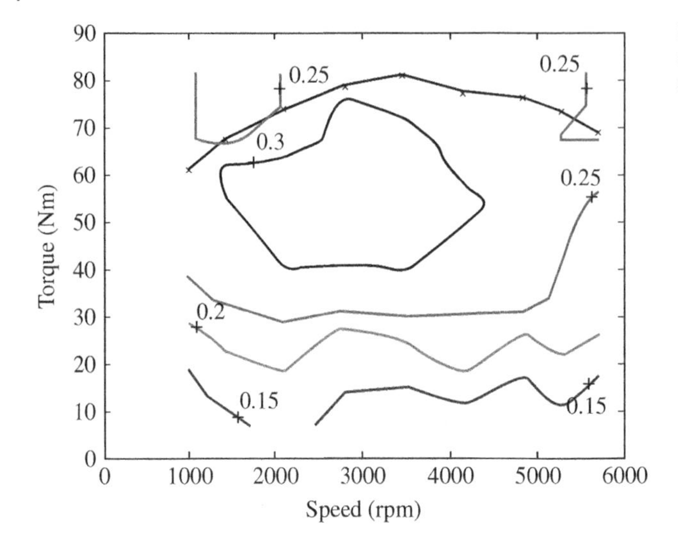

Figure 14.2 Geo 1.0 l (43 kW) spark ignition engine efficiency map.

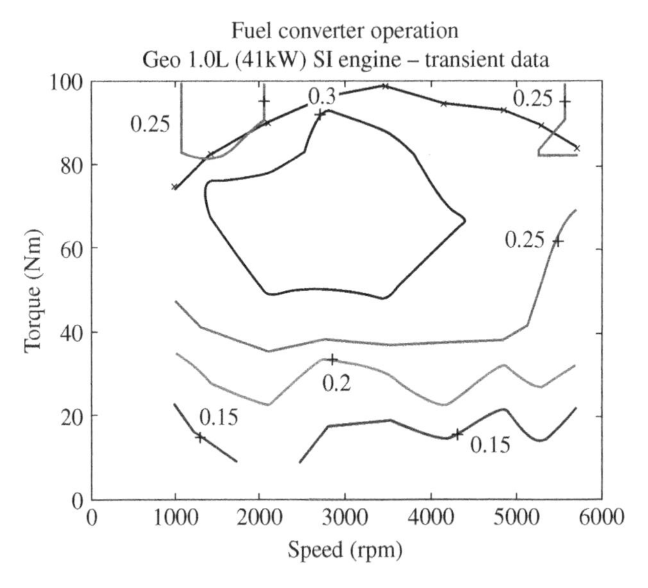

Figure 14.3 Geo 1.0 l engine scaled to give a maximum power of 50 kW by linear alteration of the torque characteristics.

The program also allows for the linear scaling of components. For an ICE, this means linear scaling of the torque so as to provide the required maximum power. This type of scaling is valid only in the neighborhood near the actual parameter where the efficiency map for a slightly larger or smaller component would not change drastically. Scaling of the Geo ICE is shown in Figure 14.3, so that the ICE gives a maximum power of 50 kW instead of the nominal 43 kW.

In the latest version of ADVISOR, the functionality of the software was improved by allowing links to other software packages such as Ansoft Simplorer [17] and Synopsys Saber [18]. These powerful packages allow for a more detailed look at the electric systems of the vehicle.

As an application example, ADVISOR is used to simulate a hybrid battery–ultracapacitor energy storage system. More extensive applications can be found in [19], where ADVISOR is used to model a hybrid fuel cell/battery powertrain and hybrid fuel cell/ultracapacitor powertrain and simulate their fuel economy and performance. The concept of using a hybrid energy storage system consisting of a battery and a UC is well known and well documented in the literature [20, 21]. The UC provides and absorbs the current peaks, while the battery provides the average power required for the electric motor. This arrangement of hybrid energy storage in an HEV extends the life of the battery and allows the motor to operate more aggressively. Simulating such a system in ADVISOR allows the user to visualize the fuel economy benefit. At the same time, the program allows the user to design the best control strategy for the battery–ultracapacitor hybrid to improve the battery life and the overall system performance. Finally, the size of the components can be optimized, and thus the cost and weight of the system can be reduced.

The default battery model in ADVISOR operates by requesting a specific amount of power from the battery as decided by the vehicle control strategy. Depending on the amount of power that the battery is able to supply, the battery module will send out the power available from the battery to the other subsystems. Due to the hybrid backward/forward simulation method of ADVISOR, the amount of power that the batteries are able and required to supply in a given time step is calculated in a single iteration. From this value, the battery model calculates the battery variables such as current, voltage, and temperature.

However, hybrid battery–ultracapacitor energy storage systems cannot be modeled within ADVISOR using the above default battery model. Here, we have to replace the energy storage model with a more complex model. Fortunately, the subsystem model in ADVISOR can be altered as long as the types of inputs and outputs to the rest of the vehicle are not altered. In our simulation, we replaced the battery model by a model of the combination of a battery and a UC connected to a local control strategy unit that splits the power demand between the battery and the UC. Detailed information about the control strategy is available in [21]. The block diagram representation of the system is shown in Figure 14.4.

The use of the model described above gives the user a way to quickly and easily simulate the battery–ultracapacitor subsystem in a vehicle environment. It allows the user to observe the benefit of using the UC on the fuel economy of the vehicle as well as

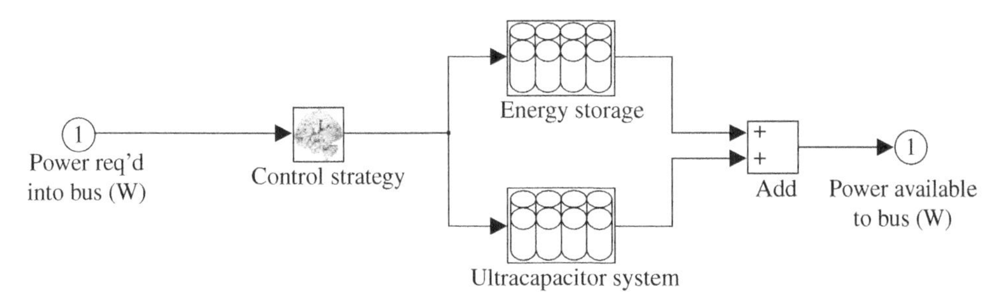

Figure 14.4 Block diagram representation of the new battery subsystem that consists of the battery and ultracapacitor. The input/output relation with the rest of the system is left unchanged.

the benefit to the battery by making the battery's state of charge more even and by reducing the peaks of the battery current that the battery has to accept. It also allows the user to validate whether the system operates as efficiently as it would if the battery size were reduced. Finally, the user can optimize the battery–ultracapacitor control strategy (in other words, how the power demand will be split) without having to think about the complexities of designing the power electronics to make this control system feasible. In addition, the system can be optimized before any system is built and the system cost and possible savings can be easily calculated at the early design stage. Once the control strategy is optimized, the actual DC–DC converter with the required control strategies can be integrated into the simulation by interfacing with Saber or Ansoft Simplorer software [20].

14.4 HEV Modeling Using PSAT

PSAT (AUTONOMIE) is state-of-the-art flexible simulation software developed by Argonne National Laboratory and sponsored by the US Department of Energy (DOE) [7]. PSAT is modeled in a MATLAB/Simulink environment and is set up with a graphical user interface (GUI) written in C#, which makes it user friendly and easy to use. Being a forward-looking model, PSAT allows users to simulate more than 200 predefined configurations, including conventional, pure electric, fuel cell, and hybrids (parallel, series, power split, and series–parallel). The large library of component data enables users to simulate light, medium, and heavy-duty vehicles.

The level of detail in component models can be flexible; for example, a lookup table model or high-fidelity dynamic model can be used for a component, depending on the user's simulation requirements. To maintain modularity, every model must have the same number of input and output parameters. The use of quasi-steady models and control strategies including propelling, braking, and shifting strategies sets PSAT apart from other steady-state simulation tools like ADVISOR. This feature allows PSAT to predict fuel economy and performance of a vehicle more accurately. Its modeling accuracy has been validated against the Ford P2000 and Toyota Prius. PSAT is designed to co-simulate with other environments and is capable of running optimization routines. HIL testing is made possible in PSAT with the help of PSAT-PRO, a control code to support the component and vehicle control [7].

As an application example, PSAT is used to optimize a parallel HEV for maximum fuel economy on a composite driving cycle. Four global algorithms – divided rectangle (DIRECT), simulated annealing (SA), genetic algorithm (GA), and particle swarm optimization (PSO) – are used in the model-based design optimization [22]. Details can be found in Chapter 13.

14.5 Physics-Based Modeling

PSAT and ADVISOR are based on experiential models in the form of lookup tables and efficiency maps. The accuracy of these tools may not be good enough for vehicles operating under extreme conditions. For detailed dynamic modeling and simulation of an HEV system, physics-based modeling is needed. VTB, PSIM, Simplorer, and V-Elph are good

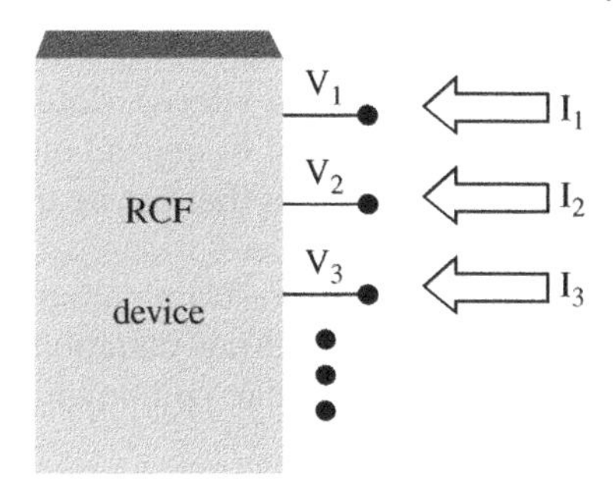

Figure 14.5 Physics-based resistive companion form (RCF) modeling technique.

examples of physics-based modeling tools, where the state variables of a component or subsystem are modeled according to the physical laws representing the underlying principles. The resulting model is a function of device parameters, physical constants, and variables. Such physics-based models can facilitate high-fidelity simulations for dynamics at different timescales and also for controller development.

In this section, the physics-based modeling technique is explored – resistive companion form (RCF) modeling in particular [23]. The RCF method originates from electrical engineering, but is suitable for multidisciplinary modeling applications such as the hybrid powertrain.

14.5.1 RCF Modeling Technique

The RCF method has been used successfully in a number of industry-standard electronic design tools such as SPICE [24] and Saber. Recently, it has also been applied in VTB [9, 23], which is recognized as the leading software for prototyping of large-scale, multitechnical dynamic systems. Using the RCF modeling technique, we can obtain high-fidelity physics-based models of each component in a modular format. These models can be seamlessly integrated to build a system simulation model suitable for design. Just as a physical device is connected to other devices to form a system, the device can be modeled as a block with a number of terminals through which it can be interconnected to other component models, as shown in Figure 14.5. Each terminal has an associated across variable and a through variable. If the terminal is electrical, these variables are the terminal voltage with respect to a common reference and the electric current flowing into the terminal, respectively. Note that the concept of across and through variables in RCF is similar to the effort–flow concepts used in ADVISOR and PSAT.

The general form of the RCF model can be expressed as follows, which is obtained by numerically integrating the differential–algebraic equations describing the dynamics of the component:

$$\begin{bmatrix} \mathbf{i}(t) \\ \mathbf{0} \end{bmatrix} = \mathbf{G}[\mathbf{v}(t), \mathbf{v}(t-h), \mathbf{i}(t), \mathbf{i}(t-h), \mathbf{y}(t), \mathbf{y}(t-h), t] \\ \times \begin{bmatrix} \mathbf{v}(t) \\ \mathbf{y}(t) \end{bmatrix} - \begin{bmatrix} \mathbf{b}_1[\mathbf{v}(t), \mathbf{v}(t-h), \mathbf{i}(t), \mathbf{i}(t-h), \mathbf{y}(t), \mathbf{y}(t-h), t] \\ \mathbf{b}_2[\mathbf{v}(t), \mathbf{v}(t-h), \mathbf{i}(t), \mathbf{i}(t-h), \mathbf{y}(t), \mathbf{y}(t-h), t] \end{bmatrix} \tag{14.1}$$

where $\mathbf{i}$ is a vector of through variables; $\mathbf{v}$ is a vector of across variables; $\mathbf{y}$ is a vector of internal state variables; h is the numerical integration time step; $\mathbf{G}$ is a Jacobian matrix; $\mathbf{b}_1$, $\mathbf{b}_2$ are vectors depending in general on past history values of through and across variables and internal states, and on values of these quantities at time instant t. Note that $\mathbf{G}$, $\mathbf{b}_1$, and $\mathbf{b}_2$ depend on the chosen integration method. The most common integration methods that can be used are the trapezoidal rule and second-order Gear's method.

After all the powertrain components are modeled in RCF, they can be integrated into one set of algebraic equations by applying the connectivity constraints between neighboring modular components, which can then be solved to get system state variables.

14.5.2 Hybrid Powertrain Modeling

Modeling examples for powertrain components are given for a DC machine, a DC–DC boost power electronics converter, and vehicle dynamics. Through these modeling examples, the principles of physics-based modeling techniques are demonstrated. Extensive coverage of models for all the powertrain components is not intended for this chapter.

14.5.3 Modeling of a DC Machine

An equivalent circuit model of a DC machine is illustrated in Figure 14.6, where R and L are the armature resistance and inductance, respectively. The DC machine has two electrical terminals (0,1) and one mechanical terminal (2).

The through variables are: $\mathbf{i} = [i_0, i_1, T_{sh}]^t$, where T_{sh} ($= i_2$) is the mechanical torque at the machine shaft; the superscript t indicates matrix transpose. The across variables are: $\mathbf{v} = [v_0, v_1, \omega]^t$, where ω ($= v_2$) is the rotational speed of the machine shaft.

The differential algebraic equations describing the machine dynamics are

$$\begin{cases} i_0 = -\dfrac{L}{R}\dfrac{di_0}{dt} + \dfrac{1}{R}(v_0 - v_1) - \dfrac{k_e\phi}{R}v_2 \\ i_1 = -i_0 \\ i_2 = -(k_T\phi)i_0 + J\dfrac{dv_2}{dt} + d \bullet v_2 \end{cases} \tag{14.2}$$

where J is shaft inertia, d is the drag coefficient, and ϕ is the flux per pole. Applying the trapezoidal integration rule, we get the following RCF model:

$$i(t) = G(h) \bullet v(t) - b(t-h) \tag{14.3}$$

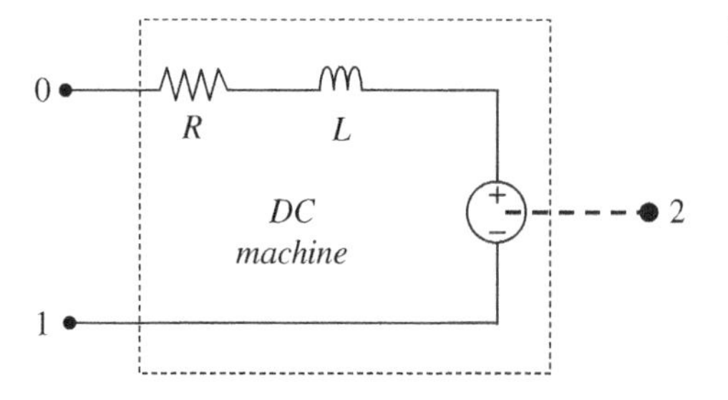

Figure 14.6 DC machine modeling.

where

$$G(h)=\begin{bmatrix} \frac{h}{hR+2L} & \frac{-h}{hR+2L} & \frac{-hk_e\phi}{hR+2L} \\ \frac{-h}{hR+2L} & \frac{h}{hR+2L} & \frac{hk_e\phi}{hR+2L} \\ \frac{-hk_T\phi}{hR+2L} & \frac{hk_T\phi}{hR+2L} & \left(\frac{hk_e\phi k_T\phi}{hR+2L}+\frac{2J}{h}\right) \end{bmatrix} \tag{14.4}$$

$$b(t-h)=\begin{bmatrix} b_0(t-h) \\ -b_0(t-h) \\ b_2(t-h) \end{bmatrix} \tag{14.5}$$

$$\begin{aligned} b_0(t-h)=&\frac{hR-2L}{hR+2L}i_0(t-h)-\frac{h}{hR+2L}v_0(t-h) \\ &+\frac{h}{hR+2L}v_1(t-h)+\frac{hk_e\varphi}{hR+2L}v_2(t-h) \end{aligned} \tag{14.6}$$

$$b_2(t-h)=-k_T\phi b_0(t-h)+k_T\phi i_0(t-h)+i_2(t-h)+\frac{2J}{h}v_2(t-h) \tag{14.7}$$

14.5.4 Modeling of DC–DC Boost Converter

An equivalent circuit model of the DC–DC boost converter is illustrated in Figure 14.7. The DC–DC boost converter has three electrical terminals (0, 1, and 2). Here, we derive the average state space model, based on the two states of the circuit when the switch is on or off.

When the switch Q is ON, we have the following state space dynamic equations:

$$\begin{aligned} \frac{di_0}{dt}&=\frac{1}{L}(v_0-v_1) \\ \frac{d(v_2-v_1)}{dt}&=\frac{1}{C}i_2 \end{aligned} \tag{14.8}$$

When the switch Q is OFF, we have the following state space dynamic equations:

$$\begin{aligned} \frac{di_0}{dt}&=\frac{1}{L}(v_0-v_2) \\ \frac{d(v_2-v_1)}{dt}&=\frac{1}{C}(i_0+i_2) \end{aligned} \tag{14.9}$$

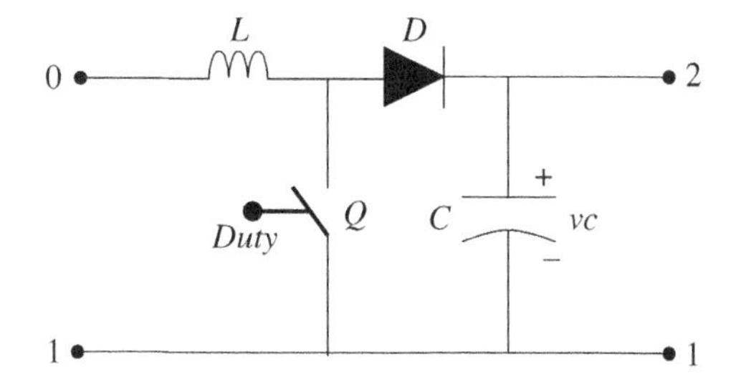

Figure 14.7 DC–DC boost converter modeling.

Hence, the Middlebrook state space averaging model is (d = duty)

$$\begin{aligned}\frac{di_0}{dt} &= \frac{d}{L}(v_0 - v_1) + \frac{(1-d)}{L}(v_0 - v_2)\\ \frac{d(v_2 - v_1)}{dt} &= \frac{d}{C}i_2 + \frac{(1-d)}{C}(i_0 + i_2)\end{aligned} \tag{14.10}$$

Applying the trapezoidal integration rule, we get the following RCF model for the boost power converter:

$$i(t) = G(h) \bullet v(t) - b(t-h) \tag{14.11}$$

where

$$G(h) = \begin{bmatrix} \frac{h}{2L} & \frac{-hd}{2L} & \frac{-h(1-d)}{2L} \\ \frac{-hd}{2L} & \frac{hd^2}{2L} + \frac{2C}{h} & \frac{hd(1-d)}{2L} - \frac{2C}{h} \\ \frac{-h(1-d)}{2L} & \frac{hd(1-d)}{2L} - \frac{2C}{h} & \frac{hd(1-d)^2}{2L} + \frac{2C}{h} \end{bmatrix} \tag{14.12}$$

$$b(t-h) = \begin{bmatrix} b_0(t-h) \\ -b_0(t-h) - b_2(t-h) \\ b_2(t-h) \end{bmatrix} \tag{14.13}$$

$$b_0(t-h) = -i_0(t-h) - \frac{h}{2L}v_0(t-h) + \frac{hd}{2L}v_1(t-h) + \frac{h(1-d)}{2L}v_2(t-h) \tag{14.14}$$

$$\begin{aligned} b_2(t-h) = &-(1-d)b_0(t-h) + (1-d)i_0(t-h) + i_2(t-h) \\ &- \frac{2C}{h}v_1(t-h) + \frac{2C}{h}v_2(t-h) \end{aligned} \tag{14.15}$$

14.5.5 Modeling of Vehicle Dynamics

The vehicle dynamic model can be derived from Newton's second law considering all the forces applied to the vehicle. The driving force comes from the powertrain shaft torque, which can be written as the wheel torque:

$$T_{wh} = R_g \; \eta_{trans} \; T_{sh} \tag{14.16}$$

where R_g and η_{trans} are the transmission gear ratio and transmission efficiency, respectively. This wheel torque provides the driving force to the vehicle:

$$F_d = \frac{T_{wh}}{r} \tag{14.17}$$

where r is the wheel radius.

The total resistance force consists of rolling resistance, aerodynamic resistance, and gravitational force. Hence, the vehicle dynamic equation can be obtained:

$$\begin{aligned} F_d &= F_{gxt} + F_{roll} + F_{ad} + ma \\ &= mg\,\sin(\alpha) + mg\left(C_0 + C_1 v\right)\text{sgn}(v) \\ &\quad + \frac{1}{2}\rho C_d\, A_F\, \left(v + v_0\right)^2 \text{sgn}(v) + \left(m + \frac{J_{wh}}{r^2}\right)\frac{dv}{dt} \end{aligned} \tag{14.18}$$

where F_{gxt} is the gravitational force on a grade, F_{roll} is the rolling resistance, F_{ad} is the aerodynamic resistance, m is the vehicle mass, g is the acceleration due to gravity, α is the angle of grade, C_0 and C_1 are the rolling coefficients, ρ is the air density, C_d is the aerodynamic drag coefficient, A_F is the vehicle frontal area, v_0 is the wind speed, v is the vehicle linear speed, and J_{wh} is the wheel inertia.

Similarly, applying the trapezoidal integration rule, we get the following RCF model for the vehicle dynamics:

$$i(t) = G(h) \bullet v(t) - b(t-h) \tag{14.19}$$

where the through variable is $i(t) = F_d$ and the across variable $v(t) = v$ (vehicle velocity).

Note that Equation 14.18 is a nonlinear model, requiring an iterative Newton–Raphson solution procedure at each simulation time step; the Jacobian $G(h)$ is

$$G(h, v(t)) = mgC_1 \operatorname{sgn}(v) + \rho C_d \ A_F (v(t) + v_0) \operatorname{sgn}(v) + \frac{2}{h}\left(m_{veh} + \frac{J_{wh}}{r^2} \right) \tag{14.20}$$

Other RCF models for induction machine, batteries, UCs, and so on, can be found in [23, 25, 26]. Based on the same principles, an ICE model and fuel cell model can be developed.

Finally, as an example of employing RCF techniques for HEVs, a hybrid fuel cell–ultracapacitor–battery vehicle model is modeled in VTB, as shown in Figure 14.8. Upon simulation, variables that are of interest can be plotted, as shown in Figure 14.9, where the reference vehicle speed, battery, UC, and DC motor currents are plotted. Details of how to use VTB can be found in [9].

14.5.6 Wheel Slip Model

In simulations involving vehicle dynamics, the wheel slip model must be implemented. Figure 14.10 shows the one-wheel model of an HEV. Applying a driving torque or a braking force F_m to a pneumatic tire produces tractive (braking) force F_d at the tire–ground contact patch due to the wheel slip. The slip ratio λ is defined as

$$\lambda = \frac{V_\omega - V}{\max\{V, V_\omega\}} \tag{14.21}$$

where V is the speed of the vehicle and V_ω is the linear speed of the wheel.

The wheel speed can be expressed as

$$V_\omega = \omega r \tag{14.22}$$

where ω is the angular speed of the wheel and r is the radius of the wheel.

During normal driving, $\lambda > 0$, there exists a friction force on the wheel in the direction of the forward motion. This friction force, also known as the *traction force*, is caused by the slip between the road surface and the tire. This force contributes to the forward motion of the vehicle during normal driving. During braking, external forces are applied to the wheel so that the wheel linear speed is less than the vehicle speed; that is, $\lambda < 0$. Therefore, there exists a braking force opposite to the forward motion.

The traction force, or braking force in the case of braking, as shown in Figure 14.10, can be expressed as

$$F_d(\lambda) = \mu(\lambda) mg \tag{14.23}$$

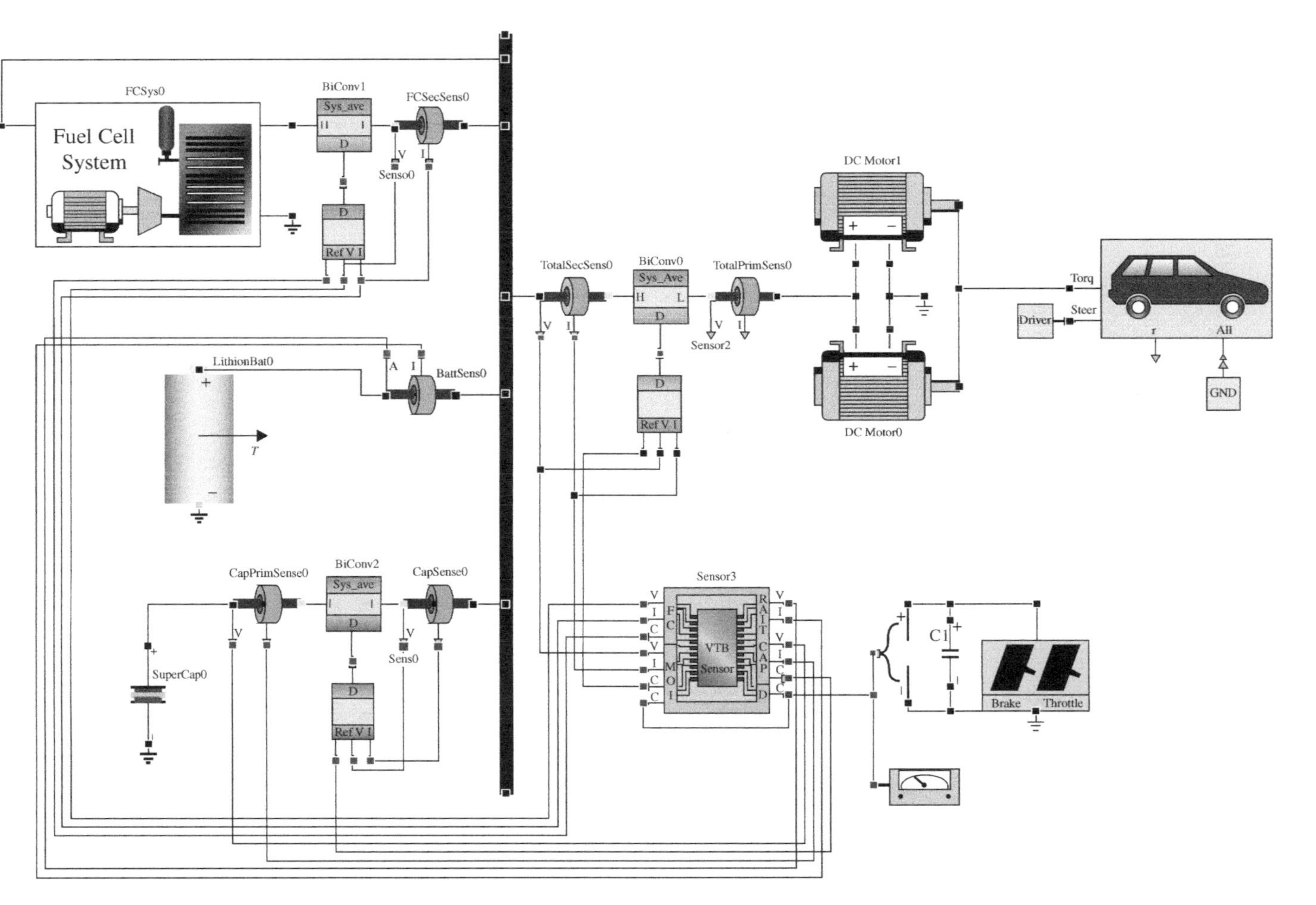

Figure 14.8 Modeling a hybrid fuel cell/ultracapacitor/battery vehicle in VTB [9]. *Source:* University of South Carolina Board of trustees.

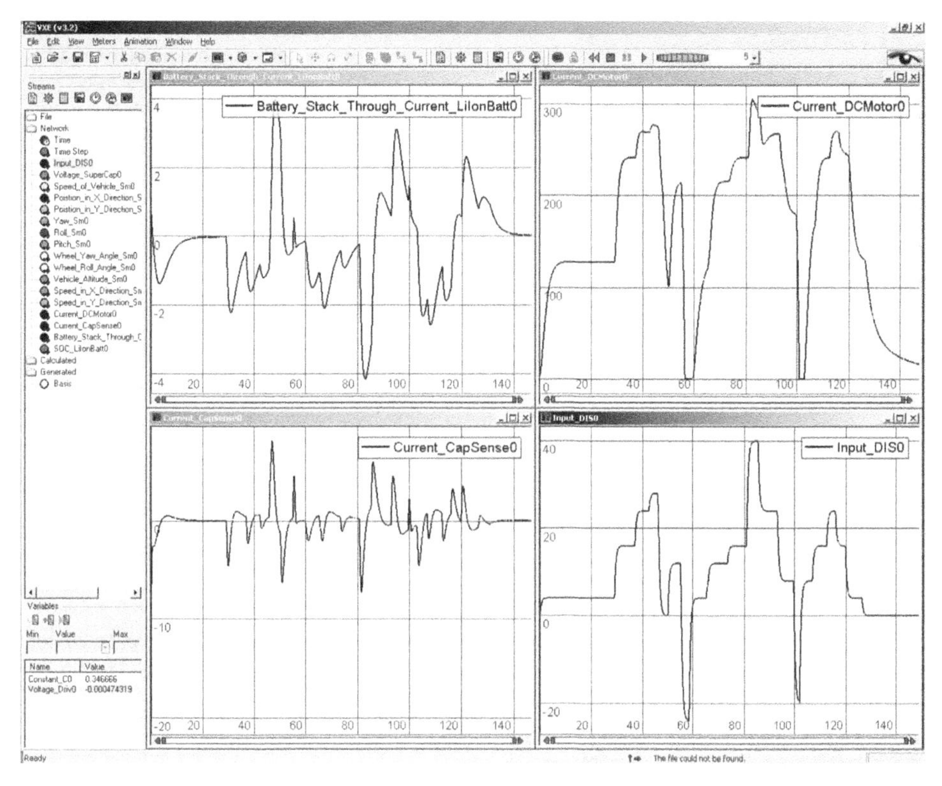

Figure 14.9 Simulation results of hybrid vehicle in VTB.

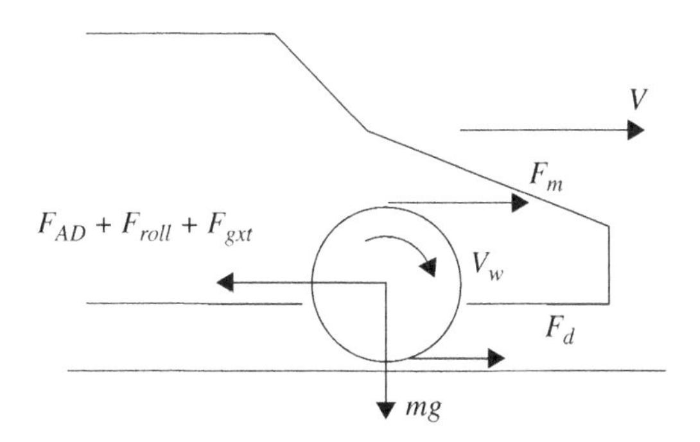

Figure 14.10 One-wheel model of vehicles, where F_m is the force applied to the wheel by the powertrain, F_d is the tractive force caused by tire slip, m is the vehicle mass.

where $\mu(\lambda)$ is the adhesive coefficient between the road surface and the tire. $\mu(\lambda)$ is a function of slip ratio λ and is a function of tire condition and road condition as shown in Figure 14.11.

The equation of the vehicle motion can be expressed as

$$m\frac{dV}{dt} = F_d(\lambda) - (F_{gxt} + F_{AD} + F_{roll}) \tag{14.24}$$

This equation is similar to Equation 14.18 but the tractive force is linked with the slip ratio.

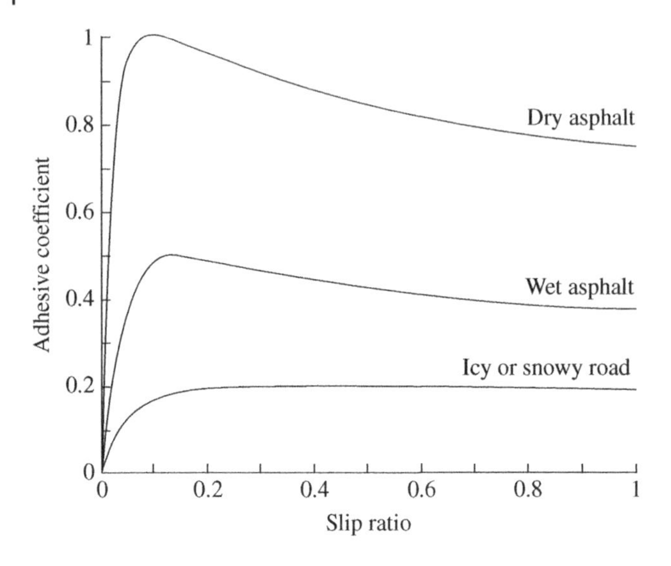

Figure 14.11 Typical adhesive coefficient between the road surface and the tires, as a function of slip ratio and road surface conditions.

During normal driving, the external torque applied to the wheels is positive. In order to enter braking mode, an external torque must be applied to the wheel to slow it down. In an HEV, this torque is the sum of the motor regenerative braking torque and additional braking torque provided by the mechanical braking systems, in case the motor torque is not enough to provide effective braking.

During normal driving, the powertrain torque tries to accelerate the wheel while the tractive force will try to slow it down. During braking, the powertrain torque applied to the wheel is in the opposite direction of the wheel's rotation and slows down the wheel. The tractive force, on the other hand, is in the same direction as the wheel rotation and therefore will accelerate the wheel, as shown in Figure 14.10.

Therefore, the equation of motion of the wheel can be expressed as

$$J_\omega \frac{d\omega}{dt} = T_m - rF_d(\lambda) \tag{14.25}$$

where J_ω is the inertia of the wheel and T_m is the total braking torque, $T_m = F_m r$.

14.6 Bond Graph and Other Modeling Techniques

14.6.1 Bond Graph Modeling for HEVs

Created by H.M. Paynter in 1959, bond graphs are a graphical tool used to describe and model subsystem interactions involving power exchange. This formulation can be used in hydraulics, mechatronics, thermodynamics, and electrical systems. Bond graphs have been proven effective for the modeling and simulation of multidomain systems including automotive systems [27–38].

In a bond graph model, a physical system is represented by basic passive elements that are able to interchange power: *resistances* (*R*), *capacitances* (*C*), and *inertias* (*I*). Although these names suggest a direct application in electrical systems, they are used in

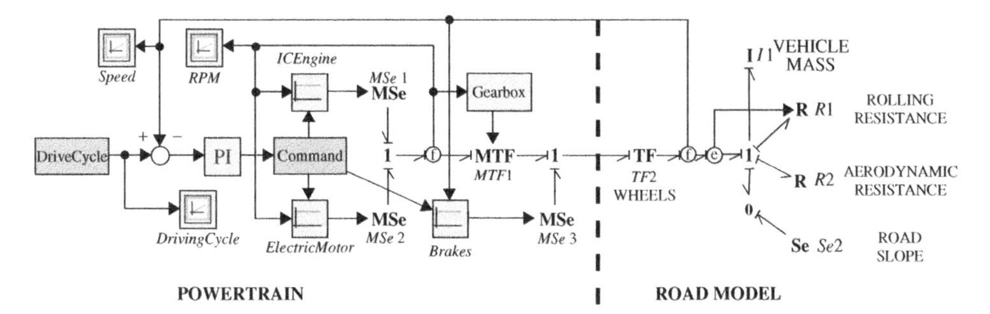

Figure 14.12 A bond graph modeling example: an HEV powertrain model connected to a road model.

other domains as well; for example, friction as a mechanical resistance, a compressible fluid as a capacitance, and a flywheel as an inertial element.

Each element has one or more ports where power exchange can occur. This *power* (*P*) is expressed as a product of two variables: *effort* (*e*) and *flow* (*f*). These names are used extensively in all domains, but have a unique name in each domain: force and speed in mechanical systems, voltage and current in electrical systems, pressure and flow in hydraulics, and so on. Additional variables are defined: *momentum* (*p*) as the time integral of effort and *displacement* (*q*) as the time integral of flow.

Additional elements are needed to fully describe a system: *sources of effort* (S_e) and *sources of flow* (S_f) are active elements that provide the system with effort and flow, respectively; transformers (TFs) and gyrators (GYs) are two-port elements that transmit power, but scale their effort and flow variables by their modulus; and *one junction* (*1*) elements are multiport elements that distribute power sharing equal flow, while *zero junction* (*0*) elements distribute power, having equal effort among all ports.

Bond graph elements are linked with half arrows (bonds) that represent power exchange between them. The direction of the arrow indicates the direction of power flow when both effort and flow are positive. Full arrows are used when a parameter is to be passed between elements, but no power flow occurs.

A bond graph can be generated from the physical structure of the system. For example, the HEV powertrain connected to the road load model can be drawn as shown in Figure 14.12, where the road load is described by Equation 14.18.

Causality in bond graph models is indicated with a vertical stroke at the start or end of the bond arrow. This causal stroke establishes the cause and effect relationships between elements. Causality in bond graphs enables the extraction of system dynamics equations. It also provides an insight into the dynamic behavior of the model and is useful to predict modeling problems such as algebraic loops, differential causality, and causal loops.

Modeling presented in [37, 38] demonstrated that bond graph modeling is an appropriate tool for the modeling and simulation of hybrid and electric vehicles.

14.6.2 HEV Modeling Using PSIM

PSIM is a user-friendly simulation package that was originally developed for simulating power electronics converters and electric machine drives. Its new version allows interactive simulation capability and provides magnetic and thermal models for more flexible and accurate analysis of automotive mechatronics design. However, with a few additional customer-built models, it can also be used to model and simulate electric and hybrid vehicles.

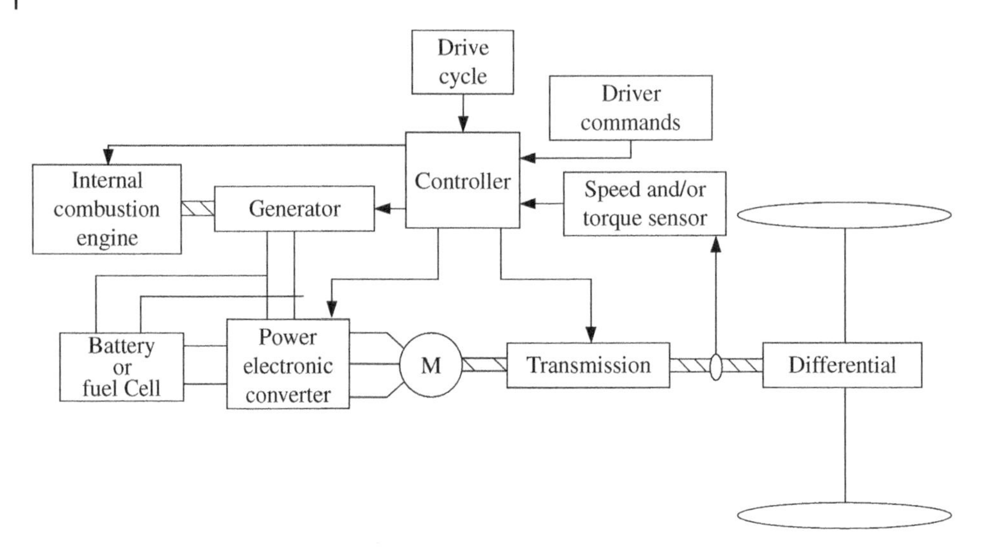

Figure 14.13 Series HEV configuration.

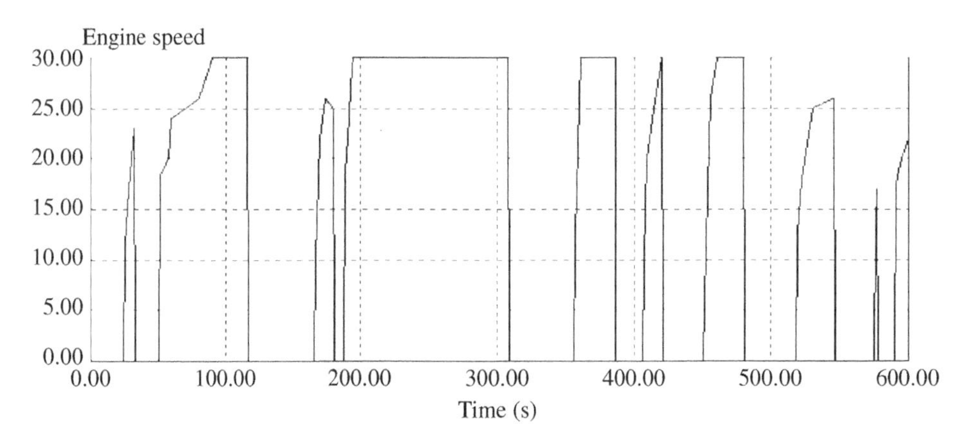

Figure 14.14 Engine speed (×100 rpm) vs. time in seconds.

Module boxes for necessary electrical systems and also for mechanical, energy storage, and thermal systems are created. These modules include ICEs, fuel converters, transmissions, torque couplers, and batteries. Once these modules are made and stored in the PSIM model library, the user can build an electric or a hybrid vehicle model. As an example, a series hybrid configuration, shown in Figure 14.13, is modeled in PSIM [39].

Since load torque is imposed only on the propulsion motor, the ICE can be operated at its optimal efficiency all the time, regardless of the load torque. Therefore, the energy management strategy is simple. The main design task is focused on how and where to operate the ICE in an optimum region [39, 40].

Simulation results of the engine speed for the UDDS drive cycle are presented in Figure 14.14.

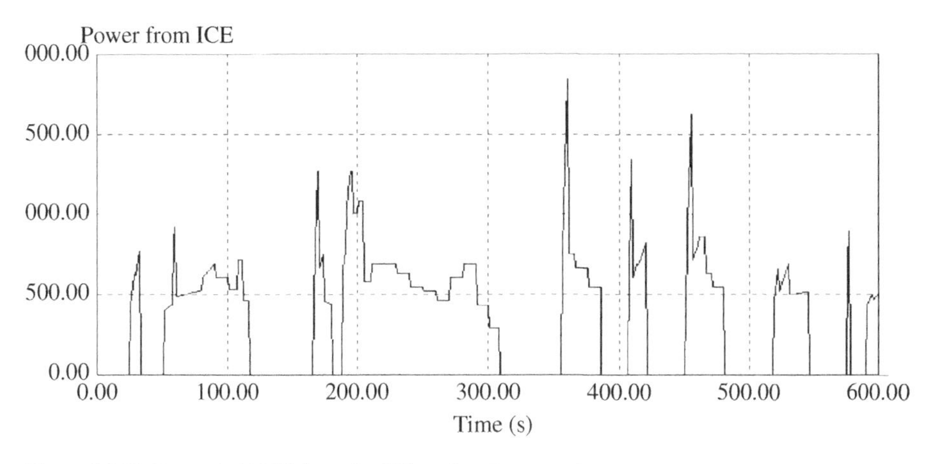

Figure 14.15 Power (×100W) from the ICE vs. time in seconds.

This simulation model assumed that the power produced by the engine is equal to the power generated by the generator and stored directly in the battery. It can be observed that power is produced when the engine is on (Figure 14.15).

14.6.3 HEV Modeling Using Simplorer and V-Elph

Simplorer is a multidomain simulation package that can be used for system-level HEV modeling. It has a comprehensive automotive component library, including batteries, fuel cells, wires (conductors), fuses, lamps, electric motors, alternators, engine models, and relays, in addition to the electronics, power electronics, and controller models. Further, Simplorer can be linked for co-simulation with a finite element-based electromagnetic field simulation package, Maxwell [17]. This capability provides even greater modeling and simulation accuracy for automotive electronics and machine design. In [41], a series hybrid electric HMMWV is modeled in Simplorer. The vehicle model consists of an ICE/generator, a PM DC motor, a DC–DC converter, a battery and battery management system, a PI controller, and a vehicle model. The simulation facilitates the development and functional verification of controller and battery management. Dynamic/transient responses of battery voltage, motor torque, and motor voltage under different drive cycles can be simulated. Also, the vehicle's response to the incline of road grades can be obtained to predict overall system performance.

V-Elph [12] is a system-level MATLAB/Simulink-based modeling, simulation, and analysis tool developed at Texas A & M University. This package uses detailed dynamic models of electric motors, ICEs, batteries, and the vehicle. The dynamic performance and fuel economy, energy efficiency, emissions, and so on, can be predicted for hybrid and electric vehicles.

In addition, software packages such as Modelica [42, 43] are also used in the physics-based modeling and simulation of hybrid and electric vehicles.

14.7 Consideration of Numerical Integration Methods

Numerical integration of differential equations or state equations is essential for performing dynamic system simulation. Therefore, discussion of numerical integration methods is an essential part of a study focusing on modeling and simulation. There are several varieties of numerical integration methods: backward Euler's, trapezoidal, Simpson's, Runge–Kutta, Gear's method, and so on. Of these methods, trapezoidal integration is the most popular one in dynamic modeling and simulation, due to its merits of low distortion and absolute stability (A-stable). For example, the trapezoidal integration rule is used in EMTP, Spice, and VTB. However, numerical oscillations are often encountered, especially in the simulation of power electronics circuits, which are very often used in hybrid powertrains. Specifically, the numerical values of certain variables oscillate around the true values. In other words, only the average values of the simulated results are correct. The magnitude and frequency of these numerical oscillations are directly related to the parameters of the energy storage elements and the simulation time step. Sometimes, this problem is so severe that the simulation results are erroneous.

Two techniques can be used to mitigate the problem for this kind of numerical oscillation: the trapezoidal with numerical stabilizer method and Gear's second-order method. Elimination of numerical oscillations is of great significance in performing a meaningful simulation for power electronics circuits in which switching of semiconductor devices cause current interruptions.

14.8 Conclusion

This chapter has presented an overview of the modeling and simulation of HEV, with specific emphasis on physics-based modeling. Methods for the mitigation of numerical oscillations in dynamic digital simulations were briefly discussed. Additional simulation techniques, such as bond graph modeling, provide added flexibility in HEV modeling and simulation [44].

With the advent of powerful computing, development of computational methods, and advances in software-in-the-loop (SIL) and HIL modeling and simulations, it is now possible to study numerous iterations of different designs with combinations of different components and different topology configurations. HIL is becoming increasingly important for rapid prototyping and development of control systems for new vehicles such as X-by-Wire [45].

Modeling and simulation will play important roles in the success of HEV design and development.

References

1 Muta, K., Yamazaki, M., and Tokieda, J. (2004) Development of new-generation hybrid system THS II – drastic improvement of power performance and fuel economy. SAE World Congress, March 8–11, Detroit, MI, SAE paper no. 2004–01-0064.
2 Horie, T. (2006) Development aims of the new CIVIC hybrid and achieved performance. SAE Hybrid Vehicle Technologies Symposium, February 1–2, San Diego, CA.
3 Struss, P. and Price, C. (2004) Model-based systems in the automotive industry. *AI Magazine*, 24 (4), 17–34.

4 Gao, W., Neema, S., Gray, J., et al. (2005) Hybrid powertrain design using a domain-specific modeling environment. Proceedings of the IEEE Vehicle Power and Propulsion Conference, September, Chicago, pp. 6–12.
5 Wipke, K.B., Cuddy, M.R., and Burch, S.D. (1999) ADVISOR 2.1: a user-friendly advanced powertrain simulation using a combined backward/forward approach. *IEEE Transactions on Vehicular Technology*, 48 (6), 1751–1761.
6 Markel, T., Brooker, A., Hendricks, T., et al. (2002) ADVISOR: a systems analysis tool for advanced vehicle modeling. *Journal of Power Sources*, 110 (2), 255–266.
7 PSAT Documentation, http://www.transportation.anl.gov/software/PSAT/ (accessed 2006).
8 PSIM web site, http://www.powersimtech.com/ (accessed February 2011).
9 VTB web site, http://vtb.ee.sc.edu/ (accessed February 2011).
10 Powell, B., Bailey, K., and Cikanek, S. (1998) Dynamic modeling and control of hybrid electric vehicle powertrain systems. *IEEE Control Systems Magazine*, 18 (5), 17–22.
11 Lin, C.C., Filipi, Z., Wang, Y., et al. (2001) Integrated, feed-forward hybrid electric vehicle simulation in Simulink and its use for power management studies. Proceedings of SAE 2001 World Congress, March, Detroit, MI.
12 Butler, K.L., Ehsani, M., and Kamath, P. (1999) A Matlab-based modeling and simulation package for electric and hybrid electric vehicle design. *IEEE Transactions on Vehicular Technology*, 48 (6), 1770–1778.
13 Rizzoni, G., Guzzella, L., and Baumann, B.M. (1999) United modeling of hybrid electric vehicle drivetrains. *IEEE Transactions on Mechatronics*, 4 (3), 246–257.
14 He, X. and Hodgson, J.W. (2002) Modeling and simulation for hybrid electric vehicles, I. Modeling. *IEEE Transactions on Intelligent Transportation Systems*, 3 (4), 235–243.
15 He, X. and Hodgson, J. (2002) Modeling and simulation for hybrid electric vehicles – Part II. *IEEE Transactions on Transportation Systems*, 3 (4), 244–251.
16 Fritzson, P. (2004) *Principles of Object Oriented Modeling and Simulation with Modelica 2.1*, IEEE Press, Piscataway, NJ.
17 Ansoft Simplorer web site, http://www.ansoft.com/ (accessed February 2011).
18 Saber web site, http://www.synopsys.com/saber/ (accessed 2006).
19 Gao, W. (2005) Performance comparison of a hybrid fuel cell–battery powertrain and a hybrid fuel cell–ultracapacitor powertrain. *IEEE Transactions on Vehicular Technology*, 54 (3), 846–855.
20 Baisden, A.C. and Emadi, A. (2004) An ADVISOR based model of a battery and an ultra-capacitor energy source for hybrid electric vehicles. *IEEE Transactions on Vehicular Technology*, 53 (1), 199–205.
21 Bose, B.K., Kim, M.H., and Kankam, M.D. (1996) Power and energy storage devices for next generation hybrid electric vehicle. Proceedings of the 31st Intersociety Energy Conversion Engineering Conference, August, Washington, DC, vol. 3, pp. 1893–1898.
22 Gao, W. and Porandla, S. (2005) Design optimization of a parallel hybrid electric powertrain. Proceedings of the IEEE Vehicle Power and Propulsion Conference, September, Chicago, pp. 530–535.
23 Gao, W., Solodovnik, E., and Dougal, R. (2004) Symbolically-aided model development for an induction machine in Virtual Test Bed. *IEEE Transactions on Energy Conversion*, 19 (1), 125–135.
24 SPICE web site, http://bwrc.eecs.berkeley.edu/Classes/IcBook/SPICE/ (accessed February 9, 2011).

25 Gao, L., Liu, S., and Dougal, R.A. (2002) Dynamic lithium-ion battery model for system simulation. *IEEE Transactions on Components and Packaging Technologies*, 25 (3), 495–505.

26 Gao, L., Liu, S., and Dougal, R.A. (2003) Active power sharing in hybrid battery/capacitor power sources. Applied Power Electronics Conference and Exposition, APEC'03, February, Miami Beach, FL, vol. 1, pp. 497–503.

27 Xia, S., Linkens, D.A., and Bennett, S. (1993) Automatic modeling and analysis of dynamic physical systems using qualitative reasoning and bond graphs. *Intelligent Systems Engineering*, 2, 201–212.

28 Gissinger, G.L., Chamaillard, Y., and Stemmelen, T. (1995) Modeling a motor vehicle and its braking system. *Journal of Mathematics and Computers in Simulation*, 39, 541–548.

29 Suzuki, K. and Awazu, S. (2000) Four-track vehicles by bond graph-dynamic characteristics of four-track vehicles in snow. 26th Annual Conference of the IEEE Industrial Electronics Society, IECON 2000, October, Nagoya, Japan, vol. 3, pp. 1574–1579.

30 Kim, J.-H. and Cho, D.D. (1997) An automatic transmission model for vehicle control. IEEE Conference on Intelligent Transportation System, ITSC'97, November, Boston, MA, pp. 759–764.

31 Nishijiri, N., Kawabata, N., Ishikawa, T. et al. (2000) Modeling of ventilation system for vehicle tunnels by means of bond graph. 26th Annual Conference of the IEEE Industrial Electronics Society, IECON 2000, October, Nagoya, Japan, vol. 3, pp. 1544–1549.

32 Kuang, M.L., Fodor, M., Hrovat, D., et al. (1999) Hydraulic brake system modeling and control for active control of vehicle dynamics. Proceedings of the 1999 American Control Conference, June, San Diego, CA, vol. 6, pp. 4538–4542.

33 Khemliche, M., Dif, I., Latreche, S., et al. (2004) Modeling and analysis of an active suspension 1/4 of vehicle with bond graph. Processing of the First International Symposium on Control, Communications and Signal Processing, March, Hammamet, Tunisia, pp. 811–814.

34 Truscott, A.J. and Wellstead, P.E. (1990) Bond graphs modeling for chassis control. IEE Colloquium on Bond Graphs in Control, April, London, UK, pp. 5/1–5/2.

35 Coudert, N., Dauphin-Tanguy, G., and Rault, A. (1993) Mechatronic design of an automatic gear box using bond graphs. International Conference on Systems, Man, and Cybernetics, October 17–20, Le Touquet, France, pp. 216–221.

36 Jaume, D. and Chantot, J. (1993) A bond graph approach to the modeling of thermics problems under the hood. International Conference on Systems, Man and Cybernetics, October 17–20, Le Touquet, France, pp. 228–233.

37 Hubbard, G.A. and Youcef-Toumi, K. (1997) Modeling and simulation of a hybrid-electric vehicle drivetrain. Proceedings of the 1997 American Control Conference, June 4–6, Albuquerque, NM, vol. 1, pp. 636–640.

38 Filippa, M., Mi, C., Shen, J., and Stevenson, R. (2005) Modeling of a hybrid electric vehicle test cell using bond graphs. *IEEE Transactions on Vehicular Technology*, 54 (3), 837–845.

39 Onoda, S. and Emadi, A. (2004) PSIM-based modeling of automotive power systems: conventional, electric, and hybrid electric vehicles. *IEEE Transactions on Vehicular Technology*, 53 (2), 390–400.

40 Juchem, R. and Knorr, B. (2003) Complete automotive electrical system design. Proceedings of the 2003 Vehicular Technology Conference, October 6–9, Orlando, FL, vol. 5, pp. 3262–3266.

41 Ducusin, M., Gargies, S., Berhanu, B., et al. (2005) Modeling of a series hybrid electric high mobility multipurpose wheeled vehicle. Proceedings of the IEEE Vehicle Power and Propulsion Conference, September, Chicago, pp. 561–566.
42 Otter, M. and Elmqvist, H. (2001) Modelica Language, Libraries, Tools, Workshop, and EU-Project RealSim, The Modelica Organization, http://www.modelica.org/documents/ModelicaOverview14.pdf (accessed June 2001).
43 Glielmo, L., Natale, O.R., and Santini, S. (2003) Integrated simulations of vehicle dynamics and control tasks execution by Modelica. IEEE/ASME International Conference on Advanced Intelligent Mechatronics, July 20–24, Kobe, Japan, vol. 1, pp. 395–400.
44 Gao, W., Mi, C., and Emadi, A. (2007) Modeling and simulation of electric and hybrid vehicles. *Proceedings of the IEEE,* Special Issue on Electric and Hybrid Fuel Cell Vehicles, 95 (4), 729–745.
45 Chu, L., Wang, Q., Liu, M., et al. (2005) Control algorithm development for parallel hybrid transit bus. Proceedings of the IEEE Vehicle Power and Propulsion Conference, September, Chicago, pp. 196–200.

15

HEV Component Sizing and Design Optimization

15.1 Introduction

One key research and development topic for hybrid electric vehicles (HEVs) is an innovative hybrid powertrain, whose parameters must be tuned for the desired performance of the hybrid vehicle [1]. A hybrid powertrain comprises electric motors, power electronics converters, energy storage devices such as batteries and ultracapacitors, and sophisticated controllers, in addition to such classical components as internal combustion engines (ICEs), transmissions, clutches, drive shafts, and differentials. Therefore, a hybrid powertrain is much more complicated than a conventional powertrain. The component sizing and system prototyping of a hybrid powertrain is difficult because of the many design options and the rapidly developing technologies in the automotive industries [2]. The cost and performance of the designed hybrid powertrain are determined by the chosen configuration and hundreds of design variables and parameters. A parametric design method can be used to determine hybrid vehicle component sizes. Example design variables include the power ratings of the fuel converter (ICE), motor controller, number of battery cells, final drive ratio, and control strategy parameters. The parametric design can yield a hybrid vehicle with a better performance than the baseline vehicle. The overall design is unlikely to be optimal, but it can be used as a rough design for further optimization so as to maximize fuel economy and minimize emissions, weight, and cost. At the same time, vehicle performance requirements must be satisfied [3, 4].

There are a variety of optimization algorithms available for HEV design. They can be categorized in different ways: for example, the local optimization algorithm versus the global optimization algorithm; deterministic versus stochastic; or gradient-based versus derivative-free. A good selection of optimization algorithms for the application of hybrid powertrain design is not very obvious. In this chapter, four optimization algorithms are thoroughly discussed. Design optimization examples are given for a parallel HEV and a series HEV. Since an analytical expression for the objective function does not exist, a vehicle simulation model is used for function evaluations. PSAT (Powertrain System Analysis Toolkit) and ADVISOR are chosen as HEV modeling tools for the optimization study.

This chapter explores the feasibility of different global optimization algorithms by comparing their performance and accuracy. Section 15.2 reviews the principles and procedures of four global optimization algorithms. Section 15.3 presents the methodology of the model-in-the-loop design process used for this study. In Section 15.4, the

Hybrid Electric Vehicles: Principles and Applications with Practical Perspectives,
Second Edition. Chris Mi and M. Abul Masrur.

constrained HEV design optimization problem is set up and the results are presented with the associated comparison. Section 15.5 presents another design example of a series HEV using the non-dominated sorting genetic algorithm (NSGA). Finally, conclusions are given in Section 15.6.

15.2 Global Optimization Algorithms for HEV Design*

For HEV design, the performance and design objectives, such as fuel economy, can be evaluated from a vehicle model and computer simulation. The value is also called the *response function*, which is typically multimodal (involving many local minima), and sometimes noisy and discontinuous [3]. Gradient-based algorithms such as sequential quadratic programming (SQP) [5] use the derivative information to find the local minima. The major disadvantage of local optimization algorithms is that they do not search the entire design space and cannot find the global minimum. Derivative-free algorithms such as divided rectangles (DIRECT) [6, 7], simulated annealing (SA) [8], genetic algorithm (GA) [9], and particle swarm optimization (PSO) [10, 11] do not rely on the derivatives and can therefore work exceptionally well when the objective function is noisy and discontinuous. Derivative-free methods are often the best global algorithms because they must often sample a large portion of the design space to be successful. A comparison of the gradient-based and derivative-free algorithms for the optimization of an HEV is given in [3, 12]. Note here that even though DIRECT, SA, GA, and PSO algorithms search the design space globally, the main difference is that DIRECT is a deterministic algorithm, whereas SA, GA, and PSO are stochastic algorithms.

15.2.1 DIRECT

DIRECT is a sampling algorithm, developed by Donald R. Jones [6]. This global optimization algorithm is a modification of the standard Lipschitz approach that eliminates the need to specify the Lipschitz constant [7]. This constant is a weighing parameter, which determines the different emphasis on the global or local search [7]. The use of the Lipschitz constant is eliminated in [6] by searching all possible values for it, thus putting a balanced emphasis on both the global and local search.

A brief introduction to the DIRECT algorithm is presented here, but a detailed explanation of it can be found in [6]. DIRECT is a modification of a one-dimensional Lipschitz algorithm by Shubert [13] and an extension to multidimensional problems. The approach proposed by Shubert is given as follows.

A function $f(x)$ defined in the closed interval $[l, u]$ is said to have a lower bound such that there exists a positive Lipschitz constant K that satisfies the following condition

$$|f(x)-f(x')| \leq K|x-x'| \text{ for all } x,x' \in [a,b] \tag{15.1}$$

From Equation 15.1, the change in the value of function $f(x)$ should be less than the change between x and x' multiplied by a constant K. Further, let us substitute a or b for x' in Equation 15.1; then the following inequality constraints are obtained for $f(x)$:

$$f(x) \leq f(a)+K(x-a) \tag{15.2}$$

$$f(x) \leq f(b)-K(x-b) \tag{15.3}$$

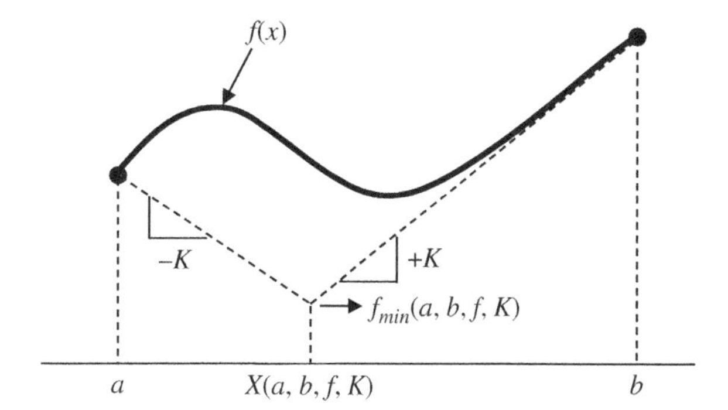

Figure 15.1 Lower bound of a function $f(x)$ using the Lipschitz constant.

These two equations correspond to two lines with slopes $-K$ and $+K$ as shown in Figure 15.1. These two lines form a V shape and the smallest value that $f(x)$ can possibly be is at the bottom of the V shape; that is, $X(a,b,f,K)$. So, $f_{\min}(a,b,f,K)$ corresponds to the lower bound of $f(x)$:

$$X(a,b,f,K) = (a+b)/2 + [f(a) - f(b)]/2K \tag{15.4}$$

$$f_{\min}(a,b,f,K) = [f(a) + f(b)]/2 - K(b-a) \tag{15.5}$$

where $f(a)$ and $f(b)$ are the function values of f at a and b.

This minimum point is taken as x_1, which is used to divide the search space into two intervals. Then one of the two intervals with the least function value f_{min} is selected for further division. This division is continued until some pre-specified tolerance of the final solution is met. Equation (15.5) explains the local and global search of the algorithm. The first term leads to the local search and the second term leads to the global search. The relative weight between the local and global search is determined by the Lipschitz constant. The larger the value of K, the higher the emphasis on the global search.

However, the Lipschitz approach followed by Shubert has the following two main disadvantages:

1) The need to specify the correct value of the Lipschitz constant K.
2) The need for 2^n function evaluations for an n-dimensional design space.

These problems are fixed in the DIRECT algorithm, in which the sampling is done at the center point of the interval rather than at the endpoints to reduce the number of function evaluations [6]. The balance between the local and global search in the DIRECT algorithm is achieved by selecting the optimal intervals (optimal rectangles) assuming all possible values for the Lipschitz constant. For example, assume that Figure 15.1 is divided into 10 intervals so that 10 center points are obtained. The function values at the center points of the intervals are evaluated. A plot showing these 10 points with the width of the interval on the x axis and the corresponding function value is given in Figure 15.2. If a line with a slope given by the Lipschitz constant K is drawn from a point, then the y intercept is the local bound for the function. Instead of fixing one value of K, various possible values of K are taken. This gives the lowest lower bound intervals represented by the lower convex hull of points shown in Figure 15.2. The same procedure will be followed to select the optimal rectangles in the DIRECT algorithm for the n-dimensional problem.

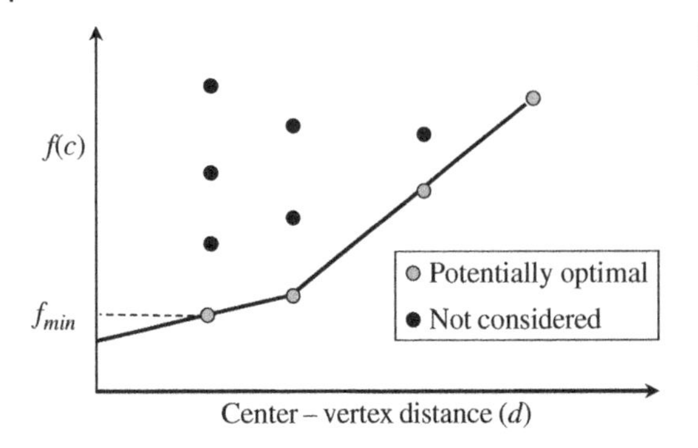

Figure 15.2 Rectangle selection using all possible K.

The DIRECT algorithm is based on the above theoretical background, and is basically a sampling algorithm. At the beginning of the algorithm, the n-dimensional design space box is scaled to an n-dimensional unit hypercube. To initiate the search, the design objective function is evaluated at the center point of the hypercube. DIRECT then trisects this hyperrectangle and samples the center points of the two resulting hyperrectangles. From here, DIRECT selects the optimal hyperrectangles using various values of the Lipschitz constant and trisects them. An example selection of the optimal hyperrectangles is shown by the lower convex hull of dots in Figure 15.2. The selection of optimal hyperrectangles selects the larger rectangles (global search) as well as smaller rectangles (local search). This division process continues until pre-specified function evaluations are reached or convergence is achieved. The division of rectangles in the first three iterations of a two-dimensional problem is illustrated in Figure 15.3. In this figure, d denotes the distance between the center and vertex, and each center point is labeled with a numeral for the sake of explanation.

In the first iteration, the unit hypercube is trisected into three rectangles. The objective function value is evaluated at the center points of the three resulting rectangles. The objective function values are plotted versus the center–vertex distance as shown in Figure 15.4a. Then the rectangle with the least objective value in each column of dots is selected as the optimal rectangle. In the first iteration, there is only one column of dots, so rectangle 1 is selected as the optimal rectangle and trisected in the second iteration. Similarly, in the second iteration, rectangle 4 and rectangle 2 have the least objective function values, as shown in Figure 15.4b. These two rectangles are selected as optimal rectangles and further trisected in the third iteration. This process is continued until the maximum number of function evaluations is exhausted.

The inequality constraints are handled by an auxiliary function given in [6], which combines the information of the objective and constraint functions. The auxiliary equation is given in Equation 15.6 and is a weighted sum of the violated constraints and the deviation of the objective function value from a projected global minimum:

$$\max(f_r - f^*, 0) + \sum_{m}^{j=1} c_j \max(g_{rj}, 0) \tag{15.6}$$

In this equation, f_r is the objective function value at the center point of the rectangle r, f^* is the assumed global minimum, m is the number of inequality constraints, c_j are the positive weighing coefficients, and g_{rj} is the constraint violation of the jth constraint at the mid-point of rectangle r.

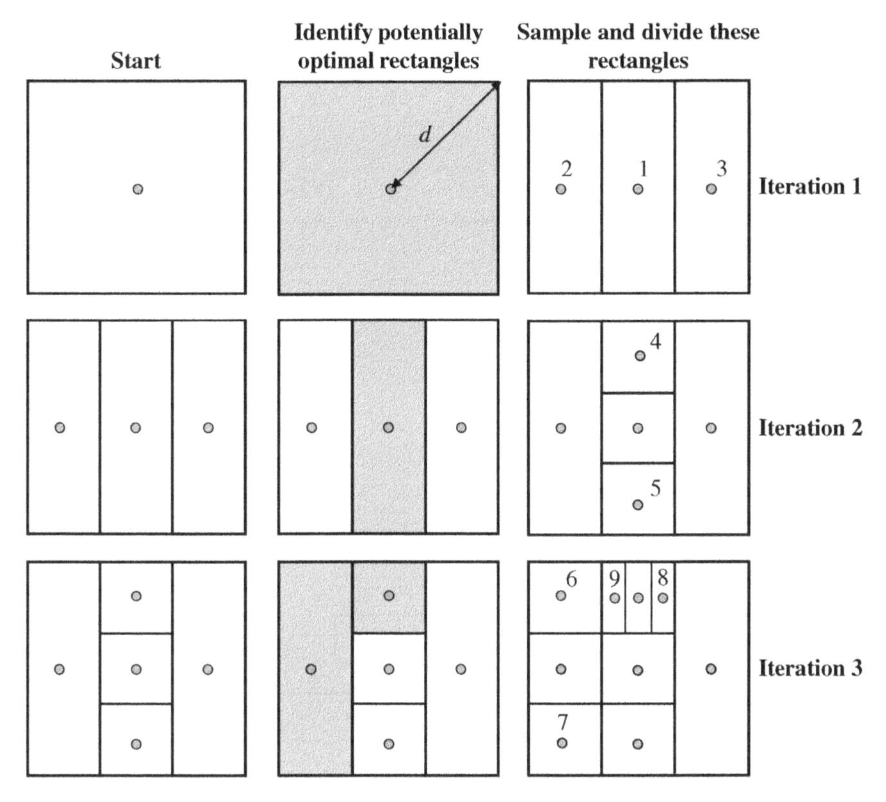

Figure 15.3 First three iterations of a two-dimensional problem.

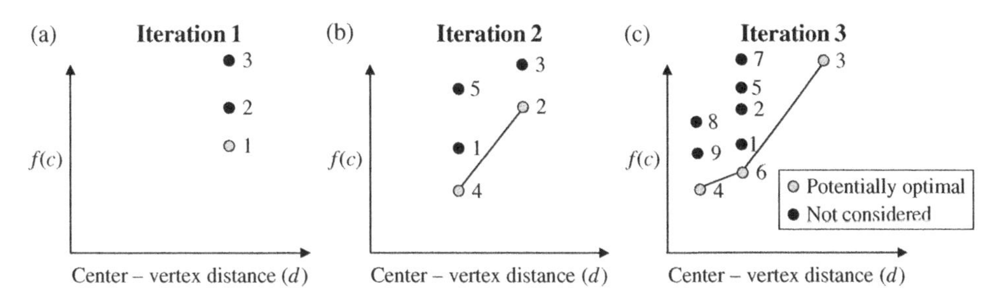

Figure 15.4 Selection of optimal rectangles in each iteration.

A flowchart explaining the DIRECT algorithm is shown in Figure 15.5.

Initially, DIRECT converts the n-dimensional design space into an n-dimensional unit hypercube. It samples and evaluates at the center point of this rectangle. This function value is assigned to a variable $f_{\min}$ which holds the minimum function objective value. Then a set of optimal rectangles is selected, assuming various possible values for the Lipschitz constant. These rectangles form the lower convex hull of dots, as shown in Figure 15.4. However, in the first iteration, the only present rectangle is selected as the optimal rectangle. Each rectangle in the optimal rectangle set is trisected to give two more rectangles (left and right rectangles). The objective function is evaluated at the

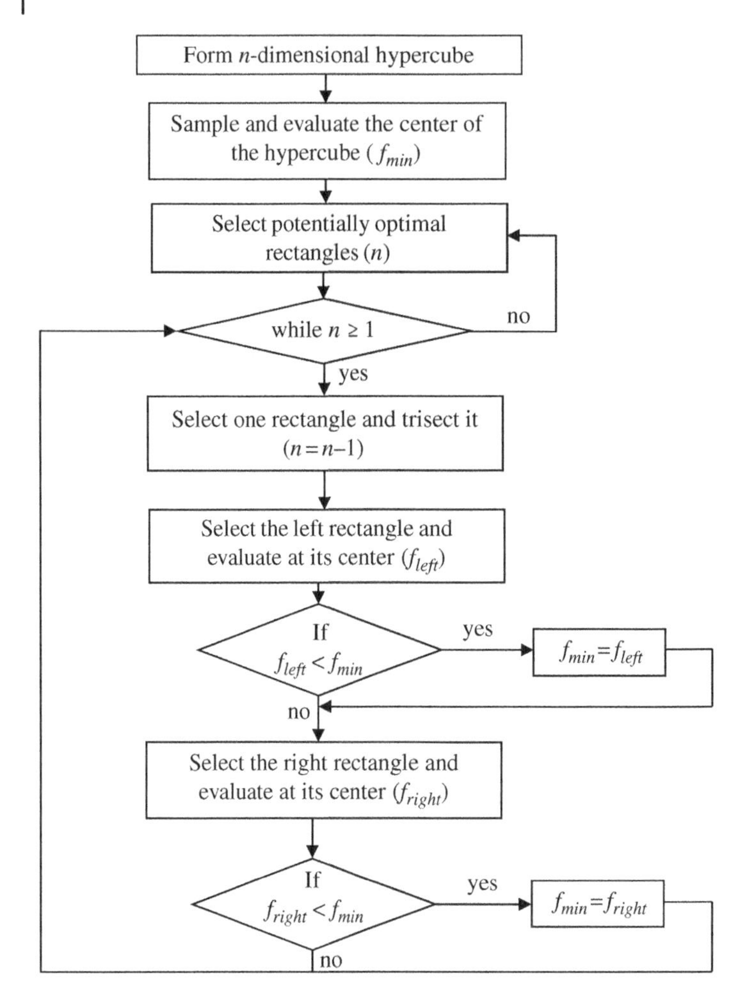

Figure 15.5 Flowchart showing the DIRECT algorithm.

center points of the left and right rectangles and $f_{\min}$ is updated if there is an improvement in the objective function. The whole process is continued until a pre-specified number of function evaluations is reached.

15.2.2 Simulated Annealing

Simulated annealing (SA) belongs to the class of stochastic hill-climbing algorithms, which means that they follow a random path in every searching process for the global optimum. SA has been presented by a large number of authors but Kirkpatrick et al. in 1983 [8] started using it in various combinatorial problems. SA is based on the Metropolis et al. Monte Carlo simulation proposed in [14].

15.2.2.1 Algorithm Description

The SA algorithm is analogous to the physical annealing process occurring in metals and function minimization. This analogy is explained briefly here. When metals are at high temperatures, the atoms in the lattice structure can move relatively freely in the higher energy states due to heating, and as the temperature is decreased slowly, the atoms have time enough to move and begin adopting the most stable orientation by taking the lowest possible energy state. However, if the temperature is decreased rapidly, the atoms become frozen at a high energy state. Attaining the lowest possible state can be thought of as reaching a global minimum in the optimization process. The temperature is decreased slowly (cooling) so that the design will find the global minimum (lowest energy state). At the beginning of the algorithm, the objective function value is evaluated at a random feasible design point. From this design point, a new random design point is selected, and the objective function value and feasibility of the new design point are evaluated. If the new neighbor is better than the previous one, then the new point is accepted as the optimal point; otherwise, acceptance or rejection of the new design point depends on the Metropolis et al. probability criterion,

$$P(f,T)=e^{-[(f_{new}-f_{current})/T]} \tag{15.7}$$

From this equation, it can be seen that the new point is more likely to be accepted if the new design point function value is close to the current design point function value. Also, the probability of accepting a design point is higher when the temperature is high. Note here that a new design point might be accepted even when it is not as good as (i.e. the function value at the new point is higher) the current design point. However, this feature will help the method avoid becoming trapped in a local minimum. So, initially, when the temperature is high, the algorithm does a global search where even the worse design points are more likely to be accepted, and then switches to a local search when the temperature is decreased, where worse design points are less likely to be selected. Thus, switching from the global search to local search depends on the value of the temperature; this is called *gradual cooling* – to gradually reduce the probability of accepting a neighbor design point with worse performance. Another parameter that is responsible for switching from the global to local search is the maximum step size. This variable is reduced as the algorithm progresses, forcing it to search more locally. If either the specified number of temperature reduction cycles or the predetermined number of function evaluations is exceeded, then the algorithm will stop.

15.2.2.2 Tunable Parameters

SA has many parameters that need to be tuned to improve the efficiency of the algorithm. So, particular attention should be given to these parameters. The parameters used in this algorithm are

- *num_steps*: number of steps before reducing the temperature and maximum step size
- T_0*_init*: initial temperature
- V_0*_init*: initial step size (maximum)
- *temp_red*: temperature reduction factor; the temperature for the next cycle is reduced by this factor
- V_0*_red*: this factor is multiplied by the initial step size to get the step size for the next design point.

15.2.2.3 Flow Chart

A flow chart showing the core of SA is given in Figure 15.6. SA starts by initializing a temperature t. Next, a random design point x_n is selected, such that it satisfies all the constraints and all the design variables are within the bounds (i.e. a feasible point). This point is passed as the current point to the algorithm core. SA is carried out in two loops. The outer loop defines the number of function evaluations that the algorithm must run before terminating. This parameter is defined by the *max_funevals* parameter. The inner loop defines the number of steps after which the temperature and the step size for x_n to jump to x_{n+1} are reduced. This parameter is defined by the *num_steps* parameter. As discussed above, this parameter is responsible for switching from the global to local search. As shown in the flow chart, the feasible point found initially is made the current point x_n. Then, it makes a random step to x_{n+1} and the objective function value and the constraints are calculated in the next step. The current step size is used in finding the new design point. In the next step, the function value and the constraints at this design point are evaluated. Then the new design point is compared to the old design point to see whether it is better or not. This comparison is done by a penalty method. The following quadratic penalty function is used:

$$Penalty(x_i) = f(x_i) + \sum_{n}^{j=1} \left\{ f(x_i) \left[\frac{\max(0, g_j(x_i))}{boundingvalue_j} \right] \right\}^2 \tag{15.8}$$

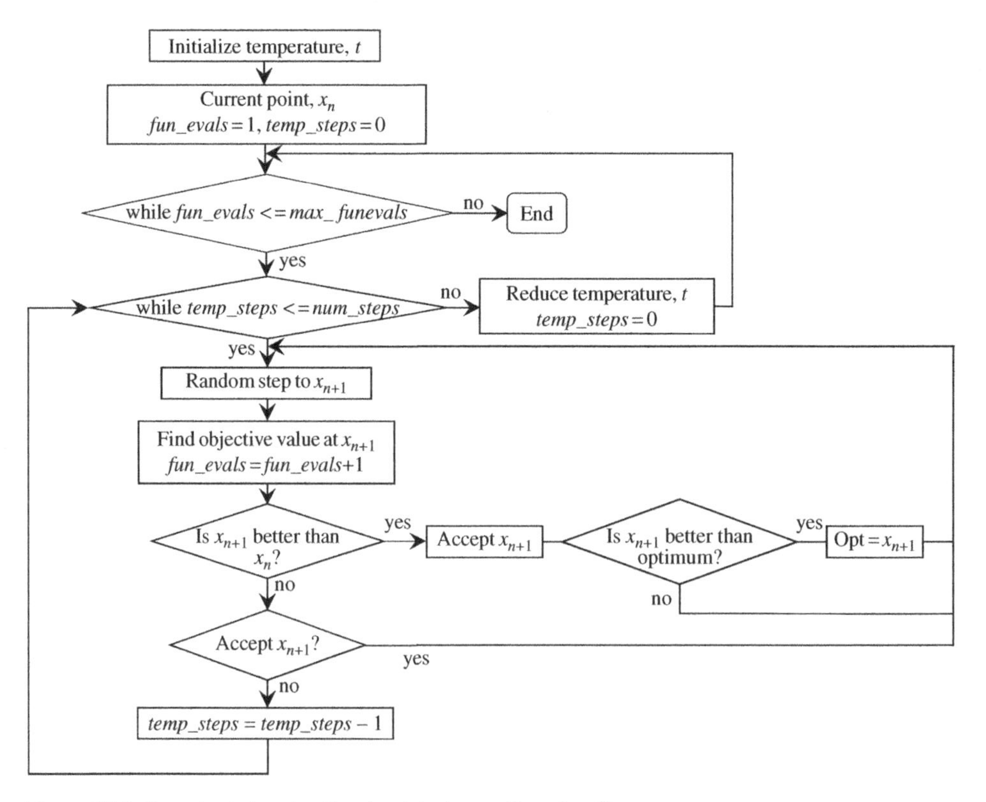

Figure 15.6 Flow chart showing the simulated annealing algorithm.

where g_j indicates the design constraint and n is the total number of design constraints. This function gives a higher penalty value to the design points which violate the design constraints more. So, if the penalty value of the current point x_{n+1} is less than the penalty for the design point x_n then it is accepted and tested to see if it is better than the current optimum. If it is better, then it is assigned as the current optimum and this point is fed back to generate a new design point. If the penalty of the current point x_{n+1} is more than the penalty for the design point x_n then the decision of its acceptance is taken using the Metropolis criterion given in Equation (15.7). If the current point is accepted then a new design point is generated based on the design point x_{n+1}. If it is not accepted then *temp_step* is increased by 1 and then a new design point is generated from the previous design point x_n.

15.2.3 Genetic Algorithms

Genetic algorithms (GAs) [9] are based on evolutionary processes and Darwin's concept of natural selection. In this selection, only the fittest populations survive and are allowed to produce offspring while the bad populations are weeded out. During the process, several natural processes like crossover, mutation, and natural selection are used to select the best fit population. The same concept is extended to mathematical optimization problems where only good design points are selected while bad design points are neglected. In this context, the objective function is usually referred to as a *fitness function*, and the process of "survival of the fittest" implies a maximization procedure. The design constraints can be incorporated as penalties in the fitness function.

15.2.3.1 Flow Chart

The flow chart for the GA is given in Figure 15.7. The GA begins by randomly generating or seeding an initial population of candidate solutions (design points). Starting with the initial random population, the GA then applies a sequence of operations like the design crossover where two individuals (parents) from the initial population are selected randomly and reproduced to get two new individuals (children), and mutation where one individual from the initial population is slightly changed to get a new individual. If the newly generated individuals created by the crossover and the mutation operators are better than the parents used, then the parents are replaced by the newly created individuals. Again, at the end, the worst design points are weeded out from the population in order to improve the fitness function. The entire process can be termed as *one generation* and continued for multiple generations or until the maximum number of function evaluations is exhausted in order to further improve the fitness function.

15.2.3.2 Operators and Selection Method

There are numerous operators for the GA. The *arithmetic crossover* operator is described as follows. Two parents reproduce to generate two new individuals (children). The parent individuals are selected randomly. The newly generated individuals can be represented as a linear combination of the parents as

$$\begin{aligned} X_{ind1} &= r.X_{par1} + (1-r)\cdot X_{par2} \\ X_{ind2} &= (1-r).X_{par1} + r\cdot X_{par2} \end{aligned} \tag{15.9}$$

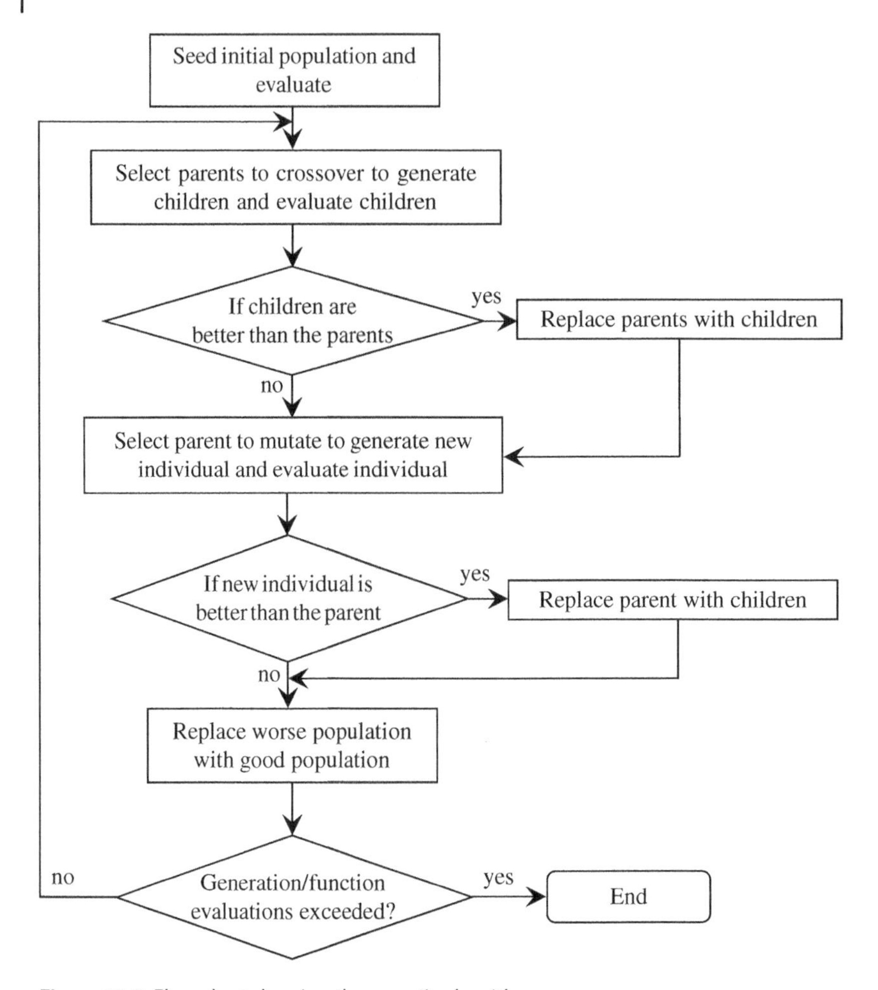

Figure 15.7 Flow chart showing the genetic algorithm.

where X_{ind1}, X_{ind2} represent the two newly created individuals, X_{par1}, X_{par2} represent the parents, and r represents a random variable between 0 and 1.

In mutation, a parent is selected and is altered to get a new individual.

A selection method is needed to choose the best fit individuals. A normalized geometric selection method, which is a ranking-type method, is used as the selection method. The ranking method was chosen because of the presence of negative fitness. With this selection method, a probability is assigned to each individual of the population using Equation (15.10), where q denotes the probability of choosing the best design, r denotes rank of the individual, and P the population size:

$$\text{Each individual probability} = \frac{q}{1-(1-q)^P}(1-q)^{r-1} \tag{15.10}$$

15.2.3.3 Tunable Parameters

Like SA, the GA has many parameters that need to be tuned for a better performance, although the GA has fewer tunable parameters than SA. The tunable parameters and their description are

- *pop_size*: number of individuals in a generation
- *xoverfns*: number of times the crossover operation is to be done
- *mutFNs*: number of times the mutation operation is to be done.

15.2.4 Particle Swarm Optimization

Particle swarm optimization (PSO) was developed by Kennedy and Eberhart, and is an evolution-based stochastic global optimization technique [10, 11]. The idea of PSO came from the swarm intelligence found in many natural systems with group behavior. Such systems are typically made up of a population (swarm) of simple agents or particles interacting locally with their neighbors and with their group organization. Ant colonies, bird flocks, and animal herds are a few of the examples of such natural systems. In these systems, the agents interact locally and this can result in a global behavior. For example, individuals can change position or velocity locally and a global behavior pattern can be observed. The underlying principle is used to develop a technique to find global maxima in optimization problems.

15.2.4.1 Algorithm Description

Just like the GA, PSO is a population-based search procedure. At the beginning, random solutions called *particles* are initialized in the multidimensional design space. In a PSO system, each particle flies in the multidimensional design space seeking the global maximum. Each particle in the PSO is defined by a point in the design space called *position* and its flight speed is called *velocity*. Also, each particle is aware of its best position reached so far (*pbest*) and the best position of the group so far (*gbest*). During flight, each particle adjusts its position according to its own experience (*pbest* value), and also according to the experience of its neighboring particles (*gbest* value). The position is modified using the concept of velocity. The velocity of each particle is updated as follows:

$$v_i^{n+1} = kv_i^n + \alpha_1\ rand_1(pbest_i - p_i^n) + \alpha_2\ rand_2(gbest - p_i^n) \tag{15.11}$$

where v_i^{n+1} is the velocity of particle i at iteration $n+1$, k is the weighting function, α_1 and α_2 are the weighing factors, $rand_1$ and $rand_2$ are two random numbers between 0 and 1, p_i^n is the position of particle i at iteration n, $pbest_i$ is the best position of particle i, and *gbest* is the best position of the group (the best of all *pbests*). For the study example in Section 15.4, $k = 0.6$ and $\alpha_1 = \alpha_2 = 1.7$ are taken for better convergence [11]. Similarly, the position is updated as follows:

$$p_i^{n+1} = p_i^n + v_i^{n+1} \tag{15.12}$$

The velocity and position updating for particle i is illustrated in Figure 15.8. Note that this figure represents a two-dimensional problem.

From Figure 15.8, it can be seen that the position of the particle is adjusting itself toward the *gbest* position. This is because the velocity has changed its direction toward the *pbest* and *gbest* values.

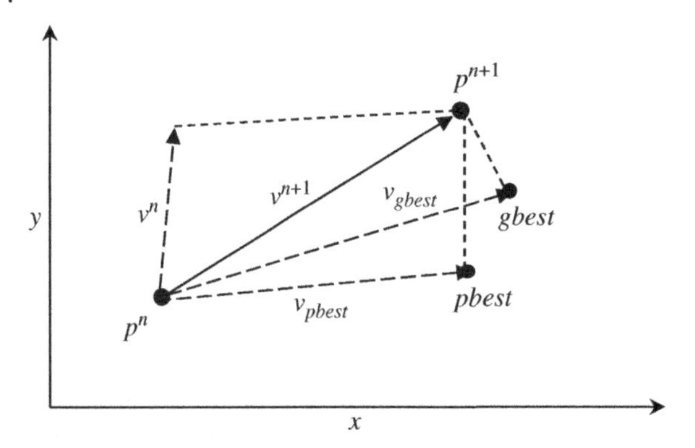

Figure 15.8 Velocity and position updating in particle swarm optimization.

The constraints in PSO are treated the same way as in SA and the GA. A penalty is assigned to each design point using Equation (15.8). The penalty is used to update *pbest* and *gbest* for each particle. For a particle *i*, the *pbest* value is updated if the penalty of the particle is less than the previous best penalty. The same is done when *gbest* is updated. This makes sure that the objective function is maximized.

15.2.4.2 Flow Chart

The flow chart representing the PSO algorithm is shown in Figure 15.9. The algorithm starts by initializing a population (particles) of random design points. In the case of PSO, the random solutions are normalized for a better performance. The population size (*n*) is defined by the user. The random design points are evaluated to give the next step. Initially, the best position values (*pbest*) and the best objective values (*pbestval*) of each particle are assigned to starting position values (*pos*) and the best starting values (*out*).

The best value of all the particles (*gbestval*) is the least value of all the *pbest*. The variable *gbestval* holds the current best global maximum of the objective function. Next, a while loop is run for a specified number of function evaluations. The while loop updates the *pbest* and *pbestval* values if there is any improvement in the current *pbestval* value. It also updates *gbestval* if there is an improvement. The updated *pbest* and *gbest* are used in determining the new velocity (*vel*) and position (*pos*). The function evaluations (*fevals*) are incremented by the population size every iteration.

15.2.5 Advantages/Disadvantages of Different Optimization Algorithms

In the following, the advantages and disadvantages of the above four different optimization algorithms are discussed.

15.2.5.1 DIRECT

DIRECT has no tuning parameters in order to get good algorithmic performance. Also, the user is not required to specify the starting point since DIRECT uses the center point of the design space as its initial point. Therefore, it eliminates the problem of choosing a good starting point. Note here that the local optimizers require a good initial point to reach the optimum value. Another advantage is that DIRECT covers the entire design space, avoiding any chance of missing the global optimum.

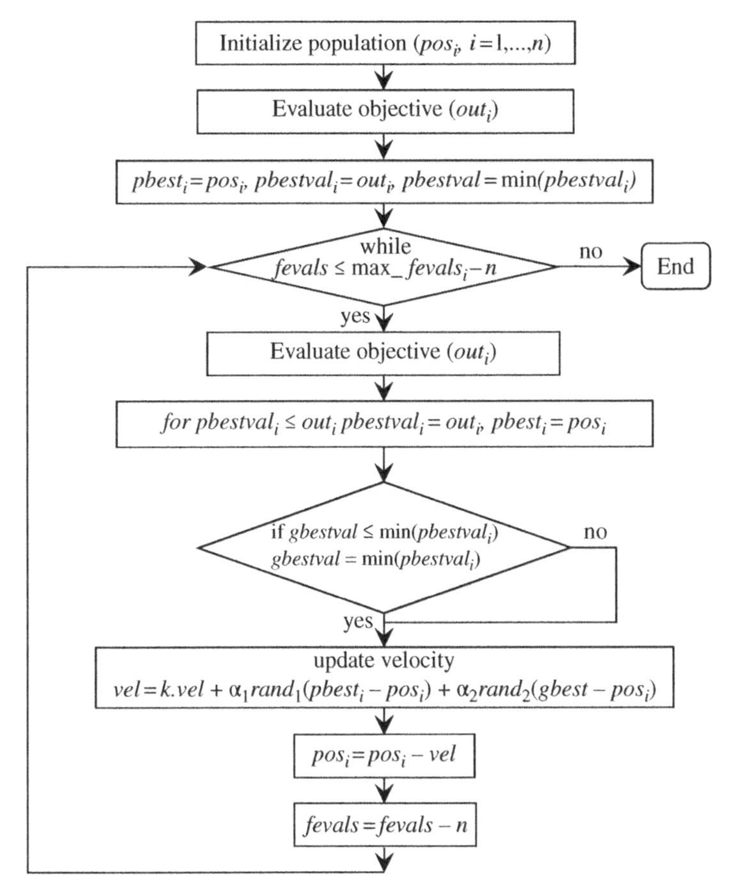

Figure 15.9 Flow chart of particle swarm optimization.

DIRECT can converge to a global optimal region after fewer function evaluations, but in order to actually reach the global optimum point more function evaluations are often required. The algorithm can be implemented for paralleling computing.

15.2.5.2 SA

The main advantage of SA is that it is a very efficient algorithm for finding global minima. It accepts some worse points during the process in the expectation of finding global minima.

The disadvantage of SA is the tuning parameters discussed above. The right set of parameters is needed to improve the efficiency of the algorithm. This tuning in turn becomes an optimization problem. Similar to DIRECT, SA is efficient in finding the region of global minima, but in order to reach the true global minima more function evaluations are often required.

15.2.5.3 GA

The initial population generated primarily determines a good starting point for the GA. So, it can be seeded by some good design points to improve its efficiency. According to the nature of the design problem, different types of selection method and operators for crossover and mutation can be chosen in using GA.

The major disadvantage is the tuning of the parameters. Also, the initial population is randomly chosen. This random initial population may not cover the entire design space uniformly. It was observed that most of the design points generated by the crossover and mutation operators were not able to meet the constraints (infeasible points) because the operators have no knowledge of the constraints. This resulted in many infeasible points. This was the reason why the algorithm had difficulty in improving the best objective function value found in the initial random population.

15.2.5.4 PSO

The major advantage of the PSO algorithm is that fewer parameters need to be adjusted compared to SA and GA. The constants for updating velocity are very critical in obtaining better performance. Many sets of constants are available and suitable for specific problems. Moreover, no natural operators such as crossover, mutation, and selection are needed in PSO. Also, PSO is easier to understand with simpler equations. The disadvantage lies in the selection of the constants for updating the velocity. If inappropriate constants are chosen then the problem may not converge to the optimum.

15.3 Model-in-the-Loop Design Optimization Process

The approach for HEV design optimization is typically a model-in-the-loop design optimization process, as illustrated in Figure 15.10. In the middle of the diagram, the vehicle is modeled in a simulation tool such as PSAT [15], ADVISOR [16], or VTB [17], and this model is embedded in a computational loop. Initially, the vehicle model is simulated using the initial values of the design variables, and we get the numerical values of the objective function, in this case the composite fuel economy measured in mpgge (miles per gallon gasoline equivalent). At the same time, the constraint functions – in this case the vehicle performance – are evaluated. Then these simulated results are fed back to the optimization algorithm, which generates a new set of values for the design variables. Subsequently, the vehicle model is simulated again to get the values for the objective function and the constraint functions. The simulation results are fed back to the optimization algorithm again to generate yet another new set of design variables. This iteration process goes on and on until some stopping criteria are

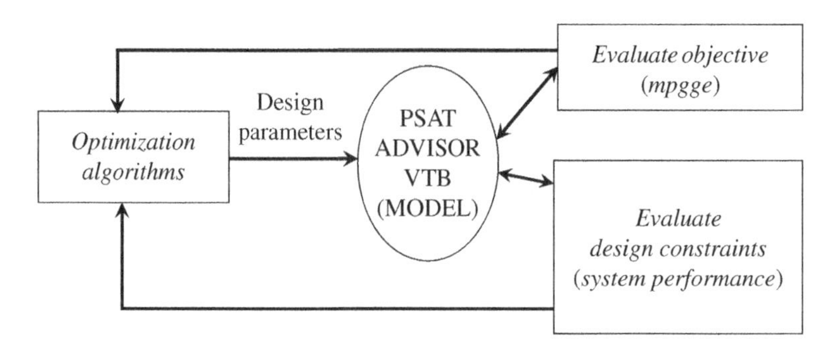

Figure 15.10 Model-in-the-loop design optimization process.

reached for the optimization process. Note that the design variables are restricted within their bounds during this process.

For illustration purposes, in this chapter PSAT is used as the modeling and simulation tool. PSAT was developed by Argonne National Laboratory, sponsored by the US Department of Energy (DOE) [15]. It can help a vehicle designer to size components and develop a realistic hybrid powertrain and its control system. PSAT can accurately simulate vehicle performance, fuel economy, and emissions. In using PSAT, we mainly need to select powertrain topology, define component sizes, and construct the control strategy. Note that the component sizing is automated by the model-in-the-loop process.

15.4 Parallel HEV Design Optimization Example*

As an application example, PSAT is used to optimize a parallel HEV for the maximum fuel economy on a composite driving cycle. Four global algorithms, DIRECT, SA, GA, and PSO, are used in the model-based design optimization. The main focus of the example is to show the comparison of different optimization algorithms for optimal vehicle design rather than the result. The vehicle model gui_par_midsize_cavalier_ISG_in (available in the PSAT model library) has been chosen for this optimization study. This vehicle is a two-wheel-drive starter–alternator parallel configuration with a manual transmission. The basic configuration of the parallel HEV used for the simulation study is illustrated in Figure 15.11, and the main components of the HEV are listed in Table 15.1.

The objective is to maximize the composite fuel economy, which is computed based on city fuel economy and highway fuel economy. For example, composite fuel economy can be computed as the weighted average of the state of charge (SOC) balanced fuel economy values during the city drive cycle and highway drive cycle using [18]

$$\text{Composite fuel economy} = \frac{1}{\dfrac{0.55}{\text{City\_FE}} + \dfrac{0.45}{\text{Hwy\_FE}}}$$

where City_FE and Hwy_FE denote the city and highway fuel economy values, respectively.

The driving cycles selected are Federal Test Procedure city driving cycle FTP-75 and Highway Fuel Economy Test drive cycle HWFET. The two drive cycles are shown in Figures 15.12 and 15.13, respectively.

The design problem's constraints come from the following required vehicle performance:

- acceleration time 0–60 mph (0–96 km/h) $\leq$ 18.1 seconds
- acceleration time 40–60 mph (64–96 km/h) $\leq$ 7 seconds
- acceleration time 0–85 mph (0–136 km/h) $\leq$ 35.1 seconds
- maximum acceleration $\geq 3.583\,\text{m/s}^2$

Table 15.2 shows the six design variables used in this study. The first two define the power ratings of the fuel converter (the engine) and the motor controller. The third, fourth, and fifth variables define the number of battery modules, minimum battery SOC

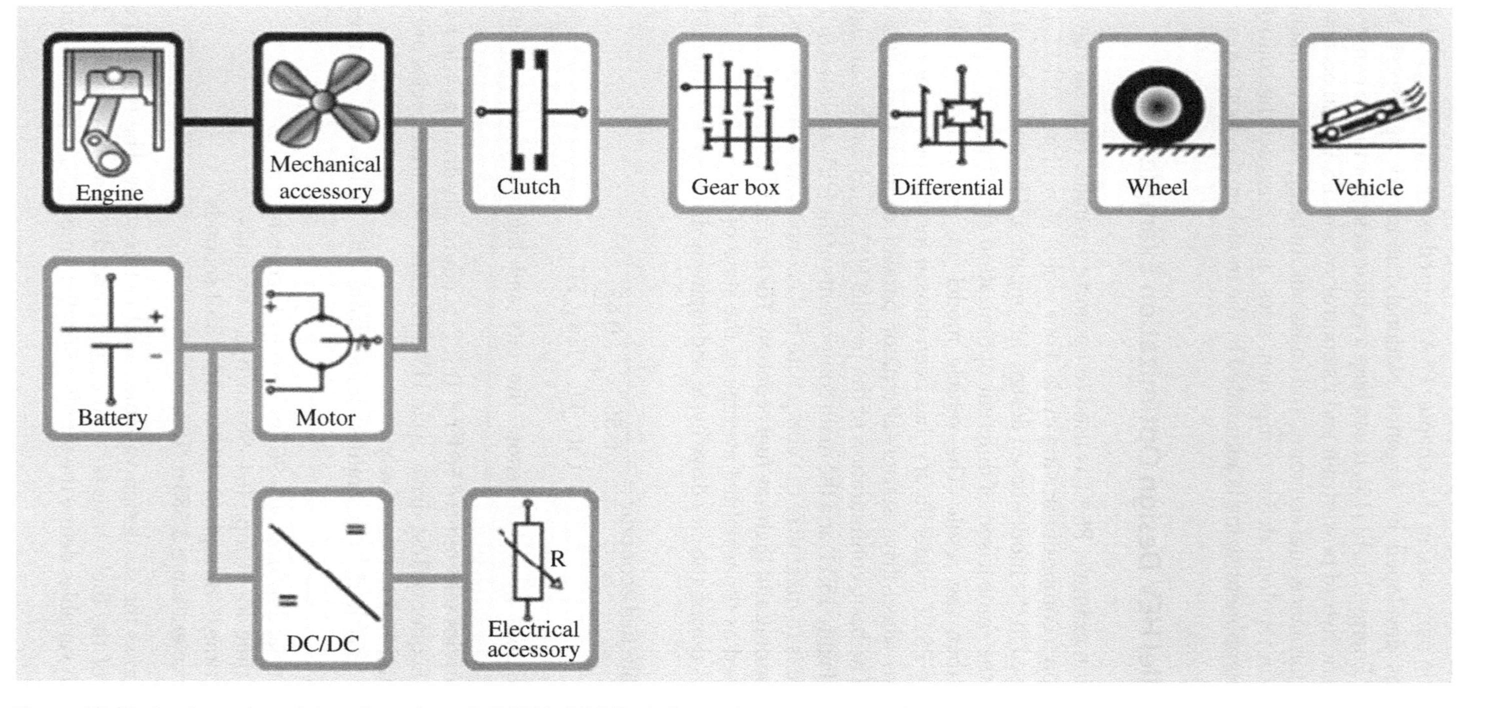

Figure 15.11 Configuration of the selected parallel HEV in PSAT [15]. *Source:* Transportation.anl.gov

Table 15.1 Parallel HEV components.

Component	Description
Fuel converter	84 kW and 2.2 l Cavalier gasoline engine
Motor	ECOSTAR motor model with continuous power of 33 kW and peak power of 66 kW
Battery	Panasonic NiMH battery with a capacity of 6.5 Ah and 240 cells
Transmission	Four-speed manual gearbox with final drive ratio 3.63
Control strategy	Default propelling, shifting, and braking strategies

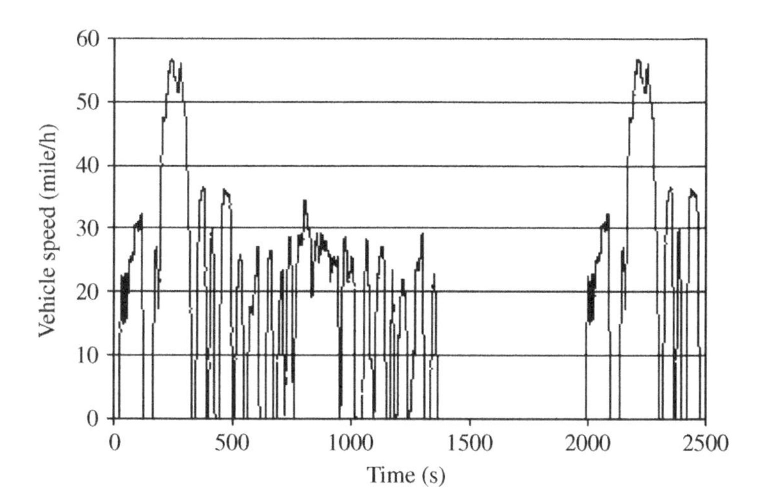

Figure 15.12 The FTP-75 drive cycles.

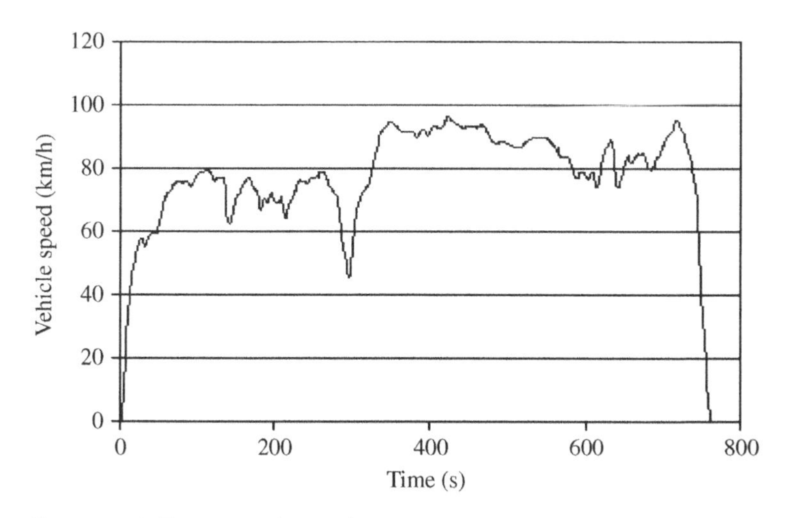

Figure 15.13 The HWFET drive cycles.

Table 15.2 Upper and lower bounds of design variables.

Design variable	Description	Lower bound	Upper bound
eng.scale.pwr_max_des	Fuel converter power rating	40 kW	100 kW
mc.scale.pwr_max_des	Motor controller power rating	10 kW	80 kW
ess.init.num_module	Battery number of cells	150	350
ess.init.soc_min	Minimum SOC allowed	0.2	0.4
ess.init.soc_max	Maximum SOC allowed	0.6	0.9
fd.init.ratio	Final drive ratio	2	4

Table 15.3 Initial design variable values.

Design variable	Initial value
eng.pwr_max_des	86 kW
mc.pwr_max_des	65.9 kW
ess.init.num_module	240
ess.init.soc_min	0
ess.init.soc_max	1
fd.init.ratio	3.63

Table 15.4 Comparison of fuel economy.

Fuel economy				
Before optimization	After optimization			
	DIRECT	SA	GA	PSO
35.1 mpg	39.64 mpg	40.37 mpg	37.6 mpg	37.1 mpg

allowed, and maximum battery SOC allowed. Note that the SOC values are part of the control strategy parameters. Although they are not related to component sizing, they have a direct impact on fuel economy of an HEV design. The sixth design variable defines final drive ratio. Each design variable is also restricted within lower and upper bounds.

The problem now becomes quite challenging, since this is a constrained multivariable optimization problem.

First, the default vehicle is simulated in PSAT. The design variables and their initial values are listed in Table 15.3. The fuel economy was observed to be 35.1 mpg (6.7 l/100 km), as given in Table 15.4 under the first column.

Second, the optimization algorithms, DIRECT, SA, GA, and PSO, are looped with the PSAT vehicle simulator and the optimization is carried out. For this step, the same default vehicle configuration given in Figure 15.12 and Table 15.1 is taken, and the

Table 15.5 Final design variable values.

Design variable	Initial value	Final value			
		DIRECT	SA	GA	PSO
eng.pwr_max_des (kW)	86	83.1	82.4	95.5	87.1
mc.pwr_max_des (kW)	65.9	20.2	21.9	24.2	14.8
ess.init.num_module	240	245	311	300	238
ess.init.soc_min	0	0.25	0.22	0.34	0.26
ess.init.soc_max	1	0.84	0.78	0.89	0.78
fd.init.ratio	3.63	3.9	4.0	3.49	3.42

Table 15.6 Comparison of the HEV performance.

Constraint	Constraint value	Before optimization	After optimization			
			DIRECT	SA	GA	PSO
0–60 mph	≤18.1 s	18.1 s	15.5 s	10.8 s	11.9 s	11.1 s
40–60 mph	≤7 s	7 s	6.8 s	5 s	4.4 s	4.9 s
0–85 mph	≤35.1 s	35.1 s	30.6 s	20.7 s	21.2 s	20 s
Maximum acceleration	≥3.583 m/s^2	3.583 m/s^2	3.97 m/s^2	4.07 m/s^2	3.94 m/s^2	3.99 m/s^2

bounds for the design variables are given in Table 15.2. The four algorithms are allowed to run for 400 function evaluations. Using the same number of function evaluations will allow us to compare the performance of the different optimization algorithms. A comparison of the fuel economy before and after the optimization is given in Table 15.4. A significant improvement in the fuel economy is seen due to optimization (to a lesser extent in the case of PSO and the GA, though). Of all the four algorithms, SA performs well with an approximate improvement of 5 mpg.

Table 15.5 shows the final values of the six design variables after optimization. Note that the rating of the electric motor is greatly reduced, implying that downsizing of the electric motor has been achieved. On the other hand, the engine is downsized to a lesser extent in the DIRECT and SA cases, while upsized by GA and PSO. Given the vehicle performance constraints, the tradeoff of engine downsizing and motor downsizing can be realized by adjusting the lower and upper bounds of the design variables.

Table 15.6 shows the performance results of the hybrid powertrain after optimization. Essentially, all the optimization algorithms resulted in improved vehicle performance.

The vehicle mass changes as the design variables change because the mass of the vehicle depends directly on some design variables. In particular, of the chosen six design variables, three (power ratings of engine and motor, and energy modules) affect the vehicle mass. The vehicle mass before and after the optimization is given in Table 15.7.

Table 15.7 Mass of HEV before and after optimization.

Mass of the vehicle (kg)				
	After optimization			
Before optimization	DIRECT	SA	GA	PSO
1683	1635	1656	1694	1690

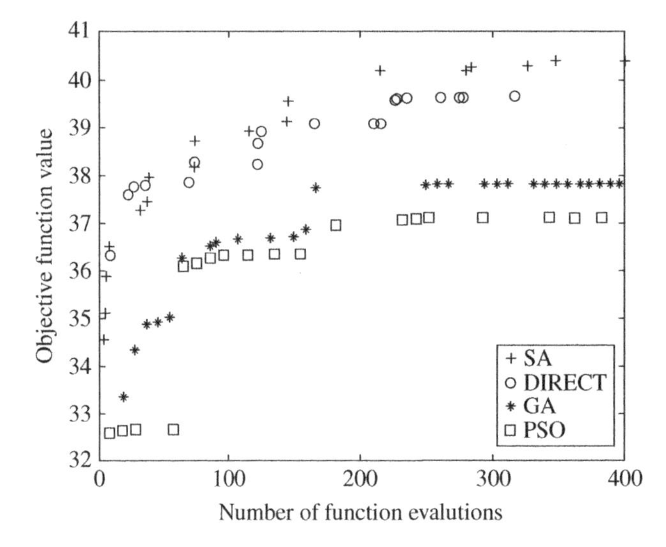

Figure 15.14 Performance comparison of DIRECT, SA, GA, and PSO.

The vehicle mass decreased for the DIRECT and SA cases and increased slightly in the case of the GA and PSO.

Figure 15.14 shows how the objective function (fuel economy) value improves versus the design iteration number. The cross curve is for the SA case; the circle curve is for the DIRECT case; the star curve is for the GA case; and the square curve is for the PSO case. We can see that fuel economy improvement with the SA and DIRECT algorithms is very close until about 125 function evaluations, after which SA leaped ahead of DIRECT. The GA is slow to catch SA and DIRECT initially because it takes some function evaluations to generate the initial populations. After about 50 function evaluations, the GA did not find any good design point to get further improvement in the fuel economy. The performance of PSO is similar to that of the GA. Overall, SA performed the best for this particular design optimization problem.

15.5 Series HEV Design Optimization Example

In this section, non-dominated sorting genetic algorithm NSGA-II is applied for the design optimization of a series HEV [19, 20]. NSGA-II is one of the most efficient evolutionary algorithms. Three different operators (crowded tournament selection, simulated binary

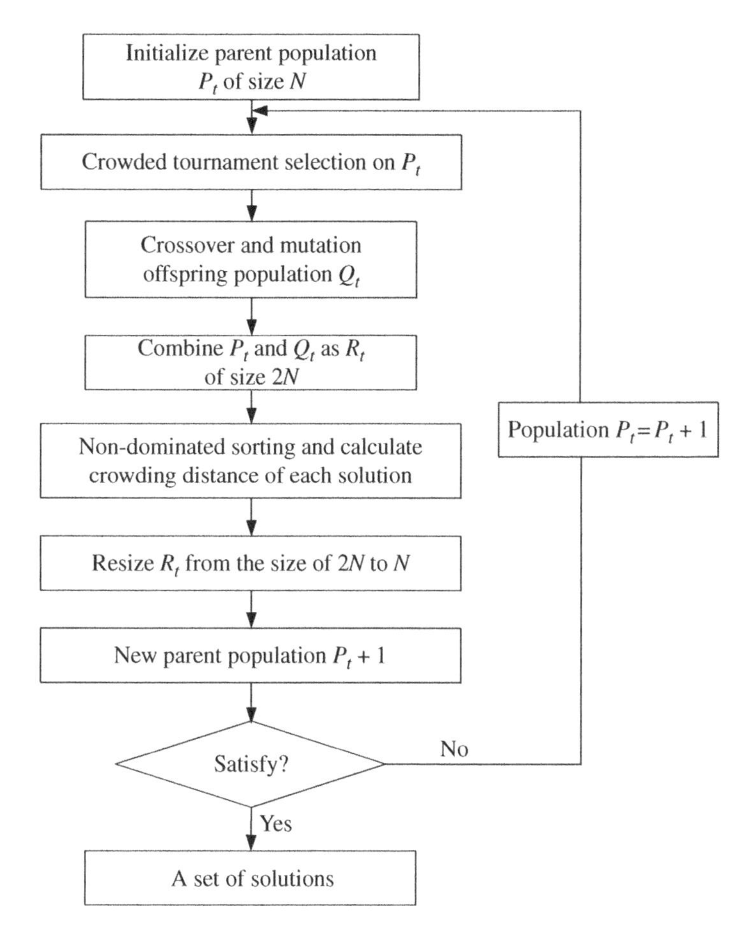

Figure 15.15 Computational steps in NSGA-II.

crossover, and polynomial mutation) are applied to the population to yield a better population at each iteration. The main task of a selection operator is to emphasize good solutions of the population by making multiple copies of them to replace bad solutions of the population. The task of a crossover operator is to exchange partial information between two or more reproduced solutions and to create new offspring solutions. The task of a mutation operator is to locally perturb the offspring solutions. Successive applications of such iterations have been demonstrated to converge close to the true optimal solution of the problem both theoretically and computationally [21, 22].

Figure 15.15 illustrates the major computational steps in NSGA-II. The parent population P_t is first initialized based on the range and constraints of the design variables. Through crowded tournament selection, simulated binary crossover, and polynomial mutation, the offspring population Q_t is generated. Then the parent population P_t and the offspring population Q_t are combined together to form R_t of size $2N$. After that, a non-dominated sorting procedure is applied to classify the entire population R_t into a number of hierarchical non-dominated fronts. Once the non-dominated sorting is completed, the crowding distance of each solution is also calculated. The new parent population P_{t+1}

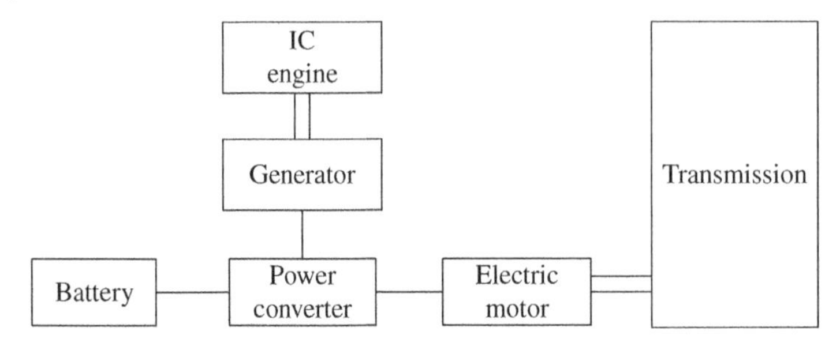

Figure 15.16 Powertrain configurations of series HEV.

of size N is derived from R_t of size $2N$ through the following calculation [19, 20]. The calculation starts with the best non-dominated front and continues with the solutions of the second non-dominated front, and so on. When the last allowed front is being considered and there are more solutions in the last front than the remaining slots in the new population, the crowding distance has to be considered to choose the members of the last front, which reside in the least crowed region in the front. The iteration of generating new populations repeats until some terminating conditions are met. As the generation progresses, the solutions in the new populations are optimized.

15.5.1 Control Framework of a Series HEV Powertrain

A typical powertrain configuration of a series HEV is shown in Figure 15.16. The vehicle is propelled by a traction motor, which is powered by a battery pack and/or an engine/generator unit. The engine/generator unit either helps the batteries to power the traction motor when load power demand is high, or charges the batteries when load demand is low. The electric motor can also be used as a generator to charge batteries during regenerative braking. The motor controller is to control the traction motor to produce the power required by the vehicle.

The series thermostat control strategy uses the generator and engine to generate electrical energy for use by the vehicle. The powertrain control strategy is described as follows [23]:

- To maintain charge in the battery, the engine turns on when the SOC reaches the lower limit.
- The engine turns off when the SOC reaches the higher limit.
- The engine operates at the most efficient speed and torque levels.

The major vehicle parameters and main components of the series HEV are given in Tables 15.8 and 15.9.

15.5.2 Series HEV Parameter Optimization

ADVISOR is used to evaluate the vehicle performance. ADVISOR is a vehicle simulation package developed on the MATLAB/Simulink software platform, which adopts forward and backward modeling methods. Figure 15.17 shows the test drive cycle used in our simulation. The test drive cycle is composed of city driving represented by UDDS and highway driving represented by HWFET. Table 15.10 shows that the optimization is initially limited to eight design variables, three component parameters, and five control

Table 15.8 Vehicle parameters.

Gross mass (kg)	Full load mass (kg)	Wheelbase (m)
1373	1659	2.6
Windward area (m^2)	Rolling coefficient	Aerodynamic coefficient
2.0	0.015	0.335

Table 15.9 Series HEV main components.

Component	Description
Fuel converter	Geo 1.0 l SI 41 kW
Motor	75 kW Westinghouse AC induction motor
Battery	Hawker Genesis 12 V, 26 Ah, 10EP sealed valve-regulated lead acid (VRLA) battery

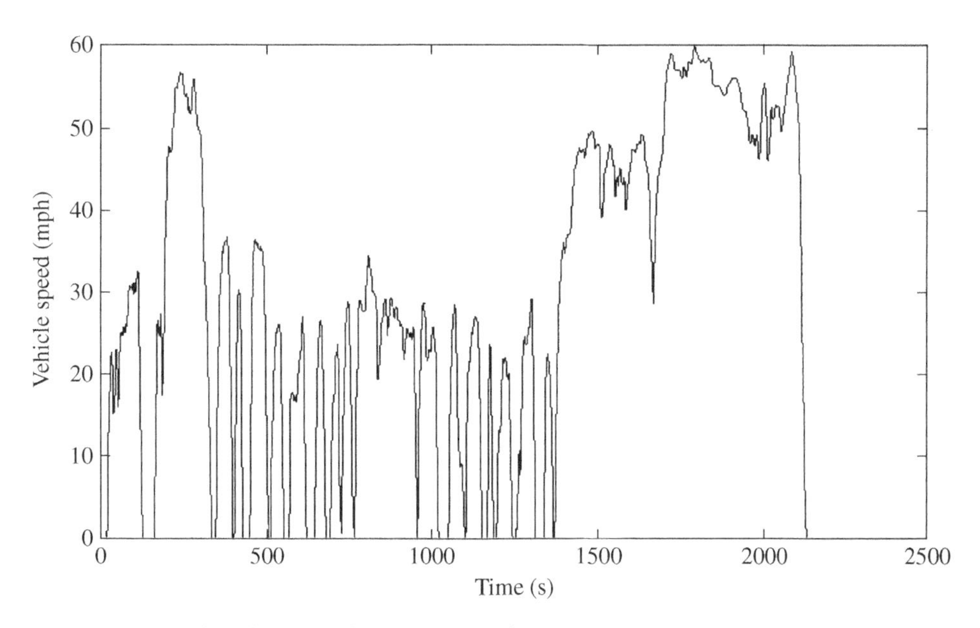

Figure 15.17 Combined UDDS and HWFET drive cycle.

strategy parameters. The initial default values and the boundaries of design parameters are also given in the table.

Before optimization, the vehicle model has the following dynamic performances:

- 0–60 mph: 10.5 seconds
- 40–60 mph: 5.6 seconds
- 0–85 mph: 24.6 seconds
- Gradability: 6.8% grade at 55 mph (88 km/h)

Table 15.10 Design variables.

Design variable	Description	Default value	Lower bound	Upper bound
eng_pwr	Engine power	41 kW	25 kW	53 kW
mc_pwr	Motor power	75 kW	38 kW	112 kW
ess_cap	Capacity of batteries	26 Ah	13 Ah	39 Ah
cs_high_soc	Highest SOC allowed	0.8	0.7	0.85
cs_low_soc	Lowest SOC allowed	0.6	0.3	0.5
cs_max_pwr	Maximum power commanded of fuel converter	30 kW	25 kW	40 kW
cs_low_pwr	Minimum power commanded of fuel converter	20 kW	5 kW	20 kW
cs_off_time	The shortest allowed period of fuel converter off	Inf.	10 s	1000 s

Since the dynamic performance must be maintained during optimization, the above performances are imposed constraints in the optimization process. At the same time, in order to eliminate the influence of initial battery energy on the fuel consumption and emissions, the SOC correction has to be selected, so the initial and final SOC can be set at almost the same level, and the delta SOC tolerance is within [−0.5%, +0.5%]. We can consider that the entire output energy for the cycle is from the engine alone.

For each solution in a population, which contains eight design variables, the algorithm calls ADVISOR to run a simulation using these parameters specified for the test drive cycle to obtain the fuel consumption and emission data. The optimization algorithm calculates the fitness value for each solution, and then generates a new set of solutions by the crossover and mutation operations. The non-dominated sorting algorithm is applied to the newly generated solutions to select the next population. This iterative process continues until the terminating condition (such as the maximum number of generations) is satisfied.

15.5.3 Optimization Results

The NSGA-II algorithm shown in Figure 15.15 is implemented in ADVISOR. The initial population is 40 randomly selected individuals of the design parameters given in Table 15.10 from the solution space. The terminating condition is set to 80 generations. For each individual, the drive cycle simulation for the evaluation including the objective function and constraints takes an average of about 1 minute on a 3.4 GHz Pentium computer, and it takes about four days for the whole program to run. Figure 15.18 shows the fuel consumption and emission data generated by the 40 solutions after 80 generations of optimization, along with the result generated by the default controller in ADVISOR. All 40 solutions consumed less fuel than the default controller, and they all generated less emission in CO, NO_X, and HC than the default controller.

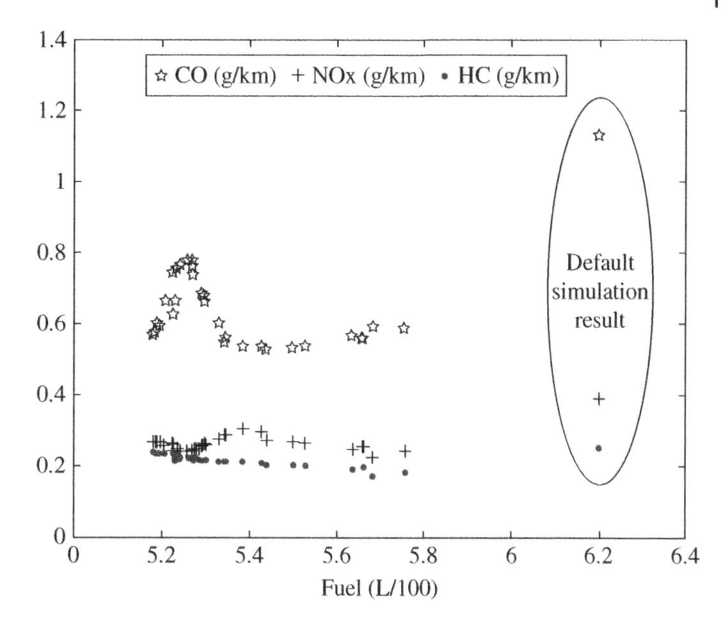

Figure 15.18 Fuel consumption and emission data generated by the tradeoff solutions generated by the optimization algorithm.

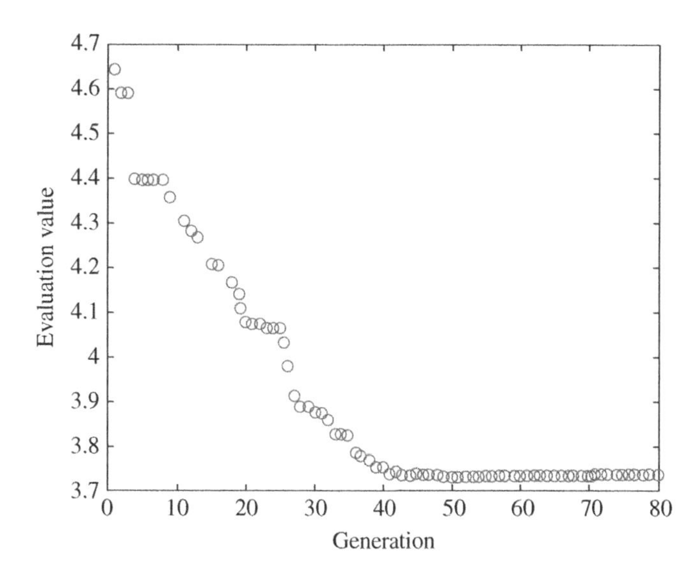

Figure 15.19 Evaluation of generation progress during optimization.

In order to illustrate progress through the generations during the optimization, we use the following equation as an evaluation function to select the best one from the 40 solutions at each generation:

$$F = 0.7 \cdot \text{fuel} + 0.1 \cdot \text{HC} + 0.1 \cdot \text{CO} + 0.1 \cdot \text{NO}_x$$

The results are illustrated in Figure 15.19. The data in the figure shows that the algorithm converges to the evaluation value of 3.72 after 50 generations.

Table 15.11 lists the optimization results generated by the five top-ranked solutions in the final population based on the non-dominated sorting method. Take solution 1, for

Table 15.11 Optimization results generated by the top five solutions selected by the non-dominated sorting.

No.	Fuel (l/100 km)	HC (g/km)	CO (g/km)	NO_X (g/km)
1	5.268	0.213	0.779	0.231
2	5.681	0.170	0.593	0.223
3	5.439	0.201	0.526	0.272
4	5.183	0.238	0.568	0.270
5	5.758	0.184	0.589	0.242

Table 15.12 Value of design variables and performance after optimization.

No.	1	2	3	4	5
eng_pwr	25 kW	25.1 kW	25 kW	25.1 kW	25 kW
mc_pwr	82.8 kW	81.4 kW	86.7 kW	80.9 kW	82.9 kW
ess_cap	39 Ah	38.2 Ah	38.9 Ah	38.3 Ah	38 Ah
cs_high_soc	0.73	0.75	0.75	0.74	0.74
cs_low_soc	0.49	0.48	0.50	0.48	0.49
cs_max_pwr	33.1 kW	33.5 kW	32.2 kW	33 kW	31.9 kW
cs_low_pwr	13.2 kW	6.7 kW	9.8 kW	5.6 kW	12.4 kW
cs_off_time	750 s	538 s	598 s	519 s	711 s
0–60 mph	10.2 s	10 s	10 s	10.3 s	9.9 s
40–60 mph	5.4 s	5.3 s	5.3 s	5.5 s	5.2 s
0–85 mph	23.8 s	23.4 s	23.4 s	24 s	22.8 s
Gradability at 55 mph	8.1%	7.9%	7.9%	8.0%	7.2%
Delta SOC	0.21%	0.26%	0.31%	−0.17%	−0.24%

example: the fuel consumption and the three emissions are reduced by 15, 14, 31, and 40%, respectively, in comparison to the performances generated by the default controller. This suggests that if the design parameters suggested by this solution are used in the series HEV design, an online controller can be optimized to reach the same or better performances than those generated by this optimal solution.

Corresponding to the five solutions in Table 15.11, Table 15.12 shows the value of the design variables and performance after optimization. Although the engine power is scaled down, by increasing the motor power and battery capacity, the acceleration ability and gradability of vehicle performance are still improved.

Table 15.13 shows the optimization results generated by the top five solutions selected by the evaluation function from the final population. We can see that the solutions in Table 15.13 are different from those shown in Table 15.10. We use this example to show the importance of the proposed optimization algorithm: it generates a population of multiple tradeoff optimal solutions. Vehicle designers and control engineers can use their own evaluation criteria to select the optimal solutions from this population of tradeoff solutions.

Table 15.13 Optimization results generated by the top five solutions selected by the evaluation function.

No.	Fuel (l/100 km)	HC (g/km)	CO (g/km)	NO_X (g/km)	Evaluation value
1	5.183	0.238	0.568	0.270	3.7357
2	5.188	0.234	0.578	0.268	3.7396
3	5.193	0.238	0.582	0.268	3.7439
4	5.194	0.237	0.601	0.266	3.7462
5	5.199	0.238	0.592	0.267	3.7490

15.6 Conclusion

Based on the optimization results of the example parallel HEV, the following observations can be made. The fuel economy of the parallel HEV is increased from 35.1 to 39.64 MPG with the DIRECT algorithm, and from 35.1 to 40.37 MPG with the SA algorithm. The performance of the optimized HEV shows a great improvement. The power rating of the traction motor is significantly reduced [24]. The second example uses the evolutionary algorithm, NSGA-II, for the multiobjective optimization problem in a series HEV. The algorithm has the capability of simultaneously optimizing fuel economy as well as three emissions. At the end of the optimization process, the algorithm generates 40 optimal tradeoff solutions and performances. All of these solutions are better, in all four categories, than the performances generated by the default controller in ADVISOR. We also demonstrated that vehicle designers and control engineers can derive their own tradeoff criteria to select the solutions that best suit their specific needs.

In these studies, only global optimization algorithms are tested for a hybrid vehicle design, which generally have slower convergence. On the other hand, derivative-based algorithms are known for their faster convergence. In fact, a hybrid optimization algorithm can be used that combines the benefits of both global and local algorithms. The global algorithm can reach a design point near the global optimum region after a certain number of optimization steps. Then a local algorithm kicks in and the process is continued until a global optimum is found.

For the parallel HEV example, the design optimization takes about 100 hours running PSAT on a single PC. This long design time necessitates the development of a more efficient optimization methodology such as using parallel and distributed computing.

References

1 Moore, T.C. (1996) Tools and Strategies for Hybrid-electric Drive System Optimization. SAE Technical Paper Series, No. SP-1189.

2 Miller, J. (2003) *Propulsion Systems for Hybrid Vehicles*, Peter Peregrinus, Stevenage.

3 Fellini, R., Michelena, N., Papalambros, P., et al. (1999) Optimal design of automotive hybrid powertrain systems, in *Proceedings of EcoDesign 99 – First International Symposium on Environmentally Conscious Design and Inverse Manufacturing* (ed. H. Yoshikawa, et al.), Tokyo, pp. 400–405.

4 Gao, W. and Porandla, S. (2005) Design optimization of a parallel hybrid electric powertrain. Proceedings of the IEEE Vehicle Power and Propulsion Conference, Chicago, September, pp. 530–535.
5 Schittkowski, K. (1985) NLQPL: a FORTRAN-subroutine solving constrained nonlinear programming problems. *Annals of Operations Research*, 5, 485–500.
6 Jones, D.R. (2001) DIRECT global optimization algorithm, in *Encyclopedia of Optimization*, Kluwer Academic, Dordrecht.
7 Jones, D.R., Perttunen, C.D., and Stuckman, B.E. (1993) Lipschitz optimization without Lipschitz constant. *Journal of Optimization Theory and Applications*, 79 (1), 157–181.
8 Kirkpatrick, S., Gelett, C., and Vechhi, M. (1983) Optimization by simulated annealing. *Science*, 220 (4598), 671–680.
9 Holland, H.J. (1975) *Adaptation in Natural and Artificial Systems*, The University of Michigan.
10 Kennedy, J. and Eberhart, R. (1995) Particle swarm optimization. Proceedings of the IEEE International Conference on Neural Networks, Perth, Australia, November–December, Vol. IV, pp. 1942–1948.
11 Trelea, I.C. (2003) The particle swarm optimization algorithm: convergence analysis and parameter selection. *Information Processing Letters*, 85 (6), 317–325.
12 Wipke, K. and Markel, T. (2001) Optimization techniques for hybrid electric vehicle analysis using ADVISOR. Proceedings of the ASME, International Mechanical Engineering Congress and Exposition, New York, November.
13 Shubert, B. (1972) A sequential method seeking the global maximum of a function. *SIAM Journal of Numerical Analysis*, 9, 379–388.
14 Metropolis, N., Rosenbluth, A.W., Rosenbluth, M.N., et al. (1958) Equations of state calculations by fast computing machines. *Journal of Chemical Physics*, 21, 1087–1092.
15 PSAT Documentation, http://www.transporation.anl.gov/software/PSAT (accessed 2006).
16 Wipke, K.B., Cuddy, M.R., and Burch, S.D. (1999) ADVISOR 2.1: A User-Friendly Advanced Powertrain Simulation using a Combined Backward/Forward Approach. NREL/JA-540-26839, September.
17 Gao, W., Solodovnik, E., and Dougal, R. (2004) Symbolically-aided model development for an induction machine in Virtual Test Bed. *IEEE Transactions on Energy Conversion*, 19 (1), 125–135.
18 Wipke, K., Markel, T., and Nelson, D. (2001) Optimizing energy management strategy and a degree of hybridization for a hydrogen fuel cell SUV. 18th Electric Vehicle Symposium (EVS-18), Berlin, Germany, October 20–24.
19 Deb, K. (2001) *Multi-Objective Optimization Using Evolutionary Algorithms*, John Wiley & Sons, Ltd, Chichester.
20 Deb, K., Pratap, A., Agarwal, S., et al. (2002) A fast and elitist multiobjective genetic algorithm: NSGA-II. *IEEE Transactions on Evolutionary Computation*, 6 (2), 182–197.
21 Deb, K., Jain, P., Gupta, N.K., et al. (2004) Multiobjective placement of electronic components using evolutionary algorithms. *IEEE Transactions on Components and Packaging Technologies*, 27 (3), 480–492.
22 Vose, M. (2001) *Simple Genetic Algorithm: Foundation and Theory*, MIT Press, Cambridge, MA.
23 ADVISOR (2004) Documentation, http://www.avl.com (accessed January 27, 2011).
24 Gao, W. and Mi, C. (2007) Hybrid vehicle design using global optimisation algorithms. *International Journal of Electric and Hybrid Vehicles*, 1 (1), 57–70.

16

Wireless Power Transfer for Electric Vehicle Applications

Wireless power transfer (WPT) is the technology which could set people free from annoying wires. In fact, WPT adopts a basic theory that has been developed for at least 30 years under the term inductive power transfer (IPT). WPT technology has been developed rapidly in recent years. At kilowatts power level, the transfer distance increases from several millimeters to several hundred millimeters with a grid to load efficiency above 90%. These advances make WPT very attractive for use in electric vehicle (EV) charging applications in both stationary and dynamic charging scenarios. This chapter reviews the technologies in the WPT area applicable to EV wireless charging. By introducing WPT in EVs, the obstacles of charging time, range and cost can be easily mitigated. Furthermore, battery technology is no longer relevant in the mass market penetration of EVs. It is hoped that researchers, encouraged by state-of-the-art achievements, will continue to push forward and further the development of WPT, as well as the expansion of EVs.

The Sections 16.1–7 introduce the basics of WPT with a focus on the IPT technology, while Section 16.8 discusses a double-sided LCC topology. Finally, capacitive wireless power transfer is introduced and discussed briefly in Section 16.9.

16.1 Introduction[1]

Owing to energy, environmental, and many other concerns, the electrification of transportation has been carried out for many years. In railway systems, electric locomotives have been well developed for a long time. A train runs on a fixed track, and it is easy to get electric power from a conductor rail using pantograph sliders. However, for electric vehicles, their high flexibility makes it difficult to get power in a similar way. Instead, a high power, large capacity battery pack is usually used as an energy storage unit to make the EV operate over a satisfactory distance.

Until now, EVs have not been very attractive to consumers, even with the many government incentive programs. Government subsidies and tax incentives are one key

---

1 Reprint, Copyright IEEE. S. Li and C. C. Mi, "Wireless Power Transfer for Electric Vehicle Applications," in IEEE Journal of Emerging and Selected Topics in Power Electronics, vol. 3, no. 1, pp. 4–17, March 2015. doi: 10.1109/JESTPE.2014.2319453

Hybrid Electric Vehicles: Principles and Applications with Practical Perspectives,
Second Edition. Chris Mi and M. Abul Masrur.

to increasing the market share of EVs today. The only problem with an electric vehicle is that its electric storage technology requires a battery, which is the bottleneck today due to its unsatisfactory energy density, limited lifetime, and high cost.

In an EV, the battery is not so easy to design because all of the following requirements – high energy density, high power density, affordable cost, long cycle lifetime, safety, and reliability – need to be met simultaneously. Lithium-ion batteries are recognized as the most competitive solution to be used in electric vehicles [1]. However, the energy density of the commercialized lithium-ion battery in EVs is only 90~100 Wh/Kg for a finished pack [2][2]. This figure is extremely low compared to that of gasoline, which is about 12,000 Wh/Kg. To challenge the 300-mile range of an internal combustion engine powered vehicle, a pure EV needs a large number of batteries, which would be too heavy and too expensive. The cost of lithium-ion batteries is currently about 500 $/kWh. Considering the vehicle's initial investment, maintenance and energy costs, owning a battery-powered electric vehicle will make the consumer spend an extra 1000 $/year on average compared with owning a gasoline-powered vehicle [1]. Besides the cost issue, the long charging time of EV batteries also makes the EV not acceptable to many drivers. For a single charge, it takes from half an hour to several hours, depending on the power level of the attached charger, which is many times longer than the gasoline refueling process. EVs cannot be ready immediately if they have run out of battery energy. To overcome this, what owners would most likely do is find any possible opportunity to plug-in and charge the battery; however, this can also be problematic, as people may forget to plug in and find themselves out of battery energy later on. The charging cables on the floor are also tripping hazards. Leakage from cracked old cables, particularly in cold zones, can produce additional hazardous conditions for the owner. Moreover, people may have to brave the wind, rain, ice, or snow to plug in, at the risk of an electric shock.

WPT technology, which can eliminate all the aforementioned charging troubles is desirable for EV owners. By wirelessly transferring energy to the EV, charging becomes a simple task. For a stationary WPT system, the drivers just need to park their car and leave it. For a dynamic WPT system, which means the EV could be powered while driving, it is possible to run the EV forever without stopping. Also, the battery capacity could be reduced by 80% or more.

Although the market demand is huge, the question is whether WPT can be realized efficiently at a reasonable cost. A research team from Massachusetts Institute of Technology (MIT) published a paper in *Science* [3], in which 60 W power was transferred at a 2-meter distance with the so-called strongly coupled magnetic resonance theory. The result surprised academia, and WPT quickly became a hot research area. A lot of interesting work has been accomplished with different kinds of innovative circuits, as well as system analysis and control [4–9]. The power transfer path can even be guided using the domino-form repeaters [10, 11]. In order to transfer power more efficiently and further, the resonant frequency is usually selected at the MHz level, and air-core coils are used.

---

2 Although a lithium-ion battery can achieve up to 250 Wh/kg for individual cells, the battery pack requires structure design, cooling, and battery management systems. The overall energy density of a battery pack is much lower than the cell density.

When WPT is used in EV charging, it is hard for the MHz frequency operation to meet the power and efficiency criteria. It is inefficient to convert a few watts to a few hundred kilowatts power at MHz frequency level using state-of-the-art power electronics devices. Moreover, air-core coils are too sensitive to surrounding ferromagnetic objects. When an air-core coil is attached to a car, the magnetic flux will go inside the chassis, causing high eddy current loss as well as a significant change in the coil parameters. To make it more practical for EV charging, ferrite as a magnetic flux guide and an aluminum plate as a shield are usually adopted in the coil design [12]. With the frequency lowered to less than 100 kHz, and the use of ferrite, the WPT system is no different from the inductive power transfer (IPT) technology that has been developed for many years [13–39]. In fact, since WPT is based on non-radiative and near-field electromagnetics, there is no difference with the traditional IPT, which is based on magnetic field coupling between the transmitting and receiving coils. The IPT system has already been proposed and applied to various applications, such as underwater vehicles [32–34], mining systems [16], cordless robots in automation production lines [36–39], and the charging of electric vehicles [13, 14, 25–27].

Recently, as the need for EV charging grows and technology has progressed, the power transfer distance has increased from several millimeters to a few hundred millimeters at kilowatts power level [12, 14, 40–60]. As a proof-of-concept of a roadway inductively powered EV, the PATH (Partners for Advance Transit and Highways) program was conducted at UC Berkeley in the late 1970s [14, 54]. A 60 kW, 35-passenger bus was tested along a 213 m long track with two powered sections. The bipolar primary track was supplied with 1200 A, 400 Hz AC current. The distance of the pickup from the primary track was 7.6 cm. The attained efficiency was around 60%, due to limited semiconductor technology. During the past 15 years, researchers at Auckland University have focused on the inductive power supply of movable objects. Their recent achievement in designing pads for stationary charging of EVs is worth noting. A 766 × 578 mm pad that delivers 5 kW of power with over 90% efficiency for distances of approximately 200 mm was reported [48, 55]. The achieved lateral and longitudinal misalignment tolerance is 250 mm and 150 mm, respectively. The knowledge gained from the On-Line Electric Vehicle (OLEV) project conducted at the Korea Advanced Institute of Science and Technology (KAIST) has also contributed to WPT design. Four generations of OLEV systems have been built: a light golf cart as the first generation, a bus for the second and fourth, and an SUV for the third. The accomplishments of the second and the third are noteworthy: 60 kW power transfer for the buses and 20 kW for the SUVs with efficiency of 70% and 83%, respectively, with allowable vertical distance and lateral misalignment up to 160 mm and up to 200 mm, respectively [56, 57]. In the USA, more and more public attention has been drawn to WPT since the publication of the 2007 Science paper [3]. The WiTricity Corporation, with technology from MIT, released their WiT-3300 development kit, which achieved 90% efficiency over a 180 mm gap at 3.3 kW output. Recently, a wireless charging system prototype for EVs was developed at Oak Ridge National Laboratory in the USA. The tested efficiency is nearly 90% for 3 kW power delivery [53]. Researchers at the University of Michigan-Dearborn achieved a 200 mm distance, 8 kW WPT system with DC–DC efficiency as high as 95.7% [61]. From the functional aspects, it can be seen that WPT for EVs is ready in both stationary and dynamic applications. However, to make it available for large-scale commercialization, there is still a lot of work to be done on performance optimization, setup of the industrial standards, cost effectiveness, and so on.

16.2 Fundamental Theory

A typical wireless EV charging system is shown in Figure 16.1. It includes several stages to charge an EV wirelessly. First, the utility alternating current (AC) power is converted to a direct current (DC) power source by an AC–DC converter with power factor correction (PFC). Then, the DC power is converted to a high frequency AC to drive the transmitting coil through a compensation network. Considering the potential insulation failure of the primary side coil, a high frequency isolated transformer may be inserted between the DC–AC inverter and primary side coil for extra safety and protection. The high frequency current in the transmitting coil generates an alternating magnetic field, which induces an AC voltage on the receiving coil. By resonating with the secondary compensation network, the transferred power and efficiency are significantly improved. Finally, the AC power is rectified to charge the battery. Figure 16.1 shows that a wireless EV charger consists of the following main parts:

1) detached (or separated, loosely coupled) transmitting and receiving coils. Usually, the coils are built with ferrite and a shielding structure. In the later sections, the term magnetic coupler is used to represent the entirety, including the coil, ferrite and shielding
2) compensation network
3) power electronics converters.

The main difference between a wireless charger and a conventional conductive or wired charger is that the transformer is replaced by a set of loosely coupled coils. To

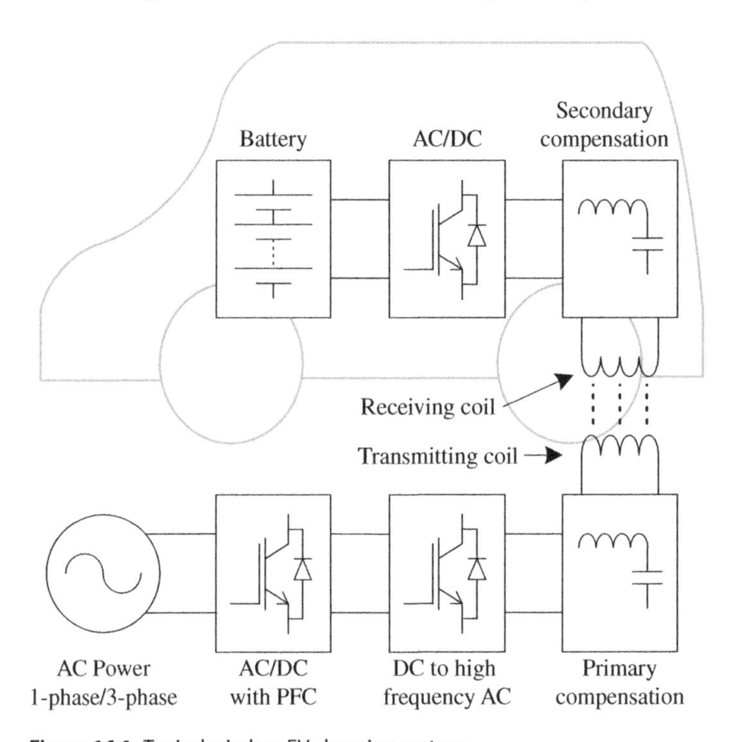

Figure 16.1 Typical wireless EV charging system.

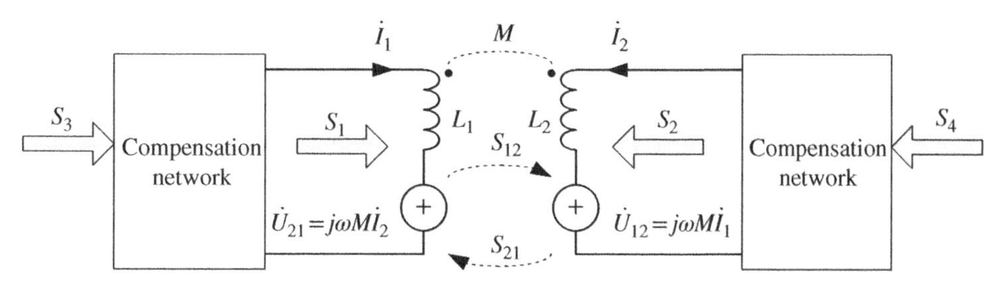

Figure 16.2 A general two-coil wireless power transfer system.

give a quick idea of the wireless power transfer principle, the coil and the compensation network are pulled out separately, as shown in Figure 16.2, where L_1 represents the self-inductance of the primary side transmitting coil and L_2 represents the self-inductance of the receiving coil; $\dot{I}_1$ and $\dot{I}_2$ are the current in the two coils; $\dot{U}_{12}$ is the voltage in the secondary coil that is induced by the current in the primary side coil. $\dot{U}_{21}$ is the voltage in the primary coil that is induced by the current in the secondary side coil due to coupling, or mutual inductance, between the primary and secondary coils. S_1 and S_2 are the apparent power that goes into L_1 and L_2 respectively. S_3 and S_4 are the apparent power provided by the power converter. S_{12} and S_{21} represent the apparent power exchange between the two coils. The form of the compensation network is not specified. The characteristics of the compensation network will be discussed later.

From Figure 16.2, neglecting the coil resistance and magnetic losses, we can calculate the simplified form of exchanged complex power from L_1 to L_2:

$$\begin{aligned}\dot{S}_{12} &= -\dot{U}_{12}\dot{I}_2^* = -j\omega M\dot{I}_1\dot{I}_2^* \\ &= \omega M I_1 I_2 \sin\varphi_{12} - j\omega M I_1 I_2 \cos\varphi_{12}\end{aligned} \tag{16.1}$$

$$\begin{aligned}\dot{S}_{21} &= -\dot{U}_{21}\dot{I}_1^* = -j\omega M\dot{I}_2\dot{I}_1^* \\ &= -\omega M I_1 I_2 \sin\varphi_{12} - j\omega M I_1 I_2 \cos\varphi_{12}\end{aligned} \tag{16.2}$$

where I_1 and I_2 are the root mean square (RMS) values, φ_{12} is the phase difference between $\dot{I}_1$ and $\dot{I}_2$. The active power transfer from the primary side to the secondary side can be expressed as

$$P_{12} = \omega M I_1 I_2 \sin\varphi_{12} \tag{16.3}$$

The system shown in Figure 16.2 can transfer active power in both directions. In the analysis below, we assume that the power is transferred from L_1 to L_2. When $\varphi_{12} = \pi/2$, which means $\dot{I}_1$ leads $\dot{I}_2$ by a quarter cycle, the maximum power can be transferred from L_1 to L_2.

The total complex power that goes into the two-coil system is

$$\begin{aligned}\dot{S} &= \dot{S}_1 + \dot{S}_2 \\ &= j\left(\omega L_1\dot{I}_1 + \omega M\dot{I}_2\right)\dot{I}_1^* + j\left(\omega L_2\dot{I}_2 + \omega M\dot{I}_1\right)\dot{I}_2^* \\ &= j\omega\left(L_1 I_1^2 + L_2 I_2^2 + 2M I_1 I_2 \cos\varphi_{12}\right)\end{aligned} \tag{16.4}$$

Therefore, the total reactive power that goes into the two-coil system is

$$Q = \omega\left(L_1 I_1^2 + L_2 I_2^2 + 2MI_1 I_2 \cos\varphi_{12}\right) \tag{16.5}$$

For a traditional transformer, the reactive power represents the magnetizing power. Higher magnetizing power brings higher copper and core loss. In order to increase the transformer efficiency, the ratio between the active power and reactive power should be maximized. The ratio is defined by

$$\begin{aligned} f(\varphi_{12}) &= \frac{|P_{12}|}{|Q|} = \left| \frac{\omega M I_1 I_2 \sin\varphi_{12}}{\omega L_1 I_1^2 + \omega L_2 I_2^2 + 2\omega M I_1 I_2 \cos\varphi_{12}} \right| \\ &= \frac{k\sqrt{1-\cos^2\varphi_{12}}}{\sqrt{\frac{L_1}{L_2}\frac{I_1}{I_2}} + \sqrt{\frac{L_2}{L_1}\frac{I_2}{I_1}} + 2k\cos\varphi_{21}} = \frac{k\sqrt{1-\cos^2\varphi_{12}}}{x + \frac{1}{x} + 2k\cos\varphi_{12}} \end{aligned} \tag{16.6}$$

where $\pi/2 < \varphi_{12} < \pi$, $x = \sqrt{\frac{L_1}{L_2}\frac{I_1}{I_2}} > 0$, and k is the coupling coefficient between L_1 and L_2.

To achieve the maximum value of $f(\varphi_{12})$, we solve the following equations:

$$\frac{\partial}{\partial\varphi_{12}} f(\varphi_{12}) = 0, \frac{\partial^2}{\partial^2\varphi_{12}} f(\varphi_{12}) < 0 \tag{16.7}$$

And the solutions are

$$\cos\varphi_{12} = -\frac{2k}{x + \frac{1}{x}}, \ \sin\varphi_{12} = \sqrt{1 - \frac{4k^2}{\left(x + \frac{1}{x}\right)^2}} \tag{16.8}$$

When k is close to 1, it is a traditional transformer. In this case, if $\dot{I}_2$ is a current induced by $\dot{I}_1$, x will be close to 1. Thus, $\cos\varphi_{12} \approx -1$, and the phase difference between $\dot{I}_1$ and $\dot{I}_2$ is nearly 180°. While for wireless power transfer, k is close to 0. $f(\varphi_{12})$ is maximized at $\sin\varphi_{12} = 1$, at which point the transferred power is also maximized, and the phase between $\dot{I}_1$ and $\dot{I}_2$ is around 90° instead of 180°. Hence we can see the difference between the tightly and the loosely coupled coils.

In addition, the degree of coupling affects the design of the compensation network. Taking the series-series topology as an example, there are two ways to design the resonant capacitor. One way is to design the capacitor to resonate with the leakage inductance [46, 62], which could achieve a higher $f(\varphi_{12})$. Another way is to resonate with the coil self-inductance [27, 41, 63] which could maximize the transferred power at a certain coil current. When the coupling is tight with a ferrite, say $k > 0.5$, it is important to increase $f(\varphi_{12})$ to achieve better efficiency. In this case, resonating with the coil's self-inductance – which makes $\varphi_{12} = \pi/2$ and lowers $f(\varphi_{12})$ – is not recommended. Otherwise the magnetizing loss may significantly increase. When the capacitor resonates with the leakage inductance, it is as though the leakage inductance were being compensated. This makes the transformer perform as a traditional one and increases $f(\varphi_{12})$. However, the overall system does not work at a resonant mode. When the coupling is loose, say $k < 0.5$, which is the case for EV wireless charging, usually the capacitor is tuned with its self-inductance to make the system work at a resonant mode to achieve maximum transferred power at a certain coil current. In this case, most of the magnetic field energy is stored in the large air gap between the two coils. The hysteresis loss in the ferrite is not particularly high compared to some other magnetic materials. However, the loss in the copper wire is proportional to the square of the conducting current. In order to efficiently transfer more power at a certain coil current, the induced current $\dot{I}_2$ should lag $\dot{I}_1$ by 90°. Since the induced voltage $\dot{U}_{12}$ on the receiving coil lags $\dot{I}_1$ by 90°, $\dot{U}_{12}$ and $\dot{I}_2$ should be in phase. The secondary side should have a pure resistive characteristic seen from $\dot{U}_{12}$ at the frequency of the current waveform of $\dot{I}_1$. In the meanwhile, the primary side input apparent power S_3 should be minimized. At $\cos\varphi_{12} = 0$, the complex power $\dot{S}_1$ is

$$\dot{S}_1 = j\omega L_1 I_1^2 + \omega M I_1 I_2 \tag{16.9}$$

Ideally, the primary side compensation network should cancel the reactive power and make $S_3 = \omega_0 M I_1 I_2$, where ω_0 is the resonant frequency. From the above analysis, we see that for a certain transferred power, it is necessary to make the secondary side resonant to reduce the coil VA rating, which reduces the loss in the coils, and to make the primary side resonant to reduce the power electronics converter VA rating, which in turn reduces the loss in the power converter. This is why we transfer power at the magnetic resonance.

Based on the above analysis, we can calculate the power transfer efficiency between the two coils at the resonant frequency. We have

$$U_{12} = I_2\left(R_2 + R_{Le}\right) = \omega M I_1 = \omega k\sqrt{L_1 L_2}\, I_1 \tag{16.10}$$

where R_2 is the secondary winding resistance and R_{Le} is the equivalent load resistance.

By defining the quality factor of the two coils, $Q_1 = \omega L_1/R_1$, $Q_2 = \omega L_2/R_2$, the transferred efficiency can be expressed as

$$\eta = \frac{I_2^2 R_{Le}}{I_1^2 R_1 + I_2^2 R_2 + I_2^2 R_{Le}} = \frac{R_{Le}}{\dfrac{(R_2 + R_{Le})^2}{k^2 Q_1 Q_2 R_2} + R_2 + R_{Le}} \tag{16.11}$$

By defining $a = R_{Le}/R_2$, we obtain the expression of efficiency as a function of a:

$$\eta(a) = \frac{1}{\dfrac{a + \dfrac{1}{a} + 2}{k^2 Q_1 Q_2} + \dfrac{1}{a} + 1} \tag{16.12}$$

The maximum efficiency is obtained by solving the following equations:

$$\frac{\partial}{\partial a}\eta(a) = 0, \quad \frac{\partial^2}{\partial^2 a}\eta(a) < 0 \tag{16.13}$$

The maximum efficiency $\eta_{\max} = \dfrac{k^2 Q_1 Q_2}{\left(1 + \sqrt{1 + k^2 Q_1 Q_2}\right)^2}$ is achieved at $a_{\eta\max} = \sqrt{1 + k^2 Q_1 Q_2}$.

In [64], the maximum efficiency is also derived based on several different kinds of compensation networks. The results are identical and are in accordance with the above

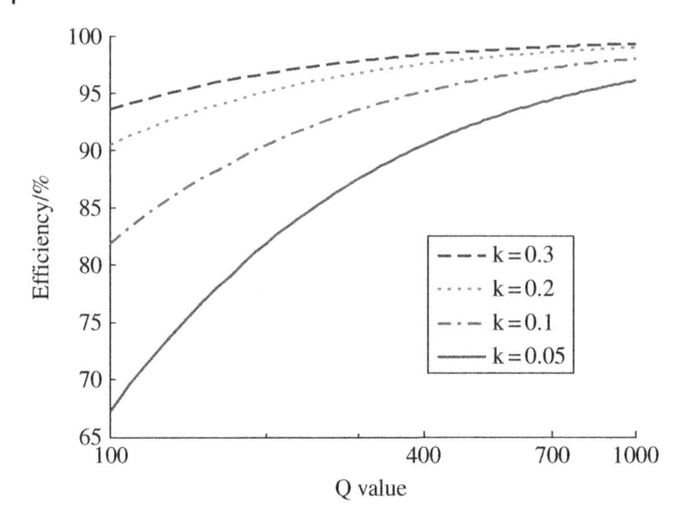

Figure 16.3 Theoretical maximum transfer efficiency between two coils.

results. The analysis here does not specify a particular compensation form. Instead, it can be regarded as a general formula to evaluate the coil performance and to estimate the highest possible power transfer efficiency.

In EV wireless charging applications, the battery is usually connected to the coil through a diode-bridge rectifier. Most of the time, there is some reactive power required. The reactive power can be provided by either the coil or the compensation network like a unit-power-factor pickup. The battery can be equivalent to a resistance $R_b = U_b/I_b$, where U_b and I_b are the battery voltage and current respectively. If the battery is connected to the rectifier directly in a series-series compensation form, the equivalent AC side resistance can be calculated by $R_{ac} = 8/\pi^2 \cdot R_b$. Thus, a battery load could be converted to a resistive load. The R_{ac} equation is different for different battery connection styles, such as with or without DC/DC converter, parallel or series compensation. Most of the time, the equivalent R_{ac} can be derived. Some typical equivalent impendences on the primary side are given in [42]. By calculating the equivalent AC resistances, the above equations could also be applied to a battery load with a rectifier.

For stationary EV wireless charging, the coupling between the two coils is usually around 0.2. If both the sending and receiving coils have a quality factor of 300, the theoretical maximum power transfer efficiency is about 96.7%. More efficiency calculations under different coupling and quality factors are shown in Figure 16.3.

16.3 Magnetic Coupler Design

In order to transfer power wirelessly, there must be at least two magnetic couplers in a WPT system. One is at the sending side – the primary coupler. The other is at the receiving side – the pickup coupler. Depending on the application scenario, the magnetic coupler in a WPT for an EV could take either a pad or a track form. For higher efficiency, it is important to have a high coupling coefficient k and a high quality factor Q. Generally, for a given structure, the larger the size-to-gap ratio of the coupler, the higher the k; the thicker the wire and the larger the ferrite section area, the higher the Q. By increasing the dimensions and improving the materials, a higher efficiency can be achieved, but this is not a good engineering approach. It is preferred to have a higher k and Q with the minimum dimensions and cost. Since Q equals $\omega L/R$, a high frequency is usually adopted to increase the value *of* Q. Researchers at MIT used a frequency at around 10 MHz and the coil's Q value reached nearly 1000 [3]. In high power EV WPT applications, the frequency is also increased in order to have these benefits. In Bolger's early design, the frequency was only 180 Hz [13]. A few years later, a 400 Hz frequency EV WPT system was designed by System Control Technology [14]. Neither 180 Hz nor 400 Hz is high enough for a loosely coupled system, and huge couplers were employed in the two designs. Modern WPT systems use a frequency of at least 10 kHz [15], and 100 kHz could be achieved [65] at high power levels. The WiTricity Company, with the technology from MIT, uses a frequency of 145 kHz in their design. In recent research and applications, the frequency adopted in an EV WPT system is between 20 kHz and 150 kHz to balance efficiency and cost. In this frequency range, to reduce the AC loss of copper coils, Litz wire is usually adopted.

Besides the frequency, the coupling coefficient k is significantly affected by the design of the magnetic couplers, which is considered one of the most important factors in a WPT system. With similar dimensions and materials, different coupler geometry and configuration will have a significant impact on the coupling coefficient. A better coupler design may lead to a 50~100% improvement compared with some non-optimal designs [48].

16.3.1 Coupler for Stationary Charging

In stationary charging, the coupler is usually designed in a pad form. Very early couplers were just like a simple split-core transformer [19, 38, 56]. Usually this kind of design could only transfer power through a very small gap. To meet the requirements for EV charging, the deformations from spilt-core transformers and new magnetic coupler forms are presented for large gap power transfer [12, 31, 37, 42, 47–50, 66–71]. According to the magnetic flux distribution area, the couplers can be classified as the double-sided or single-sided type. For the double-sided type, the flux goes to both sides of the coupler [12, 31, 67]. A flattened solenoid inductor form is proposed in [12, 67]. Because the flux goes through the ferrite as through a pipe, it is also called a flux-pipe coupler. To prevent the eddy current loss in the EV chassis, an aluminum shield is usually added, which brings a loss of 1~2% [12]. When the shielding is added, the quality factor of a flux-pipe coupler is reduced from 260 to 86 [48]. The high shielding loss makes the double-sided coupler a non-optimal choice. For the single-sided coupler, most of the flux exists on only one side of the coupler. As shown in Figure 16.4, the main flux path flows through the ferrite in a single-sided coupler. Unlike the double-sided coupler, which has half of the main flux in the back, the single-sided coupler has only a leakage flux in the back. This makes the shielding effort of a single-sided type much less.

Two typical single-sided flux type pads are shown in Figure 16.5. One is a circular unipolar pad [47]. The other one is a rectangular bipolar pad proposed by the University of Auckland, which is also named the DD pad [48]. Besides the mechanical support material, a single-sided pad is composed of three layers. The top layer is the coil. Below the coil, a ferrite layer is inserted for the purpose of enhancing and guiding the flux.

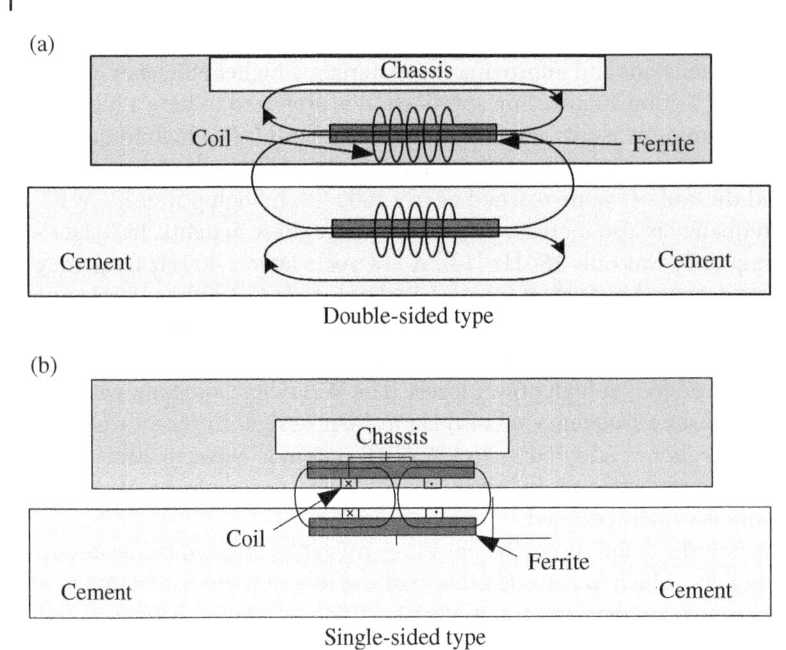

Figure 16.4 Main flux path of double-sided and single-sided couplers.

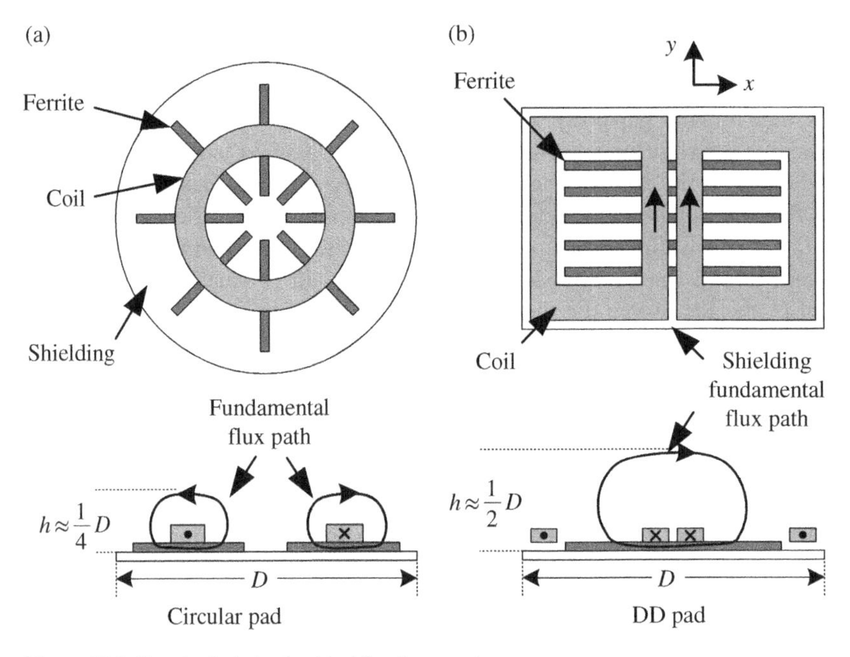

Figure 16.5 Two typical single-sided flux type pads.

At the bottom is a shielding layer. To transfer power, the two pads are put close together, coil to coil. With the shielding layer, most of the high frequency alternating magnetic flux can be confined in the space between the two pads. A fundamental flux path concept was proposed in the flux-pipe paper [67]. The flux path height of a circular pad is about ¼ of the pad's diameter, while for a DD pad, the height is about ½ of the pad's length. For a similar size, a DD pad has a significant improvement in the coupling. The charge zone for a DD pad could be about twice the area of a circular pad with similar material cost. The DD pad has a good tolerance in the y direction, making the DD pad a potential solution for dynamic charging when the driving direction is along the y axis. However, there is a null point for the DD pad in the x direction at about 34% misalignment [48]. To increase the tolerance in the x direction, an additional quadrature coil, named the Q coil, is proposed to work together with the DD pad, which is called the DDQ pad [48, 49, 68]. With a DDQ receiving pad on a DD sending pad, the charge zone is increased to five times that of the circular configuration. Due to the additional Q coil on the receiver side, the DDQ over DD configuration uses almost two times the copper compared with the circular one [48]. A variant of a DDQ pad, which is called a new bipolar pad, was also proposed by the University of Auckland [49, 50]. By increasing the size of each D pad and having some overlap between the two D coils, the new bipolar pad could have a performance similar to that of a DDQ pad with 25% less copper. With all these efforts, at a 200 mm gap, the coupling between the primary and secondary pads could reach 0.15~0.3 with an acceptable size for an EV. As shown in Figure 16.3, at this coupling level, efficiency above 90% could possibly be achieved.

16.3.2 Coupler for Dynamic Charging

Dynamic charging, also called online electric vehicles (OLEV) [56] or roadway powered electric vehicles (RPEV) [14], is a way to charge an EV while it is driving. It is believed that dynamic charging can solve the EV range anxiety, which is the main factor limiting the market penetration of EVs. In a dynamic charging system, the magnetic components are composed of a primary-side magnetic coupler, which is usually buried under the road, and a secondary-side pickup coil, which is mounted under an EV chassis. There are two main kinds of primary magnetic couplers in dynamic charging. The first is a long track coupler [26, 31, 57, 70, 72–76]. When an EV with a pickup coil is running along the track, continuous power can be transferred. The track can be as simple as just two wires [37, 77], or an adoption of ferrites with U-type or W-type [26, 56] to increase the coupling and power transfer distance. Further, a narrow-width track design with an I-type ferrite was proposed by KAIST [72, 73]. The differences between the W-type and I-type are shown in Figure 16.6. For the W-type configuration, the distribution area of the ferrite W determines the power transfer distance, as well as the lateral displacement. The total width of the W-type should be about four times the gap between the track and the pickup coil. For the I-type configuration, the magnetic pole alternates along with the road. The pole distance W_1 is optimized to achieve a better coupling at the required distance. The width of the pickup coil W_2 is designed to meet the lateral misalignment requirement. The relation between track width and transfer distance is decoupled, and the track can be made very narrow. The widths for the U-type and W-type are 140 cm and 80 cm respectively [73]. For the I-type, it can be reduced to just 10 cm with a similar power transfer distance and misalignment capacity.

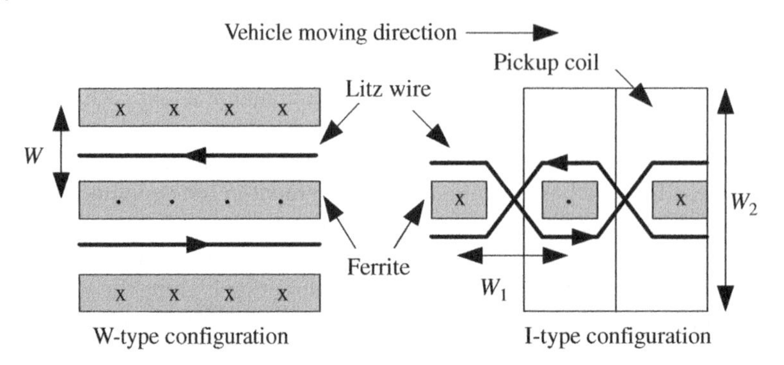

Figure 16.6 Top view of W-type and I-type track configurations.

A power of 35 kW was transferred at a 200 mm gap and 240 mm displacement using the I-type configuration [73]. With the narrowed design, the construction cost could be reduced. Also, since the track is far away from the roadside, the electromagnetic field strength that pedestrians are exposed to could also be reduced.

The problem with the track design is that the pickup coil covers only a small portion of the track, which makes the coupling coefficient very small. The poor coupling produces efficiency and electromagnetic interference (EMI) issues. To reduce the EMI issue, the track is built in segments [52, 70, 75] with a single power converter and a set of switches to power the track. The excitation of each segment can be controlled by the switches' on-off states, so the electromagnetic field above the inactive segments is significantly reduced or eliminated. However, there is always a high frequency current flowing through the common supply cables, which lowers the system efficiency. The published system efficiency is about 70~80%, which is much lower than the efficiency achieved in stationary charging.

If each segment is short enough, the track becomes like a pad in stationary charging, which is the other kind of primary magnetic coupler. Each pad can be driven by an independent power converter. In this way, the primary pads can be selectively excited without a high frequency common current. Also, the energized primary pad is covered by the vehicle. The electromagnetic field is shielded to have a minimum impact on the surrounding environment. The efficiency and EMI performance could be as good as that in a stationary charging application, but the cost to build a power converter for each pad is too high. It is desirable to use only one converter to drive a few pads, for the current in each pad to be controllable. A double coupled method was proposed with each pad configured with an intermediary coupler and a bidirectional switch [78]. The intermediary couplers are coupled to one primary coil on the converter side and perform like a high frequency current source. By controlling the on-off time of the switch, the current in each pad can be controlled. However, even if the corresponding pad is shut down by the switches, the high frequency current is always circulating in all the intermediary couplers, which may lower the efficiency. A reflexive field containment idea by North Carolina State University has also been proposed [79], in which three pads are driven from only one power converter. By carefully designing the primary and pickup parameters, the reflexive field of the pickup pad could enhance the current in the primary pad. The current in each primary pad is sensitive to the coupling condition and could be automatically built up when the pickup pad is coupled. Furthermore, the current decreases very quickly when the pickup pad moves away. The relation between the primary pad current and coupling coefficient should be carefully designed. For dynamic charging, the EV runs freely on the road, which makes the coupling vary over a wide range. To make this method more practicable, the system characteristics under coupling variation caused by the lateral misalignment, vehicle forward movement, and vehicle types need to be studied further.

16.4 Compensation Network

In a WPT system, the pads are loosely coupled with a large leakage inductance. The analysis in Section 16.2 shows that it is necessary to use a compensation network to reduce the VA rating in the coil and power supply. In early inductive charging designs, the compensation was set on the primary or secondary side only [18]. When the coupling coefficient is reduced to less than 0.3 in the EV WPT, compensation on both the primary and secondary sides is recommended to have more flexible and advanced characteristics. To compensate for a leakage inductance, the simplest way is to add a capacitor on each side. As shown in Figure 16.7, depending on how the capacitors are connected to the coils, there are four basic compensation topologies, which are series-series (SS), series-parallel (SP), parallel-parallel (PP), and parallel-series (PS) [21, 23, 27, 80–82]. If the primary side is series compensated, a voltage source converter can be connected directly to the coil. If the primary side is parallel compensated, usually an inductor is inserted to change the converter to a current source. The secondary side capacitor C_2 is usually designed to resonate with L_2 to reduce the VA capacity of the coils. When the primary side coil has a constant current, a series compensation on the

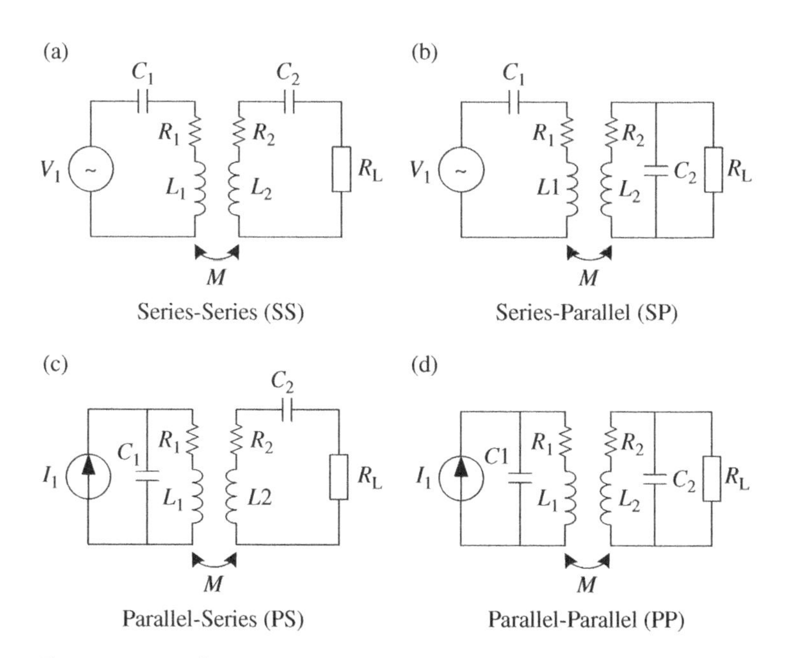

Figure 16.7 Four basic compensation topologies.

Table 16.1 Primary compensation capacitance.

Topology	Primary capacitance C_1
SS	$\frac{C_2 L_2}{L_1}$
SP	$\frac{C_2 L_2}{L_1} \cdot \frac{1}{1-k^2}$
PS	$\frac{C_2 L_2}{L_1} \cdot \frac{1}{Q_s^2 k^4 + 1}$
PP	$\frac{C_2 L_2}{L_1} \cdot \frac{1-k^2}{Q_s^2 k^4 + \left(1-k^2\right)^2}$

secondary side makes the output behave like a voltage source, while a parallel compensation makes the output behave like a current source [27]. However, not all of the design has a constant primary side current, and different output characteristics can exist for a series or parallel compensation on the secondary side.

To reduce the power converter VA rating, the primary side capacitor is usually tuned to make the input voltage and current be in phase under certain coupling and load conditions, which is called the zero-phase-angle (ZPA) method. To realize soft-switching for power electronics converters, the primary side compensation network is often tuned to make the primary side have a small portion of reactive power to reach zero voltage switching (ZVS) or zero current switching (ZCS) condition [22, 44, 83]. Since the tuned reactive power is relatively small, the parameters for realizing ZVS and ZCS are close to the parameters designed by ZPA method.

To calculate the primary side capacitance, a secondary load quality factor is defined. For series compensated secondary side, $Q_s = \frac{\omega_0 L_2}{R_L}$. For parallel compensated secondary side, $Q_s = \frac{R_L}{\omega_0 L_2}$, where ω_0 is the resonant frequency. The load quality factor is the ratio between the reactive and active power.

To achieve ZPA on the primary side, the primary capacitances for different types of compensation are listed in Table 16.1 [27]. From this table we can see the primary compensation capacitance is a constant value for the SS method, regardless of the coupling and load conditions. For the SP method, the capacitance varies when the coupling changes. For PS and PP, the capacitance is affected by both the coupling and load conditions.

When the secondary side is at resonant frequency, the reflected load on the primary side can be calculated from

$$R_{r\_ss} = R_{r\_ps} = \frac{\omega_0^2 M^2}{R_L} \tag{16.14}$$

$$R_{r\_sp} = R_{r\_pp} = \frac{M^2 R_L}{L_2^2} \tag{16.15}$$

For the SS structure, from (16.14) we see that when the coupling reduces, the reflected resistance on the primary side also reduces. This will increase the output power when the primary side is connected to a voltage source. For the PS structure, the reflected resistance changes in the same way as the change of coupling. However, the PS structure should be connected to a current source. The output power will reduce when the coupling reduces. In order to maintain a constant output power when the coupling changes, the SPS compensation method was proposed in [84]. It can be regarded as a combination of SS and PS. By designing a proper ratio between the two primary side capacitors, the characteristics of SS and PS are mixed. Thus, constant output power is achieved at a high misalignment tolerance without adjusting the primary power supply.

By introducing an LC compensation network, a primary side LCL current source structure is widely used in inductive heating and wireless power transfer applications [44, 85–88]. The advantage of the LCL structure is obvious. At the resonant frequency, an LCL network performs like a current source. The current in the primary side coil is controlled by the high frequency square wave voltage from the power converter, regardless of the coupling and load conditions. This makes the control on the primary side much easier. Moreover, by tuning the LCL parameters, the reactive power can be fully compensated. The power converter provides the active power only, so the required VA rating for the power converter can also be minimized.

Together with the LCL primary compensation, compensation using parallel form on the secondary side is adopted in many designs [55, 69]. The power control and decoupling method for a parallel structure has been well developed [21]. However, a parallel compensated system has a large reactive current in the pickup coil and the reactive power is reflected to the primary side. To overcome the disadvantage of the parallel pickup compensation, a unity power factor pickup has been proposed by the University of Auckland [89–91]. By introducing an LCL form pickup coil, the circulating current in the pickup coil can be minimized and only the active load is reflected to the primary side. To cancel the nonlinear effect of the rectifier diodes, another capacitor is introduced to form an LCLC compensation form, which can achieve an exact unity power factor under a predetermined load condition. From the comparisons, the difference between an LCL and an LCLC is insignificant. Both of LCL and LCLC structures achieve a significant efficiency improvement compared with the traditional LC parallel structure.

16.5 Power Electronics Converters and Power Control

In a WPT system, the function of the primary side power electronics converter is to generate a high frequency current in the sending coil. To increase the switching frequency and efficiency, usually a resonant topology is adopted [21–23, 55, 85, 87, 88, 92–96]. On the secondary side, a rectifier is adopted to convert the high frequency AC current to DC current. Depending on whether a secondary side control is needed, an additional converter may be employed. The primary side converter may be a voltage or a current source converter. As a bulky inductor is needed for the current source converter, the most common choice on the primary side is a full bridge voltage source resonant converter. A typical wireless power circuit schematic is shown in Figure 16.8. In the primary side, the full bridge converter outputs a high frequency square wave voltage.

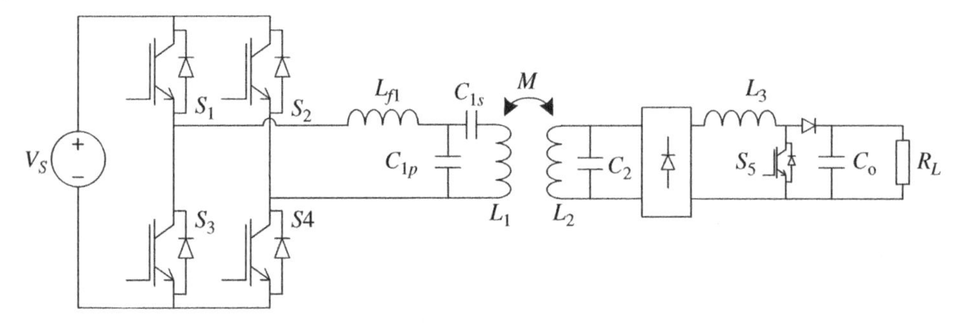

Figure 16.8 Circuit schematic of a typical WPT configuration.

By adopting the LC compensation network, a constant high frequency current can be maintained in L_1. An additional capacitor C_{1s} is introduced here to compensate part of the reactive power of L_1. Thus, the power rating on L_{f1} can be reduced. The system design flexibility can also be improved. On the secondary side, the parallel compensation is adopted. With a constant primary coil current and parallel secondary side compensation, the output behaves like a current source. At a certain coupling, the current in L_3 is almost constant. By changing the duty ratio of switch S_5, the output power can also be controlled.

Many different control methods have been proposed to control the transferred power. Depending on where the control action is applied, the control method can be classified as primary side control [92, 95, 97], secondary side control [27, 30, 45] or dual side control [55]. In most cases, the primary side and dual side control is only suitable for power transfer from one primary pad to one pickup pad. The secondary side control can be used in the scenario where multiple pickup pads are powered by one primary pad or track.

The control on the primary side can be realized by changing the frequency, duty cycle, and phase between the two legs. Since the characteristic of a resonant converter is related to the operating frequency, a frequency control at the primary side is adopted in some designs [53, 98]. When adjusting the frequency, the bifurcation phenomenon in the loosely coupled systems should be noted [23]. Power vs. frequency is not always a monotonic function. Also, the frequency control method takes up a wider radio frequency bandwidth, which may increase the risk of electromagnetic interference. When the switching frequency is fixed, the control can be carried out by a duty cycle or phase shift [99]. The problem with duty cycle or phase shift control, however, is that there is a high circulating current in the converter. Also, the ZVS or ZCS switching condition may be lost. To ensure ZVS, an alternative way to control the system output power is to adjust the input DC voltage V_S [73]. An asymmetrical voltage cancellation method, which uses an alternative way to change the duty cycle, was proposed to increase the ZVS region [88]. A discrete energy injection method, which could achieve ZCS and lower the switching frequency under light load condition, was proposed in [43, 92, 94].

On the secondary side, as shown in Figure 16.8, with parallel compensation, a boost converter is inserted after the rectifier for the control. Correspondingly, with series compensation, a buck converter can be used. When the control is after the rectifier, an additional DC inductor, as well as a diode on the current flow path, should be introduced. The University of Auckland proposed a control method on the AC side before the rectifier. By doing so, the DC inductor and additional diodes could be saved. Due to the resonance in the AC side, ZVS and ZCS could be achieved. The detailed designs for series compensation, as well as an LC compensation network, are presented in [91, 100, 101]. The dual side control is a combination of both primary and secondary side control [55]. The system complexity and cost may increase, but the efficiency can be optimized by a dual side control.

16.6 Methods of Study

Wireless power transfer involves multiple disciplines, including magnetics, power electronics, communications, mechanical and electrical engineering. The study of a WPT system can be very complex due to its multidisciplinary nature and uncertainties. For example, the magnetic field is at a high frequency and low density, and varies with the gap distance, misalignment, and power levels. The resonance in the system is key to the high efficiency power transfer, but that could also be affected by the coupling between the two coils, and the surrounding media (i.e., wet or dry environment). Typically, the study of WPT systems involves

1) an analytical method, including circuit analysis and calculation of mutual inductances through analytical approaches
2) field analysis using numerical tools such as finite element methods, finite boundary method, high frequency structured system analysis, etc.
3) simulation of a lumped model involving parametric analysis, i.e. coupling coefficient change versus efficiency, etc.
4) an experimental study involving the use of network analyzers and field measurements and parameter identification of the WPT system and its resonant characteristics
5) soft switching of the power converters in a WPT system that involves the various methods for studying power electronic circuits.

In particular, two-port network theory can be an efficient tool for the study of WPT systems [102]. Figure 16.9 shows a WPT system with an SP resonant topology and its generic two-port network representation.

The impedance matrix, transfer matrix, and scattering matrix can be defined as

Impedance Matrix – Z

$$\begin{bmatrix} V_1 \\ V_2 \end{bmatrix} = \begin{bmatrix} Z_{11} & Z_{12} \\ Z_{21} & Z_{22} \end{bmatrix} \begin{bmatrix} I_1 \\ -I_2 \end{bmatrix} \tag{16.16}$$

Transfering Matrix – T

$$\begin{bmatrix} V_1 \\ I_1 \end{bmatrix} = \begin{bmatrix} a & b \\ c & d \end{bmatrix} \begin{bmatrix} V_2 \\ -I_2 \end{bmatrix} \tag{16.17}$$

Scattering Matrix – S

$$\begin{bmatrix} b_1 \\ b_2 \end{bmatrix} = \begin{bmatrix} S_{11} & S_{12} \\ S_{21} & S_{22} \end{bmatrix} \begin{bmatrix} a_1 \\ a_2 \end{bmatrix} \tag{16.18}$$

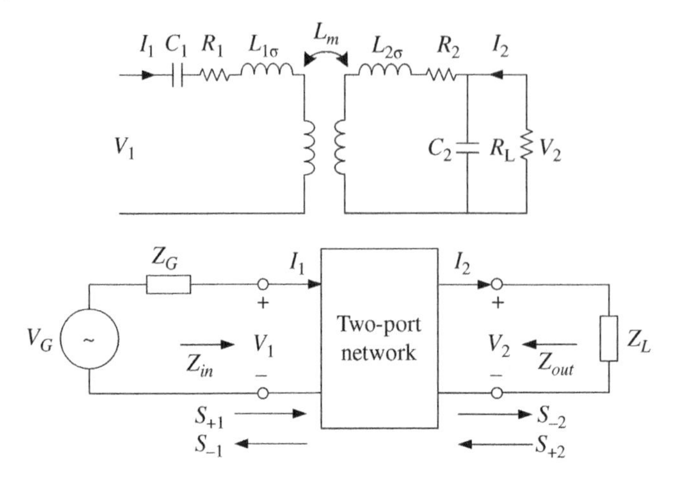

Figure 16.9 A WPT system with an SP resonant topology and its representation as a two-port network: top – SP topology; bottom – two-port network of a WPT system.

By rearranging the equations above, we can obtain the following relationships between the impedance matrix and the transfer matrix:

$$Z = \frac{1}{c}\begin{bmatrix} a & -(ad-bc) \\ 1 & -d \end{bmatrix}; T = \frac{1}{Z_{21}}\begin{bmatrix} Z_{11} & -(Z_{11}Z_{22}-Z_{12}Z_{21}) \\ 1 & -Z_{22} \end{bmatrix} \tag{16.19}$$

Similarly, we can obtain the following relationships between the impedance matrix and the scattering matrix:

$$Z = \frac{Z_0}{D_S}\begin{bmatrix} (1+S_{11})(1-S_{22})+S_{12}S_{21} & 2S_{12} \\ 2S_{21} & (1-S_{11})(1+S_{22})-S_{12}S_{21} \end{bmatrix} \tag{16.20}$$

$$S = \frac{1}{D_Z}\begin{bmatrix} (Z_{11}-Z_0)(Z_{22}+Z_0)-Z_{12}Z_{21} & 2Z_{12}Z_0 \\ 2Z_{21}Z_0 & (Z_{11}+Z_0)(Z_{22}-Z_0)-Z_{12}Z_{21} \end{bmatrix} \tag{16.21}$$

where

$$\begin{aligned} D_Z &= (Z_{11}+Z_0)(Z_{22}+Z_0)-Z_{12}Z_{21} \\ D_S &= (1-S_{11})(1-S_{22})-S_{12}S_{21} \end{aligned} \tag{16.22}$$

Input power to the network:

$$P_1 = \frac{|V_S|^2}{|Z_S+Z_{in}|^2}R_{in} = \frac{|V_S|^2}{4Z_0}\frac{\left(1-|\Gamma_{in}|^2\right)|1-\Gamma_S|^2}{|1-\Gamma_{in}\Gamma_S|^2} \tag{16.23}$$

Output power:

$$\begin{aligned} P_2 &= \frac{|Z_{21}|^2|V_S|^2}{|(Z_{in}+Z_S)(Z_{22}+Z_L)|^2}R_L \\ &= \frac{|V_S|^2}{4Z_0}\frac{(1-|\Gamma_L|^2)|1-\Gamma_S|^2|S_{21}|^2}{|(1-S_{11}\Gamma_S)(1-S_{22}\Gamma_L)-S_{12}S_{21}\Gamma_S\Gamma_L|^2} \end{aligned} \tag{16.24}$$

Efficiency:

$$\eta = \frac{P_2}{P_1} = \frac{1}{1-|\Gamma_{in}|^2} \frac{1-|\Gamma_L|^2}{|1-S_{22}\Gamma_L|^2} |S_{21}|^2 \tag{16.25}$$

where Γ_{in} is the input reflection coefficient, Γ_L is the load reflection coefficient, and Z_{in} is the input impedance.

For SP topology, we can derive the transfer parameters as

$$\begin{aligned}
a &= \frac{1}{L_m}\left(L_1 - \frac{1}{\omega^2 C_1} + R_1 C_2 R_2 + \frac{L_2 C_2}{C_1} \right) + \frac{1}{L_m}(-\omega^2 C_2 (L_1 L_2 - L_m{}^2)) \\
&\quad + \frac{j}{L_m}\left(-\frac{R_1}{\omega} \right) + \frac{j}{L_m}\left(\omega C_2 R_2 L_1 - \frac{C_2 R_2}{\omega C_1} + \omega C_2 R_1 L_2 \right) \\
b &= \frac{1}{L_m}\left(L_1 R_2 - \frac{R_2}{\omega^2 C_1} + R_1 L_2 \right) + \frac{j}{L_m}\left(-\frac{R_1 R_2}{\omega} + \omega (L_1 L_2 - L_m{}^2) - \frac{L_2}{\omega C_1} \right) \\
c &= \frac{1}{L_m}\left(C_2 R_2 + j\left(\frac{-1}{\omega} + \omega C_2 L_2 \right) \right) \\
d &= \frac{1}{L_m}\left(L_2 + j\left(-\frac{R_2}{\omega} \right) \right)
\end{aligned} \tag{16.26}$$

From the above equations, we can easily study the system performance by adjusting the parameters of the system.

16.7 Additional Discussion

16.7.1 Safety Concerns

Wireless power transfer avoids the risk of electrocution associated with the traditional contact charging method, but when charging an EV battery wirelessly, there is a high frequency magnetic field between the transmitting and receiving coils. The large air-gap between the two coils causes a high leakage field, but the magnetic flux coupled between the two coils is the foundation for wireless power transfer, which cannot be shielded. The frequency and amplitude of the leakage magnetic field should be elaborately controlled to meet safety regulations.

A safe region should always be defined for a wireless charging EV. We should ensure that the magnetic flux density meets the safety guidelines when people are in normal positions, such as standing outside a car or sitting inside a car. Fortunately, cars are usually made of steel, which is a very good shielding material.

The guidelines published by the International Commission on Non-Ionizing Radiation Protection (ICNIRP) are the most referenced standard to ensure human safety. There are two versions of the ICNIRP standards. The first was published in 1998. In ICNIRP 1998, there are two reference levels for occupational and general public exposure respectively. At a frequency of 0.8–150 kHz, which covers most of the EV WPT frequency, the limit for general public exposure is 6.25 μT. For occupational exposure, it is a little different. At frequency 0.82–65 kHz, the limit is 30.7 μT, while at 65 kHz–1 MHz, the limit is 2.0/*f*, where *f* is the frequency measured in MHz. Under the ICNIRP 1998 guidelines, the safety evaluation for a 5 kW stationary EV WPT system was conducted [55]. The average magnetic field exposure of up to a 1500 mm

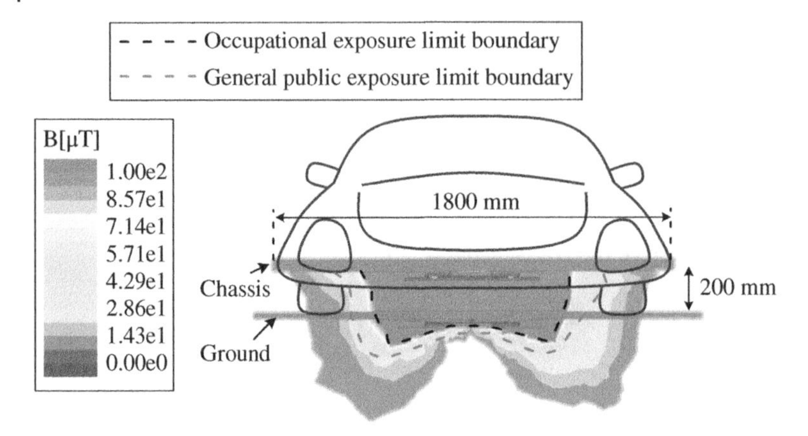

Figure 16.10 Exposure limit boundary for a 8 kW WPT system.

height body was 4.36 μT. For a 35 kW dynamic EV WPT system, the magnetic flux density at 1 meter from the center of the road is 2.8 μT [72]. Both the stationary and dynamic WPT system designs met the ICNIRP 1998 safety guidelines. A good thing for EV WPT is that, after another ten years of research on the health effects of time-varying electromagnetic field, the ICNIRP revised its guidelines in 2010 and increased the reference level significantly. For occupational exposure, the reference level was relaxed to 100 μT. For the general public, the value changed from 6.25 μT to 27 μT. The increase in the reference level is because the former guidelines were too conservative. There is another standard about electromagnetic field safety issues, IEEE Std. C95.1-2005, presented by the IEEE International Committee on Electromagnetic Safety. In this standard, the maximum permissible exposure for the head and torso is 205 μT for general public, and 615 μT for occupational. The maximum permissible exposure for the limbs is even higher, which is 1130 μT for both the general public and occupational. Compared with the IEEE Std., the ICNIRP 2010 standard is still conservative. According to ICNIRP 2010, the exposure safety boundaries of our 8 kW EV WPT system for both occupational and general public exposure are shown in Figure 16.10. Together with the chassis, the safety zone is quite satisfactory. On the premise of safety, a higher power WPT system could be developed according to ICNIRP 2010.

Besides the safety issue, the emission limit for industrial, scientific, and medical (ISM) equipment is also regulated by the Federal Communications Commission (FCC) in Title 47 of the Code of Federal Regulations (CFR 47) in part 18 in the USA. According to FCC part 18, ISM equipment operating in a specified ISM frequency band is permitted unlimited radiated energy. However, the lowest ISM frequency is at 6.78 MHz, which is too high for EV WPT. When the WPT operates at a non-ISM frequency, the field strength limit should be subjected to §18.305. The Society of Automotive Engineers (SAE) has already formed a committee, J2954, to look into many issues related to EV WPT systems. One of their goals will be creating the safety standards. It is anticipated that the J2954 SAE standard on EV WPT systems will soon be released by this committee. More standards and regulations from different regions are summarized in a paper from Qualcomm Incorporated [103].

16.7.2 Vehicle to Grid Benefits

As the development of EVs continues, the vehicle to grid (V2G) concept, which studies the interaction between mass EV charging and the power grid, has also become a hot research topic in smart grid and EV areas. It is recognized that if the EV charging procedure could be optimized, it could have many benefits for the grid. The EV could balance the loads by valley filling and peak shaving. The batteries in EVs are like an energy bank, thus some unstable new energy power supply, such as wind power, could be connected to the grid more easily. When the secondary rectifier diodes are replaced by active switches, a bidirectional WPT function is realized [104–112]. The bidirectional WPT could provide advanced performance in V2G applications. Studies show that by introducing WPT technology, drivers are more willing to connect their EVs to the grid [113], which could maximize the V2G benefits.

16.7.3 Wireless Communications

In a WPT system, it is important to exchange information between the grid side and vehicle side wirelessly to provide feedback. Thus, the power flow could be controlled by the methods mentioned in Section 16.5. The communication design can be classified by whether the signal is modulated on the power carrier or uses a separate frequency band. The Qi standard for wireless low power transfer modulates a 2 kHz communication signal onto the power carrier frequency [114], therefore the communication signal is transmitted through the power coils. The 2 kHz signal is very easy to process even by using the existing microcontroller in the device. In this way, the extra antennas and control chips used for the communication signal can be eliminated. In an EV WPT system, due to the high voltage across the power coils, isolation is required for the communication control circuit, which may increase the cost. For advanced information exchange, general wireless communication protocols, like Bluetooth and near field communication (NFC), could be adopted. In the EV WPT prototype from Oak Ridge National Laboratory (ORNL), the dedicated short range communications (DSRC) Link is used [95]. DSRC is a technology based on global position system (GPS) and IEEE 802.11p wireless fidelity (Wi-Fi), which could realize the connection between vehicle-to-vehicle (V2V) and vehicle-to-infrastructure (V2I) [115]. The FCC already allocates a 75 MHz band at 5.9 GHz for DSRC, which is being committed to use by the US Department of Transportation in the Intelligent Transportation System (ITS). As the IEEE and SAE standards have already been published, the DSRC could provide an easier way to implement the smart grid functionalities and maximize the vehicle to grid benefits.

16.7.4 Cost

An important factor that affects the future of WPT is its cost. Actually, from Figure 16.8, we see that a WPT charger is only slightly different from a wired charger. The extra cost of a WPT charger is mainly due to the magnetic coupler. For our 8 kW stationary WPT design [61], the material cost of the two magnetic couplers is about $400. This is the rough cost increase of an 8 kW wireless charger compared with a wired charger, which is quite acceptable, considering all the convenience brought by WPT and the long-term operation cost savings and reduction of battery size. For the dynamic WPT design, the infrastructure cost, including the converter and the track for a 1 km one-way road, is less than $400,000 [57]. The investment of electrification is much lower than the construction cost of the road itself. With the road electrification, the EV onboard batteries could be reduced by 80%. The savings on batteries might be much more than the investment on the infrastructure. Studies also show that with only 1% electrification of the urban road, most EVs could meet a 300-mile range easily [60, 116]. The era of road electrification is coming.

16.8 A Double-Sided LCC Compensation Topology and its Parameter Design[3]

The compensation network and the corresponding control method are the most important and difficult aspects in the design of a wireless charging system. In this section, the compensation network design is focused on a double-sided LCC compensation topology and its parameter design. The topology consists of one inductor and two capacitors on both the primary and secondary sides. With the proposed method, the resonant frequency of the compensated coils is independent of the coupling coefficient and the load condition. Moreover, the wireless power transfer system can work at a constant frequency, which eases the control and narrows the occupied frequency bandwidth. Near unity power factor can be achieved for both the primary side and the secondary side converters in the whole range of coupling and load conditions, thus a high efficiency for the overall WPT system is easily achieved. A parameter tuning method is also proposed and analyzed to achieve ZVS operation for the MOSFET-based inverter. The proposed method is more attractive in an environment where the coupling coefficient keeps changing, like the electric vehicle charging application. Also, due to its symmetrical structure, the proposed method can be used in a bidirectional WPT system. Simulation and experimental results verified the analysis and validity of the proposed compensation network and tuning method. A prototype with 7.7 kW output power for electric vehicles was built, and 96% efficiency from a DC power source to a battery load was achieved.

16.8.1 The Double-Sided LCC Compensation Topology

The proposed double-sided LCC compensation network and corresponding power electronics circuit components are shown in Figure 16.11. S_1~S_4 are four power MOSFETs in the primary side. D_1~D_4 are the secondary side rectifier diodes. L_1 and L_2 are the self-inductances of the transmitting and receiving coils, respectively. L_{f1}, C_{f1}, and C_1 are the primary side compensation inductor and capacitors. L_{f2}, C_{f2}, and C_2 are the secondary side compensation components. M is the mutual inductance between the two coils. Here, u_{AB} is the input voltage applied to the compensated coil and u_{ab} is the output voltage before the rectifier diodes. Variables i_1, i_2, i_{Lf1}, i_{Lf2} are the currents

3 Reprint, Copyright IEEE. S. Li, W. Li, J. Deng, T. D. Nguyen and C. C. Mi, "A Double-Sided LCC Compensation Network and Its Tuning Method for Wireless Power Transfer," in IEEE Transactions on Vehicular Technology, vol. 64, no. 6, pp. 2261–2273, June 2015. doi: 10.1109/TVT.2014.2347006

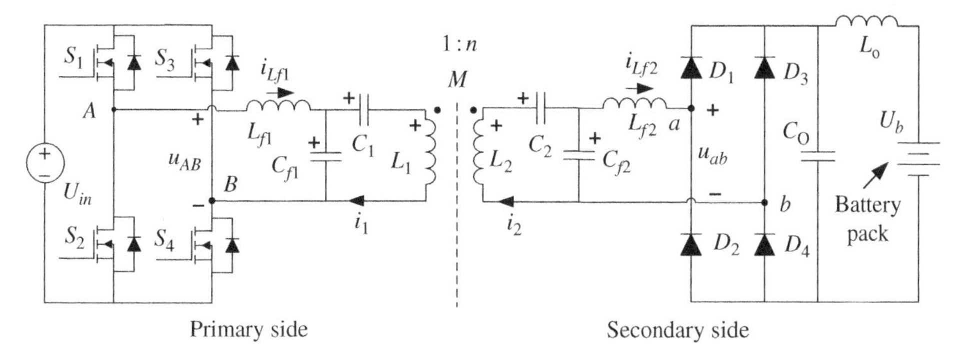

Figure 16.11 Double-sided LCC compensation topology for wireless power transfer.

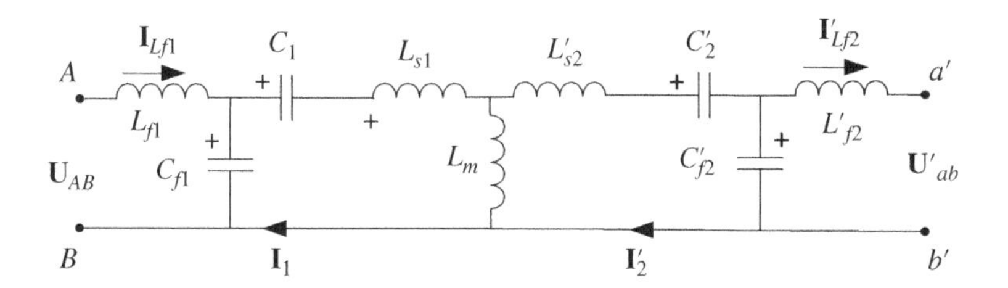

Figure 16.12 Equivalent circuit referred to the primary side of the proposed topology.

on L_1, L_2, L_{f1}, and L_{f2}, respectively. In the following analysis, $\mathbf{U}_{AB}$, $\mathbf{U}_{ab}$, $\mathbf{I}_1$, $\mathbf{I}_2$, $\mathbf{I}_{Lf1}$, $\mathbf{I}_{Lf2}$ are adopted to represent the phasor form of the corresponding variables.

For the first step, a concise characteristic of the proposed compensation network will be given by analyzing the first-order harmonics of the square voltage waveform at the switching frequency. The resistance on all the inductors and capacitors is neglected for simplicity of analysis. The accuracy of the approximations will be verified by circuit simulation and experiments in the later sections of this chapter. The equivalent circuit of Figure 16.11 referred to the primary side is derived as shown in Figure 16.12. We define the turns-ratio of the secondary to primary side as

$$n = \sqrt{L_2 / L_1} \tag{16.27}$$

The variables in Figure 16.12 can be expressed as

$$\begin{aligned}
L_m &= k \cdot L_1, \\
L_{s1} &= (1-k) \cdot L_1, \\
L'_{s2} &= (1-k) \cdot L_2 / n^2, \\
L'_{f2} &= L_{f2} / n^2, \\
C'_2 &= n^2 \cdot C_2, \\
C'_{f2} &= n^2 \cdot C_{f2}, \\
U'_{ab} &= U_{ab} / n.
\end{aligned} \tag{16.28}$$

where L_m is the magnetizing inductance referred to the primary side. The primed symbols, which also appear in the equations that follow, mean the converted values referred to the primary side.

For a high-order system in Figure 16.11, there are multiple resonant frequencies. In this section, we do not focus on the overall frequency domain characteristics. Instead, only one frequency point, which can be tuned to a constant resonant frequency, is studied. Here, resonance means that the input voltage $\mathbf{U}_{AB}$ and current $\mathbf{I}_{Lf1}$ of the compensated coil system are in phase. The circuit parameters are designed using the following equations to achieve a constant resonant frequency for the topology:

$$\begin{aligned} L_{f1} \cdot C_{f1} &= \frac{1}{\omega_0^2}, \\ L_{f2} \cdot C_{f2} &= \frac{1}{\omega_0^2}, \\ L_1 - L_{f1} &= \frac{1}{\omega_0^2 C_1}, \\ L_2 - L_{f2} &= \frac{1}{\omega_0^2 C_2}, \end{aligned} \tag{16.29}$$

where ω_0 is the constant resonant angular frequency, which is relevant only to inductors and capacitors in the system, independent of the coupling coefficient k and load conditions.

Under the above rules, the circuit characteristics will be analyzed at resonant frequency ω_0. According to the superposition theory, the effects of $\mathbf{U}_{AB}$ and $\mathbf{U}'_{ab}$ can be analyzed separately, as shown in Figure 16.13. The dashed line means that there is no current on that path. Figure 16.13a is used to analyze the effects of $\mathbf{U}_{AB}$. The additional subscript AB indicates that current is contributed by $\mathbf{U}_{AB}$. To make the analysis clear,

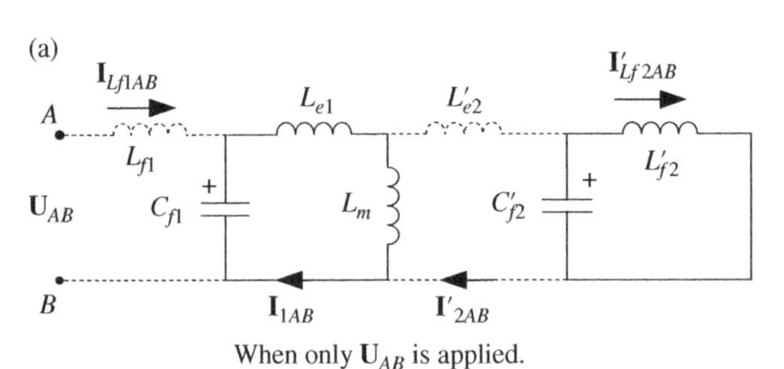

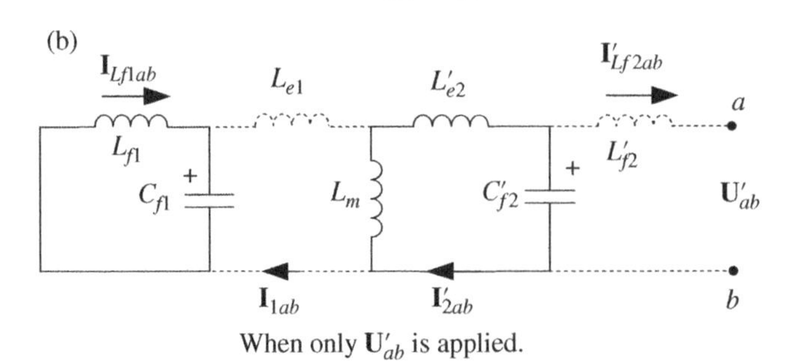

Figure 16.13 Circuit status at resonant frequency.

the series connected capacitor and inductor branches, C_1, L_{s1} and C_2', L_{s2}' are expressed using equivalent inductances L_{e1} and L_{e2}', where

$$L_{e1} = \frac{1}{j\omega_0} \cdot \left(\frac{1}{j\omega_0 C_1} + j\omega_0 L_{s1} \right) = L_{f1} - k \cdot L_1 \tag{16.30}$$

$$L_{e2}' = \frac{1}{j\omega_0} \cdot \left(\frac{1}{j\omega_0 C_2'} + j\omega_0 L_{s2}' \right) = L_{f2}' - k \cdot L_1 \tag{16.31}$$

L_{f2}' and C_{f2}' form a parallel resonant circuit, which can be regarded as an open circuit at the resonant frequency ω_0. Thus, $\mathbf{I}_{2AB}' = 0$. In the left part of the circuit, L_{e1}, L_{m} are connected in series, and

$$L_{e1} + L_m = L_{f1} - k \cdot L_1 + k \cdot L_1 = L_{f1} \tag{16.32}$$

L_{f1} forms another parallel resonant circuit with C_{f1} at the same resonant frequency ω_0. Thus, $\mathbf{I}_{Lf1AB} = 0$. Because there is no current through L_{f1} and L_{e2}', the voltage on C_{f1} equals $\mathbf{U}_{AB}$, and the voltage on C_{f2} equals the voltage on L_m. $\mathbf{I}_{1AB}$ and $\mathbf{I}_{Lf2AB}'$ can be easily solved:

$$\mathbf{I}_{1AB} = \frac{\mathbf{U}_{AB}}{j\omega_0 L_{f1}} \tag{16.33}$$

$$\mathbf{I}_{Lf2AB}' = \frac{k\mathbf{U}_{AB} L_1}{j\omega_0 L_{f1} L_{f2}'} \tag{16.34}$$

When $\mathbf{U}_{ab}'$ is applied, the analysis is similar to that of when $\mathbf{U}_{AB}$ is applied. An additional subscript ab is added to indicate that the current is contributed by $\mathbf{U}_{ab}$.

The solutions are $\mathbf{I}_{Lf2ab}' = 0$, $\mathbf{I}_{1ab} = 0$ and

$$\mathbf{I}_{2ab}' = -\frac{\mathbf{U}_{ab}'}{j\omega_0 L_{f2}'} \tag{16.35}$$

$$\mathbf{I}_{Lf1ab} = -\frac{k\mathbf{U}_{ab}' L_1}{j\omega_0 L_{f1} L_{f2}'} \tag{16.36}$$

$\mathbf{U}_{ab}'$ is a passive voltage generated according to the conduction modes of diodes $D_1 \sim D_4$. It should be in phase with $\mathbf{I}_{Lf2}'$. Since $\mathbf{I}_{Lf2ab}' = 0$, $\mathbf{U}_{ab}'$ is in phase with $\mathbf{I}_{Lf2AB}'$. If we take $\mathbf{U}_{AB}$ as the reference, $\mathbf{U}_{AB}$ and $\mathbf{U}_{ab}'$ can be expressed as

$$\mathbf{U}_{AB} = U_{AB} \angle 0° \tag{16.37}$$

$$\mathbf{U}_{ab}' = \frac{U_{ab}'}{j} = U_{ab}' \angle \varphi = U_{ab}' \angle -90° \tag{16.38}$$

where φ is the phase by which $\mathbf{U}_{ab}$ leads $\mathbf{U}_{AB}$. From (16.37) and (16.38), we can see that $\mathbf{U}_{ab}'$ lags $\mathbf{U}_{AB}$ by 90°. We substitute (16.28), (16.37), (16.38) into (16.33–16.36), and sum up the current generated by $\mathbf{U}_{AB}$ and $\mathbf{U}_{ab}'$ to get

$$\mathbf{I}_{Lf1} = \mathbf{I}_{Lf1ab} = \frac{kL_1 U_{ab}'}{\omega_0 L_{f1} L_{f2}'} \angle 0° = \frac{k\sqrt{L_1 L_2} U_{ab}}{\omega_0 L_{f1} L_{f2}} \angle 0° \tag{16.39}$$

$$\mathbf{I}_1 = \mathbf{I}_{1AB} = \frac{U_{AB}}{j\omega_0 L_{f1}} = \frac{U_{AB}}{\omega_0 L_{f1}} \angle -90° \tag{16.40}$$

$$\mathbf{I}_2 = \frac{\mathbf{I}_2'}{n} = \frac{\mathbf{I}_{2ab}'}{n} = \frac{U_{ab}'}{n \cdot \omega_0 L_{f2}'} \angle 0° = \frac{U_{ab}}{\omega_0 L_{f2}} \angle 0° \tag{16.41}$$

$$\mathbf{I}_{Lf2} = \frac{\mathbf{I}_{Lf2}'}{n} = \frac{\mathbf{I}_{Lf2AB}'}{n} = \frac{k\sqrt{L_1 L_2} \cdot U_{AB}}{\omega_0 L_{f1} L_{f2}} \angle -90° \tag{16.42}$$

From (16.37) and (16.39), we can see that the input voltage and current are in phase, and thus unity power factor for the converter is achieved. From (16.38) and (16.42), we can see that the output voltage and current are in phase, hence unity power factor for the output rectifier is also achieved. In addition, the phase relation does not rely on the coupling coefficient and battery voltage. Thus, the resonant condition can be achieved regardless of the coupling and load condition. The transferred power can be calculated by

$$P = \mathbf{U}_{AB} \cdot \mathbf{I}_{Lf1} = \frac{\sqrt{L_1 L_2}}{\omega_0 L_{f1} L_{f2}} \cdot k U_{AB} U_{ab} \tag{16.43}$$

It can be seen that the output power is proportional to the coupling coefficient k, and the input voltage U_{AB}, and the output voltage U_{ab}. Thus, a buck or boost converter can be inserted either before the primary side inverter or after the secondary side rectifier to control the output power. For some applications, such as opportunity charging or dynamic charging for electric vehicles, an accurate continuous power is not necessary. In this case, the charging power can be controlled by switching the system between maximum and zero output power. Thus, the buck or boost converter can be omitted.

16.8.2 Parameter Tuning for Zero Voltage Switching

If the coils and compensation network parameters are designed exactly according to the above rules, all the MOSFETs will be turned on and off under a zero current switching (ZCS) condition. However, ZCS is not a perfect soft switching condition in converters containing MOSFETs and diodes. To minimize the switching loss, it is better that all switches are turned on and off at a zero voltage switching (ZVS) condition. The parasitic output capacitance of the MOSFET holds the voltage close to zero during the turn-off transition, so the turn-off switching loss is very small [118]. However, in the turn-on transition, the ZVS operation is required to prevent both body diode reverse recovery and parasitic output capacitance from inducing switching loss. To realize ZVS for a MOSFET, the body diode should conduct before the MOSFET does. It is essential that the MOSFET be turned on at a negative current. For a full-bridge converter, this means the input impedance of the resonant network should be inductive. In this case, the resonant current lags the resonant voltage, which forms the ZVS operation condition for all MOSFETs.

In this section, the ZVS operation condition refers to ensuring the turn-off current is positive to realize the ZVS turn-on of another MOSFET in the same arm. There can be several ways to tune the system parameters to ensure that the MOSFETs turn off at a positive current. Here, one simple way is introduced and analyzed.

Figure 16.14 Equivalent circuit referred to the primary side.

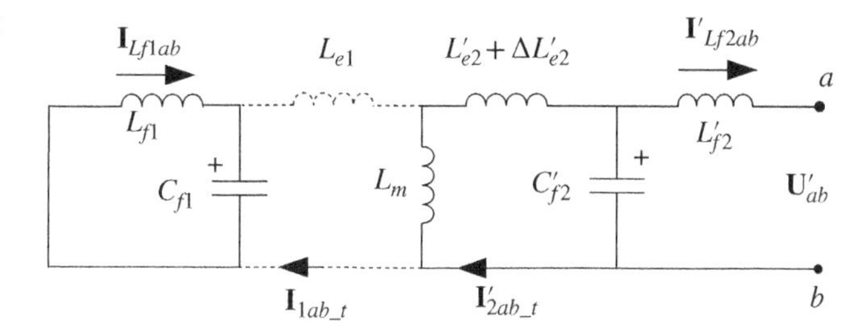

To achieve ZVS, we just slightly increase the value of L_{e2}. As shown in Figure 16.14, the change of L_{e2} is $\Delta L'_{e2}$. In addition, the superposition method is used to analyze the tuned circuit. When $\mathbf{U}_{AB}$ is applied to the circuit, the equivalent circuit is the same as in Figure 16.13a and $\mathbf{I}_{Lf1AB}$ is zero. When $\mathbf{U}'_{ab}$ is applied, $\mathbf{I}'_{Lf2ab}$ is not zero anymore. We solve the circuit again with the variation of $\Delta L'_{e2}$, and the following equations can be derived:

$$\mathbf{I}'_{Lf2ab\_1st} = -j\cdot\frac{\mathbf{U}'_{ab}}{\omega_0\cdot L'_{f2}}\cdot\frac{\Delta L'_{e2}}{L'_{f2}} \tag{16.44}$$

$$\begin{aligned}\mathbf{I}'_{Lf2\_1st} &= \mathbf{I}'_{Lf2AB}+\mathbf{I}'_{Lf2ab\_1st}\\ &= \frac{kU_{AB}L_1}{j\omega_0 L_{f1}L'_{f2}} - j\cdot\frac{U'_{ab}\cdot(\cos\varphi + j\cdot\sin\varphi)}{\omega_0\cdot L'_{f2}}\cdot\frac{\Delta L'_{e2}}{L'_{f2}}\\ &= \frac{U'_{ab}\cdot\sin\varphi\cdot\Delta L'_{e2}}{\omega_0\left(L'_{f2}\right)^2} - j\cdot\left(\frac{U'_{ab}\cdot\cos\varphi\cdot\Delta L'_{e2}}{\omega_0\left(L'_{f2}\right)^2}+\frac{kU_{AB}L_1}{\omega_0 L_{f1}L'_{f2}}\right)\end{aligned} \tag{16.45}$$

$$\begin{aligned}\mathbf{I}_{Lf2\_1st} &= \frac{\mathbf{I}'_{Lf2\_1st}}{n}\\ &= \frac{U_{ab}\cdot\sin\varphi\cdot\Delta L_{e2}}{\omega_0 L_{f2}^2} - j\cdot\frac{kU_{AB}\sqrt{L_1L_2}}{\omega_0 L_{f1}L_{f2}} - j\cdot\frac{U_{ab}\cdot\cos\varphi\cdot\Delta L_{e2}}{\omega_0 L_{f2}^2}\end{aligned} \tag{16.46}$$

In these equations, the subscript 1st is used to indicate the first harmonic component of the corresponding item.

The phase of $\mathbf{U}_{ab}$ is φ. Equation (16.38) shows that φ is $-90°$ when ΔL_{e2} is zero. When we increase L_{e2} to realize ZVS, the change of L_{e2} is relatively small, so φ is still close to $-90°$. Therefore, we have $\sin\varphi \approx -1$, $\cos\varphi \approx 0$, $\Delta L_{e2} < L_{f2}$. Usually, the additional inductors, L_{f1} and L_{f2}, used as reactive power compensators, are designed such that they are much smaller than the main coils. The following approximation can be obtained from (16.46):

$$\cos\varphi_1 \approx -\cot\varphi_1 \approx -\frac{U_{ab}}{U_{AB}}\cdot\frac{\Delta L_{e2}\cdot L_{f1}}{L_{f2}\cdot k\sqrt{L_1L_2}} \tag{16.47}$$

where φ_1 ($-90° < \varphi_1 \ll -180°$) is the phase by which $\mathbf{I}_{Lf2\_1st}$ leads $\mathbf{U}_{AB}$.

To reduce the switching loss, we prefer to achieve the ZVS condition at a minimum turn-off current. This means that at the switching point the current is close to zero, and the current slew rate at the switching point is high. A small phase error in the analysis will bring a relatively large current error. If the phase error falls into the inductive region, it means a higher turn-off current and higher switching loss. If the phase error falls into the capacitive region, the turn-off current may be negative which means that ZVS is lost and severe reverse recovery in the MOSFET diode will occur. This will bring high switching loss and cause an electromagnetic interference (EMI) problem. Therefore, the accuracy of analysis of the turn-off current is very important. The high-order harmonics of the square voltage should also be considered, but the inductor–capacitor network from the primary side to the secondary side is a high-order filter. For the high-order harmonics, the interaction between the primary and secondary sides can be neglected. Thus, the high-order current on L_{f2} can be roughly calculated as

$$
\begin{aligned}
\mathbf{I}_{Lf2\_3rd} &\approx -\frac{\mathbf{U}_{ab\_3rd}}{j\cdot 3\omega_0 L_{f2} + \dfrac{1}{j\cdot 3\omega_0 C_{f2}}} = j\frac{3\mathbf{U}_{ab\_3rd}}{8\omega_0 L_{f2}} \\
\mathbf{I}_{Lf2\_5th} &\approx -\frac{\mathbf{U}_{ab\_5th}}{j\cdot 5\omega_0 L_{f2} + \dfrac{1}{j\cdot 5\omega_0 C_{f2}}} = j\frac{5\mathbf{U}_{ab\_5th}}{24\omega_0 L_{f2}} \\
&\cdots \\
\mathbf{I}_{Lf2\_(2k+1)th} &\approx -\frac{\mathbf{U}_{ab\_(2k+1)th}}{j\cdot (2k+1)\omega_0 L_{f2} + \dfrac{1}{j\cdot (2k+1)\omega_0 C_{f2}}} \\
&= j\frac{(2k+1)\mathbf{U}_{ab\_(2k+1)th}}{((2k+1)^2-1)\omega_0 L_{f2}} \\
&\cdots
\end{aligned}
\tag{16.48}
$$

According to (16.48), the phase difference between $\mathbf{U}_{ab\_mth}$ and $\mathbf{I}_{Lf2\_mth}$ is 90°, so when $\mathbf{U}_{ab}$ jumps at the time that $i_{Lf2}=0$, i_{Lf2_mth} reaches its peak, which can be calculated as

$$
\begin{aligned}
\max\left\{\sum i_{Lf2\_mth}\right\} &= \sqrt{2}\cdot\sum_{k=1}^{\infty} I_{Lf2\_(2k+1)th} \\
&= \sqrt{2}\cdot\sum_{k=1}^{\infty}\frac{1}{((2k+1)^2-1)}\frac{U_{ab}}{\omega_0 L_{f2}} \\
&= \frac{\sqrt{2}}{4}\cdot\frac{U_{ab}}{\omega_0 L_{f2}}
\end{aligned}
\tag{16.49}
$$

The phase φ_2, by which $\mathbf{U}_{ab}$ leads $\mathbf{I}_{Lf2\_1st}$, is close to 0 ($0<\varphi_2\ll 90°$) and can be easily obtained from

$$
\sin\varphi_2 = \frac{\sqrt{2}\cdot\sum_{k=1}^{\infty} I_{Lf2\_(2k+1)th}}{\sqrt{2}\cdot I_{Lf2\_1st}} \approx \frac{1}{4}\cdot\frac{U_{ab}}{U_{AB}}\cdot\frac{L_{f1}}{k\sqrt{L_1 L_2}}
\tag{16.50}
$$

Figure 16.15 shows the effect of the high-order harmonic currents. The sign of u_{ab} is determined by i_{Lf2}, which is a composition of both the first- and high-order harmonic currents.

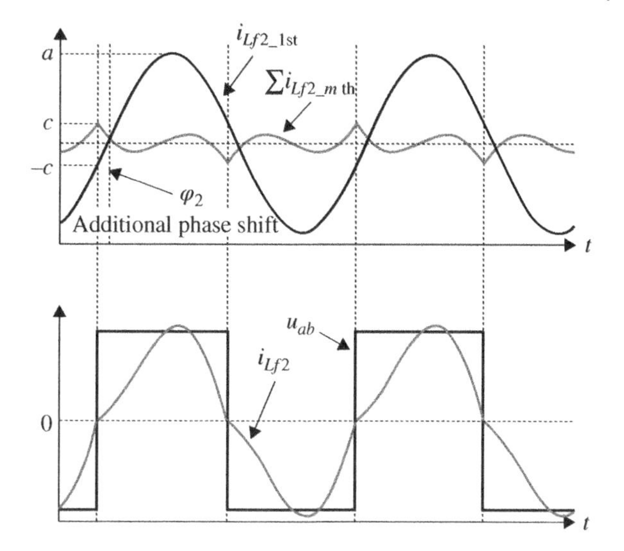

Figure 16.15 Effect of all high-order currents. $a=\sqrt{2}\cdot\frac{k\sqrt{L_1L_2}U_{AB}}{\omega_0L_{f1}L_{f2}}$, $c=\sqrt{2}\cdot\frac{U_{ab}}{4\omega_0L_{f2}}$.

From (16.47) and (16.50) we can obtain

$$\begin{aligned}\cos\varphi &= \cos(\varphi_1+\varphi_2)=\cos\varphi_1\cos\varphi_2-\sin\varphi_1\sin\varphi_2\\ &\approx \cos\varphi_1+\sin\varphi_2\\ &\approx -\frac{U_{ab}}{U_{AB}}\cdot\frac{L_{f1}}{k\sqrt{L_1L_2}}\cdot\left(\frac{\Delta L_{e2}}{L_{f2}}-\frac{1}{4}\right),\end{aligned} \tag{16.51}$$

$$\begin{aligned}\mathbf{I}_{Lf1\_1st} &= -\frac{k\mathbf{U}'_{ab}L_1}{j\omega_0L_{f1}L'_{f2}}=-\frac{kU'_{ab}(\cos\varphi+j\cdot\sin\varphi)L_1}{j\omega_0L_{f1}L'_{f2}}\\ &= -\frac{kU'_{ab}\sin\varphi L_1}{\omega_0L_{f1}L'_{f2}}-\frac{kU'_{ab}\cos\varphi L_1}{j\cdot\omega_0L_{f1}L'_{f2}}\\ &\approx \frac{kU'_{ab}L_1}{\omega_0L_{f1}L'_{f2}}+\frac{\frac{U^2_{ab}}{U_{AB}}\left(\frac{\Delta L_{e2}}{L_{f2}}-\frac{1}{4}\right)}{j\cdot\omega_0L_{f2}}\\ &= \frac{kU_{ab}\sqrt{L_1L_2}}{\omega_0L_{f1}L_{f2}}+\frac{\frac{U^2_{ab}}{U_{AB}}\left(\frac{\Delta L_{e2}}{L_{f2}}-\frac{1}{4}\right)}{j\cdot\omega_0L_{f2}}\end{aligned} \tag{16.52}$$

From (16.52), we can see that an additional reactive current item is introduced. This current will increase the MOSFET turn-off current to achieve ZVS. For the primary side, similar to the analysis of the secondary side for the high-order harmonics effect, we can obtain

$$\begin{aligned}\max\left\{\sum i_{Lf1\_mth}\right\} &= \sqrt{2}\cdot\sum_{k=1}^{\infty}I_{Lf1\_(2k+1)th}\\ &= \sqrt{2}\cdot\sum_{k=1}^{\infty}\frac{1}{((2k+1)^2-1)}\frac{U_{AB}}{\omega_0L_{f1}}\\ &= \frac{\sqrt{2}}{4}\cdot\frac{U_{AB}}{\omega_0L_{f1}}\end{aligned} \tag{16.53}$$

The MOSFET turn-off current is a composition of both the first-order and the high-order harmonic currents. From (16.52) and (16.53), the MOSFET turn-off current can be calculated as

$$I_{OFF} = \sqrt{2}\left(\frac{\frac{U_{ab}^2}{U_{AB}}\left(\frac{\Delta L_{e2}}{L_{f2}} - \frac{1}{4}\right)}{\omega_0 L_{f2}} + \frac{U_{AB}}{4\omega_0 L_{f1}}\right) \tag{16.54}$$

According to the MOSFET parameters, a minimum turn-off current to achieve ZVS can be determined [119]. Then a suitable ΔL_{e2} can be designed to ensure that there is enough turn-off current to achieve ZVS for the whole operation range. To ensure that I_{OFF} is greater than a certain positive value, the following equation should be satisfied:

$$\frac{\Delta L_{e2}}{L_{f2}} \geq \frac{1}{4} \tag{16.55}$$

According to (16.54) and (16.55), the lower the output voltage, the smaller the turn-off current. The minimal turn-off current can be derived as

$$I_{OFF\_min} = \frac{\sqrt{2} \cdot U_{ab\_min}}{\omega_0 \cdot L_{f2}} \sqrt{\frac{\Delta L_{e2}}{L_{f1}} - \frac{1}{4}\frac{L_{f2}}{L_{f1}}} \tag{16.56}$$

where U_{ab_min} is the minimum RMS value of the output voltage before the rectifier.

The minimum turn-off current is reached when

$$U_{AB} = U_{ab\_min} \cdot \sqrt{4 \cdot \left(\frac{\Delta L_{e2}}{L_{f2}} - \frac{1}{4}\right) \cdot \frac{L_{f1}}{L_{f2}}} \tag{16.57}$$

Once the minimum MOSFET turn-off current is obtained, ΔL_{e2} can be calculated as

$$\Delta L_{e2} = \frac{1}{4} L_{f2} + \frac{I_{OFF\_min}^2 \cdot \omega_0^2 \cdot L_{f1} \cdot L_{f2}^2}{U_{ab\_min}^2} \tag{16.58}$$

When the parameters are tuned for ZVS, the change of active power transferred is very limited. The conclusion can be obtained by analyzing the change of $\mathbf{I}_{Lf2\_1st}$ and the phase between the output current and voltage. In (16.46), the first term has one relatively small item, whose contribution to the amplitude is negligible. The last term is the product of two relatively small items, which can also be neglected. Compared with (16.39), the amplitude of $\mathbf{I}_{Lf2\_1st}$ is almost the same. From (16.50), we can see that the phase between the output current and voltage has no relation to ΔL_{e2}. Thus, ΔL_{e2} does not have an obvious impact on the active power transferred. However, if the harmonics are considered, the output power in (16.43) should be revised to

$$P = \frac{\sqrt{L_1 L_2}}{\omega_0 L_{f1} L_{f2}} \cdot k U_{AB} U_{ab} \cos\varphi_2 \tag{16.59}$$

where $\cos\varphi_2$ can be calculated from (16.50). If U_{AB} is not too low, $\cos\varphi_2$ will be close to 1. The output power can be calculated by both (16.43) and (16.59).

The decomposition of current harmonics adopted in this section is a refinement of a similar approach that was used in the analysis of a ZCS LCC-compensated resonant converter [117]. In the referenced paper, the battery load was considered to be an equivalent impedance, which is a kind of first-harmonics approximation. Although the high-order harmonics are hardly transferred between the primary and secondary side coils, they do affect the reactive power on the secondary side, which could change the primary side turn-off current. Here, the current harmonics are considered on both the primary and secondary side. The secondary side phase difference between the voltage and current fundamental harmonic can be solved, meaning that the reactive power of the battery load with a rectifier can also be solved. In this way, equation (16.54) gives a direct method to calculate the turn-off current by the input and output voltage instead of an equivalent impedance of the battery load, which could be used to optimize the parameters easily.

16.8.3 Parameter Design

In this section, an 8 kW WPT system is designed according to the above principle. A comparison between the simulation results and the analytical results will be given in the next section to verify the effectiveness of the above analysis.

The specifications of the wireless battery charger are listed in Table 16.2. Since the ratio between the input voltage and the output voltage is around 1, the transmitting and

Table 16.2 Wireless battery charger specifications.

Spec/parameter	Value
Input DC voltage	<425 V
Output battery voltage	300 V~450 V
Nominal gap	200 mm
Coupling coefficient[1]	0.18~0.32
Transmitting coil inductance[2]	350~370 μH
Transmitting coil AC resistance	~500 mΩ
Receiving coil inductance[2]	350~370 μH
Receiving coil AC resistance	~500 mΩ
Switching frequency	79 kHz
Maximum power[3]	~8 kW
Maximum efficiency[3]	~97.1%

1 The coupling coefficient varies because of misalignment. Its value is related to the coil design, which is not the focus of this section.
2 The main coil inductance changes a little bit when the position of the two coils changes. A middle value 360 μH was selected in the design stage.
3 The rated output power is designed as 8 kW with a maximum efficiency of 97.1%. Due to the parameter variations, a maximum efficiency of 96% was reached at 7.7 kW.

receiving coils are designed to have the same size. Thus, from (16.29), we should design $L_{f1} = L_{f2}$. From Table 16.2 and (16.43), we can get

$$L_{f1} = L_{f2} = \sqrt{\frac{k_{max} U_{AB} U_{ab}}{\omega_0 P_{max}} \cdot L_1} = \sqrt{\frac{0.32 \times \frac{2\sqrt{2}}{\pi} \times 425 \times \frac{2\sqrt{2}}{\pi} \times 450}{2\pi \times 79 \times 10^3 \times 8 \times 10^3} \cdot 360 \times 10^{-6}\,\text{H}} \approx 67\,\mu\text{H} \tag{16.60}$$

The values of C_{f1} and C_{f2} can be calculated from (16.29),

$$C_{f1} = C_{f2} = \frac{1}{\omega_0^2 L_{f1}} \approx 60.6\text{nF} \tag{16.61}$$

C_1 and C_2 can also be calculated from (16.29)

$$C_1 = C_2 = \frac{1}{\omega_0^2 (L_1 - L_{f1})} \approx 14\text{nF} \tag{16.62}$$

Then, a variation of ΔL_{e2} should be designed to increase the turn-off current for MOSFETs to achieve ZVS. Once the minimum turn-off current for ZVS is obtained, ΔL_{e2} can be designed using (16.58). The minimum turn-off current varies for different types of MOSFETs and dead-time settings. For a 8 kW system, usually a MOSFET with 80 A continuous conduction capability can be adopted. The Fairchild FCH041N60E N-Channel MOSFET was chosen as the main switches. The switches are rated at 600 V, 48 A (75°C) to 77 A (25°C). According to the parameters of the MOSFET, the calculated dead-time is 600 ns. In order to guarantee ZVS in this mode, the turn-off current must be large enough to discharge the junction capacitors within the dead-time, which can be represented by [3]

$$I_{OFF} \geq \frac{4C_{oss} U_{AB,\max}}{t_d} \tag{16.63}$$

where $U_{AB,\max}$ is the maximum input voltage, C_{oss} is the junction capacitance and t_d is the dead-time. By using the MOSFET parameters, we can calculate the turn-off current, which should be larger than 2 A to realize ZVS. Thus, we design the minimum turn-off current I_{OFF_min} to be 3 A.

By substituting (16.60–16.62), and I_{OFF_min} into (16.58), we can get

$$\Delta L_{e2} = \frac{1}{4} L_{f2} + \frac{I_{OFF\_min}^2 \cdot \omega_0^2 \cdot L_{f1} \cdot L_{f2}^2}{2 \cdot U_{ab\_min}^2} = \left(\frac{67}{4} + \frac{3^2 \cdot \left(2\pi \cdot 79 \times 10^3\right)^2 \cdot \left(67 \times 10^{-6}\right)^2 \cdot 67}{2 \cdot \left(\frac{2\sqrt{2}}{\pi} \cdot 300 \right)^2} \right) \mu\text{H} \approx 21\,\mu\text{H} \tag{16.64}$$

Table 16.3 Compensation network parameters.

Parameter	Design value
L_{f1}	67 μH
L_{f2}	67 μH
C_{f1}	60.6 nF
C_{f2}	60.6 nF
C_1	14 nF
C_2	15.1 nF

The equivalent inductance L_{e2} is determined by C_2 and the leakage inductance L_{s2}. Because L_{s2} is related to the self-inductance and coupling, it is more complicated if we tune L_{s2} to change the value of L_{e1}. It is easier to tune C_2 instead to change the value of L_{e2}. From (16.29), we know that

$$\Delta L_{e2} = \frac{1}{\omega_0^2 C_2} - \frac{1}{\omega_0^2 (C_2 + \Delta C_2)} \tag{16.65}$$

Then, the variation of C_2 can be calculated as

$$\Delta C_2 = \frac{\omega_0^2 \cdot \Delta L_{e2} \cdot C_2^2}{1 - \omega_0^2 \cdot \Delta L_{e2} \cdot C_2} \approx 1.1\text{nF} \tag{16.66}$$

Thus, to achieve ZVS, the value of C_2 should be tuned such that it is 1.1 nF larger than the value calculated by (16.62). All the designed values for the compensation network are listed in Table 16.3.

Since the system is a high-order circuit, there might be another resonant point near ω_0 with a sharp change of the system characteristics. The additional resonant point may significantly affect the performance in a real system because of the parameter variations. The frequency characteristics of the circuit shown in Figure 16.12 were given to check whether there is a sharply changing resonant point around ω_0. In the simulation, the value of C2 was 14 nF to verify the analysis results in Section 16.2. To show the load influence, a load resistor $R_{ac} = 8/\pi \cdot R_b$, where R_b is the equivalent resistance when the battery is charging, was connected between a' and b' in Figure 16.12. When the battery charging power was between 5% and 100%, the range of R_{ac} was roughly 10~200 Ω. The frequency characteristics of the input impedance were analyzed under a coupling coefficient of 0.18~0.32. Two typical bode diagrams are shown in Figure 16.16 with different load conditions at coupling coefficients 0.18 and 0.32. From Figure 16.16, we can see that there is a constant resonant frequency at about 79 kHz. There are also some other resonant frequencies. The lowest and highest resonant points do not change when the load changes, but they do change with the coupling coefficient. It should be noticed that around ω_0, another resonant point may exist under certain conditions. However, the change from ω_0 to the neighboring resonant point is quite smooth, which means there will be no sudden change when the working frequency drifts slightly from ω_0.

(a)

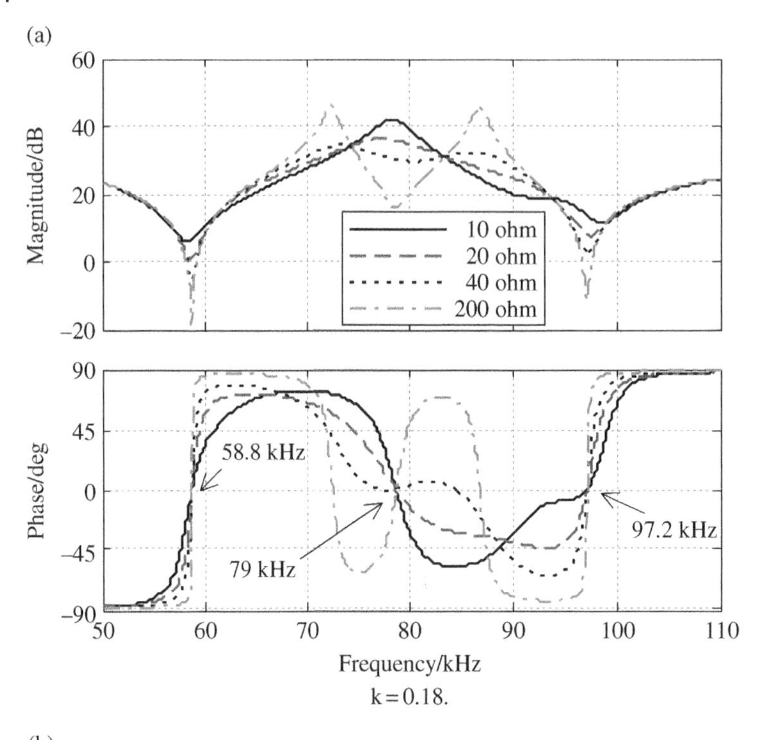

k = 0.18.

(b)

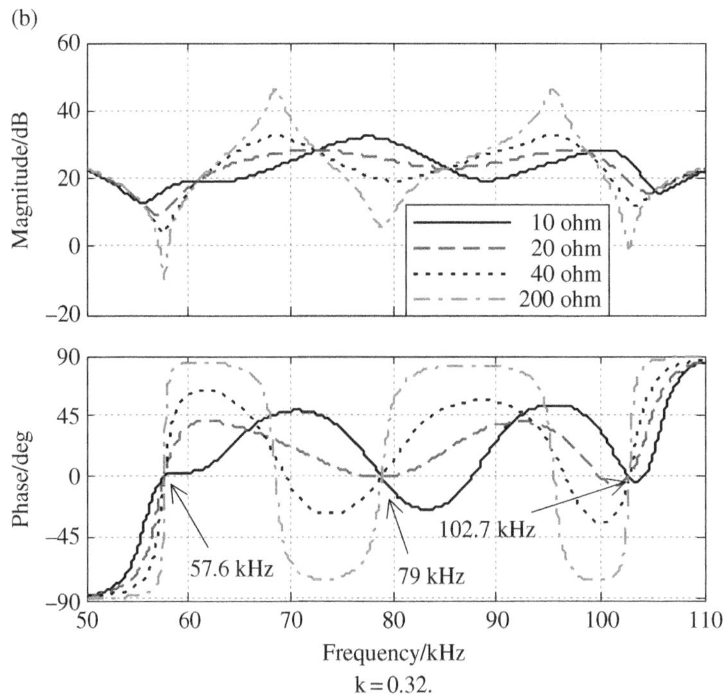

k = 0.32.

Figure 16.16 Frequency characteristics of the input impedance.

16.8.4 Simulation and Experiment Results

Both simulation and experiments are undertaken to verify the proposed double-sided LLC compensation network and its tuning method. The circuit parameters have been shown in Table 16.1 and Table 16.2. We define two kinds of misalignments, i.e. X-misalignment (door-to-door or right-to-left), and Y-misalignment (front-to-rear), as shown in Figure 16.17. When parking a car, the X-misalignment is much harder for the driver to adjust, so we chose X-misalignments for the simulation and experiments. Various misalignments are reflected by the different coupling coefficients. In this section, three coupling coefficients, namely, $k = 0.18$, 0.24, 0.32, corresponding to $X = 310$ mm, 230 mm, and 0 mm respectively, as well as three output voltages, $U_b = 300$ V, 400 V, 450 V, are chosen as case studies. The switching frequency is fixed at 79 kHz for all the cases.

16.8.4.1 Simulation Results

A model was built in LTspice to simulate the performance of the proposed topology. The simulation results for different coupling coefficients, input voltages, and output voltages were obtained. Figure 16.18 shows the comparisons between simulated and calculated output power for various conditions. As can be seen from Figure 16.18, the output power varies linearly with the input voltages for different coupling coefficients and output voltages.

For high input voltage and high coupling coefficients, the simulation and theoretical analysis agree well with each other, but for low input voltage and low coupling coefficients, the simulation does not agree well with the analytical results. This inconsistency is because at low input voltage and low coupling coefficient, the diodes on the secondary side do not conduct all the time between $t_x + n \cdot T/2$ and $t_x + (n+1) \cdot T/2$, which is shown in Figure 16.19. A similar situation can also be found in the comparison results of turn-off current, as shown in Figure 16.20.

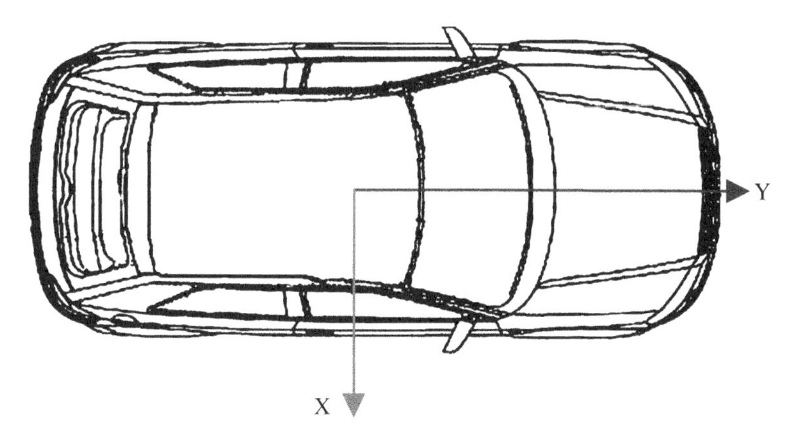

Figure 16.17 Two definitions of misalignments: X-misalignment is the door-to-door or right to left alignment, and Y-misalignment is the front-to-rear alignment.

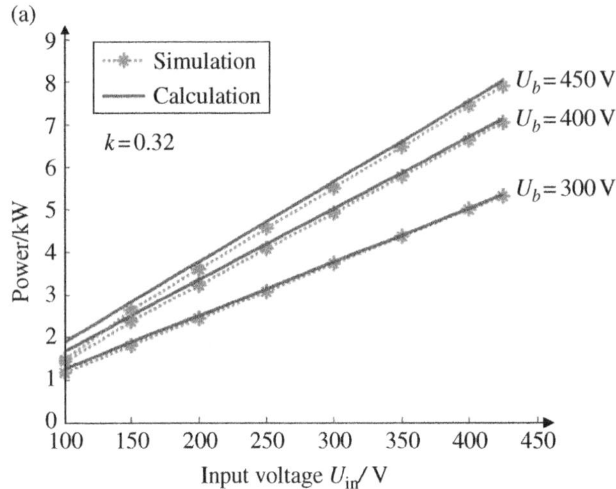

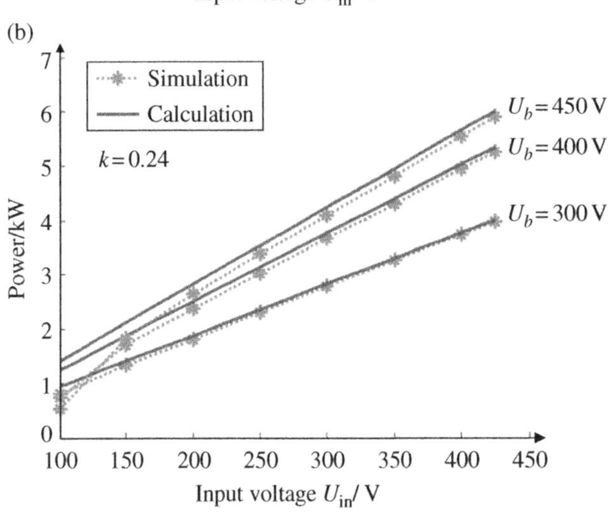

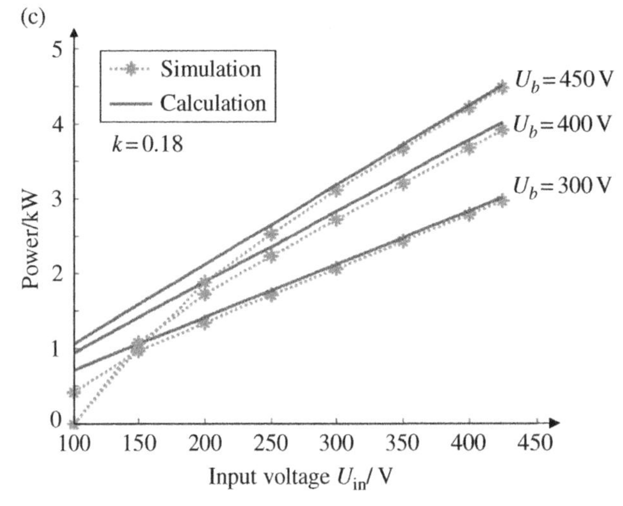

Figure 16.18 Simulation and theoretical calculation results of the power levels for the designed system. (a) $k = 0.32$. (b) $k = 0.24$. (c) $k = 0.18$.

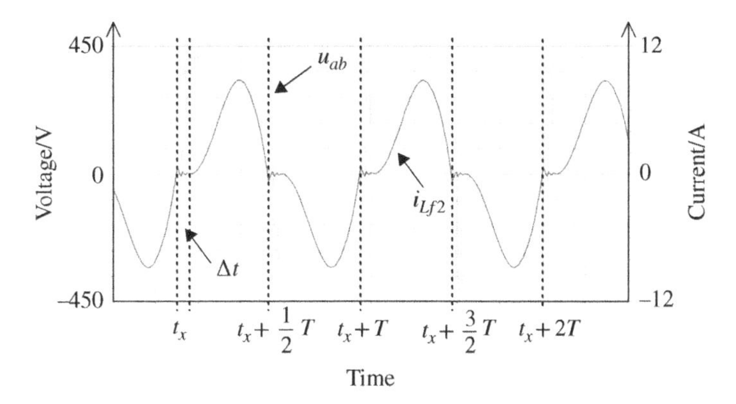

Figure 16.19 Simulation waveforms of output voltage u_{ab} and current through diodes i_{Lf2} when $U_{in} = 150$ V, $U_b = 450$ V.

16.8.4.2 Experimental Results

Figure 16.21a shows the experimental setup. The coil dimension is 800 mm in length and 600 mm in width. The gap between the two coils is 200 mm. A 10 μF capacitor (C_o) and 10 μH inductor (L_o) are selected as the output filter. The coils are connected to the input inverter and output rectifier through the LCC compensation network. An electronic load in constant voltage mode is adopted to take the position of a real battery for easy voltage adjustment. Figure 16.21b shows the efficiency screen shot from the power meter, Yokogawa WT1600, at 7.7 kW output power. U_{dc1} and I_{dc1} are the input DC voltage and current, while U_{dc2} and I_{dc2} are the output DC voltage and current. P_1 and P_2 are the input and output power, and η is the efficiency from DC power supply to the electronic load. I_{ac2} is the output current ripple into the electronic load. Figure 16.22 shows the comparison of experimental and simulation output power as a function of input voltage for three coupling coefficients and three output voltages. Different coupling coefficients are obtained by adjusting the gap and misalignment between transmitting and receiving coils. The output power matches well between simulation and experimental results, and they vary linearly with the input voltage. The same inconsistency phenomenon happens under the low input voltage and low coupling conditions, as mentioned earlier. The calculated and simulated maximum efficiency are both 97.1%. Because of the resistance and parameter variations in the real system, the maximum efficiency of 96% was reached at 7.7 kW output power, which is a little lower than the simulated result.

Figure 16.23 shows the comparison of experimental and theoretical calculated turn-off currents of the MOSFETs. The experimental results agree well with the analytical results. Figure 16.23 also verifies a good characteristic of the proposed tuning method. From (16.54), we can see the turn-off current is not a function of the coupling coefficient k. Once the parameters are designed and tuned, the ZVS condition can be achieved for all coupling conditions easily. The primary side waveforms and secondary side waveforms are shown in Figure 16.24 when the system operates at steady state, delivering 7.2 kW to the load. At this operating point, input voltage $U_{in} = 400$ V, output voltage $U_b = 450$ V, coupling coefficient $k = 0.32$. The results indicate a good ZVS condition with $I_{OFF} = 5.6$ A. The turn-off current is held higher than required, while it is quite small relative to the peak current.

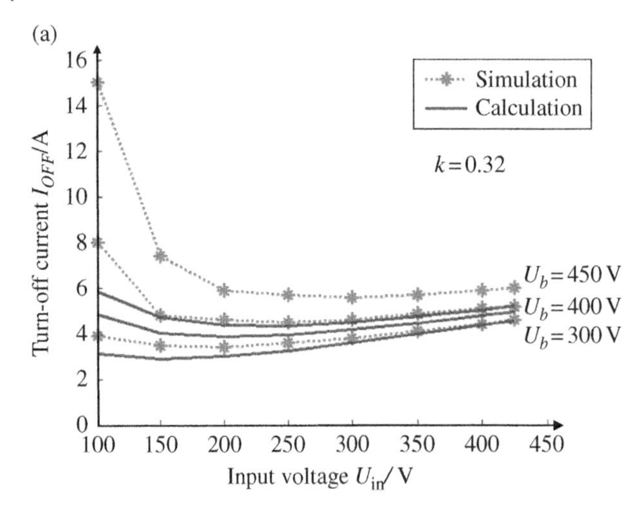

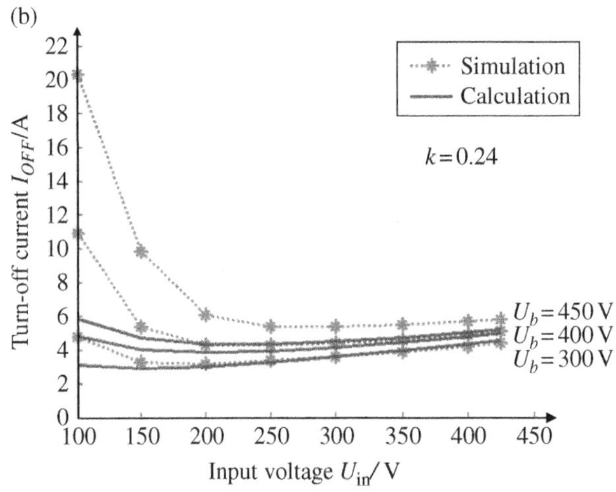

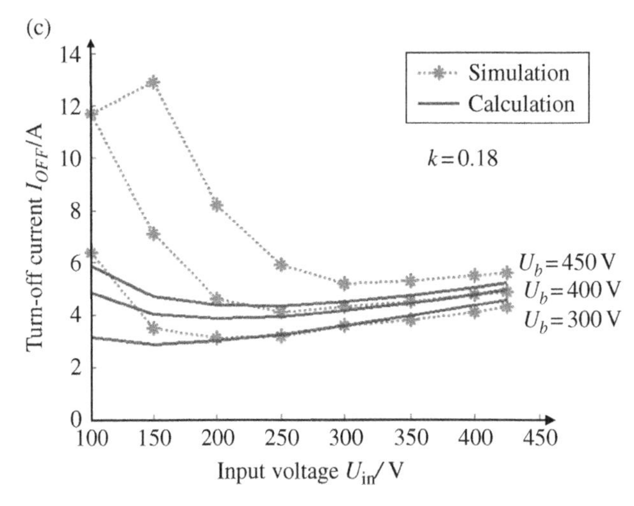

Figure 16.20 Simulation and theoretical results of the MOSFETs turn-off current I_{OFF}. (a) $k=0.32$. (b) $k=0.24$. (c) $k=0.18$.

(a)

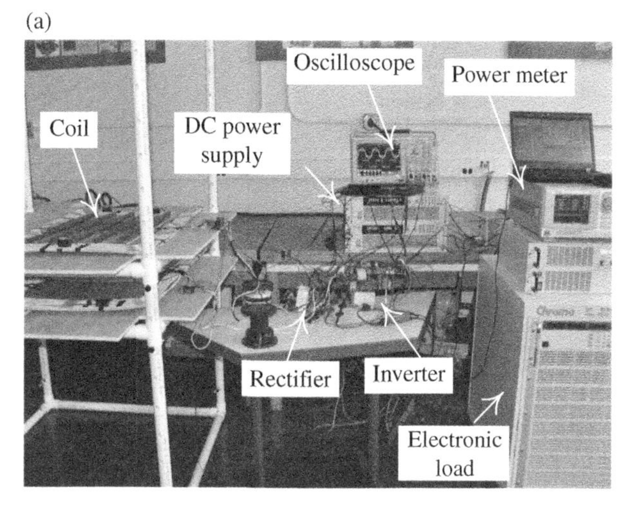

(b)

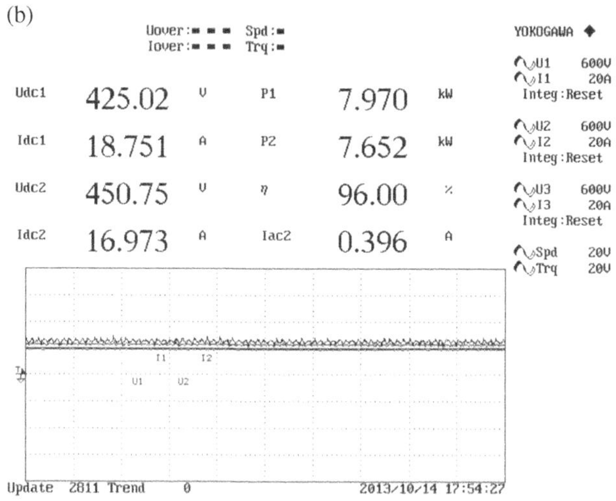

Figure 16.21 Experiment setup: (a) Physical setup of the WPT system. (b) Capture of efficiency from power meter at 7.7 kW output.

Figure 16.25 shows the simulation and experimental efficiencies from the DC power source to the battery load for the proposed double-sided LCC compensation network for a wireless power transfer system. The experiment uses a constant voltage mode electronic load to represent the battery pack for flexible voltage adjustment. A power meter, WT1600 from Yokogawa, was connected in the system to calculate the efficiency by measuring the output power from the DC source and the input power to the electronic load. From Figure 16.25c, we find that the efficiency is very high even under a large X-misalignment condition. The maximum simulated efficiency is 97.1%, while the maximum measured efficiency is about 96% when $U_{in} = 425$ V, $U_b = 450$ V and $k = 0.32$ as shown in Figure 16.25a. Table 16.4 gives a rough loss distribution at 7.7 kW output power through different parts in the system. The large voltage and current dynamic range, as well as the fast transient switching procedure, makes it hard to measure the loss of MOSFETs and diodes accurately. The loss of MOSFETs and diodes is estimated by SPICE model based simulation, whereas the losses of all the other passive

(a)

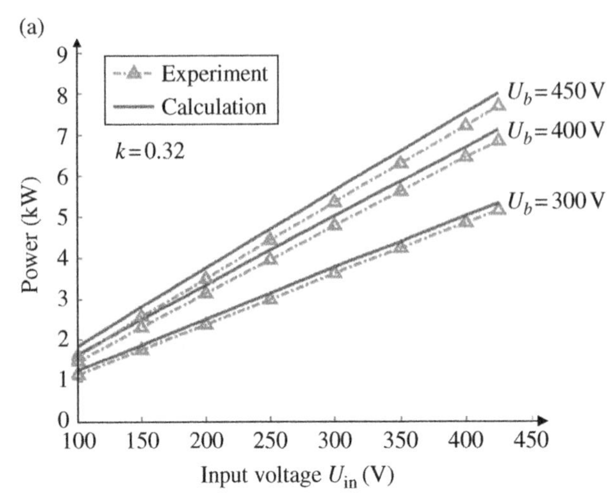

(b)

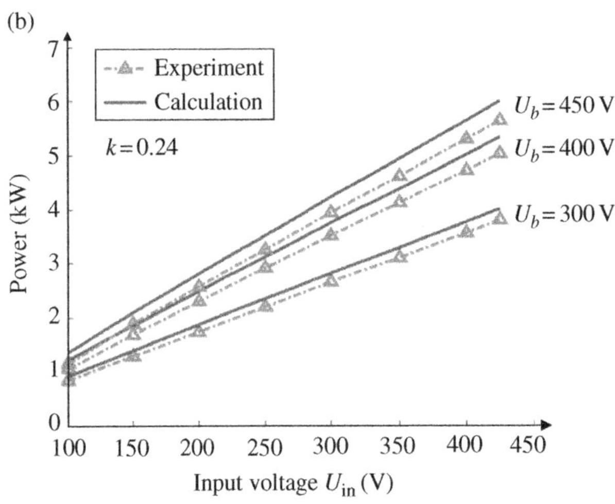

(c)

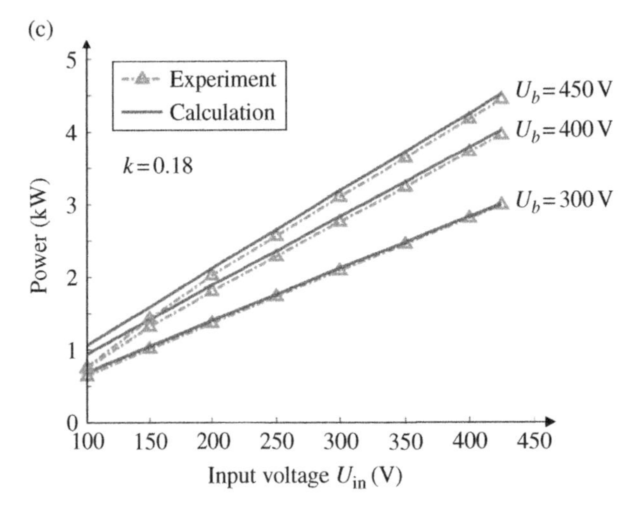

Figure 16.22 Experimental and theoretical calculation results of the power levels for the wireless charger system. (a) $k=0.32$. (b) $k=0.24$. (c) $k=0.18$.

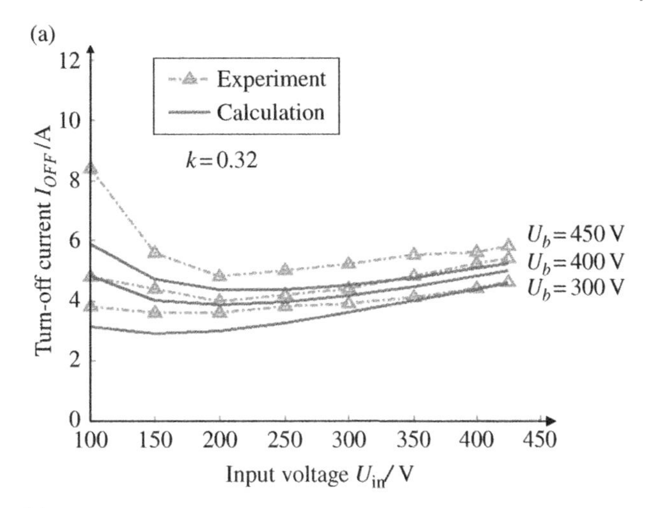

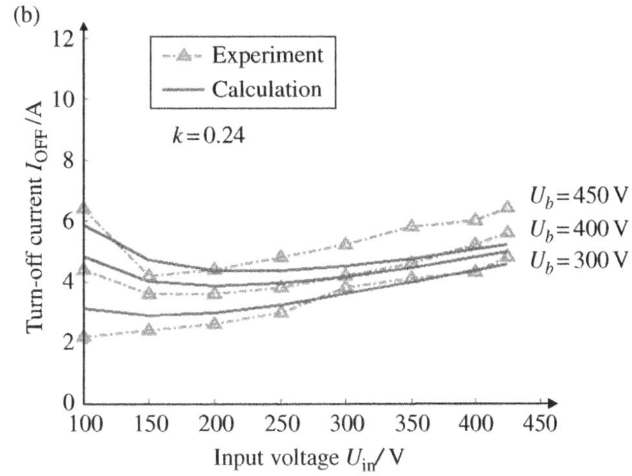

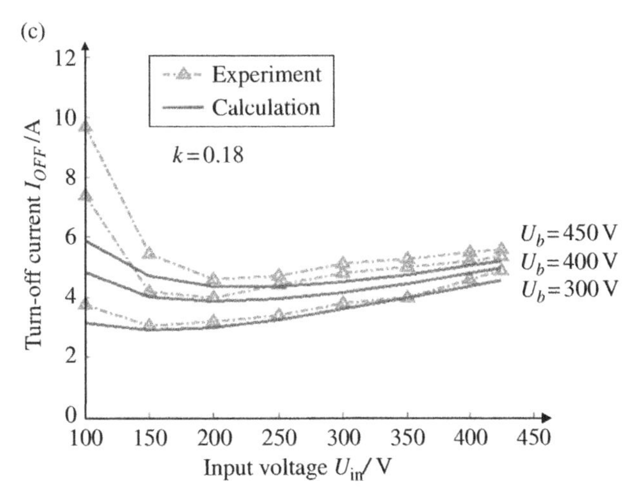

Figure 16.23 Experimental and theoretical calculation results of the MOSFETs turn-off current I_{OFF}. (a) $k=0.32$. (b) $k=0.24$. (c) $k=0.18$.

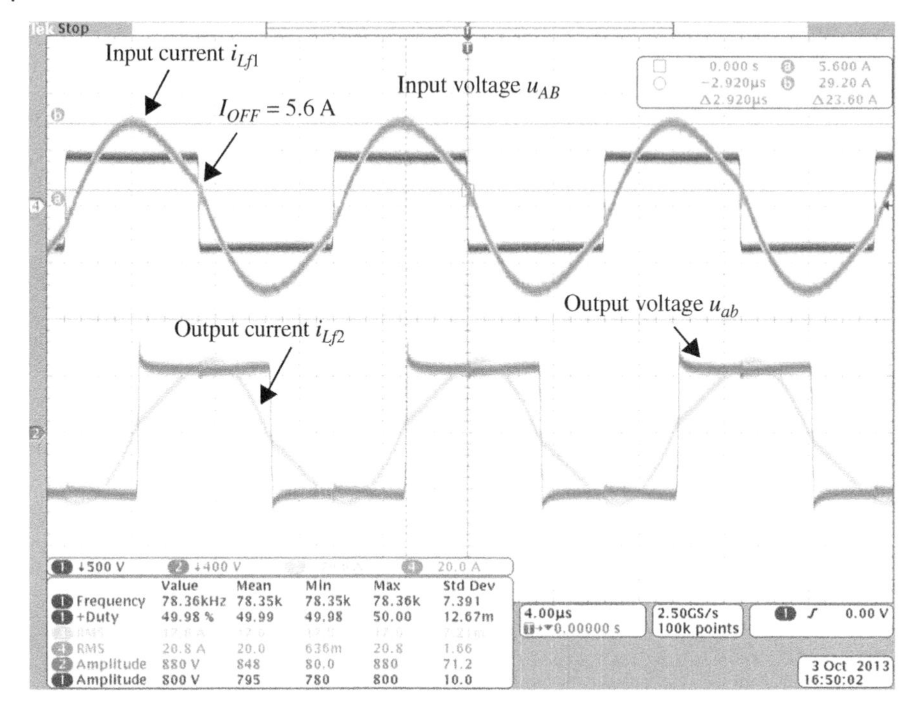

Figure 16.24 Waveforms of the input voltage u_{AB} and current i_{Lf1} and output voltage u_{ab} and current i_{Lf2} when 7.2 kW of power is delivered. $U_{in} = 400$ V, $U_b = 450$ V.

components are calculated by the current rms value and AC resistance from the experiment. From Table 16.6, we can see almost half of the loss is brought by the main coils, which means if we would like to increase the efficiency further, the optimization of the coils is still the most important approach.

The topology and tuning method discussed above can ensure that the resonant frequency is independent of the coupling coefficient and load conditions, as well as that the ZVS condition for the MOSFETs is realized.

16.9 An LCLC Based Wireless Charger Using Capacitive Power Transfer Principle[4]

Capacitive power transfer (CPT) is another effective method of transferring power wirelessly [120]. The CPT technology uses high-frequency alternating electric field to transfer power without direct electric connection, while the IPT system discussed in the previous sections uses magnetic field to transfer power. The IPT technology has

4 Reprint, copyright IEEE. F. Lu, H. Zhang, H. Hofmann, and C. Mi, "A Double-Sided LCLC-Compensated Capacitive Power Transfer System for Electric Vehicle Charging," in IEEE Transactions on Power Electronics, vol. 30, no. 11, pp. 6011–6014, Nov. 2015. doi: 10.1109/TPEL.2015.2446891

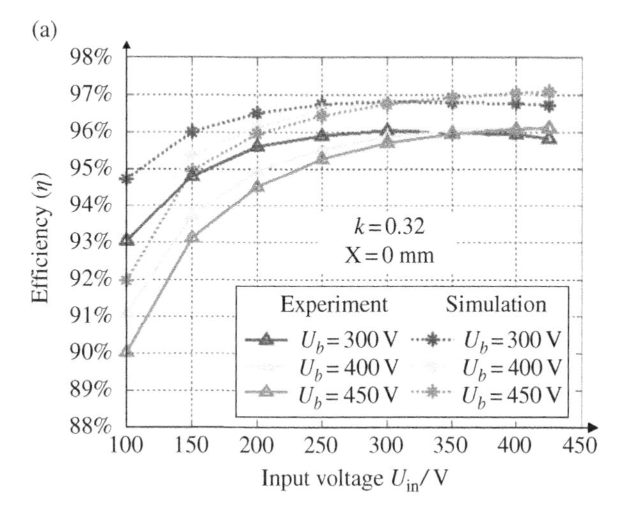

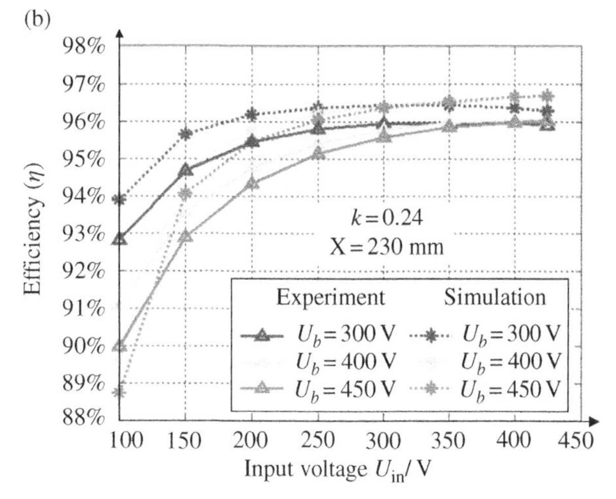

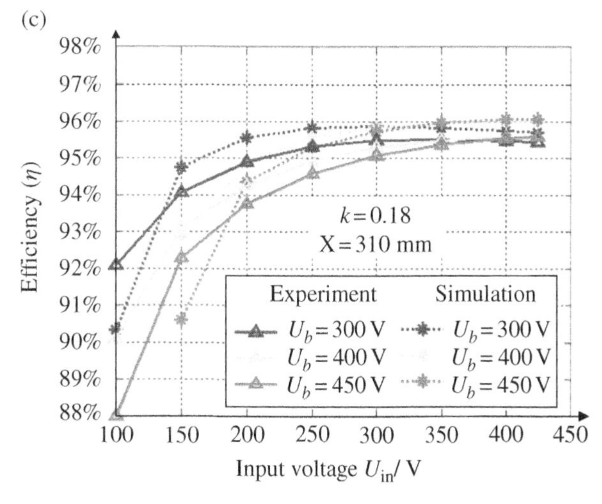

Figure 16.25 Simulation and experimental efficiencies of the system when output voltages are 300 V, 400 V, and 450 V at different X-misalignments: (a) $X=0$ mm. (b) $X=230$ mm. (c) $X=310$ mm.

Table 16.4 A rough loss distribution within the system.

Parts	Loss ratio
Power MOSFETs	11.9%
Main coils	52.3%
LCC compensation network	16.7%
Rectifier diodes	16.8%
Output filter	2.2%

already been widely used in many applications, such as portable electronic devices, biomedical devices, and electric vehicle charging [121].

Compared with the IPT system, the CPT system has many advantages. For example, magnetic fields are sensitive to nearby metal objects and the system efficiency drops quickly with this interference [122]. It can generate eddy current losses, and hence generates heat in a conductive object – a potential fire hazard. However, the electric field in the CPT system does not generate significant losses in the metal objects.

The recent CPT system can be classified by the matching network topology. The most popular topology is a single inductor resonating with the capacitor to form a simple series-resonant circuit [123, 124]. The second topology is the LCL structure at the front-end to step up the voltage for the coupling capacitor. However, there is also an inductor directly connected with the capacitor to form a series resonance [125]. In these two topologies, the series inductance is too large, because of the small value of capacitance. The voltage pressure on the capacitor is also too large. The third topology is the resonant class E converter or the non-resonant PWM converter used to replace the compensation inductor [126, 127]. All of these systems require very high capacitance values, at tens or hundreds of nF range. So the transferred distance is usually around 1 mm.

Compared with previous work [120], this section focuses on 150 mm distance power transfer by capacitive coupling. It is essentially designed for use in electric vehicle charging. At this large distance, the coupling capacitance is typically in the pF range. The series resonance topology is no longer suitable. This section proposes a double-sided LCLC-compensated topology (Figure 16.26), and its design process is discussed. The circuit design procedure, the capacitance simulation by Maxwell, the design of a 2.4 kW prototype, and the system experiment are discussed in detail.

16.9.1 Circuit Topology Design

LCLC Compensation topology: Since the challenge in the CPT system is the small coupling capacitance value, the best solution is to connect an extra capacitor in parallel with the coupling capacitor. The LCLC network can help achieve unity power factor at both the input and output. The reactive power in the circuit is therefore eliminated, and so the system efficiency is high.

Circuit working principle: The fundamental harmonic approximation (FHA) method can be used to simplify the system, as shown in Figure 16.27. The superposition theory can be used to analyze the circuit. Figure 16.27b shows components excited by the input voltage at resonance. The circuit parameters are designed to achieve

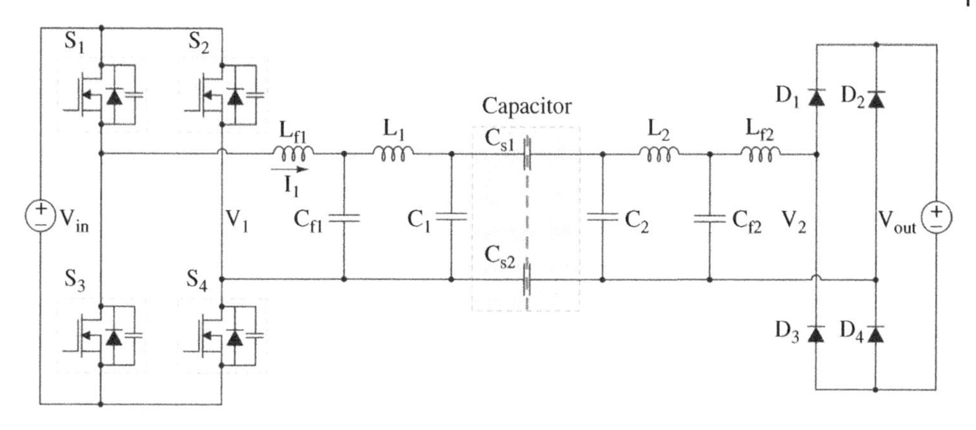

Figure 16.26 Circuit topology of the proposed capacitive power transfer system.

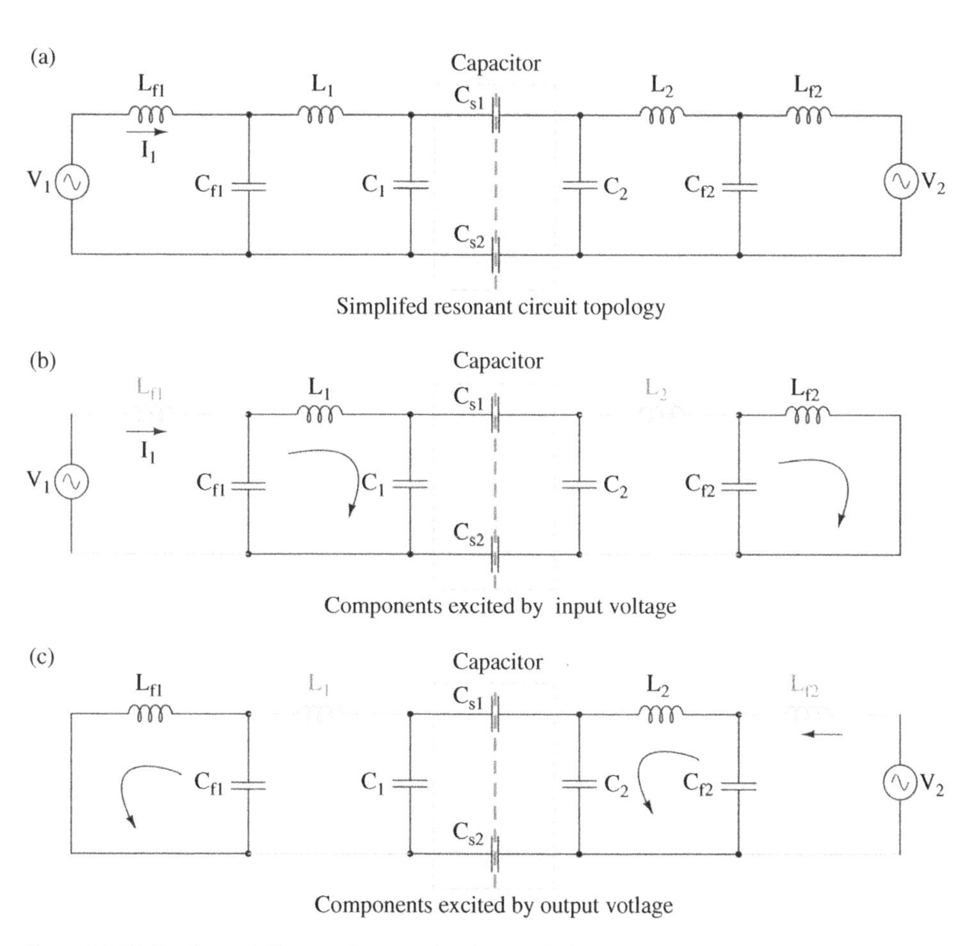

Figure 16.27 Fundamental harmonic approximation analysis.

resonance at the same frequency. This shows that the input current is independent of the input voltage. Figure 16.27c shows the components excited by the output voltage. Similarly, the output current does not depend on the output voltage. The parameter values should satisfy (16.67), where f_{sw} is the switching frequency.

$$\begin{cases} L_{f1} = 1/(\omega_0^2 C_{f1}), L_{f2} = 1/(\omega_0^2 C_{f2}) \\ L_1 = 1/(\omega_0^2 C_{p1}) + L_{f1}, L_2 = 1/(\omega_0^2 C_{p2}) + L_{f2} \\ C_{p1} = C_1 + C_s \cdot C_2/(C_s + C_2), C_{p2} = C_2 + C_s \cdot C_1/(C_s + C_1) \\ C_s = C_{s1} \cdot C_{s2}/(C_{s1} + C_{s2}), \omega_0 = 2\pi \cdot f_{sw} \end{cases} \tag{16.67}$$

Figure 16.27b shows that the output current leads the input voltage by 90°, and Figure 16.27c indicates that the input current lags the output voltage by 90°. Because an H-bridge rectifier is used at the output side, a the output voltage and current are in phase. Therefore, the current and voltage are also in phase at the input side. As a result, the input power is

$$P_{in} = V_1 \cdot I_1 = \omega_0 C_s \cdot \frac{C_{f1} C_{f2}}{C_1 C_2 + C_1 C_s + C_2 C_s} \cdot V_1 \cdot V_2 \tag{16.68}$$

Since C_1 and C_2 are much larger than C_s, equation (16.68) can be approximated as in (16.69):

$$P_{in} = V_1 \cdot I_1 \approx \omega_0 C_s \cdot \frac{C_{f1} C_{f2}}{C_1 C_2} \cdot V_1 \cdot V_2 \tag{16.69}$$

It can be seen from (16.69) that the out power is proportional to frequency, input voltage, out voltage, coupling capacitance, as well as proportional to C_{f1}, C_{f2}, and inversely proportional to C_1 and C_2. Therefore, it is essential to achieve the desired output power by adjusting C_{f1}, C_{f2}, C_1 and C_2.

16.9.2 Capacitance Analysis

Four metal plates were used to form two coupling capacitors to transfer power. Each plate size was selected as 610 × 610 mm. The nominal distance d was 150 mm. The separation d_1 was set to be 500 mm to eliminate inter-coupling between the two pairs of plates. The dimensions of the coupling capacitors are provided in Figure 16.28a,b.

Considering the edge effects [9], a single capacitor can be calculated as in (16.70),

$$C_{s1} = [1 + 2.343 \cdot (d/l)^{0.891}] \cdot (\varepsilon \cdot l^2/d) = 36.7\,\text{pF} \tag{16.70}$$

where $\varepsilon = 8.85 \times 10^{-12}$ F/m is the permittivity of free space.

Maxwell software from ANSYS was used to simulate a single capacitor at different misalignment and distance conditions. As shown in Figure 16.29, the capacitance is not highly sensitive to misalignment and distance variations.

16.9.3 A 2.4 kW CPT System Design

A 2.4 kW CPT system is designed using the procedure in Section 16.2. The system parameters are shown in Table 16.5.

It needs to be emphasized that the inductor L_2 is designed to be 5% larger than L_1 to achieve zero-voltage turn-on for the input inverter. Figure 16.30 shows the simulated input and output waveforms.

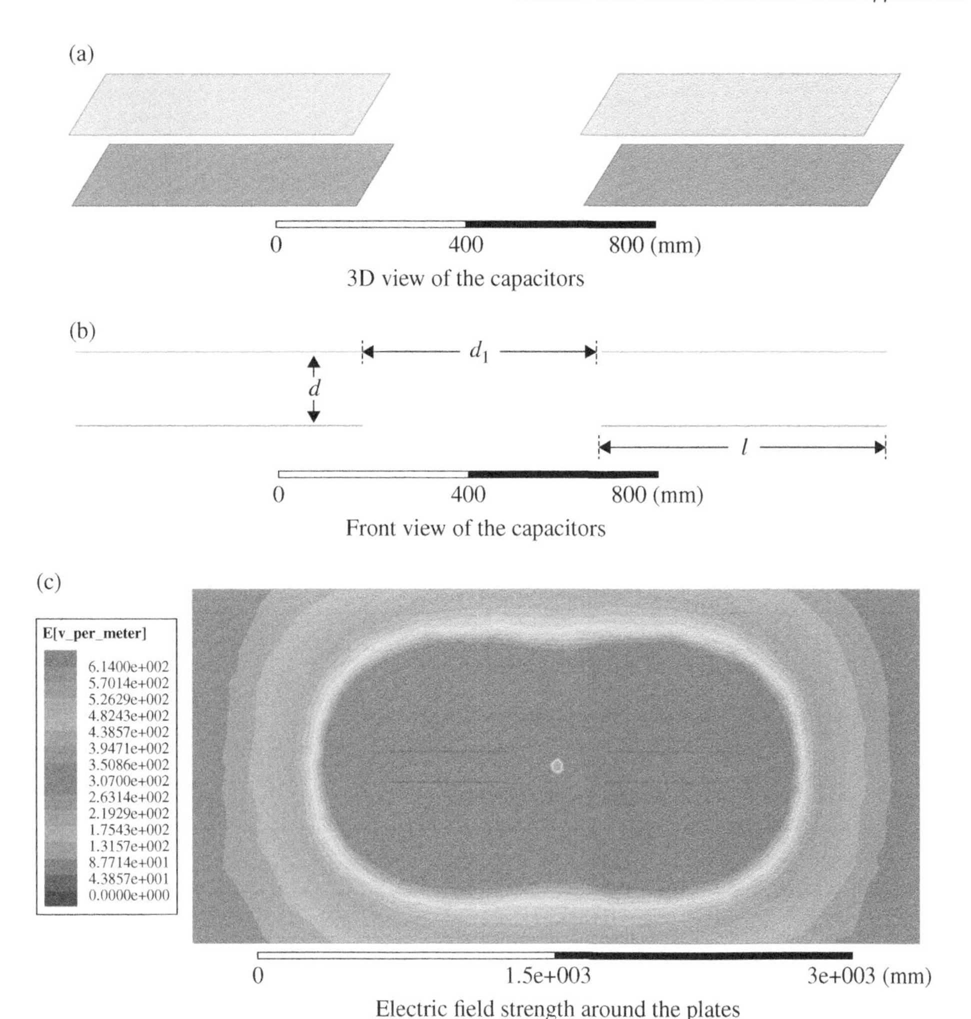

Figure 16.28 Three-dimension of the coupling capacitors.

The LTSpice-simulated output power is 2.4 kW, which is the desired value. Figure 16.30 shows that the input current is positive at the turn-on transient, which means the anti-body diode conducts current. Therefore, the switching loss is reduced. Table 16.6 shows the components peak voltage/current stress.

Figure 16.3c shows the Maxwell simulation result of the electric field strength. According to IEEE standard, the safe area for this system is about 0.6 m away from the plates, which has a field strength lower than 614 V/m.

16.9.4 Experiment

Using the parameters in Table 16.1, a 2.4 kW prototype was constructed as shown in Figure 16.31. Four 610 × 610 mm aluminum plates are used to form two capacitors, and the nominal distance is 150 mm. Table 16.2 shows that the peak voltage is 3.2 kV. Since the breakdown voltage of air is 3 kV/mm, there is no concern with arcing. For capacitors C_1 and C_2, 10 film capacitors with 2000 V DC voltage rating are connected in series.

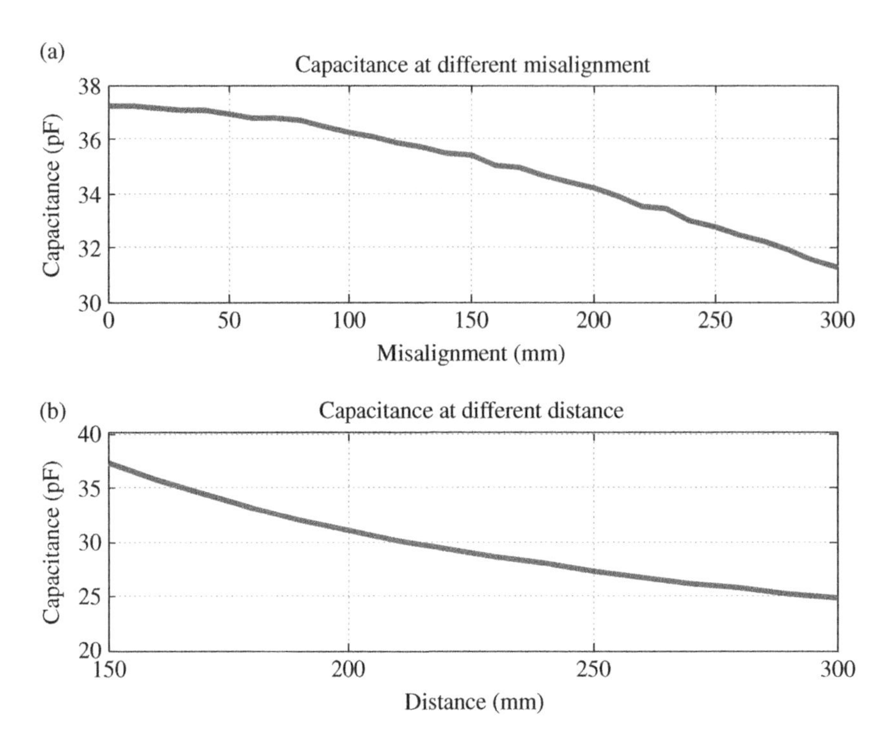

Figure 16.29 Capacitance variations with misalignment and distance.

Table 16.5 System specifications and parameter values.

V_{in}	V_{out}	f_{sw}	L_{f1} (L_{f2})	C_{f1} (C_{f2})	C_1 (C_2)	L_1	L_2
265 V	280 V	1 MHz	11.6 μH	2.18 nF	100 pF	231 μH	242 μH

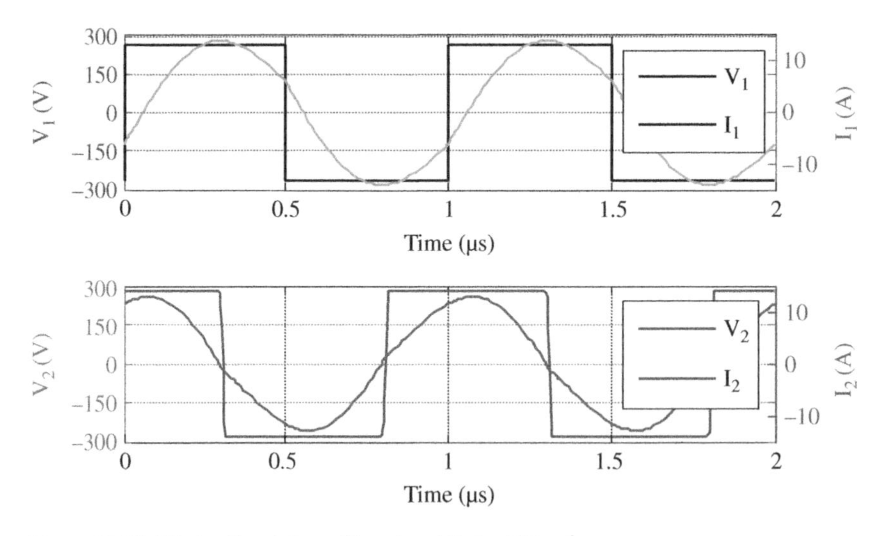

Figure 16.30 LTSpice Simulation of Input and Output Waveforms.

Table 16.6 Peak voltage/current stress on components.

Components	L_{f1} (L_{f2})	C_{f1} (C_{f2})	C_1 (C_2)	L_1(L_2)	Plates
Voltage	1.0 kV	1.0 kV	7.2 kV	7.2 kV	3.2 kV
Current	15.5 A	15.0 A	4.8 A	5.2 A	0.7 A

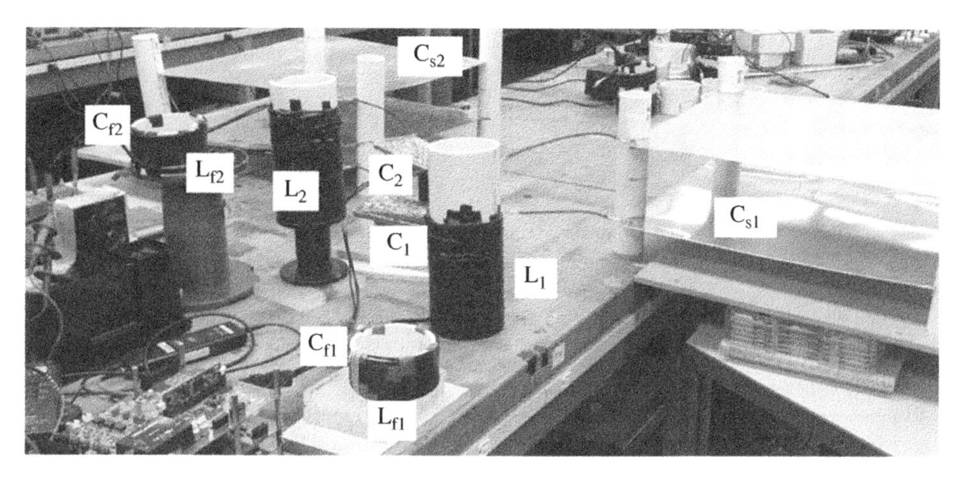

Figure 16.31 A 2.4 kW prototype of CPT system.

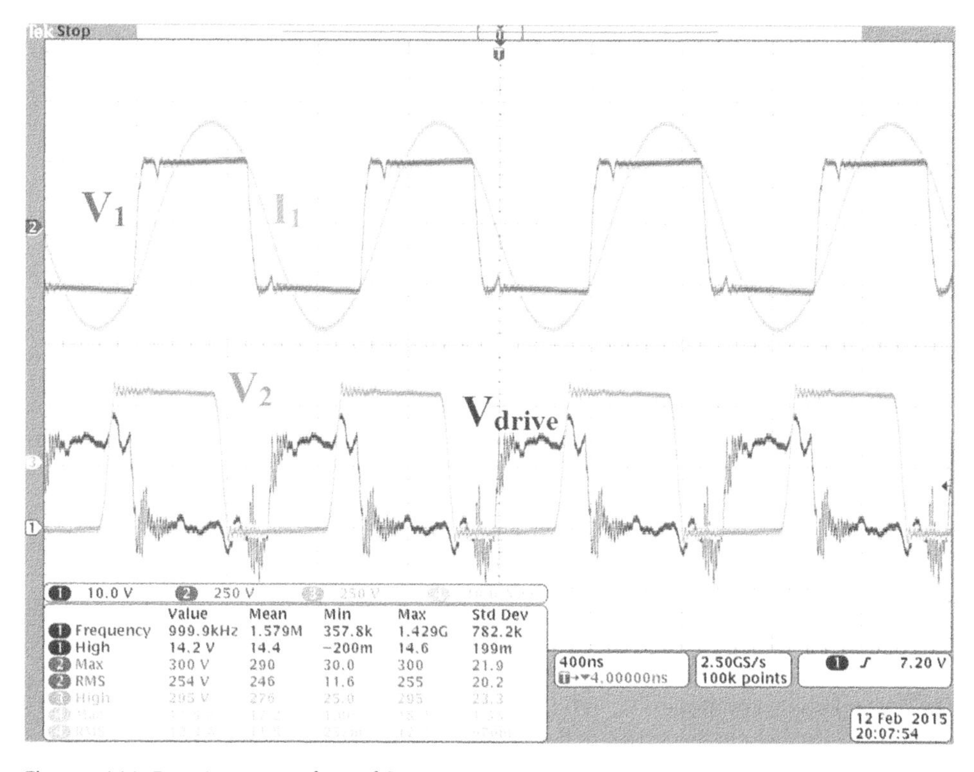

Figure 16.32 Experiment waveform of CPT system.

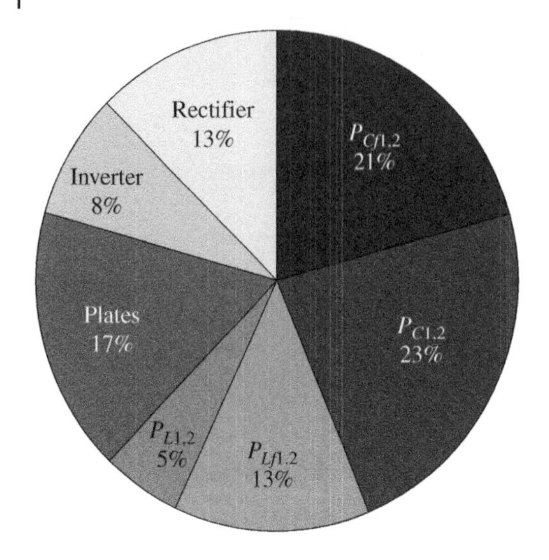

Figure 16.33 Power loss distribution of each component.

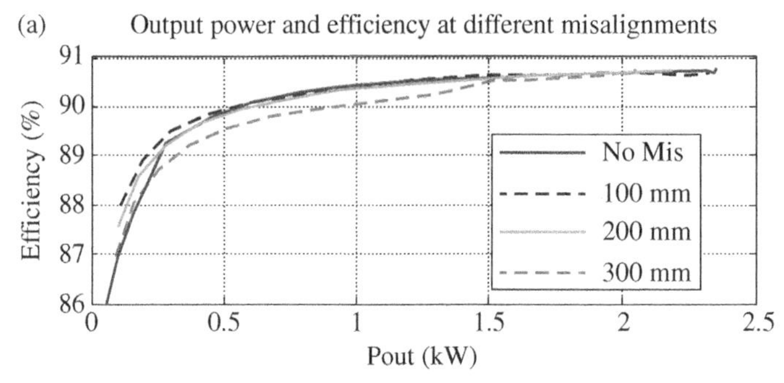

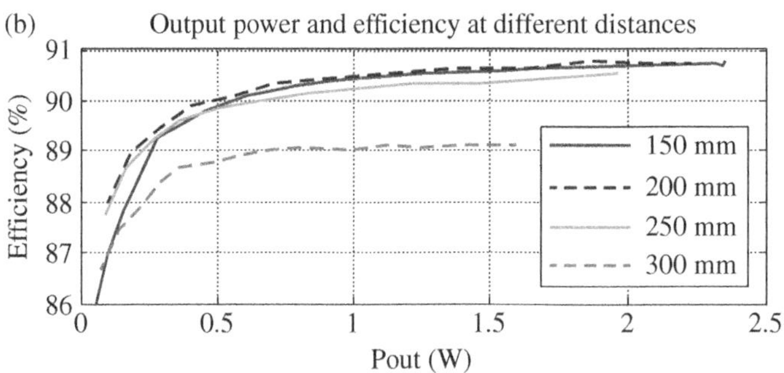

Figure 16.34 System performance under different conditions.

The inductors are wound in multi-turn shapes, so the voltage pressure between turns is limited and the insulation between turns can prevent arcing.

Experimental results are shown in Figure 16.32. The zero-voltage turn-on is achieved at the input side, and the waveform is similar to the simulation in Figure 16.30. The output power is 2.4 kW with 90.8% efficiency at 150 mm distance and no misalignment. The power loss on components can be estimated by the parasitic resistance of the component and also the current stress in Table 16.2. The power loss distribution is shown in Figure 16.33.

The system performance under different conditions is provided in Figure 16.34, which shows that the CPT system can maintain 2.1 kW output with 90.7% efficiency at 300 mm misalignment. Also, it can maintain 1.6 kW output with 89.1% efficiency at 300 mm distance. So the CPT system is robust to misalignment and distance variations compared to IPT systems.

Many progresses in capacitive wireless power transfer have been made since the writing of this book. Interested readers can refer to references [130–138].

16.10 Summary

This chapter has presented an overview of wireless charging of electric vehicles. It is clear that vehicle electrification is unavoidable due to environment and energy related issues. Wireless charging will provide many benefits compared to wired charging. In particular, when the roads are electrified with wireless charging capability, they will provide the foundation for mass market penetration for EVs regardless of battery technology. As the technology continues to develop, the wireless charging of EVs can be brought to fruition.

References

1 S.J. Gerssen-Gondelach and A.P.C. Faaij, "Performance of batteries for electric vehicles on short and longer term," *Journal of Power Sources*, vol. 212, pp. 111–129, 2012.
2 V. Etacheri, R. Marom, R. Elazari, G. Salitra, and D. Aurbach, "Challenges in the development of advanced Li-ion batteries: a review," *Energy & Environmental Science*, vol. 4, pp. 3243–3262, 2011.
3 A.K. Andre Kurs, Robert Moffatt, J.D. Joannopoulos, Peter Fisher, Marin Soljacic, "Wireless Power Transfer via Strongly Coupled Magnetic Resonances," *Science*, vol. 317, pp. 83–86, 2007.
4 A.P. Sample, D.A. Meyer, and J.R. Smith, "Analysis, Experimental Results, and Range Adaptation of Magnetically Coupled Resonators for Wireless Power Transfer," *Industrial Electronics, IEEE Transactions on*, vol. 58, pp. 544–554, 2011.
5 B.L. Cannon, J.F. Hoburg, D.D. Stancil, and S.C. Goldstein, "Magnetic Resonant Coupling As a Potential Means for Wireless Power Transfer to Multiple Small Receivers," *Power Electronics, IEEE Transactions on*, vol. 24, pp. 1819–1825, 2009.
6 A. Kurs, R. Moffatt, and M. Soljacic, "Simultaneous mid-range power transfer to multiple devices," *Applied Physics Letters*, vol. 96, pp. 044102–044102–3, 2010.
7 C. Sanghoon, K. Yong-Hae, S.-Y. Kang, L. Myung-Lae, L. Jong-Moo, and T. Zyung, "Circuit-Model-Based Analysis of a Wireless Energy-Transfer System via Coupled Magnetic Resonances," *Industrial Electronics, IEEE Transactions on*, vol. 58, pp. 2906–2914, 2011.
8 C. Kainan and Z. Zhengming, "Analysis of the Double-Layer Printed Spiral Coil for Wireless Power Transfer," *Emerging and Selected Topics in Power Electronics, IEEE Journal of*, vol. 1, pp. 114–121, 2013.
9 Z. Yiming, Z. Zhengming, and C. Kainan, "Frequency Decrease Analysis of Resonant Wireless Power Transfer," *Power Electronics, IEEE Transactions on*, vol. 29, pp. 1058–1063, 2014.

10 L. Chi Kwan, W.X. Zhong, and S.Y.R. Hui, "Effects of Magnetic Coupling of Nonadjacent Resonators on Wireless Power Domino-Resonator Systems," *Power Electronics, IEEE Transactions on*, vol. 27, pp. 1905–1916, 2012.
11 W.X. Zhong, L. Chi Kwan,, and S.Y. Hui, "Wireless power domino-resonator systems with noncoaxial axes and circular structures," *Power Electronics, IEEE Transactions on*, vol. 27, pp. 4750–4762, 2012.
12 Y. Nagatsuka, N. Ehara, Y. Kaneko, S. Abe, and T. Yasuda, "Compact contactless power transfer system for electric vehicles," in *Power Electronics Conference (IPEC), 2010 International*, 2010, pp. 807–813.
13 J.G. Bolger, F.A. Kirsten, and L.S. Ng, "Inductive power coupling for an electric highway system," in *Vehicular Technology Conference, 1978*. 28th IEEE, 1978, pp. 137–144.
14 M. Eghtesadi, "Inductive power transfer to an electric vehicle-analytical model," in *Vehicular Technology Conference, 1990 IEEE 40th*, 1990, pp. 100–104.
15 A.W. Green and J.T. Boys, "10 kHz inductively coupled power transfer-concept and control," in *Power Electronics and Variable-Speed Drives, 1994. Fifth International Conference on*, 1994, pp. 694–699.
16 K.W. Klontz, D.M. Divan, D.W. Novotny, and R.D. Lorenz, "Contactless power delivery system for mining applications," *Industry Applications, IEEE Transactions on*, vol. 31, pp. 27–35, 1995.
17 J.M. Barnard, J.A. Ferreira, and J.D. Van Wyk, "Sliding transformers for linear contactless power delivery," *Industrial Electronics, IEEE Transactions on*, vol. 44, pp. 774–779, 1997.
18 N.H. Kutkut and K.W. Klontz, "Design considerations for power converters supplying the SAE J-1773 electric vehicle inductive coupler," in Applied Power Electronics Conference and Exposition, 1997. APEC '97 Conf*erence Proceedings 1997, Twelfth Annual*, 1997, pp. 841–847 vol.2.
19 D.A.G. Pedder, A.D. Brown, and J.A. Skinner, "A contactless electrical energy transmission system," *Industrial Electronics, IEEE Transactions on*, vol. 46, pp. 23–30, 1999.
20 H. Abe, H. Sakamoto, and K. Harada, "A noncontact charger using a resonant converter with parallel capacitor of the secondary coil," *Industry Applications, IEEE Transactions on*, vol. 36, pp. 444–451, 2000.
21 J.T. Boys, G.A. Covic, and A.W. Green, "Stability and control of inductively coupled power transfer systems," *Electric Power Applications, IEE Proceedings*, vol. 147, pp. 37–43, 2000.
22 A.P. Hu, J.T. Boys, and G.A. Covic, "ZVS frequency analysis of a current-fed resonant converter," in *Power Electronics Congress, 2000. CIEP 2000. VII IEEE International*, 2000, pp. 217–221.
23 W. Chwei-Sen, G.A. Covic, and O.H. Stielau, "Power transfer capability and bifurcation phenomena of loosely coupled inductive power transfer systems," *Industrial Electronics, IEEE Transactions on*, vol. 51, pp. 148–157, 2004.
24 J.T. Boys, G.A. Covic, and X. Yongxiang, "DC analysis technique for inductive power transfer pickups," *Power Electronics Letters, IEEE Transactions on*, vol. 1, pp. 51–53, 2003.
25 J. Hirai, K. Tae-Woong, and A. Kawamura, "Study on intelligent battery charging using inductive transmission of power and information," *Power Electronics, IEEE Transactions on*, vol. 15, pp. 335–345, 2000.

26 S. Byeong-Mun, R. Kratz, and S. Gurol, "Contactless inductive power pickup system for Maglev applications," in *Industry Applications Conference, 2002. 37th IAS Annual Meeting. Conference Record of the* 2002, pp. 1586–1591, vol. 3.
27 W. Chwei-Sen, O.H. Stielau, and G.A. Covic, "Design considerations for a contactless electric vehicle battery charger," *Industrial Electronics, IEEE Transactions on*, vol. 52, pp. 1308–1314, 2005.
28 O.H. Stielau and G.A. Covic, "Design of loosely coupled inductive power transfer systems," in *Power System Technology, 2000. Proceedings. PowerCon 2000. International Conference on*, 2000, pp. 85–90, vol. 1.
29 A.P. Hu and S. Hussmann, "Improved power flow control for contactless moving sensor applications," *Power Electronics Letters, IEEE Transactions on*, vol. 2, pp. 135–138, 2004.
30 J.T. Boys, C.I. Chen, and G.A. Covic, "Controlling inrush currents in inductively coupled power systems," in *Power Engineering Conference, 2005. IPEC 2005. The 7th International, 2005*, pp. 1046–1051 vol. 2.
31 G.A.J. Elliot, J.T. Boys, and G.A. Covic, "A Design Methodology for Flat Pick-up ICPT Systems," in *Industrial Electronics and Applications, 2006 1ST IEEE Conference on*, 2006, pp. 1–7.
32 A.M. Bradley, M.D. Feezor, H. Singh, and F.Y. Sorrell, "Power systems for autonomous underwater vehicles," *Oceanic Engineering, IEEE Journal of*, vol. 26, pp. 526–538, 2001.
33 H. Singh, J.G. Bellingham, F. Hover, S. Lemer, B.A. Moran, K. von der Heydt, and D. Yoerger, "Docking for an autonomous ocean sampling network," *Oceanic Engineering, IEEE Journal of*, vol. 26, pp. 498–514, 2001.
34 S.P. Keith W. Klontz, M. Deepakraj M. Divan, M. Donald W. Novotny, and M. Robert D. Lorenz, "Submersible Contactless Power Delivery System," *US Patent* 5,301,096, 1994.
35 W. Kyung-il, P. Han Seok, C. Yun Hyun, and K. Kyung Ho, "Contactless energy transmission system for linear servo motor," *Magnetics, IEEE Transactions on*, vol. 41, pp. 1596–1599, 2005.
36 A. Esser and H.C. Skudelny, "A new approach to power supplies for robots," *Industry Applications, IEEE Transactions on*, vol. 27, pp. 872–875, 1991.
37 G.A.J. Elliott, G.A. Covic, D. Kacprzak, and J.T. Boys, "A New Concept: Asymmetrical Pick-Ups for Inductively Coupled Power Transfer Monorail Systems," *Magnetics, IEEE Transactions on*, vol. 42, pp. 3389–3391, 2006.
38 A. Kawamura, K. Ishioka, and J. Hirai, "Wireless transmission of power and information through one high-frequency resonant AC link inverter for robot manipulator applications," *Industry Applications, IEEE Transactions on*, vol. 32, pp. 503–508, 1996.
39 S.I. Adachi, F. Sato, S. Kikuchi, and H. Matsuki, "Consideration of contactless power station with selective excitation to moving robot," *Magnetics, IEEE Transactions on*, vol. 35, pp. 3583–3585, 1999.
40 J. Sallan, J.L. Villa, A. Llombart, and J.F. Sanz, "Optimal Design of ICPT Systems Applied to Electric Vehicle Battery Charge," *Industrial Electronics, IEEE Transactions on*, vol. 56, pp. 2140–2149, 2009.
41 J.L. Villa, J. Sallán, A. Llombart, and J.F. Sanz, "Design of a high frequency Inductively Coupled Power Transfer system for electric vehicle battery charge," *Applied Energy*, vol. 86, pp. 355–363, 2009.

42 H. Chang-Yu, J.T. Boys, G.A. Covic, and M. Budhia, "Practical considerations for designing IPT system for EV battery charging," in *Vehicle Power and Propulsion Conference, 2009. VPPC '09. IEEE*, 2009, pp. 402–407.

43 L. Hao Leo, H. Aiguo, and G.A. Covic, "Development of a discrete energy injection inverter for contactless power transfer," in *Industrial Electronics and Applications, 2008. ICIEA 2008. 3rd IEEE Conference on*, 2008, pp. 1757–1761.

44 T. Chun Sen, S. Yue, S. Yu Gang, N. Sing Kiong, and A.P. Hu, "Determining Multiple Steady-State ZCS Operating Points of a Switch-Mode Contactless Power Transfer System," *Power Electronics, IEEE Transactions on*, vol. 24, pp. 416–425, 2009.

45 J.U.W. Hsu, A.P. Hu, and A. Swain, "A Wireless Power Pickup Based on Directional Tuning Control of Magnetic Amplifier," *Industrial Electronics, IEEE Transactions on*, vol. 56, pp. 2771–2781, 2009.

46 A.J. Moradewicz and M.P. Kazmierkowski, "Contactless Energy Transfer System With FPGA-Controlled Resonant Converter," *Industrial Electronics, IEEE Transactions on*, vol. 57, pp. 3181–3190, 2010.

47 M. Budhia, G.A. Covic, and J.T. Boys, "Design and Optimization of Circular Magnetic Structures for Lumped Inductive Power Transfer Systems," *Power Electronics, IEEE Transactions on*, vol. 26, pp. 3096–3108, 2011.

48 M. Budhia, J.T. Boys, G.A. Covic, and H. Chang-Yu, "Development of a Single-Sided Flux Magnetic Coupler for Electric Vehicle IPT Charging Systems," *Industrial Electronics, IEEE Transactions on*, vol. 60, pp. 318–328, 2013.

49 G.A. Covic, M.L.G. Kissin, D. Kacprzak, N. Clausen, and H. Hao, "A bipolar primary pad topology for EV stationary charging and highway power by inductive coupling," in *Energy Conversion Congress and Exposition (ECCE), 2011 IEEE*, 2011, pp. 1832–1838.

50 A. Zaheer, D. Kacprzak, and G.A. Covic, "A bipolar receiver pad in a lumped IPT system for electric vehicle charging applications," in *Energy Conversion Congress and Exposition (ECCE), 2012 IEEE*, 2012, pp. 283–290.

51 N. Shinohara, "Wireless power transmission progress for electric vehicle in Japan," in *Radio and Wireless Symposium (RWS), 2013 IEEE*, 2013, pp. 109–111.

52 T.E. Stamati and P. Bauer, "On-road charging of electric vehicles," in *Transportation Electrification Conference and Expo (ITEC), 2013 IEEE*, 2013, pp. 1–8.

53 N. Puqi, J.M. Miller, O.C. Onar, and C.P. White, "A compact wireless charging system for electric vehicles," in *Energy Conversion Congress and Exposition (ECCE), 2013 IEEE*, 2013, pp. 3629–3634.

54 I. Systems Control Technology, *Roadway Powered Electric Vehicle Project: Track Construction and Testing Program. Phase 3D: California PATH Program*, Institute of Transportation Studies, University of California, Berkeley, 1994.

55 H.H. Wu, A. Gilchrist, K.D. Sealy, and D. Bronson, "A High Efficiency 5 kW Inductive Charger for EVs Using Dual Side Control," *Industrial Informatics, IEEE Transactions on*, vol. 8, pp. 585–595, 2012.

56 L. Sungwoo, H. Jin, P. Changbyung, C. Nam-Sup, C. Gyu-Hyeoung, and R. Chun-Taek, "On-Line Electric Vehicle using inductive power transfer system," in *Energy Conversion Congress and Exposition (ECCE), 2010 IEEE*, 2010, pp. 1598–1601.

57 H. Jin, L. Wooyoung, C. Gyu-Hyeong, L. Byunghun, and R. Chun-Taek, "Characterization of novel Inductive Power Transfer Systems for On-Line Electric Vehicles," in *Applied Power Electronics Conference and Exposition (APEC), 2011 Twenty-Sixth Annual IEEE*, 2011, pp. 1975–1979.

58 F. Musavi, M. Edington, and W. Eberle, "Wireless power transfer: A survey of EV battery charging technologies," in *Energy Conversion Congress and Exposition (ECCE), 2012 IEEE*, 2012, pp. 1804–1810.

59 G.A. Covic and J.T. Boys, "Modern Trends in Inductive Power Transfer for Transportation Applications," *Emerging and Selected Topics in Power Electronics, IEEE Journal of*, vol. 1, pp. 28–41, 2013.

60 S. Lukic and Z. Pantic, "Cutting the Cord: Static and Dynamic Inductive Wireless Charging of Electric Vehicles," *Electrification Magazine, IEEE*, vol. 1, pp. 57–64, 2013.

61 T.-D. Nguyen, S. Li, W. Li, and C. Mi, "Feasibility Study on Bipolar Pads for Efficient Wireless Power Chargers," in *Applied Power Electronics Conference and Exposition, Fort Worth, Texas*, 2014.

62 S. Valtchev, B. Borges, K. Brandisky, and J.B. Klaassens, "Resonant Contactless Energy Transfer With Improved Efficiency," *Power Electronics, IEEE Transactions on*, vol. 24, pp. 685–699, 2009.

63 S. Jaegue, S. Seungyong, K. Yangsu, A. Seungyoung, L. Seokhwan, J. Guho, J. Seong-Jeub, and C. Dong-Ho, "Design and Implementation of Shaped Magnetic-Resonance-Based Wireless Power Transfer System for Roadway-Powered Moving Electric Vehicles," *Industrial Electronics, IEEE Transactions on*, vol. 61, pp. 1179–1192, 2014.

64 K.V. Schuylenbergh and R. Puers, Inductive Powering – Basic Theory and Application to Biomedical Systems: *Springer*, 2009.

65 R. Mecke and C. Rathge, "High frequency resonant inverter for contactless energy transmission over large air gap," in *Power Electronics Specialists Conference, 2004. PESC 04. 2004 IEEE 35th Annual*, 2004, pp. 1737–1743 Vol.3.

66 J.T. Boys, G.A.J. Elliott, and G.A. Covic, "An Appropriate Magnetic Coupling Co-Efficient for the Design and Comparison of ICPT Pickups," *Power Electronics, IEEE Transactions on*, vol. 22, pp. 333–335, 2007.

67 M. Budhia, G. Covic, and J. Boys, "A new IPT magnetic coupler for electric vehicle charging systems," in *IECON 2010 – 36th Annual Conference on IEEE Industrial Electronics Society*, 2010, pp. 2487–2492.

68 M. Budhia, G.A. Covic, J.T. Boys, and H. Chang-Yu, "Development and evaluation of single sided flux couplers for contactless electric vehicle charging," in *Energy Conversion Congress and Exposition (ECCE), 2011 IEEE*, 2011, pp. 614–621.

69 M. Chigira, Y. Nagatsuka, Y. Kaneko, S. Abe, T. Yasuda, and A. Suzuki, "Small-size light-weight transformer with new core structure for contactless electric vehicle power transfer system," in *Energy Conversion Congress and Exposition (ECCE), 2011 IEEE*, 2011, pp. 260–266.

70 S. Choi, J. Huh, W.Y. Lee, S.W. Lee, and C.T. Rim, "New Cross-Segmented Power Supply Rails for Roadway-Powered Electric Vehicles," *Power Electronics, IEEE Transactions on*, vol. 28, pp. 5832–5841, 2013.

71 M. Kiani and M. Ghovanloo, "A Figure-of-Merit for Designing High-Performance Inductive Power Transmission Links," *Industrial Electronics, IEEE Transactions on*, vol. 60, pp. 5292–5305, 2013.

72 H. Jin, L. Sungwoo, P. Changbyung, C. Gyu-Hyeoung, and R. Chun-Taek, "High performance inductive power transfer system with narrow rail width for On-Line Electric Vehicles," in *Energy Conversion Congress and Exposition (ECCE), 2010 IEEE*, 2010, pp. 647–651.

73 J. Huh, S.W. Lee, W.Y. Lee, G.H. Cho, and C.T. Rim, "Narrow-Width Inductive Power Transfer System for Online Electrical Vehicles," *Power Electronics, IEEE Transactions on*, vol. 26, pp. 3666–3679, 2011.
74 J. Young Jae and K. Young Dae, "System architecture and mathematical model of public transportation system utilizing wireless charging electric vehicles," in *Intelligent Transportation Systems (ITSC), 2012 15th International IEEE Conference on*, 2012, pp. 1055–1060.
75 J. Young Jae, K. Young Dae, and J. Seungmin, "Optimal design of the wireless charging electric vehicle," in *Electric Vehicle Conference (IEVC), 2012 IEEE International*, 2012, pp. 1–5.
76 S. In-Soo and K. Jedok, "Electric vehicle on-road dynamic charging system with wireless power transfer technology," in *Electric Machines & Drives Conference (IEMDC), 2013 IEEE International*, 2013, pp. 234–240.
77 J.T. Boys and S. Nishino, "Primary inductance pathway," *US Patent* 5619078, 1997.
78 G.R. Nagendra, J.T. Boys, G.A. Covic, B.S. Riar, and A. Sondhi, "Design of a double coupled IPT EV highway," in *Industrial Electronics Society, IECON 2013 – 39th Annual Conference of the IEEE*, 2013, pp. 4606–4611.
79 K. Lee, Z. Pantic, and S. Lukic, "Reflexive Field Containment in Dynamic Inductive Power Transfer Systems," *Power Electronics, IEEE Transactions on*, vol. PP, pp. 1–1, 2013.
80 A. Khaligh and S. Dusmez, "Comprehensive Topological Analysis of Conductive and Inductive Charging Solutions for Plug-In Electric Vehicles," *Vehicular Technology, IEEE Transactions on*, vol. 61, pp. 3475–3489, 2012.
81 W. Zhang, S.-C. Wong, C.K. Tse, and Q. Chen, "Analysis and Comparison of Secondary Series and Parallel Compensated Inductive Power Transfer Systems Operating for Optimal Efficiency and Load-Independent Voltage-Transfer Ratio," *Power Electronics, IEEE Transactions on*, vol. 29, pp. 2979–2990, 2014.
82 C. Duan, C. Jiang, A. Taylor, and K. Bai, "Design of a zero-voltage-switching large-air-gap wireless charger with low electric stress for electric vehicles," *Power Electronics, IET*, vol. 6, pp. 1742–1750, 2013.
83 Z. Pantic, B. Sanzhong, and S. Lukic, "ZCS LCC-Compensated Resonant Inverter for Inductive Power Transfer Application," *Industrial Electronics, IEEE Transactions on*, vol. 58, pp. 3500–3510, 2011.
84 J.L. Villa, J. Sallan, J.F. Sanz Osorio, and A. Llombart, "High-Misalignment Tolerant Compensation Topology For ICPT Systems," *Industrial Electronics, IEEE Transactions on*, vol. 59, pp. 945–951, 2012.
85 S. Dieckerhoff, M.J. Ruan, and R.W. De Doncker, "Design of an IGBT-based LCL-resonant inverter for high-frequency induction heating," in *Industry Applications Conference, 1999. Thirty-Fourth IAS Annual Meeting. Conference Record of the 1999 IEEE*, 1999, pp. 2039–2045 vol. 3.
86 G.L. Fischer and H. Doht, "An inverter system for inductive tube welding utilizing resonance transformation," in *Industry Applications Society Annual Meeting, 1994., Conference Record of the 1994 IEEE*, 1994, pp. 833–840 vol.2.
87 M.L.G. Kissin, H. Chang-Yu, G.A. Covic, and J.T. Boys, "Detection of the Tuned Point of a Fixed-Frequency LCL Resonant Power Supply," *Power Electronics, IEEE Transactions on*, vol. 24, pp. 1140–1143, 2009.
88 B. Sharp and H. Wu, "Asymmetrical Voltage-Cancellation control for LCL resonant converters in Inductive Power Transfer systems," in Applied Power Electronics Conference and Exposition *(APEC), 2012 Twenty-Seventh Annual IEEE*, 2012, pp. 661–666.

89 N.A. Keeling, G.A. Covic, and J.T. Boys, "A Unity-Power-Factor IPT Pickup for High-Power Applications," *Industrial Electronics, IEEE Transactions on*, vol. 57, pp. 744–751, 2010.

90 N. Keeling, G.A. Covic, F. Hao, L. George, and J.T. Boys, "Variable tuning in LCL compensated contactless power transfer pickups," in *Energy Conversion Congress and Exposition*, 2009. *ECCE 2009. IEEE*, 2009, pp. 1826–1832.

91 C.Y. Huang, J.T. Boys, and G.A. Covic, "LCL Pickup Circulating Current Controller for Inductive Power Transfer Systems," *Power Electronics, IEEE Transactions on*, vol. 28, pp. 2081–2093, 2013.

92 H.L. Li, A.P. Hu, G.A. Covic, and T. Chunsen, "A new primary power regulation method for contactless power transfer," in Industrial Technology, 2009. ICIT *2009. IEEE International Conference on*, 2009, pp. 1–5.

93 L. Zhen Ning, R.A. Chinga, T. Ryan, and L. Jenshan, "Design and Test of a High-Power High-Efficiency Loosely Coupled Planar Wireless Power Transfer System," *Industrial Electronics, IEEE Transactions on*, vol. 56, pp. 1801–1812, 2009.

94 L. Hao Leo, A.P. Hu, and G.A. Covic, "A Direct AC-AC Converter for Inductive Power-Transfer Systems," *Power Electronics, IEEE Transactions on*, vol. 27, pp. 661–668, 2012.

95 J.M. Miller, C.P. White, O.C. Onar, and P.M. Ryan, "Grid side regulation of wireless power charging of plug-in electric vehicles," in *Energy Conversion Congress and Exposition (ECCE), 2012 IEEE*, 2012, pp. 261–268.

96 B. Nguyen Xuan, D.M. Vilathgamuwa, and U.K. Madawala, "A matrix converter based inductive power transfer system," in *IPEC, 2012 Conference on Power & Energy*, 2012, pp. 509–514.

97 Q. Wei, W. Guo, X. Sun, G. Wang, X. Zhao, F. Li, and Y. Li, "A New Type of IPT System with Large Lateral Tolerance and its Circuit Analysis," in *Connected Vehicles and Expo (ICCVE), 2012 International Conference on*, 2012, pp. 311–315.

98 S. Krishnan, S. Bhuyan, V.P. Kumar, W. Wenjiang, J.A. Afif, and L. Khoon Seong, "Frequency agile resonance-based wireless charging system for Electric Vehicles," in *Electric Vehicle Conference (IEVC), 2012 IEEE International*, 2012, pp. 1–4.

99 M. Borage, S. Tiwari, and S. Kotaiah, "Analysis and design of an LCL-T resonant converter as a constant-current power supply," *Industrial Electronics, IEEE Transactions on*, vol. 52, pp. 1547–1554, 2005.

100 H.H. Wu, J.T. Boys, and G.A. Covic, "An AC Processing Pickup for IPT Systems," *Power Electronics, IEEE Transactions on*, vol. 25, pp. 1275–1284, 2010.

101 H.H. Wu, G.A. Covic, J.T. Boys, and D.J. Robertson, "A Series-Tuned Inductive-Power-Transfer Pickup With a Controllable AC-Voltage Output," *Power Electronics, IEEE Transactions on*, vol. 26, pp. 98–109, 2011.

102 S.J. Orfanidis. (2013). Electromagnetic Waves and Antennas [Online]. Available: http://eceweb1.rutgers.edu/~orfanidi/ewa/

103 K.A. Grajski, R. Tseng, and C. Wheatley, "Loosely-coupled wireless power transfer: Physics, circuits, standards," in *Microwave Workshop Series on Innovative Wireless Power Transmission: Technologies, Systems, and Applications (IMWS), 2012 IEEE MTT-S International*, 2012, pp. 9–14.

104 U.K. Madawala and D.J. Thrimawithana, "Current sourced bi-directional inductive power transfer system," *Power Electronics, IET*, vol. 4, pp. 471–480, 2011.

105 U.K. Madawala and D.J. Thrimawithana, "A Bidirectional Inductive Power Interface for Electric Vehicles in V2G Systems," *Industrial Electronics, IEEE Transactions on*, vol. 58, pp. 4789–4796, 2011.

106 D.J. Thrimawithana and U.K. Madawala, "A three-phase bi-directional IPT system for contactless charging of electric vehicles," in *Industrial Electronics (ISIE), 2011 IEEE International Symposium on*, 2011, pp. 1957–1962.

107 T.P.E.R. Joy, K. Thirugnanam, and P. Kumar, "Bidirectional Contactless Charging System using Li-ion battery model," in *Industrial and Information Systems (ICIIS), 2012 7th IEEE International Conference on*, 2012, pp. 1–6.

108 M.J. Neath, A.K. Swain, U.K. Madawala, D.J. Thrimawithana, and D.M. Vilathgamuwa, "Controller Synthesis of a Bidirectional Inductive Power Interface for electric vehicles," in *Sustainable Energy Technologies (ICSET), 2012 IEEE Third International Conference on*, 2012, pp. 60–65.

109 A.K. Swain, M.J. Neath, U.K. Madawala, and D.J. Thrimawithana, "A Dynamic Multivariable State-Space Model for Bidirectional Inductive Power Transfer Systems," *Power Electronics, IEEE Transactions on*, vol. 27, pp. 4772–4780, 2012.

110 C. Tang, X. Dai, Z. Wang, Y. Su, and Y. Sun, "A bidirectional contactless power transfer system with dual-side power flow control," in *Power System Technology (POWERCON), 2012 IEEE International Conference on*, 2012, pp. 1–6.

111 D.J. Thrimawithana and U.K. Madawala, "A Generalized Steady-State Model for Bidirectional IPT Systems," *Power Electronics, IEEE Transactions on*, vol. 28, pp. 4681–4689, 2013.

112 D.J. Thrimawithana, U.K. Madawala, and M. Neath, "A Synchronization Technique for Bidirectional IPT Systems," *Industrial Electronics, IEEE Transactions on*, vol. 60, pp. 301–309, 2013.

113 H. Xueliang, Q. Hao, H. Zhenchen, S. Yi, and L. Jun, "The Interaction Research of Smart Grid and EV Based Wireless Charging," in *Vehicle Power and Propulsion Conference (VPPC), 2013 IEEE*, 2013, pp. 1–5.

114 D. van Wageningen and T. Staring, "The Qi wireless power standard," in *Power Electronics and Motion Control Conference (EPE/PEMC), 2010 14th International*, 2010, pp. S15–25-S15–32.

115 J.B. Kenney, "Dedicated Short-Range Communications (DSRC) Standards in the United States," *Proceedings of the IEEE*, vol. 99, pp. 1162–1182, 2011.

116 S.M. Lukic, M. Saunders, Z. Pantic, S. Hung, and J. Taiber, "Use of inductive power transfer for electric vehicles," in *Power and Energy Society General Meeting, 2010 IEEE*, 2010, pp. 1–6.

117 Z. Pantic, B. Sanzhong, and S. Lukic, "ZCS LCC-Compensated Resonant Inverter for Inductive-Power-Transfer Application," Industrial Electronics, IEEE Transactions on, vol. 58, pp. 3500–3510, 2011.

118 R.W. Erickson and D. Maksimovic, Fundamentals of Power Electronics, 2nd ed. New York: Kluwer Academic Publishers, 2001.

119 L. Bing, L. Wenduo, L. Yan, F.C. Lee, and J.D. Van Wyk, "Optimal design methodology for LLC resonant converter," in Applied Power Electronics Conference and Exposition, 2006. APEC '06. Twenty-First Annual IEEE, 2006, pp. 533–538.

120 J. Dai and D. Ludois, "A Survey of Wireless Power Transfer and a Critical Comparison of Inductive and Capacitive Coupling for Small Gap Applications", *IEEE Trans. Power Electron.*, vol. PP, pp. 1–14, 2015.

121 S. Li, W. Li, J. Deng, and C.C. Mi, "A Double-Sided LCC Compensation Network and Its Tuning Method for Wireless Power Transfer", *IEEE Trans. Veh. Tech.*, vol. PP, pp. 1–12, 2014.

122 D.J. Graham, J.A. Neasham, and B.S. Sharif, "Investigation of Methods for Data Communication and Power Delivery through Metals", *IEEE Trans. Ind. Electron.*, vol. 58, 4972–4980, 2011.

123 C. Liu, A.P. Hu, G.A. Covic and N.C. Nair, "Comparative Study of CCPT Systems With Two Different Inductor Tuning Positions," *IEEE Trans. Power Electron.*, vol. 27, pp. 294–306, 2012.

124 D.C. Ludois, M.J. Erickson, and J.K. Reed, "Aerodynamic Fluid Bearing for Translational and Rotating Capacitors in Noncontact Capacitive Power Transfer Systems," *IEEE Trans. Ind. Appl.*, vol. 50, pp. 1025–1033, 2014.

125 M.P. Theodoridis, "Effective Capacitive Power Transfer," *IEEE Trans. Power Electron.*, vol. 27, pp. 4906–4913, 2012.

126 D. Schmilovitz, A. Abramovitz, and I. Reichman, "Quasi Resonant LED Driver with Capacitive Isolation and High PF," *IEEE Jour. Emerg. Select. Topics Power Electron.*, vol. PP, pp. 1–20, 2015.

127 J. Dai and D.C. Ludois, "Single Active Switch Power Electronics for Kilowatt Scale Capacitive Power Transfer," *IEEE Jour. Emerg. Select. Topics Power Electron.*, vol. 3, pp. 315–323, 2015.

128 H. Nishiyama and M. Nakamura, "Form and Capacitance of Parallel-Plate Capacitor", IEEE Trans. Compen. Packag. Munufact. Tech.–Part A, vol 17, pp. 477–484, 199.

129 *IEEE Standard for Safety Levels with Respect to Human Exposure to Radio Frequency Electromagnetic Fields, 3 kHz to 300 GHz*, C95.1, 2005.

130 Fei Lu, Hua Zhang, and C. Mi, "A Two-Plate Capacitive Wireless Power Transfer System Using Earth Ground as Current Returning Path with Reduced Chassis Voltage for Electric Vehicle Charging Applications," Revisions, Submitted to IEEE Transactions on Power Electronics – Letters, June 16, 2017.

131 Fei Lu, Hua Zhang, Heath Hofmann, and C. Mi, "A Dual-Coupled LCC-Compensated IPT System with a Compact Magnetic Coupler," to appear, IEEE Transactions on Power Electronics, Decision on March 9, 2017.

132 Fei Lu, Hua Zhang, H. Hofmann, and C. Mi, "A Double-sided LC Compensation Circuit for Loosely-Coupled Capacitive Power Transfer," January 5, 2017. IEEE Transactions on Power Electronics. Early Access, DOI: 10.1109/TPEL.2017.2674688

133 Hua Zhang, Fei Lu, Heath Hofmann, Weiguo Liu, and C. Mi, "A Six-Plate Capacitive Coupler to Reduce Electric Field Emission in Large Air-Gap Capacitive Power Transfer," IEEE Transactions on Power Electronics, Early Access, 10.1109/TPEL.2017.2662583.

134 Hua Zhang, Fei Lu, Heath Hofmann, Weiguo Liu, and C. Mi, "An LC Compensated Electric Fields Repeater for Long Distance Capacitive Power Transfer," IEEE Transactions on Industry Applications, 10.1109/TIA.2017.2697846.

135 Fei Lu, Hua Zhang, Heath Hofmann, and C. Mi, "An Inductive and Capacitive Integrated Couper and Its LCL Compensation Circuit Design for Wireless Power Transfer," IEEE Transactions on Industry Applications, 10.1109/TIA.2017.2697838.

136 Yunlong Shang, Yunlong Shang, Bing Xia, Chenghui Zhang, Naxin Cui, and C. Mi, "An Automatic Equalizer Based on Forward and Flyback Conversion for Series-Connected Battery Strings," IEEE Transactions on Industrial Electronics, early access, DOI: 10.1109/TIE.2017.2674617

137 Fei Lu, Hua Zhang, Heath Hofmann, and C. Mi "A Dynamic Charging System with Reduced Output Power Pulsation for Electric Vehicles," IEEE Transactions on Industrial Electronics, 10.1109/TIE.2016.2563380, VOL. 63, NO. 10, OCTOBER 2016.

138 Fei Lu, Hua Zhang, Heath Hofmann, and C. Mi, "An Inductive and Capacitive Combined Wireless Power Transfer System with LC-Compensated Topology for Electric Vehicle Charging Application," IEEE Transactions on Power Electronics, 10.1109/TPEL.2016.2519903, VOL. 31, NO. 12, DECEMBER 2016.

17

Vehicular Power Control Strategy and Energy Management

17.1 A Generic Framework, Definition, and Needs

The terms power and energy management are "almost" synonymous when used in the context of vehicular systems. To be more precise, power is an instantaneous quantity, whereas energy involves a period of time over which the power is applied. Although the meaning will become clearer as we progress through this chapter, normally when power is "managed" in general energy will be managed as well.

The question of management of power or energy arises primarily for the following reasons (although the term "management" has some additional implications, which will be described later). Consider Figure 17.1 which represents a very generic system involving several power/energy sources and loads, including distribution of power to the loads.

In this diagram we have several sources [1]. Each of these sources could be of any particular type. For example, it could be an ICE, which is a power converter implying that it is an energy converter which translates chemical energy into mechanical power. It could also be a battery, which is an electrical energy storage device, or it could be a fuel cell, which converts chemical energy into electrical power, or it could be a flywheel, which stores kinetic energy in mechanical form. In the case of the flywheel, energy could be injected into the flywheel by running through an electric motor and speeding it up. The sources could also be hydraulic or pneumatic power sources.

It is apparent from the above discussion that the sources could be of different natures, but those could be similar as well, or even identical. For example, sources 1 and 2 in Figure 17.1 could both be ICEs. Or, one could be a diesel and the other a gasoline engine. Similarly, one source could be a battery, the other a fuel cell.

Loads could also be of different types. For example, in a hybrid or electric vehicle the load could be a propulsion motor type of electric (or rather electromechanical) load, or it could be an electric light or heater, i.e. some kind of resistive load. It could also include auxiliary motors like pumps, air-conditioners, and so on. Thus it is obvious that in a vehicle there are various types of loads depending upon needs. But the question is: why should one have multiple sources in a vehicular system?

Even in a conventional non-hybrid vehicle we have the ICE and also a battery (and of course, the battery is charged through a generator driven by the ICE). Here we have at least two separate sources to cater for two different types of loads, i.e. mechanical propulsion and electrical loads. In an HEV we also have two sources at least. The only difference is that in an HEV the battery can be much larger (actually consisting of a

Hybrid Electric Vehicles: Principles and Applications with Practical Perspectives,
Second Edition. Chris Mi and M. Abul Masrur.

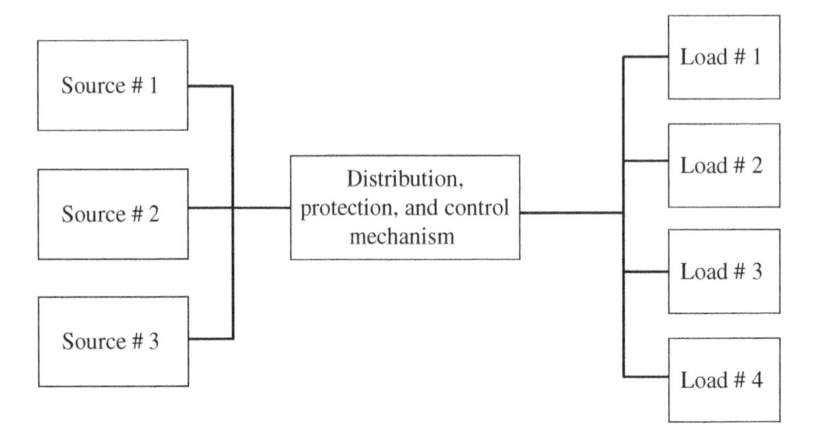

Figure 17.1 A block diagram schematic showing the source and load distribution system.

number of separate battery modules combined together) than in a regular non-hybrid vehicle. The simple answer as to why we have different types of sources is that in a conventional vehicle we have a heavy mechanical propulsion load which can be catered for by means of an ICE. Then of course we can use the battery to deliver smaller loads such as electric lights or pumps. But a more complex answer regarding the need for multiple sources is that the load demand (amount of power) varies with time, and the characteristics of the load (i.e. torque–speed characteristics of a mechanical load like propulsion) also change with time. Sources also have characteristics in terms of their power delivery profile, and efficiency. In other words, to deliver a particular load at a particular moment, if it is found that a particular source is capable of doing so at the best efficiency compared to several choices of sources, then obviously we should select the best source. Here we have to first see if one or more sources (individually or collectively) can meet the particular load demand in terms of load performance. If there is more than one source that could do the job, then we have to choose the most efficient source (or a combination of the most efficient ones) to do the job.

The reason why the above situation arises is that the ICE has better efficiencies in a certain torque–speed region of its engine efficiency map. The same applies to an electric motor, but electric machines have much higher efficiencies over a wider range of torque and speed, than an ICE. Similarly, if, for the sake of discussion, we assume that a system has both a diesel and a gasoline engine (multiengine, multifuel system), then their efficiency characteristics will be different, and at a certain torque–speed point a diesel engine will be better than a gasoline engine, and vice versa.

This idea forms the main rationale behind power management from performance and efficiency points of view. This can also be termed energy management since this rationale also increases overall system efficiency and reduces energy consumption.

In addition, other things that fall within the scope of power management involve the overall means through which the power is distributed to various loads. For example, the mechanism used to connect the loads to the source and the protection mechanism used for problems like overload, short circuits, and over voltage, in an electrical system are also within the scope of power management. Also within the scope of power management are a few additional items, such as controlling and coordinating the overall power system and various subsystems and components in a manner such that the battery charging and discharging that are controlled with the objective of enhancing battery life.

17.2 Methodology to Implement

We will treat power management and control from two different perspectives: optimization, and distribution and control. Optimization deals with things like energy efficiency and maintaining the battery's state of charge (SOC) within some threshold values, and it has more to do with generation control and management, based on load demand. Although the second perspective relates to distribution and control, there is a good interrelationship between the two as will be evident in the following.

First let us consider the characteristics of power/energy sources and look at the characteristics of: gasoline ICE, diesel engine, alternator, battery, and fuel cell.

For these engines, the curves shown in Figure 17.2 should actually contain a family of curves corresponding to different throttle positions or fuel input (additional discussion can be found in the web article cited under further reading section at the end of this chapter). We could also have another set of curves showing the relationship between speed and efficiency at different throttle positions (or equivalently the fuel input amount, or any other form of energy input amount) for each of the sources. Both of these sets – speed–torque and speed–efficiency curves – together constitute the engine map.

Similarly, for the electrical components, such as motors and generators, we could have another set of curves showing the relationship between speed and efficiency for a given power output (or equivalently the speed versus power input at a given efficiency). For static components like battery and fuel cells, we could plot voltage versus current at a given SOC (or equivalently the power input versus current). These characteristics for

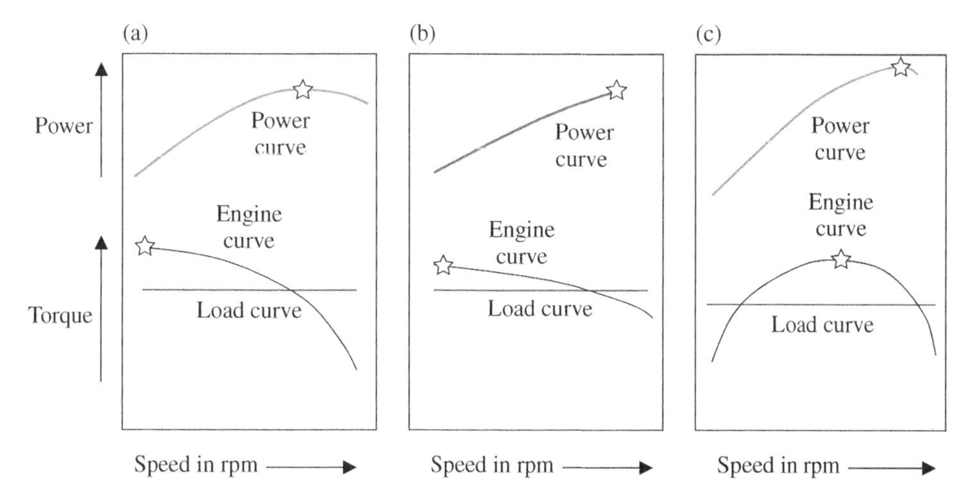

Figure 17.2 (a) Diesel engine with relatively high torque rise (slope); (b) diesel engine with relatively low torque rise (slope); (c) gasoline engine. (Maximum torque and power points are shown by a five-point star.)

components such as a generator and battery can constitute the required information in terms of the operating points of these devices.

For the characteristic curves in Figure 17.2, the engine curves and the load curves are indicated. As noted previously, the curves involving an ICE are for a specific throttle position corresponding to engine fuel intake. The intersection points between the load curves and the source (e.g. engine, battery, etc.) curves dictate the exact operating point of the load and source. When we want to drive a load, this means that we want to generate a specific speed at a specific torque. The load and its characteristics are dependent on things outside of the engine (or source). To meet the load speed and torque requirements, it is usually necessary to include a gear train in the system, since engines in general cannot handle the wide range of load demand, even after controlling the throttle. Once the throttle position and gear ratio are chosen, those define the precise engine operating point. At that point, the engine operates at a given efficiency or fuel economy.

The electrical system characteristics and their nature are shown in Figure 17.3. For convenience of comparison we have made the curves identical in form to show what variables change in each case. The variables are voltage and current on the y and x axis respectively. The parameter that is fixed is the SOC in the case of a battery, and fuel (could be hydrogen) flow rate in the case of a fuel cell. For the generator curve, the speed is held constant while the load changes. Some electrical load can be applied to each of

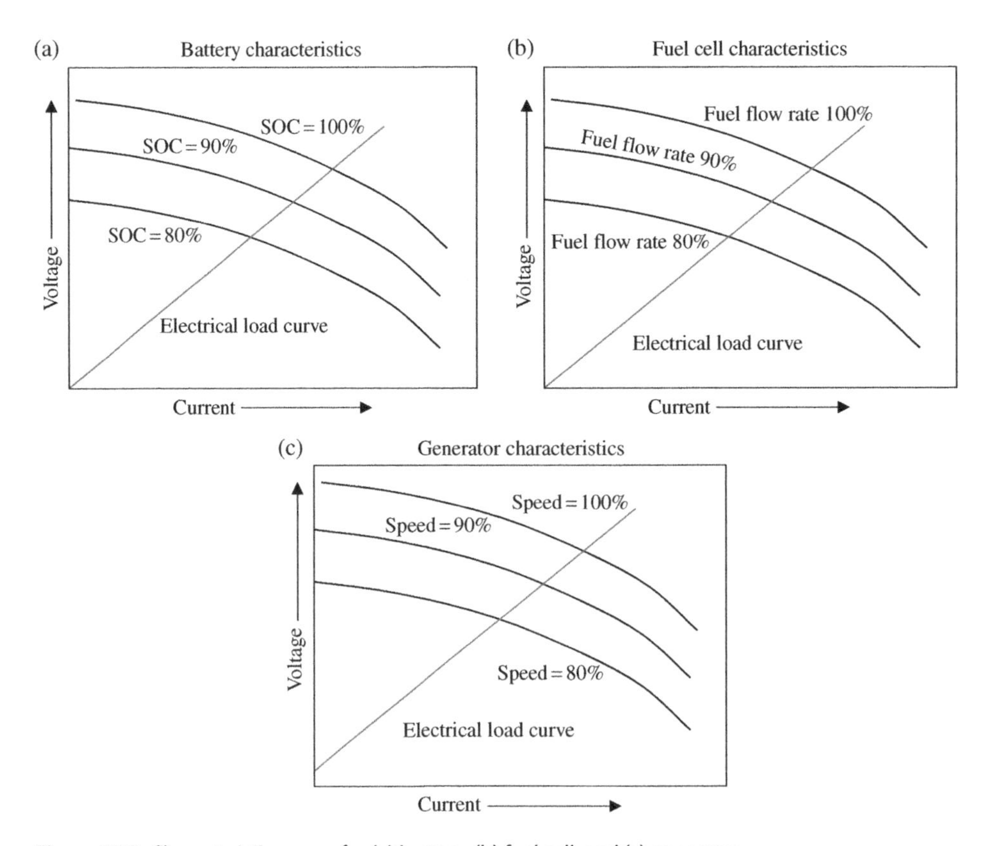

Figure 17.3 Characteristic curves for (a) battery, (b) fuel cell, and (c) generator.

these electrical sources, as depicted by the respective load curves. For example, an electrical resistance (which could be a lamp) characteristic could be the straight lines shown in the figures. In each of these graphs, the operating point is defined by the intersection point between the curves. Corresponding to each of these diagrams, a companion diagram showing the actual power input needed at the operating points might be drawn, which could be used to compute the energy used and would depend on the efficiency at each operating point. That (efficiency) might be one of the ingredients in defining the cost of power production at each point.

At this point we can define the above situation mathematically as

$$\text{Engine}: T_{e1} = f_{e1}(\omega_{e1}, \Gamma_{e1}) \tag{17.1}$$

which is the torque corresponding to a specific engine $e1$, where ω_{e1}, Γ_{e1} are speed and throttle positions.

As noted earlier, the engine will also have another equation involving its power (or energy consumption), or equivalently the efficiency, which can be written as

$$\eta_{e1} = \zeta_{e1}(\omega_{e1}, \Gamma_{e1}) \tag{17.2}$$

Similarly we can write an equation for another engine:

$$T_{e2} = f_{e2}(\omega_{e2}, \Gamma_{e2}) \tag{17.3}$$

which can have different characteristics from $e1$. Together, the engines can deliver the total load.

The second engine can have an efficiency defined by

$$\eta_{e2} = \zeta_{e2}(\omega_{e2}, \Gamma_{e2}) \tag{17.4}$$

and the load is assumed to have the characteristics

$$T_L = f_L(\omega_L, \Gamma_L) \tag{17.5}$$

where T_L is the torque, ω_L is the load speed, and Γ_L depends on the road profile and external environmental issues and driver demand. Γ_L also depends on any auxiliary load demand within the vehicle due to the direct action of the driver (e.g. increased electrical load, which can lead to a higher torque on the generator), or due to indirect control actions taken within the vehicle automatically, for example, some mechanical or electrical pump being activated. Even if the quantity T_L is dependent on only the driver input in terms of desire to accelerate, the functional nature of the torque, that is, f_L, will depend on the road characteristics and other factors outside the vehicle. Hence this function f_L will change from one road to another or from one drive cycle to another.

There may be a gear ratio between the load and the individual engine. In other words,

$$\omega_L / \omega_{e1} = g_1 \tag{17.6}$$

and

$$\omega_L / \omega_{e2} = g_2 \tag{17.7}$$

Assume, for the sake of discussion, that the gears are ideal and lossless. Under the above scenario, if the loads are driven by the two engines together, and assuming conservation of power, the power equation will be

$$T_L \ \omega_L = T_{e1} \ \omega_{e1} + T_{e2} \ \omega_{e2} \tag{17.8}$$

So, the total fuel consumption in this case will correspond to the total input power to the two engines taken together:

$$\begin{aligned} P_{in} &= \{\text{power of engine 1/efficiency of engine 1}\} \\ &\quad + \{\text{power of engine 2/efficiency of engine 2}\} \\ &= T_{e1} \ \omega_{e1} / \eta_{e1} + T_{e2} \ \omega_{e2} / \eta_{e2} \\ &= \{f_{e1}(\omega_L/g_1, \Gamma_{e1}) \times (\omega_L/g_1)\} / \{\zeta_{e1}(\omega_L/g_1, \Gamma_{e1})\} \\ &\quad + \{f_{e2}(\omega_L/g_2, \Gamma_{e2}) \times (\omega_L/g_2)\} / \{\zeta_{e2}(\omega_L/g_2, \Gamma_{e2})\} \end{aligned} \tag{17.9}$$

and of course

$$T_L = \{f_{e1}(\omega_L/g_1, \Gamma_{e1})/g_1)\} + \{f_{e2}(\omega_L/g_2, \Gamma_{e2})/g_2)\} \tag{17.10}$$

It can be immediately seen from this equation that fuel input is dependent only on the gear ratios and the throttle position. The strategy in this case will be: given the load torque T_L and load speed ω_L, choose gear ratios g_1, g_2, and throttle positions Γ_{e1}, Γ_{e2}, such that the fuel consumption rate corresponding to P_{in} is a minimum, while satisfying the torque and speed demand. It is of course understood that the various function expressions for torque and efficiency are available from the engine specifications or manufacturer's data. If this objective can be met, it will deliver the correct torque and speed to the load at a minimum fuel intake, which is very much desirable. In a real system these actions are coordinated through the fuel/throttle controller and transmission controller. If the transmission is not a continuously variable type, then the choice for g_1 and g_2 will be limited. In addition, the engine controller and other sensor information in the vehicle will be involved in the above decision process. But the point here is that, in principle, the best strategy can be realized to meet the load demand with minimum fuel consumption.

On top of the equations indicated above, it may be necessary to impose certain constraints. For example, we can impose the requirement that the engine speed, for either $e1$ or $e2$, should not exceed a certain limit, due to the engine's structural design, physical size, and other factors. This can be written in the form

$$\omega_{e1} \leq \omega_{threshold1} \tag{17.11}$$

$$\omega_{e2} \leq \omega_{threshold2} \tag{17.12}$$

and similarly for the torques,

$$T_{e1} \leq T_{threshold1} \tag{17.13}$$

$$T_{e2} \leq T_{threshold2} \tag{17.14}$$

In other words, we have to not only observe certain engine characteristics defined previously through an engine map or equivalent information, but also to make sure that we are within the maximum threshold limits of the devices, based on the manufacturer's specifications. Hence our optimization problem has to take into account these constraints, and then try to minimize the fuel economy, and so on.

We will present the methodology of optimization after further discussion on other multisource systems.

Continuing with the previous plots (torque vs. speed etc.) showing the characteristics of various sources, the idea can be extended as follows. Just as we have done for the engines, we can write similar equations for the other sources. For example, for the electrical source or storage elements such as the generator, battery, fuel cell, and so on, we can write the following for their electrical characteristics based on the previous diagrams showing their respective characteristics:

$$I_{G1} = f_{G1}\left(V_{G1}, \omega_{G1}\right) \tag{17.15}$$

(relationship showing current as a function of voltage and speed of the generator)

$$\eta_{G1} = \zeta_{G1}\left(V_{G1}, \omega_{G1}\right) \tag{17.16}$$

(relationship showing efficiency as a function of voltage and speed of the generator)

$$I_{B1} = f_{B1}\left(V_{B1}, \mathrm{SOC}_{B1}\right) \tag{17.17}$$

(relationship showing current as a function of voltage and SOC for the battery)

$$\eta_{B1} = \zeta_{G1}\left(V_{B1}, \mathrm{SOC}_{B1}\right) \tag{17.18}$$

(relationship showing efficiency as a function of voltage and SOC for the battery)

$$I_{FC1} = f_{FC1}\left(V_{FC1}, \mathrm{FR}_{FC1}\right) \tag{17.19}$$

(relationship showing current as a function of voltage and fuel rate for the fuel cell)

$$\eta_{FC1} = \zeta_{FC1}\left(V_{FC1}, \mathrm{FR}_{FC1}\right) \tag{17.20}$$

(relationship showing efficiency as a function of voltage and fuel rate for the fuel cell).

In the equations above, the subscripts *G*, *B*, and *FC* stand for generator, battery, and fuel cell, respectively. SOC indicates the state of charge, and FR indicates the fuel rate. The equations imply that the current in the generator is dependent on its speed and output voltage. Similarly the battery current is dependent on the battery voltage for a given SOC, and the fuel cell current is dependent on the fuel cell voltage for a given fuel rate. The efficiency of each of these devices depends on the voltage (or equivalently the current) and the other parameters such as speed, SOC, or FR, which are also included above. Again, we might have some constraints imposed on these devices:

$$I_{G1} \le I_{G1threshold} \tag{17.21}$$

$$I_{B1} \le I_{B1threshold} \tag{17.22}$$

$$I_{FC1} \leq I_{FC1threshold} \tag{17.23}$$

$$\mathrm{SOC}_{B1} \leq \mathrm{SOC}_{B1\_high\_threshold} \tag{17.24}$$

$$\mathrm{SOC}_{B1} \geq \mathrm{SOC}_{B1\_low\_threshold} \tag{17.25}$$

Here the upper bounds of current limits in the devices, and upper and lower bounds for the SOC of the battery, are shown as constraints.

For the sake of simplicity we can write

$$\mathrm{SOC}_{B1} = f_{SOC\_B1}\left(V_{B1}\right) \tag{17.26}$$

which indicates that the SOC is related to voltage. This is a simple equation for the sake of illustration only, since SOC equations can be quite complex and a method to evaluate them can include various nonlinearities, and also involve other variables such as temperature.

We can also assume that there is a total electrical load current and load voltage in the system, which can be written as

$$I_L = f_L\left(V_L\right) \tag{17.27}$$

$$I_L = I_{G1} + I_{B1} + I_{FC1} \tag{17.28}$$

These are just examples. In a similar manner, additional constraints, depending on the needs and specifications, could be included. Once all these constraints and equations are in place, which can involve a large number of equations, some formal method has to be introduced to deal with them. We will now briefly touch on some of these methods, since there are quite a number of methodologies available nowadays, as well as computer-based modeling, design, and development, which are often necessary and effective.

17.2.1 Methodologies for Optimization

One stage in the process of dealing with the above is the development of a "cost function" where a single function is developed, such that the equations and the constraints are all included so that if the energy consumption in a particular device increases, the cost function increases. Similarly, if a constraint is violated, the cost function increases. The word "cost" in this context implies that some "undesirable" thing happens somewhere in the system when the "cost" increases. It can mean more fuel intake indicating fuel cost, or it can be an overcurrent or overvoltage in a component, or exceeding the SOC in a battery, and so on, all ultimately somehow related to something undesirable and in some sense having the connotation of actual cost in terms of expense or money.

As an example, we can develop a cost function with the following philosophy.

We can directly include the number $1-\eta$ (i.e. inefficiency of the device) as a "cost." Similarly, we can say that if the current I in a device exceeds the threshold, then $|(|I| - |I_{threshold}|)|$, that is, absolute value of the amount exceeded, multiplied by some constant, can form part of a cost function. The constant multiplier can serve as a weighting function, that is, how important this threshold item is, and taken into account through this multiplier. Say, for example, that the "cost" of exceeding the SOC in a battery by 5% may be much higher than the "cost" of going below the lower threshold.

The physical meaning of this is that the "cost" or damage to the battery will be much higher if the upper SOC limit is exceeded compared to the lower one. Similar judgment has to be exercised in developing the individual ingredients of the cost function corresponding to other devices. As we indicated earlier, it is obvious that the word "cost" in cost function can be construed as system inefficiency (which is tantamount to actual cost of fuel), or damage done in terms of cost, or it could be in terms of poor performance of a device which can lead to undesirable ramifications and hence is costly in some sense. Thus, a proper development of the cost function is very important in terms of system optimization.

In a real system it may be necessary to implement the above in real time. Hence, depending on the complexity of the system, a significant amount of computational power might be necessary. It is therefore beneficial to develop a cost function that serves the needs of optimization and at the same time it is not too complex.

Considering the above system, with two engines (for the sake of generality), which need not necessarily be ICEs (one could be an ICE and the other an electric propulsion system), and with one battery, one fuel cell, and one generator, we could set up an overall cost function as

$$\begin{aligned} C = {} & P_{in} + K_{prop} \times |P_{prop\_actual} - P_{prop\_demand}| \\ & + \eta_{G1} \times (V_{G1} \times I_{G1}) + \eta_{B1} \times (V_{B1} \times I_{B1}) + \eta_{FC1} \times (V_{FC1} \times I_{FC1}) \\ & + K_{B\_high} \times |\mathrm{SOC}_{B1} - \mathrm{SOC}_{B1\_high\_threshold}| + K_{B\_low} \times |\mathrm{SOC}_{B1} - \mathrm{SOC}_{B1\_low\_threshold}| \\ & + K_{FC\_high} \times |I_{FC1} - I_{FC1\_threshold}| + K_{G1} \times |(|I_{G1}| - |I_{G1threshold}|)| \\ & + K_{B1} \times |(|I_{B1}| - |I_{B1threshold}|)| \end{aligned} \tag{17.29}$$

In this equation the power input to the engine, P_{in}, is dependent on the propulsion power demand and also any auxiliary power, such as the power needed to drive a generator or pumps. Other terms in C involve the efficiency of individual items such as the generator, battery, and any fuel cell. The remaining terms involve the exceeding of current and SOC thresholds, which is penalized. Since P_{in} is the total input fuel to the engine, obviously therefore, the higher its value, the higher the value of C. The term $K_{prop} \times |P_{prop\_actual} - P_{prop\_demand}|$ tells us that if the actual power output from the engine does not meet the demand for some reason, then we incorporate a penalty in C. Here this discrepancy in the actual power output from the demand can be an indicator of lack of performance by the engine, and not necessarily as inefficiency. The values of various constant multipliers can be chosen depending on their importance. For example, if engine performance is very important, we should use a relatively high value for the multiplier K_{prop}. Similarly, if an item is less important, it should be given a lower value.

Once again, the items above are all included for the sake of a general discussion and merely to give an idea of what a cost function can imply. A real system might include more of these items or fewer, depending on the situation. In other words, the cost function is an indication of the "goodness" (or rather the lack thereof) of the system.

Although the function C above is already quite involved, it should be noted that the full expression for each of the individual functions, based on all the previous equations, should be inserted to make it complete. Once that is done, the issue will be to minimize this cost function C, which will, in our illustrative example, already have taken into account the various constraints.

As noted above, the input from the driver is provided to the vehicular system through the accelerator pedal and brake pedal positions. The input basically serves as the desire of the driver to either increase or decrease speed, which is implemented through the accelerator and brake pedals. This is compared against the actual speed, and a control action is invoked. The situation on the road and the environment outside the vehicle are also taken into account through various sensors. The power management algorithm includes all of these and integrates them together to come up with a control action. This mechanism is shown in Figure 17.4.

The diagram is self-explanatory. The only input to the system comes through the accelerator and brake pedals. In addition, depending on certain actions by the driver or through the various controllers, different electrical or mechanical loads might be engaged. Examples could be the window lift motor, which is due to an action by the driver; or the activation of the air-conditioner, depending on the cabin temperature; or perhaps the water pump motor, depending on the engine temperature. All these inputs, together with the environmental and road conditions, dictate what the real fuel intake will be or the battery current will be. The outcome of the above situations (due to driver action etc.) will be transmitted through various sensor data, or certain indirectly computed information like the battery SOC will be generated, which will then be fed into the power management algorithm. The algorithm looks at the cost function (as in the equation for C above), created through the use of all this information. The algorithm then tries to optimize C, and its end result could be the updated choice of a particular gear ratio, throttle position, or battery current value. These updated values will be used to tell the various controllers to command a new status for the respective quantities. In consequence, information on torque and speed request to the engine, and other entities such as the electric propulsion system are also updated – which depend on the road and

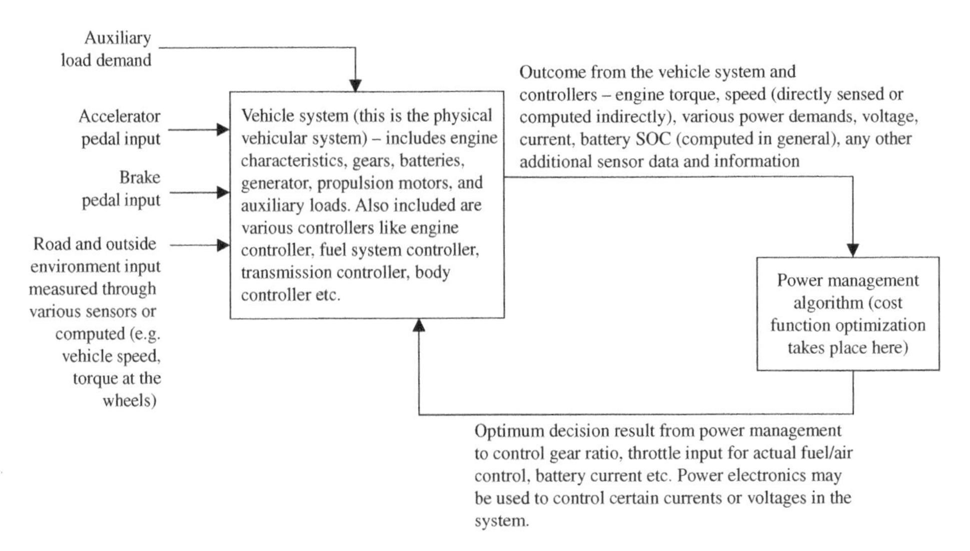

Figure 17.4 Schematic diagram of the overall power management control implementation interface for a vehicular system.

environmental conditions and the control actions. Note that the road and outside environment are outside the vehicle, so information about those has to be acquired through various sensors, or otherwise derived through computation.

It is apparent, therefore, that the ultimate outcome in Figure 17.4 on power management control is due to cost function optimization within the power management algorithm block. This optimization can be done as follows.

17.2.2 Cost Function Optimization

In our example above, pertaining to Equation (17.29), the only things that can be controlled are the throttle, gear ratios in the two engines, and current control in the alternator or battery through some power electronics. It is assumed that the propulsion and non-propulsion load powers have already been decided, based on the driver demand and external conditions. Hence the optimization is involved with trying to minimize the function C while observing various constraints. The idea here will be to achieve a global optimization (minimization in our case) so that cost C is minimized. In essence, it involves identifying various local minima in the system, and then trying to find the one which gives the global minimum. There are certain numerical techniques available to do this.

Assume the situation shown in Figure 17.5, with two engines in a vehicle. Let us say that the vehicle is at point A and intends to go to point B, where A and B are defined in time. Between these two points the path or drive cycle is defined by the load power demand of the vehicle $L(t)$, which depends on the speed between A and B, the actual road surface, and other external environmental inputs. Sometimes an alternative way to define a drive cycle is by means of vehicle speed as a function of time, along with information on the type of road (where "type" can imply gradability). But ultimately, the end result of the drive cycle will be mapped to some power demand $L(t)$ which will be used by the power management process to generate the optimization decision. The statement of the problem is as follows: given $L(t)$, we have to find the individual power allocations between the two engines, as discussed in more detail below.

In our example, where we have two engines, the objective is to select the torque and speed of the engines, and also the gear ratios as a function of time, in a manner such that

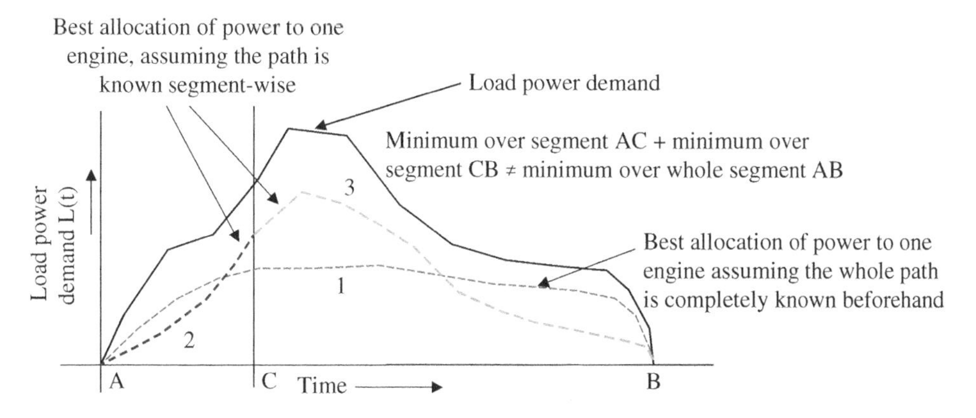

Figure 17.5 Load demand cycle segments as a function of time.

the "cost" C (which can be equivalent to fuel cost, but not necessarily so) is minimized. As long as the total power demand is met, we have the freedom to choose any combination of individual engine torques and speeds along with the corresponding gear ratios. In meeting the power demand we may be meeting the various constraints in terms of a certain quantity being less than or equal to some threshold. Thus it is very possible that in going from A to B, we have some unique (torque–speed) profile for the individual engines which gives the minimum C. Let us call this minimum C equal to C_{AB} for the sake of specificity, with the subscripts AB indicating the path. Then let us split the span AB into AC and CB. If we just focus on AC and come up with a sequence of engine (torque–speed) profiles to keep the cost C_{AC} to a minimum, and then do the same for the segment CB. Then the question is whether or not

$$C_{AC}|_{\min} + C_{CB}|_{\min} = C_{AB}|_{\min} \tag{17.30}$$

The answer is: not necessarily in general.

In Figure 17.5, the dashed line labeled (1) shows the path $C_{AB}|_{\min}$, assuming the whole path is known beforehand; the dashed line labeled (2) shows the path $C_{AC}|_{\min}$ without knowing anything after AC; and similarly the dashed line labeled (3) shows $C_{CB}|_{\min}$ within the segment CB without knowing anything prior to CB.

Before we introduce additional mathematical terminology, the reason why preemptive knowledge of the whole path can lead to a different result when minimizing, compared to segmented minimization, can be understood through a simple numerical example, which is very hypothetical but provides a suitable illustration.

Consider that in Figure 17.5 we have at our disposal two different engines running on two different fuels. Let us consider the following figures.

Assume that path AC needs 40 gallons (151 l) of fuel regardless of which engine, 1 or 2, is used. Also assume that path CB needs 80 gallons (303 l) of fuel regardless of which engine is used. In addition, assume that we have 42 gallons (159 l) of fuel in engine 1 and 80 gallons in engine 2 to begin with. Let us also assume some numerical values for the cost of fuels as indicated below. These numerical values will give a simple picture of the illustration for ease of understanding the idea.

Let us say that C_{AC1} is the cost of fuel over segment AC for engine 1 = \$2/gal for engine 1 and, as indicated above, we have 42 gallons of fuel for engine 1. Similarly, let us say that C_{AC2} is the cost of fuel over segment AC for engine 2 = \$5/gal for engine 2 and we have 80 gallons of fuel for engine 2.

These different dollar values for fuels can be due to the type of engines and other reasons, i.e. in principle the values can be different.

Assume that for the second segment CB, for whatever reason, the hypothetical engines need some special additives or whatever to make them go, perhaps due to some slope conditions etc., and that the cost of fuel increases as follows: C_{CB1} = \$5/gal for engine 1 and C_{CB2} = \$20/gal for engine 2. (These figures are chosen for ease of clearly showing the distinction.)

Say we know nothing about segment CB and we are asked to find the minimum cost over AC. Obviously we will (or, rather, will be tempted to) use only engine 1, since the fuel cost of engine 1 is cheaper over AC, and use 40 gallons at a cost of

$$\$2 \times 40 = \$80$$

Now, after we have completed our journey through AC, assume that we are told that we have to go to another segment, CB. Our engine 1 has already used 40 gallons and has only 2 gallons left, so we have no choice, and our cost for segment CB will be

$$\$5 \times 2 = \$10$$

since we have only 2 gallons of fuel for engine 1, and from engine 2 we have

$$\$20 \times 78 = \$1560$$

Hence the total cost of traversing the distance AC + CB, or AB, is

$$\$80 + 10 + 1560 = \$1650$$

Now, consider that we are told beforehand that we will have to traverse the whole path AB and that the price of fuel will vary between AC and CB for special reasons as indicated earlier. Then we may decide as follows.

For AC, we use engine 2 fully, and the cost is \$5 × 40 = \$200. Notice here that even though using engine 1 is cheaper, we decided to withhold its use. So, engine 2 will be left with 40 gallons at the end of AC. During segment CB we can use engine 1 fully for a cost of \$5 × 42 = \$210. For the remaining 38 gallons we use engine 2 for a cost of \$20 × 38 = \$760. Hence the total cost of the path AC + CB or AB will now be \$200 + 210 + 760 = \$1170.

This artificial numerical example illustrates the point that forward knowledge of the path, i.e. knowledge of the load and various cost items ahead during the whole path, allows us to make a decision in a more prudent way so that the absolute minimum is not necessarily equal to the individual minima over different segments added together.

If we know load $L(t)$ over the whole path for a period of time t, along with the different cost values of different entities such as engines, then with a system involving only two entities (engines in our case), this can be written mathematically as follows.

Min J = minimum of the integrated cost function over the individual segments t_1 and t_2 will be

$$\underbrace{\int_0^{t_1} \left\{C_1\left(p_1(t)\right) + C_2\left(L(t) - p_1(t)\right)\right\} dt}_{\min} + \underbrace{\int_{t_1}^{t_2} \left\{C_1\left(p_1(t)\right) + C_2\left(L(t)\right) - p_1(t)\right\} dt}_{\min}$$

which is not necessarily equal to

$$\underbrace{\int_0^{t} \left\{C_1\left(p_1(t)\right) + C_2\left(L(t) - p_1(t)\right)\right\} dt}_{\min} \tag{17.31}$$

where $t_1 + t_2 = t$, $p_1(t)$ is the power allocation for engine 1 as a function of time t, and the minimization is done subject to certain constraints. The constraints, in the case of entities like engines, can be in the form of expressions indicating that the torque and/or speed of the engine cannot exceed a given threshold. If the particular items related to constraints involve sources such as a battery or generator, then those thresholds could take the form of overcurrent, overvoltage, SOC, and so on. For the equation above, $C_1(p_1(t))$ is a cost function corresponding to entity 1 (which can be an engine or anything else) and similarly for entity 2. In this case it is assumed that the cost function C_1 for entity 1 depends on how many units $p_1(t)$ of some item (which can be energy or power, or other attributes as well) are delivered by this entity 1.

Once again it is important to emphasize that, although we gave a numerical example using figures to illustrate the point, the term "cost" can imply anything in the sense of penalty due to violation of some constraints like overcurrent and so on, and hence cost should not necessarily be construed as something only in the sense of fuel economy and such like. The minimization process of the cost function normally will penalize the cost function if the constraints are violated in some way. For example, in the case of an engine, exceeding the torque above its rating of 100% to a value of 110% might be assigned some numerical increase in cost through some penalty value, and if it is between 110 and 120%, the penalty value may be much higher. Similarly for the SOC of a battery, if it is overcharged the penalty might be quite a bit higher compared to if it is undercharged; this is due to the damage done to a battery during overcharging.

The point above can also be explained by introducing the term "incremental cost" which is the additional cost of the cost function C for producing one additional unit of power (in our case), where C is dependent on the operating power of the various entities (ICE, motor, etc.) involved in producing it. In addition, C is dependent on various other items or parameters which can be gear ratio, throttle position, or anything else. Hence, the final incremental cost will depend on both the operating power and the parameters. This can be expressed mathematically as follows, where i is an index which indicates the various controllable variables corresponding to different resources:

$$\begin{aligned}\text{Total incremental cost} &= \Delta C \\ &= \sum_{i=1,n} \left[\partial C / \partial(\text{parameter}_i)\right] \times \left[\text{change in the parameter}_i \text{ value}\right]\end{aligned} \tag{17.32}$$

To minimize ΔC at a particular moment, it is necessary to choose the best parameter set in the above equation to achieve this. The operating power itself can be considered a parameter, and hence the incremental cost ΔC will depend, in our example, on the allocation of power to the various engines or devices delivering it.

In general, the incremental cost will change with time, depending on various factors, including, but not limited to, the possibility of running out of fuel, battery SOC too low, and so on. If the incremental cost remained constant for each of the resources involved, independent of operating conditions, and regardless of which zone of operation they are in, and if resource quantity were unlimited, then the answer to the question posed earlier (i.e. whether minimization of cost function C over individual segments leads to the same result as minimization over the total amount of all the segments) would be affirmative. In other words, if we know the complete path AB beforehand, we can decide exactly what the individual power generation (or something else) should be throughout the journey, to make the total C a minimum at the end of the journey. This can be considered to be a preemptive minimization effort. The discussion tells us that knowing the whole path ahead beforehand allows us to choose the best combination of resources to make C a minimum at the end of the journey.

In a complex system, with multiple resources and constraints, the above issue of minimization of a complex function can be quite laborious. Numerical and mathematical methods like dynamic programming (DP) [2] can help in such situations (details are abundant in the literature [2]). It should be noted that DP depends on knowing beforehand the "cost" of crossing from section to section while traversing the path from the origin to the destination. There are recursive methods (e.g. backward and forward induction methods) [3] which can be used to find the most optimal "cost", which depends on the optimality condition developed by Bellman [4].

In a real system, however, often it is difficult to know the path (and the cost of each section on the way) ahead. Hence people have to resort to minimizing the cost based on what is known now, that is, instantaneous minimization in real time. In a real-time scenario, this can mean moving forward incrementally in little segments. In particular, for the equation related to ΔC above, as time progresses, we would see which particular parameter i, when adjusted, helps minimize ΔC of the cost function. Intuitively speaking, the algorithm might just take one parameter at a time and minimize it as much as it can, then take the next one, and so on, and then come back and iterate the process until there is no further benefit. But there is no guarantee that this process will yield the best or absolute minimum for the cost function, nor will it ensure that this method will help achieve the absolute minimum for the complete path, if we had known the path ahead beforehand. Some methods also tend to predict the path forward to come up with hopefully something better than knowing nothing at all about the path ahead. In addition, there are methods like artificial neural networks and fuzzy algorithms that have been used [5] to come up with more optimal solutions to a cost function minimization problem.

There are some practical details we can state, related to the previous paragraph: In a real-time scenario, we can create tables that can be stored in a microprocessor memory, based on the information of different resources as a function of operating point, which can be engine torque–speed and efficiency–speed maps at different throttle positions, or information about generator voltage as a function of speed and current drawn, or of a battery voltage vs. current at different SOC. At different operating points, the data can be extracted from the memory at nearby points to calculate $\partial C/\partial(\text{parameter } i) \times [\text{change in parameter } i]$, and thus find ΔC, the incremental cost of C at an operating point. By doing so we can find which resource has the minimum cost increment and choose the one to deliver the next unit of power or energy. If a single particular resource cannot provide it all, then an additional next best resource has to be brought in. Since the incremental cost changes with time, the priority list indicating which resource should be delivering how much power at a particular moment will also change. As discussed earlier, if the future needs of system loads can be predicted beforehand, then we can make decisions differently. Thus, even if the cost of one resource in terms of its immediate cost is higher than another at a particular moment, the costlier resource can still be engaged to deliver the load, notwithstanding the fact that it is not the best option at that particular moment. In this case, this apparently costlier decision is taken in anticipation that cost reduction due to "non-engagement" of this "presently higher" cost device at a future operating point will be more than offseting the loss encountered now.

It can be seen from the above that to obtain the optimal numerical solution to the cost function minimization problem, mathematical manipulations and computations are necessary. Normally this calls for the incorporation of processors with significant amounts of computational power, especially if calculations are to be done in real time. With the artificial neural network method it is possible to train a system through various drive scenarios and then use the trained network to come up with a real-time solution based on various inputs from different sensor signals. These may not be the best solutions in a rigorous mathematical sense, but could practically be very good with significant benefits. By properly training a neural network, we can derive a simple matrix of constant numbers, which are normally used to multiply different variables, so the computational demand is not too high once the network is trained. Similar reasoning holds for fuzzy algorithms too.

The above discussion ultimately relates to the power management block shown in Figure 17.4, where the goal is to deliver the desired load under different constraints while minimizing some cost, which can be a real cost in the sense of fuel economy, or a penalty factor corresponding to violation of some SOC or current thresholds, which are minimized.

17.3 Benefits of Energy Management

Based on the previous discussions, it is easily appreciated that energy management has multiple goals. It obviously involves the distribution of energy (implying instantaneous power as well) depending on load demand. It also implies protection of the system in case some threshold is exceeded, for example, voltage, current, or SOC. Further, since the resource is not unlimited, it involves the best allocation of resources, with the objective of minimizing fuel consumption while observing various constraints. Thus good energy management leads to better fuel economy and/or lower emissions, and it also leads to enhanced life of devices. For example, by controlling the SOC properly, it may be possible to better maintain the battery's health, thus requiring less replacement. To be more precise, power management has the goal of taking a holistic view of the system, not just from the point of view of fuel economy; or rather, it is to achieve a minimum life cycle cost of the system, from the point of view of operation, maintenance, and longevity, all taken together.

Note also that power and energy management is a post-design situation. It is an operational-level strategy after the system has already been built and deployed. Whereas the design optimization and component sizing are done before the vehicle is made, as discussed in Chapter 14 and 15. Knowing beforehand what kind of power management strategy and algorithm will be used can help one make a better choice of devices in terms of size of engine, battery, and so on, during the design phase, so that the same performance can be achieved by using smaller equipment and making best or optimum use of it. Thus it is important to coordinate the power management process during the design phase, which can help design a more compact system. Once a system has been designed, however, little further changes can be done, though it is still possible to improve the power management algorithm from time to time in an adaptive manner. This may be required because various parameters within the system might change (i.e. due to aging and other reasons) with time, and hence the algorithm should adaptively be able to change accordingly to take care of this.

References

1 Murphey, Y., Chen, Z., Kiliaris, L., and Masrur, M. (2010) Intelligent power management in a vehicular system with multiple power sources. *Journal of Power Sources*, 196 (2), 835–846.

2 Bertsekas, D. (2007) *Dynamic Programming and Optimal Control*, 3rd edn., Athena Scientific, Nashua, NH.
3 http://web.mit.edu/15.053/www/AMP-Chapter-11.pdf (accessed September 2015)
4 Bellman, R. (2003), *Dynamic Programming*, Dover Publications.
5 Murphey, Y. (2008) *Intelligent Vehicle Power Management: An Overview*, Studies in Computational Intelligence (SCI), vol. 132, Springer, Berlin, pp. 169–190.

Further Reading

Power and Torque – Essential Concepts: Torque is measured; Power is calculated, http://www.epi-eng.com/piston_engine_technology/power_and_torque.htm (accessed September 2015).

18

Commercialization and Standardization of HEV Technology and Future Transportation

18.1 What Is Commercialization and Why Is It Important for HEVs?

By commercialization we normally mean that a product is produced in volume quantities and whose technology has been developed to a reasonable degree of maturity such that systematic large-scale production is viable and sustainable. One important reason behind commercialization is that it allows mass production, hence the price can come down, and more people can afford the product. Without commercialization, a technology may be good yet remain unused. In connection with HEVs, the situation is no different. It involves various technologies and with the commercialization of HEVs all these technologies will benefit.

18.2 Advantages, Disadvantages, and Enablers of Commercialization

There are several benefits of commercialization. As noted above, it allows mass production of the product where the whole manufacturing process can be streamlined. This leads to a fall in price, and more people can then afford the product. As more people use it, its price can fall further – but only, of course, up to a certain point. In HEVs, various technologies are involved, such as electric motor technology, power electronics, control system technology, microcomputer technology, high power cables, and electromagnetic interference (EMI), to name but a few, which have to be addressed. Hence to commercialize HEVs, for mass production it is necessary for all the constituent ingredients and the corresponding technologies to be mature as well. Once a product is commercialized, additional incentives arise for the manufacturer and designer to improve the various technologies further and invest accordingly.

While discussing benefits, it is also instructive to see if there are any disadvantages to commercialization. In fact, there are. When something is commercialized, there is the potential that individual developers of a technology will try to keep it as a proprietary item and not share it. Thus outsiders may not have the chance to provide any worthwhile input to the technology which could have been of benefit. This situation is comparable to commercial software versus open-source software. In the latter,

Hybrid Electric Vehicles: Principles and Applications with Practical Perspectives,
Second Edition. Chris Mi and M. Abul Masrur.

everyone can provide input and thus enrich the software, even though it is not financially profitable for any particular company or entity. The same holds for HEV related technology. Thus commercialization can deprive the technical community and society of further enrichment of a technology, some of which might not even be conceivable until someone proposes it. Another disadvantage of commercialization is the fact that a particular manufacturer can make things in such a way that maintenance is very difficult without proprietary OEM parts, and no generic replacement can be used. While commercialization can lead to a fall in the price of a commodity, it can also lead to monopolization by a manufacturer, especially if the technology is proprietary and not shared, leading to a higher price. Another issue related to commercialization, especially in the case of HEVs, is that many smaller industries and businesses cannot participate in the whole process except as suppliers. Even though sometimes small businesses can participate as suppliers, big businesses cannot normally afford to use too many relatively smaller suppliers to get the same product. Another disadvantage of commercialization is that even if a very promising new technology suddenly becomes available, it may be very difficult to incorporate it in an existing product; this is due to the expensive change required in the manufacturing base of the existing components and OEMs may be very reluctant to take the risk of trying a new technology.

Thus we see that there are both advantages and disadvantages of commercialization, and we need to reckon with these.

Some of the enablers of commercialization can be investments put toward technology development. In addition, partnerships between various industries can lead to better technology development at a lower cost. One example of such a collaboration is that which was between GM, BMW, and (previously) Daimler-Chrysler, which formed a consortium to develop HEV technology and improve it. Once the job was accomplished, the partnership did not need to exist anymore, and the partners got on with their own products. In the case of HEVs, it can undoubtedly be said that the US Department of Energy plays a major role in terms of funding various projects jointly with industry. This has paved the way to the very commendable objective of developing a viable HEV product.

One important item in the process of commercialization of HEVs and EVs relates to government incentives and legislation. This applies globally and should be coordinated with the various countries that are involved in production and usage of such vehicles. Such government-level proactivity in different countries can lead to more interest among people to own such vehicles, which can then affect the production volume, thus leading to better commercialization at an affordable price. One such example is the lithium ion battery price drop in the past few years due to the large deployment of EVs in China. The price of lithium ion batteries was close to US $1000/KWh in 2011 and is under US $300/kWh in 2017.

18.3 Standardization and Commercialization

Commercialization of HEVs obviously calls for cooperation between various manufacturers, since there are multiple technologies involved. These technologies – batteries, power electronics, motors, charging mechanism of batteries in the case of plug-in hybrid vehicles – all need to conform to certain standards. These standards can concern safety, but also the need to conform to some commonality between what one technology can give, and what another can receive. Regardless of the manufacturer, it is very likely that power electronics are purchased from only a handful of suppliers, so battery and motor related design must match the basis of what is available from the power electronics, since one technology feeds another. Then there is the EMI issue, for which there are various standards. It is very important to observe these standards to make sure there is no problem with any microcontroller system due to EMI, and that it does not disrupt anything in a nearby vehicle. In connection with plug-in vehicle infrastructure, although there are battery charging stations in certain countries, such stations are still not as abundant as regular gas (i.e. gasoline or diesel refuelling) stations. It is very important that the charging interfaces between vehicles of different brands and the charging stations follow the recommendations of standards. Other issues of standardization pertain to maintenance of the vehicles. Just like their ICE counterparts, HEVs will need appropriate testing equipment, which need not necessarily be procured from the vehicle manufacturer. In general, it is likely that diagnostic instruments from the HEV manufacturer will be more costly than generic ones. Hence some standardization, such as OBD II, will be called for. From the customer's point of view perhaps the perfect standardization is one in which a component for the vehicle, including motors, batteries, and power electronics, could be replaced by generic counterparts readily available on the market. To make this a reality, the generic components must adhere to some minimum safety and other standards. In addition, there should be some kind of packaging standards (analogous, for example, to computer motherboards with various footprints when fitted to the computer case) – that is, in terms of how various items are fitted into the vehicle. It means that if a power electronics box from a generic source is to be installed in a vehicle, this can materialize only if the OEMs and the generic manufacturers follow some standard in terms of packaging footprints. All the above issues could also potentially cause difficulty in terms of warranty. A manufacturer might say that if a particular part is replaced within the vehicle using generic components, then the warranty will be void. All of the above are legitimate concerns, and need the participation of the manufacturer, consumer, various professional societies, organizations like the ISO, IEEE, SAE, and the government(s), in order to achieve successful standardization.

18.4 Commercialization Issues and Effects on Various Types of Vehicles

Commercialization may not have the same effect on all types of vehicles. That is, the effect might be different for a regular small passenger vehicle compared to heavy trucks, military vehicles, delivery trucks, refuse/garbage trucks, utility vehicles, construction vehicles and equipment, locomotives, ships, and so on. Commercialization ultimately has to do with profit and money. Thus it depends on the cost per unit and the volume produced. The impact of commercialization and its implementation need to take all of the above into account.

It appears that the effect of commercialization at this time is more important for smaller vehicles, since a larger number of the population are affected in that case, in terms of vehicle cost per unit. Heavier vehicles, especially those with less production volume, may sometimes be developed based on need (or customized) and use very special designs developed by a few niche industries. Some of these industries might need flexible manufacturing processes in order to meet low volume production which can be quickly adapted to produce items with different specifications. Military vehicles in particular may be less affected by commercialization in the sense that the military can afford to select multiple suppliers for the same thing. However, they can benefit from commercialization as well, since it is always more cost effective to use commercial-off-the-shelf (COTS) items to help ease maintenance and provide overall life cycle benefits. Locomotives, ships, and similar vehicles, that are very expensive items and produced in relatively low volumes, and sometimes custom ordered and built according to requirements, can benefit from the commercialization of various constituent components used to build the vehicle, rather than commercialization of the finished product, which does not happen in a true sense for such high-cost, low-volume items.

18.5 Commercialization of HEVs for Trucks and Off-Road Applications

As indicated earlier, HEV and EV for trucks and other off-road, and military applications involve relatively smaller volume. However, these applications can save quite a large amount of fuel if HEV and EV technologies are included for such purposes. According to the Hybrid Commercial Vehicles Report, major truck manufacturers and system developers in the USA are participants in the usage of the technology. DOE is also supporting various efforts related to this. The above report indicates that the hybrid truck industry started with only 200 vehicles in 2006 and was expected to grow to about 5000 in 2010. This seems to be quite realistic. According to a report from Pike Research, the worldwide market for medium and heavy-duty HEV trucks and buses should grow from around 9000 vehicles sold in 2010 to more than 100,000 vehicles in 2015. Regardless of these specific numbers, it appears that the overall interest among various truck industries is quite positive. That can lead to commercialization of the various constituent systems, subsystems, and components. In terms of motors and power electronics, if such needed components for trucks can be manufactured by the same suppliers who are already for catering the regular automotive industry, perhaps the risk can be reduced, thus leading to a quicker and easier path towards commercialization in the truck industry.

In addition to trucks, various off-road vehicles and construction equipment using HEV technology, could possibly be better commercialized if those are coordinated with the system developers for regular automotive industry. Quite a few construction and off-road vehicle manufacturers have actually deployed such equipment, e.g. in mining vehicles and excavators. Even though the latter (excavator) does not fall under the category of a vehicle proper, the technology used is similar to HEV. Military vehicles also fall into similar categories. However, the ruggedization issue is unique to military and must be addressed, for both adoption of the technology and also commercialization. Unlike regular commercial applications, military applications involve very difficult operational conditions and many times involve life-and-death issues. Hence highly reliable components and systems must be used. Also, the issue of low volume applications can be a barrier towards commercialization of such systems. But as discussed earlier, utilization of HEV and EV technology can lead to significant fuel savings in military vehicles and can have major consequences in terms of indirect cost reduction through lower logistic support costs.

18.6 Commercialization and Future of HEVs and Transportation

It is apparent from the previous discussion that commercialization in general can lead to the delivery of better value for customers who want HEVs. With commercialization and standardization, prices will most likely be cheaper, since commercialization can lead to better competition and provide more options to users. This same applies to suppliers of various components. If commercialization of a particular vital item such as a battery becomes wider, that will lead to a fall in the cost of the battery. As is well known, the cost, size, and weight of a battery are now real challenges that stand in the way of the overall popularity of HEVs. The cost issue is one item, but the size and weight will also depend on the technology. With more competition in the industry, various manufacturers might invest in developing new and better technologies, which in essence can be the deciding factor in the success or failure of an item, ultimately leading to people wanting to own an HEV rather than a regular ICE vehicle. Currently the fundamental issues on batteries, relating to their chemistry, remain big challenges and cannot claim to have been fully addressed. In the future, if battery technology can be really made to improve through some quantum jump, then that can lead to the next stage of development of pure electric vehicles (EVs). In addition, serious research is presently being done on fuel cells. Fuel cell technology exists in a relatively mature form for very high-power fuel cells, but for portable vehicular applications it still has some more way to go. If the fuel used in the fuel cell can directly convert the high-specific-energy content fuel to hydrogen (rather than carrying hydrogen in cylinders) then this will help significantly toward increasing the range of a pure EV. In an HEV, having both ICE and electric propulsion leads to a complexity of manufacturing, control, and maintenance. In a pure EV these can be significantly simplified. Hence a future goal of commercialization should have the objective of achieving a pure EV. This applies equally even to high-power construction equipment. In some of these vehicles, it is possible to have ultracapacitor-type high-specific-power devices to supplement the fuel cell. Combining a fuel cell and an ultracapacitor can lead to both longer range and overall longevity of dispensable items, that is, an ultracapacitor instead of a battery. Thus it appears that the HEV is an intermediate step between the ICE and pure EV, and will depend very much on the technology of the fuel cell, ultracapacitor, and battery. Other technologies such as power electronics and motors will benefit from commercialization in terms of cost reduction since they are already very efficient devices. But thermal management still remains a challenge for power electronics, and high-temperature devices based on silicon carbide, and perhaps something else in the future can be important enablers toward future EV transportation.

Further Reading

Britton, D. (2010) Barriers to Commercialization Still Exist for Hybrid Truck Industry, http://www.truckinginfo.com/article/story/2010/01/barriers-to-commercialization-still-exist-for-hybrid-truck-industry.aspx (accessed October 2015).

Browning, L. and Unnasch, S. (2001) Hybrid electric vehicle commercialization issues. The 16th Annual Battery Conference on Applications and Advances, August, Long Beach, CA, p. 45.

Kramer, F. (2009) Commercializing Plug-In Hybrids, The California Cars Initiative, http://www.calcars.org/calcars-photos.pdf (accessed Oct 4, 2015).
Van Amburg, B. (2007) Heavy Hybrid Vehicle Commercialization Progress and Directions: Clean Transportation Solutions, http://files.harc.edu/Sites/TERC/About/Events/ETAC200705/HybridCommercialization.pdf (accessed Oct 4, 2015).
Chhaya, S (2008) Commercializing Medium Duty Plug-in Hybrid Electric Vehicles, Electric Power Research Institute, Palo Alto CA.
Alam, S. (2013) Key Barriers to the Profitable Commercialization of Plug-in Hybrid and Electric Vehicles, Advances in Automobile Engineering.
Profozich, R. (1980) Electric Utility Regulatory Aspects of Electric Vehicle Commercialization.
International Energy Agency (2015) Hybrid and Electric Vehicles The Electric Drive Delivers, www.ieahev.org (accessed October 2015).

19

A Holistic Perspective on Vehicle Electrification

19.1 Vehicle Electrification – What Does it Involve?

As we come to the end of this book, it is reasonable to ask about the whole purpose of developing hybrid vehicles and electric vehicles, or in a wider sense about electrification of vehicles. We need to take an unbiased view in considering these questions and be realistic and take a scientific stand, rather than following a particular trend, which can be very short term. We should also look at the complete picture in this connection. This involves whether a vehicle should be fully electrified in terms of its propulsion – leading towards a pure EV – or should the vehicle be partially electrified, leading towards an HEV, or mild HEV, or should we just electrify some non-propulsion related components or systems within the vehicle? As is obvious from the previous chapters of the book, these issues pertain to drive cycles and applications involved. They also relate to how many vehicles we are talking about and whether fuel economy and pollution control are the foremost issues, or are there other things that we should consider? In addition, we need to look at a complete global system point of view rather than at the vehicle only: if we solve one problem of fuel economy or pollution, then are we creating some problems elsewhere? The point is that developing of HEVs, EVs etc. involves not just the automotive industry, but also other industries related to power electronics, electrical machines, and batteries to name a few, which are in turn related to other industries involved in developing constituent items, such as mining industries, various chemical industries and so on. From a socio-political viewpoint, it can have other consequences. For example, in developing electrified vehicles, are we creating opportunities (e.g. jobs) in one area and then creating problems in some other areas? These are not easy questions to answer, but we have to face them sooner or later. If a very myopic viewpoint is adopted on the above, then in the long run they will lead to unforeseen, and in some cases unpleasant, consequences. Hence this chapter will try to discuss these issues to some extent.

19.2 To What Extent Should Vehicles Be Electrified?

The previous section has set the tone on issues related to vehicle electrification. Let us now take to look at the extent to which vehicles should be electrified.

Hybrid Electric Vehicles: Principles and Applications with Practical Perspectives,
Second Edition. Chris Mi and M. Abul Masrur.

In a legacy vehicle, there are various items, such as ICE and transmission, pertaining to propulsion. But to support these systems there are other items such as various pumps and fans, in addition to items that are not directly connected to propulsion. For example, an air conditioner is a comfort item, but its inclusion demands power and hence affects the power demand from the engine and in turn affects the propulsion system, though in an indirect manner. Also, depending on the type of vehicle – regular passenger, heavy trucks, military vehicles, construction equipment, garbage trucks, all-terrain vehicles, or industry utility vehicles – we have to decide whether it should be electrified or not, and if so to what extent. Here we also need to think about the market volume and the cost involved in manufacturing and also the cost to the end customer.

Let us first take the case of regular commercial passenger vehicles – the regular cars we drive. Since these are high volume items with a significant amount of city driving cycles, these apparently have higher impact on the overall fuel economy, and environmental pollution control. So, for full electrification or full hybrid considerations, these vehicles seem to be most important potential candidates, and that is the direction the industry is currently taking.

Next comes the heavy trucks. These are generally the diesel vehicles that run long distances. An example of electrification of such vehicles is the effort by DOE on Super Truck. According to DOE (http://energy.gov/eere/articles/supertruck-making-leaps-fuel-efficiency) this Super Truck can run at about 10.7 mpg as opposed to the equivalent legacy counterpart at about 5.8 mpg. This could give a huge benefit in terms of fuel economy and pollution reduction. It should be noted, however, that the Super Truck involves not just hybridization of the propulsion, but also taking a holistic view of many other details within the vehicle in terms of cooling fans, aerodynamic design, radiator grill adjustment based on speed and allowing better cooling, plus many other lesser details. Taken together this brings the benefit noted earlier. So it is reasonably safe to say that hybridization of such trucks may be a justifiable thing to do. However, full electrification of such trucks may not be justified or even viable, due to the very long mileage range that is needed, and also because of the very high life cycle mileage involved in such vehicles. Having a battery system and charging infrastructure for that could be a challenge as well.

Using electrification for military vehicles is under serious consideration in various countries. Here again, due to the type of operation drive cycle, full electrification does not seem reasonable. However, full hybridization, or hybridization in part, may be justified.

In the area of construction vehicles some electrification is already being done. In mining vehicles, electrification of the propulsion has already been done as indicated in earlier chapters. In such applications, the vehicle has been essentially transformed with an electric power train – the ICE generates mechanical power, which is then converted into electrical power by using an electric generator, which in turn drives the electric motors to run the wheels. Without a battery source, it cannot be termed a hybrid vehicle in a strict sense; rather it is an electrified vehicle. In the area of excavators Komatsu has demonstrated hybridization of its excavator system, where the main motion is actuated by using an electric motor. Regenerative energy capture is done by using an ultracapacitor system. Construction vehicles like the excavator have different kinds of application cycles and cannot be directly compared or classified alongside regular passenger vehicle drive cycles. However, such electrification has been shown to have improved fuel economy significantly, as noted in earlier chapters. Due to the high power involved in excavator and other construction vehicles, sometimes a partially hydraulic system is used; in these applications full electrification or full hybrid method (without hydraulic) may be difficult within the limited packaging size and weight. A similar reasoning also holds in the case of refuse collection vehicles. In the case of courier service type of applications, such as postal mail, the drive cycle is short repetitive cycles, where regenerative energy capture can be important, but it may be more prudent to use partially hybrid system rather than full hybrid. For industrial utility vehicles with indoor applications, there may not be much choice other than using fully electric vehicles due to pollution considerations. For indoor applications, however, since electric charging is easier to obtain compared to outdoor applications, it makes sense to use full electric vehicles anyway, given that indoor application type of vehicles are not particularly large vehicles.

Thus it is apparent that the degree of electrification of vehicles will depend on the drive cycle and other considerations, as indicated above. But it will also depend on the volume quantity involved. For low volume applications, the effort and cost needed to electrify may not justify such transition.

19.3 What Other Industries Are Involved or Affected in Vehicle Electrification?

Both HEVs and EVs obviously entail interdisciplinary fields, involving engineering disciplines of electrical, mechanical, control, computer and microelectronics, and chemical in particular. Hence this situation calls for cooperation between various manufacturers, since there are multiple technologies involved. As already noted in the previous chapters, these technologies, such as batteries, power electronics, motors, charging mechanism of batteries in the case of plug-in hybrid vehicles, all need to conform to certain standards. These standards can concern the perspective of safety, but also the perspective of conforming to some commonality between what one technology can give, and what another can receive. For example, power electronics items are purchased from only a few selected suppliers; interfacing technologies like battery and motor related design must match properly with power electronics, and similarly for other interfaces. The EMI issue, which is very important in HEV and EV, relates not only to proper functioning of the microcontroller system, but they can eventually affect the power electronics, and even low power electrical system and control system, and can also disrupt the functionality of a nearby vehicle. So, proper standards and mitigation of these issues involve the appropriate material industry and also the packaging process. In connection with plug-in electric vehicle infrastructure, nowadays a very important topic relates to wireless power transfer. That industry may well also play a significant role in the future in connection with EV and HEV. Some additional industries, though they may be a subset of the ones indicated above, are fuel cell (including the fuel system in the complete fuel cell system package), flywheel, and ultracapacitor related industries. The technology in these is not yet fully mature in terms of specific energy and specific weight. However, substantial progress has been seen, and it is expected that viable technology will be available in the near future for all of these, which may significantly impact vehicle electrification. On the issue of maintenance of the vehicles, industries that manufacture appropriate testing equipment will also be involved. As noted in the earlier chapter, in general it is likely that diagnostic instruments from the HEV or EV manufacturers will be more costly than generic ones. Hence some standardization, similar to OBD II, will be called for. It is therefore seen from the above that quite a few industries are involved in vehicle electrification. Finally, if deregulation of electricity happens in the utility industry, then that will bring other players such as various utility industries into the scenario, along with proper standardization related to vehicle to grid interface.

19.4 A More Complete Picture Towards Vehicle Electrification

In order to get a more unbiased and complete picture related to electrified vehicles, whether fully electric or hybrid, it is important to look at both the positive and negative sides of these vehicles. This involves looking into the full life cycle of the vehicle from several viewpoints – cost (from both manufacturing and end user perspectives), fuel economy, pollution and environmental aspects – not just from the immediate perspective of the vehicle itself, but also from the broader viewpoint including the manufacturing process of such vehicles, along with any indirect ramifications, both positive and negative.

On such issues, some literature review and discussion can be helpful. Nordelöf et al. (see the Further Reading section) and others (in particular a study by The Norwegian University of Science and Technology) have indicated that for plug-in hybrid and pure electric vehicles that require charging from external entities, such as the utility industry, the environmental benefit will depend on the type of fuel that the utility industry uses. For example, if the utility industry uses coal or any other fuel with high carbon emissions, then the benefit of using the electric vehicle may in fact be worse than its conventional counterpart. These studies suggest that the benefit from vehicle electrification can materialize better in terms of pollution and environmental effects if the energy used to charge the vehicles comes from renewable energy or other forms of energy with low carbon footprints. Since electric and hybrid vehicles are more fuel efficient, from that perspective it leads to less pollution overall, but if we take into account the method of charging the vehicle then the above scenario in terms of a complete perspective may not necessarily be beneficial.

Let us now look at the issue of manufacturing the electrified vehicle, fully electric or hybrid. Electrified vehicles need several items such as battery, power electronics, electric motor, and ultracapacitor (if one is used). Depending on the type of vehicle, some reduction in certain items may take place in the electrified vehicle. A (not necessarily complete) summary of these is given in Table 19.1.

The table gives a basic idea of certain items that can be eliminated in an electrified vehicle, and some items that will be additions, plus some items that can be derated. In electrified vehicles, manufacturing the additional items (e.g. battery, and permanent magnets in motors) can be sometime very intensive in terms of environmental sensitivity. It has been suggested in various papers (e.g. Nordelöf et al.) that manufacturing of lithium-based batteries is very environmentally intensive. A simplified view of the

Table 19.1 Comparison of component and manufacturing needs in electrified vehicles.

Vehicle type ↓ / Items used →	ICE	Propulsion battery	Propulsion motor	Power electronics	Ultracapacitor	Auxiliary items
Conventional	Full power rated	None	None	None for propulsion	None	Mechanical A/C, water pump, fuel pump, oil pump, brakes, steering
Fully electric	None	Full power rated	Full power rated	Full power rated	Reduced power rated possible, if at all used.	Mechanical auxiliaries not needed – electrified A/C
Plug-in hybrid (parallel)	Reduced power rated	Reduced power rated	Reduced power rated	Reduced power rated possible	Reduced power rated possible, if at all used.	Mechanical auxiliaries can be of reduced size or eliminated, electrified A/C.
Full hybrid (parallel)	Reduced power rated	Reduced power rated	Reduced power rated	Reduced power rated possible	Reduced power rated possible, if at all used.	Mechanical auxiliaries can be of reduced size or eliminated, electrified A/C.
Series HEV	Full power rated (if no propulsion battery is used)	Reduced power rated can be used	Full power rated	Full power rated	Reduced power rated possible, if at all used.	Mechanical auxiliaries can be of reduced size or eliminated, electrified A/C.

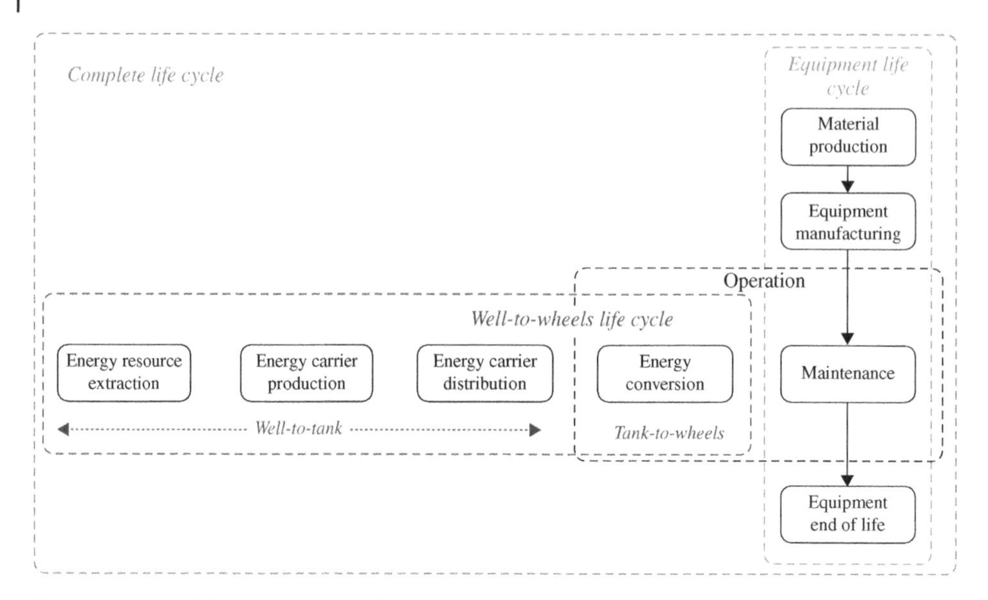

Figure 19.1 Simplified view of the well-to-wheels and equipment flows (a more detailed view would include, for example, recycling options). *Source:* Courtesy Anders Nordelöf et al.

manufacturing process and well-to-wheels life cycle has been provided by Nordelöf et al. indicated in their paper and shown in Figure 19.1.

Given the above scenario, it can be concluded that using electrified vehicles will lead to better fuel economy from the user perspective. In the process, regenerative braking will lead to longer life for the brakes. Maintenance can be done at longer intervals for the ICE in HEVs. All these are user-related issues. While certain items are added in an EV or HEV, certain items such as various pumps or fans could possibly be eliminated in some cases, or electrified, and perhaps derated in certain situations.

During the total life cycle of electrified vehicles, it is important to introduce recycling of various items, both at the end of the life cycle, and also during the running life when maintenance or replacement of components is done. Obviously 100% recycling can never be realized, and hence there will always be some items that will end up as environmental pollution, and also some of the energy consumption process during manufacturing that cannot be reversed.

The previously indicated Norwegian study (also the paper by Hawkins et al.) suggests that the production phase of electric vehicles is highly environmentally intensive, compared to the manufacturing process involved in conventional gasoline and diesel vehicles. It also suggests that the global warming potential from electric vehicle production is about double that of conventional vehicles. Furthermore, manufacture of batteries and electric motors (including those using permanent magnet and rare earth materials) can use significant amounts of toxic minerals such as nickel, copper, and aluminum. All of these can also lead to acidification of the environment, which can be much more than conventional automotive production.

Another issue related to electrified vehicles is that their benefit can be quite a bit reduced, depending on the life mileage being considered and also whether it is compared against conventional gasoline or diesel vehicles. For example, if a vehicle has a lifetime of 200,000 km then benefits in terms of global warming can be about 27–29% for electric vehicles compared to gasoline based vehicles, and 17–20% relative to diesel based vehicles. If the vehicle is assumed to have a lifetime of only 100,000 km, the benefit will be reduced for electric vehicles to only 9–14% when compared to gasoline vehicles and almost no benefit when compared to a diesel vehicle. This is based on the various references at the end of this chapter. It has been indicated by a top Daimler official that a battery electric vehicle, with the utility industry power generation mix currently used in Europe, gives a benefit of about 10% or so, compared to the conventional equivalent diesel vehicle. Apparently this benefit is in terms of overall environmental carbon footprint, regardless of the user fuel economy benefits.

The advantages and disadvantages of EV and HEV can be summarized as follows:

Advantages – (a) no gasoline for EV and lower amounts of gasoline in HEV needed, (b) no or lower carbon footprint, (c) with more popularity and higher volume production, cost can come down significantly, (d) maintenance needs are lower with no oil changes needed for EV, and less frequent changes for HEV, (e) reduced audible noise.

Disadvantages – (a) Charging infrastructure needed, (b) electricity is not free and if the demand for electricity becomes very high, the utility industry may have to increase their rates in order to to meet the higher demand, (c) low mileage range for pure EV, (d) long charging times, assuming that charging facilities are available, (e) batteries will need replacement about 3–10 years and this is very costly, (f) maintenance facilities may not be available everywhere and most likely dealerships will be the place for that, at least initially, (g) generic components may not be usable, unless proper standardization is in place, (h) low volume will lead to higher cost in the foreseeable future, thus it will be several years before the end user can recover the extra cost of such vehicles, thus discouraging users from buying such vehicles.

Overall, it seems that not every detail and benefit, or lack thereof, of electric and hybrid vehicles, will be very clear from either the user perspective or in terms of environmental benefit, that is from a complete holistic perspective, until probably close to a couple of decades have passed and the volume usage has increased substantially and become comparable to conventional vehicles. This is true of any technology and not just vehicles, but by taking a careful look and an unbiased perspective from both user and environment viewpoints, overall industry and the general population will benefit and be able to address various issues more realistically.

19.5 The Ultimate Issue: To Electrify Vehicles or Not?

Based on the discussion in the previous sections we can consider the ultimate issue of whether to work towards electrification of the vehicle or not. As we saw, the answer is not straightforward and there are two sides to the story. What items should we consider in answering this? From the user perspective it is the fuel economy and life cycle cost, including maintenance and disposal. From the environmental perspective it is the overall pollution and carbon footprint and similar items, considering the manufacturing process involved in these electrified vehicles, including the various components related to such vehicles. In addition, we need to consider the overall volume of vehicle usage. Similar considerations are also needed for vehicles used for off-highway applications, such as construction vehicles, military applications, ships, and locomotives.

It is true that electrified vehicles will be more economic in terms of energy efficiency, primarily because the ICE is fundamentally limited in efficiency and the efficiency percentage figure is rather low, in the order of middle twenties for gasoline vehicles and perhaps middle forties in diesel vehicles. On the other hand, power electronics and electrical machine efficiencies can be in the middle to high nineties, certainly the high eighties. The battery efficiencies can also be in middle to high eighties. With all the electrical components taken together, the overall system efficiency can still be somewhere in the seventies in percentage for electrified vehicles. This is significant. So, from a user perspective there is definitely a benefit in fuel economy if electrified vehicles are used. The additional cost of electrified vehicles may take several years to recover. Economics can sometimes – maybe in most cases – become the deciding factor in usage of these vehicles. Another factor is mileage range, which, for fully electric vehicles, is still very limited and stands in the way of popularity. For HEVs which are standalone, this issue or range is somewhat mitigated, but then the system becomes more complex. In terms of maintenance, it is also true that electrified vehicles are quite beneficial compared to their conventional counterpart.

We have discussed environmental issues pertaining to EV and HEVs in the earlier sections. This issue is quite complex, due to the manufacturing situation. For PHEVs and EVs, the benefit to the environment can depend quite a bit on the type of fuel used by the utility industry that provides the energy for vehicle charging. For a completely standalone HEV, there is enhanced fuel economy and as soon as there is fuel economy, there is benefit to the environment from that perspective alone. But if we take into account the manufacturing process, then some of the benefits of a pure standalone HEV may be diminished to some extent.

To answer the final question, therefore, whether the world should go after vehicle electrification or not, is not really straightforward. There have been many studies on this, and some are noted in the reference section at the end of this chapter. But again, these studies often depend on various assumptions, and some of the assumptions may be correct, and others may not. In addition to automotive applications, there are other applications like ships and locomotives, where electrification is likely to be more beneficial. In locomotives, electrification has been in place for many years, and experience is positive. For ships, although it has not taken place on a large scale yet, it appears that there are likely to be benefits. Ships and locomotives have rather predictable drive cycles and hence the design of the vehicles could be quite optimized at the outset. The same comments apply to construction and similar off-road vehicles. In fact, nowadays these construction vehicles are being electrified to some extent. These vehicles, like ships and locomotives, fall into the category of power train electrification without storage (i.e. without having batteries) in many cases. For construction vehicles there are both – electrified power trains with and without storages – and some benefit in terms of fuel economy has been achieved in both cases. But information on the additional manufacturing burden has not been thoroughly studied, in terms of their ramifications on the environment. This needs to be undertaken.

It is the opinion of the authors of this book, that perhaps the best way to answer the question posed earlier is to introduce electrification in vehicles in stages, and that is what is taking place now. Popular acceptance will automatically dictate the future, guided by economics. The customer may not necessarily be guided by environmental considerations, rather by economics, in the long run. A pure electric vehicles volume market may not be there unless there is adequate incentive, primarily due to the fact that the range of these vehicles is rather low. For standalone or PHEV, range may not be an issue, but the total life cycle cost definitely will be. Another thing that should be done is that various professional organizations should initiate environment data collection in a very organized and scientific manner and then use that to advise manufacturers accordingly. Vehicle electrification decision-making should not be an emotional issue, rather be guided by thorough scientific reasoning and economic considerations, and then an optimum point will automatically be reached where the technical community will know or rather get a reasonable idea about the best mix of conventional vehicle, EV, and HEV which should be manufactured. For now it seems that from the user perspective HEV is a good interim solution, provided cost can be reduced. Pure EV may still have some way to go, from both cost and environmental perspectives, due to the manufacturing issues.

Further Reading

A. Nordelöf, M. Messagie, A. Tillman, M. Söderman, and J. Mierlo, "Environmental impacts of hybrid, plug-in hybrid, and battery electric vehicles – what can we learn from life cycle assessment?", Int J Life Cycle Assess (2014) 19:1866–1890.

P. Egedea, T. Dettmera, C. Herrmanna, and S. Karab, "Life Cycle Assessment of Electric Vehicles – A Framework to Consider Influencing Factors", The 22nd CIRP conference on Life Cycle Engineering, Procedia CIRP 29 (2015) 233–238.

T. Hawkins, B. Singh, G. Majeau-Bettez, and A. Strømman, "Comparative Environmental Life Cycle Assessment of Conventional and Electric Vehicles", Journal of Industrial *Ecology* vol. 17, no. 1, 2012, pp. 53–64.

M. Messagie, F. Boureima, T. Coosemans, C. Macharis, and J. Mierlo, "A Range-Based Vehicle Life Cycle Assessment Incorporating Variability in the Environmental Assessment of Different Vehicle Technologies and Fuels", Energies 2014, 7, 1467–1482.

P. Lebeau, C. Macharis, J. Mierlo, and K. Lebeau, "Electrifying light commercial vehicles for city logistics? A total cost of ownership analysis", EJTIR 15(4), 2015, pp. 551–569. http://www.bbc.com/news/business-19830232 (last visited – Nov. 2015).

http://onlinelibrary.wiley.com/doi/10.1111/j.1530-9290.2012.00532.x/abstract (last visited – Nov. 2015).

http://www.conserve-energy-future.com/advantages-and-disadvantages-of-electric-cars.php (last visited – Nov. 2015).

http://greenliving.lovetoknow.com/Hybrid_Vehicles_Negative_Environmental_Impact (last visited – Nov. 2015).

http://journalistsresource.org/studies/environment/transportation/comparative-environmental-life-cycle-assessment-conventional-electric-vehicles (last visited – Nov. 2015).

http://environment.yale.edu/yer/article/electrified-vehicles-a-solid-choice (last visited – Nov. 2015).

http://www.hybridcars.com/why-an-electric-car-is-greener-cradle-to-grave/(last visited – Nov. 2015).

http://environment.yale.edu/yer/article/electrified-vehicles-a-solid-choice (last visited – Nov. 2015).

http://www.worldwatch.org/node/579 (last visited – Nov. 2015).
http://nlpc.org/stories/2011/06/21/more-bad-news-chevy-volt (last visited – Nov. 2015).
http://www.dailymail.co.uk/sciencetech/article-2876552/Your-electric-car-not-green-Researchers-say-electricity-generated-coal-plants-make-air-DIRTIER.html (last visited – Nov. 2015).

Index

C

d

S

W

X

Z

Printed and bound by CPI Group (UK) Ltd, Croydon, CR0 4YY

16/04/2025

14658388-0004